VOLCANIC AND TECTONIC HAZARD ASSESSMENT FOR NUCLEAR FACILITIES

Geoscientists worldwide are developing and applying methodologies to estimate geologic hazards associated with the siting of nuclear facilities, including nuclear power plants and underground repositories for long-lived radioactive wastes. Understanding such hazards, particularly in the context of the long functional lifetimes of many nuclear facilities, is a challenging task. This book documents the current state of the art in volcanic and tectonic hazard assessment for proposed nuclear facilities, which must be located in areas where the risks associated with geologic processes can be quantified and are demonstrably low.

Specific topics include overviews of volcanic and tectonic processes, the history of development of hazard assessment methodologies, description of current techniques for characterizing hazards, and development of probabilistic methods for estimating risks and uncertainties. Examples of hazard assessments are drawn from around the world, including the United States, Great Britain, Sweden, Switzerland and Japan.

This volume will promote much interest and debate about this important topic among research scientists and graduate students actively developing methods in geologic hazard assessment, geologists and engineers charged with assessing the safety of nuclear facilities and those with regulatory responsibility to evaluate such assessments.

CHARLES B. CONNOR is Professor and Chairman of the Geology Department at the University of South Florida. He has worked on assessment of volcanic hazards at nuclear facilities since 1992, in association with the US Nuclear Regulatory Commission, the International Atomic Energy Agency and the Nuclear Waste Organization of Japan. These professional activities have included developing the US Nuclear Regulatory scientific program for assessment of volcanic hazards at Yucca Mountain, Nevada, chairing of the committee to develop IAEA safety guidance for nuclear power plants and developing safety guidelines for nuclear installations in Japan. In addition, he served on the US National Research Council commission to review the US Geological Survey volcanic hazards program for the National Academy of Sciences.

NEIL A. CHAPMAN is Chairman of the ITC School of Underground Waste Storage and Disposal, Switzerland; Research Professor of Environmental Geology, Department of Engineering Materials, University of Sheffield, UK; Programme Director, Arius Association, Switzerland; an independent consultant. He has worked for more than 30 years on the scientific and strategic issues of the nuclear industry and radioactive wastes, for industrial, governmental and international organizations and agencies worldwide. This has involved participation in many national and international advisory committees, in the management of internationally funded projects and as a visiting expert. He is currently chairman of the INSITE site investigation overview group for the Swedish regulatory authority, SSM, and was recently a member of the International Technical Advisory Committee (ITAC) of the Japanese radioactive waste management organization (NUMO).

LAURA J. CONNOR is a computational scientist and research associate in the Department of Geology at the University of South Florida. Her work has focused on computational methods in geologic hazard assessment and geophysical research, which have highlighted new methods for optimization of volcanic hazard models, uncertainty assessment for volcanic hazard models and applications in real-time monitoring of geophysical processes. She has authored numerous codes, including the probabilistic volcanic hazard assessment codes currently in use by the US Nuclear Regulatory Commission and the Nuclear Waste Organization of Japan. She is co-editor of *Statistics in Volcanology*, recently published by the Geological Society of London.

VOLCANIC AND TECTONIC HAZARD ASSESSMENT FOR NUCLEAR FACILITIES

Edited by

C. B. CONNOR,[1] N. A. CHAPMAN,[2] L. J. CONNOR[1]

[1] *Department of Geology, University of South Florida*
[2] *MCM Consulting, Switzerland*

CAMBRIDGE
UNIVERSITY PRESS

CAMBRIDGE UNIVERSITY PRESS
Cambridge, New York, Melbourne, Madrid, Cape Town, Singapore, São Paulo, Delhi

Cambridge University Press
The Edinburgh Building, Cambridge CB2 8RU, UK

Published in the United States of America by Cambridge University Press, New York

www.cambridge.org
Information on this title: www.cambridge.org/9780521887977

First published 2009

Printed in the United Kingdom at the University Press, Cambridge

A catalog record for this publication is available from the British Library

ISBN 978-0-521-88797-7 hardback

Contents

The colour plates appear between pages 306 and 307.

Contributors

W. R. Alexander Bedrock Geosciences, Veltheimerstrasse 18, CH-5105 Auenstein, Switzerland.

J. Andersson JA Streamflow AB, Vinodlargatan 6, SE-117 59 Stockholm, Sweden.

W. P. Aspinall Aspinall & Associates, 5 Woodside Close, Beaconsfield HP9 1JQ, UK.

K. Berryman GNS Science – Te Pu Ao, 1 Fairway Drive, Avalon, Lower Hutt 5010, New Zealand.

J. Beavan GNS Science – Te Pu Ao, 1 Fairway Drive, Avalon, Lower Hutt 5010, New Zealand.

C. Bonadonna Section des Sciences de la Terre, Université de Genève, 13, rue des Maraîchers, CH-1205 Genève, Switzerland.

M. L. Caplinger Department of Geology, University of South Florida, 4202 E. Fowler Ave, Tampa FL, 33620, USA.

N. A. Chapman MCM Consulting, Täfernstrasse 11, CH 5405 Baden-Dättwil, Switzerland.

M. Cloos Department of Geological Sciences, Jackson School of Geosciences, University of Texas at Austin, Austin TX, 78712, USA.

C. B. Connor Department of Geology, University of South Florida, 4202 E. Fowler Ave, Tampa FL, 33620, USA.

L. J. Connor Department of Geology, University of South Florida, 4202 E. Fowler Ave, Tampa FL, 33620, USA.

K. J. Coppersmith Coppersmith Consulting, Inc., 2121 North California Blvd, Suite 290, Walnut Creek CA, 94596, USA.

M. Díez Department of Geology, University of South Florida, 4202 E. Fowler Ave, Tampa FL, 33620, USA.

G. Downes GNS Science – Te Pu Ao, 1 Fairway Drive, Avalon, Lower Hutt 5010, New Zealand.

S. J. Fowler Department of Earth Science, University of California, Santa Barbara CA, 93106, USA.

A. R. Godoy ESS/NSNI, P.O. Box 100, Wagramer Strasse 5, A-1400 Vienna, Austria.

J. Goto Nuclear Waste Management Organization of Japan (NUMO), Mita MN building, 1-23, Shiba 4-chome, Minato-ku, Tokyo 108-0014, Japan.

A. Hasegawa Research Center for Prediction of Earthquakes and Volcanic Eruptions, Tohoku University, Sendai 980-8578, Japan.

B. E. Hill US Nuclear Regulatory Commission, NMSS/HLWRS, EBB 2-02, Washington DC, 20555-0001, USA.

S. S. Hughes Department of Geosciences, Idaho State University, 921 S. 8th Avenue – Stop 8072, Pocatello ID, 83209-8072, USA.

D. Inoue Central Research Institute of Electric Power Industry, 1646 Abiko, Abiko-shi, Chiba-ken 270-1194, Japan.

O. Jaquet In2Earth Modelling Ltd., c/o Wirtschafts-Treuhand AG, Arnold Böcklin-Strasse 25, CH-4051 Basel, Switzerland.

K. E. Jenni Insight Decisions, LLC, 1616 Seventeenth St., Suite 268, Denver CO, 80202, USA.

K. Kitayama Nuclear Waste Management Organization of Japan (NUMO), Mita NN building, 1-23, Shiba 4-chome, Minato-ku, Tokyo 108-0014, Japan.

S. Kodaira Institute for Research on Earth Evolution, Japan Agency for Marine–Earth Science and Technology, Yokohama 236-0001, Japan.

J. -C. Komorowski Institut de Physique du Globe de Paris (IPGP) – CNRS (UMR 7154), Equipe de Géologie des Systèmes Volcaniques, 4, Place Jussieu, B 89, 75252 Paris cedex 05, France.

H. Kondo Nuclear Fuel Cycle Backend Research Center, Central Research Institute of Electric Power Industry, 1646 Abiko, Abiko-shi, Chiba-ken 270-1194, Japan.

C. Lantuéjoul MinesParisTech, Équipe Géostatistique, 35, rue Saint Honoré, F-77305 Fontainebleau, France.

A. -M. Lejeune Laboratoire de Pétrologie, Modélisation des Matériaux et Processus, Universite, Pierre et Marie Curie, Case 110 - 4 place Jussieu, 75252 Paris cedex 05, France.

N. Litchfield GNS Science – Te Pu Ao, 1 Fairway Drive, Avalon, Lower Hutt 5010, New Zealand.

B. Lund Department of Earth Sciences, Uppsala University, Villavägen 16, 752 36 Uppsala, Sweden.

S. H. Mahony Department of Earth Sciences, University of Bristol, Wills Memorial Building, Queen's Road, Bristol BS8 1RJ, UK.

R. McCaffrey GNS Science – Te Pu Ao, 1 Fairway Drive, Avalon, Lower Hutt 5010, New Zealand.

T. McEwen McEwen Consulting, Cobblestones, Main Street, Hickling, Melton Mowbray LE14 3AJ, UK.

I. G. McKinley McKinley Consulting, Täfernstrasse 11, CH-5405 Baden/Dättwil, Switzerland.

T. Menand Centre for Environmental and Geophysical Flows, Department of Earth Sciences, University of Bristol, Queens Road, Bristol BS8 1RJ, UK.

D. Merritts Department of Geosciences, Franklin and Marshall College, Lancaster PA, 17603-3003, USA.

S. Miura Research Center for Prediction of Earthquakes and Volcanic Eruptions, Tohoku University, Sendai 980-8578, Japan.

S. Nakada Earthquake Research Institute, University of Tokyo, Yayoi 1-1-1, Bunkyo-ku, Tokyo 113-0032, Japan.

J. Nakajima Research Center for Prediction of Earthquakes and Volcanic Eruptions, Graduate School of Science, Tohoku University, Sendai, 980-8578, Japan.

J. O. Näslund Swedish Nuclear Fuel and Waste Management Company, Blekholmstorget 30, Stockholm, Sweden.

Y. Ota Yokohama National University, Tokyo, 145-0063, Japan.

S. C. P. Pearson Department of Geology, University of South Florida, 4202 E. Fowler Ave, Tampa FL, 33620, USA.

R. C. Perman AMEC Geomatrix, Inc., 2101 Webster St., 12th Floor, Oakland CA, 94612, USA.

F. V. Perry Earth and Environmental Sciences Division, Los Alamos National Laboratory, Los Alamos NM, 87545, USA.

J. C. Phillips Centre for Environmental and Geophysical Flows, Department of Earth Sciences, University of Bristol, Queens Road, Bristol BS8 1RJ, UK.

W. Power GNS Science – Te Pu Ao, 1 Fairway Drive, Avalon, Lower Hutt 5010, New Zealand.

L. Reiter Consultant, 1960 Dundee Road, Rockville MD, 20850, USA.

W. D. Smith GNS Science – Te Pu Ao, 1 Fairway Drive, Avalon, Lower Hutt 5010, New Zealand.

R. S. J. Sparks Centre for Environmental and Geophysical Flows, Department of Earth Sciences, University of Bristol, Queens Road, Bristol BS8 1RJ, UK.

F. J. Spera Department of Earth Science, University of California, Santa Barbara CA, 93106, USA.

M. Stirling GNS Science – Te Pu Ao, 1 Fairway Drive, Avalon, Lower Hutt 5010, New Zealand.

Y. Tamura Institute for Research on Earth Evolution, Japan Agency for Marine-Earth Science and Technology, Yokosuka, 237-0061, Japan.

H. Tsuchi Nuclear Waste Management Organization of Japan (NUMO), Mita NN building, 1-23, Shiba 4-chome, Minato-ku, Tokyo, 108-0014, Japan.

G. A. Valentine Department of Geology, University at Buffalo, 876 Natural Sciences Complex, Buffalo NY, 14260-3050, USA.

A. C. M. Volentik Department of Geology, University of South Florida, 4202 E. Fowler Ave, Tampa FL, 33620, USA.

L. M. Wallace GNS Science – Te Pu Ao, 1 Fairway Drive, Avalon, Lower Hutt 5010, New Zealand.

P. H. Wetmore Department of Geology, University of South Florida, 4202 E. Fowler Ave, Tampa FL, 33620, USA.

A. W. Woods BP Institute, Cambridge University, Madingley Rise, Madingley Road, Cambridge CB3 OEZ, UK.

R. R. Youngs AMEC Geomatrix, Inc., 2101 Webster St., 12th Floor, Oakland CA, 94612, USA.

Preface

Worldwide, geoscientists are exploring and developing methodologies to estimate volcanic and tectonic hazards associated with the siting of nuclear facilities, including nuclear power plants and proposed long-lived geological repositories of radioactive wastes. Understanding such geological hazards, particularly in the context of long-lived nuclear facilities, is a challenging task. This book presents the current state of the art in volcanic and tectonic hazard assessment for nuclear facilities, with the goal of promoting interest and debate in this important topic.

Nuclear energy has been a source of power for a little over fifty years. Today, 30 countries utilize nuclear power plants to generate 16% of the world's electricity. By 2015, world energy demand is set to double from its 1980 figure. Nevertheless, in the early years of this century it would not have been possible to forecast the renewed worldwide interest in nuclear energy that is now evident. Low carbon emission requirements and need for security of energy supply have caused many countries to take steps to renew or increase their existing nuclear power capacity. Other countries may soon embark upon nuclear power programs for the first time. It is not inconceivable that within the next twenty years a dozen additional countries will have nuclear power plants or associated nuclear fuel cycle facilities.

One reason for the slow development of nuclear power during the last two decades has been concern, and sometimes controversy, about the safety of nuclear installations. Although much of this concern revolves around the safe management and operation of nuclear power plants, the possibility that natural events could jeopardize facilities has attained increasing significance among those charged with regulating safety. Many current and potential future nuclear power countries lie in regions that are tectonically active. These regions will inevitably experience volcanic eruption, earthquake and tsunami in the future. Site assessment and hazard analysis are essential in order to understand and account for potential natural hazards in the siting, design and operation of nuclear facilities.

Sites hosting operational, surface-based nuclear facilities can today be envisaged to have operational lifetimes of the order of one hundred years. Underground repositories for the disposal of radioactive wastes have to provide isolation and containment for thousands of years. That any such facility might be vulnerable to the forces of nature was appreciated from the earliest days of nuclear power. During the course of writing this book, two very significant events occurred that highlight the importance of the issues we address. The world's largest nuclear power generating complex, the Kashiwazaki–Kariwa nuclear power plant in Japan, was struck by a major earthquake in 2007. Although its safety systems were not compromised, it remains closed for extensive checks, at tremendous economic cost. Several chapters in this book address lessons learned in Japan and elsewhere from experience in seismic hazard assessment for nuclear facilities. In 2008, a license application was finally submitted for the proposed United States' geological disposal facility for spent fuel and high-level waste, Yucca Mountain, after decades of research and discussion. The susceptibility of the Yucca Mountain site to future volcanism will be a central issue for the regulatory appraisal that will take place over coming years. Several chapters address specific issues in tectonic and volcanic hazard assesment of the proposed Yucca Mountain repository.

In this book, we begin by looking at the nature of tectonic and volcanic hazards with respect to nuclear facilities. Chapters 1–3 provide essential background on the nature of hazard assessment for nuclear facilities and progress in understanding volcanic and tectonic processes. In Chapters 4–8, the reader will find rich details about the physical conditions that give rise to natural hazards, the rates of tectonic and volcanic processes, and the geological and geophysical observations that make it possible to understand them. Translating observations into hazard models is a complex area of research, and the focus of Chapters 9–18. Techniques of probabilistic seismic hazard analysis have been available for many years and, indeed, owe much to the requirement for seismic analysis of the early generations of nuclear power stations. They are, however, only recently being adapted and applied to volcanic hazard and the reader will see that we have concentrated many of our examples on the latter, as this is an emerging area. Tsunami hazard is also highlighted in this section, as the area of probabilistic tsunami hazard assessment is only now receiving the attention it deserves, following the 2004 Great Sumatran earthquake and resulting global disaster. All hazard models must be based on good understanding of geological processes, basic observations and scientific deduction. The practical aspects of sifting alternative models of causal mechanisms must not be overlooked. Chapters 19–26 address the development of risk-informed approaches to site hazard assessment and the nature of regulation in light of our improving, and increasingly complex, understanding of natural phenomena.

The concept for this book arose from a long-running project initiated by the Nuclear Waste Management Organization of Japan (NUMO). The tectonically active nature of the Japanese islands means that NUMO is naturally concerned with evaluating possible risks to the isolation of potential geological repository sites that emerge from its voluntarist siting process. With the likelihood of future volcanic and rock deformation impacts varying widely from location to location across Japan, NUMO brought together a small team of experts from Japan and around the world to devise techniques for assessing the nature and probability of volcanic and tectonic hazards. Some chapters in this volume describe the initial results of this ongoing project, which has furthered development of hazard models for nuclear facilities generally. This illustrates not only the central nature of tectonic risk assessment in Japan, but also the far-sighted approach that is being adopted in that country to the progressive development and testing of hazard assessment techniques. Similarly, several chapters in this volume reflect the efforts of the International Atomic Energy Agency to foster methods in seismic and volcanic hazard assessment.

We would like to acknowledge the support of NUMO – not only in encouraging us to produce this book and thereby, we hope, promoting interest in this important topic – but also for their financial assistance in producing the color illustrations. Through their active encouragement, the authors have begun to dig more deeply into the subject than has ever been done before. Numerous individuals have helped bring this work to fruition. We thank Hideki Kawamura, Akira Chigama, Raymond Munier, Peter LaFemina, Ivan Savov, Gordan Woo, Diana Roman and Chris Newhall for their efforts. Special thanks go to Susan Francis and colleagues at Cambridge University Press for their guidance and enthusiasm.

Chuck Connor and Laura Connor, Tampa, USA
Neil Chapman, Remigen, Switzerland
July 2008

1

Tectonic events and nuclear facilities

N. A. Chapman, H. Tsuchi and K. Kitayama

Nuclear power had its origins over half a century ago, during the Cold War. Some eight years after the first nuclear reactors for plutonium production had begun operation in the USA, as part of the Manhattan Project, the first reactor to produce electricity entered service in late 1951 (EBR-1, in Idaho, USA). Just two years later, in 1953, President Eisenhower made his famous "Atoms for Peace" proposal, which effectively launched commercial nuclear power generation and led to the formation of the International Atomic Energy Agency (IAEA).

The spread of nuclear power was slow during the early 1950s, with only the USA, the Soviet Union and the UK having operating power reactors by 1958. In 1959, France and Germany began their nuclear power operations. Nuclear power plants (NPPs) began real commercial development in the early 1960s, led by the Pressurized Water Reactor design (PWR, originally developed for submarine propulsion units), and there was a rapid spread worldwide during the 1970s and 1980s (Figure 1.1). By the mid 1980s, although the number of NPPs being put into operation was at its peak (in 1985, when 42 NPPs were brought into operation), nuclear power was actually entering a marked decline. In 1986, further development of the nuclear industry essentially stopped in many European countries, primarily caused by reaction to the Chernobyl accident in the former Soviet Union (Ukraine).

However, other nations continued expansion, particularly in the Asia–Pacific region and, although the average number of NPPs commissioned each year since 1990 has only been about five, what has been called a worldwide "nuclear renaissance" was considered to be underway in the early years of the present century. Countries that had not ordered NPPs for decades were showing a new interest in nuclear power and the growth of nuclear power in Asia continued, especially in India and China. This resurgence is seen by many as partly a response to the drive to reduce greenhouse gas emissions and partly as a desire by nations to ensure security of electricity supply, independent of the political uncertainties of fossil fuel imports. By the beginning of 2008, there were 439 NPPs in operable condition worldwide, 34 under construction, 93 planned and 222 proposed (World Nuclear Association, 2008). The NPPs that are operating, under construction, or already closed down are located on 237 sites spread around the globe. In a period of about fifty years, nuclear power has reached a point where it is generating about 370 GW of electrical power, around 16% of the world's electricity supply.

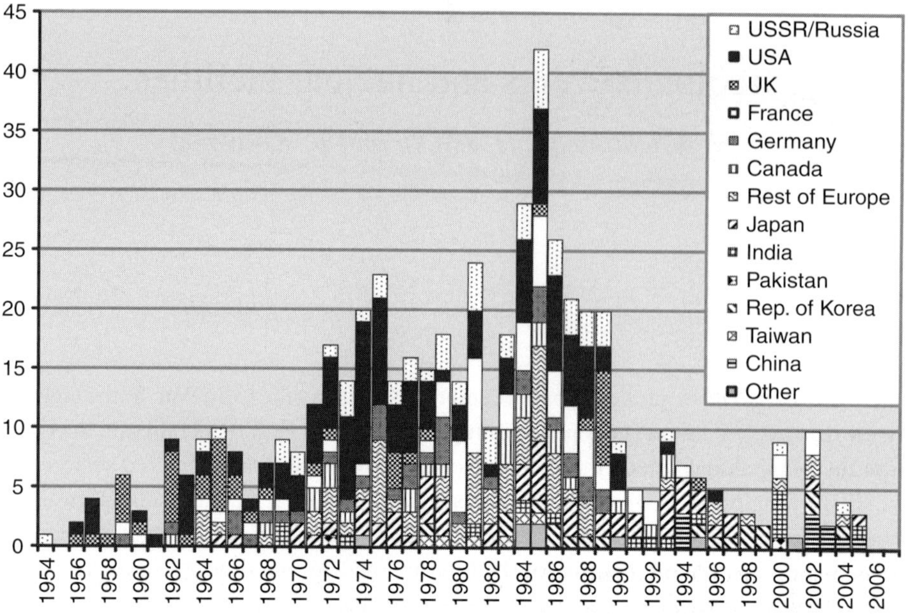

Fig. 1.1 Temporal and geographical spread of nuclear power by country. Shown are the number of nuclear power plants coming into operation each year, from the dawn of nuclear power until the end of 2007. The rapid growth in the 1970s and 1980s is evident, as is the even more marked decline in the early 1990s. Not all countries are shown individually; "Rest of Europe" shows NPPs in Europe excluding France, Germany and the UK (the early developers). Data taken from the World Nuclear Association database.

The widespread use of small nuclear reactors for research purposes or isotope production is often overlooked when considering the global distribution of nuclear reactors. Around 280 research reactors exist today, in 56 countries, although there were more in the 1970s; their distribution includes many more countries than have NPPs, including several small and developing countries (e.g. Bangladesh, Algeria, Colombia, Ghana, Jamaica, Libya, Thailand and Vietnam all currently have research reactors). There is a trend now to decommission many research reactors and repatriate the fuel to the countries that provided them. More than 360 reactors have been closed over recent years.

1.1 Tectonics and nuclear power plant location

The same half-century also saw the dawn of our current understanding of global tectonic processes, with the explosion in knowledge and research into "seafloor spreading"; then the development of plate tectonic theory, beginning in the early 1960s at about the same time that the first nuclear electricity was being generated. Strikingly, it was another aspect of nuclear energy that pushed forward our ability to build our present understanding of tectonic processes. The 1963 nuclear test-ban treaty proscribed the use of above-ground nuclear

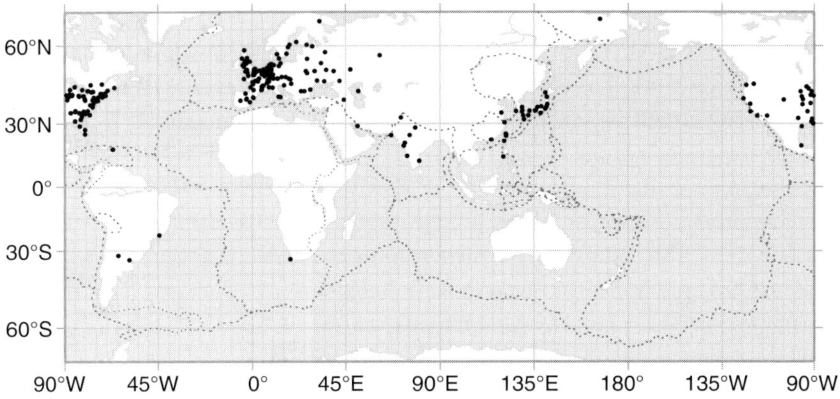

Fig. 1.2 Locations of NPPs (black dots) are shown with respect to the outlines of Earth's major active plate boundaries (gray dotted lines).

weapons testing. In order to monitor compliance, the Worldwide Standardized Seismograph Network (WWSSN) was set up and the greatly improved data that it provided allowed the precise mapping of global earthquake zones that was to underpin plate tectonic concepts.

Figure 1.2 shows the locations of operating NPPs worldwide with respect to their global-scale tectonic setting. It can be seen that NPP sites are preponderantly located in relatively "quiet" regions of the world, in Europe, Russia and North America. However, it is apparent that many NPPs and several nuclear power nations also lie in highly active regions, close to active plate margins.

Regions of the world that most clearly need to consider elevated probabilities of tectonic impacts on existing or proposed NPPs include the western USA, southern Europe, Iran, Turkey, Pakistan, Taiwan, the Philippines, Indonesia, China, Japan and Korea. Ironically, it is in these latter regions that we are currently seeing the most rapid actual expansion of nuclear power, or interest in developing new nuclear power programs.

That NPPs, with their requirement to be fail-safe in the event of accidents, could be at risk from tectonic events, most specifically from earthquakes, was realized early on, but the first NPPs were located with relatively little consideration of possible tectonic impacts. In the first decade of nuclear power, those constructed in Europe were anyway in what is generally regarded tectonically as a relatively quiet region of the world, although two of the first five power plants built in the initial stage of nuclear power in the USA were in California, in the seismically more active western part of the country. This led to several problems that are discussed later in this chapter and by Reiter (Chapter 20, this volume). The first reactor to be built in Japan came into operation in 1965.

Seismic hazard evaluation and seismic design of NPPs were both common by the early 1970s and have become increasingly well specified and internationalized since that time. All countries have seismic hazard and design codes for NPPs and the IAEA issues general guidance in its Safety Series and Safety Guide reports. These cover techniques for evaluating seismic hazard (IAEA, 2002), the evaluation, or reevaluation of seismic hazard at existing NPPs (IAEA, 2003a), the seismic design of NPPs (IAEA, 2003b) and the site characterization work that is needed to evaluate seismic hazard when planning a NPP (IAEA, 2004). Nevertheless, seismic events have caused problems for NPPs in the past and the July 2007 experience of the Niigata–Chuetsu–Oka earthquake in Japan (discussed later in this chapter) shows that they will continue to pose problems in the future.

In plate margin regions, seismicity is not the only tectonic hazard that needs to be considered with respect to NPPs. Sites located on low-lying land near the coast have to consider the likelihood and possible impacts of tsunamis, especially those generated by offshore, ocean trench earthquakes (Power and Downes, Chapter 11, this volume). The December 26, 2004 Great Sumatran earthquake that resulted in widespread and catastrophic tsunami impacts and loss of life around the Indian Ocean caused the automatic shut-down of the Kalpakkam NPP on the east coast of India, which was restarted six days later. The potential for such impacts and means of evaluating them are also discussed in IAEA safety guidance (IAEA, 2003d).

Proximity to active volcanoes and the possible impacts of ash fall, lahars, pyroclastic and lava flows, and other volcanic phenomena also need to be assessed. The concept of volcanic hazard assessment of NPPs has developed more slowly and patchily, even though around a quarter of IAEA member states have Holocene volcanoes within their territories. In the USA, the need to consider volcanism was recognized in NPP siting regulations in the early 1970s. The 1980 eruption of Mount St. Helens strengthened concerns, leading to a number of evaluations of possible volcanic impacts on, and siting guidelines for, NPPs (e.g. Hoblitt *et al.*, 1987). The proximity of a NPP in the Philippines to an active volcano was one of the reasons why it was never put into operation in the mid 1980s. The IAEA issued a provisional safety standard in 1997 and referred to volcanic hazards in its safety guide on "external events other than earthquakes" (IAEA, 2003c). A full Safety Guide on volcanic hazards is currently in preparation (Hill *et al.*, Chapter 25, this volume).

Nuclear power plant designers endeavor to mitigate the impacts of any type of accident or adverse event (i.e. a malfunction), or an event that is natural or operational in origin, by a system known as Defense in Depth (DID). The DID philosophy involves the use of diverse, redundant and reliable safety systems, with two or more systems performing key functions independently, such that, if one fails, another will back it up, providing continuous protection. The systems include both static components of the NPP (physical barriers) and dynamic operational, control and response systems (such as cooling systems, emergency action measures). The static components of the multiple physical DID barriers are the ceramic fuel pellets, the metallic fuel cladding, the reactor pressure vessel, the reactor containment and the surrounding building.

The first level of DID aims at prevention of occurrence of hazardous events. Clearly, this is not possible for tectonic events, other than by locating an NPP in an area where the event is essentially extremely unlikely or impossible during the (geologically short) operational lifetime of the plant. The second level of DID aims at preventing propagation of the impacts and the third level at mitigating the impacts. In both cases, seismic design of reactor systems

(Section 1.5.1) is a clear example of DID in practice. Generally, the DID expectations on the performance of NPP static barriers are very high.

Defense in Depth was developed early in the history of nuclear power as a conceptually simple, "belt-and-braces" design philosophy, before risk-based, probabilistic techniques were available to quantify the impacts of events (Sorensen *et al.*, 1999). Now that such methods are available and well tested, through integrated analysis of complete NPP systems, it is possible to use quantitative estimates of risk to be more specific about the requisite functions of the static and dynamic DID components for various accident/event scenarios. Risk analysis does not totally supplant the original DID philosophy, however, owing both to uncertainties in risk estimates and the need to provide robust safety systems to reassure the public. Specifically, in the context of the subject of this book, later chapters illustrate the constraints on probabilistic evaluations of tectonic events. Whilst we are now able to develop soundly based estimates of tectonic event likelihood for a particular area or site, the range of agreed values from expert elicitation is often wide and there is generally some uncertainty about the exact nature of impacts on nuclear facilities. We return to this issue later, when considering waste repositories.

1.2 Other nuclear facilities

Nuclear power plants are the most widespread, but not the only types of facility that are required by a nuclear power program. The nuclear fuel cycle also involves fuel fabrication plants and, in some countries, facilities for reprocessing spent fuel once it has come out of the reactors at the end of its useful life. These are major industrial complexes, especially reprocessing plants, which are currently a part of the fuel cycle in France, Japan, Russia and the UK. For nuclear weapons states, such facilities have sometimes been closely linked with weapons production and Russia, the UK and the USA, in particular, have a legacy of old military nuclear facilities (plutonium production reactors and fuel processing plants) that have been, or will need to be, decommissioned.

However, evaluation of the susceptibility of fuel cycle facilities to seismic hazard is generally less advanced than for NPPs and their susceptibility to other tectonic hazards, such as volcanism, has not been widely considered. Chung *et al.* (1990) looked at volcanic hazard to the Idaho National Laboratory, USA, and is one of the few such studies. At the time of writing, an IAEA Safety Guide on seismic hazards to existing nuclear facilities (to parallel IAEA (2003b) for NPPs) was in preparation. The hazard potential in case of seismic impacts varies greatly from one type of facility to another, but most fuel manufacturing and fuel reprocessing facilities include potentially vulnerable components for the movement or storage of gaseous or liquid radioactive materials. Many facilities worldwide are now old and the emphasis today is generally on reevaluation and back-fitting design features to improve robustness and ensure they are at modern levels of standards.

The nuclear fuel cycle center of Rokkasho in northern Honshu, Japan, contains a major new reprocessing facility, which has a seismic design that is intended to withstand an earthquake of magnitude M 8.25. Of course, many countries are not involved in the fuel cycle, only being users of nuclear fuel. Consequently, they only possess NPPs and the necessary storage facilities for radioactive wastes. Even in a seismically "quiet" region such as the Netherlands, the recently constructed HABOG 100-year, passively cooled storage facility for spent fuel and high-level waste contains a number of engineering features (e.g. automatically triggered latches on its massive radiation isolation doors) designed to mitigate the impacts of a low-probability major earthquake if it were to occur when material was being moved in the facility.

Nuclear power production generates radioactive wastes at each step and, although a large proportion of the more radiotoxic and long-lived classes of waste produced over the fifty years of nuclear power is currently in storage, we are now beginning to see the first geological repositories being constructed and operated. The majority of countries with nuclear facilities have surface or near-surface repositories for storing or disposing of their less-active, short-lived radioactive wastes until they have decayed to levels below concern. However, almost all countries have been very slow to site and construct deep (> 300 m) geological repositories for their reactor operating wastes and spent fuel (and vitrified high-level wastes and longer-lived intermediate-level wastes, if they practice fuel reprocessing).

Unlike other fuel-cycle facilities, a deep geological repository is based upon a series of multiple barriers with no dynamic components. Once waste is emplaced, the "engineered barrier system" of solid waste-form, metallic or concrete container and rock or mineral buffer and backfill provide passive isolation, even as the system evolves and progressively degrades over tens of thousands of years. Understanding of the behavior of a geological repository far into the future requires knowledge of the geochemical environment at depth, how water moves through pores and fractures in the rock, how stress affects the stability of the barriers and the rock, and how all of these slowly change in response to external processes and events, such as changing climate and tectonic activity. Vulnerability to tectonic impacts and the hazard potential of the radioactive materials once a repository is completed and sealed are of a different character to those of other nuclear facilities.

The identification of sites that can provide adequate, long-term stability for a geological repository is one of the principal themes of this book. Importantly, the requirements for geological disposal take us much farther into the future when it comes to assessing tectonic stability. This leads us into consideration of the time periods for which potential tectonic hazards need to be evaluated.

1.3 Operational lifetimes with respect to tectonic hazards

Clearly, the main period of concern with respect to potential impacts from tectonic events is during the operational life of a NPP and as long afterwards as spent fuel might continue to be stored at the reactor site. The operational life of the early NPPs was planned to be only a few decades. Many of the earliest power reactors, which were to a large extent developmental, were typically shut down after only five to fifteen years of operation. The Calder Hall reactors in the UK are a famous exception, having operated for over forty-five years before closure, but the typical lifetime of reactors commissioned in the first decade of nuclear electricity was fifteen to twenty-five years. Progressively, a forty-year operational period became typical, then a sixty-year period. Today, considering the difficulties of finding societally acceptable new locations for nuclear facilities of any type, it is common to consider the continued development of existing nuclear power stations sites by the construction of additional or replacement NPPs, such that the lifetime of an NPP site might now stretch over at least one hundred years. Consequently, susceptibility to tectonic hazards needs to be seen over a much longer period than may originally have been envisaged.

The planned lifetimes of other fuel-cycle facilities, such as fuel fabrication or reprocessing plants, are of a similar order to that of an NPP. However, for geological repositories, which aim to isolate long-lived radioactive wastes until they have decayed at least to levels similar to natural uranium ores, we must now consider periods out to thousands or hundreds of thousands of years. Table 1.1 indicates the differences in hazard potential that we need to consider for different types of facility in response to possible tectonic impacts.

Even though a sealed geological repository is expected to provide passive isolation of the waste and containment of radionuclides in a stable deep environment, tectonic processes and events could compromise its overall performance. It is important to emphasize that such impacts are likely to have insignificant consequences in terms of radiological exposure of people, compared to those that might result from a severe damage scenario to an operating NPP or reprocessing plant, but the convention is to treat them equally seriously.

The metric for all possible radiological exposures from all nuclear facilities is that of risk. The topic of risk, expressed in a number of different ways, will be covered in depth in many of the chapters in this book. In the context of tectonic events affecting nuclear facilities, radiological health risk is broadly an expression of the likelihood of an exposure occurring (itself, a function of the likelihood of a disruptive tectonic event occurring) multiplied by the consequences, in terms of the scale and nature of radiological health impact. It is clear that, if the likelihood of an event occurring is extremely small over the vulnerable, operational lifetime of a facility, then even quite large health impacts that might be caused if it were to occur would result in a low estimated risk. Short (in a tectonic time framework) periods of vulnerability will lead to low risks. Long periods of vulnerability, such as those for geological repositories, where it takes thousands to hundreds of thousands of years for the waste to decay to natural levels of radioactivity (Figure 1.3), even if they lead to very low health exposures, can have commensurate risks.

The very much longer periods over which safety assessments are required for geological repositories mean that it is not only tectonic events such as the seismic shaking caused by earthquakes, flooding caused by tsunamis and the eruptions of nearby volcanoes that have to be evaluated. We enter a different realm of possible impacts from slower, long-term processes, along with the possibility of much more infrequent events, when we begin to look out towards 100 000 a or even 1 Ma. Factors that become important include the possibility that active faults may develop or extend into the repository volume, especially where undetected structures at depth might propagate upwards; the cumulative impact of displacements on small fractures in the repository host rock caused by repeated movements along nearby major active faults; slow uplift and erosion of the geological formations

Table 1.1. Period of concern for tectonic hazard evaluation for nuclear facilities and features representing significant risk

Years	Principal features at risk	Comments
Fuel fabrication facilities		
100	Volatile uranium hexafluoride storage, transfer and centrifuge plant	Generally low hazard. Processing of natural uranium with with low activity levels and limited potential for airborne release and transport.
Nuclear power plants		
100	Reactor management systems, including fuel handling, control rods, coolant and emergency coolant systems	Massive concrete containment, pond walls and other foundation structures give considerable structural protection against seismic and many volcanic events. Damage to emergency and control systems, from seismic shaking, volcanic ash or marine flooding, is a key issue. Water has slopped out of open ponds; items have fallen in.
	Spent fuel storage, especially if in water-filled ponds	Potential hazards are high: airborne or waterborne releases of volatile fission products from damaged reactor core or spent fuel in storage. Large areas and large populations could be exposed in severe scenarios.
Fuel reprocessing plants		
	Fuel storage ponds or dry cask stores	Some parts involve massive concrete containment (storage pond walls, fuel dismantling hot cells, vitrification hot cells). Facilities larger, more spread out, more complex than NPPs. Also likely to have much larger inventories of radioactive materials.
100	Liquid chemical extraction process systems, including transfer piping and storage reservoirs and their coolant systems	Hazards probably comparable to or significantly higher than those of NPPs. Slopping, rupture of pipes or storage tanks or loss of coolant to storage tanks could lead to airborne or waterborne releases of volatile fission products. Large areas and large populations could be exposed in severe scenarios.
	Spent fuel storage, especially if in water-filled ponds	
Geological repositories for long-lived radioactive wastes		
~ 10 000 to several 100 000s	Operational period: surface interim stores and spent-fuel encapsulation hot cells, waste transfer and handling systems (e.g. shaft hoist systems), underground power and pumping systems	Spent-fuel encapsulation plants have similar risks to those at NPPs or fuel processing plants. Many wastes will arrive at interim stores solidified, encapsulated and ready for disposal. Repositories and ancillary facilities thus less susceptible to hazards than NPPs or fuel reprocessing plant. During open, operational period (may be up to 100s of years) hazards principally to surface facilities. Problems underground may cause operational recovery difficulties but unlikely to lead to significant public radiation exposures.
	Post-closure: container–overpack–buffer systems (EBS: engineered barrier system) for spent fuel, vitrified high-level waste and other classes of long-lived waste	The requirement to assess post-closure safety over hundreds of thousands of years means impacts on the EBS become more probable and there are regulatory requirements to understand and assess them in detail. However, potential exposures and risks are generally insignificant.

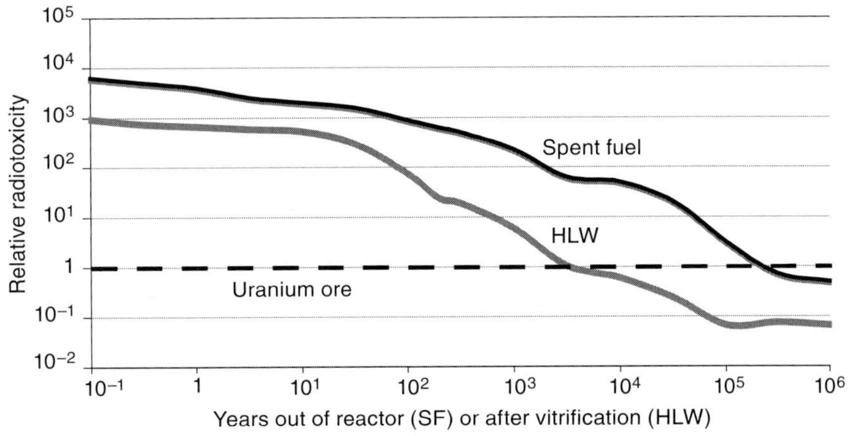

Fig. 1.3 The decline in radiotoxicity of typical spent fuel and vitrified HLW as a function of time, compared to that of an amount of uranium ore equivalent to that used to make the fuel (or the glass from reprocessing it after use). HLW toxicity approaches the toxicity of uranium ore after a few thousand years, while spent fuel takes $\sim$ 100 000 a to decline to the toxicity of uranium ore.

hosting or overlying the repository; the possibility that a new volcano might form near or even through the repository; possible exposures to larger, infrequent tsunami events caused by sector collapse of distant volcanic islands.

In addition, long-term safety assessments have to account for the impacts of a range of climate change events, the majority of which are not covered in this book, including glacial cycling and its effects on sea level, groundwater flow and chemistry and erosion. One aspect that we do, however, consider in this volume are the impacts of what are generally termed neotectonic processes, such as postglacial faulting and associated earthquakes, caused by the response of the rock mass to loading and then unloading by kilometer-thick ice sheets (Lund and Näslund, Chapter 5, this volume).

1.4 Tectonic problems for early nuclear power plants

We should say at the outset of this section that, while NPPs have certainly suffered damage from earthquakes, there have been no accidents leading to serious loss of containment of radioactive materials from nuclear facilities that are attributable to tectonic processes. Nevertheless, the risk of tectonic impacts has been a serious concern in a number of instances (e.g. the 2007 earthquake impacts on the Kashiwazaki–Kariwa NPP in Japan, discussed later) and has indeed led to the abandonment of some planned facilities.

Possibly the best-known examples are associated with the early development of nuclear power in California. In the 1960s, the Pacific Gas and Electric (PG&E) Company began development of an NPP at Bodega Head, a site located only about 300 m from the edge of the active zone of the San Andreas fault. Faulting observed in the granitic rocks of a shaft constructed for the NPP foundations resulted in an analysis of possible seismic impact.

A study by the USGS (Schlocker and Bonilla, 1964) concluded that, although the most recent fault displacement in the foundations was probably $\sim$ 42 ka, this was not certain, and there had been $\sim$ 7 m total displacement along the fault over $\sim$ 400 ka. The report concluded that the site was "almost certain" to experience a severe earthquake in the next fifty years (see also, Reiter, Chapter 20, this volume).

The PG&E Company proposed a design for the reactor that would accommodate fault movement but the Atomic Energy Commission (AEC) was unconvinced that other parts of the NPP would be protected and concluded that the site was not suitable; this led to the abandonment of the project in late 1964. The foundation excavations (disparagingly known locally as the "hole in the head") filled with freshwater and are now a coastal wildlife habitat.

The PG&E Company looked elsewhere and selected a site at Diablo Canyon (see cover illustration), where it was thought that there was no evidence of active faulting. The subsequent discovery of a major offshore active fault led to years of investigations and hearings, with revised NPP design and seismic back-fitting; it was one of the main causes of delays that resulted in the first reactor not becoming operational until 1984, fifteen years after the first work on site.

Seismic hazards continued to pose problems to power companies in California throughout the 1970s and 1980s. An NPP was proposed for the Malibu site near Los Angeles in the mid 1960s and eventually abandoned in the early 1970s owing to the likely difficulties of designing and then obtaining a license with respect to seismic impacts. The Vallecitos facility run by General Electric hosted an old fuel test reactor, whose "precautionary decommissioning" took place in 1977 when it was found to lie on the splayed Verona thrust fault. The USGS again carried out detailed studies of the site (Herd and Brabb, 1980), producing evidence for late Quaternary surface rupture caused by the Verona fault. The history of these famous cases is described in depth by various authors, including Novick (1969), Meehan (1984) and Walker (1990) and is summarized by Reiter (Chapter 20, this volume).

The Metsamor NPP in Armenia was closed for four years following the M 7.2 Spitak earthquake in 1988, before one of the units was restarted in 1995. The Kozloduy NPP in Bulgaria suffered slight damage to two old reactor units in 1977, as a result of an earthquake that occurred some 400 km away in Romania. This resulted in design changes to units then under construction. Many countries are now back-fitting seismic protection to NPPs as it has often been found that, on recent evaluation, original seismic designs have underestimated ground motions.

1.5 The current situation with NPPs

As noted earlier in this chapter, the principal focus of developments in evaluation of tectonic hazards to nuclear facilities has been on seismic impacts. The evaluation of volcanic and other tectonic impacts on NPPs, whilst under evaluation, has lagged far behind. The disparity between consideration of seismic and volcanic risks can only be accounted for by a general, seemingly "risk-uninformed," perception of the frequency of earthquakes compared to major volcanic eruptions. Even so, it has been known for decades that volcanism can be of real concern. The Bataan NPP in the Philippines was completed in 1984 but never fueled and commissioned, in part as a result of its possible susceptibility to volcanism (and, more so, to seismicity). This risk had been known since it was originally sited and was demonstrated in fact in 1991, when ash from the Mt. Pinatubo eruption fell on the site (Volentik *et al.*, Chapter 9, this volume). By this time, it had already been decided some years previously that the location was not safe. Using current evaluation techniques such as those described by Hill *et al.* (Chapter 25, this volume) many sites may prove vulnerable to some type of volcanic impact.

There have been numerous instances over the last thirty years where the resilience of NPPs and other nuclear facilities to seismic shaking has been tested, the more widely known instances being in Japan, the USA, Taiwan and Armenia (Hore-Lacy, 2007). Japan has experienced many large earthquakes in the vicinity of NPPs. Between 1984 and 2004 none of these caused automatic reactor shut-downs, triggered by seismic motion detectors installed at the NPP, although there have been several instances of tripped closures since 2004, with the July 2007 shut-down of the Kashiwazaki–Kariwa NPP being not only the most widely publicized but perhaps the most seriously perceived incident.

An important consideration is that, to date, in all cases known worldwide where major earthquakes have occurred in the vicinity of NPPs, there has been no serious damage to the reactors themselves leading to releases of radioactivity from the reactor containment. Generally, it has been possible for most facilities to go back on stream within a few days, after inspections.

Even countries not normally associated in the public mind with tectonic hazards have become increasingly occupied in assessing seismic hazards to NPPs. One of the most recent major studies, PEGASOS (Abrahamson *et al.*, 2002), was a probabilistic seismic hazard assessment (PSHA) that focused on updating risk estimates for four NPP sites in Switzerland (Coppersmith *et al.*, Chapter 26, this volume). The study was requested by the regulatory authorities and looked at annual probabilities of events as low as 10^{-7}. One of the key aspects emerging from this study revolved around the problems encountered in propagating the inherent uncertainties in data interpretation and expert opinion in forward modeling while still being able to arrive at a definitive result.

With more countries looking to develop nuclear power over the next decade, the issue of tectonic hazards will become increasingly important. At the time of writing, the prospect of new or increased nuclear power was being considered in several countries located in tectonically active regions. Chile, Indonesia, Iran, the Philippines and Turkey are examples of potentially new power programs. Pakistan, China, Taiwan and Japan are examples of countries where nuclear power is in use and likely to increase significantly.

1.5.1 Mitigating the potential impacts of earthquakes on NPPs

An important aspect of seismic design philosophy for NPPs is the requirement that, in addition to the safety of the structure and the occupants, the facility must continue to meet its safety functions during and after an earthquake. Owing to the geographically patchy spread of nuclear power described at the beginning of this chapter, the inclusion of seismic factors in NPP design has been at variable levels, ranging from no specific seismic requirements through the life-safety provisions normally incorporated in national building codes, to the latest seismic nuclear codes. The latter provide not only for structural robustness but also include operational requirements for continued operation of essential safety functions in the event of a major earthquake (Asmis and Eng, 2001).

In Japan, the specific measures taken against earthquake damage begin with thorough geological investigation to avoid active faults, construction on rigid rocks, design with large safety margins and tests using large-scale shaking tables. Nuclear power plants incorporate automatic shutdown systems that trip the reactor on detecting significant ground motion, larger than a criterion specified for each plant individually. In addition, the height of the plant site above sea level is based on evaluation of records of historical tsunamis and simulation analysis.

The new advanced boiling water reactor (ABWR) design in Japan incorporates several specific design measures against seismic ground motion. For example, the cooling water circulation pump that had previously been installed outside the reactor pressure vessel has been replaced with an internal pump, which has resulted in less vulnerable piping runs and a lowering of the overall center of gravity, leading to enhanced seismic resistance. The reinforced concrete containment vessel is now integrated with the reactor building, again contributing to seismic resilience (Figure 1.4).

1.5.2 Case study: Seismic hazards to NPPs in Japan

Japan is in one of the most active plate boundaries in the world and consideration of tectonic hazards is a priority in the nuclear industry. Consequently, it makes a useful case

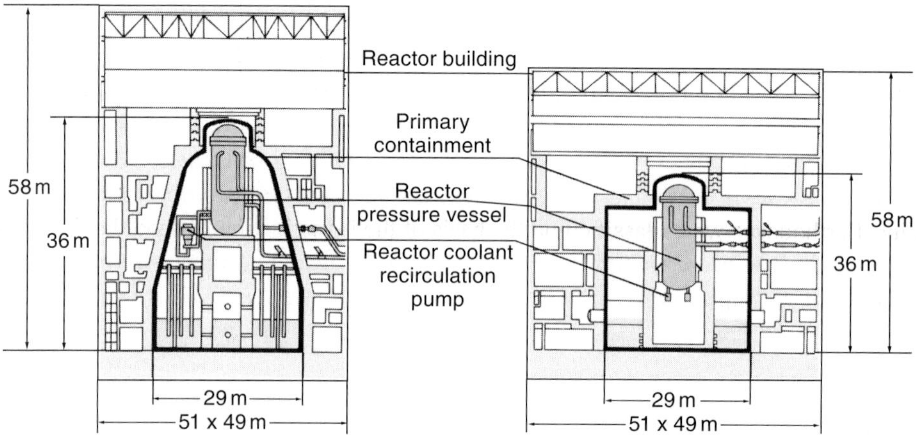

Fig. 1.4 Japan's current boiling water reactor (BWR) design (left) and the advanced ABWR design (right) are two reactors where earthquake hazard reduction was a central objective. The ABWR has high seismic resistance owing to its internal pumps, reduced piping and lowered center of gravity.

study for our current review. Historically, the siting process for NPPs in Japan endeavored first to avoid altogether locations that were thought to have the potential for igneous activity, or where active faults lie close by, then to apply seismic design principles to the facilities to be constructed. The seismic design guidelines used for the majority of operational NPPs were established from 1978 to 1981 (Committee on Examination of Nuclear Reactor Safety, Japan, 1978; JNSC, 1981) but have recently been reviewed and a revised system partially adopted. One of the most significant changes is a move from an entirely deterministic evaluation of potential earthquake ground motions at a candidate site, to a partially probabilistic hazard approach. This follows the broader, risk-informed approach to nuclear facility regulation being developed, since 2003, by the Nuclear Safety Commission (NSC). In fact, the process of updating NPP seismic guidelines originates somewhat earlier, stemming from the 1995 Hyogoken–Nanbu Earthquake (M 7.3 on the Japanese Meteorological Agency scale). Although no damage to nuclear facilities occurred, the experience led to initiation of the revision of the seismic design guidelines to reflect up-to-date knowledge on seismology and earthquake engineering and development of seismic design techniques.

When the 2000 Tottori-ken Seibu Earthquake (M 7.3) occurred, some further difficulties with the seismic assessment guidelines were pointed out. In particular, no active fault had been described on the existing national active fault map, and no distinctive ground rupture fault appeared at the surface as a result of the earthquake, although it had been thought highly likely that faults would break surface for earthquakes with M 6.5 or larger. In the 2005 Miyagiken–Oki earthquake (M 7.2), the ground motion recorded for some frequency intervals at the Onagawa NPP was greater than the calculated basic ground motion used in the seismic design. Consequently, in September 2006, the Nuclear Safety Commission decided to revise the guidelines for seismic design. The new guidelines are considered by the IAEA (IAEA, 2007) to be in line with the recommendations of their 2002 Safety Guide for NPPs (IAEA, 2002). The revised guidelines have been applied to new nuclear power plants and, for existing plants, have been used for reevaluation (Inoue, Chapter 21, this volume).

The old guidelines for NPP seismic design considered two categories of design basis earthquake: $S1$ (maximum design earthquake) and $S2$ (extreme design earthquake). The magnitude of $S1$ is identified by deterministic analysis of information on past earthquakes and on active faults in the NPP vicinity. The former assesses magnitude > 5 (Japan Meteorological Agency scale) earthquakes near the site, along with those occurring within a 200 km radius. Earthquakes were classified on the basis of the investigation method that identified them: "historical," "active fault" or "(plate) marginal." For the purposes of $S2$, active faults were defined as faults that have been active during the last 50 ka. The magnitude of $S2$ is determined by combining the active fault data with evaluation of the regional-scale seismic structure of the crust (largest possible earthquake in a broad region with generally similar tectonic structure) and considerations of the potential for shallow-focus earthquakes close to the NPP site. The $S1$ and $S2$ data are used to simulate possible ground motions (based on static vertical seismic load) on the free surface of the "base stratum" at the site.

The new, modified approach makes no distinction of earthquake categories but aims at assessing the site-specific ground motion (S_s), without specifying earthquake source. A more process-based classification of earthquake types is used: interplate (originating along the subducting and overlying plate boundary), inland (occurring within the overlying plate) and intraplate (occurring within the subducting oceanic plate). Active faults are now defined as those showing evidence of movement in the last 120–130 ka. Calculation of ground motion now looks at both static and dynamic vertical seismic load.

By early 2007, NPP operators were being asked to carry out voluntary probabilistic seismic hazard assessments. Earlier in this chapter, we discussed how probabilistic, risk-informed techniques can be used to strengthen and add versatility to the DID philosophy for NPP design. The NSC review did not result in full introduction of PSHA, but it was partially adopted in the sense that exceedance probabilities need to be referred to in determining basic earthquake ground motion. A logic-tree technique has been developed to consolidate expert opinion for probabilistic prediction of earthquakes, based on fault models and uncertainty evaluation.

The anticipated advantages of the PSHA approach include the capability to estimate quantitative values of risk to a nuclear power plant and express them as probabilistic indices (for example, the frequency of reactor core failure); the capability to manage uncertainties quantitatively, for example those associated with handling earthquake data and those associated with the marginal state of NPP systems damaged by an earthquake; the ability to identify and focus on NPP systems that have the most significant contribution to risk.

Nevertheless, it was considered by many that detailed analysis would become possible only when the construction of NPPs has been completed. The Japanese nuclear industry is currently only in an exploratory stage when it comes to implementing PSHA and assessing the implications of this relatively new approach. To date, NPP operators have developed PSHA mainly as a tool to generate reference information to help promote safety "self regulation." The industry began slowly to make efforts to solve problems associated with PSHA, with, for example, further efforts being required to improve evaluation techniques in order to reduce the associated uncertainties (see also comments above on the outcome of the Swiss PSHA project), to build consensus among experts and to accumulate experience.

The paragraphs above describe the situation of progressive adaptation and change in both guidelines and methodologies for seismic hazard analysis in Japan up to mid 2007. Shortly after 10 am of July 16, 2007 a M 6.8 earthquake occurred about 16 km offshore of the Kashiwazaki–Kariwa NPP on the west coast of Honshu, causing the three NPP units that were operating to close down automatically (a further three units were closed for periodic checks and a seventh was about to come on-line). Kashiwazaki–Kariwa is the largest nuclear power station in the world, with its seven NPPs capable of generating over 8 GW of power, up to 7% of Japan's electricity consumption. The earthquake caused widespread damage in the surrounding area, fourteen deaths, over two thousand injuries and over ten thousand people were made temporarily homeless.

Damage to the NPPs was considered light, both by the operators and a subsequent IAEA inspection mission (IAEA, 2007). Nevertheless, because a small amount of radioactivity was released to the environment and because an electrical unit caught fire and could be seen burning in television coverage, it attracted enormous attention worldwide, more so than the tragic results and fatalities caused by the earthquake itself. In objective terms, the NPPs at the site fared well. The reactors maintained their containment function and closed down as designed. A small amount of slightly contaminated water slopped out of spent fuel storage ponds and about one cubic meter of water was discharged to the sea, containing an insignificant amount of radioactivity (the estimated public exposure at around one billionth of that from annual exposure to natural background radioactivity). A second discharge occurred from a turbine condenser up one of the NPP stacks, again causing very low potential exposures. The tumbling of hundreds of stacked drums of solid waste in a storage building, with the loss of lids from some tens of them, whilst not posing a public hazard, gave a vivid visual impression of damage and disarray.

In fact, the real issue with this incident was not the risk posed by damage to the NPP, which was insignificant though immensely costly, but the fact that the event exceeded the seismic hazard design guidelines and called into question the adequacy of the data used for modeling seismic event magnitude, as well as the modeling process itself. A pre-construction offshore seismic survey for Units 7 and 8 in 1988 identified four faults around 20–40 km from the site, but concluded that three of them were not active and one of them was short enough to neglect. The design basis earthquake (*S*1) was set at M 6.5 and was exceeded. Both the survey and the assessment were, in retrospect, clearly flawed. Partly as a result of the changed NSC guidelines and partly as a result of the 2007 earthquake, detailed geophysical investigations are now planned on land and offshore to identify and characterize capable and active faults in and around the site, including the potential existence of active faults beneath the site which could cause possible surface rupturing (IAEA, 2007).

In the hazard analysis, source parameters such as fault mechanism and directivity effects may play an important role in the calculated impacts even at different locations that are close together. At Kashiwazaki–Kariwa, the site is split into two areas about 1500 m apart, with different thicknesses of Miocene and Pliocene sediments beneath each, contributing to notably different observed impacts. When active faults are located within a few kilometers of a NPP, these factors become important contributors to the hazard assessment. It is also vital to take recent seismicity into account. Japan has a dense seismic monitoring network, providing an enormous increase in the records of near-field earthquakes. Sometimes, these records have shown larger than expected accelerations compared to earlier derived seismic attenuation relationships. If active faults are present close to the site, these recent records need to be taken into account (IAEA, 2007).

It is worth quoting directly from the August 2007 IAEA report at this point:

> Both deterministic and reference probabilistic methods will be used in the re-evaluation of seismic hazard. Probabilistic seismic hazard analysis will be needed for the seismic PSA study. It is important to conduct both studies for this site in order to understand the different ways of quantifying uncertainties. There is worldwide interest in conducting seismic PSA and PSHA studies are needed for this purpose for a variety of seismotectonic settings. The faults in the near region of Kashiwazaki–Kariwa nuclear power plant site will also be of interest for the modeling of the attenuation relationship and how new methods such as empirical Green's functions can be applied within the context of a nuclear power plant seismic hazard evaluation. Source related parameters such as fault mechanism and directivity

Table 1.2. Main items evaluated for seismic hazard assessment and design of nuclear facilities in Japan

	Uranium enrichment plants for fuel manufacture	Low-level waste disposal facilities (near-surface repositories)	Vitrified high-level waste storage facilities	NPPs and spent fuel reprocessing plants
Historical earthquake record in vicinity	○	○	○	○
Proximity to active faults with high level of past activity	–	–	○	○
Proximity to active faults with low level of past activity	–	–	–	○
Regional seismotectonic crustal structure	–	–	–	○
Shallow-focus earthquakes beneath or very close to facility (Mj 6.5)	–	–	–	○
Assessment approach for ground motion	Based on standard building code	Based on standard building code	Design basis earthquake ($S1$)	Design basis earthquake ($S2$)

were observed to play an important role in the recent earthquake. It is expected that new methods may provide more information relating to these issues.

The development of such new methods is one of the main topics dealt with in this book.

The closure of Kashiwazaki–Kariwa had a huge effect on Japan's energy market and affected summertime energy use during 2007. The power company operating the site, TEPCO, estimated a $5.2 billion impact on its annual accounts. At the time of finalizing this chapter, almost two years after the earthquake, Kashiwazaki–Kariwa remains closed and undergoing further tests and inspections.

The Kashiwazaki–Kariwa experience vindicates the NSC's decision to strengthen the approach to seismic design and illustrates clearly the need for better modeling and more detailed and reliable geological information. The financial implications of this for the nuclear power industry are clearly likely to be large. Critically, the new approach in Japan requires reevaluation of existing NPPs, with possible seismic back-fitting being required. This reevaluation is already demanding additional, more detailed site characterization work, especially within a 5 km radius of the NPPs, aimed at identifying what may previously have been undetected potential earthquake sources ("hidden" active faults).

Japan has used its experience with NPP seismic hazard evaluation to develop equivalent approaches for other nuclear facilities potentially at risk from earthquake impacts. Table 1.2 shows how seismic design for different facilities utilizes information and can be compared with the preceding discussion of the data inputs for the now-superseded seismic design approach for NPPs. The only other type of facility with seismic design requirements as high as those for NPPs is the spent fuel reprocessing plant. The features at risk for these facilities were shown previously, in Table 1.1.

1.6 Tectonic hazards and geological repositories for radioactive wastes

Deep geological repositories, as discussed earlier, have considerably lower hazard potential than other fuel cycle facilities, but introduce a different timeframe into tectonic hazard considerations. The lower vulnerability is due to the passive functioning of the static multi-barrier system, which means that a geological repository, unlike a NPP, does not have even a remote possibility of failing catastrophically as a result of the gradual processes of geosphere evolution. In many safety assessments it has also been assumed that an environmentally and societally catastrophic tectonic event such as a major earthquake or volcanic eruption is only likely locally to impair barrier functions, and lead to earlier or elevated releases of radioactivity, if it occurs in the vicinity of a geological repository. However, the direct injection of magma into a repository, by what may be a relatively small volcanic event, can have serious consequences for containment (Valentine and Perry, Chapter 19, this volume; Menand *et al.*, Chapter 17, this volume). Thus, knowledge of the probability of such events becomes of central importance.

The multi-barrier system of a geological repository has been likened to the DID design philosophy for NPP protection. However, the analogy is only partial and differs in some important respects. The multiple barriers of a repository are conceptually similar to the static components of a reactor's DID system but, in a repository, there is no dynamic control component and no equivalent concept of independent defense levels or redundant barriers. Repository barriers work together, in concert, to provide isolation, with each barrier responding in different measure to the stresses of different perturbing scenarios. The role of each barrier also evolves with the passage of time, as the radioactivity in the waste declines and the engineered components progressively degrade (Figure 1.3).

The US Nuclear Regulatory Commission considered the DID parallels for NPPs and repositories and concluded that, because a repository is a much less complex system than an NPP and the comparison with NPP accidents and their consequences is inappropriate, the concept of defense in depth for repositories should be targeted more towards using the combined effects of the natural and engineered barrier systems to protect natural resources (such as groundwater supplies), where there are high uncertainties due to the very long time periods involved (USNRC, 2000). They recommended that the contribution that each individual repository safety system makes in achieving risk acceptance criteria should be determined by risk assessment with quantified uncertainty distributions.

The growth of risk-informed approaches to design and regulation of all types of nuclear facilities means that there are now closer parallels in the way that safety and performance in the case of low-probability events are evaluated for geological repositories and other nuclear installations. Much of this book is dedicated to looking at developments in this area for low-probability tectonic events, so we do not deal with it in any depth here but merely provide some pointers. Three examples are introduced below, each with quite a different type of significance to the development of a geological repository.

1.6.1 The USA's proposed spent-fuel repository: an example of post-siting concerns

The site of the proposed Yucca Mountain repository lies in the Basin and Range province of southern Nevada. It was selected as the national repository site for disposal of the spent fuel from the nation's NPPs more than twenty years ago. The story of the siting process and the investigations that have been undertaken in the years since then is long, complex and, currently, incomplete. We do not attempt to address this here. Rather, we draw an analogy with the problems encountered in the early decades of nuclear power development a little further west, in California.

The site investigations at and around Yucca Mountain began in the mid 1980s. As more became known about the site, the realization that it could be affected by both seismicity and volcanism grew. More investigation yielded more data and the developing understanding of the site coincided with (and contributed to) the growth of probabilistic seismic and volcanic hazard assessment techniques. Changes in thinking on regulatory standards pushed the period of detailed assessment further into the future and brought potential tectonic impacts into increased prominence. As a consequence, the seismic hazard potential of the site is now well understood and has been largely eliminated as a matter of concern. However, there is still considerable debate about the likelihood and the potential consequences of volcanism (Connor and Connor, Chapter 14, this volume; Valentine and Perry, Chapter 19, this volume; Coppersmith *et al.*, Chapter 26, this volume).

There seems to be a clear analogy with the story of California's nuclear experience. The issue of tectonics appeared of little (or limited) relevance to begin with, grew in significance as data were obtained and then (in some cases) had many years to mature and ferment as reports were commissioned, panels sat and licensing was progressively delayed. In the end, this resulted in the abandonment of some sites and significant upgrades and design modifications to others. Critically, the problems arose because tectonics was overlooked or poorly dealt with at the time of original site selection. This leads us to our second example.

1.6.2 Japan's HLW repository project: an example of pre-siting concerns

It was always clear to the Japanese nuclear power sector and to the government that any geological repository development program would have to address tectonics head-on. The developing saga of difficulties at Yucca Mountain during the 1990s would have been evidence of this on its own but, of course, Japan already had a long history of accounting for at least some tectonic events in operating its NPPs, even though recent events have shown the approach to be at least partially flawed. Consequently, when the Nuclear Waste Management Organization of Japan (NUMO) began its repository siting program in 2002 it already had a basis from which to begin to tackle some of the issues, notably, active faulting and seismicity.

In its methodology for finding a suitable site, NUMO began by establishing simple, deterministic exclusion criteria to remove potential sites (to emerge from a nationwide volunteering process) from consideration at the outset (NUMO, 2004). Essentially, locations very close to volcanoes or astride active faults would not be considered. This then left the problem of how to decide whether non-excluded sites will eventually prove acceptable in terms of their long-term tectonic risk and the way that this affects calculated radiological safety.

Some of the key issues are the likelihood of a new volcano developing in areas where there has been no volcanism during the Quaternary (especially in areas of monogenetic volcanism), the likelihood that an area might contain undetected active faults at depth and the likelihood that land uplift and other, slow deformation processes could affect a repository over the long term. A suite of essentially probabilistic approaches is under development to deal with each of these in an integrated fashion (Chapman *et al.*, in press). Nevertheless, the basic analysis that NUMO makes will follow a deterministic/empirical approach, with the results of probabilistic evaluations being incorporated as necessary (that is, largely depending on the exact location of volunteer sites: for example, some will lie well outside areas with any conceivable volcanic intrusion hazard over the next million years or more). How to couple the deterministic (e.g. Kondo, Chapter 12, this volume; Tamura *et al.*, Chapter 7, this volume) and probabilistic (e.g. Mahony *et al.*, Chapter 13, this volume; Jaquet and Lantuéjoul, Chapter 15, this volume; Connor and Connor, Chapter 14, this volume) approaches will be important in the eventual assessment of potential repository sites. In this respect, the repository siting studies are a close parallel to the situation with the introduction of risk-informed decision making into NPP seismic hazard evaluation, discussed earlier in this chapter. In both instances, the nuclear industry is feeling its way towards the best means of introducing probabilistic methods that will match possible future regulatory expectations on the use of risk targets.

Many of the chapters in this volume are associated with the work being carried out by NUMO to address these matters. Perhaps the most important point to note here is that, in Japan, the difficult issue of tectonic hazard has been addressed directly, at the very beginning of the geological repository program.

1.6.3 Sweden's spent-fuel repository under ice: an unexpected concern about very long-term safety

The Swedish spent-fuel repository development program is one of the longest running in the world, arguably the most consistent in its approach and consequently among the most successful. Isolation of the spent fuel in granitic bedrock is predicated on containment in a very long-lived copper container, protected from changes in the deep stress, groundwater flow and chemical environment by a thick buffer of compacted bentonite. Through several decades of safety assessment, attention was principally focused on very long-term chemical corrosion of the container. Relatively recently, attention has shifted to another possible failure mechanism, shearing of the buffer and, possibly, the container as a result of movement on fractures in the repository as they take up strain in the rock during large-magnitude earthquakes (Lund and Näslund, Chapter 5, this volume; McEwen and Andersson, Chapter 23, this volume).

Sweden does not have large-magnitude earthquakes today, so this seemingly unlikely concern arises simply because the assessment timescales for geological repositories are very long. In the aftermath of a glaciation, which can result in ice cover of several kilometers thickness in central Scandinavia, there is evidence that massive readjustment of the major bedrock blocks to ice unloading can cause earthquakes in the range up to M 6, and possibly up to M 8 (Lund and Näslund, Chapter 5, this volume). Over the next hundred thousand years or so, it is expected that glacial conditions (with all their implications for permafrost development, land uplift and subsidence, and changes in sea level) will return to the northern hemisphere, even accounting for the immediate impacts of global warming. The potential for neotectonic phenomena related to ice loading and unloading affecting a geological repository in the distant future is evident.

As a consequence of these major environmental changes, the tectonically stable Scandinavian nations of Sweden and Finland are both evaluating in detail the response of the deep bedrock to all aspects of glaciation, but especially the way in which strain is taken up on minor fractures close to major, kilometers-long, potentially seismogenic, deformation zones (Lund and Näslund, Chapter 5, this volume; Hökmark *et al.*, 2006). Relatively small amounts of cumulative strain (10–20 cm) along a fracture intersecting a deposition borehole for a waste container, some 300–500 m deep in the rock, could result in damage to the container that might accelerate releases of radionuclides into the surrounding rock. Consequently, the response of the whole fracture network between a major seismogenic deformation zone and the repository needs to be evaluated in order to establish a reasonable separation distance that will reduce radiological risks to an appropriate level. This type of analysis generally uses a combination of a deterministic description of the major structural features in the rock with a superimposed stochastic network of fractures with conservatively assumed mechanical properties.

It must be recalled that the possibility that such impacts could occur within a repository lies only many tens of thousands of years into the future. No current climate model predicts another global glaciation within less than about 50 000 a. As noted previously, the way in which such hazards are assessed, and the radiological risk targets that apply to them, need to be balanced against the shorter-term risks posed to operating nuclear facilities over the next decades.

Concluding remarks

Some readers might consider that we have just been lucky that there have been no serious radiological incidents caused by natural disasters at or close to nuclear facilities over the last half century. Others might say that this is simply a testament to the robust design of those facilities. Whichever view you might adhere to, it is abundantly clear that, with nuclear power seemingly set to grow significantly in many parts of the world that are tectonically highly active, possibly expanding into several countries on the Pacific rim that have so far not had major nuclear facilities, the importance of reliable methodologies for assessing and responding to tectonic hazards is growing. There is a long experience in several countries of evaluating seismic hazards to NPPs and of assessing its impacts on operational facilities. This is captured in a suite of national legislation and guidelines, as well as advisory documents from the IAEA that reflect a consensus of expert views.

Nevertheless, it is obvious from recent events in Japan that there is much room for improvement in data gathering and in modeling and assessment techniques. Right across the field of tectonic hazard analysis, our data gathering and interpretation is developing rapidly. We acknowledge the growing contribution made by improved process-level models of volcanism and tectonism and by relatively recently deployed remote sensing technology, whose impacts can be seen in several of the chapters in this volume. In particular, our ability to probe the deep structure of the crust and upper mantle by seismic tomography, the use of satellite global positioning systems to measure and constantly monitor millimeter changes in Earth's surface strain, the application of synthetic aperture radar to make time-series measurements of deformation around fault zones and active volcanoes, and the ability to probe the surface structure of the rock by "stripping" vegetation cover in airborne laser imaging are providing invaluable assistance to extend our understanding of tectonic processes and the scale of impacts of tectonic events.

Beyond the impacts of earthquakes, we note that other types of tectonic hazard to NPPs have been less ably dealt with to date. Tsunami hazards have only recently become widely recognized, although they have been part of seismic design code in some countries for decades. The fact that they can potentially impact coastal nuclear facilities became starkly evident in December 2004. Development of approaches to volcanic hazards lags well behind those of seismic hazards, possibly because the perceived probabilities of volcanic impacts and the number of facilities at potential risk are both smaller.

Risks to other types of nuclear facility are less widely discussed than those to NPPs, although again, in most countries, there are well-established procedures for seismic design for all types of fuel-cycle plant. Approaches to assessing tectonic risk to deep geological repositories are concerned with much longer time periods, typically hundreds of thousands of years, but are significantly less developed than approaches for other nuclear facilities. As will be seen later in this volume, experiences at the proposed Yucca Mountain repository site in the USA have shown this topic to be potentially contentious in the scientific community. This is one reason why the relatively new Japanese deep-repository program is looking so carefully into tectonic issues.

A topic that is developing slowly is the best approach to factoring deterministic and probabilistic analyses into tectonic hazard assessment. Historically, the emphasis has been on the former; but with growing experience in PSHA (in particular) and the progressive move to risk-based regulatory systems in many countries, the desire to have comprehensive probabilistic evaluations, with their capability to encapsulate wide ranges of uncertainties, is growing. We now see PVHA (probabilistic volcanic hazard analysis) being developed and applied to nuclear facilities.

Returning to the first sentence of our conclusions, we close this chapter by observing that humankind has had a tendency to disbelieve, discount or forget the magnitude of events that Earth can throw at us. Events such as the December 2004 Indian Ocean tsunami show that large regions of the planet, the home of numerous potentially hazardous facilities, can be put at risk with little or no warning. Nuclear power systems have yet to be exposed to a really large-magnitude earthquake or volcanic event on their doorstep. While the toll of such natural events in terms of human suffering may be enormous, it is still our duty to take reasonable measures to reduce any additional, consequential risks from what are anyway generally regarded by the public to be hazardous facilities. In the near term, however, it may be the smaller events that could cause the bigger hazards to nuclear facilities, particularly if siting has led to poorly understood or over-confident, but underestimated, levels of risk. Being able adequately to calculate real risks, for all scales and circumstances, is an essential underpinning for the inevitable next half century of nuclear power. That is the topic of this book.

References

Abrahamson, N. A., P. Birkhauser, M. Koller *et al.* (2002). PEGASOS – a comprehensive probabilistic seismic hazard assessment for nuclear power plants in Switzerland. Presented at the 12th International Conference on Earthquake Engineering, September 9–13, London.

Asmis, G. J. K. and P. Eng (2001). Seismic hazard assessment in intra-plate areas and backfitting. In: Seismic design considerations of nuclear fuel cycle facilities, IAEA-TECDOC-1250. Vienna: International Atomic Energy Agency, 83–90.

Chapman, N. A., J. Goto and H. Tsuchi *et al.* (in press). Development of methodologies for the identification of tectonic hazards to potential repository sites in Japan. The Kyushu case study. Conference proceedings on the stability and buffering capacity of the geosphere for long-term isolation of radioactive waste: application to crystalline rock, Tokyo, Japan: Nuclear waste organization of Japan.

Chung, D. H., D. W. Carpenter, B. M. Crowe *et al.* (1990). Assessment of potential volcanic hazards for new production reactor site at the Idaho National Engineering Laboratory, UCRL-ID-104722. Paris: OECD Nuclear Energy Agency University of California, Lawrence Livermore National Laboratory.

Committee on Examination of Nuclear Reactor Safety, Japan (1978). Guideline on licensing examination for geology and ground conditions of nuclear power plants. Tokyo: Committee on Examination of Nuclear Reactor Safety, Japan.

Herd, D. G. and E. E. Brabb (1980). Faults at the General Electric test reactor site, Vallecitos Nuclear Center, Pleasanton, California: a summary review of their geometry, age of last movement, recurrence, origin, and tectonic setting and the age of the Livermore Gravels, USGS Administrative Report, US Nuclear Regulatory Commission Docket 8006060. Menlo Park, CA: US Geological Survey.

Hoblitt, R. P., C. D. Miller and W. E. Scott (1987). Volcanic hazards with regards to siting nuclear power plants in the Pacific Northwest, USGS Open-File Report 87-297. Vancouver, WA: US Geological Survey.

Hökmark, H., B. Fälth and T. Wallroth (2006). T-H-M couplings in rock. Overview of results of importance to the SR-Can safety assessment, SKB Report SKB R-06-88. Stockholm: Svensk Kärnbränslehantering.

Hore-Lacy, I. (2007). Nuclear power plants and earthquakes. In: Cleveland, C. J. (ed.) *Encyclopedia of Earth*. Washington, DC: Environmental Information Coalition, National Council for Science and the Environment.

IAEA (1997). Volcanoes and associated topics in relation to nuclear power plant siting, Provisional Safety Standards Series 1. Vienna: International Atomic Energy Agency.

IAEA (2002). Evaluation of seismic hazards for nuclear power plants, Safety Guide NS-G-3.3. Vienna: International Atomic Energy Agency.

IAEA (2003a). Seismic evaluation of existing nuclear power plants, Safety Series Report 28. Vienna: International Atomic Energy Agency.

IAEA (2003b). Seismic design and qualification for nuclear power plants, Safety Guide NS-G-1.6. Vienna: International Atomic Energy Agency.

IAEA (2003c). External events excluding earthquakes in the design of nuclear power plants, Safety Guide NS-G-1.5. Vienna: International Atomic Energy Agency.

IAEA (2003d). Flood hazard for nuclear power plants on coastal and river sites, Safety Guide NS-G-3.5. Vienna: International Atomic Energy Agency.

IAEA (2004). Geotechnical aspects of site evaluation and foundations for nuclear power plants, Safety Guide NS-G-3.6. Vienna: International Atomic Energy Agency.

IAEA (2007). Preliminary findings and lessons learned from the 16 July 2007 earthquake at Kashiwazaki-Kariwa NPP, engineering safety review services, seismic safety expert mission report to the government of Japan, 1. Vienna: International Atomic Energy Agency.

JNSC (1981). Examination guide for aseismic design of nuclear power reactor facilities. Tokyo: Japan Nuclear Safety Commission.

Meehan, R. L. (1984). *The Atom and the Fault: Experts, Earthquakes, and Nuclear Power.* Cambridge, MA: MIT Press.

Novick, S. (1969). *The Careless Atom*. New York, NY: Dell Publishing.

NUMO (2004). Evaluating site suitability for a HLW repository site, scientific background and practical application of NUMO's siting factors, Nuclear Waste Management Organization of Japan Report NUMO-TR-04-04. Tokyo: NUMO.

Schlocker, J. and M. G. Bonilla (1964). Engineering geology of the proposed nuclear power plant on Bodega Head, Sonoma County, California, un-numbered USGS report for the AEC. Menlo Park, CA: US Geological Survey.

Sorensen, J. N., G. E. Apostolakis, T. S. Kress and D. A. Powers (1999). On the role of defense in depth in risk-informed regulation. In: *Proceedings of PSA'99, International Topical Meeting on Probabilistic Safety Assessment: Risk-informed and Performance-based Regulation in the New Millenium*, August 22–26, 1999, Washington, DC, Lagrange Park, IL: American Nuclear Society.

USNRC (2000). Use of defense in depth in risk-informing NMSS activities. Washington, DC: Advisory Committee on Reactor Safeguards, US Nuclear Regulatory Commission, http://www.nrc.gov/reading-rm/doc-collections/acrs/letters/2000/4721893.html.

Walker, J. S. (1990). Reactor at the fault: the Bodega Bay nuclear plant controversy, 1958–1964: a case study in the politics of technology. *The Pacific Historical Review*, **59**(3), 323–348.

WNA (2008). Reference Docs. WNA reactor database. London: World Nuclear Association, http://www.world-nuclear.org.

2

The nature of tectonic hazards

M. Cloos

The tectonic forces at work deep within Earth are enormous and outside the control of humans. The power of large earthquakes to devastate cities is evident from the 1906 San Francisco and 1923 Tokyo events. The 2004 mega-disaster arising from the earthquake offshore Sumatra generated tsunamis that caused destruction in areas far from where any shaking was felt. For many people, the dangers from large volcanic eruptions are even more frightening, evinced by the AD 79 eruption of Mount Vesuvius in Italy, that destroyed the cities of Pompeii and Herculaneum.

Devastating earthquakes and volcanic eruptions, infrequent on the scale of human lifetimes, are indeed commonplace tectonic phenomena over timescales of millions of years, the standard geologic time unit required to understand the evolution of Earth. But the key observation is that these kinds of events are restricted in occurrence even over multimillion-year geologic timescales (Figure 2.1). Most earthquake and volcanic activity is localized to the margins of the Pacific Ocean basin, the "Ring of Fire." Another region where large earthquakes are frequent but volcanism occurs only in scattered clusters is along the great, roughly east–west trending mountainous belt extending from the European Alps to the Himalayas. In the context of plate tectonic theory, we now understand the cause of this activity. The circum-Pacific Ring of Fire is the product of plate convergence (subduction) and plates sliding past one another (strike-slip transform faulting). The greatest mountain belt on Earth is the product of ongoing continental collision of Eurasia with Africa and India. Active deformation in these regions is indeed spectacular, but equally profound is the long-term geologic stability of most of the rest of Earth's surface. A central premise in plate tectonic theory is that nearly all of the earthquake-generating faulting and volcanic activity occurs along, and defines, the locations of plate boundaries (Figure 2.2). Vast areas of flat-lying sedimentary rocks blanket the crust and thus clearly record an absence of significant earth movements and igneous activity over timescales of tens to hundreds of millions of years. In some large areas of continental crust, the cratons (Figure 2.3), geologic stability has been maintained for more than one billion years.

Of significant importance to the selection of sites for nuclear power plants and waste repositories is the recognition that even within actively deforming areas, there are regions of the crust many tens of kilometers across that are geologically stable over timescales of many millions of years. These stable blocks are caught up in regional movements as they

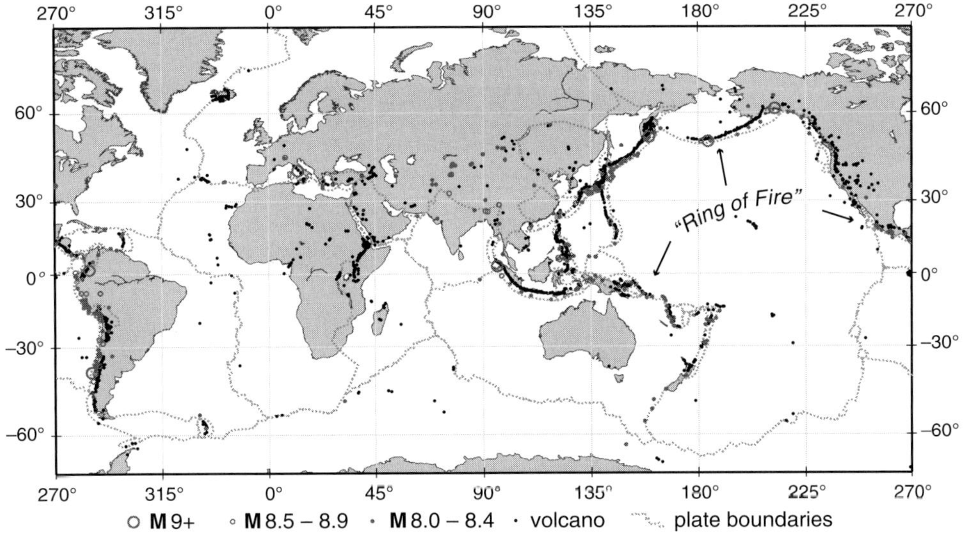

Fig. 2.1 Map showing the locations of all large (M > 8) earthquakes since AD 1900. Most of the volcanic activity and large earthquakes occur around the margins of the Pacific Ocean basin. This concentration of activity is commonly referred to as the "Ring of Fire." See color plate section.

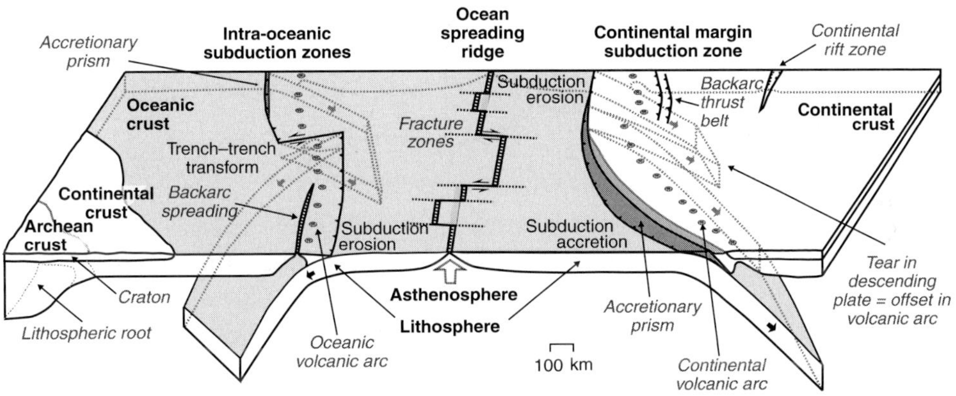

Fig. 2.2 Schematic diagram illustrating the lithosphere and asthenosphere and the three types of plate margin. Divergent plate margins occur where seafloor spreading generates new ocean-crust-capped lithosphere as the asthenosphere upwells to fill the gap left by diverging plates. Rift zones are areas of incipient divergent plate motion. Convergent plate margins occur where the process of subduction recycles oceanic lithosphere deep into Earth. Ocean trenches and arcs of explosive volcanoes are generated in the process. Transform margins occur where plates slide passed one another.

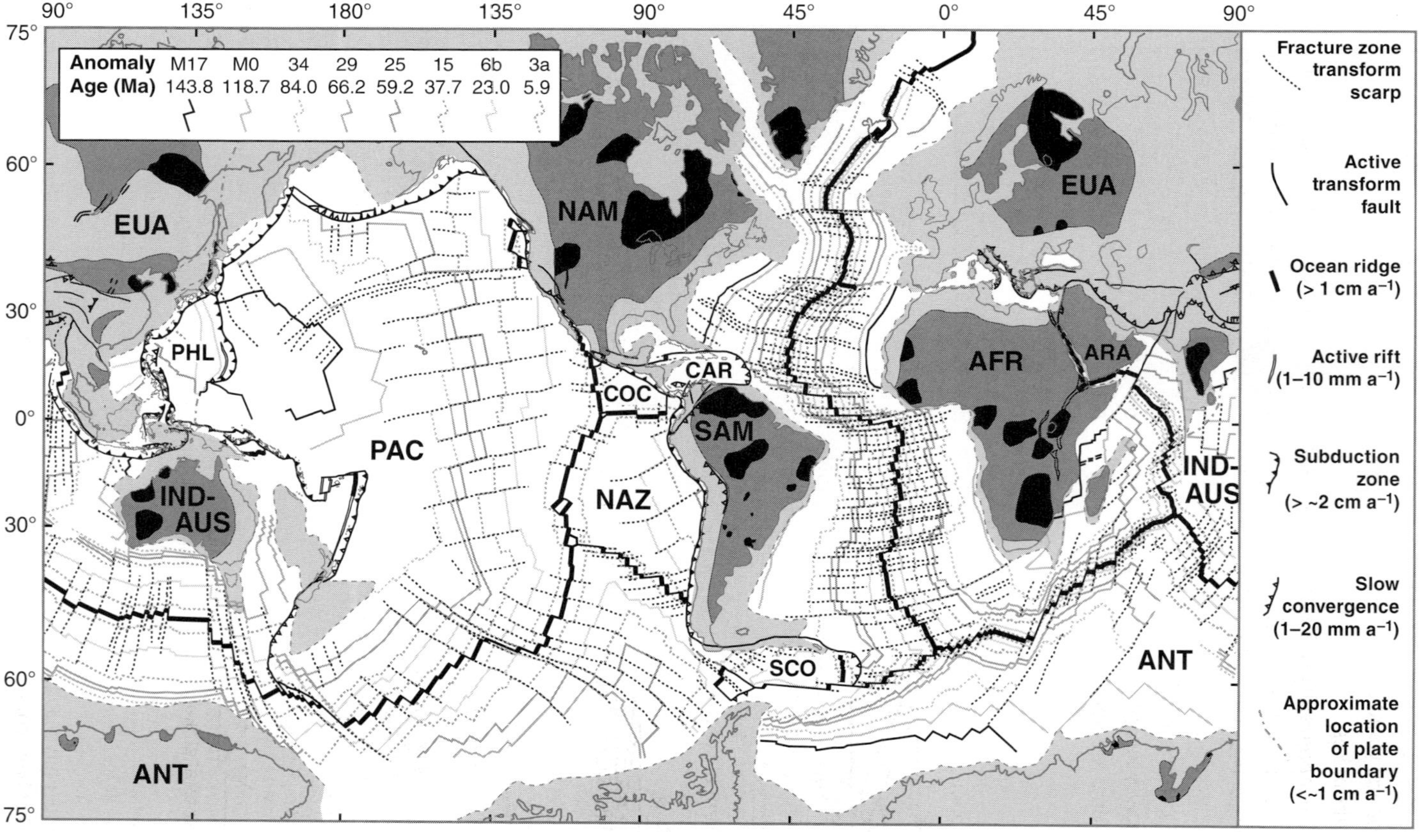

Fig. 2.3 Magnetic anomalies delineate age of ocean floor. Continental crust (light gray), Precambrian crust (dark gray), Archean crust (black). Major plates: North America (NAM), South America (SAM), Nazca (NAZ), Eurasia (EUR), India-Australian (IND-AUS), Antarctic (ANT), Africa (AFR). Minor plates: Philippine (PHL), Cocos (COC), Caribbean (CAR), Scotia (SCO), Arabia (ARA). See color plate section.

translate along and perhaps slowly warp, but for the most part their internal distortions are only temporary and minute elastic strains that come and go, as distant fault-slip events generate earthquakes.

Much of the topography of Earth's surface is produced by successive displacements along faults. Any discussion of faulting must consider the three basic classes of displacement: normal, reverse and strike-slip (Figure 2.4). Another distinguishing attribute is the dip of a fault. With respect to Earth's surface, faults inclined $> 45°$ are steeply dipping, while those inclined $< 45°$ are gently dipping. Strike-slip faults are near-horizontal offset structures that are typically steeply dipping and commonly near-vertical. Most normal-slip faults dip $\sim 60°$, but some are gently dipping structures known as detachment faults. Steeply dipping

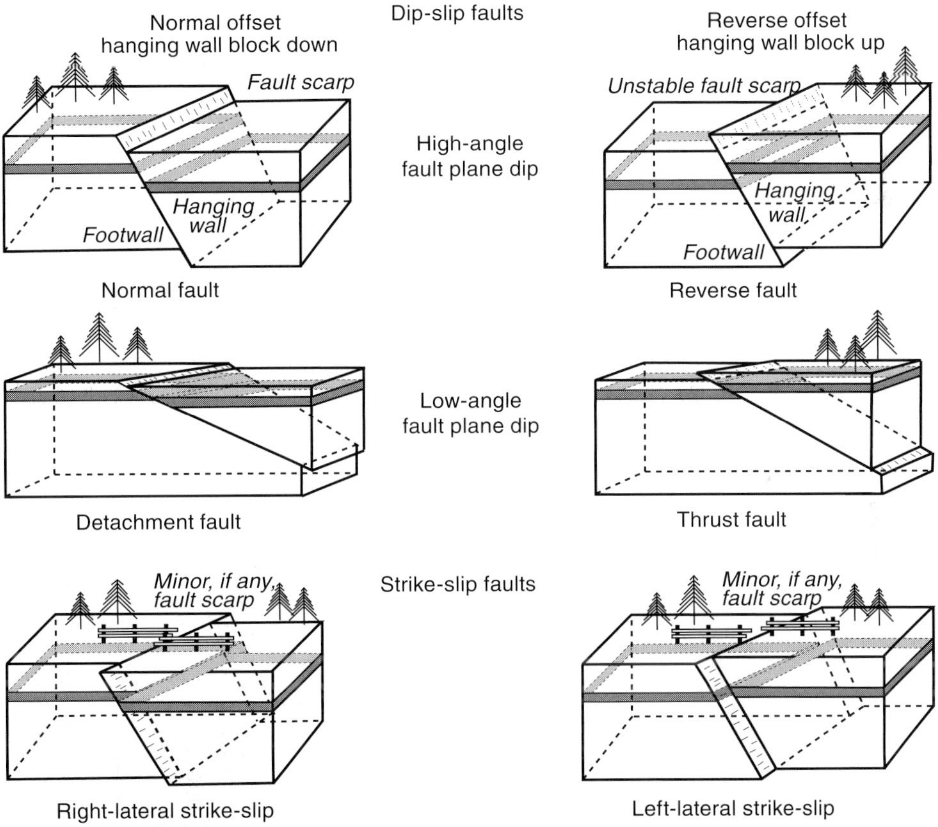

Fig. 2.4 Schematic diagram illustrating the three types of faults. Dip-slip faults have sliding movements that are largely up and down. Dip of the fault plane is a primary classification criteria with normal faults and reverse faults dipping $> 45°$ and detachment and thrust faults dipping $< 45°$. Strike-slip faults have sliding movements that are largely parallel to the surface. Many faults have movements that have components of dip-slip and strike-slip movements. Such oblique-slip faults are not illustrated.

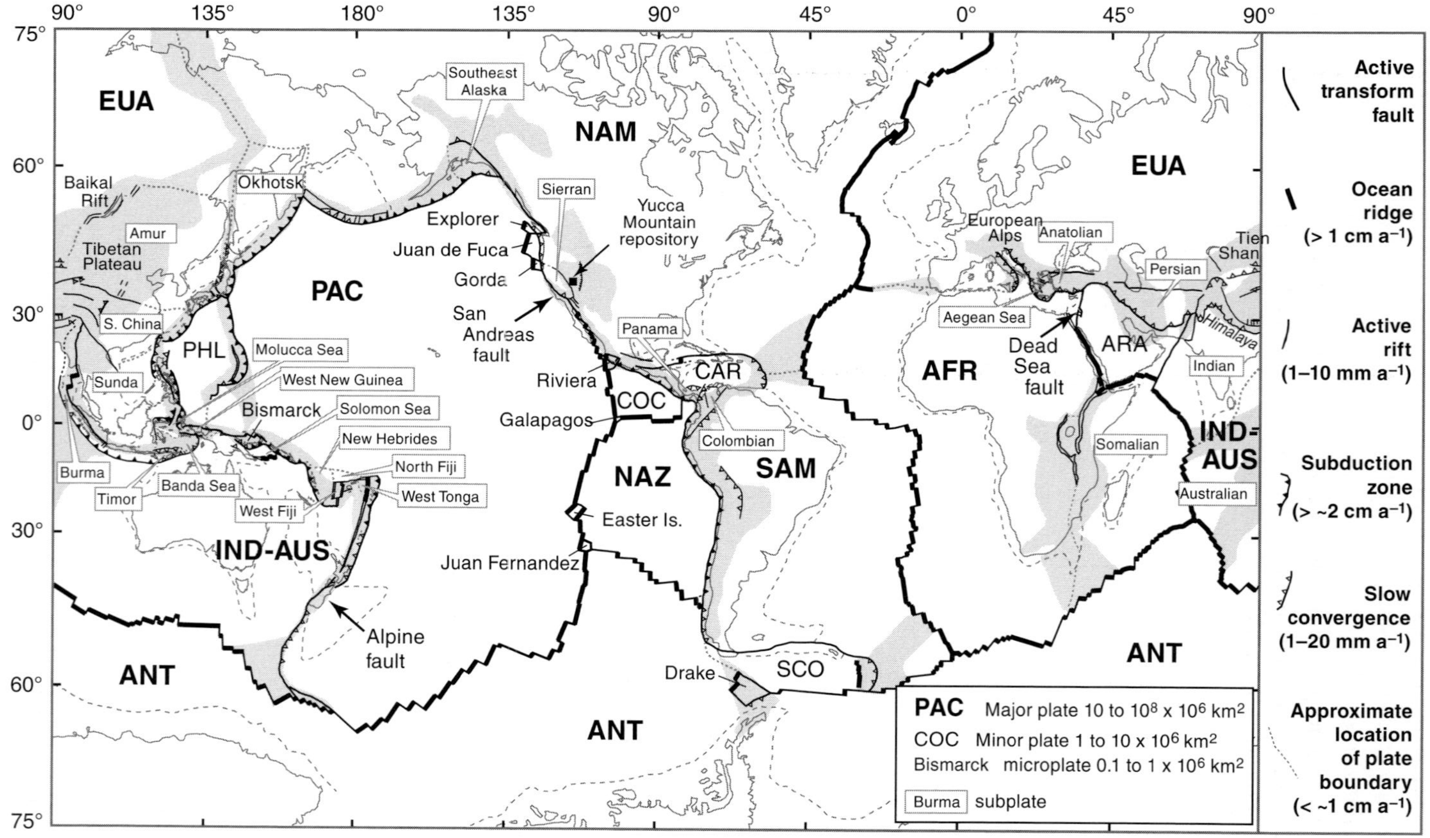

Fig. 2.5 Map showing the major and minor plates. Diffuse deformation zone limits (shaded) after Gordon and Stein (1992).

reverse faults are common, but so are gently dipping thrust faults. Movements intermediate to the vertical offsets of dip-slip faults and the horizontal offsets of strike-slip faults occur on oblique-slip faults. Dip-slip movements cause blocks of crust to move up or down. When ruptures propagate to the surface, broad regions of rock are displaced, but strain is localized near the fault surface. Where ruptures are confined to the subsurface, the strain in the overlying rock is distributed with many events, causing the slow growth of folds and other structures.

The San Andreas fault of California and the Alpine fault of New Zealand are especially well known because they are active structures and traceable on land for hundreds of kilometers (Figure 2.5). These structures are segments of the edges of the Pacific plate. In the lexicon of plate tectonics, both are transform faults: the San Andreas is the boundary with the North American plate and the Alpine is the boundary with the Australian plate. Most movement along these plate boundaries occurs episodically, with segments slipping several meters during large earthquakes. Time intervals between large-slip events are measured in many decades to millennia.

2.1 Faulting

Less appreciated is the fact that San Andreas, Alpine and other plate boundary faults are actually zones, about one kilometer across, that contain a multitude of fault splays and broken, variably altered rock. Fault zones widen over time as slip events repeat and cumulative offset increases. Since 10 Ma, cumulative offset on these plate boundaries has totaled several hundred kilometers. Major faults across which displacements of a few kilometers have occurred typically have associated zones of disruption that are $\lesssim$ 10 m across. Faults that extend for only a few kilometers are most commonly nearly planar surfaces.

During an earthquake-generating fault slip event, movement is essentially instantaneous and offset is localized along a gently curved to undulating, but overall nearly planar, surface. Scarps are ground offsets that form when fault offsets propagate to Earth's surface. One major and several subsidiary and near-parallel scarps are common. Complex fault and scarp patterns are sometimes found near fault bends or at the ends of faults.

The generalization that rock disruption is highly localized during earthquake-generating fault slip events is well documented in numerous photographs showing the effects of the 1906 San Francisco earthquake (Lawson and Reid, 1908). In this earthquake, and many others since, structures straddling the slip plane were ruptured, but buildings located just meters away remained standing. That the nature of the underlying surface materials strongly determines the amount of damage to engineered structures became well-recognized during the analysis of the destruction near San Francisco. Similar structures built on bedrock withstood the shaking much better than those built on unconsolidated materials. This was dramatically reaffirmed via engineering analysis of the selective destruction during the 1989 Loma Prieta M 7.1 earthquake in California (McNutt *et al.*, 1990), the 1995 Hanshin (Kobe) M 6.9 earthquake in southwest Japan (Schiff, 1999) and many other events. Damage is most extensive for structures constructed on weak substrates such as mud or shale, but generally minor or even undetectable in structures with foundations constructed on bedrock outcrops.

It is also important to be aware that landsliding, slumping and other forms of near-surface earth mobilization are commonplace on weathered slopes or in soft sediments during earthquakes. Ground ruptures associated with these kinds of movements are often confused as surface expressions of tectonic fault offset. In reality, very few earthquake-generating ruptures propagate to the surface.

Earthquake magnitude determines the amplitude of ground motion, but even more important for engineered structures, the larger the event the longer the timespan of shaking. Earthquakes are waves of elastic distortion that move away from a nucleation point known as the focus (or hypocenter). The epicenter is the point on Earth's surface directly above the focus. Large earthquakes ($M > 6.5$) originate at depths of 10–25 km. When these waves, both compressional and shear type, reach the surface, a large amount of their energy is converted to a type of wave that propagates along the surface. Seismic surface waves are much like those formed by a disturbance of the surface of a lake. Their passage causes most of the damage to man-made structures as well as triggering slumping, sliding and other movements in unconsolidated surface materials. Far less appreciated is the fact that the effects of shaking dissipate rapidly with depth from Earth's surface.

For some investigators, the most spectacular confirmation of the profound downward attenuation of shaking comes from the disastrous 1975 Tangshan earthquake in China (Huixian *et al.*, 2002). This earthquake, a M 7.8 event, occurred directly beneath the industrial city of Tangshan and killed at least 250 000 people. The Kailuan coal mines are directly beneath the city where more than 85% of the surface structures collapsed or had to be destroyed following the earthquake. When this cataclysm occurred, more than 10 000 miners were scattered throughout a vast network of excavated passages extending to depths of $\sim$ 1000 m. Nearly all of the underground damage in the coal mines that was directly due to the earthquake occurred along the upper 20 m of the entrance shafts (Huixian *et al.*, 2002). Very few of the working miners were even aware that a major earthquake-generating fault slip event had occurred beneath them and that the resultant surface waves were collapsing their homes and other buildings.

The record of the Tangshan earthquake vividly demonstrates that subsurface elastic displacements, even very near the nucleation sites of large earthquakes, are very small and transient. Slow earth movements over decades to millennia locally build up elastic strain in volumes of rock that can be quickly released when part of the strained rock mass becomes loaded to the point of failure. The rupture of intact masses of rock obviously occurs, but this is actually a very rare event for locations far from active fault zones. In fact, almost every historic surface rupture has occurred along pre-existing faults with demonstrable evidence of prior movement in the Quaternary (Yeats *et al.*, 1997).

Repeated earthquake-generating fault slip events, the seismic cycle, involve a period of slow elastic strain build-up that culminates with rapid elastic energy release. The volume of elastically strained rock and thus the magnitude of resultant earthquakes depends upon the rigidity of the rock in the region and the strength of the weakest anisotropies such as old faults. Even very near the slipped fault plane, the slow build-up of elastic strain and rapid unstraining leaves the rock intact. At distant points, elastic distortions associated with the passage of earthquake waves quickly come and go. Crystalline rock is unchanged, but episodes of shaking can cause compaction and even liquefaction of overlying layers of porous sedimentary strata. From the point of view of hazard assessment for many types of nuclear facilities, the faulting/earthquake problem centers on avoiding locations that become ruptured by faults during the timescale of concern.

The preceding discussion alludes to an important fact of the Earth's behavior. Faults with hundreds of meters of offset are the result of many movement events. Fault slip events of 10 m are extraordinary and only occur during great earthquakes. The largest documented historic offset was the 1855 rupture of the Wairarapa fault in New Zealand. This $M > 8.1$ event caused a surface displacement as great as 18 m (Rodgers and Little, 2006). The largest recorded earthquake, the titanic Chilean M 9.5 subduction-zone event of 1960, generated $\sim 25\%$ of all the seismic energy released in the years 1900–2000 (Scholz, 2002). The depth of earthquake nucleation was about 30 km and the average subsurface fault slip was 20–30 m (Cifuentes, 1989). There is no evidence that there has ever been a larger tectonic earthquake.

2.2 Timescale of concern

This chapter is primarily concerned with the tectonic processes that give rise to faulting and earthquakes. The rates of plate movement, and rates of faulting, are of primary concern. It is worthwhile comparing these rates of tectonic activity with the expected period of performance of nuclear facilities. For surface facilities, such as nuclear power plants, tectonic hazards are primarily of concern from the perspective of rare large-magnitude earthquakes, topics addressed by Inoue (Chapter 21, this volume), Stirling *et al.* (Chapter 10, this volume) and others. For geological high-level waste (HLW) repositories, the performance periods are much longer and the potential for faulting in the immediate area of the site is of primary concern. Many examples used in this chapter consider the potential for faulting in the context of HLW repositories.

For HLW repositories, the timescale of concern depends on whether waste is stored as canisters of spent fuel or reprocessed waste solidified as borosilcate glass. The radioactivity of spent-fuel pellets is primarily due to uranium. Without dilution, spent fuel attains a level of radioactivity similar to common uranium ore bodies over timespans of 100–300 ka (Chapman *et al.*, Chapter 1, this volume). The radioactivity of reprocessed and vitrified waste arises largely from the concentration of short-lived ^{90}Sr and ^{137}Cs and depends strongly upon dilution factors in making the glass. The level of radioactivity in vitrified waste will become comparable to that of typical uranium ore bodies much earlier, over timespans < 15 ka and perhaps as short as 3 ka (Cohen, 1989; Chapman, 2006).

Debate also remains about the timescales necessary for ensuring complete isolation of the waste. The comparison of waste radioactivity with that of uranium ore bodies is a comparison rooted in concern for the biosphere. Most uranium ore bodies are so shallow that they are extracted with surface mining methods (Plant *et al.*, 1999). In fact, roll-front-type uranium deposits are found in rock units that are so shallow and permeable that they are, or were, aquifers. Over geologic time, the biosphere has obviously co-existed with the erosion and redistribution of many uranium-rich rock masses.

There is general agreement that 10 ka is the minimal time requirement for isolation of radioactive waste. After 10 ka, the radioactivity of reprocessed HLW is actually less than that of typical ore bodies that are mined to produce nuclear fuel. In the case of spent fuel, the radioactivity has reduced to $\sim$ 50 times that of typical uranium ore. After 100 ka, the residual radioactivity of spent fuel is comparable to uranium-rich ore bodies. A timescale of 100 ka has become a common standard for agencies regulating HLW repositories, whether designed for spent fuel or reprocessed material. After 100 ka, the radiological hazard of the waste is comparable to that of naturally occurring geologic materials, but unnatural concentrations of radioactive isotopes would be easily detectable with advanced instrumentation.

To protect the biosphere, the geologic setting is selected and the engineered barriers are designed to contain the waste. Understanding the kinds and amounts of earth movements that could disrupt these barriers over timescales of 10–100 ka requires an understanding of plate tectonic processes.

2.3 Plate tectonic movements

Earth remains a dynamic planet because a significant amount of heat is still steadily produced within it from the decay of radioactive isotopes of uranium, thorium and potassium. Rock outcroppings reveal abundant evidence of crustal deformation driven by the slow release of this heat. Geochronological studies, based upon the analysis of natural radioactive decay locked into crystals (e.g. parent–daughter isotope ratios of U to Pb, K to Ar) date many geologic events and provide our firm understanding of the timescales of tectonic change (Dalyrmple, 1991). With the Global Positioning System (GPS), we now have the ability to measure precisely how the surface of the Earth is moving today (Wallace *et al.*, Chapter 6, this volume). These measurements, combined with an understanding of the local geologic history for the past few million years and a region's plate tectonic setting, enable confident prediction of what permanent movements, if any, are likely over the next 100 ka.

One of the great revolutions in our thinking about how Earth works is the now well-tested theory of plate tectonics. While the concept of continental displacement has been speculated upon by some people since the first maps of the coastlines of the Atlantic ocean were created, the notion that the continents drifted about and collided forming mountain belts was most clearly expounded upon by Alfred Wegener in 1912 (Marvin, 1973). His research culminated in a 1929 book that is now celebrated, but the concept was generally dismissed as implausible until the 1950s. The widespread change in viewpoint first arose from studies in rock magnetism that indicated continental blocks had rotated and translated with respect to one another. But it was primarily the discovery that the geology of the ocean floor was fundamentally different from that of the continents that led to the scientific revolution of plate tectonics. Oceanographic studies following World War II discovered that the sea covers a bifurcating, globally encircling mountain system, the 66 000 km-long network of ocean ridges. In the 1960s, it was realized that almost all of the deep ocean floor is probably much younger than almost all the rock that forms the basement of the continents (Figure 2.3).

The theory of plate tectonics arose in the late 1960s. It is rooted in these new discoveries about the geology of the ocean floor and a greatly improved ability to locate earthquake sources that arose from the need to detect nuclear explosions (Isacks *et al.*, 1968). The key revelation was that horizontal movements of the surface of the Earth dwarf the vertical mountain-generating movements.

In the lexicon of plate tectonics, the outermost part of Earth is a cool, strong zone, known as the lithosphere, that is broken into pieces known as plates (Figure 2.3). The plates are underlain by a hot, weak zone, the asthenosphere, that readily flows in response to lithospheric movements and internal variations in temperature or composition. Nearly all of the world's earthquake and volcanic activity occurs at plate margins (Isacks *et al.*, 1968). The ocean ridges form at sites of plate divergence (via "seafloor spreading"), and ocean trenches and nearby lines or arcs of volcanoes form at sites of plate convergence (via "subduction"). Ocean-crust-capped plate is steadily created by seafloor spreading and an equivalent amount of plate, ranging from very young to $\sim$ 200 Ma, is consumed by subduction. There is a third kind of plate boundary, the transform margin, along which plates are neither created or destroyed (Wilson, 1965). At this type of margin, plates slide past one another with movement largely localized along strike-slip faults.

Basaltic volcanism along fissures and normal-type faulting dominate the tectonic activity at divergent plate margins. The process of seafloor spreading is comparable to the movement of two diverging conveyor belts. As the plates separate, the hot asthenosphere upwells, decompresses and partly melts. The upwelling movements continuously generate mafic magma which rises to solidify as a layer of basalt and gabbro, the ocean crust, that is almost everywhere between 6 and 8 km thick.

One still poorly understood aspect of Earth behavior is that the magnetic field episodically reverses polarity. The time between reversals varies from thousands to several tens of millions of years. Evidence of the irregular oscillations in magnetic field polarity is best recorded in the magnetism locked into the basaltic layers that cap ocean crust when it forms by spreading at ocean ridges. Where the polarity frozen in the crust is parallel to the present field, the local intensity is additive. Where the magnetic polarity frozen in the crust is opposite to the present field, the local intensity is subtractive. Variations in the local strength of the magnetic field are detected with sensitive magnetometers and form a pattern of "magnetic stripes" parallel to the formative ocean ridges (Figure 2.3). The magnetic stripes are offset along fracture zones that connect ridge segments almost everywhere at 90° angles. The spreading history of the ocean floor for the past 100 Ma is well recorded in the pattern of magnetic stripes and transform faults in the seafloor. From the age–distance relationships, it was well established that plate forming and destroying tectonic motions steadily occur at speeds typically averaging in the range of 1–10 $\mathrm{cm\,a^{-1}}$.

Explosive andesitic volcanism and thrust-type earthquakes are the most obvious forms of tectonic activity at convergent plate margins (Figure 2.1). The process of subduction is the descent of ocean-crust-capped lithosphere into the mantle. This creates deep ocean trenches and volcanic arcs. Subduction zones, with an aggregate length of $\sim$ 38 000 km, are readily located from the position of the inclined seismic zones that were discovered in the 1920s by the Japanese seismologist K. Wadati. The worldwide distribution and character of these inclined seismic zones was established in the 1950s by H. Benioff.

Wadati–Benioff seismic zones extend to depths as great as 700 km. Earthquakes can occur at great depth because the old, cold subducting plate heats slowly. The deepest earthquakes occur beneath the Tonga islands where the plate is steeply dipping and descending at speeds of $\sim$ 8 cm a^{-1}. Thrust-type fault offsets are common along the interface with the bottom of the overriding plate. Why some subduction-zone segments have had enormous M 9 earthquakes (e.g. Chile, Alaska, Sumatra) and others have had no events of $M > 7.5$ (e.g. Mariana) remains a matter of great scientific debate. Subduction drags a thin layer of water-rich sediments into the mantle and substantial melting occurs at depths of $\sim$ 100 km. These water-rich magmas rise to the surface to form the lines, or arcs, of explosive composite volcanoes. They are typically spaced 50–150 km apart and commonly tower as much as 3 km above the surrounding countryside.

Transform plate boundaries are conservative overall. Neither large-scale plate creation or destruction occurs, but most transform boundaries are sites of either slightly convergent or slightly divergent strike-slip faulting. In addition, long transforms, especially those cutting continental crust (e.g. San Andreas, Alpine and Dead Sea faults; Figure 2.5), are only locally perfectly oriented to accommodate plate motions. Complex movements locally occur at bends, jogs and splays. Some bends are divergent with pull-apart motions forming basins that can be associated with igneous activity. Other bends are convergent causing blocks of rock to pop up, forming mountains. Consequently, while transform margins are regions dominated by strike-slip faulting, segments of long transform margins may be zones with abundant normal- or reverse-type faulting. As the locus of movement can change over time, blocks of rock caught within transform plate boundary zones can have exceedingly complex geologic histories.

2.3.1 Major tests support the theory

Specific predictions based upon plate tectonic theory were subjected to several tests in the late 1960s that were rigorous and definitive. By 1970, acceptance of the theory by geoscientists was so widespread that historians consider it a scientific revolution (Glen, 1982; Le Grand, 1988). Specific predictions concerned the age of the ocean floor as a function of position with respect to spreading ridges and the location and type of earthquakes. These predictions have been tested and confirmed at many locations.

Starting in 1968, the deep-sea drilling ship, the *Glomar Challenger*, was used to obtain many hundreds of cores from the ocean basins that confirmed the age of the underlying ocean crust as predicted from the patterns of magnetic stripes (Figure 2.3). The other diagnostic test arose from the fact that since 1966 it has been possible to determine remotely the type and direction of fault movement that caused earthquakes from the analysis of seismic wave arrival patterns. This work, known as focal mechanism analysis, showed that earthquakes along ocean ridges were normal fault movements and those near trenches were almost entirely thrust fault movements (Isacks *et al.*, 1968). The motion along transform faults are strike-slip displacements, which are either right- or left-lateral, depending upon how they connect two spreading ridge segments. The analysis of earthquake-generating strike-slip along transform faults connecting ocean ridge segments was a diagnostic test that confirmed a major prediction of plate tectonic theory (Sykes, 1967).

In the 1970s, Earth science textbooks were rewritten from the viewpoint that the large-scale surface features of our planet are primarily shaped by horizontal motions. Rift zones in the continents, such as East Africa, were recognized as incipient stages of ocean basin formation. The great mountains of the European Alps, Himalayas, and islands such as New Guinea and Taiwan were recognized as the sites of collisional tectonism due to the underthrusting and jamming of subduction zones by the edges of continents.

2.3.2 Hot spots – an addition to the theory

There is one distinctive anomaly to the grand scheme of plate tectonics: volcanism that occurs in the otherwise stable interior of a plate (Morgan, 1972; Perfit and Davidson, 2000). One of the best examples is Hawaii and the associated Hawaiian-Emperor chain of seamounts (Wilson, 1963). Other examples include the Galapagos, Canary and Réunion island groups. The Yellowstone area in North America and the Afar region of Africa (which has evolved into a site of continental rifting) are other examples of magmatic activity affecting the interior of the continents. These sites of anomalous intraplate magmatism are known as "hot spots." About 15 hotspots are recognized as presently active. Most workers believe they form above upwellings from very deep in the mantle that are active for tens of millions of years (Morgan, 1972). Where hot-spot magmas reach the surface, a chain or ridge of volcanoes is created. The Hawaiian hot spot is the type example and the Hawaiian-Emperor seamount chain records the movement of the Pacific plate. This 5500 km-long volcanic chain formed over the past 80 Ma and directly records horizontal plate movements at speeds of $\sim 7\,\mathrm{cm\,a^{-1}}$.

2.3.3 Rates of plate movement

When one speaks of tectonic movements at plate boundaries, one is usually referring to relative movements occurring at speeds of $\sim 1{-}10\,\mathrm{cm\,a^{-1}}$. Currently, the fastest relative plate movement is the speed of seafloor spreading along the central segment of the East Pacific Rise near 30° S where the Nazca plate is separating from the Pacific plate at velocities as fast as $15\,\mathrm{cm\,a^{-1}}$ (Figure 2.6). Relative movements everywhere else are occurring at speeds $\lesssim 10\,\mathrm{cm\,a^{-1}}$. "Fast" plate movements are those with relative motions $\gtrsim 5\,\mathrm{cm\,a^{-1}}$. Over 1 Ma, $5\,\mathrm{cm\,a^{-1}}$ movements correspond to displacements of 50 km. Over the 100 ka timescale of concern for HLW repositories, this corresponds to displacements of ~ 5 km.

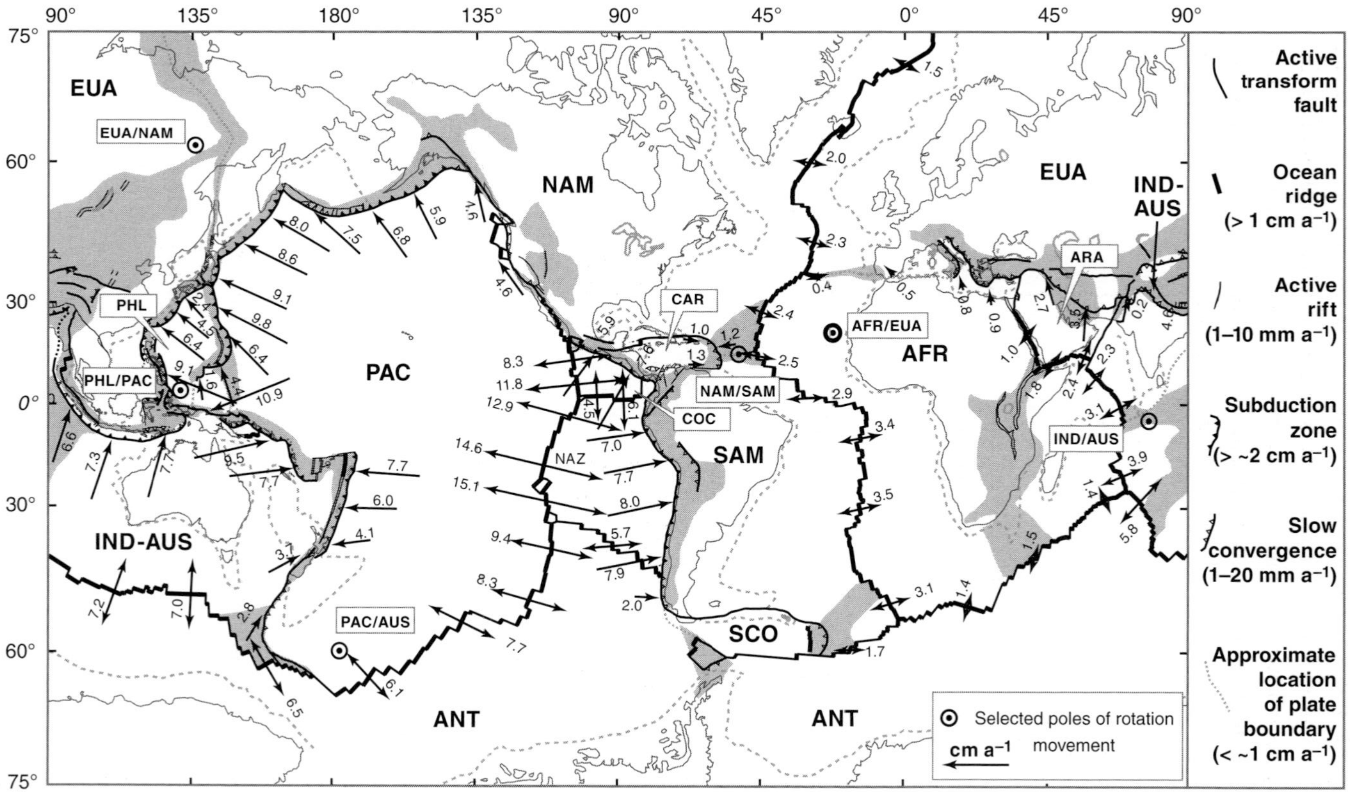

Fig. 2.6 Map showing the relative plate motions at cm a^{-1} speeds at divergent and convergent plate margins. Where movements are $\gtrsim 2$ cm a^{-1}, a plate boundary location is well defined because earthquakes are numerous. Relative plate tectonic movements are very slow near poles of rotation, the pivoting point on the Earth surface for the movement between two plates. Where two plates are juxtaposed at their pole of rotation, the local motions approach zero and the plate boundary location is not well defined. Examples include the boundaries between the Eurasian and North American plates and the North and South American plates. Diffuse deformation zones are shaded. From the perspective of crustal stability, it is very important to recognize that about 80% of Earth's surface is not detectably distorting. Plate abbreviations are the same as in Figure 2.3. Relative velocities and selected poles of rotation are taken from the NUVEL-1A model (DeMets *et al.*, 1994). Diffuse deformation zone boundaries are after Gordon and Stein (1992).

Where the relative movements are $> 2\,\mathrm{cm\,a^{-1}}$, plate boundaries are clearly evident in the local geology as areas of uplift, subsidence or strike-slip faulting and generally locatable to within a distance of 10 km, and commonly much less (e.g. the central segment of the San Andreas fault). However, where relative movements are slower, locating the plate boundary is generally difficult because earthquakes are rare and the surface effects of deformation are greatly muted, especially where they are distributed across a broad region. Over the 100 ka timescale of concern, relative tectonic movements at speeds of $1\,\mathrm{cm\,a^{-1}}$ correspond to cumulative offsets of only 1 km. Horizontal displacements averaging $< 1\,\mathrm{mm\,a^{-1}}$ generally only leave a subtle imprint on the near-surface rock record.

The analysis of plate movements from magnetic anomalies and associated transform faults, complemented by the analysis of hot-spot volcanic chains, led to the generation of "global plate motion models." Because the Earth is essentially spherical, the movements between any two points on the surface can be quantified in terms of a pole of rotation and an angular velocity (McKenzie and Parker, 1967). Locating the poles of rotation for the movement between major plates separated by spreading transform networks is a straightforward exercise (Cox and Hart, 1986). Lines drawn perpendicular to transform faults bounding a pair of plates intersect at their pole of rotation. A property of tectonics on a sphere is that relative motions are fastest at locations 90° from the pole of rotation and zero at the pole.

Most poles of rotation are located far from the relevant plate boundaries. However, where a pole of rotation is located near the corresponding plate boundaries, the plate boundary location is usually not clearly evident in the local geology because movements are very slow. The boundaries between the North American and South American or North American and Eurasian plates illustrate this problem (Figure 2.6). The pole of rotation between the American plates is located in the central Atlantic. The pole of rotation between the North American and Eurasian plates is located in Siberia (northeast Russia). Similarly slow movements near poles of rotation obscure the boundary location between the African/European and Pacific/Australian plates.

Plates are kinematically distinct entities that can be divided, based on size, into major ($> 10^7\,\mathrm{km^2}$), minor (10^6–$10^7\,\mathrm{km^2}$) and microplates ($< 10^6\,\mathrm{km^2}$) (Figure 2.5). About 95% of Earth's surface can be assigned to one of the eight major plates. About 4% of Earth's surface can be assigned to one of five minor plates. Microplates are commonly mentioned as important tectonic entities, but they constitute $< 1\%$ of Earth's surface. Nearly all of the faulting and volcanic activity is concentrated along, or very near, major, minor or microplate boundaries.

The position of plate boundaries are almost always evident where differential movements between adjacent plates are $> 1\,\mathrm{cm\,a^{-1}}$. Complications arise because several of the major plates are torn or breaking apart, forming subplates (Figure 2.5). Subplates are sections of lithosphere that are moving with a velocity that is not quite the same as the parent plate. They are kinematically distinct fragments: tears, splinters or flakes, but are still connected to the parent plate. Velocity differences between a subplate and the parent plate range from zero at the tip of a tear to speeds as fast as $10\,\mathrm{mm\,a^{-1}}$. Large subplates are found in East

Africa and East Asia. The East African rift system is the line of active divergence where the Somalian subplate slowly separates from the rest of the African plate. Eastern Asia is divisible into at least two major subplate regions (East Asia and Southeast Asia, which are commonly divided into even smaller subplates), because the great Himalaya-forming collision has caused a broad region of distributed deformation to develop in the overriding Asian plate.

The India–Australian plate is a unique case because it is capped at both ends with colliding continental blocks. The western end of this plate is involved with the Himalaya-forming collision and the eastern end is involved in collision with an oceanic arc, forming the island of New Guinea. Collisional tectonism at both ends is causing the middle of the plate, the center of the Indian Ocean, slowly to warp.

Subplate movements of 1 mm a^{-1} cause horizontal displacements of 1 km in 1 Ma. During the 100 ka timescale of concern, only 100 m of displacement will have occurred. The corresponding vertical displacements (at fractions of 1 mm a^{-1}) in areas of convergence or divergence are comparable to rates of erosion and deposition. Consequently, rising blocks of crust can be beveled by erosion and subsiding lows can be filled in by sedimentation as fast as the movements are occurring. Because such slow displacements are also widespread, many potential HLW repository sites may be near areas where differential tectonic movements are occurring at speeds of a few millimeters per year.

2.3.4 Global plate motion models

A best-fit of all magnetic and fault observations is necessary to generate a global plate motion model. This kind of analysis was first done by two independent workers, Le Pichon (1968) and Morgan (1968), with landmark papers published in 1968. During the 1970s, magnetic surveys in remote parts of the ocean led to important refinements. In 1978, Minster and Jordan (1978) published a global model that became the standard for the calculation of local plate motion movement rates. This best-fit global plate motion model was based upon the seafloor spreading and transform-fault patterns averaged over the past 3 Ma.

With the Minster and Jordan model, the local motions at any position along a plate boundary could be calculated with a precision reported to a millimeter per year. With this, the magnitude of tectonic movements recorded in the amount of fault offset and folding as quantified in map and geologic cross-sections over the past few million years could be compared to movements predicted by global plate motion model calculation. For example, motion along the San Andreas transform fault separating the Pacific and North American plates near San Francisco was calculated as 59 mm a^{-1} along central California. An anomaly was soon discovered when cross-fault trenching studies in southern California indicated offsets along the San Andreas fault over the past few thousand years are $\sim$ 34 mm a^{-1}, substantially less than predicted (Sieh, 1978; Sieh *et al.*, 1989). Thus, it became evident that a significant fraction of the movement between the Pacific and North American plates is being accommodated at locations east and west of the San Andreas fault.

2.3.5 Direct measurement of plate motion

The direct measurement of plate tectonic motions became possible in the 1980s. This required sophisticated analysis of the relative position between pairs of radio-telescopes based upon signals from quasars, very long baseline interferometry (VLBI) and by determining positions on Earth by measuring the time it takes for laser beams to reflect off special reflectors on satellites, satellite laser ranging (SLR). The VLBI and SLR studies by the late 1980s confirmed that points on Earth's surface were moving with respect to one another at the centimeter-per-year rates predicted by the Minister and Jordan motion models (Carter and Robertson, 1986; Smith *et al.*, 1990). The remarkable agreement of plate movement rates based upon a decade or so of telescopic and satellite measurements with the global plate motion model calculations based upon best-fit geological relationships (magnetic anomalies and transform fault patterns) generated over the past 3 Ma was taken by many workers as definitive proof of plate tectonic theory (Gordon and Stein, 1992). However, the number and locations of these kinds of measurements are inherently limited. Radio-telescopes are enormous installations and few exist (Herring *et al.*, 1986). Satellite laser ranging utilizes mobile units, but measurements are time intensive and the number of benchmarks that can be resurveyed over time in this manner are limited (Christodoulidis *et al.*, 1985).

The current standard global plate motion model was published by DeMets *et al.* in 1990. A revision in 1994 arose from a small recalibration of the magnetic reversal timescale. The revised model (NUVEL-1A) is the current standard of comparison. In making this model, careful consideration was given to broad regions with evidence of deformation at scattered locations. From the recognition of such regions, Gordon and Stein (1992) expounded the concept that some plate boundary regions and a few plate interior areas are "diffuse deformation zones." They concluded that $\sim$ 15% of Earth's surface consists of regions across which scattered occurrences of faulting, volcanism and crustal warping of some kind are ongoing (Figure 2.5). But more importantly, this analysis confirmed a central tenet of plate tectonic theory, $\sim$ 85% of Earth's surface is simply translating, with negligible internal distortion. For the most part, movements along and near the plate edges only generate temporary elastic distortions that are episodically released during earthquake-generating, "stick-slip" faulting events.

Plate interiors are commonly described as rigid, but this is not the best perspective to take because they do deform. Large parts of plates are not permanently deforming laterally because the push and pull forces that are sufficient to drive plate motion are less than the yield strength of the lithospheric mantle underlying the crust. However, slow vertical movements, known as isostatic adjustments, occur in response to erosion or deposition at the surface or buoyant forces exerted from below by flow in the asthenosphere. Between 6 and 20 ka, the melting of the great ice sheets caused vast areas of North America and Europe to rebound as much as 200 m. This has been called "nature's great experiment" because it directly indicates that isostatic adjustments due to crustal unloading from erosion or loading from deposition must be steady and rapid compared with rates of plate motion.

2.3.6 Diffuse deformation zones

The concept of a diffuse deformation zone provides a framework to classify regions across which slow and permanent tectonic movements are accommodated (Stein and Sella, 2002). The leading edges of the overriding plate at subduction zones and regions near long transform faults, such as the San Andreas and Alpine faults, are classified as diffuse deformation zones (Figure 2.5). Areas near subplate boundary zones are also regions of diffuse deformation.

Subduction zones between the trench axis and somewhere in the backarc, typically distances of 150–350 km wide, are regions of scattered tectonic activity (Figure 2.6). However, nearly all of the deformation from plate convergence is localized to very near the base of the trench slope and from there along a gently curving shear zone that extends to the depths of arc magmagenesis (Shreve and Cloos, 1986; Byrne *et al.*, 1988). The region near and behind the volcanic arc is commonly a site of episodic extension forming backarc basins (e.g. behind the Mariana arc) or episodic shortening forming backarc thrust-and-fold belts (e.g. central Andes). The episodes of backarc extension or shortening are slow compared to the local speed of subduction. Such episodes of deformation occur at speeds as fast as $\sim 20\,\mathrm{mm\,a^{-1}}$ but are generally much slower. Deformation is localized along the line of the volcanic arc because repeated intrusion of magma heats the region and makes it a zone of lithospheric weakness. The forearc block, on the other hand, is strengthened to some extent because it is cooled from below by the underflow of the descending plate.

The region forming the mountain belt extending from the European Alps to the Himalayas is a diffuse deformation zone across which shortening from continent–continent collision is ongoing (Figure 2.6). To the north and east, Asia can be divided into subplates that move slightly differently than the rest of Eurasia. The Tibetan plateau is largely a broad region of near-vertical uplift. Much of east China and southeast Asia is a region with many strike-slip faults and local zones of extension (e.g. Baikal rift) and shortening (e.g. Tien Shan). Parts of western North America and the belt that includes the East African rift are regions across which extension is occurring. In all of these regions, the global plate motion model of DeMets *et al.* (1994) enables precise quantification of the relative movements occurring somewhere between the boundaries of the diffuse deformation zones.

If a HLW repository were to be located in a diffuse deformation zone, then understanding how movements are distributed across the zone becomes a matter of concern. How the movements manifest themselves depends upon the nature of the local geology, specifically the distribution of zones of weakness in the igneous and metamorphic rocks that make up the crystalline crust. In eastern Asia and western North America, it is clear that there are several major fault zones across which most of the movement is accommodated over time in a step-like fashion. In the central Andes and parts of the Himalayas, folds are growing and near-surface shortening deformation is more evenly distributed across a broad region.

The key point is that the global plate motion models provide a rigorously quantitative framework to understand local geologic histories. The direct quantification of relative tectonic movements with millimeter-per-year precision for dozens of sites was a theoretician's dream in the 1980s. The dream became reality with the deployment of the GPS in the 1990s.

And with this, the direct quantification of local tectonic movements with millimeter-per-year precision became possible.

2.4 The global positioning system (GPS)

The measurement of ground motions associated with active faulting dates back to the analysis of benchmarks around San Francisco Bay after the great 1906 earthquake. The comparison of triangulation benchmark locations after the earthquake with several surveys completed before the event revealed that there were precursory movements far from the San Andreas fault. From this, it became apparent that slow and distant movements caused a build-up of elastic strain along a locked portion of the San Andreas fault. This is the key concept behind the elastic rebound theory developed by Reid (1910). Previously, most people believed earthquake energy was locally generated as in explosive detonations.

Triangulation is based upon precise measurement of angles, and errors can be as small as $\sim$ 10 ppm. In the 1970s, electronic distance measuring equipment was developed that had a precision of $\sim$ 1 ppm for line-of-sight measurements. An incredible advance came with the development of the GPS in the 1980s. By 1993, the system was fully operational with a constellation of 24 satellites equipped with transmitters sending positional and timing signals from atomic clocks. With receivers on the ground, high-precision geodetic measurements over distances of thousands of kilometers became possible. In 1991, continuous GPS measurement of relative positions began at several dozen worldwide benchmarks. The horizontal distances between stations that are out of sight of one another could be measured with errors that were proven to be on the order of a few parts per billion (Mao *et al.*, 1999). This means that over a timespan of a few years, relative horizontal positions along baseline distances of 1000+ km can be determined with a precision of a few millimeters. The precision of positional measurements increases as data are continuously acquired over longer periods of time. Note that a line length change divided by original line length is the standard measure of strain. A strain of one part per billion is a "nanostrain." By the mid 1990s, the relative position changes between the major plates were directly measured with uncertainties in horizontal position of a few millimeters per year (Larson *et al.*, 1997; Segall and Davis, 1997).

Transient elastic strains, albeit small, can be transmitted far from plate boundaries. This is spectacularly evident in the case of the great Sumatran M 9.2 earthquake on December 26, 2004 (Lay *et al.*, 2005). The movement of benchmarks (Vigny *et al.*, 2005; only detectable with GPS geodetic surveying) revealed that coseismic elastic displacements of $\sim$ 10 cm were detected $\sim$ 400 km from the epicenter. Remarkably, displacements of $\sim$ 1 mm were detected $\sim$ 3000 km from the epicenter. Differentiating earth movements that are due to temporary elastic strains from those that are permanent tectonic strains, is a rapidly developing area of research with potential application to earthquake prediction.

Most earth movements now studied with GPS surveying are tectonically driven, but the sensitivity is great and small deflections due to seasonal changes in a region's water load (liquid or snow) have proven to be detectable (Van Dam *et al.*, 2001; Elósegui *et al.*, 2003).

Even the daily vertical movements due to Earth tides have been detected (Watson *et al.*, 2006). With careful data analysis from clusters of receivers that are built on stable foundations, GPS surveys appear capable of directly measuring relative horizontal displacements as slow as 0.2 mm a^{-1} (Davis *et al.*, 2003). Movements at these rates could cause as much as 2 m of strike-slip fault offset every 10 ka. The magnitude of corresponding dip-slip movements is dependent upon fault orientation and dip (Figure 2.4). Where movements along faults are convergent, uplift will occur. Where movements along faults are divergent, subsidence will occur.

Global positioning system receivers have modest cost and thousands of receivers are now deployed in networks about the surface of Earth. They enable direct and continuous monitoring of Earth movements. The most advanced network, the GPS Earth Observation Network (GEONET) array, was established by the Japanese Geographical Survey Institute starting in 1993 (Sagiya *et al.*, 2000). The GEONET network collects high precision GPS data from more than 1200 stations deployed in a grid with about 20 km spacing across the Japanese islands.

Since the early 1990s, GPS measurements have corroborated the limited measurements of plate motion by VLBI and SLR techniques that confirmed the plate motion velocities and directions predicted with the DeMets *et al.* (1994) global plate motion model to mm a^{-1} precision (Larson *et al.*, 1997). The GPS networks have clearly demonstrated that $\sim$ 80% of Earth's surface is translating laterally with negligible permanent internal distortion (Stein and Sella, 2002; Prawirodirdjo and Bock, 2004). A global plate motion model, REVEL, that is entirely based on GPS measurements since 1993, has been developed and is now used to compare long-term and short-term tectonic movements (Sella *et al.*, 2002).

Most Earth movements are localized at plate boundaries. Global GPS networks now indicate that $\sim$ 20% of Earth's surface comprises zones of diffuse deformation. The amount of movement occurring across these zones is locally as fast as $\sim$ 40 mm a^{-1}, but most are accommodating movements $<$ 10 mm a^{-1} (Kreemer *et al.*, 2000). The fastest rates are where backarc spreading is occurring, such as behind the Mariana island arc, and where backarc shortening is occurring, such as along the central Andes.

Global positioning system networks have already provided fundamental new insights into movement patterns across many of the diffuse deformation zones (Figure 2.6). Moreover, these surveys are revealing where movements are currently localized. Local networks have already yielded detailed records of active Earth movements for East Asia (Shen *et al.*, 2000), Japanese islands (Miyazaki and Heki, 2001), Indonesia (Bock *et al.*, 2003), western New Guinea (Stevens *et al.*, 2002), eastern New Guinea (Wallace *et al.*, 2004a), New Zealand (Wallace *et al.*, 2004b), central South America (Khazaradze and Klotz, 2003), western North America (Bennett *et al.*, 1999; Hammond and Thatcher, 2004), the Aegean region (Nyst and Thatcher, 2004) and Anatolia of the eastern Mediterranean (McClusky *et al.*, 2000). Wernicke *et al.* (1998, 2004) were the first to apply GPS measurements to evaluate the geologic stability near the proposed Yucca Mountain HLW repository. Wallace *et al.* (Chapter 6, this volume) present a methodology rooted in GPS measurements to evaluate site stability in northeast Japan.

2.5 Tectonic movements and HLW repositories

Many countries are partly within a diffuse deformation zone and a few (e.g. Japan, Indonesia) are entirely within such zones (Apted *et al.*, 2004). The Yucca Mountain repository site is within the diffuse deformation zone along the western margin of North America. Evaluating the tectonic hazard for many potential HLW repository sites will center on evaluating a region for internal movements at speeds slower than a few millimeters per year.

Japan adopted a three-tiered classification for fault activity for the purpose of analyzing seismic hazards near critical facilities (Table 2.1; Inoue, Chapter 21, this volume). Class A active faults have high rates of slip, but this slip is episodic, with periods of quiescence often lasting centuries. Class B active faults experience quiescence between slip events commonly lasting for many centuries to millennia. At the slip rates of Class C faults, evidence of activity, either seismic or geomorphic, is almost always very sparse. Nevertheless, on timescales of 10–100 ka, these rates of motion may result in substantial displacements (Table 2.1).

Table 2.2 compares deformation at these rates in 100 ka for movement entirely localized along a single fault and movement uniformly distributed (via penetrative strain) across a region of greater width. The wider the zone, the less the local strain. At the meter-scale of waste canisters, a penetrative strain of 0.1 corresponds to a distortive movement of 10 cm. A penetrative strain of 0.01 corresponds to a distortive movement of 10 mm. Distortions at this scale are so small that strain near a waste canister should be absorbed by flowage in the enveloping bentonite clay pack. The extreme case occurs when all of the tectonic movement is localized as slip along a single planar fault. How do these rates of movement correspond to fault slip during earthquakes?

2.5.1 Earthquakes, elastic strains and fault offset

There are quantitative relationships between earthquake magnitude (M), the size of rupture areas and the amount of fault slip (Sibson, 1986). The most recent assessments of average global relationships (Wells and Coppersmith, 1994; Hanks and Bakun, 2002) are summarized in Table 2.3. The database is limited for events $M > 8$ and it appears there may be some systematic differences among tectonic environments (Fujii and Matsu'ura, 2000). New Zealand strike-slip earthquake ruptures are large, relatively numerous and well studied since the 1848 Marlborough event. These events appear to depart significantly from global averages (Stirling *et al.*, 2002). The earthquakes have rupture characteristics that correspond to events ~ 0.4 magnitude units smaller than global averages (Dowrick and Rhoades, 2004), possibly as a result of variations in the rigidity of the deforming rocks and fault zone properties. The scarcity of events $M > 8.6$ precludes proper statistical characterization of great earthquakes. For a given magnitude, the average slip and rupture-dimension relationships listed in Table 2.3 appear to be accurate within about a factor of two.

Most earthquakes that are not aftershocks are nucleated at depths of 10–25 km. A region's geothermal gradient is very important. Where thermal gradients are low, earthquake-generating fault slip events can be larger and deeper. Figure 2.7 is a schematic cross-sectional diagram illustrating normal, thrust and strike-slip fault ruptures associated

Table 2.1. Tectonic displacement rates

Speed	Relative motions	Movement (100 ka)	Japanese seismic hazard classification
Rates of plate motion			
Fast	> 5 cm a^{-1}	> 5000 m	
Slow	1–5 cm a^{-1}	1000–5000 m	
Rates on individual faults			
Very slow	1–10 mm a^{-1}	100–1000 m	A
Extremely slow	0.1–1 mm a^{-1}	10–100 m	B
Limits of detection	0.01–0.1 mm a^{-1}	1–10 m	C

Table 2.2. Localized offset or diffuse strains per 100 ka

Displacement rate	Localized offset on faults	Strain (DI^a/I_0^b) if distributed			
		0.1 km	1 km	10 km	100 km
10 mm a^{-1}	1000 m	10	1	0.1	*0.01*
1 mm a^{-1}	100 m	1	0.1	*0.01*	*0.001*
0.1 mm a^{-1}	10 m	0.1	*0.01*	*0.001*	*0.0001*
0.01 mm a^{-1}	1 m	*0.01*	*0.001*	*0.0001*	*0.00001*

[a] DI = change in distance between benchmarks. [b] I_0 = initial separation of benchmarks. Italics indicate negligible strains at the scale of waste containers.

with M 4–8 earthquakes. It is very rare for earthquakes of M < 6 to have ruptures that extend to Earth's surface and uncommon for events of M < 7 (Table 2.4).

Figure 2.8 schematically illustrates what happens during an earthquake-generating fault slip event for a simple rupture confined to the subsurface. Most of the elastic strain energy that makes up an earthquake in a simple rupture event arises from very near the earthquake focus. In principle, this focus can be located from the analysis of seismic-wave arrival times at only four seismographs. In practice, the focus is located from averaging arrival-time data from many stations. In a simple rupture, the maximum amount of slip occurs at the focus point and decreases to zero at the end of the rupture. Earthquake energy is also released from the nearby region of rock that was not loaded to the point of failure, but becomes less strained during the slip event.

Detailed analysis of earthquake waveforms recorded as seismograms reveals most large earthquake ruptures are complex events with varied amounts of energy release along the rupture plane (Mai and Beroza, 2000). In many cases, the sites of maximum energy release and inferred maximum slip are not exactly the earthquake focus. It is evident that a slip event

Table 2.3. Earthquake magnitude and average slip and rupture dimensions

M	Mean slip	Max. slip	Rupture dimension (km)	Rupture area (km^2)	Focus zone width (km)	Comments
9	10 m	15 m	~ 750 × 40	~ 30 000	34	Rare, complex subduction zone events
8	5.2 m	7.8 m	~ 170 × 30	~ 5000	11	Potential surface rupture hazard
7	1.1 m	1.7 m	~ 35 × 25	~ 890	3.4	
6	22 cm	33 cm	~ 10	100	1.1	Rupture to surface very rare
5	4 cm	6 cm	~ 3	11	0.34	
4	9 mm	14 mm	~ 1	1.4	0.11	
3	2 mm	3 mm	~ 0.4	0.2	0.03	
2	0.4 mm	0.6 mm	~ 0.15	0.02	0.001	

Mean slip, rupture dimensions and area for M $<$ 6 after Wells and Coppersmith (1994). Rupture dimensions for M 7–8 after Hanks and Bakun (2002). Maximum slip at an asperity are calculated assuming 1.5 times the average slip. Width of focus zone calculated following Ohnaka (2000).

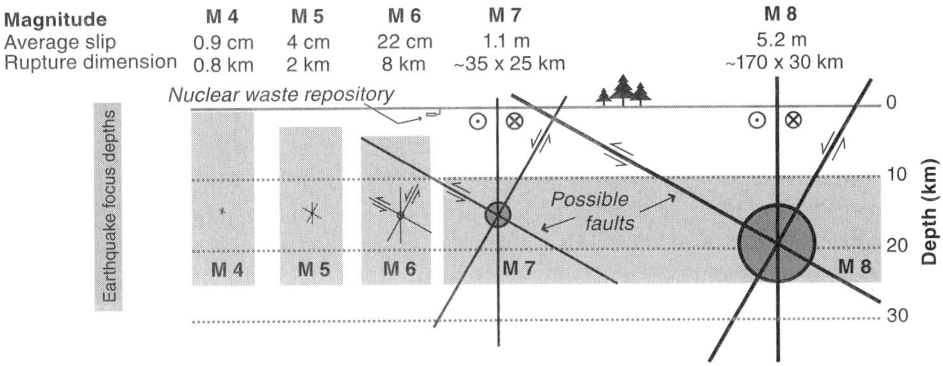

Fig. 2.7 Schematic diagram showing the location of a nuclear waste repository with respect to the length of fault ruptures associated with earthquakes of M 4–8. Note that few earthquakes of M $\lesssim$ 7 are likely to cause ruptures that would disrupt a waste repository constructed at depths of 300–1000 m. The circular volume at the intersection point of the possible faults denotes the equivalent spherical volume of elastically strained rock that is the source of seismic waves for an event of given magnitude. Source-volume estimate calculated assuming stress decrease of 1000 bar and bulk modulus of 1000 kbar.

can rapidly cause strained regions along and near the slip plane to become loaded to the point of failure. When this occurs, the release of a larger pulse of elastic strain energy becomes the locus of the mainshock; the magnitude of which is taken to characterize the earthquake event. The analysis of complex ruptures is an active area of research (Abercrombie *et al.*, 2006).

Table 2.4. Earthquake magnitude and land surface ruptures

Earthquake magnitude	Historic surface ruptures[a]	Global occurrence (per year[b])	Likelihood of surface rupture
M 4.0–4.9	4	13 000	Extremely rare
M 5.0–5.9	33	1320	Very rare
M 6.0–6.9	113	134	Uncommon
M 7.0–7.9	125	17	Common
M 8.0–8.9	20	1	Very common

[a] Yeats *et al.* (1997, Appendix). [b] USGS (2008).

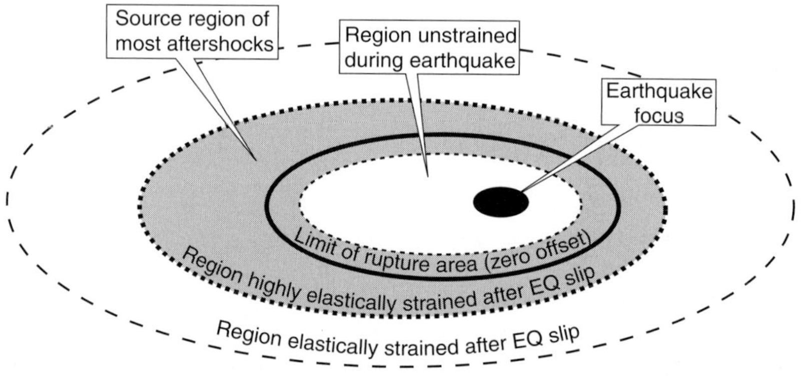

Fig. 2.8 Schematic diagram illustrating the unstrained and strained regions near an earthquake-generating fault slip event. The earthquake focus is where fault rupture begins. Rupture propagates outwards into less elastically strained rock. As this occurs, the highly strained rock becomes unstrained, generating seismic waves in the process, and the less strained rock absorbs motion by becoming strained. Where these strains are large, aftershocks result. $M > 5$ earthquakes are likely to have aftershocks.

The details of earthquake nucleation, energy-release patterns and rupture propagation are matters which exceed the scope of the problem at hand. For the sake of this discussion, the scaling of mainshock ruptures is the issue of concern. The amount of fault offset at the point of maximum energy release is ~ 1.5 to perhaps as much as 2 times the average slip along the fault (Table 2.3; Scholz, 2002).

The elasticity of rocks is so minute that it is considered infinitesimal in most forms of tectonic analysis. Rocks obey Hooke's law: the elastic strain is directly proportional to the magnitude of the change in the applied stress. Until the advent of GPS surveying, regional elastic strains associated with earthquake-generating fault ruptures were undetectable except in a few cases where benchmark networks had been very precisely surveyed two or more times before the event. This is because the elastic strains in crystalline rocks are so small,

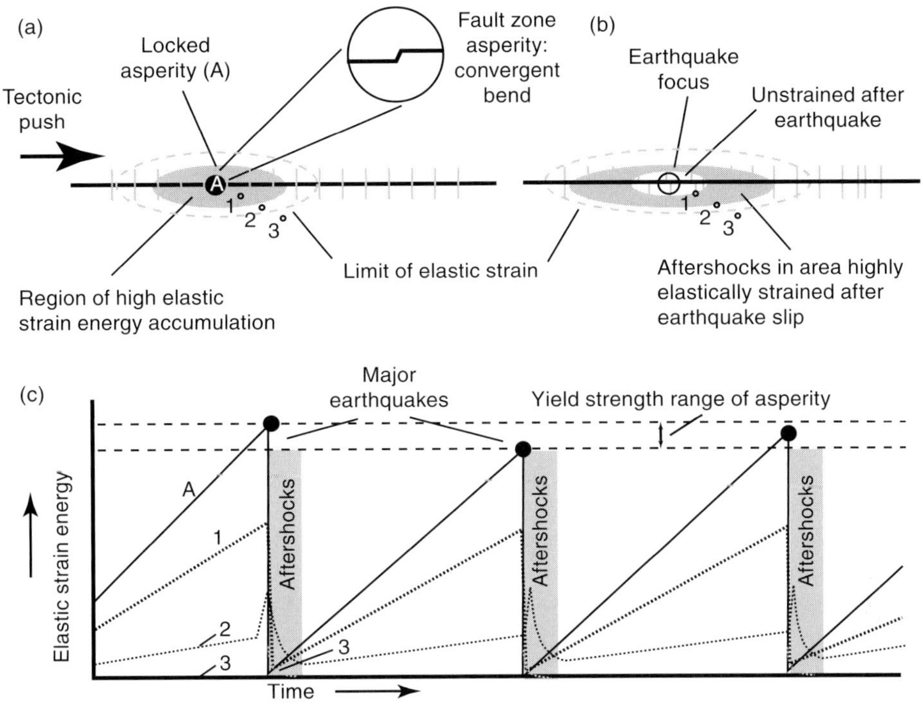

Fig. 2.9 Schematic diagrams illustrating what happened during an earthquake-generating fault slip event (conditions at the asperity (A); positions (1, 2, 3) away from the asperity). (a) An elastic strain builds up near a sticking point along a fault such as a mechanical irregularity. (b) After the earthquake, the elastic strains increase away from the slip plane and near the edges of the rupture. (c) The elastic strain versus time history near an idealized fault-plane asperity that slips in a time-systematic manner because the seismogenic asperity has a similar yield strength between these events. The slow build-up and rapid release of elastic strain near the asperity is known as the seismic, or earthquake, cycle.

$< 0.01\%$. Elastic-strain histories associated with episodic stick-slip offset near a mechanical asperity are schematically illustrated in Figure 2.9.

As fault slip movement propagates outward into less-strained rock, some of the released elastic strain energy driving displacement becomes absorbed by the elastic distortion of surrounding rock that was little strained before the event. Near the line of zero offset in moderate and larger earthquakes ($M > 5$), it is typical that the rapid increase in elastic strain locally loads the rock to the point of failure causing aftershocks (Kisslinger, 1996). These lesser earthquake-generating fault slip events are actually predictable, in a statistical sense, because they are located near and soon after larger events (Shcherbakov *et al.*, 2004). One statistical relationship, Omori's law, accounts for the decreasing frequency of aftershocks after a mainshock. Another statistical relationship, Bath's law, accounts for the numbers of aftershocks. For example, one M 7 event will trigger $\sim$ 1 M 6 aftershock, $\sim$ 10 M 5 aftershocks, $\sim$ 100 M 4 aftershocks and so forth.

The typical M 5 event is associated with fault slip over a distance of $\sim$ 3 km, with an average displacement of $\sim$ 4 cm (Table 2.3). The typical M 6 event causes a fault to slip over a distance of $\sim$ 10 km, with an average displacement of $\sim$ 33 cm. Only where crystalline igneous and metamorphic basement rock extends to near the surface is it possible for mainshock events of M $<$ 6 to cause a rupture that reaches the surface. While earthquake-generating fault slip events from small earthquakes are unlikely directly to affect repository integrity, it is possible that cumulative offsets from small events can cause penetrative strains (e.g. folding or stretching) to develop in an overlying blanket of sedimentary strata. Where this occurs, the effects are muted because strains at repository depths would become broadly distributed.

About 17 M 7–8 earthquakes occur each year somewhere within Earth (Table 2.4). Scaling relations are such that a M 7 event will have a rupture extending for $\sim$ 30 km with an average slip of $\sim$ 1 m. Events of M $<$ 7 are numerous along almost all plate boundaries where movements are faster than a few centimeters per year. They are scattered in occurrence within diffuse deformation zones. As Earth's surface commonly becomes ruptured by earthquakes of M $>$ 7, the likelihood of such events near potential HLW repository sites must be carefully evaluated.

Earth averages about one M 8–8.5 event per year and shallow events of this magnitude can be catastrophic. A M 8 event will have a rupture length extending $\sim$ 200 km with an average slip of $\sim$ 5 m. M 9 events are much larger and catastrophic. They have rupture lengths extending $>$ 500 km and an average fault slip of $\sim$ 10 m. Fortunately only five M 9 events have occurred since instrumental recording began around 1900 (Figure 2.1). During this time there also have been seven events of M 8.6–8.9. All of the M 8.6+ great earthquakes are due to large thrust-fault movements along subduction-zone boundaries where relative movements are at velocities $\gtrsim$ 4 cm a^{-1}. Many of these titanic events caused displacements of the seafloor that generated large tsunamis.

No HLW repository would be constructed within a subduction shear zone, the source of all historic great earthquake fault ruptures, but a repository site could be constructed near to where M 8–M 8.5 earthquake-generating fault ruptures are seemingly possible. For the most part, the primary concern is to evaluate the likelihood that potential repository sites may have earthquakes in the range of $\sim$ M 6–8.

2.5.2 *Infrequent fault ruptures and HLW repositories*

There are rare cases of large earthquakes occurring within otherwise seemingly stable plate interiors. The best known examples are the New Madrid earthquakes of Missouri, located in the interior of the North American plate (Figure 2.1). Three large, $\sim$ M 7.5–8 earthquakes occurred over the winter of 1811–1812. Trenching studies have revealed clear evidence for at least two similar events at AD 900 $\pm$ 100 a and AD 1450 $\pm$ 150 a (Tuttle *et al.*, 2002). The recurrence interval appears to be about 500 a. Infrequent large events are of special concern because the historical record is short and the geologic evidence for their past occurrence in areas of slow Earth movement may be very subtle.

Table 2.5. Potential numbers of fault-slip-generating earthquakes

Displacement rate	Movement (10^5 a^{-1})	Number			Hazard class
		M 6	M 7	M 8	
10 mm a^{-1}	1000 m	3030	590	130	A
1 mm a^{-1}	100 m	300	60	13	B
0.1 mm a^{-1}	10 m	30	6	1–2	B
0.01 mm a^{-1}	1 m	3	1	–	C

Assumes displacements at earthquake nucleation sites of 33 cm for M 6, 1.7 m for M 7 and 7.8 m for M 8 events.

Table 2.6. Potential frequency of earthquake fault slip events for Class B faults

M	Focus slip	Rupture width	Recurrence (1 mm a^{-1})	Recurrence (0.1 mm a^{-1})
8	7.8 m	~ 170 km	7800 a	78 000 a
7	1.7 m	~ 30 km	1700 a	17 000 a
6	33 cm	~ 10 km	330 a	3300 a
5	6 cm	~ 3 km	60 a	1600 a

With knowledge of the maximum slip (earthquake magnitude relationships in Table 2.3), we can deduce the approximate numbers of earthquake-generating fault slip events during the 100 ka timescale of concern for different imposed displacement rates (Table 2.5). In the unlikely event of a repository being sited near a fault with a Class A hazard, ~ 60–600 M 7 fault slip events, or ~ 13–130 M 8 fault slip events, could potentially occur during this performance period. Avoiding the main zone of such fault activity should be straightforward, but branch faults (discussed in the following) are common near such regions and a careful assessment of their potential is necessary when evaluating a repository site.

Similarly, for a repository located near a Class B fault, ~ 6–60 M 7 fault slip events, or ~ 1–13 M 8 fault slip events, are possible during the 100 ka timescale of concern. The approximate frequency of potential earthquake ruptures of different magnitude events is shown in Table 2.6.

Assuming careful site selection, the most likely potential tectonic hazard for a repository site is Class C fault activity. Less than six M 7 slip events, or perhaps one M 8 event, are possible during the 100 ka timescale of concern; M 6 events are not likely to extend up into the near-surface rocks hosting a repository. They are included in Table 2.5 primarily because it is possible that the deformation from many small movements in the seismogenic basement could accumulate as some form of distributed strain in overlying strata.

Of course, it is most likely that the movements over time will be accommodated by slip events in a variety of sizes with proportionate numbers of larger and smaller fault slip events. The primary value of Table 2.5 is to provide a prediction of the approximate time between events of a given size for a given displacement rate across a diffuse deformation zone. This information is needed to analyze fault movement recurrence times in the context of the historical record of seismicity or from offsets revealed in trenching studies (see McCalpin, 1996 for a review of techniques).

Globally, the instrumental record has located almost all $M > 7$ earthquakes since 1900 (Engdahl and Villanseñor, 2002). It is complete for events of $M > 6$ since 1953 and for $M > 5$ since 1964. Relative movement directions (focal mechanisms) have been determined for all earthquakes of $M > 5.5$ since 1976.

The length of a region's historical records varies greatly. In some countries, such as China and Japan, the record of $M \gtrsim 7$ earthquakes extends back several centuries. Where earthquake locations are documented near potential repository sites, the faulting hazard is evident. The problem then becomes one of evaluating the potential for large events where the recurrence time is many centuries. Why the size of earthquake-generating fault ruptures varies from place to place, as well as why some regions undergo much more diffuse deformation such as folding, is problematic. Different parts of the crust respond to tectonic movements in different ways. Geologic mapping and structural analysis can reveal how a region has responded in the recent past and this is the basis for predicting a region's behavior over the next 100 ka.

Understanding why deformation can be concentrated in some areas and not in other areas requires an understanding of how the continental crust and the lithosphere behave in response to the tectonic forces that drive plate motion.

2.6 How the crust and lithosphere deform

With a few exceptions, such as beneath the Himalayas and the Andes mountains, the thickness of continental crust is 30–50 km. Earthquake ruptures propagate outward and mostly upward towards the surface. Earthquakes are generated from the rapid release of stored elastic strain energy. To build up large elastic strains, temperatures must be low enough that rocks do not flow, preventing the accumulation of elastic strain energy.

2.6.1 Peak strength: fracture and flow strengths

Except where thermal gradients are greatly perturbed by the rapid subduction of cold lithosphere deep within Earth, most large earthquakes nucleate in the depth range of $\sim$ 10–25 km because of two competing factors: fracture strength and flow strength (Figure 2.10; Brace and Kohlstedt, 1980). Faulting is a mechanical process that involves friction-resisting sliding. There are two situations: the reactivation of old faults, which is commonplace, and the creation of entirely new faults, which is rare. The frictional resistance to sliding and the

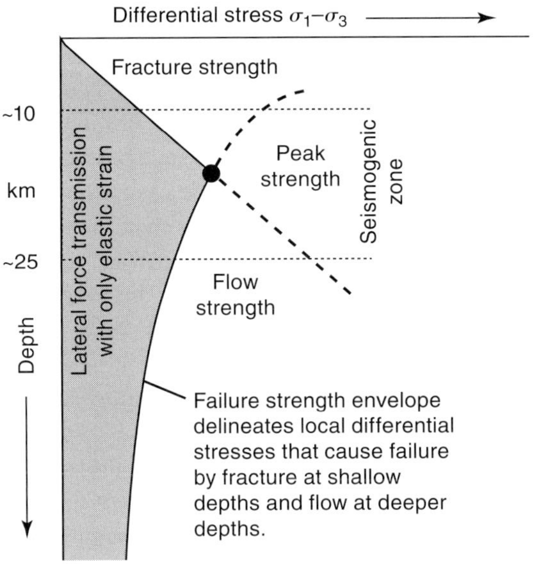

Fig. 2.10 Schematic diagram illustrating how fracture strength increases with depth and flow strength sharply decreases with depth. Fracture strength is controlled by increased friction, which arises from increased normal stress due to higher confining pressure. Flow strength decreases exponentially with depth because increasing temperature weakens the crystals that make up rocks and enables their solid-state distortion (flow). Maximum differential stresses can be supported, and thus elastic strains can attain their largest values, at the cross-over depth, $\sim$ 10–25 km. This is the depth at which the largest earthquakes are nucleated. If, as is generally the case in the interior of the plates, the imposed differential stress is less than the fracture or flow strength, the rock mass translates in response to the push or pull forces that drive plate tectonic motions.

rupture strength of intact rock depends upon the magnitude of the normal stress (σ) acting on potential slip surfaces (Sibson, 1985).

Experiments show that the frictional resistance to sliding increases linearly with confining pressure (Byerlee, 1978). These experiments also reveal that the ambient pore fluid pressure conditions are very important because fluid pressure acts to reduce the effect of normal stress. This is the well proven "effective stress concept," clearly recognized in the 1930s in the field of soil mechanics by Terzaghi and first applied to tectonic faulting by Hubbert and Rubey in 1959.

At high enough temperatures, crystals in rocks will distort by penetrative and cohesive flow. The viscosity, or flow strength, depends strongly upon the temperature and the rock type. The flow strength of crystalline continental crust is largely determined by the content of quartz and feldspar. The ability of quartz and feldspar to flow increases rapidly between temperatures of 300 to 450 °C (Tsenn and Carter, 1987). As geothermal gradients are typically in the range of 25–35 °C km^{-1}, flow processes in continental crust typically start at depths of $\sim$ 10 km and dominate at depths $>$ 25 km. Temperature has a profound (exponential) effect on the flow strength of rocks, which increases dramatically upwards towards the cold surface of Earth.

The lithosphere has its greatest strength (i.e. supports the largest differential stress: $\sigma_1 - \sigma_3$) and thus the largest earthquake-generating elastic strains, near the depth where the fracture-strength line intersects the flow-strength curve. At depths shallower than the cross-over point, common crustal materials fracture more easily than they distort by flowage. At greater depths, common rock materials flow rather than fracture. In the rock mechanics literature, the term flow refers to two end-member behaviors: plastic and viscous. A plastic material has the ability to support and transmit some differential stress before yielding. A viscous material flows in response to any differential stress. The mantle portion of the lithosphere has some long-term strength and behaves as a plastic material. The asthenosphere is mantle that is hot enough to flow viscously in response to very small differential stresses.

The fracture strength of almost all rock materials is about the same and depends only upon local confining-pressure and fluid-pressure conditions. Hence, fracture strength with depth profiles are everywhere about the same. The local flow strength with depth profile is controlled by the rock type and local geothermal gradient (Figure 2.11a). Where gradients are higher than normal in continental crust, the peak strength can approach 10 km depth. Where

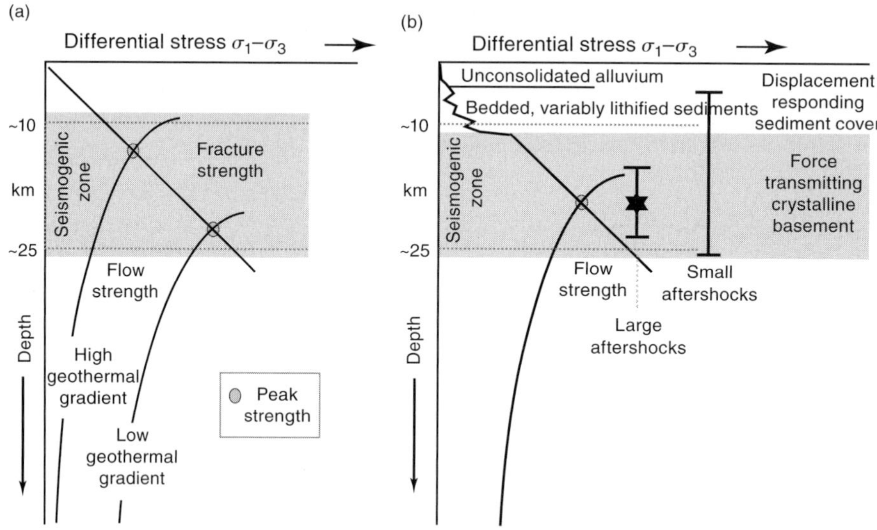

Fig. 2.11 The fracture strength of crustal materials systematically increases with depth, if fluid pressures in the rocks are low, and is similar for all rock types. The flow strength decreases with depth at a rate that depends upon the local geothermal gradient. (a) Typical crustal strength–depth profiles in areas with high and low geothermal gradients. (b) The crustal strength–depth profile in areas with thick covers of sedimentary strata. Sediments are weak rock that cannot laterally transmit substantial differential tectonic stresses. For the most part, sediments deform in response to the failure of the underlying crystalline basement.

gradients are lower than normal, the peak strength is nearer 25 km. This understanding of rock behavior leads to the common generalization that continental crust has two layers: an upper "brittle" layer and a lower "ductile" layer. The depth of the "brittle–ductile" transition zone varies across regions primarily in response to differences in geothermal gradient.

The part of the crystalline continental crust that can generate large earthquakes when it fractures has significant ability to transmit laterally tectonically generated forces. As discussed, old faults are planes of weakness that can reactivate time and again. Many new faults in sedimentary rocks are located above old faults in the basement that have reactivated in response to tectonic forces. New faults rarely form in intact crystalline rock remote from active faults. Most new faults are offshoots (splays) or propagations of fault tips during slip events along old faults.

Episodic movements occur because faults lock up where mechanical irregularities (asperities) hinder movement or where sliding friction is higher than elsewhere because of lower fluid pressure, greater cementation or other factors. Near locked regions, steady tectonic movements cause differential stresses and resultant elastic strains to increase. When the local differential stresses become large enough over timespans of decades to millennia to overcome the rupture strength of the locked region, elastic strain energy is rapidly released by the earthquake-generating fault slip.

2.6.2 Behavior of the sedimentary cover

The two-fold mechanical division of continental crust into an upper brittle layer and a lower ductile layer is an appropriate description of behavior where crystalline rock (igneous and metamorphic) extends to the surface. However, in many areas, the crystalline basement is blanketed by sedimentary strata; in some areas as much as 10 km thick. It is probable that some HLW repositories will be constructed in sedimentary rock. Sediments are porous and most types are very weak materials compared to crystalline igneous and metamorphic rocks. Even where strongly lithified strata are present, some layers are weaker than others. Shale beds and other types of rock that are weak at low temperatures, such as salt, commonly accommodate structural movements more by flowage than by fracture.

The mechanical behavior of sediments is highly variable and depends upon the mineralogy, degree of lithification (cementation), water content (porosity) and fluid pressure. Much strain can be accommodated in sediments by the closing of pores (compaction). Layers of sediment have a very limited ability to transmit tectonic stresses laterally without penetratively distorting, bending or breaking. In most areas the sediment cover is a passive blanket that deforms in direct response to movements in the underlying crystalline basement. This response can range from slow draping over rising or subsiding basement blocks, to rapid rupture when M 7+ earthquake-generating fault slip events propagate upward from the strong seismogenic zone. Where continental crust is blanketed by a thick cover of sedimentary materials, there are three mechanical layers: an upper layer of weak strata and lower crystalline layer that is weak because it is hot enough to flow. In between there is a strong member of igneous and metamorphic rock at temperatures $\lesssim 400\,°C$ (Figure 2.11b).

2.6.3 *Push and pull of plate tectonic forces*

The lithosphere is the strong outer layer of the Earth that consists of crust (with or without sediment cover) in addition to the uppermost mantle that is sufficiently cool that it has strength and the ability to laterally transmit the push and pull forces that drive plate motion. The strength profile of the ocean-crust-capped lithosphere is much simpler than that of the continental lithosphere. Peak strength in the ocean-crust-capped lithosphere resides in the upper mantle (Figure 2.12a). The continent-capped lithosphere has three or four (depending upon sediment cover) mechanical layers (Figure 2.12b) and two levels of high strength. Peak strength in the continental-crust-capped lithosphere resides in the crust, but the cool uppermost lithospheric mantle has significant flow strength. In both continental and oceanic areas, the push and pull forces that drive plate movement are laterally transmitted in the uppermost cool mantle.

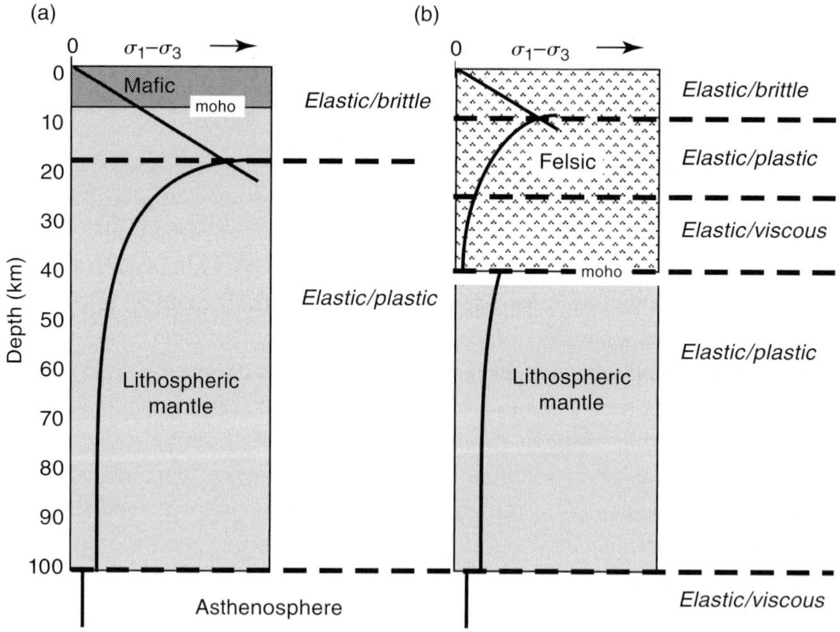

Fig. 2.12 The strength profiles of the lithosphere capped by (a) oceanic and (b) continental crust. The peak strength in the oceanic lithosphere occurs in the uppermost mantle. The continent-capped lithosphere has two zones of peak strength; an upper zone of maximum strength that is within the crust and a lower zone of high flow strength that is in the uppermost mantle. Where ambient differential stresses from the push and pull forces of plate tectonics are less than the maximum strength of the lithosphere, the lithosphere simply translates without internal deformation.

The forces that drive plate motion arise because the uppermost mantle lithosphere is so cool ($< 1200\,°C$) that it is more dense than the hotter underlying asthenosphere (Cloos, 1993). While gravity acts to pull the more dense lithospheric mantle downward, the temperature structure is such that the lithosphere usually has sufficient strength to resist the downward pull. This property, combined with the fact that downward movement requires concurrent displacement of the asthenosphere, restricts the wholesale sinking of the lithosphere (subduction) to places where it bends downward. Rift zones form where plates pull apart and the lithosphere thins. Where two parts of a torn plate are moving away from one another, new ocean crust is generated because new asthenosphere upwells to fill the gap and partly melts as fast as it widens. All evidence indicates the current plate tectonic mode of deformation has been the way Earth has behaved for at least the past 1 Ga (Windley, 1995).

The sinking of the plates at subduction zones generates a pull force and the topography at the spreading ocean ridges generates a push force; these forces are laterally transmitted for long distances. Where the differential stresses generated by the push and pull of plate tectonic forces are less than the plastic yield strength of the uppermost cool lithospheric mantle, the typical behavior, then the overlying crust rides passively along. This is the situation for $\sim 80\%$ of Earth's surface.

Belts of deformation along ocean spreading ridges are very narrow (typically < 20 km) because the steady ascent of magma maintains extremely high geothermal gradients along the line of weakness that extends to very near the surface. Deformation zones within rifting continents (e.g. East Africa) are typically a few tens of kilometers wide but can be hundreds of kilometers wide (e.g. the Great Basin of western North America). Deformation zones are diffuse and wide at subduction zones because the generation and ascent of magma from depth warms the overriding lithosphere along the line of volcanoes forming the arc and makes it a zone of weakness. The forearc block, on the other hand, is strengthened because it has low geothermal gradients due to the underthrusting of cold oceanic lithosphere. This block, which is part of the overriding plate, actually sits directly atop the descending plate and thus is directly affected by subduction movements.

The broadest deformation zones occur where plate movements cause continents or large island arcs to enter subduction zones. The positive buoyancy of continental crust resists subduction, and collisional tectonism can cause substantial differential stress to become transmitted laterally in the brittle crystalline crust (Cloos *et al.*, 2005). As evident in East Asia, collisional tectonism at convergent plate margins can cause crustal movements many hundreds of kilometers from the plate boundary. Where differential movements are localized in the otherwise strong upper crust (by the reactivation of ancient faults), strata overlying movement zones will become deformed and tectonically driven flowage will be induced in the ductile lower crust.

2.6.4 Upper-crust response to tectonic movements

The brittle upper crystalline crust deforms primarily by faulting. In areas of crustal shortening or extension, steeply dipping faults in the crystalline basement are commonly

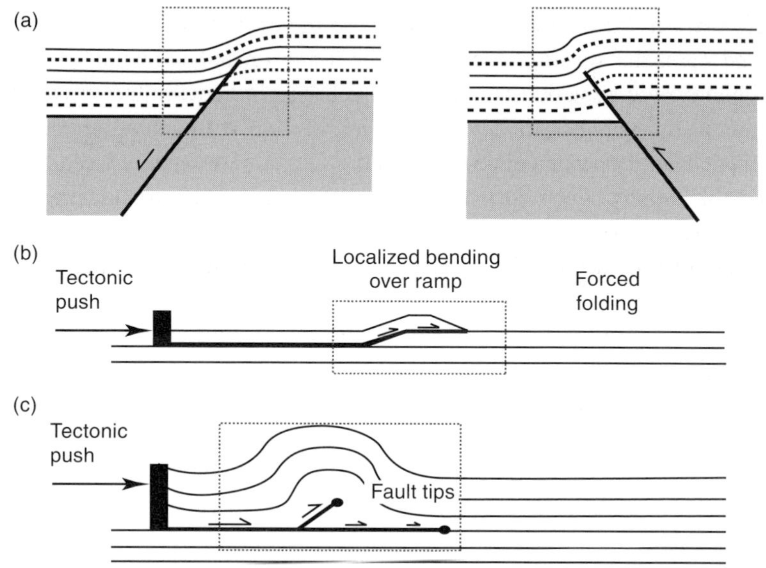

Fig. 2.13 Schematic diagram illustrating diffuse folding strain in sedimentary strata due to fault movements in the underlying basement: (a) above high-angle normal and reverse faults; (b) above a thrust fault; and (c) above a décollement.

reactivated. Sediment cover becomes draped over rising fault blocks (Figure 2.13a). In some areas of crustal shortening, low-angle thrust faulting occurs and folding is common near faults that emerge at the surface (Figure 2.13b), or have tips that end at shallow depth (Figure 2.13c). Where the strain is distributed as in the case of folding, a steeply dipping, commonly near-vertical fabric can develop as elongate minerals become aligned (cleavage) in the distorted sediments. In areas of crustal extension, there is stretching and penetrative strain can occur in the form of a near-horizontal fabric which is largely parallel to bedding. This phenomenon generally cannot be differentiated from the flattening fabric (fissility) generated as compaction reduces porosity.

If a repository is sited in a diffuse deformation zone, the question of whether tectonic deformation of the crust is localized to slip along one or more faults is the fundamental concern. Deformation by localized slip along faults is commonplace, but some regions deform in a more distributed manner with folding and tilting movements. Most importantly, it is critical to recognize that within broad zones of diffuse deformation, large blocks of crust can be translating along for millions of years without internal distortions that would affect the integrity of a HLW repository. Demonstrating the stability of crustal blocks within such regions is the challenge.

2.6.5 Stable blocks in diffuse deformation zones

Diffuse deformation zones that include or bound subplates can be many hundreds of kilometers across (Figure 2.6). If movements were distributed uniformly, the magnitude of the tectonic deformation that would occur within the region near a potential repository site would depend upon the speed of the relative movements and the width of the deformation zone. The tectonic behavior of every region is unique because the crystalline crust is the product of differing subduction, collision, and rifting histories. As a result, continental crust is mechanically heterogeneous with scattered ancient faults and shear zones. One generalization is clear: uniform penetrative strain does not occur across broad regions of Earth because zones of weakness localize movement. Consequently, all diffuse deformation zones contain fault-bounded blocks of crust that translate along without permanent internal distortion.

Tectonic hazard assessment for potential repository sites within diffuse deformation zones requires identification of the number, location and activity levels of faults, or fault zones that bound "stable" blocks. Consider a 100 km-wide region that is shortening or extending at a rate of 10 mm a^{-1}. If there are five faults that accommodate all of the movement, each one could accommodate $\sim$ 2 mm a^{-1} offset. More likely, the distribution of movement will be varied. Perhaps, two of the faults are Class A with movements averaging 6 and 3 mm a^{-1} and three are Class B with movements $<$ 1 mm a^{-1}. The spacing between faults is also critical. In the unlikely situation that faults are equally spaced, stable blocks would be $\sim$ 20 km wide. Most likely, the distribution is such that one of the fault-bounded stable blocks is $>$ 40 km across. And of course, not all of the differential motion has to be accommodated by faulting. Distributed deformation can locally occur in the form of folding or by the stretching of the crust forming a zone of subsidence toward which surface water will flow and sediments will accumulate.

The most straightforward way to identify the boundaries of potential stable blocks within a diffuse deformation zone is to identify the most extensive areas with horizontal strata. Flat-lying Miocene strata, for example, indicate no significant tilting has occurred since at least six million years ago.

2.6.6 Rock fracture: faulting, fault zones and joints

Major fault zones can be traced for tens to hundreds of kilometers. The width of disturbed rock is greater the longer the fault and the greater the amount of cumulative offset. While ruptures during M 7–8 earthquakes extend for tens to hundreds of kilometers, the associated fault offset in a given event is only a few meters and mostly occurs on planar slip surfaces. Cracking and brecciation (damage or process zone) are restricted to areas very near (meters) the fault and its splays (Shipton *et al.*, 2006). Granulation (cataclasis) is localized along the slip surface. Wide zones of disruption that characterize major faults result from many movement events.

When a trench is excavated for geologic site characterization (and when a HLW repository is excavated), every fracture encountered is characterized. There are two common types of fracture: extension and shear (Figure 2.14a–b) that occur within more complex zones (Figure 2.14c–e).

Extension fractures include joints and veins. Joints form naturally near the surface as rock units are unloaded by erosion. Joint formation can also be induced by man-made

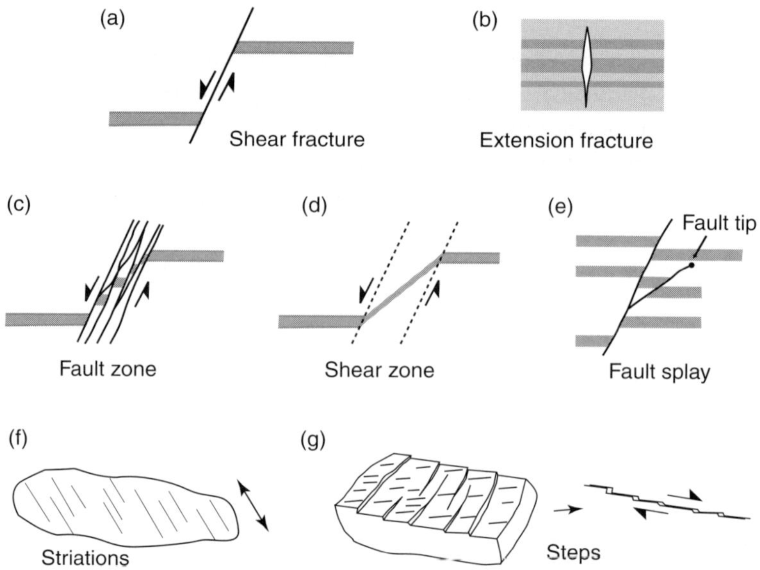

Fig. 2.14 Neotectonic analysis of areas around potential repository sites requires analysis of broken rock in outcrops and excavations. There are two fundamental types of fracture: (a) shear and (b) extension. Shear fractures are faults and offset can occur along planes or (c) across zones with many faults. A shear zone (d) forms where rock flowage causes the penetrative distribution of permanent strain. Fault zones commonly grade into shear zones at depth, and (e) form splays as they propagate toward the surface. Fault planes commonly have (f) striations and (g) steps that record the direction of slip.

excavations. The differentiation of joints and faults is usually straightforward because rocks contain internal markers at scales ranging from grains to layers. Joints have a small opening (generally < 1 mm) with opening movement perpendicular to their length. Veins are openings commonly millimeters to a few centimeters wide that become mineralized because formation fluids are commonly saturated in quartz or calcite.

Faults are shear fractures and thus they offset rock markers. Fault surfaces are commonly marked with striations generated by slip and zones of cataclasis, commonly millimeters in thickness that form thin layers of gouge. In some cases, fluids along fault planes precipitate minerals that are fibrous because of movement. Mechanical striations or mineral fibers record the direction, but not the sense of fault slip (Figure 2.14f). Some fault planes contain systematic steps that uniquely record the direction, or sense of slip (Figure 2.14g). Where striations and steps are present, the full kinematics of the fault slip event are known. The fault slip directions recorded in the rocks can be compared to slip directions for recent large earthquakes in the region.

In the past century, the world has averaged fewer than three earthquakes a year that caused offset of rupturing to the land surface (Yeats *et al.*, 1997; see their Appendix for a table of historic earthquakes with surface rupture). Some fault scarps are segmented with overlaps that are generally proportionate to differences in the local amount of offset near the fault tips. These differences reflect how displacement was transferred from one segment to the next. In many cases, minor faults splay from the main fault (e.g. Figure 2.14e). Where this occurs, several sub-parallel fault scarps are common. Splays that fork from the main slip plane at depth but do not reach the surface must be common. Major splays that extend from depth are branch faults. Some are reactivated and become significant structures over time. They commonly diverge from the main fault at a low angle, but they can also diverge at a high angle with similar but opposite (conjugate) dip.

Most regionally mappable faults have offsets of many tens of meters. Such faults must be the product of many episodes of movement because even M 8 events typically only cause about 5 m of offset. In eroded exposures or excavated outcrops, faults with offset measured in kilometers are observed to be zones of interconnected sub-parallel, anastomosing and cross-cutting faults. With each movement event, old slip planes are reactivated and new fractures form as minor splays locally veer off into the foot and hanging wall blocks, widening the zone (Figure 2.4). Rock that is broken can become infiltrated by water from the surface or fluids (formation, magmatic) from below. Infiltrating fluids can cause mineralization (veining) and/or rock alteration.

Faults are planes of weakness that are commonly reactivated. With this understanding it is evident that the simplest way to minimize the likelihood that a waste canister in a repository is ever ruptured is to mandate that no canister is ever interred in a position that cross-cuts a fault.

2.6.7 Branch-fault hazards

In some earthquakes, significant movements can occur on faults that branch as major splays extending hundreds of meters to many kilometers from the main slip plane (Figure 2.15a–c). The likelihood that a branch fault could impact a repository depends strongly upon the mode of active faulting near the site.

Branch faults near strike-slip zones are usually located very near the main fault. Minor normal and reverse faults are common near overlapping segments along strike-slip faults. Branch offshoots near normal faults commonly have dips opposite to the main fault. Their formation can generate grabens. Some branch offshoots near thrust faults have opposite dip to the main fault. These faults are commonly known as backthrusts and near the surface they typically bound anticlines. Branch faults are common in areas of dip-slip faulting and they are associated with the passive draping of the sediment cover to form folds. For the most part, branch faults are structures that form and reactivate as subsidiary movements during the main slip event. As with the main slip zone, substantial offsets only develop with many episodes of movement.

2.7 Identification of active tectonism: neotectonic analysis

Over million-year timescales, new faults locally cut intact rock far from locations of active faulting, but on timescales of millennia, this is a rare phenomenon and almost certainly

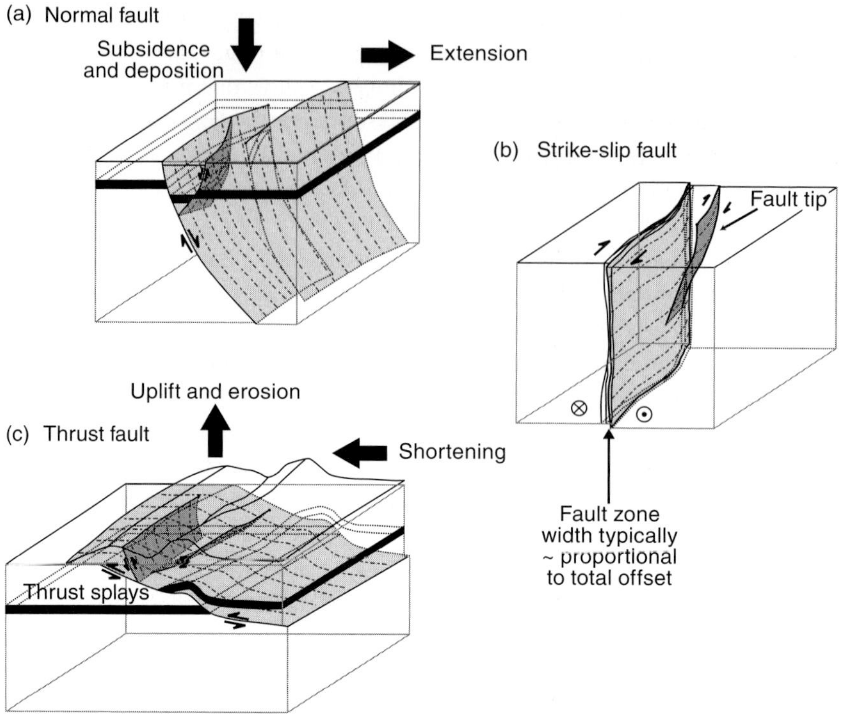

Fig. 2.15 Schematic diagram of branch faults in (a) normal, (b) strike-slip, and (c) thrust-fault settings. Branch faults are splay faults that are offshoots to the main displacement surface. They are commonplace as comparatively minor, secondary structures near major faults. Some earthquake-generating fault slip events have associated secondary or branch-fault offsets that are recognized from ground breakages away from the main rupture surface or from aftershock distribution patterns. They form by the accumulation of small offsets during large offsets on the main fault plane or by periodic activation during episodes of slip on the main fault plane.

localized to areas where active tectonism is evident. But what is the definition of active tectonism?

Active tectonics can be defined as deep-seated Earth movements that are expected to continue in the near future. Neotectonics is the study of Earth movement phenomena that have occurred in the past few million years and that are continuing in a similar mode today. There are various perspectives on what constitutes an active fault. Essentially all workers agree than any fault with movement during the past 10 ka should be considered active. Wherever differential movements are occurring faster than several millimeters per year, offsets and distortions in the land surface are generally evident. In such areas the historical record of earthquakes usually provides evidence of active tectonism. The lack of historical events does not mean no fault movement is possible, however, because the repose intervals between large events can be millennia. Van Dissen and Berryman (1996) show that intervals between the last five surface ruptures of the Wairarapa fault of New Zealand range from 1160 to 1880 a. This fault zone is accommodating a long-term average slip rate of $\sim 7\,\mathrm{mm\,a^{-1}}$.

Where horizontal movements are slower than a few millimeters per year, the geomorphic effects can be very subtle, especially if local topography is irregular. This is because the corresponding vertical movements are at speeds less than a few tenths of a millimeter per year, and rising land can be planed by erosion as fast as it rises, as subsiding areas can be infilled with sediment as fast as the rate of subsidence.

Techniques to understand and quantify geomorphic changes due to tectonic movements are steadily improving (see Burbank and Anderson (2001) and Bull (2007) for excellent reviews). Wave-cut marine terraces commonly provide extensive horizontal reference frames that are sensitive recorders of local sea-level change. Lake shores are similar and their deposits commonly also record changes in a region's sedimentation patterns due to tectonic movements. River terraces also provide horizontal reference frames to evaluate changes in tilt and vertical displacement. Because of the 120 m change in sea level due to glaciation and deglaciation during the last 120 ka, major rivers record a history of downcutting in downstream reaches during the sea-level fall followed by backfilling with terrace aggradation since the sea-level rise. Understanding these patterns can provide insight into tectonic movements far from the coastline (e.g. Berryman *et al.*, 2000). Alluvial fans, especially in desert climates, have morphologies that are so distinctive that cross-cutting faults are commonly easily recognizable. Lava flows, debris flows and landslides can provide geomorphologic marker horizons that commonly contain datable material. The development of regional depositional surfaces such as pediments or erosional surfaces such as peneplains indicate long-term tectonic stability. Strike-slip faults cause particularly distinctive offsets in river networks. Much neotectonic analysis is rooted in geomorphic characterization of a region. All calculations of deformation rates deduced from changes in landforms hinge on the dating of offset or tilted features.

Trenches are excavated in an effort to expose all active faults and tilted strata near potential nuclear facilities. Datable materials are sought in offset or tilted layers. With these kinds of data, rates of fault movement can be quantified. Excellent examples of this kind of analysis have been performed in California (Sieh and Jahns, 1984), New Zealand (Villamor and Berryman, 2001; Nicol *et al.*, 2006) and Japan (Inoue, Chapter 21, this volume). Geophysical surveys such as seismic reflection, ground-penetrating radar, magnetotellurics and electrical resistivity methods are conducted to image subsurface rock layers. Rock cores are drilled to obtain samples at depth that test cross-sectional relationships projected from the surface or deduced from geophysical surveys. These techniques and others are effectively utilized to discover, precisely locate and quantify the offset on faults near potential repository sites. For excavated sites, the construction of underground passages enable another systematic subsurface search for faults. If offsets are found in the underground passageways that cannot be correlated with known faults, additional trenching and geophysical surveys, and drill coring should be performed to determine their nature and linkage to structures in the region.

2.8 Regional strain rate and deformation budget

If a tectonically active area is selected as a repository site, neotectonic analysis should reveal much evidence about recent movements. But complete quantification of the movements is unlikely and assessment of potential activity is the matter of concern. The quantitative prediction of potential tectonic activity over the next 100 ka is now possible because regional movements can be quantified as a regional "strain rate" (e.g. Stirling *et al.*, Chapter 10, this volume). A *deformation budget* can be constructed which assigns movement rates to the active structures in the region.

Strain rate is the amount of displacement that occurs along a length of line per unit time. Deformation is the net sum of translations, rotations, distortions and dilations. Distortions are penetrative changes in shape. Many types of rocks record the effects of penetrative (grain-scale) distortion. Dilations are changes in volume, which are negligible in crystalline rocks, but can be substantial in compacting sedimentary rocks.

The geology of a region records its recent deformation history to varying degrees. In areas blanketed by distinctive sequences including datable young strata with cross-cutting relationships and geomorphic features, a very detailed record of recent movements is attainable. But in most areas, the cover of young strata is sparse and only portions of the recent deformation history may be directly observed.

Around the world, many nuclear facilities are found within regions classified as diffuse deformation zones. Knowledge of plate motion patterns enables calculation of the movements that must be accommodated across a region (Figure 2.16). Combined with the local

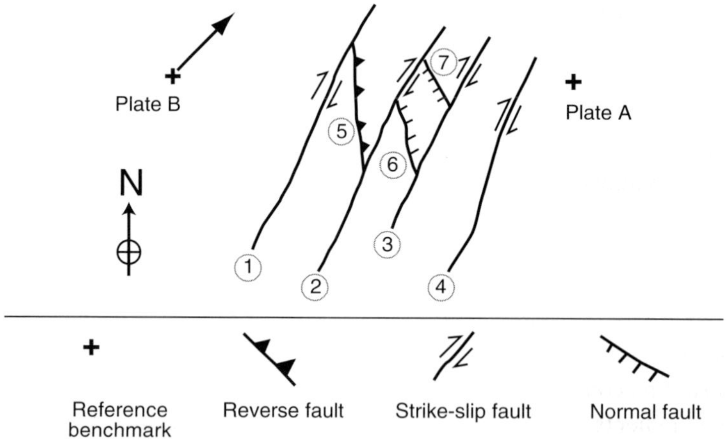

Fig. 2.16 Schematic illustration of a diffuse deformation zone at the boundary between two plates, A and B. The relative velocities are known from the plate motion model and confirmed with GPS measurements. The regional strain rate is the velocity of plate B with respect to A divided by the distance between the two stable benchmarks. The strain rate can be quantified in terms of components in the north–south and east–west directions. The deformation budget is an accounting of the movements along faults 1–7 in terms of the directions and amount of slip.

width of the deformation zone, the region's strain rate can be calculated (e.g. Kreemer *et al.*, 2000). On this basis alone, the tectonic activity in the area surrounding potential sites can be compared. A deformation budget accounts for what fraction of the region's movement occurs by slip along different faults or by distributed movements such as tilting and folding.

Using deformation budgets, the stability of crustal blocks can be assessed with confidence for timescales of 10 ka. The strain rate/deformation budget approach is rooted in the fact that the amount of tectonic movement across a region can be predicted based upon calculation from plate motion models and comparison to earthquake energy release (termed the seismic moment), fault slip and surface deformation rates, as well as current movements that are directly measured with GPS surveying (Ward, 1998). Examples of this approach to the problem of siting nuclear facilities is presented by Wallace *et al.* (Chapter 6, this volume), who focus on the analysis of regional movements as measured with GPS, and Stirling *et al.* (Chapter 10, this volume), who focus on the analysis of regional movements as indicated by strain rate derived from various analyses. The agreement between predicted and measured motions across a region and the nature of any internal movements that are detected during neotectonic analysis are cornerstones in the evaluation of tectonic hazards at specific sites.

2.8.1 Assessment of crustal stability: horizontal motions

With GPS-based geodesy, it is possible to measure directly the present-day surface movements in the region around any potential site. The Japanese GEONET is a network of GPS stations deployed in a grid with $\approx$ 20 km spacing, and measurements since 1995. With this network, movements as slow as 1 mm a^{-1} are currently detectable (Table 2.7). In most countries, GPS networks are more limited, but improving. A network of GPS stations may be deployed around any potential site. The GPS stations in a grid centered on the potential site (Figure 2.17) would enable the local displacement field to become apparent in a few years and well characterized within a decade.

If all GPS benchmark stations are moving in unison, they are translating with no local strain and the block is stable (Figure 2.17). Site stability assessment is a smaller-scale application of the same basic approach used to prove that negligible lateral strains (< 1 mm a^{-1}) are occurring within GPS networks having 2000+ km-long baseline distances across the interior of the Pacific plate (Beavan *et al.*, 2002) and most of the North American plate (Calais *et al.*, 2006). The GPS has confirmed the long-held belief that nearly all of the Australian continent is very stable (Tregoning, 2003).

If there are gradients in the displacement field, movements are occurring (Figures 2.18). The strain field can be monitored over time and related to the location of active faults and the regional geology. Differential movements between stations can be quantified as a local strain rate that is compared to the regional rate. Areas of faster and slower than average movements can be identified. One possibility is that the region is permanently deforming by pervasive shortening or extensional strain. The other possibility is that the region is elastically distorting and that all detected differential movements will recover following

Table 2.7. Detection of horizontal displacements with GPS

Relative motion	10 a displacement	Resolution needed for 25 km station spacing[a]
$10\,mm\,a^{-1}$	100 mm	4 ppm
$1\,mm\,a^{-1}$	10 mm	0.4 ppm
$0.1\,mm\,a^{-1}$	1 mm	0.04 ppm

[a] The resolution of the Japanese GPS network is $\sim 0.1\,ppm\,a^{-1}$ (Sagiya *et al.*, 2000).

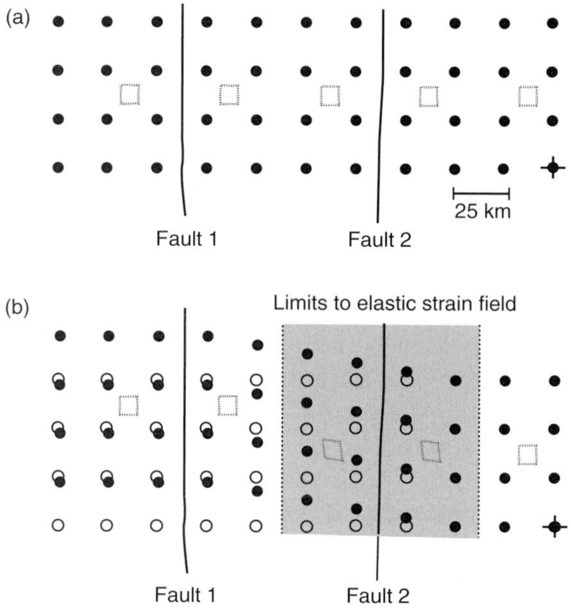

Fig. 2.17 Schematic diagrams illustrating the detection of fault activity with GPS networks. (a) A GPS station network around two faults, 1 and 2. (b) Over time, the relative changes in GPS station location indicate differential movements about fault 2 but not fault 1 because it is inactive. The width and amount of differential motion about fault 2 is a direct indicator of the depth of locking and the size of potential earthquake-generating fault slip events.

fault slip during the next large earthquake. Developing techniques to differentiate between permanent and temporary elastic strains is a rapidly expanding area of research.

The magnitude of the potential fault hazard can be estimated from the width of the gradient in the displacement field about active faults. The ability of GPS to resolve tectonic movements is demonstrated with the analysis of the diffuse deformation zone around the San Andreas fault in California. Earthquakes of M $\sim$ 8 with multi-meter surface ruptures

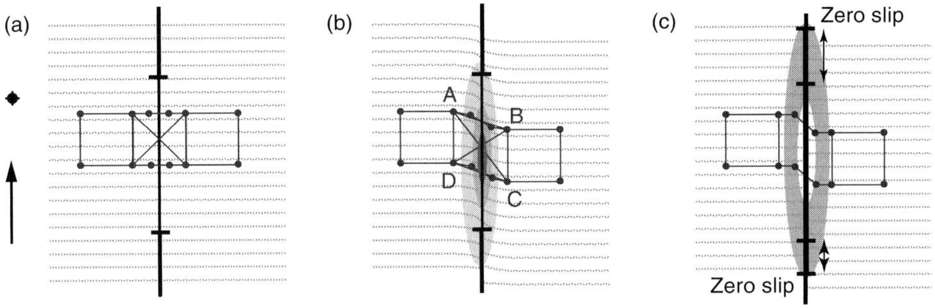

Fig. 2.18 Schematic diagram showing a possible elastic strain history near an active fault: (a) the initial benchmark network; (b) distortion of the benchmark grid over time; and (c) the rebound of the grid after an earthquake. The recurrence of earthquakes is never perfectly periodic because the local strain history depends upon movements along the fault extending above and below the line of benchmark grid.

are expected in the future, given past activity, such as the great 1906 San Francisco and 1857 Los Angeles area events. The GPS networks have revealed major differences in the displacement fields near San Francisco, near Los Angeles, and the intervening central segment. Displacement gradients extend for tens of kilometers on either side of the fault near San Francisco (Savage *et al.*, 2004) and near Los Angeles (Meade and Hager, 2005). The width of the region across which displacement gradients are detected is much smaller in central California where the San Andreas fault is steadily creeping (Argus and Gordon, 2001).

2.8.2 Assessment of crustal stability: vertical motions

Vertical motions are a natural result of predominantly horizontal plate motions. The largest vertical motions occur where all movement is localized along a single fault. The type of faulting (normal, reverse (thrust) or strike-slip) is a critical factor. Strike-slip faults by themselves do not cause changes in land elevation. But many regions dominated by strike-slip faulting, such as California and New Zealand, are either slightly convergent or divergent. Convergent regions uplift. Divergent regions subside. In both cases, the magnitude of strike-slip-related vertical motions are less than those that arise from pure normal- and reverse-fault motions caused by similar amounts of horizontal plate motion.

Relative convergent or divergent plate motions of 1 mm a^{-1} cause horizontal displacements of 100 m in 100 ka, or 10 m in 10 ka. If all the movement occurs along a fault dipping 60°, the local land surface elevation will change by $\sim$ 170 m in 100 ka, or $\sim$ 17 m in 10 ka. These elevation changes are muted by isostatic adjustments together with erosion of the rising areas and deposition in low areas. The issue of analysis of vertical motions from the geologic record is discussed by Litchfield *et al.* (Chapter 4, this volume). The potential impacts of vertical movements on repository safety is discussed by McKinley and Chapman (Chapter 24, this volume).

Concluding remarks

Overall, then, measured regional and local strain rates may be evaluated in the context of the deformation budget that accounts for the magnitude of movement along specific faults, growing folds and other active structures. The deformation budget across the region can be assessed and compared in three very different ways: geodetic measurement, the history of seismogenic movements and regional geologic analysis. Table 2.8 lists the types of data sets that will be integrated during the analysis of potential HLW repository sites in Japan. Where the long-term and short-term records of movement are in agreement, the regional deformation pattern is well understood.

Our understanding of Earth's behavior is rooted in more than two hundred years of study of the continents. The first studies were of natural outcrops supplemented by excavations made for mines, canals and roadways. The search for hydrocarbons since the 1920s brought deep drilling and spurred the invention of geophysical surveying techniques to image the subsurface. The study of the ocean floor has been an avid pursuit for more than fifty years. The worldwide deployment of the systematic seismic network in the early 1960s led to the realization that most earth movements are localized in narrow zones. The theory of plate tectonics arose in the late 1960s from the new observations about the ocean floor and the remote detection of active movements via the analysis of seismograms. The fundamental tenets of the theory, that the crust is simply a passenger forming the top of a mechanical layer, the lithosphere, which is broken into pieces, the plates, which diverge, converge and slide past one another, have been tested and confirmed in many ways.

Active tectonic movements around the world are now continuously monitored with GPS technology and these movements have been, and will be, routinely compared to the recent geologic history across broad regions. It is now evident that $\sim$ 80% of Earth's surface is simply translating along, internally undistorted for many millions years, and $\geq$ 1 Ga in the interior of some continents. From the point of view of tectonic stability, there are large areas suitable for siting nuclear power plants or hosting HLW repositories. Nevertheless, some countries will have to locate nuclear facilities in and near zones of active deformation. Parts of plates have become classified as subplates, or diffuse deformation zones, because their internal movements are now measurably different with GPS technology from the bulk of the plate. The problem of site selection in such areas is a challenge, but solvable because within these zones are blocks of crust with dimensions of tens of kilometers and more across, which are highly unlikely to undergo permanent internal distortion over the next 100 ka.

The demonstration that a site is tectonically stable and thus suitable to host a nuclear power plant or a HLW repository arises from the integration of traditional geologic analyses that indicate negligible local deformational activity in the past million years or more with confirmatory GPS measurements. The prediction of future stability is bolstered by GPS measurements across the region because these measurements are sensitive to current tectonic movements. Tectonically suitable sites for nuclear facilities are widespread. A potentially larger problem in the matter of tectonic hazards and nuclear facilities is the issue of increasing the public awareness of how the crust of the Earth actually behaves.

Table 2.8. Detection of active tectonic movements in Japan

	Methodology	Comments	Timescale
Geodetic			
(1)	Global Positioning System (GPS) Geographical Survey Institute (GSI) GPS Earth Observation NETwork (GEONET)	Initial stations in 1993; 1200 station grid since 1997	10 a
(2)	Precise geodetic network with electronic distance meter (EDM)	2760 stations, 7715 lines, 950 angles, precision 1–2 ppm; established 1973	~ 30 a
(3)	Geodetic Triangulation Network	300 stations spaced ~ 45 km; established 1883; angular precision is 1 arc s	~ 125 a
Seismic			
(4)	Historical seismicity	M 7+ events since 1581; first instrumental record 1880; first systematic felt reports 1884;	420 a
(5)	Seismic network *(old)*	M 5+ events since 1920±; locations ±30 km; focal mechanisms for M 6+ since 1930	~ 80 a
(6)	Seismic network *(improved)* Japan Meteorological Agency (JMA)	M 4+ events since 1961; locations ±15 km	~ 40 a
(7)	Seismic network *(modern)*	M 4+ events since 1985; locations ±10 km; focal mechanisms	~ 20 a
Geologic			
(8)	Geologic mapping/trenching	Active faults; type movement/location	~ 10 ka
(9)	Geomorphic analysis	Active uplift/subsidence/tilt; type movement/location	1+ ka
(10)	Geological mapping/structural analysis	Strata < 10 Ma	Ma stability

Further reading

The fundamentals of structural geology are well covered in Suppe (1985) and Davis and Reynolds (1996). For further information about faulting and earthquakes see Yeats *et al.* (1997) and Scholz (2002).

Acknowledgments

I thank the Nuclear Waste Management Organization of Japan (NUMO) for organizing the International Tectonic Meetings, which focused on many of the matters in this report. Over the past years, I have appreciated numerous discussions about nuclear waste issues with K. Kitayama, H. Tsuchi and many other staff at NUMO as well as K. Berryman and his colleagues at Geological and Nuclear Sciences (GNS) in New Zealand. M. J. Apted, N. A. Chapman, C. B. Connor, L. J. Connor, K. Berryman and B. Wagner are also thanked for discussions and review comments that improved this chapter.

References

Abercrombie, R., A. McGarr, G. Di Toro and H. Kanamori (eds.) (2006). *Earthquakes: Radiated Energy and the Physics of Faulting*, AGU Geophysical Monograph 170. Washington, DC: American Geophysical Union.

Apted, M., K. Berryman, N. Chapman *et al.* (2004). Locating a radioactive waste repository in the Ring of Fire. *Eos, Transactions of the American Geophysical Union*, **85**, 465–471.

Argus, D. F. and R. G. Gordon (2001). Present tectonic motion across the Coast Ranges and San Andreas fault system in central California. *Geological Society of America Bulletin*, **113**, 1580–1592.

Beavan J., P. Tregoning, M. Bevis, T. Kato and C. Meertens (2002). Motion and rigidity of the Pacific plate and implications for plate boundary deformation. *Journal of Geophysical Research*, **107**(B10), 2261, doi:10.1029/2001JB000282.

Bennett, R. A., J. L. Davis and B. P. Wernicke (1999). Present-day pattern of Cordilleran deformation in the western United States. *Geology*, **27**, 371–374.

Berryman, K., M. Marden, D. Eden *et al.* (2000). Tectonic and paleoclimatic significance of Quaternary river terraces of the Waipaoa River, east coast, North Island, New Zealand. *New Zealand Journal of Geology and Geophysics*, **43**, 229–245.

Bock, Y., L. Prawirodirdjo, J. F. Genrich *et al.* (2003). Crustal motion in Indonesia from global positioning system measurements. *Journal of Geophysical Research*, **108**(B8), 2367, doi:10.1029/2001JB000324.

Brace, W. F. and D. L. Kohlstedt (1980). Limits on lithospheric stress imposed by laboratory experiments. *Journal of Geophysical Research*, **85**, 6248–6252.

Bull, W. B. (2007). *Tectonic Geomorphology of Mountains: A New Approach to Paleoseismology*. Malden, MA: Blackwell.

Burbank, D. W. and R. S. Anderson (2001). *Tectonic Geomorphology*. Malden, MA: Blackwell.

Byerlee, J. D. (1978). The friction of rocks. *Pure and Applied Geophysics*, **116**, 615–626.

Byrne, D. E., D. M. Davis and L. R. Sykes (1988). Loci and maximum size of thrust earthquakes and the mechanics of the shallow region of subduction zones. *Tectonics*, **7**, 833–857.

Calais, E., J. Y. Han, C. DeMets and J. M. Nocquet (2006). Deformation of the North American plate interior from a decade of continuous GPS measurements. *Journal of Geophysical Research*, **111**, B06402, doi:10.1029/2005JB004253.

Carter, W. E. and D. S. Robertson (1986). Studying the Earth by very-long-baseline interferometry. *Scientific American*, **255**(5), 46–54.

Chapman, N. A. (2006). Geological disposal of radioactive wastes – concept, status, and trends. *Journal of Iberian Geology*, **32**, 7–14.
Christodoulidis, D. C., D. E. Smith, R. Kolenkiewicz *et al.* (1985). Observing tectonic plate motions and deformations from satellite laser ranging. *Journal of Geophysical Research*, **90**, 9249–9263.
Cifuentes, I. L. (1989). The 1960 Chilean earthquakes. *Journal of Geophysical Research*, **94**, 665–680.
Cloos, M. (1993). Lithospheric buoyancy and collisional orogenesis: subduction of oceanic plateaus, continental margins, island arcs, spreading ridges, and seamounts. *Geological Society of America Bulletin*, **105**, 715–737.
Cloos, M., B. Sapiie, A. Quarles van Ufford *et al.* (2005). *Collisional Delamination in New Guinea: The Geotectonics of Subducting Slab Breakoff*, Special Paper 400. Boulder, CO: Geological Society of America.
Cohen, B. L. (1989). Risk analysis of buried wastes from electricity generation. In: Paustenbach, D. J. (ed.) *The Risk Assessment of Environmental and Human Health Hazards*. New York, NY: John Wiley and Sons, 561–576.
Cox, A. and R. B. Hart (1986). *Plate Tectonics, How It Works*. Boston, MA: Blackwell.
Dalrymple, G. B. (1991). *The Age of the Earth*. Stanford, CA: Stanford University Press.
Davis, G. H. and S. J. Reynolds (1996). *Structural Geology of Rocks and Regions*. New York, NY: John Wiley and Sons.
Davis, J. L., R. A. Bennett and B. P. Wernicke (2003). Assessment of GPS velocity accuracy for the Basin and Range Geodetic Network (BARGEN). *Geophysical Research Letters*, **30**(7), 1411, doi:10.1029/2003/GL016961.
DeMets, C., R. Gordon, D. F. Argus and S. Stein (1990). Current plate motions. *Geophysical Journal International*, **101**, 425–478.
DeMets, C., R. Gordon, D. F. Argus and S. Stein (1994). Effect of recent revisions to the geomagnetic reversal time scale on estimates of current plate motion. *Geophysical Research Letters*, **21**, 2191–2194.
Dowrick, D. J. and D. A. Rhoades (2004). Relations between earthquake magnitude and fault rupture dimensions: how regionally variable are they? *Seismological Society of America Bulletin*, **94**, 776–788.
Elósegui, P., J. L. Davis, J. X. Mitrovica, R. A. Bennett and B. P. Wernicke (2003). Crustal loading near Great Salt Lake, Utah. *Geophysical Research Letters*, **30**(3), 1111, doi:1029/2002GL016579.
Engdahl, E. R. and A. Villaseñor (2002). Global seismicity: 1900–1999. In: Lee, W. H. K., H. Kanamori, P. C. Jennings and C. Kisslinger (eds.) *International Handbook of Earthquake and Engineering Seismology*, **81A**(24). San Diego, CA: Academic Press, 665–690.
Fujii, Y. and M. Matsu'ura (2000). Regional difference in scaling laws for large earthquakes and its tectonic implication. *Pure and Applied Geophysics*, **157**, 2283–2302.
Glen, W. (1982). *The Road to Jaramillo: Critical Years of the Revolution in Earth Science*. Stanford, CA: Stanford University Press.
Gordon, R. G. and S. Stein (1992). Global tectonics and space geology. *Science*, **256**, 333–342.
Hammond, W. C. and W. Thatcher (2004). Contemporary tectonic deformation of the Basin and Range province, western United States: 10 years of observation with the global positioning system. *Journal of Geophysical Research*, **109**, B08403, doi:10.1029/2003JB002746.

Hanks, T. C. and W. H. Bakun (2002). A bilinear source-scaling model for M−log A observations of continental earthquakes. *Seismological Society of America Bulletin*, **92**, 1841–1846.

Herring, T. A., I. I. Shapiro, T. A. Clark *et al.* (1986). Geodesy by radio interferometry: evidence for contemporary plate motion. *Journal of Geophysical Research*, **91**, 8341–8347.

Hubbert, M. K. and W. W. Rubey (1959). Mechanics of fluid-filled porous solids and its application to overthrust faulting. *Geological Society of American Bulletin*, **70**, 115–166.

Huixian, L., G. W. Housner, X. Lili and H. Duxin (eds.) (2002). The great Tangshan earthquake of 1976. Technical Report CaltechEERRL:2002.001. Pasadena, CA: Earthquake Engineering Research Laboratory, California Institute of Technology.

Isacks, B., J. Oliver and L. R. Sykes (1968). Seismology and the new global tectonics. *Journal of Geophysical Research*, **73**, 5855–5899.

Khazaradze, G. and J. Klotz (2003). Short- and long-term effects of GPS measured crustal deformation rates along the south central Andes. *Journal of Geophysical Research*, **108**(B6), 2289, doi:10.1029/2002JB001879.

Kisslinger, C. (1996). Aftershocks and fault-zone properties. *Advances in Geophysics*, **38**, 1–36.

Kreemer, C., J. Haines, W. E. Holt, G. Blewitt and D. Lavallee (2000). On the determination of a global strain rate model. *Earth Planets Space*, **52**, 765–770.

Larson, K. M. (2001). Crustal displacements due to continental water loading. *Geophysical Research Letters*, **28**, 651–654.

Larson, K. M., J. T. Freymueller and S. Philipsen (1997). Global plate velocities from the global positioning system. *Journal of Geophysical Research*, **122**, 9961–9981.

Lawson, A. C. and H. F. Reid (1908). *The California Earthquake of April 18, 1906: Report of the State Earthquake Investigation Commission in Two Volumes and Atlas*. Washington, DC: Carnegie Institution of Washington.

Lay, T., H. Kanamori, C. J. Ammon *et al.* (2005). The great Sumatra–Andaman earthquake of 26 December 2004. *Science*, **308**, 1127–1133.

Le Grand, H. E. (1988). *Drifting Continents and Shifting Theories: The Modern Revolution in Geology and Scientific Change*. Cambridge: Cambridge University Press.

Le Pichon, X. (1968). Sea-floor spreading and continental drift. *Journal of Geophysical Research*, **73**, 3661–3697.

Mai, P. M. and G. C. Beroza (2000). Source scaling properties from finite-fault-rupture models. *Seismological Society of America Bulletin*, **90**, 604–615.

Mao, A., C. G. A. Harrison and T. H. Dixon (1999). Noise in GPS coordinate time series. *Journal of Geophysical Research*, **104**, 2797–2816.

Marvin, U. B. (1973). *Continental Drift: The Evolution of a Concept*. Washington, DC: Smithsonian Institution Press.

McCalpin, J. P. (ed.) (1996). *Paleoseismology, International Geophysical Series*, 62. San Diego, CA: Academic Press.

McClusky, S., S. Balassanian, A. Barka *et al.* (2000). Global positioning system constraints on plate kinematics and dynamics in the eastern Mediterranean and Caucasus. *Journal of Geophysical Research*, **105**, 5695–5719.

McKenzie, D. P. and R. L. Parker (1967). The North Pacific: an example of tectonics on a sphere. *Nature*, **216**, 1276–1279.

McNutt, S. R., R. H. Sydnor and California Division of Mines and Geology (1990). *The Loma Prieta (Santa Cruz Mountains), California, Earthquake of 17 October 1989*. Sacramento, CA: Department of Conservation, Division of Mines and Geology.

Meade, B. J. and B. H. Hager (2005). Block models of crustal motion in southern California constrained by GPS measurements. *Journal of Geophysical Research*, **110**, B03403, doi:10.1029/2004JB003209.

Minster, J. B. and T. H. Jordan (1978). Present-day plate motions. *Journal of Geophysical Research*, **83**, 5331–5354.

Miyazaki, S. and K. Heki (2001). Crustal velocity field of southwest Japan: subduction and arc–arc collision. *Journal of Geophysical Research*, **106**, 4305–4326.

Morgan, W. J. (1968). Rises, trenches, great faults, and crustal blocks. *Journal of Geophysical Research*, **73**, 1959–1982.

Morgan, W. J. (1972). Plate motions and deep mantle convection. *Geological Society of America Memoir*, **132**, 7–22.

Nicol, A., J. Walsh, K. Berryman and P. Villamor (2006). Interdependence of fault displacement rates and paleoearthquakes in an active rift. *Geology*, **34**, 865–868.

Nyst, M. and W. Thatcher (2004). New constraints on the active tectonic deformation of the Aegean. *Journal of Geophysical Research*, **109**, B11406, doi:10.1029/2003JB002830.

Ohnaka, M. (2000). A physical scaling relation between the size of an earthquake and its nucleation zone size. *Pure and Applied Geophysics*, **157**, 2259–2282.

Perfit, M. R. and J. P. Davidson (2000). Plate tectonics and volcanism. In: Sigurdsson, H. *et al.* (eds.) *Encyclopedia of Volcanoes*. San Diego, CA: Academic Press, 89–113.

Plant, J., P. R. Simpson, B. Smith and B. F. Windley (1999). Uranium ore deposits: products of the radioactive Earth. *Mineralogy Society of America Reviews in Mineralogy and Geochemistry*, **38**, 255–319.

Prawirodirdjo, L. and Y. Bock (2004). Instantaneous global plate motion model from 12 years of continuous GPS observations. *Journal of Geophysical Research*, **109**, B08405, doi:10.1029/2003JB002944.

Reid, H. F. (1910). The mechanics of the earthquake: the California earthquake of April 18, 1906, Report of the State Investigation Commission, **2**. Washington, DC: Carnegie Institution of Washington.

Rodgers, D. W. and T. A. Little (2006). World's largest coseismic strike-slip offset: the 1855 rupture of the Wairarapa fault, New Zealand, and implications for displacement–length scaling of continental earthquakes. *Journal of Geophysical Research*, **111**, B12408, doi:10.1029/2005JB005JB004065.

Sagiya, T., S. Miyazaki and T. Tada (2000). Continuous GPS array and present-day crustal deformation of Japan. *Pure and Applied Geophysics*, **157**, 2303–2322.

Savage, J. C., W. Gan, W. H. Prescott and J. L. Svarc (2004). Strain accumulations across the Coat Ranges at the latitude of San Francisco, 1994–2000. *Journal of Geophysical Research*, **109**, B03413, doi:101029/2003JB002612.

Schiff, A. J. (1999). *Hyogoken–Nanbu (Kobe) Earthquake of January 17, 1995: Lifeline Performance*. Technical Council on Lifeline Earthquake Engineering Monograph No. 14. Washington, DC: American Society of Civil Engineers.

Scholz, C. H. (2002). *The Mechanics of Earthquakes and Faulting*. Cambridge: Cambridge University Press.

Segall, P. and J. L. Davis (1997). GPS application for geodynamics and earthquake studies. *Annual Reviews in Earth and Planetary Science*, **25**, 301–336.

Sella, G. F., T. H. Dixon and A. Mao (2002). REVEL: a model for recent plate velocities from space geodesy. *Journal of Geophysical Research*, **107**(B4), 2081, doi:10.1029/2000JB000033.

Shcherbakov, R., D. L. Turcotte and J. B. Rundle (2004). A generalized Omori's law for earthquake aftershock decay. *Geophysical Research Letters*, **31**, L11613.

Shen, Z. S., C. Zhao, A. Yin *et al.* (2000). Contemporary crustal deformation in east Asia constrained by Global Positioning System measurements. *Journal of Geophysical Research*, **105**, 5721–5734.

Shipton, Z. K., A. M. Soden, J. D. Kirkpatrick, A. M. Bright and R. J. Lunn (2006). How thick is a fault? Fault displacement–thickness scaling revisited. *American Geophysical Union Monograph*, **170**, 193–198.

Shreve, R. L. and M. Cloos (1986). Dynamics of sediment subduction, melange formation, and prism accretion. *Journal of Geophysical Research*, **91**, 10229–10245.

Sibson, R. H. (1985). A note on fault reactivation. *Journal of Structural Geology*, **7**, 751–754.

Sibson, R. H. (1986). Earthquakes and rock deformation in crustal fault zones. *Annual Reviews of Earth and Planetary Sciences*, **14**, 149–175.

Sieh, K. E. (1978). Prehistoric large earthquakes produced by slip on the San Andreas fault at Pallett Creek, California. *Journal of Geophysical Research*, **83**, 3907–3939.

Sieh, K. E. and R. Jahns (1984). Holocene activity of the San Andreas fault at Wallace Creek, California. *Geological Society of America Bulletin*, **95**, 883–896.

Sieh, K. E., M. Stuiver and D. Brillinger (1989). A more precise chronology of earthquakes produced by the San Andreas fault in southern California. *Journal of Geophysical Research*, **94**, 603–623.

Smith, D. E., R. Kolenkiewicz, P. J. Dunn *et al.* (1990). Tectonic motion and deformation from satellite laser ranging to LAGEOS. *Journal of Geophysical Research*, **95**, 22 013–22 041.

Stein, S. and G. F. Sella (2002). Plate boundary zones: concepts and approaches. *American Geophysical Union Geodynamics Series*, **30**, 1–26.

Stevens, C. W., R. McCaffrey, Y. Bock *et al.* (2002). Evidence for block rotations and basal shear in the world's fastest slipping continental shear zone in NW New Guinea. *American Geophysical Union Geodynamics Series*, **30**, 87–99.

Stirling, M. W., G. H. McVerry and K. R. Berryman (2002). A new seismic hazard model for New Zealand. *Bulletin of the Seismological Society of America*, **92**, 1878–1903.

Suppe, J. (1985). *Principles of Structural Geology*. Englewood Cliffs, NJ: Prentice-Hall.

Sykes, L. R. (1967). Mechanism of earthquakes and nature of faulting on the mid-oceanic ridges. *Journal of Geophysical Research*, **72**, 2131–2153.

Tregoning, P. (2003). Is the Australian plate deforming? A space geodetic perspective. *Geological Society of Australia Special Publication*, **22**, 41–48.

Tsenn, M. C. and N. L. Carter (1987). Upper limits of power law creep of rocks. *Tectonophysics*, **136**, 1–26.

Tuttle, M. P., E. S. Schweig, J. D. Sims, R. H. Lafferty, L. W. Wolf and M. L. Haynes (2002). The earthquake potential of the New Madrid seismic zone. *Seismological Society of America Bulletin*, **92**, 2080–2089.

USGS (2008). Earthquake Hazards Program: earthquake facts and statistics. US Geological Survey, http://neic.usgs.gov/neis/eqlists/eqstats.html.

Van Dam, T., J. Wahr, P. C. D. Milly *et al.* (2005). Crustal displacements due to continental water loading. *Geophysical Research Letters*, **28**(4), 651–654.

Van Dissen, R. J. and K. R. Berryman (1996). Surface rupture earthquakes over the last ~ 1000 years in the Wellington region, New Zealand, and implications for ground shaking hazard. *Journal of Geophysical Research*, **101**, 5999–6019.

Vigny, C., W. J. F. Simons, S. Abu *et al.* (2005). Insight into the 2004 Sumatra–Andaman earthquake from GPS measurements in southeast Asia. *Nature*, **436**, 201–206.

Villamor, P. and K. Berryman (2001). A late Quaternary extension rate in the Taupo volcanic zone, New Zealand, derived from fault slip data. *New Zealand Journal of Geology and Geophysics*, **44**, 243–269.

Wallace, L. M., C. Stevens, E. Silver *et al.* (2004a). GPS and seismological constraints on active tectonics and arc–continent collision in Papua New Guinea: implications for mechanics of microplate rotations in a plate boundary zone. *Journal of Geophysical Research*, **109**, B05404, doi:10.1029/2003JB002481.

Wallace, L. M., J. Beavan, R. McCaffrey and D. Darby (2004b). Subduction zone coupling and tectonic block rotations in the North Island, New Zealand. *Journal of Geophysical Research*, **109**, B12406, doi:10.1029/2004JB003241.

Ward, S. N. (1998). On the consistency of earthquake moment rates, geological fault data, and space geodetic strain: the United States. *Geophysical Journal International*, **134**, 172–186.

Watson, C., P. Tregoning and R. Coleman (2006). Impact of solid Earth tide models on GPS coordinate and tropospheric time series. *Geophysical Research Letters*, **33**, L0803, doi:10.1029/2005GL025538.

Wegener, A. (1929). *The Origin of Continents and Oceans*. Translated from the 4th German edition by John Biram. New York, NY: Dover Publications.

Wells, D. L. and K. J. Coppersmith (1994). New empirical relationships among magnitude, rupture length, rupture width, rupture area, and surface displacement. *Seismological Society of America Bulletin*, **84**, 974–1002.

Wernicke, B., J. L. Davis, R. A. Bennett *et al.* (1998). Anomalous strain accumulation in the Yucca Mountain area, Nevada. *Science*, **279**, 2096–2100.

Wernicke, B., J. L. Davis, R. A. Bennett *et al.* (2004). Tectonic implications of a dense continuous GPS velocity field at Yucca Mountain, Nevada. *Journal of Geophysical Research*, **109**, B12404, doi:10.1029/2003JB002832.

Wilson, J. T. (1963). A possible origin for the Hawaiian islands. *Canadian Journal of Physics*, **41**, 863–970.

Wilson, J. T. (1965). A new class of faults and their bearing on continental drift. *Nature*, **207**, 343–347.

Windley, B. F. (1995). *The Evolving Continents*, 3rd edn. Chichester: John Wiley and Sons.

Yeats, R. S., K. Sieh and C. R. Allen (1997). *The Geology of Earthquakes*. New York, NY: Oxford University Press.

3

The nature of volcanism

C. B. Connor, R. S. J. Sparks, M. Díez, A. C. M. Volentik and S. C. P. Pearson

Few people living in the town of Armero, Colombia, realized the immediate danger they faced in the autumn of 1985. Nevado Del Ruiz volcano, 65 km from Armero, had reawakened after more than one hundred years of repose. During the previous year new magma had risen beneath this ice-capped Andean volcano (Figure 3.1) to within a few kilometers of the surface. Intermittent explosions showered the summit glacier with pyroclasts, fragments of rock propelled by the sudden expansion of volcanic gases within the ascending magma. After months of this intermittent explosive activity, a much larger explosive eruption sent pyroclastic flows, hot gaseous clouds loaded with pyroclastic rock fragments, across the summit glaciers. The ice melted rapidly and water mixed with pyroclasts swept into river channels that source high on the volcano.

When these flows, termed lahars, reached the rainforest on the flanks of the volcano, their energy was sufficient to completely strip the channel banks of vegetation (Figure 3.2). The lahars that descended the Rio Lagunilla canyon toward Armero had incredible momentum. The river takes a sharp bend 1 km upstream from the mouth of the canyon, where the flow debouched onto the alluvial plain and through the town of Armero. Here, the momentum of the flow, created by a loss of elevation of about 5 km along the 65 km flow path, carried the flow up and over the river's steep embankment. Based on the resulting pattern of destruction, the mean velocity of the flow was estimated to be $\sim$ 12 m s^{-1}, its depth in the center of the channel $\sim$ 45 m and its discharge $\sim$ 50 000 m^3 s^{-1} (Lowe *et al.*, 1986). More than 23 000 people in Armero died as a result of these flows, and nearly every building in the town was destroyed (Figure 3.3).

This disaster provides key lessons for volcanic hazard assessment in general. Many of these lessons are directly applicable to volcanic hazard assessments for the siting of nuclear facilities. First, volcanic eruptions are rare phenomena, even in volcanically active parts of the world. Many people in Armero were not even aware that Nevado del Ruiz was an active volcano. This is completely understandable, as the volcano expressed none of the overt traits of volcanic activity that are commonly imagined, such as spewing lava flows or spectacular fire fountains. Indeed many active volcanoes can be dormant for centuries, or even much longer. Clearly, volcanic hazard assessments are required for nuclear facilities even in regions that appear to have exceedingly low rates of volcanism.

Fig. 3.1 Nevado Del Ruiz volcano, Colombia, an Andean composite volcano. This volcano erupted in 1985, resulting in the melting of part of the summit glacier, and the generation of debris flows (lahars) that inundated towns low on the flanks of the volcano (photo by C. Connor.)

Fig. 3.2 The November, 1985, lahar flowed > 65 km from Nevado Del Ruiz to where it debouched into the Rio Magdellena river valley. Along the flow path the lahar completely stripped the river banks of vegetation, adding volume to the flow as it descended the volcano (photo by C. Connor).

Fig. 3.3 The 1985 lahar flow discharged at a rate of $\sim 50\,000\ \mathrm{m^3\ s^{-1}}$ as it reached the town of Armero. Most of the town was destroyed (photo by C. Connor).

Some types of volcanic phenomena are extremely destructive and are capable of destroying man-made structures, including nuclear facilities. Although the 1985 Nevado del Ruiz eruption was devastating in terms of loss of life, the eruption itself was not particularly energetic or voluminous. This tragedy had much to do with the vulnerability of the town of Armero, sited within a major river valley sourced on the volcano (Voight, 1990). Apart from volcanic lahars, destructive volcanic phenomena include: pyroclastic flows that travel at speeds of tens of meters per second and sustain high impact pressures on structures; volcanic debris avalanches that may have very large ($> 10^8\ \mathrm{m^3}$) volumes and which may bury hundreds of square kilometers in thick aprons of debris; slow-moving lava flows that can be tens of meters thick and which envelop and destroy structures as they advance; and falls of chemically and physically corrosive volcanic ash that can result in failure of advanced engineering systems. Nuclear facilities are vulnerable to such phenomena. Of course, many types of surface flows, such as the Nevado del Ruiz debris flow, would not likely destroy a nuclear facility outright. But, even if nuclear facilities survive such events due to good engineering design, these facilities can become isolated in a sea of catastrophic destruction. It is vital to consider how such facilities in volcanically active regions would function in such circumstances over long periods of time.

Despite the rarity and complexity of volcanic eruptions, volcanic hazard assessment is feasible, and probabilistic volcanic hazard assessments can yield practical quantitative results. Prior to the eruption of Nevado Del Ruiz, volcanologists had prepared reasonably accurate hazard maps. These maps depicted the reach and potential consequences of eruptions. Specifically, the maps indicated that, given an eruption of Nevado Del Ruiz, lahars would potentially inundate Armero and the surrounding region. In the last decades, volcanological models have improved dramatically (Sparks, 2003). Volcano monitoring has increased in sophistication, and yields ever-improving forecasts of volcanic activity. These developments lend credence to the idea that volcanic hazard assessment for nuclear facilities can be robust, despite the rarity and complexity that is inherent to volcanic activity.

The best hazard assessment, however, is useless if no one is listening. The tragedy at Armero was in large part a result of poor communication (Voight, 1990). Decision makers simply did not act upon information provided by volcanologists. Perhaps information was not provided to them in a sufficiently clear or timely manner (Voight, 1990). Today there are examples of nuclear power plants in operation, and some closed, which were constructed in volcanically active regions (Table 3.1, Figure 3.4). Some nuclear facilities were constructed without adequate characterization of the potential or consequences of volcanic eruptions. One example is the closed nuclear power plan at the Bataan site in the Philippines. This power plant, constructed on the flanks of Natib volcano and which received tephra (volcanic ash) fallout from the 1991 eruption of Pinatubo volcano, was cited by the Union of Concerned Scientists as being at particular risk from volcanic activity (D'Amato and Engel, 1988). Another example is the Metsamor nuclear power plant in Armenia, which was constructed within a region of many small basaltic volcanoes that will inevitably experience future eruptions (Figure 3.5a). Yet, at the time of construction in the mid 1970s, no volcanic hazard assessment was considered and hazard assessments have only been done retrospectively (Kharakhanian *et al.*, 2003; Weller *et al.*, 2006). Similarly, recent volcano seismic data collected near Tatun volcano, Taiwan, and nearby operating nuclear power plants suggest that this area is more volcanically active than previously thought (Lin *et al.*, 2005). Such facilities require thorough volcanic hazard studies (Figure 3.5b). Of course, it is best to conduct such studies in the earliest stages of site evaluation, exemplified by assessments at the proposed site of a nuclear power plant on the Muria Peninsula, Indonesia (McBirney *et al.*, 2003; Figure 3.5c) and at the proposed high-level radioactive waste site at Yucca Mountain, USA (Figure 3.5d). One purpose of this volume is to provide readers with a sense of the scope and strategy for such investigations.

In light of these lessons, the primary goals of volcanic hazard assessments at nuclear facilities are to: (i) quantify the potential of volcanic activity in the region of a planned or existing nuclear facility, where the region is defined based on the magnitude of potential volcanic phenomena that might impact the site; (ii) characterize the potential magnitudes and nature of these volcanic phenomena, and their probability of affecting the site and surrounding environs; and (iii) estimate the impact of volcanic events, both on the integrity of the facility itself and on the surrounding region. A hazard assessment that achieves these goals can provide an adequate basis for a hazard and risk assessment for a specific site. Volcanology provides perspectives and techniques that make these assessments possible.

In this chapter we provide an overview of the nature of volcanism from the perspective of making such hazard and risk assessments. This overview encompasses the distribution

Table 3.1. Some nuclear facilities / nuclear power plants (NPP), where volcanic hazard has arisen as an issue in site selection or safety evaluation

Nuclear facility	Country	Volcanic hazard	References
Metsamor NPP	Armenia	Near Quaternary monogenetic volcanoes; pyroclastic flows from nearby composite volcanoes	Figure 3.5a; Karakhanian *et al.*, 2003; Weller *et al.*, 2006
Bataan NPP	Philippines	Near Natib composite volcano (erupted ≈ 14 ka); tephra fall from Mt. Pinatubo (4 cm of tephra fallout, 1991)	Volentik *et al.*, Chapter 9, this volume; D'Amato and Engel, 1988
Trojan NPP	USA	Near Mt. St. Helens volcano; debris flows mainly associated with 1980 eruption	Hoblitt *et al.*, 1987
Chin-shan 1 NPP, Chin-shan 2 NPP	Taiwan	Near Quaternary Tatun volcano group; active hydrothermal zones; volcanic seismicity; magmatic gases	Lin *et al.*, 2005
Mülheim–Kärlich NPP	Germany	Near Quaternary Eifel monogenetic volcanic field; risk of pyroclastic deposits damming Rhine river and flooding site	Figure 3.5b; Jaquet and Carniel, 2006
Idaho National Laboratory (INL) nuclear facility	USA	Located in monogenetic volcanic field; associated with hot-spot track across East Snake river plain	Wetmore *et al.*, Chapter 16, this volume
Muria site evaluation for NPP	Indonesia	Near Quaternary Muria volcano complex; Holocene maar volcanism; magmatic gas	Figure 3.5c; McBirney *et al.*, 2003
Yucca Mountain site evaluation for HLW repository	USA	Near monogenetic volcanoes (Pliocene–Quaternary in age)	Figure 3.5d; Valentine and Perry, Connor and Connor, Chapters 19, 14, this volume
HLW facility	Japan	Nuclear Waste Organization of Japan using volcanic hazard assessment as one criterion for proposed HLW site selection; site not yet identified	Apted *et al.*, 2004

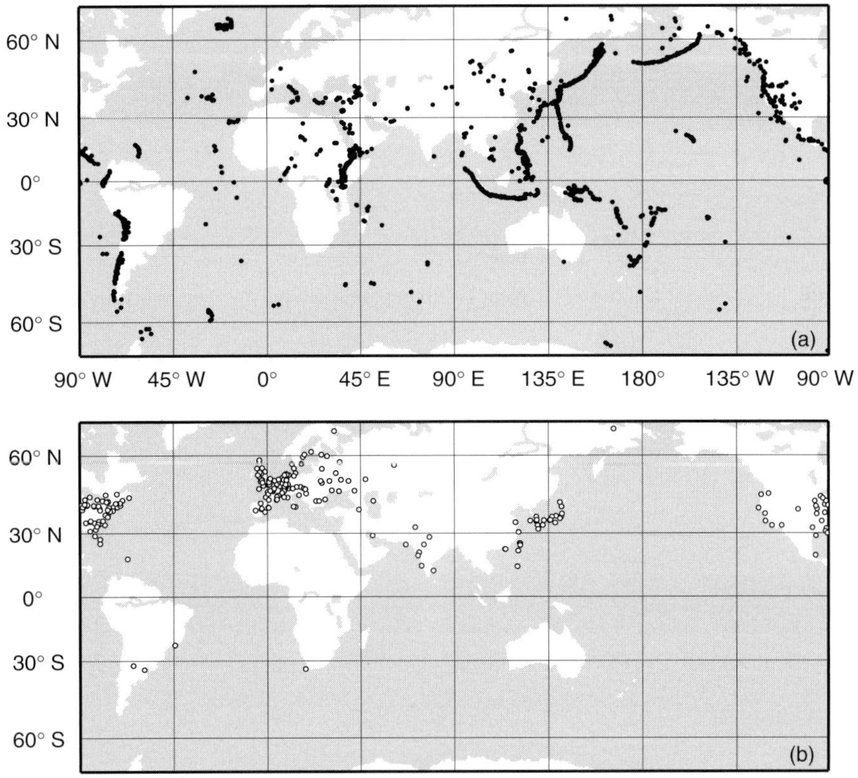

Fig. 3.4 Map (a) shows the global distribution of subaerial volcanoes, and some submarine volcanoes, active within the last 10 ka (black dots) and the boundaries of tectonic plates (dotted gray lines). Map (b) shows the distribution of 250 nuclear facilities from the International Atomic Energy Agency database, relative to tectonic plate boundaries.

of volcanism, rates of volcanic activity, physical nature of magmas, products of volcanic eruptions, monitoring and hazard assessment strategies. This chapter is only an overview. Specific topics are developed in detail in subsequent chapters and references therein.

3.1 Where are volcanoes in the world?

Volcanism is a natural manifestation of the high temperature of Earth's interior. Much of Earth's interior is solid. In many places the temperature of Earth's mantle is so close to the solidus that it partially melts when perturbed by tectonic processes. The solidus is the temperature at which rocks under specific pressure start to melt. Very high temperatures, which are rarely achieved, are required to completely melt rocks, so most magmas are generated by partial melting at depths less than 200 km beneath the surface. During volcanism, partial melts separate from the hot mantle and the resulting magmas ascend from the interior of

Fig. 3.5 Nuclear facilities are often sited in volcanically active regions. (a) The Metsamor nuclear power plant in Armenia is located in close proximity to Quaternary monogenetic volcanoes. (b) The Mülheim–Kärlich nuclear power plant in Germany, currently shut down, is located in the Eifel volcanic field and constructed on Late Pleistocene reworked pyroclastic surge facies. (c) A digital elevation model of the Holocene Muria volcano complex in Indonesia shows the proposed Muria nuclear power plant site (outlined in black). (d) In Nevada, USA, monogenetic volcanoes lie west of Yucca Mountain, the proposed site of a high-level-waste geologic repository. (Photos by C. Connor.)

Earth and erupt at the surface. In the process, heat is transferred from the interior to the exterior of the planet.

Volcanism results mostly from plate tectonic processes (Figure 3.4a), when lithospheric plates, $\sim$ 100 km thick, move with respect to Earth's deep interior. Volcanism on Earth occurs at divergent plate margins, where ocean crust is constructed from the products of volcanism and intrusive magmatism. Magmas in these environments are produced by decompression melting, which is the consequence of the solidus temperature decreasing at low pressures. Mantle rising from depth below diverging plates toward the surface initiates partial melting and the generation of basaltic magmas. Worldwide, there is much variability in this process and the details can be complex, but, in general, basaltic magmas are poor in volatiles and produce volcanism that is dominantly effusive rather than explosive in nature. Most divergent plate margins are on the seafloor at depths of 2–2.5 km, and thus are of little interest to hazard assessments for nuclear facilities. Exceptions occur at divergent plate boundaries in continental settings at the earliest stages of rift development, or where rifts have failed to progress to the stage of actual ocean crust development. Nuclear facilities have been located in some of these areas, such as the Rhine Graben in central Europe (Figure 3.5b).

Volcanism also occurs where lithospheric plates collide and one lithospheric plate is thrust under, or subducted beneath, another plate. In these areas partial melting of the mantle is triggered by the introduction of volatile compounds from the subducted plate into the overlying mantle asthenosphere. Volcanism in such regions occurs in volcanic arcs. These volcanic arcs are linear chains or groups of volcanoes that roughly parallel the margins between the two plates and, typically, an ocean trench that marks the subduction zone. Earthquakes often accompany subduction-zone volcanism. Thus, seismic and volcanic hazards are linked on a regional scale in these environments. The Pacific rim is literally a rim of subduction zones, colloquially called the "Ring of Fire." As population densities increase in this economically vital region, and global concerns about climate change and use of fossil fuels for energy increases, nuclear facilities will likely proliferate around the Pacific rim (Chapman *et al.*, Chapter 1, this volume).

Hot-spot volcanism is thought to occur where hot mantle plumes ascend from deep within Earth's mantle, causing partial melting of the mantle asthenosphere and sometimes the lithosphere. Many models locate the source of the largest mantle plumes at the core–mantle boundary. The largest mantle-plume systems are long-lived, persisting for many millions of years, and may occur in intraplate settings, such as Hawaii or Yellowstone, USA; or at plate margins, such as Iceland. Although more rare, some nuclear facilities, such as the Idaho National Laboratory (Wetmore *et al.*, Chapter 16, this volume), are located in areas affected by hot-spot volcanism. Not all such hot spots are clearly related to mantle plumes and the exact origin of many hot spots is debated. There are many smaller volcanic fields within the boundaries of lithospheric plates and the interiors of continents where mantle-plume origin is not obvious. Here, it is possible that smaller-scale processes in the asthenosphere, for example incipient rifting, are causing volcanism.

Within any one of these plate tectonic settings, the distribution of volcanoes can be quite complex and non-uniform. Active volcanoes occur in several segments along the west coast of South America, separated by zones that have no recent volcanic activity (Figure 3.6). The segmented nature of this volcanic arc is caused by variations in the angle of the subducting ocean lithosphere and changes in the composition of the subducted ocean crust. Interestingly, seismic hazards on the west coast of South America are highest in

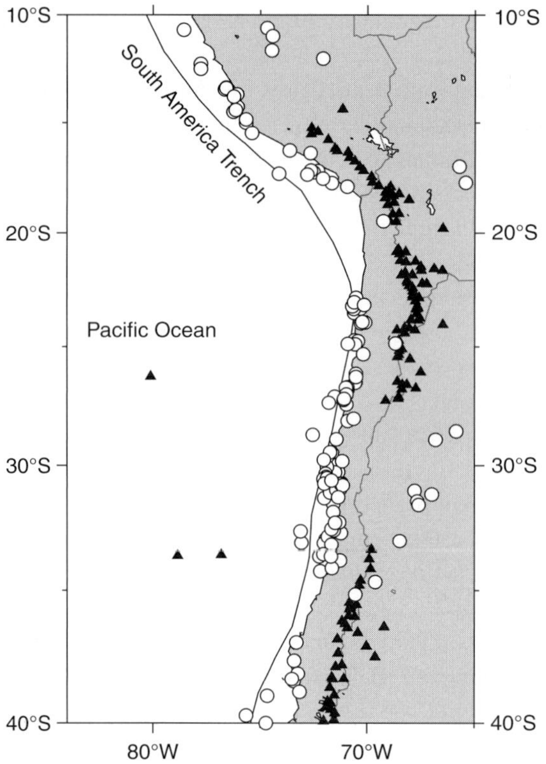

Fig. 3.6 This map illustrates the segmented nature of volcanism along the western margin of South America. Holocene volcanoes are shown as solid triangles, earthquake epicenters (Mw > 6.0) as white circles and the ocean trench, marking the subduction zone, by the thick line. Although subduction is continuous along the entire length of South America, volcanism is discontinuous. This distribution is thought to relate to the angle of the subducting slab, which is shallower in regions lacking Holocene volcanic activity.

segments where the angle of subduction is shallow, partial melts are not generated, and volcanism does not occur (Allmendinger *et al.*, 1997).

As another example, consider the Tohoku volcanic arc, northern Honshu, Japan. Several authors have noted that, although this a classic subduction-zone environment, volcanoes cluster in the Tohoku volcanic arc, rather than being distributed uniformly or randomly along the arc (this volume, Tamura *et al.*, Chapter 7; Kondo, Chapter 12; Mahony *et al.* Chapter 13; Jaquet and Lantuéjoul, Chapter 15). This clustering of volcanoes is associated with several geophysical anomalies, such as the occurrence of seismic tomographic slow velocity regions in the upper mantle and gravity anomalies in the crust. Both are indications of the persistence of clusters as a fundamental geophysical feature of the subduction zone (Zhao, 2001). The Tohuku case shows that the occurrence of volcanism is commonly not uniform along a given volcanic arc. In some parts of the arc volcanic hazards will be higher than elsewhere. Conversely, there may be locations in any volcanic arc where nuclear facilities might be sited safely.

Volcanism is not limited to these generic tectonic environments. In the western USA, volcanism is widely distributed throughout physiographic provinces known as the Basin and Range and the Colorado Plateau. These regions are characterized by low-volume, infrequent eruptions of basaltic magmas within volcanic fields. Volcanism throughout much of the region is attributed to decompression melting, resulting from extension of the lithosphere of the western USA. This partial melting is thought to be enhanced in the region because the area was the location of subduction over a period of tens of millions of years which chemically modified the mantle and made it more susceptible to partial melting (Farmer *et al.*, 1989). Nuclear facilities throughout much of the western USA, such as the proposed repository at Yucca Mountain, Nevada (this volume, Spera and Fowler, Chapter 8; Connor and Connor, Chapter 14; Valentine and Perry, Chapter 19) must take this type of volcanism into account in hazard assessments (Figure 3.5d). Similarly, subduction ceased millions of years ago in the Anatolian region of Turkey, Armenia and Iran, yet active volcanism persists today in this region (e.g. Keskin, 2003). Furthermore, widespread but diffuse volcanism in continental Asia, particularly in China, is not directly related to an active convergent or divergent plate boundary (Liu, 1999).

Volcanoes within continents have added complexity because the crust as well as the mantle can melt to generate magmas. Crustal magmas can be formed by a variety of melting mechanisms, but heating of the crust by intruding hot basalt, fluxing of hot crust by volatiles and internal heating of sedimentary rocks by radioactive decay are the most commonly proposed (England and Thompson, 1984; Annen and Sparks, 2002). These melts are typically very silica rich and form the great granite intrusions of Earth's continental crust. The volcanic equivalent of silica-rich magmas is called rhyolite. These volcanic systems can erupt in huge explosive eruptions which disgorge hundreds to thousands of cubic kilometers in single catastrophic events (Self, 2005). There are many volcanic regions that will experience several such eruptions over the lifetime of high-level radioactive repositories (e.g. Japan). Disruption to the affected human societies by such events might be so severe that records of the whereabouts of repositories could well be lost.

3.2 What types of volcanoes are there?

The term volcano refers to any place where magma has reached a planet's surface at any time in the past. Clearly this term encompasses a broad spectrum of processes, and volcanoes have tremendously varied forms and dimensions. The largest volcanoes are shield volcanoes, characterized by low topographic slopes constructed from the eruptions of tremendous volumes of comparatively low-viscosity basaltic lavas (Figure 3.7a). In contrast, composite volcanoes are steep-sided conical volcanoes built by the effusion of lava flows and domes, and by the explosive eruption of pyroclastic materials that form density currents and tephra falls (Figure 3.7b). Many individual eruptive episodes lead to the formation of both shield and composite volcanoes, which collectively are termed polygenetic.

Fig. 3.7 Volcanoes have widely varying morphologies, reflecting the processes that lead to their construction. (a) Mauna Kea, a shield volcano in Hawaii, is constructed predominantly from effusion of basaltic lava flows (photo by A. Leonard–Hintz). (b) San Cristobal, a composite volcano in Nicaragua, consists of interbedded lavas and tephra fallout deposits (photo by C. Connor). (c) Quilotoa Caldera is 1.3 km in diameter and formed during an explosive eruption 0.8 ka (photo by A. Volentik). (d) Monogenetic cinder cones in the Yerevan basin, Armenia formed along a major strike-slip fault in the basin. The large Ararat composite volcano appears in the background (photo by C. Connor). See color plate section.

Although some composite volcanoes have patterns of past activity that can be used to assess the likelihood of hazardous phenomena occurring, composite volcanoes show considerable variation in their development, and appropriate consideration should be given to a broad range of potentially hazardous explosive and effusive phenomena. The geologic record at many composite volcanoes shows that abrupt changes in composition or eruptive character are common, and that eruptive centers can suddenly emerge kilometers away from the central (summit) vent (Perry *et al.*, 2001).

Composite and shield volcanoes are characterized by many eruptions that gradually build the edifice of the volcano. Individual eruptions at polygenetic volcanoes can last for months, or in some cases for years. In some circumstances the eruption is so voluminous that the overlying edifice is completely destroyed by the eruption, and the ground over the eruptive vent may actually subside as a result of the eruption of magma from a shallow magma chamber within the crust. Depressions formed as a result of voluminous volcanic activity are termed calderas (Figure 3.7c). Some eruptions are so large and violent that an edifice never forms; the volcano is the caldera depression. Calderas are commonly greater than 10 km in diameter and produce eruptive products that devastate regions greater than 100 km from the volcano.

Not all volcanism occurs from the central vents of existing composite or shield volcanoes, or calderas. In many circumstances, volcanism is distributed, and renewed volcanic activity results in the formation of new monogenetic volcanoes. These volcanoes are characterized by single episodes of volcanic activity. On large volcanoes, such as Mt. Fuji in Japan, or Mt. Etna in Italy, the process of new vent formation is shown by hundreds of scoria cones and related volcanic features that dot the landscape for tens of kilometers around the volcano. In other areas, volcanism builds volcanic fields, sometimes consisting of hundreds of individual vents distributed over hundreds or thousands of square kilometers, each of which opened separately as an individual batch of magma ascended to the surface (Figure 3.7d). Parícutin, Mexico is a famous example of such a volcano where, in 1943, the new volcano formed in the fields of a surprised farmer (Luhr and Simkin, 1993). Volcanic hazards from such distributed sources is of major concern in siting many nuclear facilities (Table 3.1) because of the potential for new volcanic vents to form in the site vicinity (e.g. Connor *et al.*, 2000; Martin *et al.*, 2004; Jaquet and Carniel, 2006; Weller *et al.*, 2006).

3.3 How frequently do volcanoes erupt?

To our knowledge, sites for nuclear facilities are simply not considered near the world's most active volcanoes. Rather, when volcanic hazard issues arise they are typically associated with long-dormant volcanoes or very low productivity volcanic systems. For example, McBirney *et al.* (2003) discuss the problem of determining whether the long-dormant Muria volcano on Java, Indonesia (Figure 3.5c) poses a hazard to a proposed nuclear power plant site, located $\approx$ 30 km from the volcano. Is this volcano extinct or is it merely dormant? There is unfortunately no clear-cut method of distinguishing long-dormant from extinct volcanoes (Szakács, 1994).

Based on the Holocene geologic record of volcanism (Siebert and Simkin, 2007), we know that volcanoes often experience long periods of repose between episodes of volcanic activity. Holocene volcanoes have mean repose intervals of $\sim$ 0.45 ka prior to moderately explosive eruptions. The mean repose interval increases to $\sim$ 1.5 ka before the most explosive eruptions (Connor *et al.*, 2006). Furthermore, although the record is thin, volcanoes can remain dormant for thousands of years and even tens of thousands of years can separate eruptions. A good example is the $\sim$ 16 ka repose interval between caldera-forming eruptions at Santorini, Greece (Druitt *et al.*, 1999). Long repose intervals have also been documented at Mt. Lamington in New Guinea (6.8 ka) and for Tongariro, in New Zealand (7.2 ka) (Siebert and Simkin, 2007). Most recently, Chaitén volcano, in Chile, erupted in May 2008 after a reported 9.5 ka repose. Certainly, lack of Holocene activity is not clear evidence that a volcano is extinct. Repose intervals are even longer in monogenetic fields. For example, the average repose interval between eruptions of monogenetic volcanoes in the Yucca Mountain region is $\sim$ 500 ka, yet volcanism has persisted in the region at this low rate since 8 Ma (Valentine and Perry, 2006).

As these long repose intervals imply, for many of the world's volcanoes the record of past activity is inadequate to assess the likely return times of eruptions of different magnitude and style. Written records typically do not go back more than a few centuries and at most a few thousand years. Geological studies are painstaking and require long-term commitment by funding agencies. Indeed only about 20% of the world's volcanoes have had adequate geological histories constructed for even the last 10 ka (i.e. the Holocene) from an analysis of data on the Smithsonian global volcano database. Such histories become more difficult to reconstruct back in time due to erosion and burial of older stratigraphic units. Increasingly sophisticated and accurate radiometric dating methods, such as ^{14}C, $^{39}Ar/^{40}Ar$ geochronology and cosmogenic isotope dating, have emerged in the last couple of decades. However, improvement in the general dating of volcanoes remains slow because the methods are expensive and require scarce specialists and laboratories. What is axiomatic is that the accurate and comprehensive dating of young volcanic rocks in the vicinity of a nuclear facility should have a high priority because such data are essential for assessment of rates of volcanism and assessing future hazards.

3.4 Magma ascent and eruption

Magma is generated by partially melting rock through various mechanisms, such as decompression, the introduction of volatiles or the heating of crust. This partial melt accumulates along grain boundaries in the source region creating an interconnected network of fluid. Typically the melt is 5%–10% less dense than the residual matrix. Once a critical volume of melt is reached, buoyancy forces trigger its segregation from the mantle by upward, buoyant flow and the formation of fracture systems, which allow the melt to drain from regions of partial melt. Magma then ascends through the lithosphere and either crystallizes as intrusive bodies, such as dikes, sills or plutons, or reaches the surface and creates a volcanic eruption.

3.4.1 Physical properties of magmas

Magmas are multiphase mixtures of melt, crystals and bubbles with physical properties dependent on temperature, pressure, chemical composition and additional parameters, such as crystallinity and vesicularity. For purposes of understanding magma flow and to understand many types of volcanic hazard, the two most important physical properties of magmas are density and viscosity.

Small changes in density have a significant impact on the dynamics of magma ascent as density controls the buoyancy forces acting on the magma. The effect of temperature on magma density is limited because the thermal expansion coefficients of silicate compounds comprising magmas are very small. Thus, large temperature changes are needed to cause small changes in magma density. In contrast, small differences in chemical composition can produce substantial density changes. For example, basaltic, andesitic and rhyolitic melts at liquidus temperatures have densities of 2700, 2500 and 2300 kg m^{-3}, respectively. At crustal depths (e.g. < 35 km) changes in pressure do not have a significant impact on magma density except quite close to Earth's surface where pressure is low enough to allow gases to bubble out of magmas. Magmas contain variable amounts of volatiles with different compositions. The most abundant is H_2O, although CO_2 and SO_2 are also important. Water, in particular, is a low-density component and increasing water content decreases density of magmas. For example, andesite melt of intermediate silica content but with a volatile content of 4 wt.% water is less dense than dry silica-rich rhyolitic melt.

Viscosity modulates the transport of magma through the lithosphere and the mechanics of volcanic eruptions. Hazardous volcanic phenomena are largely controlled by changes in viscosity during magma ascent to the surface along the volcanic conduit. When crystals and bubbles are not present, the behavior of a magmatic melt is essentially Newtonian under a wide range of flow conditions. The relationship between melt viscosity and temperature and composition for a viscous melt is generally described using the Arrhenius equation given by

$$\mu_{\text{melt}} = A \exp\left(\frac{E}{RT}\right) \tag{3.1}$$

where μ_{melt} is the melt viscosity, A is a parameter that accounts for melt composition, E is activation energy for viscous flows, R is the gas constant and T is the melt temperature. A commonly used method to calculate the viscosity of a magmatic melt of any composition at any temperature, based on the Arrhenius model, was proposed by Shaw (1972). More recent non-Arrhenian models have been developed by Hess and Dingwell (1996), and Giordano and Dingwell (2003), and have improved viscosity estimates. Laboratory experiments demonstrate that chemical composition has a significant effect on the viscosity of silicate melts (Bottinga and Weill, 1970; Dingwell *et al.*, 2000; Dingwell *et al.*, 2004) reflecting the degree of polymerization of the liquid. A rhyolitic melt, richer in silica than a basaltic melt, contains a larger number of long chains of silicates and rings of silicates that are difficult to shear, thus giving the silicate melt a comparatively higher viscosity. The viscosity for basaltic melts ranges from roughly 10–1000 Pa s, whereas for more silicic melts, such as dacitic and rhyolitic, viscosities are several orders of magnitude higher, generally from 10^5-10^{10} Pa s for temperatures 800–1000 °C. As (3.1) suggests, changes in temperature impact the viscosity of the melt substantially. For example, by increasing the temperature of a basaltic melt $\approx$ 200 °C, viscosity can decrease one order of magnitude.

Crystals tend to increase magma viscosity. For low-crystal volume-fraction magmas, flow is expected to be Newtonian (i.e. the relationship between applied stress and the resulting strain in the magma is linear). Magma viscosity can be described by the Einstein–Roscoe equation (Marsh, 1981; Pinkerton and Stevenson, 1992; Lejeune and Richet, 1995)

$$\mu_{\text{magma}} = \mu_{\text{melt}} \left(1 - \frac{v_{\text{s}}}{v_{\text{max}}}\right)^{-2.5} \tag{3.2}$$

where v_s is the volume crystal fraction and v_{max} is the volume fraction of crystals at which maximum packing is achieved. On the other hand, high-crystal volume-fraction magmas develop yield strength, that is, they do not deform initially at low applied stress. Because such magmas have yield strength, their viscosity is non-Newtonian and different empirical relationships are proposed to calculate the viscosity (Pinkerton and Stevenson, 1992; Dingwell *et al.*, 1993; Costa, 2005). The presence of bubbles in magmas also induces non-Newtonian flow behavior (e.g. Manga *et al.*, 1998; Llewellin *et al.*, 2002; Rust and Manga, 2002). A method has been proposed to calculate the viscosity of bubbly flows along volcanic conduits that accounts for the effects of changing crystal content and bubble growth in the ascending magma (Llewellin and Manga, 2006). The sensitivity of viscosity to these factors is one of the main reasons for the wide variation in the intensity and style of volcanic eruptions (e.g. Woods *et al.*, 2002; Spera and Fowler, Chapter 8, this volume).

3.4.2 Magma ascent

Magmas ascend through the lithosphere due to buoyancy forces that result from the density contrast between the comparatively cold and rigid rock of the crust and the ascending hot magma. The viscosity of magma and the rheology of the lithospheric host rock dictate the mechanisms that govern magma ascent.

Basaltic magmas invariably ascend as vertical planar fractures filled with magma through cold and brittle crust. These structures are known as dikes. Studies of the migration of seismic swarms and deformation prior to eruptions of Kilauea volcano, Hawaii, and Krafla and Hekla volcanoes, Iceland, have revealed dike ascent velocities of 0.5–3 m s^{-1} in the shallow crust (Aki *et al.*, 1977; Einarsson and Brandsdóttir, 1980; Linde *et al.*, 1993). The study of xenoliths in alkaline lavas (e.g. Spera, 1984) has also provided minimum dike ascent velocity estimates for mantle-derived dikes in the range of 0.001–10 m s^{-1}. These magmas can feed eruptions at the surface or crystallize at shallow depths forming dikes and horizontal sheets called sills (Figure 3.8). Reported average thicknesses of basaltic dikes range from 1 m in eroded Hawaiian volcanoes to 4 m in the Tertiary swarms of Scotland

Fig. 3.8 A basaltic feeder dike and sill exposure in the San Rafael subvolcanic field, Utah. This sill formed at a depth of ~ 800 m, based on paleotopographic reconstruction and stratigraphy (photo by M. Díez). See color plate section.

and Iceland, although some Proterozoic basaltic dike swarms can have dikes which are tens of meters thick (Rubin, 1995).

Silicic magmas have viscosities that are several orders of magnitude higher than basaltic magmas. These magmas commonly form large magma bodies in the middle and upper crust, known as plutons, that can be hundreds to tens of thousands of cubic kilometers in volume. Calderas are usually associated with such large magma bodies emplaced close to the surface. A more diverse range of ascent mechanisms have been proposed for silicic magmas. Ascent as diapirs within Newtonian (Marsh, 1982) and non-Newtonian (viscoelastic) host rock (Weinberg and Podlachikov, 1994; Miller and Patterson, 1999) has been an accepted mechanism for many years, but more recent models suggest that such huge bodies are assembled incrementally *in situ* over tens of millions of years and favor dike transport from depth (e.g. Clemens and Mawer, 1992; Petford *et al.*, 1993; Glazner *et al.*, 2004). Rhyolite dikes are increasingly recognized in rock outcrops. Despite numerous laboratory experiments, numerical models and field studies, the mechanism of silicic magma ascent remains an open problem in magma transport science.

Magma intruded at shallow levels forms magma reservoirs that act as feeders of volcanic eruptions. For example, the magma reservoir feeding the most active volcano on Earth, Kilauea volcano, Hawaii, has been delineated with geophysical techniques (Ryan *et al.*, 1981). This magma reservoir extends from 2 to 6 km deep with a volume of ~ 10 km^3.

The nature of magma ascent holds key information for hazard assessment. Individual batches of magma may be spatially distributed over a broad area. The vents created by past eruptions might not be the vents that will host future eruptions. Underground facilities, such as high-level radioactive waste repositories, may be intruded and disrupted by igneous intrusions such as dikes, sills and plutons (Woods *et al.*, 2002, 2006; Dartevelle and Valentine,

2005; Menand *et al.*, Chapter 17, this volume; Lejeune *et al.*, Chapter 18, this volume). The potential impact of ascending magma is strongly influenced by magma rheology. The interplay between magma rheology and eruption dynamics is considered next.

3.4.3 Eruption dynamics

Active shallow magma reservoirs beneath many volcanoes are open systems that are repeatedly supplied by magma from deeper reservoirs during the span of activity of a particular volcanic system. The dynamics of the reservoir is complex, with convective flows caused by density contrasts related to variations in temperature, melt composition, crystal content and bubble content (Sparks *et al.*, 1984). The pressure in the reservoir results from the lithostatic load plus a magma overpressure created by the input–output budget of magma in the system and internal factors, such as crystallization within the reservoir (Tait *et al.*, 1989). When the magma reservoir is perturbed by an external factor, such as the arrival of a new magma batch or a landslide in the volcanic edifice that suddenly reduces lithostatic load, bubble nucleation can be triggered, increasing the magma overpressure in the reservoir. If this overpressure increase is greater than the strength of the host rock, an eruption can be triggered.

As magma initiates its ascent through the shallowest crust toward the surface, pressure decreases and bubbles grow by diffusion of volatiles from the surrounding melt. Gas exsolution can be accompanied by crystallization because the solidus and liquidus are increased as a consequence of degassing. In fact, the temperature of ascending magma can increase considerably as a consequence of latent heat released upon crystallization (Blundy *et al.*, 2006) impacting flow dynamics. Depending on the rheology, initial volatile content and crystallization kinetics of the ascending magma, eruptions can be either effusive or explosive.

Effusive eruptions are commonly associated with basaltic magmas. Due to the low viscosity of these magmas, bubbles can rise along the conduit escaping from the melt and reducing the driving force for a major explosion. Typical effusive basaltic activity ranges from slow bubbling in lava lakes to more vigorous lava fountains that can feed lava flows. Mass flow rates associated with this activity can be $1–10^6\,\mathrm{kg\,s^{-1}}$ with total volumes of $10^{-2}–20\,\mathrm{km}^3$, although much more voluminous eruptions are found in the geologic record (e.g. Swanson *et al.*, 1975). Basaltic magmas vary considerably in volatile content. For example, mid-ocean ridge and Hawaiian basalts usually contain rather low gas contents while some of the basalts at arcs and in continental volcanic fields can have quite high volatile contents (e.g. Roggensack *et al.*, 1997). These latter basalt volcanoes can display more vigorous explosive activity but they rarely reach the intensity of the more viscous silicic eruptions. Silicic magmas, such as rhyolites, that have lost a large amount of volatiles during their ascent through highly fractured shallow conduits, often form lava domes at the vent. Lava-dome eruptions are characterized by slow extrusion rates of about $10^2–10^4\,\mathrm{kg\,s^{-1}}$. These lava domes can grow to over 1 km in height and their collapse can produce voluminous pyroclastic flows.

Explosive eruptions are generally associated with silicic magmas having high volatile contents (typically 6–8 wt.%). Such volcanic activity is characterized by the explosive fragmentation of multiphase magma along the volcanic conduit. Magma fragmentation has been intensively studied over the last few decades and several mechanisms have been proposed (e.g. Sparks, 1978; Papale, 1999; Gonnermann and Manga, 2003; Büttner *et al.*, 2006). However, the details of magma fragmentation remains an open problem and an active area of research.

Magma, depending on its deformation rate, can fragment by two different rheological regimes (Zimanowski *et al.*, 2003). Hydrodynamic fragmentation generally involves the rapid acceleration of magma by pressurized bubbles restricted to bubble–melt interfaces and is most efficient at low viscosities and low interfacial tension between the accelerating bubble and surrounding melt. Alternatively, brittle fragmentation results from crack growth caused by large strains that exceed the elastic properties of the magma.

As a result of fragmentation, the bubbly magma is transformed into a mixture of particles and gas that accelerates towards the surface. This mixture exits the vent at high speeds $\sim 100 - 400\,\mathrm{m\,s^{-1}}$ with mass flow rates ranging from $10^5 - 10^9\,\mathrm{kg\,s^{-1}}$, forming an eruption column (Sparks *et al.*, 1997). Immediately above the vent, the ascent of the eruption column is dominated by momentum. The eruption column entrains air from the surrounding atmosphere and decreases its density with height. When the column density becomes lower than the atmospheric density, due to its air entrainment and height, the column starts to ascend buoyantly until the neutral buoyancy level is reached and the column spreads laterally as a volcanic ash cloud. Alternatively, if the vent gradually erodes during the eruption, the eruption column reaches a point where the column is too wide for sufficient air to be entrained to attain buoyancy. At this point the column collapses and forms pyroclastic density currents. Eruption column heights during explosive eruptions range from a few kilometers in small-magnitude explosions up to 40 km during large explosive eruptions, such as the Pinatubo volcano eruption in 1991 in the Philippines (Koyaguchi, 1996).

The flow of magma through conduits and eruption columns is studied with numerical models using mass, momentum and energy equations along with an equation of state for the magmatic mixture (e.g. Díez, 2006). These equations are solved to compute different flow variables along the conduit and eruption column (Figure 3.9). These models have been extensively studied during the last three decades (see Freundt and Rosi, 1998 for an excellent review) and have helped volcanologists to quantify the dynamics of volcanic eruptions. An important outcome of this modeling effort is the realization that eruption conditions, and therefore hazards, are extremely sensitive to physical conditions in the magma and volcano conduit. From a hazard perspective, this conclusion implies that it is normally necessary to consider a broad range of eruption scenarios at potentially active volcanoes.

3.4.4 Sizes of volcanic eruptions

Volcanologists have developed methods to measure the size of volcanic eruptions. A widely used method is the volcano explosivity index (VEI) (Newhall and Self, 1982), which uses an

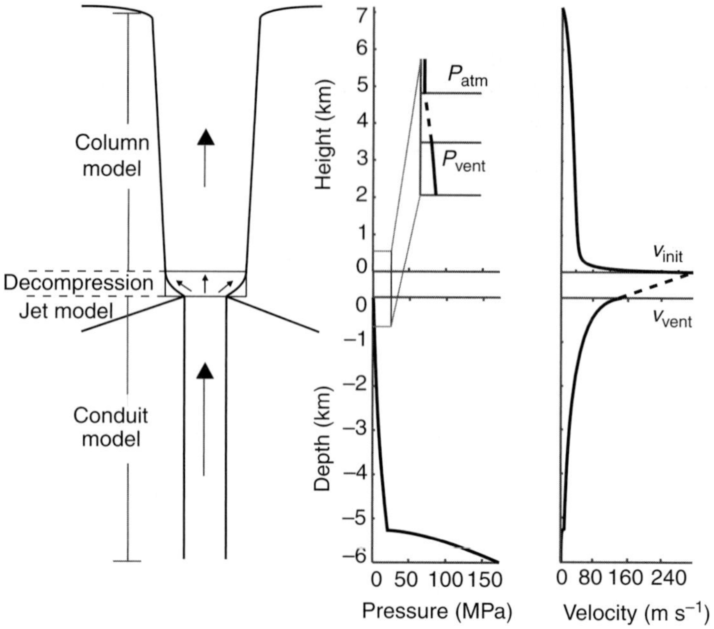

Fig. 3.9 Pressure and velocity profiles along the volcanic conduit and eruption column are computed by coupling steady-state conduit flow and eruption-column models for a magma with an initial volatile content of 4 wt.%. Such models are used to estimate the mass flow and column height of eruptions for specific magma pressures, depths and volatile contents.

integer scale from 0–8 and is based on both magnitude (erupted volume in m^3) and intensity (eruption column height in km). Unfortunately, there is not a unique relationship between VEI and the effects of volcanic eruptions (Table 3.2; Siebert and Simkin, 2007). For many nuclear facilities, eruptions of VEI 3–6 are of most concern, as such eruptions have sufficient magnitudes to effect areas tens of kilometers from the volcano, and frequencies sufficient to warrant investigation. In contrast, eruptions of VEI > 6 have regionally devastating effects, likely widely to disrupt societies, but are fortunately exceedingly rare.

The VEI assumes that the relationship between magnitude and intensity can be fully described using only one number. This index is not intended to characterize effusive eruptions, which are primarily non-explosive and therefore receive a default classification of 0 or 1. To relax the first assumption and to account for lava eruptions, Pyle (2000) proposed a method based on two logarithmic equations, one for the magnitude and another for the intensity based on erupted mass and mass eruption rate, respectively. For most of the eruptions, magnitudes defined with this scale are similar to their VEI. In terms of intensity, however, an extremely vigorous eruption will have an intensity of 10–12, whereas a comparatively mild eruption might have an intensity of 4–5.

Table 3.2. Volcano explosivity index (VEI) is a measure of eruption magnitude, with only a general relationship to expected eruption effects

VEI	Volume[a] (m^3)	Col.[b] (km)	Freq.[c] (a^{-1})	Scale of effects
1	10^4–10^6	0–1	$> 10^3$	Very localized effects. Channelized pyroclastic flows and debris flows may accompany some dome-building eruptions and affect areas < 10 km from the volcano [Cerro Negro (1999), Iwake (1863), Unzen (1990–1993)].[d]
2	10^6–10^7	1–5	50	Typically isolated explosions and short-lived eruptions producing localized effects. Channelized pyroclastic flows and debris flows may affect areas usually < 10 km from the volcano. [Stromboli (1985), Mt. Hood (1865)].
3	10^7–10^8	3–15	1	Wide variation in eruptive effects. Pyroclastic flows and debris flows generally confined to channels beyond the edifice of the volcano, but occasionally extend great distances (e.g. > 50 km). [Monte Nuovo (new vent, 1538), Miyake-jima (1983), Mayon (1968), Nevado del Ruiz (1985)].
4	10^8–10^9	10–25	0.1	Wide variation in eruptive effects, encompassing very long-lived eruptions to very brief, highly explosive eruptions. Often substantial effects < 10 km; channelized flows to < 30 km from the volcano, or occasionally to > 50 km. [Parícutin (new vent, 1943), Tolbachik (new vent, 1975), Colima (1913)].
5	10^9–10^{10}	> 25	0.02	Often substantial effects < 10 km from the volcano, specific sectors of the volcano may be devastated < 30 km; channelized flows to > 100 km. [Mount St. Helens (1980), Tarawera (1886), Fuji (1707)].
6	10^{10}–10^{11}	> 25	0.01	Areas devastated < 30 km from the volcano, substantial effects < 100 km, channelized flows to > 100 km. [Krakatau (1883), Pinatubo (1991)].
7	10^{11}–10^{12}	> 25	10^{-3}	Regions devastated on order 100 km or more from the volcano. Worldwide there are four known Holocene VEI 7 eruptions. [Aso (90 ka), Taal (1815)].
8	$> 10^{12}$	> 25	$< 10^{-4}$	No Holocene eruptions of this magnitude. These eruptions have continent-scale effects [Yellowstone Caldera (600 ka)].

[a] Total volume of tephra erupted. [b] Eruption column height (above vent for VEI 1–2; above mean sea level VEI ≥ 3. [c] Approximate global annual frequency of eruptions. [d] Example eruptions.

3.5 Volcanic products and hazards for nuclear facilities

Magma ascent and volcano eruptions are dynamic, transient processes that are difficult to observe. Much of what is known about eruptions stems from interpretation of the geologic record and from direct observations of the formation of deposits during volcanic eruptions. Certainly in order to identify, quantify and assess volcanic hazards at nuclear facilities, the products of volcanic eruptions have to be studied and their mode of emplacement understood. Furthermore, a volcano is not only hazardous during an eruptive phase of activity. Rather, a possible threat remains when the volcano is inactive, dormant or even extinct, because volcanic landscapes are commonly unstable.

As volcanic hazards for nuclear facilities are generally site-specific, volcanic phenomena can be usefully characterized in terms of distal and proximal effects. There are also issues of timescale and nature of the nuclear facility. For geological repositories subsurface intrusions are manifestly important, but surface phenomena can pose threats to reactors and also to underground repositories during construction over periods of a number of decades. Eruptions may also compromise any planned monitoring of an underground facility and eruptions can be severe enough to disrupt societies and eradicate knowledge that the facility ever existed.

We use the concept of the energy cone (Figure 3.10) described by Sheridan (1979) and Sheridan *et al.* (2004) to discriminate between distal and proximal effects of volcanic eruptions. This simple model, while not capturing the full complexity of volcanic phenomena, provides a first-order approximation of the potential runout of various flow phenomena, and is currently widely used as part of volcanic hazard assessments (e.g. Alberico *et al.*, 2002; Major *et al.*, 2004; Hubbard *et al.*, 2007; Toyos *et al.*, 2007). Potential runout of surface flows (e.g. lahars, pyroclastic flows) is based on the potential energy of the flows at their origin and the topography of the flow path, usually simply defined as radiating from the volcanic vent. In two dimensions, the energy cone is reduced to an energy line (Figure 3.10) that is defined by the vertical drop, H, of the geophysical flow (from the source point to the head of the flows) and the runout of the flow, L. For volcanic hazards at a proposed nuclear site from a specific composite volcano, one might choose an appropriately conservative ratio H/L, say 0.05–0.2, to differentiate between proximal and distal hazards (e.g. Volentik *et al.*, Chapter 9, this volume). The value of H might be chosen as the height of the volcano, or height of a potential eruption column, and the potential runout distance, L, calculated using a digital elevation model of the volcano and site region.

Although much more physically robust models for flow and transport are extant in the literature (e.g. Wadge *et al.*, 1998; Woods, 2000; Patra *et al.*, 2005; Neri *et al.*, 2007) the simplicity and wide use of the energy cone model make it a useful tool to differentiate among the types of volcanic flow phenomena that might be expected to potentially reach the site.

3.5.1 Distal volcanic hazards

Nuclear facilities throughout the world are located in regions that could be distally impacted by the products of an erupting volcano. For example, virtually all operating nuclear power

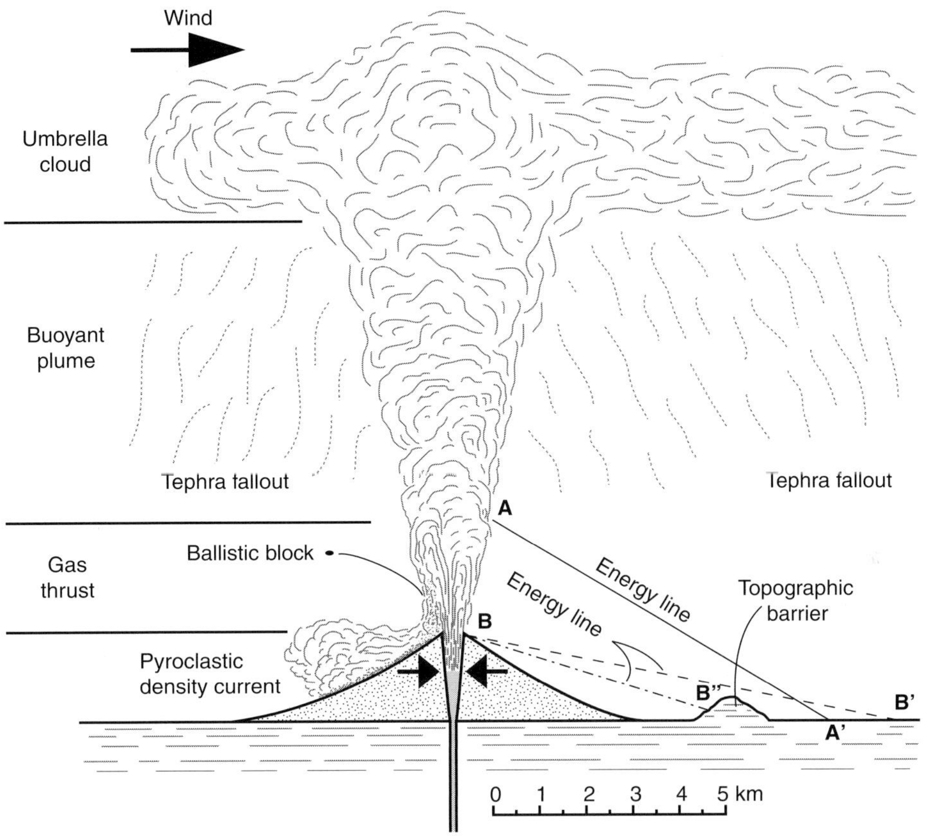

Fig. 3.10 Schematic figure representing a hypothetical Plinian eruption from a composite volcano, showing the gas-thrust, buoyant and umbrella cloud regions of the volcanic plume, as well as a pyroclastic flow sweeping down the slopes of the volcano (modified after Sheridan, 1979). The two black arrows inside the volcano are pointing toward the fragmentation level. The energy line AA' represents the potential runout of a pyroclastic density current originated from the column collapse (top of the gas-thrust region), whereas energy lines BB' and BB" represent potential runout of pyroclastic density currents generated from summit-dome collapse. Note that both AA' and BB' have the potential to overcome topographic barriers.

plants in Japan are potentially threatened by the possibly harmful impacts of tephra fallout, channelized distal lava or mudflows, or tsunami. These phenomena may easily impact sites tens of kilometers or even hundreds of kilometers from the erupting volcano. A discussion of these phenomena follows.

Distal tephra fallout

The collective term for fragmented volcanic particles is tephra and tephra fallout is the phenomena that occurs when such particles fall from eruption columns (Figures 3.10 and 3.11a) to drape the landscape (Figure 3.11b–c). The main constituents of tephra fall deposits

Fig. 3.11 Tephra fallout is one of the most widespread hazards resulting from volcanic eruptions. (a) Tephra fallout during the 1995 eruption of Cerro Negro volcano, Nicaragua (photo by B. Hill). (b) A tephra fallout deposit from the 2.45 ka Plinian eruption of Pululagua volcano, Ecuador, reveals several distinct phases of the eruption (photo by A. Volentik). The ruler is 30 cm long. (c) Interpretation of these deposits allows volcanologists to reconstruct the volume and energy of the eruption. Here, different stratigraphic layers of varying thicknesses (e.g. layer BF1, 2–3 cm) are characterized by maximum pumic size (MP) in a matrix of finer-grained material. This stratigraphic section is located $\approx$ 13 km from the vent.

are juvenile particles formed by the fragmentation of magma and accidental rock fragments and crystals (Fisher and Schmincke, 1984; Cas and Wright, 1987). The particle size, shape and physical characteristics of juvenile particles are variable. Particle size commonly varies from blocks the size of cars that are deposited near the vent to micron-sized dust which might be transported thousands of kilometers by wind-blown volcanic plumes. The thickness and mass of tephra-fall deposits are variable (Table 3.3) and depend on many factors, including eruption magnitude, degree and type of magma fragmentation, and wind conditions. Tephra deposits are thickest near the vent and thin roughly exponentially with distance (Table 3.3). Although exponential thinning is common (Pyle, 1989), the rate and style of thinning can

Table 3.3. Tephra fallout from four modern volcanic eruptions

	Tephra volume (m^3)	Thickness (cm) at 10 km	Thickness (cm) at 100 km
[1]Cerro Negro, 1995	2.8×10^6	2	< 0.1
[2]Cerro Negro, 1992	2.3×10^7	11	~ 0.1
[3]Mount St. Helens, 1980	1.1×10^9	20	3
[4]Pinatubo, 1991	$\sim 4 \times 10^9$	~ 30	3.5

[1] Hill *et al.* (1998); [2] Connor *et al.* (1993); [3] Sarna-Wojcicki *et al.* (1981); [4] Koyaguchi (1996).

vary significantly between proximal and distal areas (Bonadonna *et al.*, 1998; Bonadonna and Houghton, 2005).

Ash particles also can absorb corrosive and toxic chemical compounds that are easily remobilized by water. The compounds, including HF, HCl and a variety of others cause adverse health effects following explosive eruptions (e.g. Baxter, 2000). Similarly, a source of risk in volcanic eruptions through nuclear facilities, particularly waste repositories, is that ash might transport radionuclides, either as particles incorporated into tephra fragments or as compounds adsorbed onto the surfaces of ash.

Tsunami

As nuclear facilities are commonly located in coastal areas, tsunami threats deserve very careful consideration (Power and Downes, Chapter 11, this volume). Tsunami events can be generated by shallow submarine volcanic eruptions or when debris avalanches, landslides or large pyroclastic flows enter the sea or other large water bodies, displacing great volumes of water. The 1883 Krakatau eruption generated a tsunami with widespread distal impact (Carey *et al.*, 2001). Over 30 000 people were killed when tsunami hit the coast of Indonesia and Sumatra following this eruption.

Although a huge amount of rock can enter water bodies during volcanic eruptions, volcano slopes can collapse at any time, even during periods of no volcanic activity, due to the instability of some edifices. An example is the collapse of an old lava dome at Unzen volcano, Japan in 1792. This edifice collapse produced a tsunami that killed over 15 000 people (Hoshizumi *et al.*, 1999), many in locations far from the volcano. Such events are not easy to forecast, and hence deterministic models of tsunami inundation are the primary means of assessing these hazards.

3.5.2 Proximal phenomena

Although many nuclear facilities around the world will experience the effects of distal tephra fallout, tsunami and long-runout, channelized flows, these phenomena are readily accounted for by their design (Hill *et al.*, Chapter 25, this volume). In contrast, proximal volcanic hazards from these and a plethora of related eruptive phenomena are not necessarily mitigated through engineering design. Eruptive phenomena near active volcanic vents, including proximal tephra fallout (Table 3.3), can easily exceed reasonable design bases. Thus, such phenomena are crucial to consider in site evaluation and for the reassessment of volcanic hazards at existing facilities. Some of these proximal phenomena are described in the following.

Pyroclastic density currents

Pyroclastic density currents are high-temperature gravity-driven flows, consisting of a mixture of rock fragments, volcanic gases and ingested air (Druitt, 1998). They originate from the collapse of lava domes (both gravitational and explosive), and the gravitational collapse of eruption columns in explosive eruptions (Figure 3.12a). Thicknesses can vary from a few meters to more than 100 m (Figure 3.12b) and runouts can vary from a few kilometers to over 100 km in the largest-magnitude explosive eruptions. These flows might have temperatures of hundreds of degrees Celsius (depending on the original temperature of the magma and the degree of mixing with air) and can travel at high velocities (tens to hundreds of $m\,s^{-1}$). Two end-members of pyroclastic density currents have been identified (Wilson and Houghton, 2000): (i) dense pyroclastic density currents, called pyroclastic flows, which are usually controlled by topography, and (ii) more dilute (yet more energetic), turbulent pyroclastic density currents, referred to as surges, which are less influenced by topography than pyroclastic flows and can overcome topographic barriers or ridges. Pyroclastic density currents commonly have both high- and low-density components. Both pyroclastic flows and surges are common in, but not restricted to, explosive volcanic eruptions. Indeed, the explosive or gravitational collapse of effusive lava domes can generate hazardous pyroclastic flows and surges (Loughlin *et al.*, 2002). The impact of pyroclastic density currents on structures is significant (Valentine, 1998). Their high temperature and momentum, combined with the potential load of large, dense particles, lead to large dynamic pressures on structures, sufficient in some cases to cause complete destruction of buildings (Valentine, 1998).

Lava flows

Effusive volcanic activity leads to the formation of lava flows. Lava flows are not usually life threatening, but have a high potential for significant destruction. Their impact is mainly controlled by (i) their rheological properties (i.e. viscosity, temperature, crystallinity), (ii) their discharge rate at the source, (iii) the duration of the eruption and volume of magma emitted, and (iv) the topography. Low-viscosity lava flows can travel long distances (up to tens of kilometers) at velocities ranging from $<<1\,m\,s^{-1}$–$20\,m\,s^{-1}$ (in extreme cases). Lava-flow thicknesses vary from $<$ 1 m to $>$ 100 m. Such flows can inundate areas up to thousands of km^2 and can travel $>$ 100 km from their source.

Fig. 3.12 Pyroclastic density currents are capable of completely devastating regions about an erupting volcano. (a) The volcanic plume of May 8, 1997 from a Vulcanian eruption of Soufrière Hills Volcano, Montserrat is viewed from the NW $\approx$ 5.5 km from the volcano. Two fountain-collapse pyroclastic flows are sweeping the eastern and western slopes of the volcano (photo courtesy of C. Bonadonna). (b) The Shoshone Ignimbrite in California is part of a thick (300 m) sequence of pyroclastic flows and tephra fallout deposits that inundated the region $\sim$ 20 Ma. The glassy, welded interior of one flow creates the prominent black middle layer (photo courtesy of Sean Callihan). See color plate section.

Extrusion of high-viscosity lava usually results in more localized lava flows or domes, with lower magma-extrusion rates. This type of activity is typically associated with the generation of pyroclastic density currents. The highly viscous lava domes commonly collapse and produce pyroclastic flows or surges.

Debris flows and lahars

Debris flows and lahars are surface flows that consist of a mixture of rocks, water, delaminated soils and vegetation. These gravity-driven flows are mainly controlled by topography. Flow velocities reach speeds of 10–20 $m\,s^{-1}$, travel long distances downstream (up to 150 km) and have volumes $>10^7$ m^3 (Iverson *et al.*, 1998). As already described for Nevado del Ruiz (Figures 3.2 and 3.3), these flows gain volume as they propagate down the steep flanks of a volcano. However, as they debouch in low-lying areas, these flows lose energy, drop much of their sediment load and transform into floods, sometimes referred to as hyperconcentrated flows. When these flows reach major river systems, they can cause flooding hundreds of kilometers downstream. Thus, debris flows and lahars may pose a threat to riverine nuclear facilities located far from volcanoes.

Debris flows and lahars can occur at any stage of volcanic activity, even long after the eruption has ended (Figure 3.13). Therefore, they represent a potential hazard for nuclear facilities located near steep-sloped volcanoes.

Debris avalanches

Large composite volcanoes are often unstable edifices, due to steep slopes (as much as 35°), rock alteration (Principe and Marini, 2005), ground deformation (Voight, 2000), fault slip

Fig. 3.13 Deposits of the 2005 Panabaj, Guatemala, debris flow. This debris flow from Toliman volcano was triggered by heavy rainfall from hurricane Stan rather than directly by volcanic activity. The debris flow killed ≈ 600 people (photo by L. J. Connor).

(Vidal and Merle, 2000) and rapid erosion of the volcanic cone. They can travel at high velocities (50–70 m s^{-1}), reach distances up to 150 km and have very large volumes. A debris avalanche deposit formed by the collapse of Colima Volcano, Mexico, had an estimated volume of 30 km^3 (Ui *et al.*, 2000). Deposits resulting from debris avalanches are matrix supported and contain rock fragments, or clasts, ranging in size from a few centimeters to tens of meters in diameter. Volcanic debris avalanches are commonly accompanied by very high-energy pyroclastic density currents known as volcanic blasts (Voight, 2000).

Near-surface intrusions and new vent formation

Renewed magmatism often results in the formation of new volcanic vents. New vents form when intrusions, such as igneous dikes, transect the shallow crust and erupt magma at the surface. For subsurface nuclear facilities, such events are a potential mechanism for the release of radionuclides into the biosphere. For surface facilities, the formation of new vents must be evaluated in order to estimate accurately the potential for various volcanic phenomena to occur at the site (e.g. McBirney *et al.*, 2003; Weller *et al.*, 2006).

A wide variety of techniques are used to assess the probable locations of future vents (e.g. Connor and Connor, Chapter 14; Jaquet and Lantuéjoul, Chapter 15, this volume). These new vents may be sources of tephra, lavas and a variety of other types of flows. Thus, it becomes necessary to forecast the potential locations of future volcanic vents, and to forecast potential flows from these vents (e.g. Wadge *et al.*, 1994; Connor *et al.*, 2000, Valentine and Perry, Chapter 19, this volume).

3.5.3 Secondary impacts and eruption scenarios

Volcanic eruptions are complex and varied geophysical events. Thus, site-specific hazard assessments should address a broad range of volcanological phenomena and consider potential scenarios, or sequences of events.

An example of a complex sequence of events comes from the analysis of hazards at the Mülheim–Kärlich nuclear power plant located in Rhine Graben, Germany (Figure 3.5b). This power plant was sited close to the Eifel volcanic field and hazard assessment of the site included evaluation of the potential for new vents forming in the vicinity of the power plant. Of primary concern was the possibility of new maar formation, with attendant pyroclastic density currents. Such events occurred as recently as 12.9 ka following the eruption of Laacher See volcano (Park and Schmincke, 1997). However, the hazard assessment stopped short of considering secondary effects of eruptions that might impact the site adversely. The Laacher See eruption, for example, dammed the Rhine river downstream from the site, flooding the entire region, including the site, under $\sim$ 10 m of water. This tephra dam later burst, flooding vast areas downstream (H. Schmincke, personal communication).

These types of scenarios are virtually inevitable in volcanically active regions. Detailed understanding of site geology and volcanological processes are required in order to discern and mitigate these hazards.

3.6 Volcano monitoring

Volcano monitoring techniques are used to watch known active volcanoes for signs of imminent unrest (e.g. Scarpa and Tilling, 1996; Sparks, 2003). Monitoring can be a powerful tool for identifying potential future activity. Unfortunately, many active volcanoes are equipped with minimal monitoring networks or none at all. Thus, for many parts of the world the risk of future volcanic activity is impossible to quantify, and potentially active volcanoes may not be identified at all. Consequently, when siting a nuclear facility in a volcanically active region, a program for monitoring the current activity of nearby volcanoes and investigations into their past activity is very important.

For example, volcano seismological investigations were not conducted at the Tatun volcano, Taiwan (Table 3.1) until long after nuclear power plants were constructed within the area. These investigations revealed a higher level of seismic activity than previously anticipated. Long-period earthquakes, often associated with volcanic activity, were also identified (Lin *et al.*, 2005). Another example is the site investigation at Muria Peninsula, Java, Indonesia (Figure 3.5c and Table 3.1). Early investigations of nearby geochemical monitoring revealed the presence of mantle-derived helium by isotopic analyses of borehole gases (McBirney *et al.*, 2003). The isotopic composition of these gases signaled potential volcanic unrest. These types of data are crucial for identifying potential volcanic hazards (Hill *et al.*, Chapter 25, this volume).

3.6.1 Volcano seismology

Seismicity at volcanoes may be indicative of potential volcanic unrest and can give warning of impending eruptions. Micro-seismic networks are commonly deployed at potential nuclear sites to assess the potential impact of earthquakes. It can be cost-effective and efficient in some circumstances simply to extend these networks to include nearby volcanoes. At volcanoes, seismic investigations are used to characterize rates of seismic energy release, earthquake frequency–magnitude relationships, and to identify earthquake source regions with respect to volcanic vents. During periods of volcanic unrest or eruption the local tectonic stress field can be disturbed, resulting in earthquakes away from the volcano itself.

In addition to providing information on the frequency, magnitude and location of earthquakes, seismic waveforms of volcano-seismic events carry information about their origin that is extremely useful in site hazard assessments. Common types of volcanic seismicity are long-period earthquakes, hybrid earthquakes, volcano-tectonic earthquakes, tremor and surface signals. Explosions, rockfalls and pyroclastic flows are examples of phenomena that generate surface signals (Figure 3.14). The mechanisms causing each of these are related to a wide variety of internal processes including rock breakage related to magma ascent, degassing phenomena, movement of hydrothermal fluids and explosions.

3.6.2 Volcano deformation

Ground deformation due to volcanic activity or the instability of inactive volcanoes is not always revealed by seismic activity, as deformation can occur without earthquakes. Ground

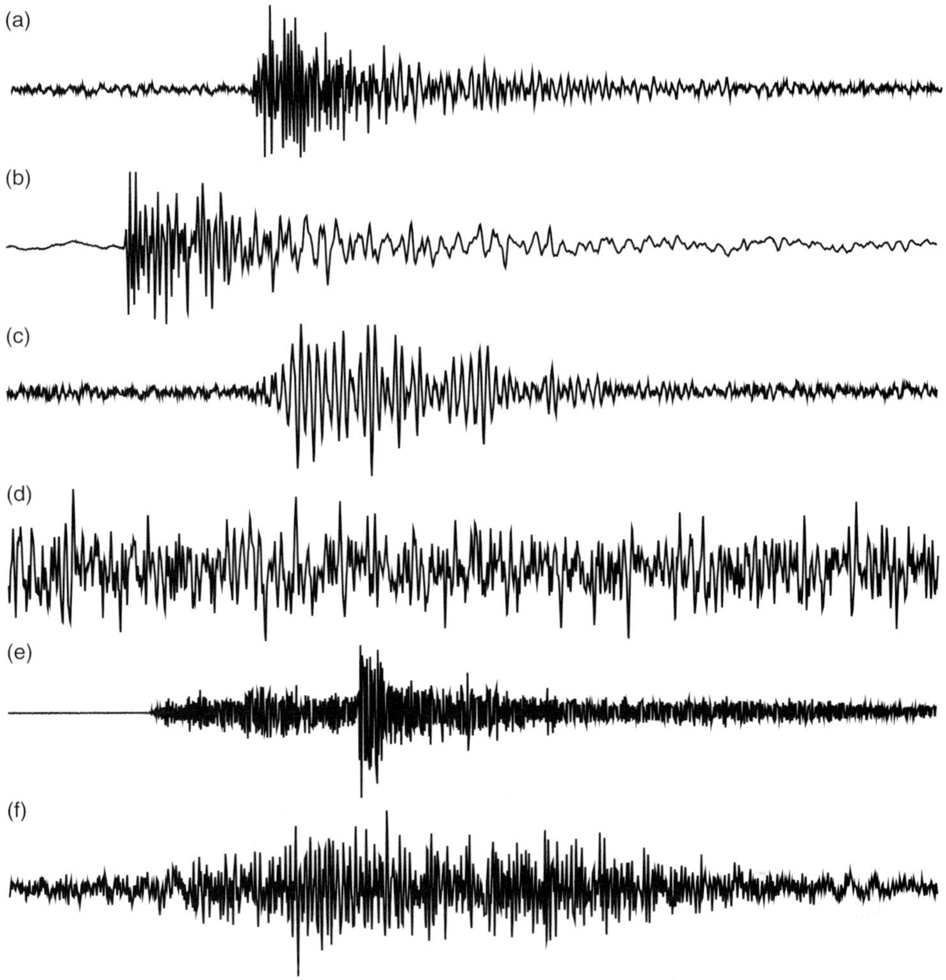

Fig. 3.14 Frequency and type of seismicity are the most common means of assessing potential volcanic activity. Examples of types of seismicity on active volcanoes include: (a) volcano-tectonic earthquake, 46 s duration, (b) hybrid event, 45 s duration, (c) long-period event lasting 66 s, (d) tremor recorded over 25 s, (e) explosion lasting 340 s, and (f) rockfall event lasting 94 s. Data provided by Silvio DeAngelis, Montserrat Volcano Observatory, and the Alaska Volcano Observatory.

deformation can be detected by several methods, including direct measurements using GPS systems (Wallace *et al.*, Chapter 6, this volume). Global positioning system measurements can be recorded either continuously or by carrying out regular campaigns. Less direct methods include tiltmeters, which measure changes in the angle of the ground, and deep borehole dilatometers, which are extremely sensitive to nanostrain-scale deformation, and hence have been used to provide early warnings of eruptions (Linde *et al.*, 1993).

Remote sensing methods, particularly interferometric synthetic aperture radar (InSar), can reveal large areas of ground motion under some circumstances. Interpreting InSar data involves calculating the phase difference between two consecutive satellite images and removing topographic effects. The resulting image reveals the difference between the distance to the ground and the satellite on successive orbits. The InSar method is particularly sensitive to vertical displacements (Wicks *et al.*, 2002). Within the Three Sisters volcanic complex in Oregon, uplift was clearly recorded by InSar beginning in 1998 and provided clues of active volcanism, although eruptions have not occurred at the Three Sisters since the eruption of rhyolite domes $\sim$ 2 ka (Scott, 1987).

3.6.3 Geochemical monitoring

Hot springs and active geothermal systems are often a sign of volcanic unrest, but do not exclusively occur in volcanic settings. Such features also occur associated with fault zones and similar structures that have little or nothing to do with potential volcanism. Often the geochemistry of waters and gases provide the only means to identify a volcanic source associated with these emissions. Although major ion geochemistry provides substantial insights, most modern investigations rely on the isotopic composition of fluids to identify the origin and mixing of geothermal fluids. For example, elevated $^3He/^4He$ is indicative of a mantle source of some helium in emissions (Rizzo *et al.*, 2006), and thus is indicative of potential unrest at volcanoes (McBirney *et al.*, 2003; Lin *et al.*, 2005).

3.6.4 Additional monitoring techniques

A host of monitoring techniques are emerging that will undoubtedly improve our understanding of the state of many volcanoes and their potential for future eruptions. Many of these techniques are now used routinely to survey volcanoes, and might be adapted to monitor changes in activity. For example, seismic tomography uses regional earthquakes and seismic networks to determine areas of high or low compressional or shear-wave velocity in the subsurface. Slow-velocity regions have been identified and mapped in many regions of active volcanism (e.g. Tamura *et al.*, Chapter 7, this volume). These velocity anomalies are attributed to areas of partial melting of the mantle. Thus seismic tomography provides essential clues about the potential for eruptions to occur, even in areas lacking obvious surface manifestations (e.g. Martin *et al.*, 2004; Umeda *et al.*, 2006). Repeated surveys might be a way of identifying potential changes in the distribution of slow-velocity regions.

Similarly, magnetotellurics and related electromagnetic methods are of increasing utility in identifying magma and magma source regions within the crust. These methods work because magma has very high electrical conductivity, but rocks that crystallize from magma have very low electrical conductivity. For example, Umeda *et al.* (2006) used magnetotelluric methods to identify partial melt zones within the crust in an area of hot springs and high $^3He/^4He$, but not characterized by magmatic activity since the Mesozoic. In addition, patterns in groundwater circulation that are affected by magma in the shallow subsurface can be mapped with a variety of electrical methods (e.g. MacNeil *et al.*, 2007).

This short summary of monitoring techniques highlights approaches to characterizing volcanic activity. It is clear from many studies of volcanism, and from a few studies dedicated to siting nuclear facilities, that detecting signs of volcanic unrest or potential activity requires detailed geophysical and geochemical investigations. Multiple techniques now exist for studying an area, and it is only by synthesis of different types of data that an area can truly be classified in terms of its potential for future volcanism (e.g. Hill *et al.*, 2002).

3.7 Probabilistic volcanic hazard assessment

Ultimately, volcanic hazard assessments for nuclear facilities require a level of detail that is proportional to the hazard. Specific approaches to these hazard assessments are described in the context of regulation by Hill *et al.* (Chapter 25, this volume). From a strictly volcanological perspective, the volcanic hazard analysis can usually be broken down into estimation of three probabilities: the probability of an eruption within some time interval, such as the performance period of the facility; the probability that eruptions will occur within a specific area; and the probability that volcanic hazards will affect the site, given that an eruption occurs. This set of three probabilities can by multiplied to estimate volcanic hazards at the site, provided that they are independent (e.g. provided the probable location of future volcanic vents does not change with time over the period of interest). Furthermore, at a specific site it may be necessary to estimate hazards associated with several volcanoes or volcanic systems separately. An additional probability, the probability that the site facility will fail and release radionuclides into the environment, can be added in a full risk assessment.

Understanding the timing of events and estimation of repose intervals, or recurrence rates, of volcanic activity is a major issue in many site evaluations. For example, for nearly every site in Table 3.1, the timing of past volcanic activity has been a major source of uncertainty in the analysis. Estimation of the recurrence rate of volcanism and the probability of eruptions within a given time period are very active areas of research. One approach is to assume that volcanic eruptions are a Poisson process in time. That is, the likelihood of an eruption in one time interval is independent of the number of eruptions during previous time intervals. For Poisson processes the distribution of repose intervals is exponential, so one can calculate a point estimate of the mean of the repose interval as:

$$\hat{\mu} = \frac{\sum_{i=1}^{n} t_i}{n} \tag{3.3}$$

where t_i is the repose interval before the ith eruption and n is the total number of repose intervals. The interval estimate (at 95% confidence) is then

$$P\left\{\frac{2\sum_{i=1}^{n} t_i}{\chi^2_{2n,\alpha=0.975}} < \mu < \frac{2\sum_{i=1}^{n} t_i}{\chi^2_{2n,\alpha=0.025}}\right\} = 0.95 \tag{3.4}$$

which states that the mean repose interval has a probability, P, of occurring within some interval, based on a χ^2 distribution with $2n$ degrees of freedom, at a given confidence level,

α, and μ is the mean repose interval. Although simple in form, the underlying assumption of a Poisson process appears to be incorrect for many volcanic systems. It is necessary to explore alternative models, such as Weibull, log–logistic, Bayesian and time-volume dependent models in a full hazard assessment (e.g. Ho, 1991; Connor *et al.*, 2003; Gelman *et al.*, 2003; Coles and Sparks, 2006). All of these models require data, and significant expense is usually devoted to resolving the distribution and timing of volcanism in the geologic record as part of site hazard assessments.

The probable location of future volcanic events is a problem that varies in complexity depending on the nature of the volcanic hazard. In some volcanic hazard assessments, the potential threat clearly originates from a specific polygenetic composite volcano. Often, however, the problem is more complex. On some polygenetic volcanoes, the location of specific vents changes with time, depending on the activity in rift zones on the volcano and related structures. Exactly where future vents form is important, because this controls the path of volcanic flows that issue from the vent. For such volcanoes it is necessary to estimate the likelihood, given some activity, that activity will occur in specific areas on the volcano. Still other volcanic systems, such as monogenetic volcanic fields and some calderas, have no central vent that is clearly the site of most likely future activity (Figure 3.15). In these areas it is necessary to model the spatial density of volcanism. This topic is covered in this volume by Kondo (Chapter 12), Mahony *et al.* (Chapter 13), Connor and Connor (Chapter 14) and Jaquet and Lantuéjoul (Chapter 15).

A volcanological perspective is essential at this stage of the hazard assessment. Significant geological investigations are required to identify all volcanic features in the region and to determine their ages and geologic properties. The spatial-density estimate has to consider geologic structures, which may in some circumstances influence vent distribution (see Wetmore *et al.*, Chapter 16, this volume), spatial shifts in the locus of volcanism through time (e.g. Condit and Connor, 1996) and geological mechanisms that give rise to vent clustering (e.g. Connor *et al.*, 2000; Mahony *et al.*, Chapter 13, this volume).

From the perspective of nuclear facilities, it simply does not matter how frequent, violent or spectacular volcanic eruptions might be, if these eruptions do not affect the site or its vicinity. Conversely, as the unfortunate example of the 1985 eruption of Nevado Del Ruiz reveals, historically, the affects of volcanism have been underestimated, occasionally with catastrophic consequences. Detailed volcanological information is required in order to assess scientifically the impacts of volcanism. This includes mapping the products of volcanic eruptions at the site and around volcanoes that may affect the site. Such mapping is often challenging because volcanic phenomena, such as pyroclastic density currents, may have devastating impact, but leave a sparse geologic record. In other words, a geologic map and site borehole logs, by themselves, are insufficient. Volcanological interpretation of these data is needed.

During the last decades volcanologists have made vast improvements in the numerical simulation of volcanic phenomena. These models may be used to forecast the areas inundated by volcanic products, given an eruption from a volcano. Examples include the modeling of tephra fallout (Bonadonna *et al.*, 2005; Volentik *et al.*, Chapter 9, this volume).

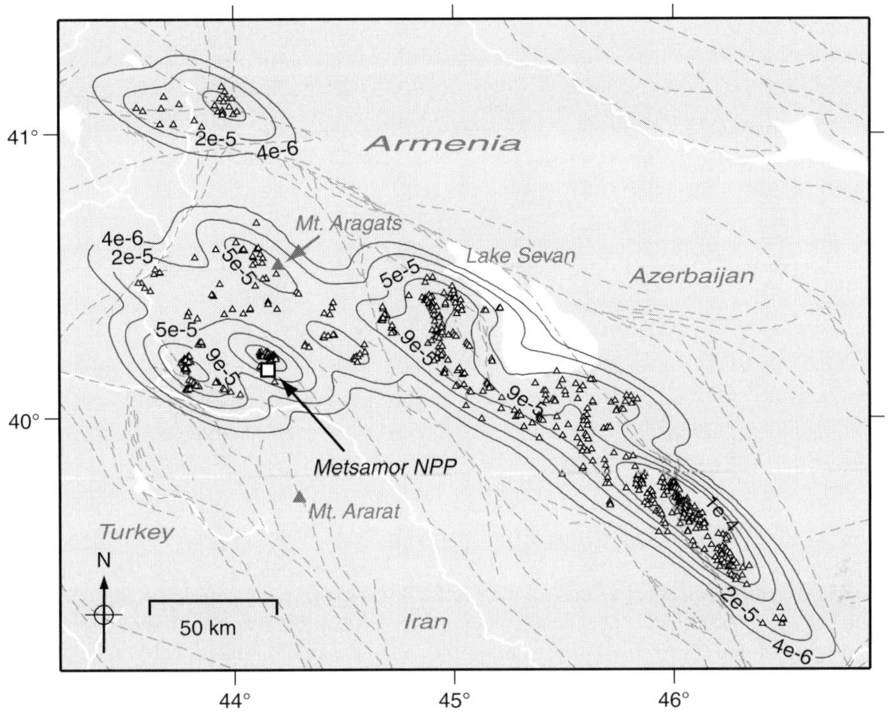

Fig. 3.15 Spatial density of monogenetic volcanoes in Armenia is estimated using non-parametric statistical techniques (described in detail in Connor and Connor, Chapter 14, this volume). Such techniques are used to estimate the probability of new vent formation during the lifetime of nuclear facilities, in this case the Metsamor nuclear power plant. The annual probability of new vent formation within 50 km^2 about the site is $\sim 6 \times 10^{-6}$ (Weller *et al.*, 2006). Contour lines are labeled such that 4e-6 is equivalent to 4×10^{-6}.

Models of tephra fallout can be used to forecast the amount of tephra that is likely to accumulate at a given location in response to a given eruption, under specific meteorological conditions. Alternatively, these models can be used to forecast the probability that tephra accumulation will exceed a specific value, given a probable range of eruption and meteorological conditions (Figure 3.16). Similar models have been developed to understand debris-flow inundation, lava flows and related phenomena. Their development offers a great deal of promise that volcanic hazard assessments will become more robust and consistent over time.

Further reading

Subsequent chapters in this volume provide a much closer look at specific techniques and case studies of volcanic hazard assessments for nuclear facilities. The *Encyclopedia of Volcanoes* (Sigurdsson *et al.* eds., 2000, Academic Press) gives a comprehensive

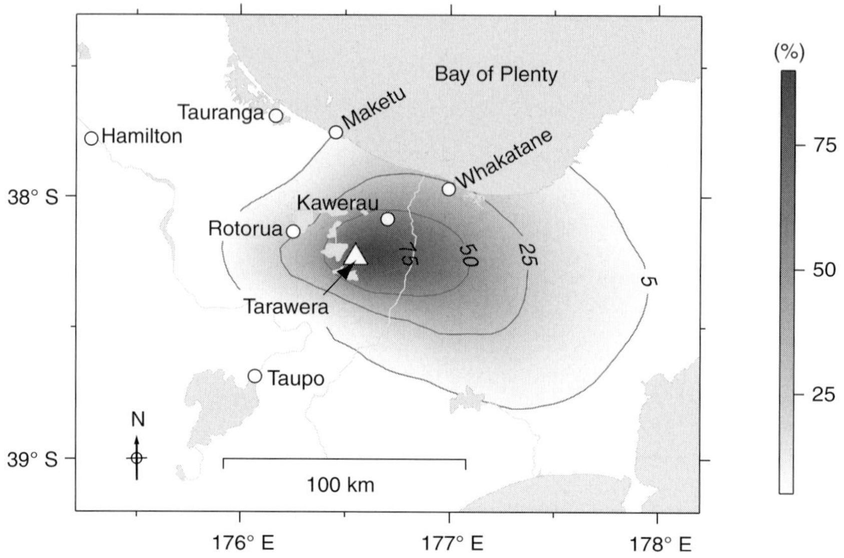

Fig. 3.16 Forecasting volcanic hazards for nuclear facilities places a high burden on the accuracy and parsimony of numerical models of volcanic phenomena. Probabilistic application of a tephra fallout model involves Monte Carlo simulation of eruptions, and determining the probable accumulation of tephra at a particular site (modified from Bonadonna, 2005). Contour interval is the percent probability that tephra accumulation will exceed $30\,\mathrm{kg\,m^{-2}}$, given an ~VEI 4 eruption at Tarawera volcano, New Zealand.

introduction to volcanic phenomena. The book, *Monitoring and Mitigation of Volcano Hazards* (Scarpa and Tilling, eds., 1996, Springer Verlag) provides a great deal of information about volcano monitoring. For more details on statistical methods in volcanology, see *Statistics in Volcanology* (Mader *et al.*, eds., 2006, Geological Society of London, IAVCEI Special Publication 1).

Acknowledgments

The authors thank Laura Connor and Diana Roman for providing assistance with several figures in this chapter. Comments by Laura Connor, Neil Chapman and Britt Hill improved the text.

References

Aki, K., M. Fehler and S. Das (1977). Source mechanisms of volcanic tremors: fluid-driven crack models and their application to the 1963 Kilauea eruption. *Journal of Volcanology and Geothermal Research*, **2**, 259–287.

Alberico, I., L. Lirer, P. Petrosino and R. Scandone (2002). A methodology for the evaluation of long-term volcanic risk from pyroclastic flows in Campi Flegrei (Italy). *Journal of Volcanology and Geothermal Research*, **116**, 63–78.

Allmendinger, R. W., T. E. Jordan, S. M. Kay and B. L. Isacks (1997). The evolution of the Altiplano–Puna Plateau of the Central Andes. *Annual Review of Earth and Planetary Sciences*, **25**, 139–174.

Annen, C. and R. S. J. Sparks (2002). Effects of repetitive emplacement of basaltic intrusions on thermal evolution and melt generation in the deep crust. *Earth and Planetary Science Letters*, **203**, 937–955.

Apted, M., K. Berryman, N. Chapman *et al.* (2004). Locating a radioactive waste repository in the Ring of Fire. *Eos, Transactions, American Geophysical Union*, **85**, 465–471.

Barca, D., G. M. Crisci, S. Di Gregorio and F. P. Nicoletta (1994). Cellular automata for simulating lava flows: a method and examples from Etnean eruptions. *Transport Theory and Statistical Physics*, **23**, 195–232.

Baxter, P. J. (2000). Impacts of eruptions on human health. In: Siggurdson, H., B. Houghton, S. R. McNutt, H. Rymer and J. Stix (eds.) *Encyclopedia of Volcanoes*. San Diego, CA: Academic Press, 1035–1043.

Blundy, J., K. Cashman and M. Humphreys (2006). Magma heating by decompression-driven crystallization beneath andesite volcanoes. *Nature*, **443**, 76–80, doi:10.1038/nature05100.

Bonadonna, C. and B. F. Houghton (2005). Total grain-size distribution and volume of tephra-fall deposits. *Bulletin of Volcanology*, **67**, 441–456.

Bonadonna, C., G. G. J. Ernst and R. S. J. Sparks (1998). Thickness variations and volume estimates of tephra fall deposits: the importance of particle Reynolds number. *Journal of Volcanology and Geothermal Research*, **81**, 173–187.

Bonadonna, C., C. B. Connor, B. F. Houghton *et al.* (2005). Probabilistic modeling of tephra dispersal: hazard assessment of a multiphase rhyolitic eruption at Tarawera, New Zealand. *Journal of Geophysical Research*, **110**, doi:10.1029/2003JB002896.

Bottinga, Y. and D. F. Weill (1970). Densities of liquid silicate systems calculated from partial molar volumes of oxide components. *American Journal of Science*, **269**, 169–182.

Büttner, R., P. Dellino, H. Raue, I. Sonder and B. Zimanowski (2006). Stress-induced brittle fragmentation of magmatic melts: theory and experiments. *Journal of Geophysical Research*, **111**, B08204, doi:10.1029/2005JB003958.

Carey, S., D. Morelli, H. Sigurdsson and S. Bronto (2001). Tsunami deposits from major explosive eruptions: an example from the 1883 eruption of Krakatau. *Geology*, **29**, 347–350.

Cas, R. A. F. and J. V. Wright (1987). *Volcanic Successions, Modern and Ancient: A Geological Approach to Processes, Products, and Successions*. Berlin: Springer Verlag.

Clemens, J. D. and C. K. Mawer (1992). Granitic magma transport by fracture propagation. *Tectonophysics*, **204**, 339–360.

Coles, S. G. and R. S. J. Sparks (2006). Extreme value methods for modeling historical series of large volcanic magnitudes. In: Mader, H. M., S. G. Coles, C. B. Connor and L. J. Connor (eds.) *Statistics in Volcanology*, Special Publications of IAVCEI 1. London: Geological Society, 47–56.

Condit, C. D. and C. B. Connor (1996). Recurrence rate of basaltic volcanism in volcanic fields: an example from the Springerville Volcanic Field, AZ, USA. *Geological Society of America Bulletin*, **108**, 1225–1241.

Connor, C. B., L. Powell, W. Strauch *et al.* (1993). The 1992 eruption of Cerro Negro, Nicaragua: an example of Plinian-style activity at a small basaltic cinder cone. *Eos, Transactions, American Geophysical Union*, **74**, 640.

Connor, C. B., J. Stamatakos, D. Ferrill *et al.* (2000). Geologic factors controlling patterns of small-volume basaltic volcanism: application to volcanic hazards at Yucca Mountain, Nevada. *Journal of Geophysical Research*, **105**, 417–432.

Connor, C. B., R. S. J. Sparks, R. M. Mason, C. Bonadonna and S. R. Young (2003). Exploring links between physical and probabilistic models of volcanic eruptions: the Soufrière Hills Volcano, Montserrat. *Geophysical Research Letters*, **30**, 1701, doi:10.1029/2003GL017384.

Connor, C. B., A. R. McBirney and C. Furlan (2006). What is the probability of explosive eruption at a long-dormant volcano? In: Mader, H. M., S. G. Coles, C. B. Connor and L. J. Connor (eds.) *Statistics in Volcanology*, Special Publications of IAVCEI 1. London: Geological Society, 39–46.

Costa, A. (2005). Viscosity of high crystal content melts: dependence on solid fraction. *Geophysical Research Letters*, **32**, L22308, doi:10.1029/2005GL024303.

D'Amato, A. and K. Engel (1988). State responsibility for the exportation of nuclear technology. *Virginia Law Review*, **74**, 1011–1066.

Dartevelle, S. and G. A. Valentine (2005). Early-time multiphase interactions between basaltic magma and underground openings at the proposed Yucca Mountain radioactive waste repository. *Geophysical Research Letters*, **32**, L22311, doi:10.1029/2005GL024172.

Díez, M. (2006). Solution and parametric sensitivity study of a coupled conduit and eruption column model. In: Mader, H. M., S. G. Coles, C. B. Connor and L. J. Connor (eds.) *Statistics in Volcanology*, Special Publications of IAVCEI 1. London: Geological Society, 185–200.

Dingwell, D. B., N. S. Bagdassarov, G. Y. Bussod and S. L. Webb (1993). Magma rheology. In: Luth, R. W. (ed.) *Handbook on Experiments at High Pressure and Applications to the Earth's Mantle*, Mineralogical Association of Canada 1. Toronto: Mineralogical Association of Canada, 131–196.

Dingwell, D. B., K.-U. Hess and C. Romano (2000). Viscosities of granitic (*sensu lato*) melts: influence of the anorthite component. *American Mineralogist*, **85**, 1342–1348.

Dingwell, D. B., P. Courtial, D. Giordano and A. R. L. Nichols (2004). Viscosity of peridotite liquid. *Earth and Planetary Science Letters*, **208**, 337–349.

Druitt, T. H. (1998). Pyroclastic density currents. In: Gilbert, J. S. and R. S. J. Sparks (eds.) *The Physics of Explosive Volcanic Eruptions*, Special Publication 145. London: Geological Society, 145–182.

Druitt, T. H., L. Edwards, R. M. Mellors *et al.* (1999). *Santorini Volcano*, Memoir 19. London: Geological Society.

Einarsson, P. and B. Brandsdóttir (1980). Seismological evidence for lateral magma intrusion during the July 1978 deflation of the Krafla volcano in NE Iceland. *Journal of Geophysical Research*, **47**, 160–165.

England, P. C. and A. B. Thompson (1984). Pressure–temperature–time paths of regional metamorphism. *Journal of Petrology*, **25**, 894–928.

Farmer, G. L., F. V. Perry, S. Semken *et al.* (1989). Isotopic evidence on the structure and origin of subcontinental lithospheric mantle in southern Nevada. *Journal of Geophysical Research*, **94**, 7885–7898.

Fisher, R. V. and H.-U. Schmincke (1984). *Pyroclastic Rocks*. Berlin: Springer Verlag.

Freundt, A. and M. Rosi (eds.) (1998). *From Magma to Tephra: Modelling Physical Processes of Explosive Eruptions*, Developments in Volcanology 4. Berlin: Springer Verlag.

Gelman, A., J. B. Carlin, H. S. Stern and D. B. Rubin (2003). *Bayesian Data Analysis*, 2nd edn. New York, NY: CRC Press.

Giordano, D. and D. B. Dingwell (2003). Non-Arrhenian multicomponent melt viscosity: a model. *Earth and Planetary Science Letters*, **208**, 337–349.
Glazner, A. F., J. M. Bartley, D. S. Coleman, W. Gray and R. Z. Taylor (2004). Are plutons assembled over millions of years by amalgamation from small magma chambers? *GSA Today*, **14**, doi:10.1130/1052-5173.
Gonnermann, H. M. and M. Manga (2003). Explosive volcanism may not be an inevitable consequence of magma fragmentation. *Nature*, **426**, 432–435, doi:10.1038/nature0213.
Hess, K.-U. and D. B. Dingwell (1996). Viscosities of hydrous leucogranitic melts: a non-Arrhenian model. *American Mineralogist*, **81**, 1297–1300.
Hill, D. P. and S. Prejean (2005). Magmatic unrest beneath Mammoth Mountain, California. *Journal of Volcanology and Geothermal Research*, **146**, 257–283, doi:10.1016/j.jvolgeores.2005.03.002.
Hill, B. E., C. B. Connor, M. S. Jarzemba *et al.* (1998). 1995 eruptions of Cerro Negro volcano, Nicaragua, and risk assessment for future eruptions. *Geological Society of America Bulletin*, **110**(10), 1231–1241.
Hill, D. P., D. Dzurisin, W. L. Ellsworth *et al.* (2002). Response plan for volcano hazards in the Long Valley caldera and Mono Craters region, California. *US Geological Survey Bulletin*, **2185**.
Ho, C.-H. (1991). Time trend analysis of basaltic volcanism at the Yucca Mountain site. *Journal of Volcanology and Geothermal Research*, **46**, 61–72.
Hoblitt, R. P., C. D. Miller and W. E. Scott (1987). Volcanic hazards with regard to siting nuclear-power plants in the Pacific northwest. US Geological Society Open-File Report 87–297.
Hoshizumi, H., K. Uto and K. Watanabe (1999). Geology and eruptive history of Unzen volcano, Shimabara Peninsula, Kyushu, SW Japan. *Journal of Volcanology and Geothermal Research*, **89**, 81–94.
Hubbard, B. E., M. F. Sheridan, G. Carrasco-Núñez, R. Díaz-Castellón and S. R. Rodríguez (2007). Comparative lahar hazard mapping at Volcán Citlaltépetl, Mexico using SRTM, ASTER and DTED-1 digital topographic data. *Journal of Volcanology and Geothermal Research*, **160**, 99–124.
Iverson, R. M., S. P. Schilling and J. W. Vallance (1998). Objective delineation of lahar-inundation hazard zones. *Geological Society of America Bulletin*, **110**, 972–984.
Jaquet, O. and R. Carniel (2006). Estimation of volcanic hazards using geostatistical methods. In: Mader, H. M., S. G. Coles, C. B. Connor and L. J. Connor (eds.) *Statistics in Volcanology*, Special Publications of IAVCEI 1. London: Geological Society, 89–104.
Karakhanian, A., R. Djrbashian and V. Trifonov (2003). Volcanic hazards in the region of the Armenian nuclear power plant. *Journal of Volcanology and Geothermal Research*, **126**, 31–62.
Keskin, M. (2003). Magma generation by slab-steepening and breakoff beneath a subduction–accretion complex: an alternative model for collision-related volcanism in Eastern Anatolia, Turkey. *Geophysical Research Letters*, **30**, 8046, doi:10.1029/2003GL18019.
Koyaguchi, T. (1996). Volume estimation of tephra-fall deposits from the June 15, 1991, eruption of Mount Pinatubo by theoretical and geological methods. In: Newhall, C. G. and R. S. Punongbayan (eds.) *Fire and Mud: Eruptions and Lahars of Mount Pinatubo, Philippines*. Seattle, WA: University of Washington Press, 583–600.

Lejeune, A.-M. and P. Richet (1995). Rheology of crystal-bearing silicate melts: an experimental study at high viscosities. *Journal of Geophysical Research*, **100**, 4215–4229.

Lin, C. H., K. I. Konstantinou, W. T. Liang *et al.* (2005). Preliminary analysis of volcanoseismic signals recorded at the Tatun Volcano Group, northern Taiwan. *Geophysical Research Letters*, **32**, L10313, doi:10.1029/2005GL022861.

Linde, A. T., I. Agustsson, S. Sacks and R. Stefansson (1993). Mechanism of the 1991 eruption of Hekla from continuous borehole strain monitoring. *Nature*, **365**, 737–740.

Liu, J. Q. (1999). *Volcanoes of China*. Beijing: Science Press.

Llewellin, E. M. and M. Manga (2006). Bubble suspension rheology and application for conduit flow. *Journal of Volcanology and Geothermal Research*, **143**, 205–217.

Llewellin, E. W., H. M. Mader and S. D. R. Wilson (2002). The rheology of a bubbly liquid. *Proceedings of the Royal Society of London – Series A*, **458**, 987–1016.

Loughlin, S. C., E. S. Calder, A. Clarke *et al.* (2002). Pyroclastic flows and surges generated by the 25 June 1997 dome collapse, Soufrière Hills Volcano, Montserrat. In: Druitt, T. H. and B. P. Kokelaar (eds.) *The Eruption of Soufrière Hills Volcano, Montserrat, from 1995 to 1999*, Memoir 21. London: Geological Society, 191–210.

Lowe, D. R., S. N. Williams, H. Leigh *et al.* (1986). Lahars initiated by the 13 November 1985 eruption of Nevado del Ruiz, Colombia. *Nature*, **324**, 51–53, doi:10.1038/324051a0.

Luhr, J. and T. Simkin (1993). *Parícut: A Volcano Born in a Mexican Cornfield*. Phoenix, AZ: Geoscience Press.

MacNeil, R. E., W. Sanford, C. B. Connor, S. Sandberg and M. Díez (2007). Investigation of the groundwater system at Masaya Caldera, Nicaragua, using transient electromagnetics and numerical simulation, *Journal of Volcanology and Geothermal Research*, doi:10.1016/j.jvolgeores.2007.07.016.

Major, J. J., S. P. Schilling, C. R. Pullinger and C. D. Escobar (2004). Debris-flow hazards at San Salvador, San Vicente, and San Miguel volcanoes, El Salvador. In: Rose, W. I., J. J. Bommer, D. L. López, M. J. Carr and J. J. Major (eds.) *Volcanic Hazards in El Salvador*, Geological Society of America, Special Paper 375, 89–108.

Manga, M., J. Castro, K. V. Cashman and M. Loewenberg (1998). Rheology of bubble-bearing magmas. *Journal of Volcanology and Geothermal Research*, **87**, 15–28.

Marsh, B. D. (1981). On the crystallinity, probability of occurrence, and rheology of lava and magma. *Contributions to Mineralogy and Petrology*, **78**, 85–98.

Marsh, B. D. (1982). On the mechanics of igneous diapirism, stoping, and zone melting. *American Journal of Science*, **282**, 808–855.

Martin, A. J., K. Umeda, C. B. Connor *et al.* (2004). Modeling long-term volcanic hazards through Bayesian inference: example from the Tohoku volcanic arc, Japan. *Journal of Geophysical Research*, **109**, B10208, doi:10.1029/2004JB003201.

McBirney, A. R., L. Serva, M. Guerra and C. B. Connor (2003). Volcanic and seismic hazards at a proposed nuclear power plant site in central Java. *Journal of Volcanology and Geothermal Research*, **126**, 11–30.

Miller, R. B. and S. R. Paterson (1999). In defense of magmatic diapirs. *Journal of Structural Geology*, **21**, 1161–1173.

Neri, A., T. E. Ongaro, G. Menconi *et al.* (2007). 4-D simulation of explosive eruption dynamics at Vesuvius. *Geophysical Research Letters*, **34**, L04309, doi:10.1029/2006GL028597.

Newhall, C. G. and S. Self (1982). The volcanic explosivity index (VEI): an estimate of explosive magnitude for historical volcanism. *Journal of Geophysical Research*, **87**, 1231–1238.

Papale, P. (1999). Strain-induced magma fragmentation in explosive eruptions. *Nature*, **397**, 425–428.

Park, C. and H.-U. Schmincke (1997). Lake formation and catastrophic dam burst during the late Pleistocene Laacher See eruption (Germany). *Naturwissenschaften*, **84**, 521–525.

Patra, A. K., A. C. Bauer, C. C. Nichita *et al.* (2005). Parallel adaptive numerical simulation of dry avalanches over natural terrain. *Journal of Volcanology and Geothermal Research*, **139**, 1–21.

Petford, N., R. Kerr and J. R. Lister (1993). Dike transport of granitoid magmas. *Geology*, **21**, 845–848.

Perry, F. V., G. A. Valentine, E. K. Desmarais and G. WoldeGabriel (2001). Probabilistic assessment of volcanic hazard to radioactive waste repositories in Japan: intersection by a dike from a nearby composite volcano. *Geology*, **29**, 255–258.

Pinkerton, H. and R. J. Stevenson (1992). Methods of determining the rheological properties of magmas at sub-liquidus temperatures. *Journal of Volcanology and Geothermal Research*, **53**, 47–66.

Power, J. A., J. C. Lahr, R. A. Page *et al.* (1994). Seismic evolution of the 1989–1990 eruption sequence of Redoubt Volcano, Alaska. *Journal of Volcanology and Geothermal Research*, **62**, 69–94.

Principe, C. and L. Marini (2005). Reaction path modeling of argillic alteration (AA) and advanced argillic alteration (AAA): consequences for debris avalanches induced by flank collapse and hydrothermal eruptions. In: Hunziker, J. C. and L. Marini (eds.) *The Geology, Geochemistry and Evolution of Nisyros Volcano (Greece): Implications for the Volcanic Hazards*, Mémoires de Géologie 44. Lausanne: Université de Lausanne, 164–178.

Pyle, D. M. (1989). The thickness, volume and grainsize of tephra fall deposits. *Bulletin of Volcanology*, **51**, 1–15.

Pyle, D. M. (2000). Sizes of volcanic eruptions. In: Sigurdsson, H. *et al.* (eds.) *Encyclopedia of Volcanoes*. Academic Press, 263–271.

Rizzo, A., A. Caracausi, R. Favara *et al.* (2006). New insights into magma dynamics during last two eruptions of Mount Etna as inferred by geochemical monitoring from 2002 to 2005, *Geochemistry, Geophysics, Geosystems*, **7**, Q06008, doi:10.1029/2005GC001175.

Roggensack, K., R. L. Hervig, S. B. McKnight and S. N. Williams (1997). Explosive basaltic volcanism from Cerro Negro volcano: influence of volatiles on eruptive style. *Science*, **277**, 1639–1642.

Rubin, A. M. (1995). Propagation of magma filled cracks. *Annual Review of Earth and Planetary Sciences*, **23**, 287–336.

Rust, A. C. and M. Manga (2002). Effects of bubble deformation on the viscosity of dilute suspensions. *Journal of Non-Newtonian Fluid Mechanics*, **104**, 53–63.

Ryan, M. P., R. Y. Koyanagi and R. S. Fiske (1981). Modeling the three-dimensional structure of macroscopic magma transport systems: applications to Kilauea volcano. *Journal of Geophysical Research*, **86**, 7111–7129.

Sarna-Wojcicki, A. M., S. Shipley, J. R. Waitt, D. Dzurisin and S. H. Wood (1981). Areal distribution thickness, mass, volume, and grainsize of airfall ash from the six major eruptions of 1980. In: *The 1980 Eruptions of Mount St. Helens*, US Geological Survey Professional Paper 1250, 577–600.

Scarpa, R. and R. Tilling (1996). *Monitoring and Mitigation of Volcano Hazards*. Berlin: Springer Verlag.

Scott, W. E. (1987). Holocene rhyodacite eruptions on the flanks of South Sister volcano, Oregon. In: Fink, J. H. (ed.) *The Emplacement of Silicic Domes and Lava Flows*, Geological Society of America Special Paper 212, 35–53.

Self, S. (2005). The effects and consequences of very large volcanic eruptions. *Meeting on Extreme Natural Hazards*. London: Royal Society.

Shaw, H. R. (1972). Viscosities of magmatic silicate liquids: an empirical method of prediction. *American Journal of Science*, **272**, 870–893.

Sheridan, M. F. (1979). Emplacement of pyroclastic flows: a review. In: Chapin, C. E. and W. E. Elston (eds.) *Ash-Flow Tuffs*. Geological Society of America Special Paper 180, 125–136.

Sheridan, M. F., B. Hubbard, G. Carrasco-Núñez and C. Siebe (2004). Pyroclastic flow hazard at Volcán Citlaltépetl. *Natural Hazards*, **33**, 209–221.

Siebert, L. and T. Simkin (2007). Volcanoes of the world: an illustrated catalog of Holocene volcanoes and their eruptions. Smithsonian Institution, Global Volcanism Program Digital Information Series GVP-3, http://www.volcano.si.edu/world/.

Sparks, R. S. J. (1978). The dynamics of bubble formation and growth in magmas: a review and analysis. *Journal of Volcanology and Geothermal Research*, **3**, 1–37.

Sparks, R. S. J. (2003). Forecasting volcanic eruptions. *Earth and Planetary Science Letters*, **210**, 1–15.

Sparks, R. S. J., H. E. Huppert and J. S. Turner (1984). The fluid dynamics of evolving magma chambers. *Philosophical Transactions of the Royal Society of London – Series A*, **310**, 511–534.

Sparks R. S. J., M. I. Bursik, S. N. Carey *et al.* (1997). *Volcanic Plumes*. New York, NY: John Wiley and Sons.

Spera, F. J. (1984). Carbon dioxide in petrogenesis III: role of volatiles in the ascent of alkaline magma with special reference to xenolith-bearing mafic lavas. *Contributions to Mineralogy and Petrology*, **88**, 217–232.

Swanson, D. A., T. L. Wright and R. T. Helz (1975). Linear vent systems and estimated rates of magma production and eruption for the Yakima Basalt on the Columbia Plateau. *American Journal of Science*, **275**, 877–905.

Szakács, A. (1994). Redefining active volcanoes: a discussion. *Bulletin of Volcanology*, **56**, 321–325.

Tait, S., C. Jaupart and S. Vergniolle (1989). Pressure, gas content and eruption periodicity of a shallow, crystallizing magma chamber. *Earth and Planetary Science Letters*, **92**, 107–123.

Toyos, G. P., P. D. Cole, A. Felpeto and J. Martí (2007). A GIS-based methodology for hazard mapping of small volume pyroclastic density currents. *Natural Hazards*, **41**, 99–112.

Ui, T., S. Takarada and M. Yoshimoto (2000). Debris avalanches. In: Sigurdsson, H., B. Houghton, S. R. McNutt, H. Rymer and J. Stix (eds.) *Encyclopedia of Volcanoes*. San Diego, CA: Academic Press, 617–626.

Umeda, K., K. Asamori, T. Negi and Y. Ogawa (2006). Magnetotelluric imaging of crustal magma storage beneath the Mesozoic crystalline mountains in a nonvolcanic region, northeast Japan, *Geochemistry, Geophysics, Geosystems*, **7**, Q08005, doi:10.1029/2006GC001247.

Valentine, G. A. (1998). Damage to structures by pyroclastic flows and surges, inferred from nuclear weapons effects. *Journal of Volcanology and Geothermal Research*, **87**, 117–140.

Valentine, G. A. and F. V. Perry (2006). Decreasing magmatic footprints of individual volcanoes in a waning basaltic field. *Geophysical Research Letters*, **33**, L14305, doi:10.1029/2006GL026743.

Vidal, N. and O. Merle (2000). Reactivation of basement faults beneath volcanoes: a new model of flank collapse. *Journal of Volcanology and Geothermal Research*, **99**, 9–26.

Voight, B. (1990). The 1985 Nevado del Ruiz volcano catastrophe: anatomy and retrospection. *Journal of Volcanology and Geothermal Research*, **44**, 349–386.

Voight, B. (2000). Structural stability of andesite volcanoes and lava domes. *Philosophical Transactions of the Royal Society of London – Series A*, **358**, 1663–1703.

Volentik, A. C. M., C. Bonadonna, C. B. Connor *et al.* (2005). A study of the 2450 BP Pululagua Plinian eruption (Ecuador): implications for models of tephra dispersal. *Eos, Transactions of the American Geophysical Union*, **86**, V53B-1543.

Wadge, G., P. A. V. Young and I. J. McKendrick (1994). Mapping lava flow hazards using computer simulation. *Journal of Geophysical Research*, **99**, 489–504.

Wadge, G., P. Jackson, S. M. Bower, A. W. Woods and E. Calder (1998). Computer simulations of pyroclastic flows from dome collapse. *Geophysical Research Letters*, **25**, 3677–3680.

Walker, G. P. L. (1973). Lengths of lava flows. *Philosophical Transactions of the Royal Society of London – Series A*, **274**, 107–118.

Weinberg, R. F. and Y. Podlachikov (1994). Diapiric ascent of magmas through power law crust and mantle. *Journal of Geophysical Research*, **99**, 9543–9559.

Weller, J. N., A. W. Martin, C. B. Connor, L. J. Connor and A. Karakhanian (2006). Modelling the spatial distribution of volcanoes: an example from Armenia. In: Mader, H. M., S. G. Coles, C. B. Connor and L. J. Connor (eds.) *Statistics in Volcanology*, Special Publications of IAVCEI 1. London: Geological Society, 77–88.

Wicks, C. W., D. Dzurisin, S. E. Ingebritsen *et al.* (2002). Magmatic activity beneath the quiescent Three Sisters volcanic center, central Oregon, Cascade Range, USA. *Geophysical Research Letters*, **29**, doi:10.1029/2001GL014205.

Wilson, C. J. and B. F. Houghton (2000). Pyroclast transport and deposition. In: Sigurdsson, H., B. Houghton, S. R. McNutt, H. Rymer and J. Stix (eds.) *Encyclopedia of Volcanoes*. San Diego, CA: Academic Press, 545–554.

Woods, A. W. (2000). Dynamics of hazardous volcanic flows. *Philosophical Transactions of the Royal Society of London – Series A*, **358**, 1705–1724.

Woods, A. W., R. S. J. Sparks, O. Bokhove *et al.* (2002). Modeling magma–drift interaction at the proposed high-level radioactive waste repository at Yucca Mountain, Nevada, USA. *Geophysical Research Letters*, **29**, 1641, doi:10.1029/2002GL014665.

Woods, A. W., O. Bokhove, A. de Boer and B. E. Hill (2006). Compressible magma flow in a two-dimensional elastic-walled conduit. *Earth and Planetary Science Letters*, **246**, 241–250.

Zhao, D. (2001). Seismological structure of subduction zones and its implications for arc magmatism and dynamics. *Physics of the Earth and Planetary Interiors*, **127**, 197–214.

Zimanowski, B., K. Wohletz, P. Dellino and R. Büttner (2003). The volcanic ash problem. *Journal of Volcanology and Geothermal Research*, **122**, 1–5.

4

Tectonic uplift and subsidence

N. Litchfield, Y. Ota and D. Merritts

Vertical movements of Earth's surface and the rocks that lie beneath it occur around many parts of the globe in response both to tectonic and non-tectonic processes (e.g. glacial unloading, igneous intrusion). Upward vertical movement (uplift) forms topography, which generally results in erosion; and downward vertical movement (subsidence) creates accommodation space, which generally results in burial. In return, particularly over long timescales, feedbacks occur such that erosion and burial may enhance uplift and subsidence, respectively. Thus, it is vitally important to understand the fundamental drivers of uplift and subsidence (e.g. tectonic or non-tectonic) and the relationship with erosion and burial.

For nuclear facilities, the impacts of uplift and erosion generally are more significant than the impacts of subsidence and burial (McKinley and Chapman, Chapter 24, this volume). Furthermore, given the rates of vertical movement (typically $< 10\,\mathrm{mm\,a^{-1}}$) and the timescales of interest for different kinds of nuclear facilities, their impacts are most significant for geological repositories for long-lived radioactive waste. The potential impacts of uplift, subsidence, erosion and burial on geological repositories are described in McKinley and Chapman (Chapter 24, this volume) and McKinley and Alexander (Chapter 22, this volume). Briefly, the key factors of interest are: (i) the average regional rate of uplift or subsidence (on a scale on the order of tens or hundreds of kilometers); (ii) the magnitude of differential uplift and erosion (on a scale of 100 m to several kilometers); (iii) the potential for extreme, localized erosion or burial; and (iv) the balance between uplift and erosion, and subsidence and burial.

This chapter reviews current scientific knowledge and understanding of the processes driving tectonic uplift and subsidence, and related erosion and burial, and currently available methods for identifying their occurrence and measuring their rates. Because of their greater importance for geological repositories, the chapter has a greater focus on uplift and erosion than on subsidence and burial. Volcanic processes are only briefly discussed as they are described further by Connor *et al.* (Chapter 3, this volume). Isostatic uplift due to glacial unloading is described by Lund and Näslund (Chapter 5, this volume).

Key issues addressed here include the importance of definition of spatial reference frames, the need to understand the type of vertical movement measured, and the importance of both spatial scales and timescales (years to millions of years). The description of the fundamental mechanics and measurement of tectonic uplift and subsidence, erosion and burial are followed by three case studies from tectonically dynamic environments around the Pacific Rim. These case studies provide examples of measurement of tectonic uplift and erosion at coastal and inland settings over a range of timescales (thousands to millions of years), and include some of the "classic" sites of studies of uplifted marine terraces. They can be viewed as "worst case" models that can be used for developing safety scenarios for geological repositories.

4.1 Mechanics of tectonic uplift and subsidence

4.1.1 Tectonic drivers at regional and local scales

Most uplift and subsidence is the result of tectonic processes that thicken, thin or flex Earth's crust (Figure 4.1). Thickening and thinning changes the density structure of the crust, leading to vertical crustal movement due to isostatic responses. Flexure of the crust may be driven by horizontal compression or loading (Figure 4.1c–d), which also result in vertical movement due to isostatic responses. These tectonic processes can be described at different spatial scales, from regional (tectonic plate) scale processes resulting in mountain building or the formation of deep ocean basins (Figure 4.1a–d), to local (kilometer) scale processes that produce differential uplift and subsidence across faults and folds (Figure 4.1e–f).

Regional uplift is generally caused by convergence of tectonic plates and concentrated largely at plate boundaries (described further by Cloos, Chapter 2, this volume). A well-studied continent–continent collision is the Alpine–Himalayan orogenic system (along the Indian and Eurasian plate collision zone). At some oblique transform margins, the component of collision is sufficient to generate regional uplift. Examples include the Southern Alps along the Alpine fault portion of the Pacific–Australian plate boundary and coastal northern California along the San Andreas fault portion of the Pacific–North American plate boundary. At subduction zones, aseismic tectonic uplift can occur due to subduction of buoyant pieces of crust such as continental fragments, oceanic plateaus or seamount chains that exist on otherwise normal oceanic crust (e.g. Corrigan *et al.*, 1990; Cloos, 1992). In some cases, uplift at subduction margins may also be driven by deeper-seated processes, such as sediment underplating (e.g. Walcott, 1987). At subduction zones characterized by volcanic arcs, uplift may occur above swelling and rising magma chambers (e.g. Kaizuka, 1992). Regional-scale coseismic uplift can occur as a result of rupture of the subduction interface, known as subduction earthquakes (e.g. Plafker, 1964). Because this uplift is often recovered (reversed) elastically during the interseismic period before the next earthquake (e.g. Savage, 1983), coseismic uplift will only be preserved as permanent uplift if some other mechanism, such as aseismic subduction processes (described above) or upper plate faulting or folding (e.g. Kelsey *et al.*, 1994), occurs.

Local-scale tectonic uplift generally is the result of faulting and folding, although local uplift can also occur above volcanic hot spots. Uplift most commonly occurs on the hanging-wall of reverse faults, and is sometimes accompanied by folding in a hanging-wall anticline

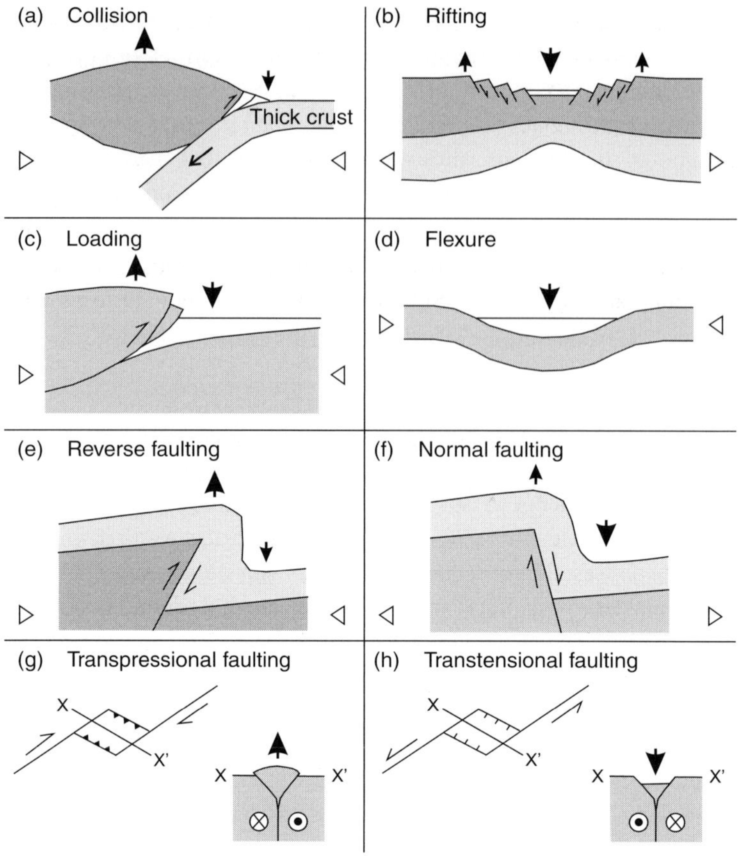

Fig. 4.1 Some of the main mechanisms of regional (a–d) and local (e–h) tectonic uplift and subsidence described in the text. The major vertical movement is shown by the large black arrows, with small arrows denoting minor vertical movement.

(Figure 4.1e). The rate of uplift is directly proportional to the convergence rate and fault dip, with steeper faults resulting in higher rates of uplift. Local uplift also occurs on the foot wall of normal faults, due to mechanical unloading of the hanging wall and isostatic rebound (Weissel and Karner, 1989), and can be accompanied by fault-related folding (Figure 4.1f). Finally, localized uplift occurs along strike-slip faults as a result of oblique slip at compressional jogs and step-overs (Figure 4.1g–h). Most active faults rupture in large earthquakes and uplift is thus often coseismic (i.e. episodic rather than steady), measuring several centimeters to meters per event.

Regional-scale tectonic subsidence is primarily the result of two processes, crustal thinning and tectonic loading. Crustal thinning occurs by rifting at divergent plate boundaries (Figure 4.1b) or by the formation of transtensional basins at a transform plate boundary. Basins formed by rifting and/or seafloor spreading initially subside due to lithospheric thinning by normal faulting and associated hanging-wall subsidence, and subsequently undergo basin-wide subsidence due to continued cooling of the underlying crust (McKenzie, 1978; Lister *et al.*, 1986). Most of Earth's large ocean basins are the result of rifting (e.g. the Atlantic Ocean basin formed along the mid-Atlantic spreading ridge). At convergent plate boundaries regional subsidence can occur as a result of tectonic loading and regional lithospheric flexure (Figure 4.1c–d), forming forearc and foreland basins (e.g. Busby and Ingersoll, 1995). Depending on the relative movement and forces between the two plates, subduction processes such as slab-rollback can result in rifting and subsidence in backarc basins (Uyeda and Kanamori, 1979). Subduction erosion is thought to be responsible for steady subsidence at some subduction margins (e.g. Wells *et al.*, 2003; Heki, 2004). Although subduction earthquakes can produce regional-scale coseismic subsidence, this sudden vertical deformation is often recovered elastically during the interseismic period (similar to coseismic uplift during earthquakes along subduction zones).

Local tectonic subsidence occurs on the downthrown side of faults and in associated foot-wall synclines (Figure 4.1e–f). In normal fault (extensional) settings, foot-wall subsidence leads to the formation of grabens or half-grabens, which may rotate due to listric or planar normal faulting. In reverse fault (compressional) settings, foot-wall subsidence occurs as a result of tectonic loading, which may also lead to the formation of a block-faulted graben or half-graben, or a foot-wall syncline. Subsidence can occur along strike-slip faults, as a result of oblique slip at tensional jogs and at extensional step-overs (Figure 4.1h). Localized subsidence may also occur as a result of magma-chamber drainage, which may form caldera collapse structures.

4.1.2 Erosion and burial

Natural consequences of tectonic uplift and subsidence are erosion and burial, respectively. Tectonic uplift forms topography (or relief), which is subjected to mechanical and chemical weathering and erosion, and tectonic subsidence creates accommodation space (or a basin), which is subject to deposition of eroded sediments (burial).

The regional-scale erosion of mountain belts, often referred to as denudation, is the general lowering of the bedrock surface that takes place over many glacial and interglacial cycles. Denudation rate is controlled primarily by relief (Ahnert, 1970) and orographic precipitation (e.g. Beaumont *et al.*, 1992), with feedbacks among tectonics, climate and erosion (Molnar and England, 1990; Willett *et al.*, 2006). Over long timescales, these controlling factors may balance, forming steady-state topography (e.g. Adams, 1980; Brandon *et al.*, 1998). The mechanisms of regional-scale denudation can be broadly grouped into hillslope (mainly bedrock landsliding), glacial and fluvial (incision) processes. The relative roles of each mechanism, and the locus of maximum erosion, may vary as the landscape matures (Hovius *et al.*, 1998).

At the local scale, erosion occurs on the upthrown side of active faults and on the crests of associated folds (anticlines). On short timescales, fault rupture creates scarps that may be eroded relatively quickly by mass wasting and fluvial incision by knickpoint retreat

(Wallace, 1977a; Arrowsmith and Rhodes, 1994). Ground shaking associated with large earthquakes can also cause "instantaneous" hillslope erosion over an area proportional to earthquake magnitude (Keefer, 1984). On a millennial timescale, tectonic uplift can affect rivers by causing them to form deeply incised canyons (antecedent drainage), or to abandon former channels and cut new ones (e.g. Schumm, 1986). Incising rivers often leave a record of incision in the form of erosional terraces (Merritts *et al.*, 1994; Burbank *et al.*, 1996).

Regional-scale burial occurs in sedimentary basins as a result of the creation of accommodation space (base-level lowering) and sediment supply. Sediment supply is a function of the erosion rate, as well as the basin environment, lithology and mechanisms for delivering sediment. For example, if a basin subsides below sea level, it will be subject to eustatic sea-level changes, resulting in deposition of marine depositional sequences (e.g. Vail *et al.*, 1977). Alternatively, if it remains terrestrial, it will be subjected to tectonically controlled alluvial fan, fluvial and delta deposition (e.g. Miall, 1981). There are also feedbacks between burial and subsidence in that sediment loading can drive further subsidence and flexure, which in turn provides more accommodation space for continued burial.

At the local scale, burial occurs on the downthrown side of faults and in associated footwall synclines. On short timescales, fault rupture and ground shaking can result in deposition of talus or colluvial wedges at the foot of fault scarps (e.g. Wallace, 1977a), rapid deposition of lacustrine deposits in fault-controlled sag ponds and lakes (e.g. Sieh, 1978), and rapid deposition of estuarine and marine deposits in coastal areas subject to coseismic subsidence (e.g. Atwater *et al.*, 1995). Ground displacements and shaking associated with large earthquakes can result in burial by tsunami deposits in coastal settings (e.g. Komatsubara and Fujiwara, 2007) and turbidites in lakes and in deep-water marine settings (e.g. Heezen and Ewing, 1952; Adams, 1990). Over millennial timescales, fault and fold-related subsidence can result in burial by stacking of alluvial-fan deposits along terrestrial basin margins (e.g. Bull, 1977; Viseras *et al.*, 2003), of fluvial deposits (e.g. Leeder and Jackson, 1983) and of sequences of lake and marine deposits.

4.1.3 Measuring uplift and subsidence rates

The measurement of tectonic uplift and subsidence rates requires definition of a spatial reference frame. The most commonly used reference frame is the geoid, or sea level corrected for eustatic changes. Differential uplift and subsidence can also be calculated across faults or folds, or across regions relative to a stable external control point.

It is also crucial to separate tectonic uplift (and subsidence) from surface uplift (and subsidence) or rock uplift (and subsidence). Surface and rock uplift are generally easier to measure than tectonic uplift (England and Molnar, 1990; Burbank and Anderson, 2001). Surface uplift is the upward displacement of Earth's surface with respect to the geoid, whereas rock uplift is the upward displacement of rocks with respect to the geoid. Tectonic uplift is the portion of rock uplift that is driven exclusively by tectonic processes. These differences are particularly important at longer timescales, when processes such as isostatic compensation due to erosion, or compaction due to burial, provide a significant component

Table 4.1. Current techniques for measuring uplift and subsidence rates

Timescale	Technique	Maximum precision	Type of rate/movement*
Short	GPS	mm	Cont, Cos
	Geodetic surveying	mm	Cont, Cos
	Tide gauges	cm	Cont, Cos
	Coral microatolls	cm	Cont, Cos
	InSAR	cm	Cont, Cos
	Geomorphology	cm to tens of cm	Cos
Medium	Marine geomorphic markers	Tens of cm to m	Cos, Ave
	Fluvial geomorphic markers	Tens of cm to m	Cos, Ave
Long	Marine geologic markers	Tens of m	Ave
	Mudstone porosity	Tens of m	Ave
	Low-T thermochronology	Hundreds of m	Ave

* Cont = continuous rate; Cos = coseismic uplift; Ave = average rate.

of the rock uplift or subsidence. In general, most of the techniques described measure rock or surface uplift and subsidence rates, but we assume that the non-tectonic component can be removed and thus refer generally to these as tectonic uplift and subsidence rates.

Many of the currently available techniques for measuring tectonic uplift and subsidence rates are listed in Table 4.1 and are briefly described in the following. Different techniques have been developed for measurements over different timescales, which we have separated into short (annual to centurial), medium (millennial) and long (million-year) timescales. In general, the precision of rates measured over longer timescales decreases and it is of vital importance to quantify these uncertainties to allow comparison between timescales.

On short timescales (annual to centurial), tectonic uplift and subsidence have traditionally been measured by geodetic surveying methods that include observations of trilateration networks (e.g. Walcott, 1987) and precise leveling of benchmarks (e.g. Kato, 1979). In coastal areas, tide gauges (e.g. Savage and Plafker, 1991) and coral microatolls (e.g. Zachariasen *et al.*, 2000) can be used to measure ongoing tectonic uplift and subsidence, and in areas affected by coseismic uplift and subsidence, emergent or drowned markers such as shorelines (beaches, watermarks), mangroves, shellfish, algae or constructions (wharves, buildings) have also been utilized (e.g. Plafker, 1964; Rajendran *et al.*, 2007). A rapidly developing technique is the use of global positioning system (GPS) measurements, either by repeated, intermittent measurements of survey points; or by continuous (daily, automated) measurements (e.g. Beavan *et al.*, 2003; Suwa *et al.*, 2006). Another rapidly developing technique is interferometric synthetic aperture radar (InSAR). This technique can be used in non-vegetated terrain to compare successive satellite images to calculate coseismic (e.g. Massonnet *et al.*, 1993) or ongoing vertical deformation (e.g. Burgmann *et al.*, 2006).

On medium timescales (millennial), tectonic uplift and subsidence rates are measured using geomorphic markers. The most commonly used markers are paleoshoreline features, such as marine terraces, coral reefs or marginal marine sediments, which can be tied to mean sea level through the use of sea-level curves. Holocene marine terraces preserved along uplifting coastlines are often interpreted to be the result of a single uplift event and can therefore record both the average tectonic uplift rate and coseismic uplift (e.g. Ota and Yamaguchi, 2004; Merritts, 1996). With the possible exception of young (30–53 ka) marine terraces at Huon Peninsula (Chappell *et al.*, 1996), Pleistocene marine terraces generally do not preserve evidence of coseismic events because of post-uplift erosion and the deposition of coverbeds, but can record longer average rates of uplift (e.g. Merritts and Bull, 1989). Buried marine and marginal marine sequences recorded by drillholes and seismic surveys can be used to calculate subsidence rates (e.g. Itoh *et al.*, 2000) and, in some cases, to identify coseismic subsidence events (e.g. Atwater 1987; Cochran *et al.*, 2006). Drowned coral reefs preserved on the shelf have also been used to calculate tectonic subsidence rates (e.g. Webster *et al.*, 2004).

Away from coastal areas, measurement of millennial-scale tectonic uplift and subsidence is more problematic, because of the lack of a datum such as mean sea level. Differential uplift across faults and folds can be measured using markers such as fluvial terraces, channels and alluvial fans, if the pre-deformation profile can be reconstructed (e.g. Rockwell *et al.*, 1988; Pazzaglia and Brandon, 2001). Flights of fluvial terraces can be used to calculate tectonic uplift rates only if they can be tied into a datum such as sea level (e.g. Yoshiyama and Yanagida, 1995) or assumptions can be made about their relative altitudes during terrace deposition (e.g. Berryman *et al.*, 2000). Buried fluvial deposits can potentially also be used to calculate subsidence rates following the same methodologies. A correlation between fluvial incision rates and rock uplift rates has also been proposed in some settings (e.g. Burbank *et al.*, 1996; Litchfield and Berryman, 2006).

Long-term (million-year) rates of tectonic uplift can be calculated using geologic markers if their relation to sea level during deposition is known, and if eustatic sea-level changes and isostatic compensation due to erosion can be taken into account (e.g. Abbott *et al.*, 1997). Long-term tectonic subsidence rates can also be calculated using geologic markers if corrections for water depth, eustatic sea-level change and compaction due to burial can be made (e.g. Allen and Allen, 1990). Exhumation rates, calculated from low-temperature thermochronology (e.g. fission track, U–Pb) and mudstone porosity, can be converted to rock uplift rates by comparison with surface uplift rates using geomorphic markers or by correcting for present-day altitude (e.g. Sobel and Strecker, 2003; Pulford and Stern, 2004).

4.1.4 Extrapolation of rates into the future

Confidence associated with extrapolating tectonic uplift and subsidence rates into the future depends on a number of factors, including the nature of the process that has been measured, the timeframe over which it has been measured, the precision of the measurement and assumptions about the stability of the tectonic and climatic regimes. For example, short-term uplift and subsidence rate measurements from GPS or InSAR studies may include both interseismic and coseismic deformation and, in the case of areas affected by subduction earthquakes, the interseismic uplift or subsidence may be in the opposite sense to the coseismic movement. Furthermore, the uplift and subsidence measurements may also be a composite of both upper-plate and subduction interface deformation, which also needs to be removed if upper-plate deformation is of interest (Wallace *et al.*, Chapter 6, this volume).

Extrapolation over millennial timescales relies upon the time-averaged behavior of tectonic processes. For example, in seismic hazard studies, active faults are usually assumed to behave characteristically, whereby similar-sized faults tend to generate the same-size earthquake (Schwartz and Coppersmith, 1984) and, hence, the same amount of uplift or subsidence. To quantify this behavior and justify this assumption, long records of large earthquakes are required.

Extrapolation of rates over millions of years fundamentally requires stability in both the tectonic and climatic regimes. Over millions of years tectonic regimes can alter, with old faults and folds abandoned and new faults and folds formed, and faults can be reactivated in the opposite sense. Furthermore, there can be feedbacks with climate and erosion; for example, changing climate regimes may alter erosion rates, which may in turn affect tectonic uplift rates.

4.2 Studies quantifying uplift, subsidence and erosion rates

In this section, examples of quantifying tectonic uplift, subsidence and erosion rates are presented from three different parts of the Pacific Rim. These examples are from some of the "classic" sites of studies of uplifted marine terraces, and the relatively rapid rates of tectonic uplift allow comparison of rates over different timescales. The first two examples demonstrate measuring tectonic uplift using marine terraces, with Section 4.2.1 focusing on millennial uplift using Holocene marine terraces to construct relative sea-level curves and Section 4.2.2 focusing on 100 ka uplift using Pleistocene marine terraces. However, both also make some comparisons between Holocene and Pleistocene uplift rates. Section 4.2.3 presents a study from an inland setting, in which fluvial terraces are used to calculate Pleistocene uplift and river incision (erosion) rates, which are then compared with uplift rates from marine terraces and with longer-term (million-year) uplift (and denudation) rates using geologic markers. Other examples of studies quantifying uplift rates in inland settings are listed in Table 4.2.

4.2.1 Holocene relative sea-level curves and marine terraces, Japan

Holocene sea-level rise (transgression) is recorded along most of the Japanese coastline and in the lower valleys of the major rivers, resulting in a complex shoreline geometry with embayments of various sizes (Figure 4.2a). The position and elevation of the

Table 4.2. Estimated rock and surface uplift rates from various tectonic settings around the world

Location	Uplift rate range ($m\,Ma^{-1}$)	Time period	Method	Tectonic setting	Sources
Appalachian Mountains; Piedmont, USA	> 5–10 (rock uplift)	$\sim$ 10 Ma–present	Apatite fission track, ages, regional stratigraphy and geodynamic models	Passive margin	Pazzaglia *et al.* (1998)
Northern coastal California, USA	$\approx$ 280–350 regionally; locally higher at faults	$\sim$ 220 ka–present	U-series dating of elevated marine terraces combined with inferred uplift estimates	Transpression along transform plate boundary	Merritts (1996)
Central Alps	300–600 (rock uplift)	6–10 Ma	Apatite fission track age	Collision zone	Schaer *et al.* (1975)
Central Alps	400–1000 (rock uplift)	10–35 Ma	Rb and K–Ar apparent ages of biotite	Collision zone	Clark and Jager (1969)
Kulu-Mandi belt, Himalayas	700 (rock uplift)	25 Ma–present	Apparent Rb–Sr ages of co-existing biotites and muscovites	Collision zone	Mehta (1980)
Huon Peninsula, Papua New Guinea	1000–3000 (surface uplift)	120 ka–present	U-series and ^{14}C dating of elevated marine terraces	Subduction collision zone	Chappell (1974)
Southern Alps, New Zealand	5000–8000 (surface uplift)	140 ka–present	Estimated ages of elevated marine terraces from inferred uplift analysis	Transpression along transform plate boundary	Bull and Cooper (1986)
Southern Alps, New Zealand	10 000 (rock uplift)	1 Ma–present	Apparent K–Ar ages of schists	Transpression along transform plate boundary	Adams (1981)

* Compiled from Summerfield (1991) and Pazzaglia *et al.* (1998).

coastline during sea-level rise have been reconstructed from geological, geomorphological and paleontological data, supplemented by data from archeological sites, excavations and research drillholes. A widespread volcanic ash, named *K-Th*, which erupted 7.3 ka from Kikai caldera in southern Kyushu, provides a useful marker for the culmination of Holocene transgression. The transgression data are summarized on a 1:2 000 000 scale map (Japan Association of Quaternary Research, 1987; Ota *et al.*, 1987; Machida and Koike, 2002; Ota, in press a) and the complete set of references are included in these publications.

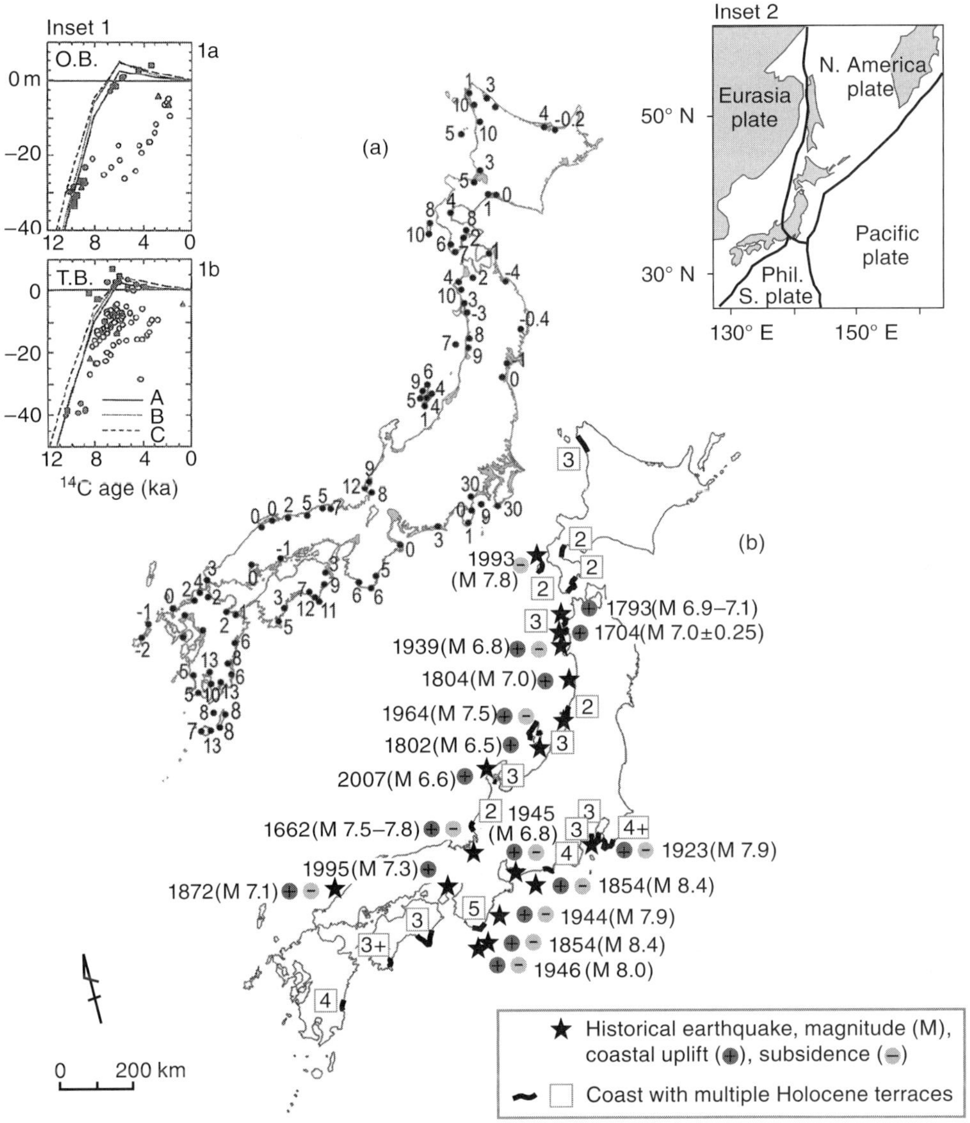

Fig. 4.2 Characteristic coastal features and amount of Holocene uplift along the coast of the Japanese Islands. Map (a) indicates altitude of the paleoshoreline at the time of the culmination of Holocene sea-level rise (5–7 ka). The numbers shown along the coast are the present altitude of the paleoshoreline in meters, which equates to the amount of uplift. Higher uplift occurs mainly on the western side of northern Japan and the eastern side of central Japan (Ota, in press a). Map (b) indicates flights of uplifted Holocene marine terraces and sites of coseismic uplift associated with historical earthquakes (Ota, in press b). Insets 1a and 1b show predicted eustatic sea-level curves taking into account the hydro-isostatic effects and relative sea-level curves from Osaka Bay (O.B.) and Tokyo Bay (T.B.) (Nakada *et al.*, 1991). The locations of earthquake epicenters are correct, but the symbols for uplift and subsidence are slightly shifted for clarity. Dots represents dated samples (solid circle: intertidal shell; open circle: subtidal shell; solid triangle: wood; solid square: peat). Inset 2 shows generalized plate boundaries of Japan.

The age and altitude of the paleoshoreline recording the culmination of Holocene sea-level rise varies regionally as well as locally, reflecting the different plate boundaries (Inset 2, Figure 4.2). The age generally ranges from ~ 5 to 7 ka and the altitude ranges from ≈ 0 to 30 m. Also, at subsiding sites, the submerged shoreline is recorded by drowned archaeological sites. The altitude of the paleoshoreline is high (> 10 m) along the northwestern coast (Sea of Japan), close to the plate boundary between the Eurasian plate and North America plate, and along the southeastern coast from southern Kanto to southern Shikoku, along the northern boundary of the subducting Philippines Sea plate. Along the southern Kanto coast, the shoreline altitude is 25–30 m. However, the age and altitude are different from the neighboring coast at the Izu Peninsula ($\sim$ 2–3 ka and 0–2 m, respectively), which is located on the subducting Philippine Sea plate, reflecting its unique tectonic setting. In southern Kyushu, the 10–15 m altitude records local uplift caused by volcanic activity. The culmination of sea-level rise is also recorded as coral terraces in the Ryukyu (Nansei) Islands along the subducting Ryukyu trench, which are up to 13 m in altitude.

Relative sea-level curves reconstructed for different parts of the coastline are highly variable, reflecting the different tectonic settings. Construction of eustatic sea-level curves, separating local tectonic effects from relative sea-level curves, is challenging. A great number of relative sea-level curves have been proposed from different tectonic areas (Ota *et al.*, 1987; Ota, in press a, b), some showing minor regression during the late Holocene. Recently hydro-isostatically predicted sea-level curves have been proposed as the standard method for establishing eustatic sea-level curves (e.g. Nakada *et al.*, 1991; Insets 1a and 1b, Figure 4.2). However, these are dependent on assumptions such as the viscosity and thickness of the crust or the melting of the Arctic ice sheet. So far, along the active Japanese coast, local tectonic effects are considered to be the first-order controlling factor on paleoshoreline altitude, which can then be used as the measure of the amount of uplift during the last ~ 7 ka.

The presence of multiple subdivided Holocene marine terraces, usually 3–5 steps, separated by clear terrace risers, is a characteristic feature of the tectonically active Japanese coast (Figure 4.2b). This is particularly so along rapidly uplifting coasts such as the Japan Sea side of northern Japan, or along the Pacific Ocean side near the Philippine Sea plate. Their coseismic origin is proposed by many workers (Ota, in press b), based on measurements of coseismic uplift resulting from historical earthquakes (Figure 4.2b) and from terrace morphology. The timing of uplift and abandonment of the terraces is constrained by dating boring shell fossils, barnacles and other microfossils, although the accuracy is sometimes poor and it is often difficult to obtain datable materials from rocky coastlines. Holocene marine-terrace data are used as the key for constraining the interval and magnitude of great earthquakes that occurred offshore, or onshore, close to the coast.

Although the pattern of coseismic uplift in a given area is usually similar throughout the Holocene, as well as being similar to the pattern of late Quaternary terrace deformation (Ota and Yamaguchi, 2004), there are some exceptions. For example, Okushiri Island, west of southern Hokkaido, consists of a flight of middle to late Quaternary and Holocene terraces, up to the top of the island (585 m altitude), recording long-term uplift. The Holocene terrace, up to 10 m altitude at its inner margin, is subdivided into two, recording intermittent

Holocene uplift. However, the 1994 Hokkaido–Nansei-oki earthquake resulted in coastal subsidence that is discordant with the Holocene uplift. This implies the pattern of tectonic movement can be variable with time, depending on the location of seismogenic fault sources. Another issue is that the recurrence interval estimated from the Quaternary terrace morphology, such as scattered raised notches, narrow terrace widths and locally attached fossils (e.g. Boso Peninsula), is long (1 ka), whereas the historical record of great subduction earthquakes suggests a shorter (< 1 ka) interval may be reasonable. For example, the 1995 Hyogo-ken Nanbu earthquake resulted in 0.6 m of coastal uplift; and the 2007 Noto Peninsula earthquake resulted in up to 0.4 m of coastal uplift, tilted to the southwest, similar to the marine isotope stage (MIS) 5e, 125 ka, terrace. However, these small amounts of uplift are not recorded in the Holocene terrace sequences. The effects of superimposition of long-interval earthquakes, with greater amount of uplift, and short-interval earthquakes, with an order of tens of centimeters of uplift, requires further work based on very precise dating and morphological characteristics.

4.2.2 Pleistocene marine terraces, northern California, USA

One of the earliest studies of wave-cut bedrock marine-terrace platforms in the USA was that of Bradley and Griggs (1976), who mapped six terrace complexes, some of which had up to three separate platforms, along ~ 32 km of the coast of central California, near the town of Santa Cruz (Figures 4.3 and 4.4b). The lowest terrace, the Davenport terrace, is correlated with MIS 5a based on high-precision U-series dates of 11 solitary corals (*Balanophyllia elegans*) from beach gravel at a site near Point Ano Nuevo, where the coral dates range from 75 800 ± 0.8 to 84 000 ± 0.6 ka (Muhs *et al.*, 2003). No higher platforms have been dated yet, but it is possible that the three platforms within the Santa Cruz terrace represent the high stands of MIS 5a, 5c and 5e.

The inner-edge altitude data can be used to estimate the amount of uplift that has occurred since the time of formation of a given marine platform, as shown in Figure 4.4c. An uncertainty in such analysis is the actual altitude of formation, or paleo-sea level, for different high stands. However, high-precision U–Th ages of coral from marine terraces at multiple locations throughout the world are providing better estimates of these past sea levels. The greater uncertainty is in estimating the age of a terrace if no independent ages are available. In cases where one terrace within a sequence is dated, as at Santa Cruz, it is possible to make the assumption that the long-term rate of uplift has been constant, and then to infer the ages of other terraces based on that assumption. The inferred uplift of a given terrace is calculated as the difference between its present altitude (at the inner edge) and altitude of formation, or paleo-sea level. Based on this approach, a plot of inferred uplift versus estimated age (i.e. time) illustrates the long-term uplift rate (slope of the best-fit line) for a given section of coast. For the Santa Cruz area, a long-term uplift rate over hundreds of thousands of years that is consistent with the dated Davenport terrace and the altitude of higher terraces is ≈ 0.28 mm a^{-1}. In general, marine terraces along the northern California coastline, including near Mendocino (Figure 4.4a), occur at similar altitudes over length

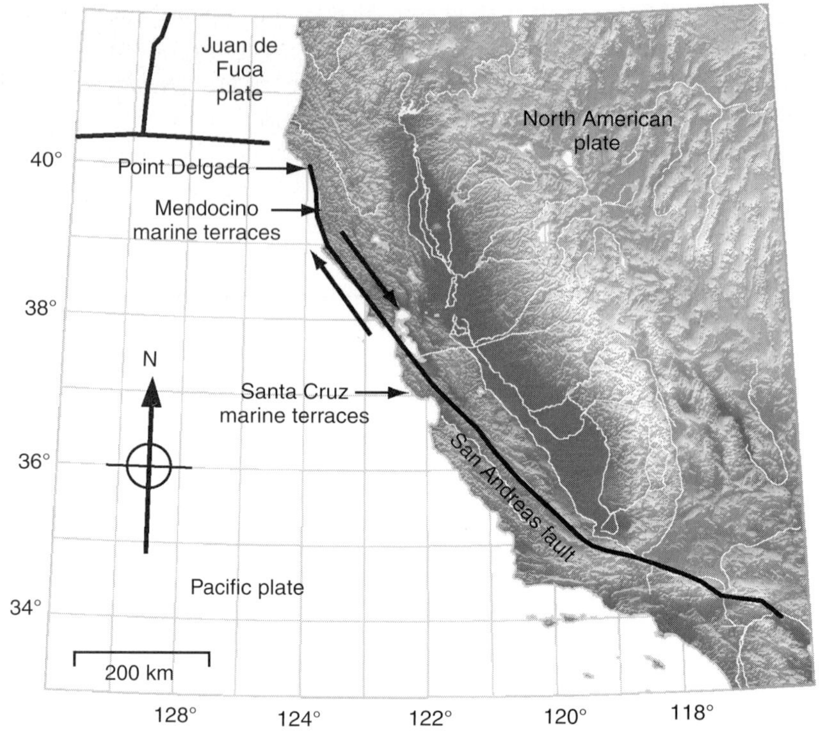

Fig. 4.3 Shaded relief map of coastal California along the San Andreas fault (topography data from GLOBE, 2008). Sites illustrated in Figure 4.4 are at Santa Cruz and Mendocino, and in Figure 4.5 at Point Delgada. The Mendocino triple junction is the region at the northern termination of the San Andreas fault where three plates, the Juan de Fuca, North American and Pacific, are adjacent to one another.

scales tens of kilometers (cf. Merritts *et al.*, 2005) and are uplifted at ≈ 0.2–$0.4\,\text{mm a}^{-1}$, rates that are interpreted to be the result of slight oblique compression along the San Andreas boundary (cf. Merritts and Bull, 1989; Anderson and Menking, 1994).

Some of the highest rates of surface uplift and seismic activity in North America occur at the northern termination of the San Andreas fault, in the vicinity of the Mendocino triple junction (Merritts, 1996). South of the triple junction, uplift rates increase by nearly an order of magnitude at $\sim$ 40 km compared to rates at Santa Cruz and Mendocino, and are sufficiently high to result in emergent Holocene marine terraces that can be traced northward to the triple junction (Figure 4.5). As the three plates, separated by the San Andreas fault, the Cascadia subduction zone and the Mendocino fracture zone, move in different directions at different rates, interplate strain accumulates and locally deforms the crust (Figure 4.3). Accumulated strain is released during large earthquakes, which uplifts the ground surface as much as several meters along the northernmost San Andreas fault (Merritts, 1996), and perhaps even more during earthquakes along the southern Cascadia subduction zone fault

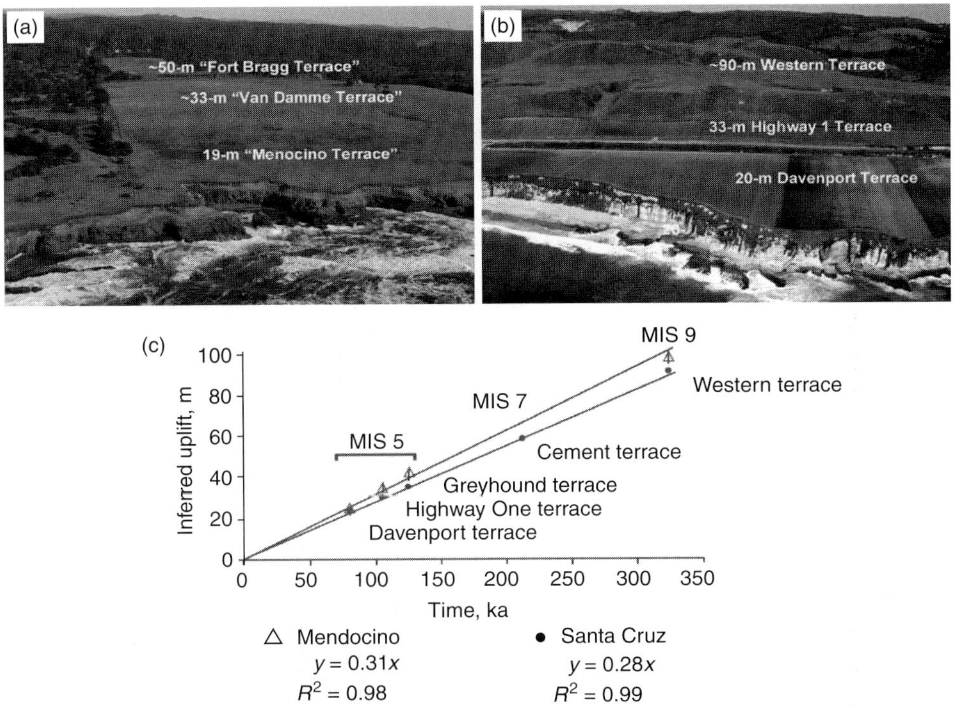

Fig. 4.4 (a) Air photo of the northern California coast south of the town of Mendocino (area of Van Damme State Park), taken November 14, 2002 (Adelman and Adelman, 2002; terrace surveying and mapping by Merritts and students, 1999–2006, unpublished data). Altitude estimates are ±1 m, but uncertainty of inner-edge estimate is ±2 m for the 2nd and 3rd terraces because of sedimentary cover on the platforms. (b) Air photo of the northern California coast near Laguna Creek, Santa Cruz taken in 1972 (Adelman and Adelman, 2002; terrace altitudes are from Bradley and Griggs, 1976). (c) Inferred uplift-rate diagram compares terrace inner-edge altitudes from the locales shown in (a) and (b). See color plate section.

(e.g. Atwater *et al.*, 1995). Long-term, permanent deformation at the terminus of the San Andreas fault has been documented from studies of uplifted Pleistocene and Holocene marine terraces. Uplift rates are as high as $\approx 4\,\mathrm{mm\,a^{-1}}$ (Merritts and Bull, 1989; Merritts, 1996). As a consequence of this range in uplift rates, the region has been studied to evaluate landscape response to tectonic forcing (Merritts and Vincent, 1989; Merritts *et al.*, 1994; Snyder *et al.*, 2000).

Within the pervasively faulted structural knot of the Mendocino triple junction, the styles of earthquake faulting and, subsequently, the rates of uplift, vary significantly over distances of several kilometers to tens of kilometers. The 1906, M 7.8 earthquake appears to have ruptured with predominantly right-lateral strike slip, and produced no discernible coastal uplift in the Shelter Cove–Point Delgada area (Lawson, 1908). Just 30 km to the north, along the southernmost Cascadia subduction zone, the 1992, Cape Mendocino M 7.1 earthquake

Fig. 4.5 Early geologic studies (Lawson, 1908) mapped the San Andreas fault onshore at Shelter Cove, but were unable to trace the fault north of Telegraph Hill due to steep topography and vegetation cover. Merritts *et al.* (2000) demonstrate that Pleistocene alluvial fans upslope (to the right) of the Shelter Cove airstrip are offset from their source at Dead Mans Gulch by the San Andreas fault and partially buried by colluvial and mass-movement deposits. Long-term slip $\geq$ 13 mm a^{-1} indicates that this is the main trace of the San Andreas fault in this region (Prentice *et al.*, 1999).

exhibited pure reverse slip and uplifted the coastline as much as 1.4 m over a coastal distance of $\sim$ 20 km (Carver *et al.*, 1994; Oppenheimer *et al.*, 1993). Along this same coastal stretch, Holocene uplift rates vary from 1.5–3.0 mm a^{-1}, with an arched pattern that matches the uplift pattern produced by the Cape Mendocino earthquake (Merritts, 1996).

The 30 km coastline extending south of the Mendocino triple junction along the northern terminus of the San Andreas fault, in contrast, has uplift rates that are even higher (1.8–4.1 mm a^{-1}) than north of the triple junction. Merritts (1996) relates the rapid uplift in this region, which preserves multiple Holocene marine terraces above a rising sea, to complex processes of deformation along the northern terminus of the San Andreas fault. Indeed, several high-angle northwest-striking reverse faults cut the uplifted Holocene platforms in this area. This zone of rapid Holocene uplift corresponds to the region where it becomes increasingly difficult to trace the 1906 San Andreas fault rupture north of Shelter Cove (using air photos, topographic data and field work; Figure 4.5). Both the San Andreas fault main trace, which lies to the east of the area of coastal uplift, and other local structures, probably mostly reverse faults, contribute to net uplift and differential tilt along the coastline.

4.2.3 Hikurangi subduction margin, New Zealand

Calculating uplift and subsidence rates away from coastal areas is problematic because of the lack of a reference frame. Recently, in inland areas of the uplifting Hikurangi subduction margin (Figure 4.6a–b), uplift rates have been calculated using flights of fluvial terraces (Berryman *et al.*, 2000; Litchfield and Berryman, 2006; Litchfield *et al.*, 2007). These studies

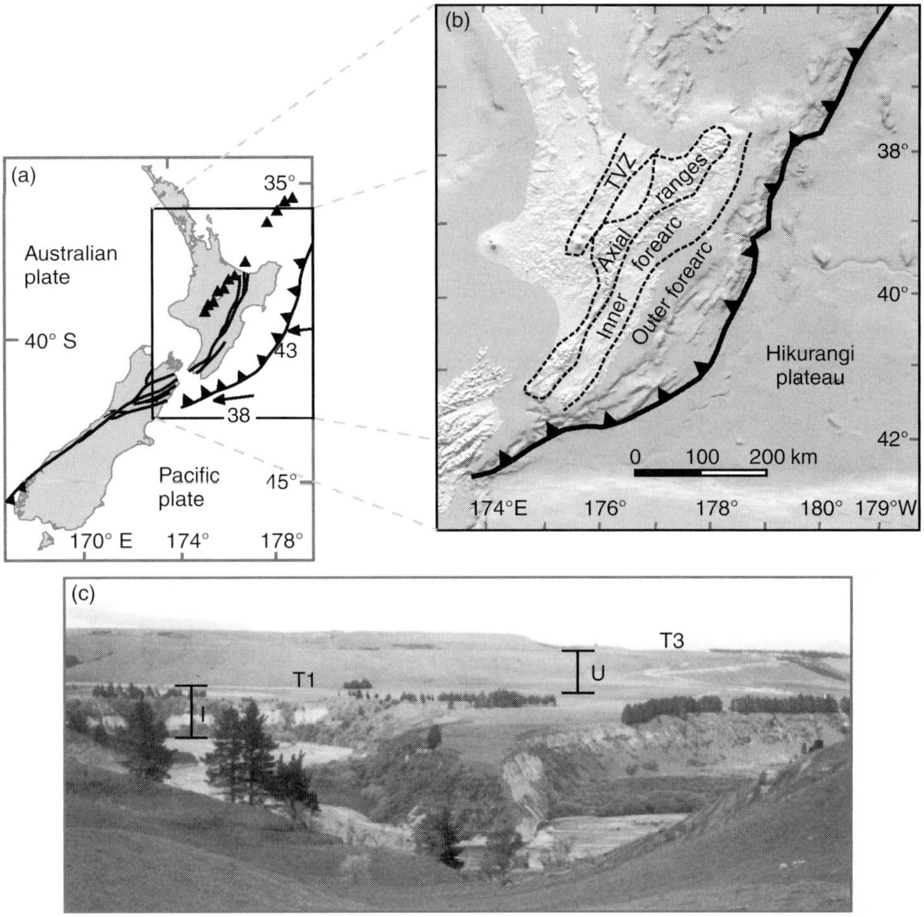

Fig. 4.6 Location maps for the Hikurangi margin, New Zealand (a) and (b), and photograph of fluvial terraces used for uplift-rate and incision-rate calculations (c). Map (a) shows the regional tectonic setting. Black triangles represent volcanoes of the Taupo volcanic zone (TVZ). Numbers by arrows are plate convergence rates (mm a^{-1}). Map (b) is a shaded digital terrain and bathymetry model showing the major structural elements of the Hikurangi margin. This study focused on uplift of the axial ranges and the inner forearc. As shown in photo (c), uplift rates were calculated from the altitude difference between T3 and T1 treads (U). Fluvial incision rates were calculated from the altitude difference between the T1 tread and the present riverbed (I).

are summarized here, focusing on the uplift-rate calculation methodologies and results over two timescales (late Quaternary and million-year) and the relationships between uplift and erosion (fluvial incision). It is important to note that the uplift rates described here are all rock uplift rates relative to the geoid or sea level, as they have not been corrected for isostatic compensation due to erosion. This overestimation is considered minor at late-Quaternary timescales, but will be greater at million-year timescales. The absence of significant (sheet) glaciation in the North Island during glacial periods means that glacial isostatic rebound is also considered negligible.

Late-Quaternary (post-55 ka) uplift rates (Figure 4.7a) were calculated from the surveyed (by GPS) altitude differences between the treads of fluvial-fill terraces (Figure 4.6c). Fill terraces in non-glaciated parts of New Zealand are primarily climate controlled, and are formed during cold stadials or glacial periods in response to increased sediment supply (Litchfield and Berryman, 2005). Thus, if the assumption can be made that aggradation during each cold period filled each valley to a similar level, then the altitude difference between successive fill terrace treads, divided by the difference in age, is equal to the rock uplift rate at that position along the river between the aggradation events (Berryman *et al.*, 2000). Alternatively, subsidence rates can be calculated by comparing the depths of buried fluvial deposits. Uplift rates were calculated using two fill terraces, T1 (18 ± 2 ka, calibrated) and T3 (55 ± 5 ka, calibrated), which were dated by radiocarbon dating and optically stimulated luminescence dating (OSL), loess stratigraphy and tephrochronology (Litchfield and Berryman, 2005). Calculated mean uplift rates are up to 4 mm a^{-1} (Figure 4.7a), with root-mean-square uncertainties of ± 0.3–0.7 mm a^{-1}, which take into account the altitude and age uncertainties. In areas close to the coast, these post-55 ka uplift rates are consistent with post-125 ka uplift rates shown in Figure 4.7a (e.g. Pillans, 1986; Berryman and Hull, 2003).

Million-year rock uplift rates (Figure 4.7b) were calculated from four data sets. The first three data sets use the present-day altitudes (from geologic and topographic maps) of geologic markers formed at or close to sea level. These markers include: (i) the 1.6 Ma marine–terrestrial contact, (ii) the equivalent age, exhumed contact forming an erosional surface on the crest of the axial ranges, and (iii) 1.8–5.7 Ma shallow-water marine limestones. Uplift-rate calculations include root-mean-square uncertainties in altitude and age and are as high as ± 0.1 mm a^{-1}. The fourth data set uses the present-day altitudes of mudstones for which maximum depth of burial has been calculated from porosity. Mudstone porosity measurements were plotted against standard porosity–depth trends from stable regions west of New Zealand (Armstrong *et al.*, 1998), allowing estimation of maximum depth of burial beneath the seafloor (and, hence, exhumation). To calculate rock uplift, the depth of deposition and present altitude was added. The rock uplift rate was calculated by dividing either by the age of the mudstone, or the time of maximum burial, whichever is younger. Uncertainties using this data set are greater ($\leq \pm 0.6$ mm a^{-1}) than for the other three data sets, and take into account altitude, deposition depth, porosity measurement, altitude uncertainty in the standard porosity curves and the age. Million-year uplift rates from these data sets are relatively low (< 1 mm a^{-1}), ~ 5 times lower than the late-Quaternary uplift rates (compare Figures 4.7a and b); despite being maximum tectonic uplift rates, as discussed above. The difference is inferred to reflect temporal variations in uplift rate on a million-year timescale, as observed from other studies (Beanland *et al.*, 1998; Chanier *et al.*, 1999; Nicol *et al.*, 2002).

The relationship between uplift and erosion was examined by comparing late-Quaternary uplift rates (Figure 4.7a) with post-glacial fluvial incision rates (Figure 4.7c). The incision rates indicate vertical erosion of the river valleys since ~ 18 ka, calculated from the altitude

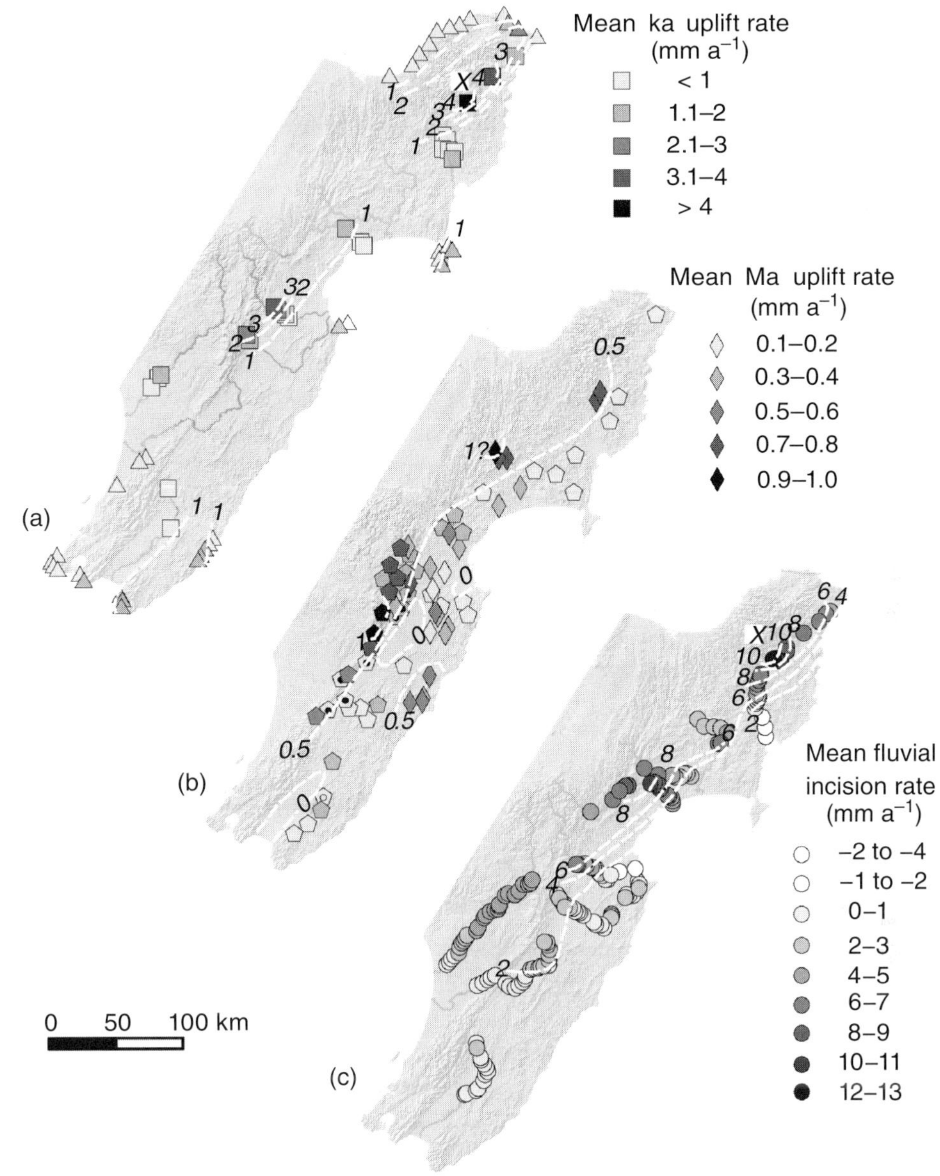

Fig. 4.7 Uplift and fluvial incision rate maps for the Hikurangi margin, New Zealand. Map (a) shows late-Quaternary (ka) mean uplift rates derived from the altitude difference between pairs of fluvial fill terraces (~ 55 ka and ~ 18 ka) (squares) and the present-day altitude of the MIS 5e (125 ka) marine terrace (triangles). Map (b) shows million-year (Ma) mean uplift rates derived from the present-day altitude of geologic markers (pentagons) and mudstones for which maximum burial depth has been calculated from porosity (diamonds). Map (c) shows mean fluvial incision rates derived from the altitude difference between T1 (~ 18 ka) and the present-day river bed (Litchfield and Berryman, 2006; Litchfield *et al.*, 2007). See color plate section.

difference between the T1 terrace tread and the present river level, divided by the time since terrace abandonment, $\sim$ 18 ka (Figure 4.7c). It is important to note that the incision rate is calculated over a half glacial cycle and thus overestimates the rock incision rate. This would be measured from the bedrock strath on which the T1 fluvial terrace deposits rest, to the present river bed. The incision rates range from $-4\,\mathrm{mm\,a^{-1}}$ (where T1 is buried) to $12\,\mathrm{mm\,a^{-1}}$, with calculated root-mean-square uncertainties of $\leq 1\,\mathrm{mm\,a^{-1}}$, taking into account surveyed altitude, projection to a mid-valley axial line, tread relief and age. Although the post-18 ka incision rates are higher (by ≈ 1.5 times) than the late-Quaternary uplift rates (compare Figures 4.7a and c), the patterns of variability are similar and, after careful consideration of the possible non-tectonic controls on incision rates (climate, base-level, bedrock lithology and stream power), Litchfield and Berryman (2006) inferred tectonics to be the main control on river incision in the eastern North Island. Thus, the post-glacial fluvial-incision rate map, a more complete data set, can also be used to infer patterns of regional-scale tectonic uplift. By examining the wavelengths of uplift and by comparison with geometries and uplift rates predicted by two-dimensional finite-element numerical models of subduction processes, Litchfield *et al.* (2007) concurred with previous workers that most of the uplift was the result of subduction processes including plateau subduction (Davy, 1992), sediment underplating (Reyners *et al.*, 1999, 2006; Upton *et al.*, 2003) and tectonic erosion (Collot *et al.*, 1996, 2001).

Concluding remarks

This chapter has summarized the current state of knowledge of the mechanics of tectonic uplift and subsidence and associated erosion and burial, and currently available techniques to measure them. The highest rates of tectonic uplift tend to occur at convergent plate boundaries, and one of the current challenges is to unravel regional-scale (subduction-related) uplift from local-scale (upper plate faults) uplift. A related challenge is to separate the short-term elastic uplift associated with ongoing build-up of strain above subduction thrusts from permanent uplift, in order to calculate robust long-term uplift rates. Some of these issues will be better understood once longer records of current rates of uplift from GPS and coseismic uplift events from contemporary earthquakes and paleoseismology records become available.

The examples described in Section 4.2 demonstrate that, for late-Quaternary timescales, well-established techniques exist for measuring regional-scale uplift in coastal areas and that techniques are currently being developed for inland areas. Table 4.2 shows that techniques for measuring rock uplift over million-year timescales in inland areas are also well established. Many of the techniques in Table 4.2 also calculate denudation rates, which are also important for geological repositories. The case studies also demonstrate comparison of rates at different timescales, which is important for understanding the variability of uplift rates over different timescales and, ultimately, to defining the level of confidence in extrapolation of rates into the future. Ultimately, extrapolation of rates to million-year timescales requires stability of the tectonic and climatic regimes or, at the very least, some understanding of changes that can be reasonably excluded. This is particularly challenging at complex plate boundaries such as subduction zones, some of which evolve rapidly.

For a site-specific analysis, consideration needs to be given to separating local- from regional-scale uplift and subsidence, which may result in different deformation effects on a nuclear facility (e.g. fault rupture versus regional uplift). Localized versus regional erosion or burial also needs to be considered, such as localized erosion on the top of a fault scarp or a fold crest, versus regional-scale denudation. On long timescales, the balance between uplift and erosion or subsidence and burial becomes increasingly important. Thus, when extrapolating past rates, it is important to separate measurements of rock uplift from exhumation rates.

Further reading

Wallace (1977b), Yeats *et al.* (1997) and Burbank and Anderson (2001) provide further information on the tectonic processes outlined here and how to measure them using geomorphic techniques. For further information about tectonic subsidence and burial in tectonic basins, see Allen and Allen (1990) and Busby and Ingersoll (1995). England and Molnar (1990) is the required reading for the importance of separating tectonic versus surface and rock uplift and exhumation. Willet *et al.* (2006), and references therein, summarize the state of the art on the interactions between regional-scale uplift and erosion of mountain belts.

Acknowledgments

The authors would like to acknowledge the efforts of T. Gardner, C. Prentice, C. Crosby, R. Walter, C. Lippincott and 15 Keck Geology Consortium students for surveying and mapping terraces around the area of Van Damme State Park from 1999–2006.

References

Abbott, L. D., E. A. Silver, R. S. Anderson *et al.* (1997). Measurement of tectonic surface uplift rate in a young collisional mountain belt. *Nature*, **385**, 501–507.

Adams, C. J. (1981). Uplift rates and thermal structure in the Alpine Fault Zone and Alpine Schists, Southern Alps, New Zealand. In: McClay, K. R. and N. J. Price (eds.) *Thrust and Nappe Tectonics*, Special Publication 9. London: Geological Society, 211–222.

Adams, J. (1980). Contemporary uplift and erosion of the Southern Alps, New Zealand. *Geological Society of America Bulletin*, **91**(1), 1–114.

Adams, J. (1990). Paleoseismicity of the Cascadia Subduction Zone: evidence from turbidites off the Oregon–Washington margin. *Tectonics*, **9**, 569–583.

Adelman, K. and G. Adelman (2002). Digital image 11645 and digital image 7220004 obtained from copyright ©2002–2006. California Coastal Records Project, http://www.californiacoastline.org.

Ahnert, B. (1970). Functional relationships between denudation, relief, and uplift in large mid-latitude drainage basins. *American Journal of Science*, **268**, 243–263.

Allen, P. A. and J. R. Allen (1990). *Basin Analysis: Principles and Applications*. Boston, MA: Blackwell Science.

Anderson, R. S. and K. M. Menking (1994). The Quaternary marine terraces of Santa Cruz, California: evidence for coseismic uplift on two faults. *Geological Society of America Bulletin*, **106**, 649–664.

Armstrong, P. A., R. G. Allis, R. H. Funnell and D. S. Chapman (1998). Late Neogene exhumation patterns in Taranaki Basin (New Zealand): evidence from offset porosity–depth trends. *Journal of Geophysical Research*, **103**, 30 269–30 282.

Arrowsmith, J. R. and D. D. Rhodes (1994). Original forms and initial modifications of the Galway Lake Road Scarp formed along the Emerson fault during the 28 June 1992 Landers, California, Earthquake. *Bulletin of the Seismological Society of America*, **84**, 511–527.

Atwater, B. F. (1987). Evidence for great Holocene earthquakes along the outer coast of Washington State. *Science*, **236**, 942–944.

Atwater, B. F., A. R. Nelson, J. J. Clague *et al.* (1995). Summary of coastal geologic evidence for past great earthquakes at the Cascadia subduction zone. *Earthquake Spectra*, **11**, 1–18.

Beanland, S., A. Melhuish, A. Nicol and J. Ravens (1998). Structure and deformational history of the inner forearc region, Hikurangi subduction margin, New Zealand. *New Zealand Journal of Geology and Geophysics*, **41**, 325–342.

Beaumont, C., P. Fullsack and J. Hamilton (1992). Erosional control of active compressional orogens. In: McClay, K. R. (ed.) *Thrust Tectonics*. London: Chapman and Hall, 1–18.

Beavan, R. J., D. W. Matheson, P. Denys *et al.* (2004). A vertical deformation profile across the Southern Alps, New Zealand, from 3.5 years of continuous GPS data. In: *Proceedings of the Workshop: The State of GPS Vertical Positioning Precision: Separation of Earth Processes by Space Geodesy, April 2-4, 2003, Hotel Parc Bell-Vue, Luxembourg, Grand-Duchy of Luxembourg*, Cahiers du Centre europêan de géodynamique et de séismologie 23. Luxembourg: Centre européen de géodynamique et de séismologie.

Berryman, K. and A. Hull (2003). Tectonic controls on late Quaternary shorelines: a review and prospects for future research. In: Goff, J. R., S. L. Nichol and H. L. Rouse (eds.) *The New Zealand Coast, Te Tai o Aotearoa*. Palmerston North, NZ: Dunmore Press, 25–58.

Berryman, K., M. Marden, D. Eden *et al.* (2000). Tectonic and paleoclimatic significance of Quaternary river terraces of the Waipaoa River, east coast, North Island, New Zealand. *New Zealand Journal of Geology and Geophysics*, **43**, 229–245.

Bradley, W. C. and G. B. Griggs (1976). Form, genesis, and deformation of central California wave-cut platforms. *Geological Society of America Bulletin*, **87**, 433–449.

Brandon, M. T., M. K. Roden-Tice and J. I Garver (1998). Late Cenozoic exhumation of the Cascadia accretionary wedge in the Olympic Mountains, northwest Washington State. *Geological Society of America Bulletin*, **110**, 985–1009.

Bull, W. B. (1977). The alluvial-fan environment. *Progress in Physical Geography*, **1**, 222–270.

Bull, W. B. and A. F. Cooper (1986). Uplifted marine terraces along the Alpine Fault, New Zealand. *Science*, **234**, 1225–1228.

Burbank, D. W. and R. S. Anderson (2001). *Tectonic Geomorphology*. Boston, MA: Blackwell Science.

Burbank, D. W., J. Leland, E. Fielding *et al.* (1996). Bedrock incision, rock uplift and threshold hillslopes in the northwestern Himalayas. *Nature*, **379**, 505–510.

Burgmann, R., G. Hilley, A. Ferretti and F. Novali (2006). Resolving vertical tectonics in the San Francisco Bay Area from permanent scatterer InSAR and GPS. *Geology*, **34**, 221–224.

Busby, C. J. and R. V. Ingersoll (1995). *Tectonics of Sedimentary Basins*. Boston, MA: Blackwell Science.

Carver, G. A., A. S. Jayko, D. W. Valentine and W. H. Li (1994). Coastal uplift associated with the 1992 Cape Mendocino earthquake, northern California. *Geology*, **22**, 195–198.

Chanier, F., J. Ferriere and J. Angelier (1999). Extensional deformation across an active margin, relations with subsidence, uplift, and rotations: the Hikurangi subduction, New Zealand. *Tectonics*, **18**, 862–876.

Chappell, J. (1974). Upper mantle rheology in a tectonic region: evidence from New Guinea. *Journal of Geophysical Research*, **79**, 390–398.

Chappell, J., Y. Ota and K. Berryman (1996). Late Quaternary coseismic uplift history of Huon Peninsula, Papua New Guinea. *Quaternary Science Reviews*, **15**, 7–22.

Clark, S. P. and E. Jager (1969). Denudation rate in the Alps from geochronologic and heat flow data. *American Journal of Science*, **267**, 1143–1160.

Cloos, M. (1992). Lithospheric buoyancy and collisional orogenesis: subduction of oceanic plateaus, continental margins, island arcs, spreading ridges, and seamounts. *Geological Society of America Bulletin*, **105**, 715–737.

Cochran, U., K. Berryman, J. Zachariasen *et al.* (2006). Paleoecological insights into subduction zone earthquake occurrence, eastern North Island, New Zealand. *Geological Society of America Bulletin*, **118**, 1051–1074.

Collot, J.-Y., J. Delteil, K. B. Lewis *et al.* (1996). From oblique subduction to intra-continental transpression: structures of the southern Kermadec–Hikurangi margin from multibeam bathymetry, side-scan sonar and seismic reflection. *Marine Geophysics Research*, **18**, 357–381.

Collot, J.-Y., K. B. Lewis, G. Lamarche and S. Lallemand (2001). The giant Ruatoria debris avalanche on the northern Hikurangi margin, New Zealand: result of oblique seamount subduction. *Journal of Geophysical Research*, **106**(B9), 19 721–19 297.

Corrigan, J., P. Mann and J. C. Ingle (1990). Forearc response to subduction of the Cocos Ridge, Panama–Costa Rica. *Geological Society of America Bulletin*, **102**, 628–652.

Davy, B. W. (1992). The influence of subducting plate buoyancy on subduction of the Hikurangi–Chatham Plateau beneath the North Island, New Zealand. In: Watkins, J. S. *et al.* (eds.) *Geology and Geophysics of Continental Margins*, AAPG Memoir 53. Tulsa, OK: American Association of Petroleum Geologists, 75–91.

England, P. and P. Molnar (1990). Surface uplift, uplift of rocks, and exhumation of rocks. *Geology*, **18**, 1173–1177.

GLOBE (2008). The Global Land One-km Base Elevation (GLOBE) Project. NOAA's National Geophysical Data Center (NGDC), http://www.ngdc.noaa.gov/mgg/topo/globe.html.

Heezen, B. C. and M. Ewing (1952). Turbidity currents and submarine slumps, and the 1929 Grand Banks earthquake. *American Journal of Science*, **250**, 849–873.

Heki, K. (2004). Space geodetic observation of deep basal subduction erosion in northeastern Japan. *Earth and Planetary Science Letters*, **219**, 13–20.

Hovius, N., C. P. Stark, M. A. Tutton and L. D. Abbott (1998). Landslide-driven drainage network evolution in a pre-steady-state mountain belt: Finisterre Mountains, Papua New Guinea. *Geology*, **26**, 1071–1074.

Itoh, Y., K. Takemura, T. Ishiyama, Y. Tanaka and H. Iwaki (2000). Basin formation at a contractional bend of a large transcurrent fault: plio-Pleistocene subsidence of the Kobe and northern Osaka basins, Japan. *Tectonophysics*, **321**, 327–341.

Japan Association for Quaternary Research (1987). *Quaternary Map of Japan*. Tokyo: University of Tokyo Press.

Kaizuka, S. (1992). Coastal evolution at a rapidly uplifting volcanic island: Iwo-Jima, western Pacific Ocean. *Quaternary International*, **15/16**, 7–16.

Kato, T. (1979). Crustal movements in the Tohoku District, Japan, during the period 1900–1975, and their tectonic implications. *Tectonophysics*, **60**, 141–167.

Keefer, D. K. (1984). Landslides caused by earthquakes. *Geological Society of America Bulletin*, **95**, 406–421.

Kelsey, H. M., D. C. Engebretson, C. E. Mitchell and R. L. Tucker (1994). Topographic form of the Coast Ranges of the Cascadia Margin in relation to coastal uplift rates and plate subduction. *Journal of Geophysical Research*, **99**(B6), 12 245–12 555.

Komatsubara, J. and O. Fujiwara (2007). Overview of Holocene tsunami deposits along the Nankai, Suruga, and Sagami troughs, Southwest Japan. *Pure and Applied Geophysics*, **164**, 493–507.

Lawson, A. C. (1908). *The California Earthquake of April 18, 1906: Report of the State Earthquake Investigation Commission*, Carnegie Institution of Washington publication 87. Washington, DC: Carnegie Institute of Washington.

Leeder, M. R. and J. A. Jackson (1993). The interaction between normal faulting and drainage in active extensional basins, with examples from the western United States and central Greece. *Basin Research*, **5**, 79–102.

Lister, G. S., M. A. Etheridge and P. A. Symonds (1986). Detachment faulting and the evolution of passive continental margins. *Geology*, **14**, 246–250.

Litchfield, N. J. and K. R. Berryman (2005). Correlation of fluvial terraces within the Hikurangi Margin, New Zealand: implications for climate and base-level controls. *Geomorphology*, **68**, 291–313.

Litchfield, N. J. and K. R. Berryman (2006). Relations between postglacial fluvial incision rates and uplift rates in the North Island, New Zealand. *Journal of Geophysical Research*, **111**, F02007, doi:10.1029/2005JF000374.

Litchfield, N., S. Ellis, K. Berryman and A. Nicol (2007). Insights into subduction-related uplift along the Hikurangi Margin, New Zealand, using numerical modeling. *Journal of Geophysical Research*, **112**, F02021, doi:10.1029/2006JF000535.

Machida, H. and K. Koike (2001). *Atlas of Quaternary Marine Terraces in the Japanese Islands*. Tokyo: University of Tokyo Press (in Japanese).

Massonnet, D., M. Rossi, C. Carmona *et al.* (1993). The displacement field of the Landers earthquake mapped by radar interferometry. *Nature*, **364**, 138–142.

McKenzie, D. (1978). Some remarks on the development of sedimentary basins. *Earth and Planetary Science Letters*, **40**, 25–32.

Mehta, P. K. (1980). Tectonic significance of the young mineral dates and the rates of cooling and uplift in the Himalaya. *Tectonophysics*, **62**, 205–217.

Merritts, D. (1996). The Mendocino triple junction: active faults, episodic coastal emergence, and rapid uplift. *Journal of Geophysical Research*, **101**(B3), 6051–6070.

Merritts, D. and W. B. Bull (1989). Interpreting Quaternary uplift rates at the Mendocino triple junction, northern California, from uplifted marine terraces. *Geology*, **17**, 1020–1024.

Merritts, D. and K. R. Vincent (1989). Geomorphic response of coastal streams to low, intermediate, and high rates of uplift, Mendocino triple junction region, northern California. *Geological Society of America Bulletin*, **101**, 1373–1388.

Merritts, D., K. Vincent and E. Wohl (1994). Long river profiles, tectonism, and eustasy: a guide to interpreting fluvial terraces. *Journal of Geophysical Research*, **99**(B7), 14 031–14 050.

Merritts, D., P. Bodin, E. Beutner, C. S. Prentice and J. Muller (2000). Active surface deformation in the Mendocino triple junction process zone. In: Bokelman, G. and R. L. Kovach (eds.) *Proceedings of the 3rd Conference on Tectonic Problems of the San Andreas Fault System*, Stanford University Publications, 21, Stanford, CA: School of Earth Sciences, Stanford University, 128–143.

Merritts, D., L. Savoy, G. Griggs and R. Walter (2005). Point Delgada to Point Arena. In: Griggs, G., K. Patsch and L. Savoy (eds.) *Living with the Changing Coast of California*. Berkeley, CA: University of California Press, 192–203.

Miall, A. D. (1981). Alluvial sedimentary basins: tectonic setting and basin architecture. In: Miall, A. D. (ed.) *Sedimentation and Tectonics in Alluvial Basins*, Geological Association of Canada Special Paper 23. Waterloo, Ontario: Geological Association of Canada, 1–33.

Molnar, P. and P. England (1990). Late Cenozoic uplift of mountain ranges and global climate change: chicken or egg? *Nature*, **346**, 29–34.

Muhs, D., C. Prentice and D. Merritt (2003). Marine terraces, sea level history and Quaternary tectonics of the San Andreas fault on the coast of California. In: Easterbrook, D. (ed.) *Quaternary Geology of the United States: INQUA 2003 Field Guide Volume*. Reno, NV: Desert Research Institute, 1–22.

Nakada, M., N. Yonekura and C. Lambeck (1991). Late Pleistocene and Holocene sea level changes in Japan: implications for tectonic histories and mantle rheology. *Paleogeography, Paleoclimatology, Paleoecology*, **85**, 107–122.

Nicol, A., R. Van Dissen, P. Vella, B. Alloway and A. Melhuish (2002). Growth of contractional structures during the last 10 m.y. at the southern end of the emergent Hikurangi forearc basin, New Zealand. *New Zealand Journal of Geology and Geophysics*, **45**, 365–385.

Oppenheimer, D. H., G. C. Beroza, G. A. Carver *et al.* (1993). The Cape Mendocino earthquakes of April 1992: subduction at the triple junction. *Science*, **261**, 433–438.

Ota, Y. (in press, a). Holocene sea level changes in the Japanese Islands and regional difference of former shoreline height. *Digital Book on Quaternary Studies in Japan*. Tokyo: Japan Association for Quaternary Research.

Ota, Y. (in press, b). Holocene marine terraces and coseismic coastal uplift in Japan. *Digital Book on Quaternary Studies in Japan*. Tokyo: Japan Association for Quaternary Research.

Ota, Y. and M. Yamaguchi (2004). Holocene coastal uplift in the western Pacific Rim in the context of late Quaternary uplift. *Quaternary International*, **120**, 105–117.

Ota, Y., Y. Matsushiima, M. Umitsu and T. Kawana (1987). Middle Holocene Shoreline map of Japan, 1:2 000 000, 1 sheet. Japan: Japanese Working Group for IGCP Project 200.

Pazzaglia, F. J. and M. T. Brandon (2001). A fluvial record of long-term steady-state uplift and erosion across the Cascadia Forearc High, western Washington State. *American Journal of Science*, **301**, 385–431.

Pazzaglia, F. J., T. W. Gardner and D. J. Merritts (1998). Bedrock fluvial incision and longitudinal profile development over geologic time scales determined by fluvial terraces. In: Tinkler, K. J. and E. Wohl (eds.) *Rivers Over Rock: Fluvial Processes in Bedrock Channels*, Geophysical Monograph Series 107. Washington, DC: American Geophysical Union, 207–235.

Pillans, B. (1986). A late Quaternary uplift map for North Island, New Zealand. In: Reilly, W. I. and B. E. Harford (eds.) *Recent Crustal Movements of the Pacific Region: Proceedings of the International Symposium on Recent Crustal Movements of the Pacific Region, Wellington, New Zealand, 9–14 February, 1984*. Wellington, NZ: Royal Society of New Zealand, 409–417.

Plafker, G. (1964). Tectonic deformation associated with the 1964 Alaska Earthquake. *Science*, **148**, 1675–1687.

Prentice, C. S., D. J. Merritts, E. Beutner *et al.* (1999). The northern San Andreas fault near Shelter Cove, California. *Geological Society of America Bulletin*, **111**, 512–523.

Pulford, A. and T. Stern (2004). Pliocene exhumation of landscape evolution of central North Island, New Zealand: the role of the upper mantle. *Journal of Geophysical Research*, **109**, F01016, doi:10.1029/2003JF000046.

Rajendran, C. P., K. Rajendran, R. Anu *et al.* (2007). Crustal deformation and seismic history associated with the 2004 Indian Ocean Earthquake: a prespective from the Andaman–Nicobar Islands. *Bulletin of the Seismological Society of America*, **97**, 174–191.

Reyners, M., D. Eberhart-Phillips and G. Stuart (1999). A three-dimensional image of shallow subduction: crustal structure of the Raukumara Peninsula, New Zealand. *Geophysical Journal International*, **137**, 873–890.

Reyners, M., D. Eberhart-Phillips, G. Stuart and Y. Nishimura (2006). Imaging subduction from the trench to 300 km depth beneath the central North Island, New Zealand, with V_p and V_p/V_s. *Geophysical Journal International*, **165**, 565–583.

Rockwell, T. K., E. A. Keller and G. R. Dembroff (1988). Quaternary rate of folding of the Ventura Avenue anticline, western Transverse Ranges, southern California. *Geological Society of America Bulletin*, **100**, 850–858.

Savage, J. C. (1983). A dislocation model for strain accumulation and release at a subduction zone. *Journal of Geophysical Research*, **88**, 4984–4996.

Savage, J. C. and G. Plafker (1991). Tide gauge measurements of uplift along the south coast of Alaska. *Journal of Geophysical Research*, **96**(B3), 4325–4335.

Schaer, J. P., G. M. Reimer and G. A. Wagner (1975). Actual and ancient uplift rate in the Gotthard region, Swiss Alips: a comparison between precise levelling and fission-track apatite age. *Tectonophysics*, **29**, 293–300.

Schumm, S. A. (1986). Alluvial river response to active tectonics. In: Wallace, R. E. (ed.) *Active Tectonics*. Washington, DC: National Academy Press, 80–94.

Schwartz, D. P. and K. J. Coppersmith (1984). Fault behavior and characteristic earthquakes: examples from the Wasatch and San Andreas fault zones. *Journal of Geophysical Research*, **89**, 5681–5698.

Sieh, K. (1978). Prehistoric large earthquakes produced by slip on the San Andreas fault at Pallett Creek, California. *Journal of Geophysical Research*, **83**, 3907–3939.

Snyder, N. P., K. X. Whipple, G. E. Tucker and D. J. Merritts (2000). Landscape response to tectonic forcing: digital elevation model analysis of stream profiles in the Mendocino triple junction region, northern California. *Geological Society of America Bulletin*, **112**, 1250–1263.

Sobel, E. R. and M. R. Strecker (2003). Uplift, exhumation and precipitation: tectonic and climatic control of Late Cenozoic landscape evolution in the northern Sierras Pampeanas, Argentina. *Basin Research*, **15**, 431–451.

Summerfield, M. A. (1991). *Global Geomorphology*. London: Longman Scientific and Technical.

Suwa Y., S. Miura, A. Hasegawa, T. Sato and K. Tachibana (2006). Interplate coupling beneath NE Japan inferred from three-dimensional displacement field. *Journal of Geophysical Research*, **111**, B04402, doi:10.1029/2004JB003203.

Upton, P., P. O Koons and D. Eberhart-Phillips (2003). Extension and partitioning in an oblique subduction zone, New Zealand: constraints from three-dimensional numerical modeling. *Tectonics*, **22**, 1068, doi:10.1029/2002TC001431.

Uyeda, S. and H. Kanamori (1979). Back-arc opening and the mode of subduction. *Journal of Geophysical Research*, **84**(B3), 1049–1061.

Vail, P. R., R. M. Mitchum and S. Thompson, III (1977). Seismic stratigraphy and global changes of sea level, part 3, relative changes of sea level from coastal onlap. In: Payton, C. E. (ed.) *Seismic Stratigraphy: Applications to Hydrocarbon Exploration*, AAGP Memoir 26. Tulsa, OK: American Association of Petroleum Geologists, 63–81.

Viseras, C., M. L. Calvache, J. M. Soria and J. Fernandez (2003). Differential features of alluvial fans controlled by tectonic or eustatic accommodation space. Examples from the Betic Cordillera, Spain. *Geomorphology*, **50**, 181–202.

Walcott, R. I. (1987). Geodetic strain and the deformational history of the North Island of New Zealand during the late Cainozoic. *Philosophical Transactions of the Royal Society of London – Series A*, **321**, 163–181.

Wallace, R. E. (1977a). Profiles and ages of young fault scarps, north-central Nevada. *Geological Society of America Bulletin*, **88**, 1267–1281.

Wallace, R. E. (1977b). *Active Tectonics*. Washington, DC: National Academy Press.

Webster, J. M., L. Wallace, E. Silver *et al.* (2004). Coralgal composition of drowned carbonate platforms in the Huon Gulf, Papua New Guinea: implications for lowstand reef development and drowning. *Marine Geology*, **204**, 59–89.

Weissel, J. K. and G. D. Karner (1989). Flexural uplift of rift flanks due to mechanical unloading of the lithosphere during extension. *Journal of Geophysical Research*, **94**(B10), 13 919–13 950.

Wells, R. E., R. J. Blakely, Y. Sugiyama, D. W. Scholl and P. A. Dinterman (2003). Basin-centered asperities in great subduction earthquakes: a link between slip, subsidence, and subduction erosion? *Journal of Geophysical Research*, **108**(B10), 2507, doi:10.1029/2002JB002072.

Willett, S. D., N. Hovius, M. T. Brandon and D. M Fisher (2006). Introduction. In: Willett, S. D. *et al.* (eds.) *Tectonics, Climate, and Landscape Evolution*, GSA Special Paper 398, Boulder, CO: Geological Society of America, vii–ix.

Yeats, R. S., K. Sieh and C. R. Allen (1997). *The Geology of Earthquakes*. New York, NY: Oxford University Press.

Yoshiyama, A. and M. Yanagida (1995). Uplift rates estimated from relative heights of fluvial terrace surfaces and valley bottoms. *Journal of Geography*, **104**, 809–826.

Zachariasen, J., K. Sieh, F. W. Taylor and W. S. Hantoro (2000). Modern vertical deformation above the Sumatran subduction zone: paleogeodetic insights from coral microatolls. *Bulletin of the Seismological Society of America*, **90**, 897–913.

5

Glacial isostatic adjustment: implications for glacially induced faulting and nuclear waste repositories

B. Lund and J. O. Näslund

The redistribution of mass associated with the growth and decay of continental ice sheets gives rise to major glacial loading and unloading effects over timescales of several tens of thousands of years. The response of Earth's crust, mantle and gravitational field is referred to as glacial isostatic adjustment (GIA). For instance, during the decay of a major ice sheet, the unloading of mass results in glacial rebound of the crust, which continues well after the disappearance of the ice. This process is well known from previously glaciated regions such as Canada and the United States, Fennoscandia, the British Isles and Siberia (e.g. Ekman, 1991); areas where this process is still active today, some 10–15 ka after the last deglaciation. In previously glaciated terrain without strong tectonism, GIA is the most significant geodynamic process governing vertical deformation of the crust (e.g. Peltier, 1994).

The downwarping and rebound processes can result in reactivation of major bedrock fracture zones with related differential crustal movement. Such glacially induced fault movements in large fracture zones may cause earthquakes of large magnitude, possibly up to M 8 (e.g. Stewart *et al.*, 2000). There is evidence of glacially induced faulting events having occurred as the ice sheet retreated at several locations in northern Fennoscandia (e.g. Lagerbäck, 1979; Olesen, 1988). The possibility that faulting may occur during future periods of glaciation is one of many scenarios that safety assessors of radioactive waste repositories are dealing with. The location of a deep geological repository is chosen to avoid it being transected by major fracture zones capable of hosting a large magnitude earthquake. However, the potential for shear on smaller fractures in the host rock volume of a repository also needs to be evaluated. As discussed in Chapman *et al.* (Chapter 1, this volume), the concern is not with seismic shaking, but rather with the possibility that spent nuclear fuel canisters might be broken. Current studies in Sweden consider that a decimeter-scale shear movement on a fracture intersecting a canister deposition hole may have the potential to jeopardize canister integrity (SKB, 2006a). Consequently, understanding the causes, location and potential magnitude of glacially induced faulting is currently of direct relevance to nuclear waste management programs in Fennoscandia, where safety assessments are looking at periods of 100 ka and more into the future (e.g. SKB, 2006a). This timescale is so long that the effects of ice-sheet conditions, and glacially induced faulting, need to be assessed regardless of the inferred global climate warming (e.g. Solomon, 2007), since the global-warming effect may taper off before the end of the assessment period (SKB, 2006b). Although this chapter is primarily concerned with potential impacts on geological repositories, glacially induced seismicity is also of concern for a variety of surface nuclear facilities.

In addition to the glacial loading and unloading effects described in this chapter, climate change and the presence of an ice sheet over a geological repository for spent nuclear fuel also result in other important changes in the geosphere and the repository. These include changes in groundwater flow and composition, and large hydrostatic pressures due to the ice load and permafrost growth. These processes are also included in the assessments of long-term repository safety (cf. SKB, 2006a), but are not covered in this book.

5.1 Glacial isostatic adjustment

During the last glacial maximum, $\sim$ 20 ka, the Laurentide ice sheet of Canada and the United States, and the Fennoscandian ice sheet, had maximum thicknesses of $\sim$ 2.5–3 km. When these large ice masses slowly formed, their weight resulted in a slow downwarping of Earth's crust. Important factors that governed both this process and the following glacial unloading include the physical properties of Earth's crust and mantle. In the downwarping process, mantle material has to be displaced and flow laterally in order to make room for the flexing crust. At times of ice-sheet decay, mantle flow is reversed and the crust rebounds. Since the mantle viscosity is high, the downwarping and subsequent rebound are slow processes. For instance, postglacial rebound in Fennoscandia has been suggested to continue for another 30 ka or more (Whitehouse, 2006), with a remaining uplift of up to $\approx$ 100 m, before the isostatic effect of the last glaciation is gone. Furthermore, it has been shown that a deglaciation of large northern hemisphere ice sheets results in deformation of the Earth's entire surface, producing a series of upwarps and downwarps away from the areas of the former ice sheets. However, the deformation is largest in the regions that were glaciated.

The GIA process manifests itself not only in the slow rebound of regions of past ice sheets. Current global warming trends and the subsequent melting of glaciers produces additional GIA effects worldwide. As an example, Iceland is currently undergoing rapid glacial rebound due to a mass loss of Vatnajökull and other smaller ice caps (e.g. Pagli *et al.*, 2007). Since the viscosity of the mantle is inferred to be two orders of magnitude lower beneath Iceland (Pagli *et al.*, 2007) than below, for example, Fennoscandia, uplift rates are on the order of 20 mm a^{-1}, in spite of the much smaller volume of ice loss.

Outside the ice-sheet margin, an uplifted forebulge, or peripheral bulge, is formed (e.g. Mörner, 1977; Fjeldskaar, 1994; Lambeck, 1995). The forebulge is caused by flexure and a lateral displacement of mantle material extending outside the ice margin, and it may stretch for several hundreds of kilometers beyond a major ice sheet. The uplift of the forebulge is considerably smaller than the downwarping of the crust beneath the central parts of the ice sheet, on the order of tens of meters. During and after deglaciation, the area of the forebulge experiences land subsidence, exemplified by the ongoing lowering of the Netherlands, southern England and the east coast of the United States. Also, the location of maximum forebulge uplift migrates towards the formerly glaciated region as the ice sheet withdraws. In addition to the formation of the forebulge, the elasticity of the lithosphere may result in a downwarping of the crust, not only under the ice sheet, but also to some extent outside the ice margin. This produces a flexural depression between the ice margin and the forebulge, a depression where lakes may form from glacial meltwater.

At present, the crust beneath the Antarctic and Greenland ice sheets are depressed in a similar way as previously occurred under the Laurentide, Fennoscandian and Siberian ice sheets. In Greenland and Antarctica, the crust would also be subject to significant glacial rebound if these areas were to be deglaciated in the future.

5.1.1 Observations

One traditional way of measuring the amount and rate of postglacial rebound has been to determine the elevation and age of raised paleoshorelines. In this context, shorelines or other geomorphological features that formed at the highest postglacial sea level, in front of the retreating ice-sheet margin, are the ones that give relevant information. However, the total amount of postglacial uplift at a site is typically larger than can be inferred from, for example, raised beaches. A large portion of the uplift takes place as the ice sheet starts to decay, prior to the actual deglaciation of a typical site situated at some distance from the maximum ice margin. The total maximum amount of glacial rebound that has occurred due to the decay of the Fennoscandian ice sheet is around 800 m (e.g. Mörner, 1979; Figure 5.1a). This may be compared to the largest Fennoscandian rebound as inferred from the highest marine limit, which is situated at $\approx$ 280 m above sea level in the Swedish coastal region of the Gulf of Bothnia. Another related method to study postglacial rebound is to analyze the amount and direction of tilt of paleoshorelines of glacial lakes that formed behind the retreating ice margin.

The rate of current postglacial uplift in previously glaciated regions can be studied by direct or indirect observations. One traditional way of measuring present-day uplift in coastal regions is by observing the associated apparent lowering of sea level. In the Baltic region, for instance, such sea-level observations have been conducted systematically for more than a century. When measuring isostatic uplift rates with this method, it is necessary to correct the result in order to exclude the effect from eustatic variations in sea level. In non-coastal areas, present postglacial rebound rates have been estimated by repeated geodetic leveling, tied to locations of sea-level observation. The results from these measurements show that the present uplift rate displays a concentric uplift pattern over Fennoscandia, with a maximum rate of slightly greater than 9 mm a^{-1} in the northernmost part of the Baltic Sea (Fredén, 1994).

Another way of making direct observations of ongoing postglacial crustal deformation is to use high-quality data from networks of continuously operating, permanent GPS receivers. Typically, such GPS stations were established within national land-survey programs with an initial aim of providing reference coordinates for other GPS measurements. However,

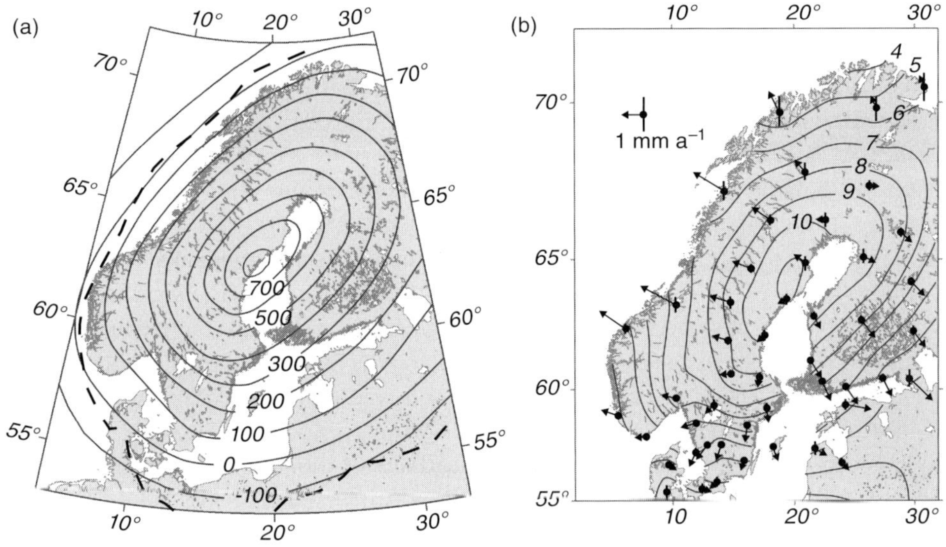

Fig. 5.1 (a) Total amount of glacial rebound, contoured in 100 m intervals, that has taken place due to the decay of the Fennoscandian ice sheet. The bold dashed line shows the maximum extent of the ice sheet. Note that the total amount of rebound is more than twice as large as the amount of rebound that may be inferred from raised beaches formed at the highest postglacial sea level (see the text). Redrawn after Fredén (1994). (b) Present-day crustal deformation over Fennoscandia as observed by continuous GPS measurements within the BIFROST project (e.g. Scherneck *et al.*, 2001). The rebound displays a concentric pattern with a maximum rebound rate of $\approx 11\ \text{mm a}^{-1}$ located approximately in the area of the former maximum ice-sheet thickness. In the postglacial upwarping process, a horizontal component of crustal deformation (arrows) is also present, directed outward from the area of maximum depression. Vertical bars represent the error ($1\,\sigma$) on vertical rebound rate. Horizontal errors in displacement are uniformly small ($< 1\ \text{mm a}^{-1}$). Map modified from Lidberg (2007).

detailed analyses of data from such permanent GPS stations have provided new insight into the processes of postglacial rebound or GIA in Fennoscandia and Canada (cf. Scherneck *et al.*, 2001; Henton *et al.*, 2006). The results provide information on both the rate of the vertical uplift component, as well as on the associated smaller horizontal component of crustal motion (e.g. Johansson *et al.*, 2002; Figure 5.1b). Uplift determined by GPS observations show the same concentric uplift patterns as those derived from sea-level and leveling measurements. The fastest rebound occurs approximately in the areas where the Laurentide and Fennoscandian ice sheets had their greatest thicknesses. The maximum vertical uplift rate measured in this way in the area of the former Laurentide ice sheet is $\approx 13\ \text{mm a}^{-1}$ (Henton *et al.*, 2006), while the corresponding value for Fennoscandia is $\approx 11\ \text{mm a}^{-1}$ (Figure 5.1b). The largest horizontal displacements are generally found in the area of the ice margins of the maximum extent of the ice sheet. The difference between the maximum vertical uplift rates as observed from sea-level/leveling measurements and from the analysis of GPS data is to a large extent covered by the uncertainty errors of the measurements, mainly in the sea-level measurements.

Another, indirect, way of studying ongoing postglacial rebound is by absolute gravity measurements (e.g. Lambert *et al.*, 2001; Mäkinen *et al.*, 2005), a method that may be used to estimate also the remaining uplift in areas where the process is not complete. Yet another method of studying postglacial rebound is by very long baseline interferometry.

5.1.2 Model studies

Numerical models have been used to study many aspects of postglacial rebound, such as changes in sea level, ice-sheet growth and decay, and the inner structure of Earth. One type of model uses inversion of observed sea-level and land uplift data in order to infer both mantle rheology parameters and ice thickness history (e.g. Lambeck *et al.*, 1998). Other GIA models couple the sea-level equation, describing the evolution of the distribution of water between oceans and continental ice sheets over time, to models of Earth's response. These GIA models (e.g. Mitrovica and Peltier, 1991; Milne *et al.*, 1999) calculate the redistribution of mass (ice and water) on the surface of Earth, taking into account resulting changes of the geoid. These changes, in turn, determine the redistribution of water in the ocean basins in the model. Present GIA models typically solve the sea-level equation for the entire globe, and also include effects of Earth's rotation.

Since the rate of glacial rebound is mainly determined by the rheology of the mantle and the ice load history, the output from GIA models is strongly dependent on how these parameters are described in the model simulations. In order to improve the description of Earth structure and ice load history, the output from GIA models may be compared with observed uplift rates from, for example, sea-level observations or GPS data (Figure 5.1b). These comparisons have made it possible to provide constraints on the viscoelastic properties of the mantle (e.g. Lambeck *et al.*, 1998; Steffen and Kaufmann, 2005) and have also shown the importance of including 3D Earth structure properties in GIA model simulations (e.g. Kaufmann *et al.*, 2000; Whitehouse *et al.*, 2006).

Misfits between observed and modeled postglacial uplift have in some cases been interpreted to reflect tectonic components in the ongoing crustal deformation in Fennoscandia (e.g. Fjeldskaar *et al.*, 2000), i.e. reflecting components that are not related to glacial loading and unloading. Nevertheless, in formerly glaciated areas that are not strongly affected by tectonics or volcanism, GIA is the main process controlling vertical displacement.

5.2 Glacially induced faulting

In this section, we use the term "glacially induced faulting" or "glacially induced fault" (GIF) for all faulting activity related to the emplacement or removal of large quantities of ice in glaciers, ice caps or continental ice sheets. The term GIF is generic in the sense that it describes the phenomenon without specifying an occurrence time related to the evolution of the ice. Other frequently used terms for GIFs are glacio-isostatic, endglacial or postglacial faults. The latter two are, in our context, subclasses of GIFs, occurring at the end of or any time after a glaciation, respectively. The usage of postglacial faulting for all types of glacially associated faults is common in the literature, although some of the faults described are, in fact, associated with glacial advance. Here, we will refer to faults as end- or postglacial only when timing is important to the discussion.

5.2.1 Observations

Most the glacially induced faults described in the literature are inferred to be either end- or postglacial. Munier and Fenton (2004) reviewed the current state of investigations of GIFs and discussed how these faults have been encountered in North America, Fennoscandia, Russia and the British Isles. The GIFs have almost exclusively been recorded in regions of low to moderate seismicity, namely passive continental margin, failed rift or intraplate/craton environments. They generally also occur only in regions where there is no evidence of surface rupture during historical time. In addition, these regions have no historical record of seismicity that approaches the magnitude thresholds for generating surface faulting. To date, all examples of postglacial faulting have involved reactivation of existing faults and fractures. Interestingly, there is no clear correlation between the location of maximum thickness or extent of a particular ice sheet and the location of GIFs within its area, although most GIFs are inferred to be endglacial, occurring as the ice retreats from the area.

Almost all of the large (kilometer-scale) faults that are currently generally accepted as being glacially induced are located in northern Fennoscandia (Kujansuu, 1964; Lagerbäck, 1979; Olesen, 1988; Munier and Fenton, 2004). Most of these GIFs strike north to northeast with dip to the east and downthrow to the west. They are almost exclusively reverse faults. The longest, the Pärvie fault, is $\sim$ 160 km long, with maximum vertical displacement of 10–15 m. The GIFs are inferred to have ruptured as one-step earthquakes, reaching M 7–8, just as the ice was retreating from the respective areas (e.g. Lagerbäck, 1979; Olesen, 1988). They mostly ruptured through old zones of weakness (shear zones), not necessarily following one zone but instead jumping to another to comply with the restraints set by the causative stress state. Although much effort has been spent on investigating the faults with both geological and geophysical methods, key questions concerning the formation and current status of the faults are still unresolved. These include fault geometry at depth, fault strength and current deformation rates. The fact that, to date, large GIFs have been identified exclusively in northernmost Fennoscandia is an intriguing problem which poses a difficult challenge to models of GIF formation. Other observations (e.g. Anda *et al.*, 2002; Kotilainen and Hutri, 2004; Mörner, 2004) suggest that glacially induced faulting has been widespread throughout Fennoscandia. When actual faults have been observed, they are, however, significantly smaller than those in northern Fennoscandia.

The influence of a large ice sheet on seismicity can be studied today in Greenland and Antarctica (e.g. Johnston, 1987). The absence of high seismic activity and large earthquakes in these areas, at least over the timespan of modern seismic instrumentation ($\approx$ 100 a), confirms similar results from GIA modeling and fault mechanics analysis. Sauber *et al.* (2000) showed that current seismic strain release in the eastern Chugach Mountains of Alaska may be controlled by glacial loading and unloading.

5.2.2 Process/mechanics

Due to uncertainties in fault geometries, the tectonic state of stress, crustal pore pressures and ice histories, the mechanics of glacially induced faulting is not well understood. There are currently two lines of reasoning suggesting different mechanisms for the formation of GIFs. These are not mutually exclusive and it is likely that the process includes both mechanisms.

The temporal evolution of an ice sheet and the subsequent deformation (GIA) of Earth was discussed previously. As pointed out by Johnston (1987), the vertical stress induced by the ice load decreases the differential stress in the crust, increasing the stability of faults. During the glaciation, tectonic strain accumulates in the crust under the ice and is later released at deglaciation when the vertical load is removed. This sudden release of stored stress is inferred to trigger end-glacial faulting.

A second approach can be taken by studying glacially induced stresses. The horizontal stresses do not simply follow the increase in vertical stress. As the lithosphere warps, compressional and tensional horizontal flexural stresses are induced which, at some locations, are significantly larger in magnitude than the actual ice load (e.g. Johnston *et al.*, 1998). Near and under the ice sheet, the horizontal stresses will be larger than the vertical stress all through the glacial cycle, and especially so towards the end of deglaciation (Figure 5.2). The isostatic rebound of the depressed elastic lithosphere is a much slower process than the retreat of the ice sheet and corresponding removal of the ice load, leaving remnant high horizontal stresses in the lithosphere, after the vertical stress induced by the ice sheet is gone. The shape of the depression and, therefore, the magnitude and distribution of induced stresses, depend critically on the ice-sheet configuration (e.g. spatial extent, surface slope and ice thickness). Without the stabilizing effect of the ice sheet, the high horizontal stresses will cause fault instability and trigger end-glacial earthquakes (e.g. Wu and Hasegawa, 1996). The tectonic stress state is very important in this approach to glacially induced faulting (e.g. Wu and Hasegawa, 1996; Lund, 2005). The glacially induced shear stresses, superimposed on an isotropic, lithostatic stress, are not large enough at seismogenic depths to induce faulting. A further source of deviatoric stress, such as tectonic stress, is necessary. The tectonic stress state largely determines which faults will be reactivated, how and when they will slip and where, in relation to the ice sheet, instabilities will occur.

Pore water pressure in bedrock fractures and pore space is a vital component in all faulting related processes. Increased pore pressure decreases the normal stress on a fault and, therefore, decreases the shear stress needed to cause movement on the fault. The ice sheet is expected to increase pore pressure in the crust below, both through the increase in mean stress by the load itself and through high water pressures at the base of the ice sheet. The increased pore pressure will tend to lower fault stability all through the glacial cycle.

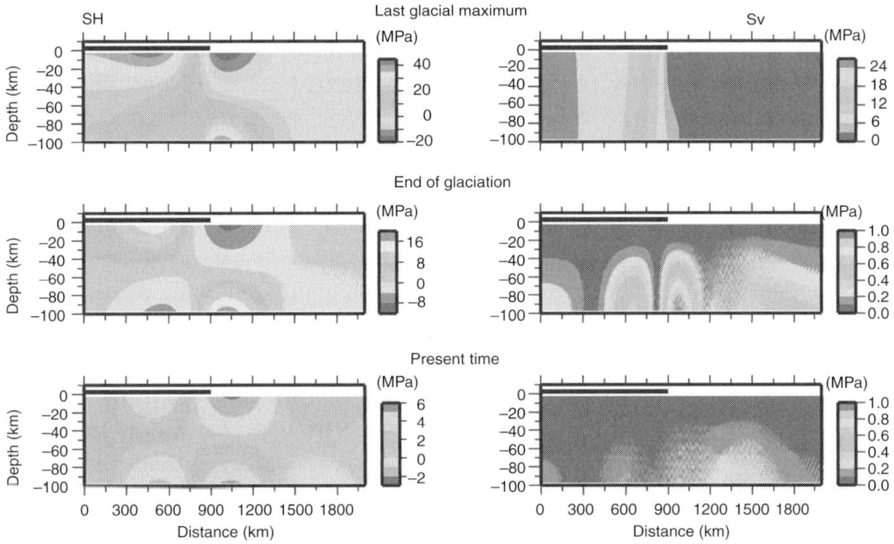

Fig. 5.2 Glacial stresses induced in a generic model of Earth. Maximum horizontal stress (SH) and vertical stress (SV) at three different times: the last glacial maximum, the end of glaciation and present. The Earth model is a simple two-dimensional model with a 100 km elastic lithosphere overlying a viscoelastic half-space subject to a generic elliptic cross-section ice load with 900 km lateral extent and 25 MPa central pressure at the last glacial maximum. See Lund (2005) for model details. See color plate section.

The effect will, however, be largest at the end of glaciation when fault stability is generally decreasing.

Glacially induced faulting is a process strongly correlated with the growth or decay of an ice sheet in a specific region. It is, however, not necessarily limited in time to the actual end of a glacial cycle, but may well occur at any rapid increase or decrease in ice volume and/or areal extent during the glacial cycle. The evolution of the ice sheet, therefore, determines the likelihood of glacially induced faulting at any specific period in time.

5.2.3 Model studies

Recent advances in algorithms and computing power allows the response of the solid Earth to a glaciation to be modeled in some detail (e.g. Martinec, 2000; Milne *et al.*, 2004; Latychev *et al.*, 2005). In order to produce realistic simulations, these algorithms require good knowledge of ice histories (i.e. the lateral extent and thickness of the ice sheet through time) and realistic models of Earth. The resulting output can be used to infer glacially induced stresses which, together with appropriate background stresses, pore pressure conditions and fault failure models, can be used to infer fault stability (e.g. Wu *et al.*, 1999; Ivins *et al.*, 2003; Lund, 2005).

Ice-sheet model simulations of the last glacial cycle are steadily improving and use a variety of modeling approaches. Näslund (2006), for example, uses glacial dynamics, geological information and climate data to simulate the Weichselian ice sheet over Fennoscandia, whereas Lambeck *et al.* (1998), for example, build a Weichselian model based on sea-level, lake-level and geological data. Although the general features of the output from these models are similar, they display enough variety in their details to warrant careful examination in the fault stability context, due to the sensitivity of the GIA-induced stresses to the shape of the ice load.

Observations of postglacial rebound can, to first order (e.g. Milne *et al.*, 2004), be modeled using very simple models of Earth with a single-layered elastic lithosphere and a two-layered viscoelastic mantle characterized by linear Maxwell rheology. The Earth models used in most GIA modeling to date have been rather simple, horizontally layered models. Elastic properties are derived from seismic studies and viscoelastic properties are inferred from the fit of the GIA models to observations such as sea-level data, GPS data and geological records, such as those discussed above. Observations at different time and length scales can be difficult to reconcile. For example, the lower crust is inferred from laboratory measurements and the lack of earthquakes to be ductile. The GIA models of cratonic regions such as Fennoscandia, however, fit the observations well with the assumption of a 100 km thick elastic layer, ignoring a ductile lower crust. Reconciling these different rheological models is an area of active research. Recently, fully three-dimensional spherical Earth models have been developed, where lateral variations in viscosity and elasticity can be studied (e.g. Martinec, 2000; Latychev *et al.*, 2005). The results from these models indicate that three-dimensional structures cause enough perturbations to one-dimensional model results to be distinguished above the uncertainty levels in current GPS observations (e.g. Whitehouse *et al.*, 2006).

The initial tectonic stress field in the models can either be taken directly from measured data or the available data are used as a basis for assumptions regarding the initial field. There is much evidence, from such diverse sources as *in situ* borehole stress measurements, earthquakes induced by fluid injection or reservoir impoundment, and earthquakes triggering other earthquakes, for an intraplate brittle crust in frictional-failure equilibrium on pre-existing faults (e.g. Zoback and Townend, 2001). Using a background stress field in frictional-failure equilibrium in the models is advantageous as it will clearly indicate when the applied GIA stresses cause fault instability. It is imperative to analyze various initial stress fields as they determine which areas, with respect to the ice sheet, and which faults of which orientation, will become most unstable. In addition, the timing of onset of fault instability is governed by the initial stress field and the ice evolution.

A specific example of this type of modeling is shown in Figure 5.3 and is based on the procedures discussed in Lund (2005). The figure shows the estimated stability of faults at various times along a two-dimensional profile that crosses northern Fennoscandia at the latitude of the largest glacially induced faults. The Earth model is a very simple 100 km thick elastic plate on a viscoelastic half-space with viscosity 10^{21} Pa s. The SCAN-2 ice model of Lambeck *et al.* (1998) is used, projected along the two-dimensional profile. The initial state of stress is a reverse faulting regime in frictional equilibrium on optimally oriented

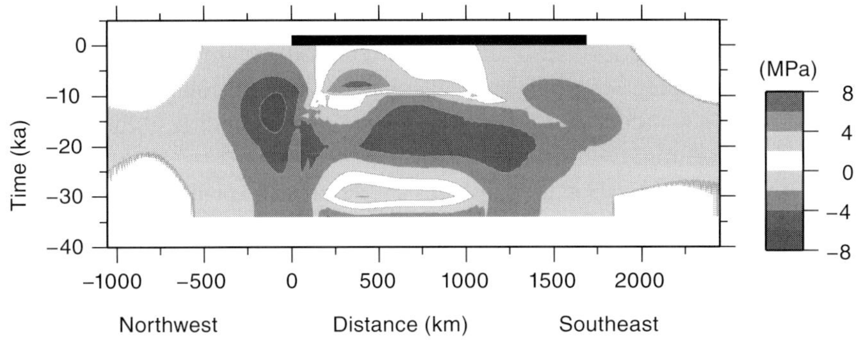

Fig. 5.3 Stability of optimally oriented faults at different times along a two-dimensional profile across northern Fennoscandia at the latitude of the large glacially induced faults. Zero along the profile corresponds to the westernmost extent of the ice sheet, off the coast of Lofoten, Norway. To the east, the profile extends well into Russia. Positive colors indicate unstable faults and negative colors indicate stable faults. The black bar across the top shows the maximum lateral extent of the ice along the profile. See the text for information about the Earth and ice models. See color plate section.

faults with a coefficient of friction of 0.6. Pore pressure is not taken into account in the model. We see in Figure 5.3 that fault instability is developed in the model during two time periods, the first at $\sim$ 30 ka and the second after deglaciation, $\approx$ 12 ka to the present. Rapid deglaciation occurred in both time periods. Instability develops under the central parts of the ice sheet, higher in magnitude to the west where the slope in the ice edge is largest. The two-dimensional profile cuts through the Pärvie fault at 260 km and the Lansjärv fault at 460 km, both in the central section of the developed instability field. The example shows that quantitative models can predict fault instability at approximately the correct locations and at the right time. There are, however, still many unresolved questions in going from the response of models such as these to actual prediction of earthquake rupture and earthquake hazard.

Most models of the kind discussed above do not include the effect of slow tectonic strain accumulation during glaciation. In Fennoscandia, the strain rates due to plate tectonics have been considered small in comparison to the glacially induced strain rates (e.g. Muir-Wood, 1995) and therefore negligible in the context of glacially induced faulting. Recently, however, Adams (2005) argued that the accumulated effect of tectonic strain rate would be sufficient to generate all the observed glacially induced faults of northern Fennoscandia. As this is a purely statistical approach, no inferences can be made about the location, orientation or type of faulting of the events.

Concluding remarks

Current models of the process of glacially induced faulting are still not mature enough that they can accurately predict the location and magnitude of future GIFs. Efforts are underway to include more complex processes in the GIF models, such as poroelastic effects, permafrost effects, strain-release effects and local geological conditions. As discussed previously, fully three-dimensional Earth models have only recently become feasible and recent ice-sheet models are detailed enough to produce the necessary loading resolution. Within the next few years, there is potential for a large increase in the understanding of the mechanics of glacially induced faulting.

The Swedish site investigation programs for a repository of spent nuclear fuel have carried out dedicated field investigations in search of glacially induced faults, but not been able positively to identify any such faults (e.g. Lagerbäck *et al.*, 2005). Nevertheless, as the possibility of glacially induced faulting at a particular repository site cannot be ruled out, current safety assessments of possible repository sites make the cautious assumption that any major fracture zone is potentially capable of reactivation in association with future glaciations. Consequently, within the Swedish program, no canister positions in the repository are allowed within a certain "respect distance" from major fracture zones (Munier and Hökmark, 2004; Fälth and Hökmark, 2006). In addition, there are several deposition hole criteria that have to be fulfilled in order for a deposition hole to be used, of which one is that the deposition hole may not be intersected by fractures larger than a certain size (Fälth and Hökmark, 2006; Fälth *et al.*, 2008).

Glacial isostatic adjustment and glacially induced faulting are important processes that need to be addressed in the assessment of long-term safety for deep geological repositories for long-lived radioactive wastes in regions that might be subject to future glaciations.

References

Adams, J. (2005). Appendix 5: On the probable rate of magnitude $\geq$ 6 earthquakes close to a Swedish site during a glacial cycle. In: Hora, S. and M. Jensen (eds.) Expert panel elicitation on seismicity following glaciation in Sweden. SSI Report 2005:20. Stockholm, Sweden: Swedish Radiation Protection Authority.

Anda, E., L. H. Blikra and A. Braathen (2002). The Berill fault – first evidence of neotectonic faulting in southern Norway. *Norsk Geologisk Tidsskrift*, **82**, 175–182.

Ekman, M. (1991). A concise history of postglacial land uplift research (from its beginning to 1950). *Terra Nova*, **3**(4), 358–365, doi:10.1111/j.1365-3121.1991.tb00163.x.

Fjeldskaar, W. (1994). The amplitude and decay of the glacial forebulge in Fennoscandia. *Norsk Geologisk Tidsskrift*, **74**, 2–8.

Fjeldskaar, W., C. Lindholm, J. F. Dehls and I. Fjeldskaar (2000). Postglacial uplift, neotectonics and seismicity in Fennoscandia. *Quaternary Science Reviews*, **19**(14–15), 1413–1422.

Fredén, C. (ed.) (1994). *National Atlas of Sweden; Geology*. Stockholm: SNA Publishing.

Fälth, B. and H. Hökmark (2006). Seismically induced slip on rock fractures. Results from dynamic discrete fracture modelling. Report R-06-48. Stockholm: Swedish Nuclear Fuel and Waste Management Company.

Fälth, B., H. Hökmark and R. Munier (2008). Slip on repository rock fractures induced by large earthquakes. Results from dynamic discrete fracture modelling. Presented at: 42nd US Rock Mechanics Symposium and 2nd US–Canada Rock Mechanics Symposium, June 29–July 2. San Francisco, CA: American Rock Mechanics Association.

Henton, J. A., M. R. Craymer, R. Ferland *et al.* (2006). Crustal motion and deformation monitoring of the Canadian landmass. *Geomatica*, **60**(2), 173–191.

Ivins, E. R., T. S. James and V. Klemann (2003). Glacial isostatic stress shadowing by the Antarctic ice sheet. *Journal of Geophysical Research*, **108**(B12), doi:10.1029/2002JB002182.

Johansson, J. M., J. L. Davis, H.-G. Scherneck *et al.* (2002). Continuous GPS measurements of postglacial adjustment in Fennoscandia: 1. Geodetic results. *Journal of Geophysical Research*, **107**, 400–428.

Johnston, A. C. (1987). Suppression of earthquakes by large continental ice sheets. *Nature*, **330**, 467–469.

Johnston, P., P. Wu and K. Lambeck (1998). Dependence of horizontal stress magnitude on load dimension in glacial rebound models. *Geophysical Journal International*, **132**, 41–60.

Kaufmann, G., P. Wu and G. Li (2000). Glacial isostatic adjustment in Fennoscandia for a laterally heterogeneous Earth. *Geophysical Journal International*, **143**(1), 262–273.

Kotilainen, A. and K.-L. Hutri (2004). Submarine Holocene sedimentary disturbances in the Olkiluoto area of the Gulf of Bothnia, Baltic Sea: a case of postglacial palaeoseismicity. *Quaternary Science Reviews*, **23**, 1125–1135.

Kujansuu, R. (1964). Nuorista siirroksista Lappissa. Summary: recent faults in Lapland. *Geologi*, **16**, 30–36.

Lagerbäck, R. (1979). Neotectonic structures in northern Sweden. *Geologiska Föreningen i Stockholms Förhandlingar*, **100**, 263–269.

Lagerbäck, R., M. Sundh, J. O. Svedlund and H. Johansson (2005). Searching for evidence of late- or postglacial faulting in the Forsmark region: results from 2002–2004. Site investigation report R-05-51. Stockholm: Swedish Nuclear Fuel and Waste Management Company.

Lambeck, K. (1995). Late Devensian and Holocene shorelines of the British Isles and North Sea from models of glacio-hydroisostatic rebound. *Journal of the Geological Society, London*, **152**, 437–448.

Lambeck, K., C. Smither and P. Johnston (1998). Sea-level change, glacial rebound and mantle viscosity for northern Europe. *Geophysical Journal International*, **134**, 102–144.

Lambert, A. N., G. S. Courtier, F. Sasagawa *et al.* (2001). New constraints on Laurentide postglacial rebound from absolute gravity measurements. *Geophysical Research Letters*, **28**, 2109–2112.

Latychev, L., J. X. Mitrovica, M. E. Tamisiea, J. Tromp and R. Moucha (2005). Influence of lithospheric thickness variations on 3-D crustal velocities due to glacial isostatic adjustment. *Geophysical Research Letters*, **32**, L01304, doi:10.1029/2004GL021454.

Lidberg, M. (2007). Geodetic reference frames in presence of crustal deformations. Unpublished Ph.D. thesis. Göteborg, Sweden: Chalmers University of Technology.

Lund, B. (2005). Effects of deglaciation on the crustal stress field and implications for end-glacial faulting: a parametric study of simple Earth and ice models. Technical Report TR-05-04. Stockholm, Sweden: Swedish Nuclear Fuel and Waste Management Company.

Martinec, Z. (2000). Spectral-finite element approach to three-dimensional viscoelastic relaxation in a spherical earth. *Geophysical Journal International*, **142**, 117–141.

Milne, G. A., J. X. Mitrovica and J. L. Davis (1999). Near-field hydroisostasy: the implementation of a revised sea-level equation. *Geophysical Journal International*, **139**, 464–482.

Milne, G. A., J. X. Mitrovica, H.-G. Scherneck *et al.* (2004). Continuous GPS measurements of postglacial adjustment in Fennoscandia: 2. Modeling results. *Journal of Geophysical Research*, **109**, B02412.

Mitrovica, J. X. and W. R. Peltier (1991). On postglacial geoid subsidence over the equatorial oceans. *Journal of Geophysical Research*, **96**, 20 053–20 071.

Muir-Wood, R. (1995). Reconstructing the tectonic history of Fennoscandia from its margins: the past 100 million years. Technical Report TR-95-36. Stockholm, Sweden: Swedish Nuclear Fuel and Waste Management Company.

Munier, R. and C. Fenton (2004). Appendix 3: Review of postglacial faulting. In: Munier, R. and H. Hökmark (eds.) Respect distances. Rationale and Means of Computation. R-report SKB R-04-17. Stockholm, Sweden: Swedish Nuclear Fuel and Waste Management Company.

Munier, R. and H. Hökmark (2004). Respect distances. Rationale and means of computation. R-report SKB R-04-17. Stockholm, Sweden: Swedish Nuclear Fuel and Waste Management Company.

Mäkinen, J., A. Engfeldt, B. G. Harsson *et al.* (2005). The Fennoscandian land uplift gravity lines 1966–2003. In: Jekeli, C., L. Bastos and J. Fernades (eds.) *Gravity, Geoid and Space Missions GGSM 2004 IAG International Symposium Porto, Portugal August 30–September 3*. International Association of Geodesy Symposia, 129. Berlin: Springer E-books, 328–332.

Mörner, N.-A. (1977). Past and present uplift in Sweden: glacial isostasy, tectonism and bedrock influence. *Geologiska Föreningen i Stockholms Förhandlingar*, **99**, 48–54.

Mörner, N.-A. (1979). The Fennoscandian uplift and Late Cenozoic geodynamics: geological evidence. *GeoJournal*, **3**(3), 287–318.

Mörner, N-A. (2004). Active faults and paleoseismicity in Fennoscandia, especially Sweden. Primary structures and secondary effects. *Tectonophysics*, **380**, 139–157.

Näslund, J.-O. (2006). Ice sheet dynamics – model studies. In: Climate and climate-related issues for the safety assessment SR-Can. TR-report SKB TR-06-23. Stockholm, Sweden: Swedish Nuclear Fuel and Waste Management Company, 33–54.

Olesen, O. (1988). The Stuoragurra fault, evidence of neotectonics in the Precambrian of Finnmark, northern Norway. *Norsk Geologisk Tidsskrift*, **68**, 107–118.

Pagli, C., F. Sigmundsson, B. Lund *et al.* (2007). Glacio-isostatic deformation around the Vatnajökull ice cap, Iceland, induced by recent climate warming: GPS observations and Finite Element Modeling. *Journal of Geophysical Research*, **112**, B08405, doi:10.1029/2006JB004421.

Peltier, W. R. (1994). Ice age paleotopography. *Science*, **265**, 195–201.

Sauber, J., G. Plafker, B. F. Molina and M. A. Bryant (2000). Crustal deformation associated with glacial fluctuations in the eastern Chugach Mountains, Alaska. *Journal of Geophysical Research*, **105**, 8055–8077.

Scherneck, H.-G., J. M. Johansson, M. Vermeer *et al.* (2001). BIFROST project: 3-D crustal deformation rates derived from GPS confirm postglacial rebound in Fennoscandia. *Earth Planets Space*, **53**, 703–708.

SKB (2006a). Long-term safety for KBS-3 repositories at Forsmark and Laxemar – a first evaluation: main report of the SR-Can project. TR-report SKB TR-06-09. Stockholm, Sweden: Swedish Nuclear Fuel and Waste Management Company.

SKB (2006b). Climate and climate-related issues for the safety assessment SR-Can. TR-report SKB R-06-23. Stockholm, Sweden: Swedish Nuclear Fuel and Waste Management Company.

Solomon, S., IPCC *et al.* (2007). *Climate Change 2007: The Physical Science Basis. Contribution of Working Group I to the Fourth Assessment Report of the Intergovernmental Panel on Climate Change*. Cambridge: Cambridge University Press.

Steffen, H. and G. Kaufmann (2005). Glacial isostatic adjustment of Scandinavia and northwestern Europe and the radial viscosity structure of the Earth's mantle. *Geophysical Journal International*, **163**(2), 801–812.

Stewart, I. S., K. Sauber and J. Rose (2000). Glacio-seismotectonics: ice sheets, crustal deformation and seismicity. *Quaternary Science Reviews*, **19**, 1367–1389.

Whitehouse, P. (2006). Isostatic adjustment and shore line migration. In: Climate and Climate-Related Issues for the Safety Assessment SR-Can. TR-report SKB TR-06-23. Stockholm, Sweden: Swedish Nuclear Fuel and Waste Management Company, 68–92.

Whitehouse, P., K. Latychev, G. A. Milne, J. X. Mitrovica and R. Kendall (2006). Impact of 3-D Earth structure on Fennoscandian glacial isostatic adjustment: implications for space-geodetic estimates of present-day crustal deformations. *Geophysical Research Letters*, **33**, L13502, doi:10.1029/2006GL026568.

Wu, P. and H. S. Hasegawa (1996). Induced stresses and fault potential in eastern Canada due to a disc load: a preliminary analysis. *Geophysical Journal International*, **125**, 415–430.

Wu, P., P. Johnston and K. Lambeck (1999). Postglacial rebound and fault instability in Fennoscandia. *Geophysical Journal International*, **139**, 657–670.

Zoback, M. D. and J. Townend (2001). Implications of hydrostatic pore pressures and high crustal strength for the deformation of intraplate lithosphere. *Tectonophysics*, **336**, 19–30.

6

Using global positioning system data to assess tectonic hazards

L. M. Wallace, J. Beavan, S. Miura and R. McCaffrey

Since the early 1990s, our understanding of plate boundary zone crustal deformation has been revolutionized by advances in global positioning system (GPS) techniques. These allow us to track directly the movement of the ground in real time, quantify the rates of crustal deformation within plate boundary zones and determine the displacement of the Earth's surface during earthquakes. The GPS measurements are taken at survey points permanently attached to the ground either by intermittent (survey-style) or continuous (daily, automated) collection of phase and pseudorange data from the constellation of GPS satellites that orbit the Earth. The GPS measurements spanning some period of time (usually longer than one year) can accurately track the movement of one point on the Earth's surface relative to others (to within a few mm a^{-1} uncertainty). Such measurements have allowed scientists to determine where and how much tectonic strain is currently accumulating within plate boundary zones (e.g. Kreemer *et al.*, 2000; McClusky *et al.*, 2000; Sagiya *et al.*, 2000; Beavan and Haines, 2001).

One of the major issues facing siting of nuclear facilities is the possibility of rapid seismic or slow aseismic strain at or near the facility. Elevated strain within a site (possibly due to a seismic event) could perturb a nuclear facility and/or jeopardize the long-term isolation of a high-level waste (HLW) repository in numerous ways, including activation/formation of faults, enhanced creep deformation of engineered barriers, flexural folding of the host rock or enhanced groundwater flow. Geological and seismological data are commonly used to assess future seismic shaking and rock deformation hazards for nuclear facilities (Stepp *et al.*, 2001). Although GPS data provide a direct measure of contemporary surface displacements, and contain important information regarding "rock deformation" hazards, they are not typically used in nuclear facility or HLW repository hazard assessments. This is undoubtedly due in part to the relative newness of the technique and uncertainties regarding how modern strain rates relate to geologic risk assessment.

Results of continuous GPS surveys in the region of the proposed Yucca Mountain HLW repository site showed evidence for ongoing tectonic strain that could jeopardize the repository's safety (Wernicke *et al.*, 1998, 2004; Hill and Blewitt, 2006). Although there are some differences in the results and interpretations of Wernicke *et al.* (1998) in comparison to the later studies of Wernicke *et al.* (2004) and Hill and Blewitt (2006), such studies do highlight the utility of geodetic/GPS data as one of the standard "tools" in the tectonic hazards toolkit. An advantage of GPS over geological and seismological studies is that it can help to detect "hidden" active faults that have not yet been recognized (e.g. Donnellan *et al.*, 1993; Stevens *et al.*, 2002). Seismological data identify regions that have been the site of earthquakes during the instrumental and/or historical record ($<$ 100–400 a) but the geologically short record suggests that many active faults with recurrence intervals $>$ 500 a may be under-represented. Geological studies of active faults (the field of paleoseismology) require an identifiable surface trace. However, many active faults do not have a well-defined surface expression; for example, a blind thrust similar to the one that produced the 1994 Mw 6.7 Northridge earthquake (e.g. Hauksson *et al.*, 1995; Wald *et al.*, 1996). Faults with very low slip rate (i.e. $< 0.1\ \text{mm a}^{-1}$) or, for example, reverse faults associated with broad folding, can be difficult to identify in the field. Such faults pose a significant rock deformation hazard in the long term. Global positioning system techniques can offer an alternative measure of contemporary tectonic strain rates and can reveal the full deformation field. However, to be useful, the contemporary strain rates measured by GPS need to be understood in terms of long-term (i.e. tens of thousands of years) tectonic strain rates.

Some of the strain being measured by GPS is transient, elastic strain that builds in the time between large earthquakes (the interseismic period) and can occur over a wide region on either side of the fault (Figure 6.1a). Thus, we cannot assume that all of the strain measured

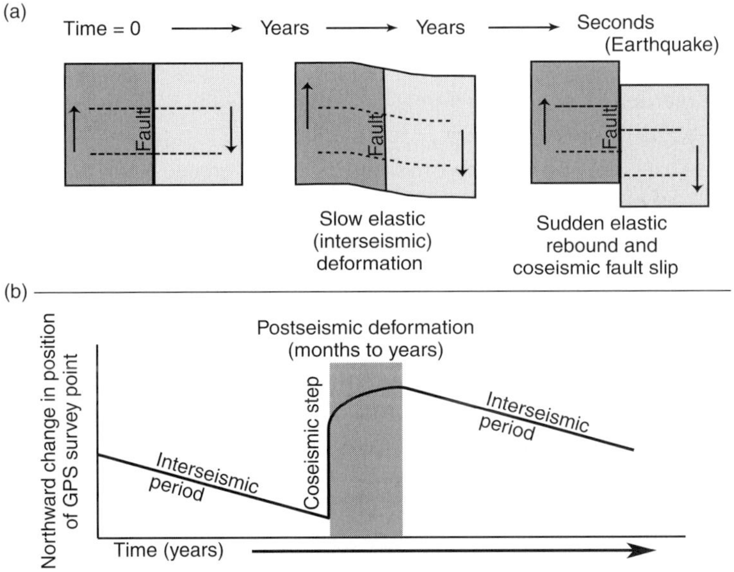

Fig. 6.1 (a) An idealized schematic of crustal deformation throughout a seismic cycle, assuming deformation of an elastic medium. The two blocks are sliding past each other along a strike-slip fault, and the dashed lines crossing the fault are drawn to illustrate the deformation occurring across the fault as time progresses. (b) Hypothetical displacement of GPS survey point (in a similar position to that shown in (a)) throughout an earthquake cycle.

by GPS will result in an earthquake in the exact location where the strain is currently accumulating. Moreover, large displacements of the surface during earthquakes (coseismic motions) and for some extended period of time following an earthquake (postseismic) can complicate the interpretation of the GPS time series.

Given that GPS commonly records transient effects (e.g. coseismic and postseismic displacements, interseismic elastic strain) in addition to the long-term tectonic deformation, a variety of techniques have been devised to extract the long-term, tectonic component from the short contemporary time series. This chapter reviews the tectonic processes that can influence GPS observations. The long-term tectonic strain signal can be extracted from a regional GPS velocity field and used in the assessment of rock deformation hazards relevant to nuclear facility siting issues. To illustrate this method, a case study uses GPS techniques to estimate where significant tectonic strain occurs within the Tohoku region of northern Japan.

6.1 What does GPS measure?

In the most general sense, repeated GPS measurements at a network of survey points measure the changes in relative positions of the sites. The position changes through time can then be interpreted in terms of strain rates within the GPS network (often due to tectonic processes, unless there are some non-tectonic effects, such as slope instability, etc.). A reference frame for the displacements can be defined by using another set of GPS points. For example, to define an Australian plate reference frame for site displacements, GPS sites in the interior of the Australian plate could be used as reference sites. A linear fit to time series of the positions (that are tied to a chosen reference frame) yields the velocity of the GPS survey point relative to that reference frame.

The motion of a tectonic plate relative to some surface reference frame can be described by an angular velocity. Since the plate moves along the surface of a sphere, the angular velocity has the center of the Earth as its origin (e.g. Cox and Hart, 1986). The intersection of the plate's angular velocity vector with the Earth's surface is known as a "pole of rotation." The GPS velocities are often used to estimate the pole and rate of rotation (the angular velocity) of a tectonic plate (e.g. Beavan *et al.*, 2002; Sella *et al.*, 2002; Prawirodirdjo and Bock, 2004). Seafloor spreading rates, oceanic transform fault azimuths and earthquake slip vectors can also be used to determine tectonic plate motions (e.g. Morgan, 1968; DeMets *et al.*, 1994) and were the main source of such knowledge prior to the development of the GPS method. Angular velocities of tectonic plates derived from geological data taken over the last 1–2 Ma (DeMets *et al.*, 1994) and those derived from GPS data (e.g. Beavan *et al.*, 2002; Sella *et al.*, 2002; Prawirodirdjo and Bock, 2004) agree very well in most cases. This agreement suggests that tectonic plate motions have been largely steady over the last 1–2 Ma. Velocities of GPS points within the interiors of major tectonic plates (i.e. away from plate boundaries) can be explained by these steady plate movements.

Accurate estimates of tectonic plate movements from GPS can also define the "plate motion budget" constraint (that is, the total motion across the deforming region) that must be accounted for within plate boundary zones when conducting tectonic hazard assessments. A key component of any tectonic hazard assessment is to ensure that all of the plate motion across a plate boundary zone is accounted for. This is a first step towards knowing how the rock deformation accommodating the total relative plate motion is distributed.

Global positioning system measurements may also comprise a variety of signals that must be understood prior to using them for tectonic/rock deformation hazard assessment, particularly in plate boundary zones. Much of the strain that contemporary GPS detects is elastic, accumulating in the Earth's crust during the time between large earthquakes (e.g. McCaffrey *et al.*, 2000; Mazzotti *et al.*, 2000; L. Wallace *et al.*, 2004). This strain arises when two pieces of crust are prevented from sliding past one another along a fault by friction on the fault's surface. The two pieces of crust become stuck together at the fault and the surrounding crust deforms as the crustal blocks continue to move in the far-field (Figure 6.1a).

Throughout this chapter, we refer to this phenomenon of frictional sticking (the "stick" part of stick-slip) as interseismic coupling, and the degree of coupling is defined as the "coupling coefficient" (ϕ), which can have a value between 0 and 1. Where $\phi = 0$, the fault creeps steadily with no elastic strain accumulation, and where $\phi = 1$, the fault is fully stuck, resulting in substantial elastic strain. Eventually, the stress becomes so large that the fault fails (an earthquake occurs; the "slip" part of stick-slip) and the blocks slide rapidly past each other along the fault. During the earthquake, the elastic strain that has built up in the surrounding crust during the preceding interseismic period is converted to permanent slip across the fault (Figure 6.1a). In some cases, the elastic strain from one fault is so large that it can mask strain build-up related to other active faults in the region. This problem is particularly prevalent near some subduction zones where elastic strain due to interseismic coupling on the subduction interface is observable hundreds of kilometers from the trench (e.g. McCaffrey *et al.*, 2000; Mazzotti *et al.*, 2000; L. Wallace *et al.*, 2004).

Deformation following a large earthquake can continue aseismically for some period of time. Continued slip on the fault or nearby faults (called "postseismic slip" or "afterslip") can occur for a period of a few months to several years following the earthquake (e.g. Heki *et al.*, 1997; Savage and Svarc, 1997; Segall *et al.*, 2000). In addition, the mantle and lower crust may relax viscoelastically as they adjust to the change in the stress that occurred during the earthquake ("postseismic relaxation," which can last for several decades following the earthquake; Pollitz, 1997, and references therein).

To correct for coseismic and postseismic offsets in the GPS time series (Figure 6.1b), many studies use time-series analysis techniques to estimate the coseismic offset as a step function, while postseismic deformation is commonly approximated by an exponential decay function (e.g. Heki *et al.*, 1997; Hsu *et al.*, 2002; Prawirodirdjo and Bock, 2004). Accounting for and removing the effect of transient coseismic and postseismic displacements that occur within a GPS network is critical because these displacements often produce large, measurable strains that are much higher than the background tectonic strain rate. Slow slip events (accelerated fault creep occurring in a matter of days to years, e.g. Hirose *et al.*, 1999, Dragert *et al.*, 2001; Ozawa *et al.*, 2002; Larson *et al.*, 2004) that have occurred at several subduction margins also contribute to the GPS time series. Transient deformation related to volcanic activity (e.g. Dixon *et al.*, 1997; Mattioli *et al.*, 1998; Miura *et al.*, 2000; LaFemina *et al.*, 2005) usually only affects GPS velocities within several kilometers of the volcano. Seasonal effects that are observed in most GPS time series can be removed (e.g. Nikolaidis, 2002; Williams *et al.*, 2004).

6.2 Extracting the component of the GPS signal relevant to rock deformation hazards

Although numerous types of transient deformation can influence contemporary GPS measurements, in many cases they can be revealed by various modeling techniques. One of the more challenging transient signals is elastic deformation due to friction on faults. Where these elastic strains occur is not always representative of where faulting will occur. Subtle straining due to minor faults (or distributed strain) may not be easily detected in the presence of the larger elastic strain rates related to nearby major faults.

To model these elastic strains, we often approximate the Earth's crust as an elastic half-space. Measurements of deformation of the crust between earthquakes and during earthquakes have shown that to first order, the crust can be approximated as an elastic material (e.g. Savage and Prescott, 1978; Beanland *et al.*, 1990; Sagiya and Thatcher, 1999; Mazzotti *et al.*, 2000; Miura *et al.*, 2004b; Meade and Hager, 2005; McCaffrey, 2005). Mathematical expressions have been derived using elastic dislocation theory to predict horizontal and vertical ground deformation due to fault slip (dislocations) in an elastic medium (e.g. Mansinha and Smylie, 1971; Savage 1983; Okada, 1985).

Some studies suggest that the elastic half-space assumption is too simplistic, and that layered models (e.g. Wald and Graves, 2001; Zhu and Rivera, 2002) or models of the crust that have an elastic layer overlying a viscoelastic layer may be more appropriate (e.g. Thatcher and Rundle, 1984). In particular, elastic, half-space models perform poorly in comparison to layered space models when dealing with coseismic displacements due to shallow earthquake ruptures, and in basin environments (the half-space models tend to underpredict the surface displacements in these situations; Wald and Graves, 2001). Viscoelastic models are often used because they may help explain discrepancies between geodetic and geological estimates of fault slip rates that may be due to possible time-dependent (earthquake-cycle-related) deformation (Thatcher and Rundle, 1984; Dixon *et al.*, 2003).

Despite some of these possible limitations of the elastic, half-space approach, it works remarkably well in characterizing most coseismic and interseismic deformation fields measured by GPS (e.g. Mazzotti *et al.*, 2000; Miura *et al.*, 2004b; Nishimura *et al.*, 2004; McCaffrey, 2005; Meade and Hager, 2005; L. Wallace *et al.*, 2004, 2007), and is widely used in crustal deformation research. For example, the interseismic coupling distribution in the source region of the 2003 Tokachi-oki earthquake and the coseismic slip distribution (both estimated using elastic, dislocation modeling techniques) match well, indicating that the coseismic slip occurred on the portion of the subduction interface that was previously coupled in the interseismic period as estimated using elastic, half-space models (e.g.

Miura *et al.*, 2004b). Moreover, the coseismic slip distributions from GPS displacements (using elastic, half-space models) in many earthquakes coincide well with the region of coseismic slip determined using seismological data (e.g. Miura *et al.*, 2004b; Subaraya *et al.*, 2006). Interpretation of GPS velocity fields in plate boundary zones that assume deformation of an elastic medium usually yield an excellent fit to geologically estimated fault slip rates (e.g. McCaffrey, 2005; Reilinger *et al.*, 2006; L. Wallace *et al.*, 2007), although there are exceptions (e.g. Dixon *et al.*, 2003; Bennett *et al.*, 2004; K. Wallace *et al.*, 2004).

Recently, many studies have shown that GPS velocities measured in zones of active faulting during the interseismic period are explained by interseismic elastic strains (Figure 6.1a), as well as long-term rotation of crustal blocks in the deforming zone (e.g. McCaffrey *et al.*, 2000; McClusky *et al.*, 2001; McCaffrey, 2002, 2005; Meade and Hager, 2005; L. Wallace *et al.*, 2004, 2007). Methods have been devised by McCaffrey (1995, 2002) and Meade and Hager (2005) to invert GPS velocities for long-term rotations of tectonic blocks, and elastic strain due to coupling on block-bounding faults. For the purposes of rock deformation/tectonic hazard assessment, if the elastic deformation from the inversion (predicted from the fault coupling parameters) is subtracted from the original GPS velocities, the resulting velocity field will be approximately free of the elastic effects of interseismic coupling on known, major active faults in the region. Gradients in the residual velocity fields may then be interpreted in terms of deformation due to other minor faults or perhaps zones of distributed deformation.

McCaffrey's (1995, 2002) method performs a non-linear inversion to estimate simultaneously the angular velocities of elastic blocks and coupling coefficients on block-bounding faults. These give the best fit to the GPS velocities, and optionally, earthquake slip vectors, and geological fault slip rates and azimuths. The data misfit, defined by the reduced chi-squared statistic (χ_n^2), is minimized. The method also allows us optimally to rotate multiple GPS velocity solutions into a common reference frame. McCaffrey's approach also has the benefit of including all of the bounding plates so that we can establish the plate motion budget that needs to be accounted for in the plate boundary zone. Once the elastic deformation effects from the major, known faults (estimated using the elastic block method) have been removed from the GPS velocity field, the residual strain can be mapped using a variety of methods (e.g. Haines and Holt, 1993; Sagiya *et al.*, 2000; Beavan and Haines, 2001; Miura *et al.*, 2004a). Although the elastic block approach inherently assumes that the plate boundary zone is composed of rigid blocks, this method is used primarily to estimate the potential elastic strain due to coupling on "known" (block-bounding) faults. Once this is accomplished, the next step is to use the residuals to see where the "rigid block" assumption does not hold (i.e. regions where strain occurs that are potentially related to "unknown" faults within the block).

We caution that the strain-rate-mapping results from GPS should only be used as a guide to identify regions where strain rates are too high for safety. Given the uncertainties inherent in GPS measurements, GPS techniques are unable reliably to detect strain rates below a certain threshold (this threshold will be dependent on the quality of the GPS network and data, and distribution of GPS sites). Thus, GPS measurements should not be used on their own to determine if a site is tectonically stable (i.e. additional geological and seismological investigations will be needed to confirm tectonic stability.)

6.3 A northern Honshu case study

Is it possible to isolate permanent crustal strain in an area of very large superimposed elastic strains due to interseismic coupling on an offshore subduction zone? In Tohoku, northern Honshu, subduction of the Pacific plate beneath northern Honshu occurs at the Japan Trench (Figure 6.2). The subduction interface has produced major earthquakes in the past (e.g. Yamanaka and Kikuchi, 2004 and references therein), and large portions of the subduction interface are currently coupled, resulting in elastic strains detected by the continuous GPS network in northern Honshu (e.g. Mazzotti *et al.*, 2000; Nishimura *et al.*, 2004; Suwa *et al.*, 2006). The GPS velocities are also influenced by an active convergent zone in the Sea of Japan where the Sea of Japan crust thrusts beneath the west coast of northern Honshu. Some of this convergence may also be accommodated by faulting on

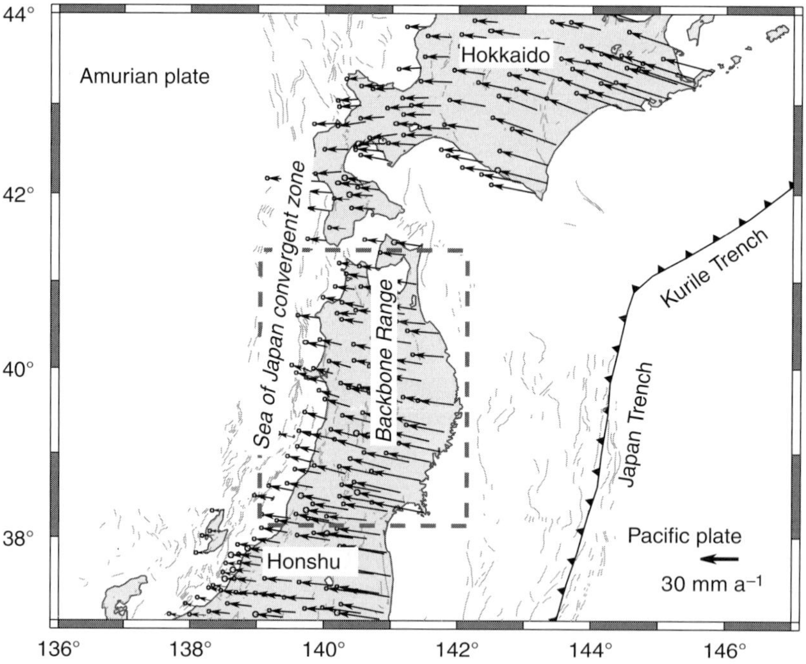

Fig. 6.2 Tectonic setting and GPS velocity field in northern Honshu (shown relative to a fixed Eurasian plate). Thinner gray lines are active fault traces. Ellipses show uncertainty at 68% confidence level. Velocities are derived from daily solutions provided by the Geographical Survey Institute of Japan (see Section 6.3.1). The dashed box refers to the area shown in Figures 6.4 and 6.5.

land, near the west coast of northern Honshu (Figure 6.2). Active reverse faulting occurs within the Backbone Range of the northern Honshu region, although the contraction related to this zone (as measured by contemporary GPS) is largely masked by the contractional strain from interseismic coupling on the subduction interface (Miura *et al.*, 2004a). Elastic strain due to interseismic coupling on the subduction interface may penetrate up to a few hundred kilometers inland from the Japan Trench.

McCaffrey's (1995, 2002) elastic block method is used to estimate the elastic strain due to coupling on the Japan Trench and possible elastic effects from interseismic coupling on offshore faults in the Sea of Japan convergent zone. The elastic effects due to interseismic coupling on the subduction zone and the offshore Sea of Japan convergent zone are removed from the GPS velocities and the residual strain is mapped. The assumption is that this residual strain will eventually result in permanent rock deformation in the upper plate, which could pose a problem if a nuclear facility is located in one of these high-strain regions. Removing "known" elastic strain signals may also provide a means to facilitate the incorporation of geodetic data into seismic hazard models.

6.3.1 GPS velocities and uncertainties

Velocities of GPS sites throughout Japan are derived from the combination of SINEX (Solution INdependent EXchange format) files provided by the Geographical Survey Institute (GSI, 2008) from their daily processing of the GEONET Continuous GPS (CGPS) network in Japan (CGPS sites $\approx$ 1200 in total). The daily network processing of the GEONET network is conducted by GSI, with Bernese GPS processing software (Rothacher and Mervart, 1996; Beutler *et al.*, 2001) using standard processing methods. The SINEX files are from one day every three months for the period 1996–2004. The GLOBK software (e.g. Herring, 2001) is used to combine the daily SINEX files to estimate velocities for the CGPS sites in Japan relative to a known terrestrial reference frame. The Japanese data set is placed in a global context by using daily solutions from Scripps Institute of Oceanography processing of the global IGS network of GPS sites (SIO, 2008), as well as SINEX files from processing a subset of Japanese sites ($\approx$ 10) and several global sites that have been submitted by GSI to the Crustal Dynamics Data Information System (NASA, 2008). Using GLOBK we estimate a rotation and translation of each data set into the International Terrestrial Reference Frame, ITRF2000 (Altamimi *et al.*, 2002) for each day. To accomplish this, we tightly constrain the coordinates of a subset of the most reliable IGS GPS stations to their known ITRF2000 values. This process is repeated for each set of daily solutions to obtain a time series of site positions in the ITRF2000 reference frame. The ITRF2000 velocities at each GPS station are calculated by a linear fit to the daily ITRF2000 coordinates.

The uncertainties in the linear fits are derived using a white-noise model which results in the uncertainties being underestimated (e.g. Zhang *et al.*, 1997; Williams *et al.*, 2004). We multiply the formal uncertainties by 5 to give more reasonable values of about 1 mm a^{-1} uncertainty in horizontal velocities for long-running stations within Japan (T. Nishimura, personal communication 2005). Ideally, the GPS velocity errors should be assessed more rigorously. This would require maximum-likelihood analysis of (probably daily, perhaps weekly) time series of GPS positions, to define the appropriate noise model for the data and to calculate a realistic velocity uncertainty (e.g. Langbein, 2004; Williams *et al.*, 2004).

Removing the effects of earthquakes and slow slip events from GPS velocities is an important step. This task is more straightforward in northern Honshu than in other parts of Japan. For example, slow slip events have not been observed in the CGPS time series in northern Honshu. Large surface displacements were observed at CGPS sites in northern Honshu due to the September 23, 2003 Tokachi-oki earthquake and July 25, 2003 northern Miyagi earthquake. Given that these earthquakes occurred late in the data time series, the daily solutions from July 2003 and later were omitted in the velocity estimation. A visual inspection of the daily position time series for all the sites in the data set was also conducted to be sure that there was no non-linear behavior recorded by the GPS sites that is not representative of steady movement during the interseismic period. Iwate volcano in central northern Honshu experienced unrest and caused measurable deformation at nearby GPS sites in 1998 (Miura *et al.*, 2000). The GPS velocities of affected sites were removed from the data set.

To interpret the GPS site velocities in a tectonically meaningful way, the velocities need to be placed into a "plate-fixed" reference frame. We use the Eurasia-fixed frame, imposed at the inversion/tectonic modeling stage by estimating a rotation of the entire data set that minimizes the velocities at sites known to be on the stable Eurasian plate. The GPS velocity field in northern Honshu is shown in a Eurasia-fixed reference frame in Figure 6.2. In addition to the GPS velocities described previously, published GPS velocities from Heki *et al.* (1999), Calais *et al.* (2003), Sella *et al.* (2002), Beavan *et al.* (2002) and Prawirodirdjo *et al.* (2004) were also used in the inversion. These data help to further place the northern Honshu velocities into a regional plate kinematic context. This step is critical because all of the possible relative motion between the various tectonic plates in the system that could be influencing the GPS measurements in Japan must be accounted for.

6.3.2 Elastic block model set-up for northern Honshu

Two clear block boundaries exist in northern Honshu, both of which crop out offshore: (i) the subduction interface of the Pacific plate beneath northern Honshu, and (ii) the convergent zone in the Sea of Japan. The Japan trench and Sea of Japan boundaries are used as the eastern and western boundaries of the northern Honshu block, respectively. For block modeling, an Okhotsk block (Figure 6.3), an Amurian plate, a Eurasian plate, a North American plate and a Pacific plate are defined. The existence of these plates has been proposed by numerous previous studies (e.g. Seno *et al.*, 1996; Wei and Seno, 1998; Heki *et al.*, 1999; Takahashi *et al.*, 1999), although whether northern Honshu, Okhotsk and North America constitute separate blocks/plates is controversial. A slightly better fit to the data is obtained where the northern Honshu and Okhotsk blocks are independent of North America; thus, for the purposes of this study, it is assumed these blocks are independent. The boundaries

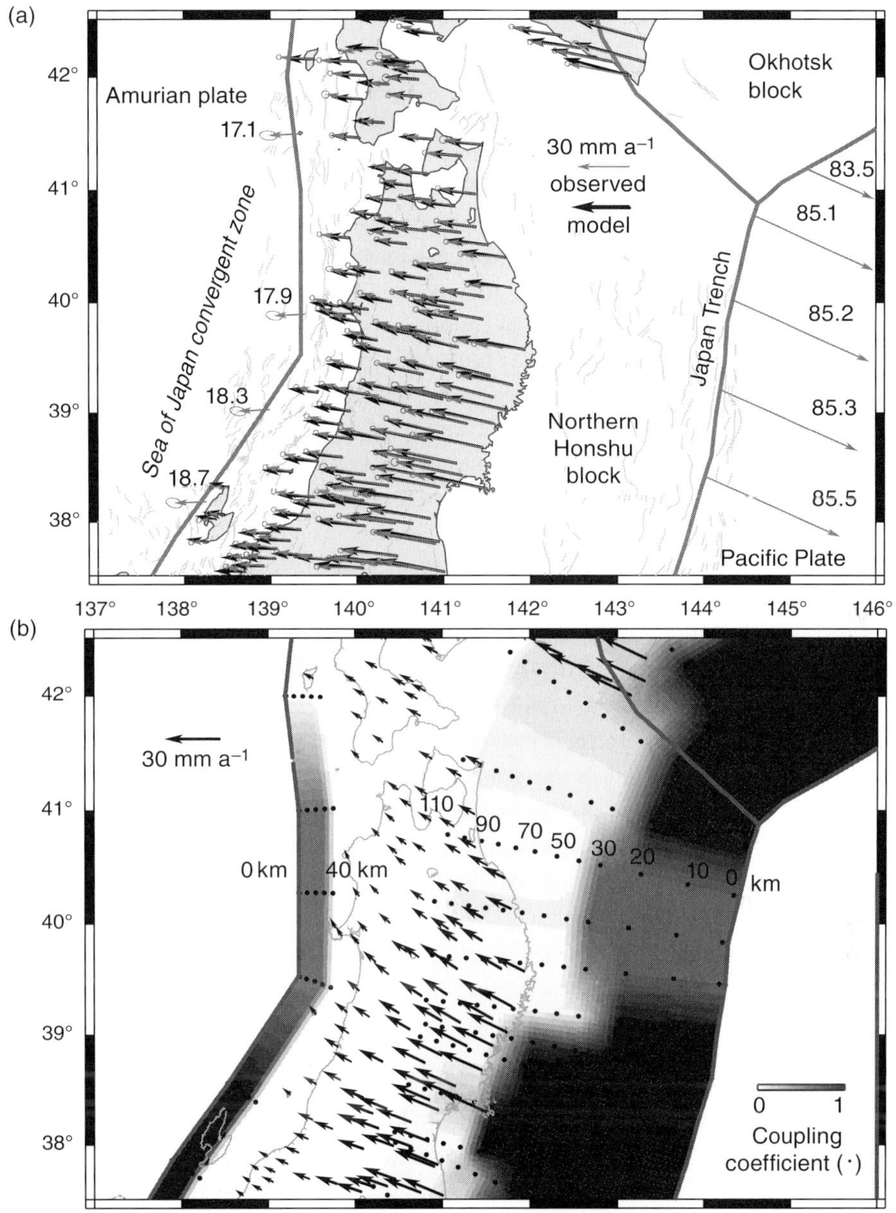

Fig. 6.3 (a) Block boundaries used in elastic block modeling are shown as heavy gray lines. Gray arrows represent GPS velocities. Corresponding black arrows show the best-fitting model. Gray arrows (and values) along the boundaries show northern Honshu block motion relative to Amurian and Pacific plates (mm a^{-1}). (b) Coupling coefficients (ϕ) were estimated for each node (block dot) along the Japan Trench and Sea of Japan fault boundaries (values indicate depth of the fault nodes in km). Degree of coupling between nodes was calculated by bilinear interpolation (gray scale). Black vectors show influence of interseismic coupling on the GPS velocity field. This influence is due to elastic (temporary) processes and is removed from the GPS velocity field to isolate long-term upper-plate strain rate in the Tohoku region (Figure 6.5). See color plate section.

of these large plates are defined based on a digital compilation of tectonic plate boundaries by Bird (2003).

The subduction interface fault (Japan Trench) is defined by approximating the configurations shown in Hasegawa *et al.* (1994) and Mazzotti *et al.* (2000) (Figure 6.3a). Individual nodes on the subduction surface are spaced $\approx$40 km apart along strike, and at 10 km depth intervals from 0 to 110 km (Figure 6.3b). The Sea of Japan convergent zone is approximated as a single fault dipping 50° E (Figure 6.3b), although alternative dips and fault configurations can be tried as well (Stirling *et al.*, Chapter 10, this volume). McCaffrey's (1995) method is used to solve for coupling coefficients at nodes on the Sea of Japan fault and the Japan Trench. To represent the change in coupling coefficient (ϕ) values between adjacent nodes, ϕ values on 5×5 km rectangular fault patches between the nodes are estimated by bilinear interpolation. Solutions for the rotation of the northern Honshu (NHON), Okhotsk (OKHO), Pacific (PACI), Amurian (AMUR) and North American (NOAM) blocks relative to Eurasia, and rotation parameters that rotate each GPS velocity data set into a Eurasia-fixed reference frame are calculated. The addition of larger, surrounding tectonic plates to the model (PACI, AMUR and NOAM) assists in balancing the plate motion budget. In addition to the GPS velocities, earthquake slip vectors are included from events on the Japan Trench and in the Sea of Japan convergent zone (obtained from Harvard CMT, 2008). These data provide information about the direction of relative motion between the NHON and PACI blocks, and the NHON and AMUR blocks, respectively.

Coupling coefficients (ϕ) at the fault nodes are constrained to decrease in value down-dip. This down-dip constraint in ϕ is necessary when using an elastic dislocation modeling approach for a subduction interface in order to avoid unrealistic extensional strain above the up-dip end of the coupled zone that is predicted by the model when the area up-dip of the coupled zone of a dipping fault slips aseismically (e.g. McCaffrey, 2002). In other words, the actual up-dip end of coupling on the subduction interface cannot be discerned when using elastic dislocation methods. No improvement in fit to the data was found after running inversions without imposing this down-dip decrease in ϕ constraint, suggesting that the data are insensitive to "uncoupled" areas up-dip of the "coupled" zone.

6.3.3 Block model results

Elastic block modeling results are summarized in Figures 6.3a and 6.3b. The best calculated fit to the measured GPS velocities gives the reduced chi-squared statistic, $\chi_n^2 = 2.3$, using 1573 data to estimate 232 free parameters. The data include 604 horizontal GPS velocities (east and north components, comprising 1208 data) and 365 earthquake slip vectors from events on the Japan Trench and in the Japan Sea. The free parameters include: rotation parameters for three tectonic blocks and six GPS data sets (three parameters for each), and coupling coefficients for 205 free nodes on the Japan Trench and Sea of Japan block boundary faults. Significant interseismic coupling on the Japan Trench and in the Sea of Japan is estimated, consistent with previous studies (Mazzotti *et al.*, 2000; Nishimura *et al.*, 2004; Suwa *et al.*, 2006). Rotation estimates of the NHON block relative to the PACI and AMUR plates suggest that convergence is accommodated at a rate $\approx$ 85 mm a^{-1} at the Japan Trench and 15–20 mm a^{-1} in the Sea of Japan, yielding a total plate motion budget $\approx$ 102–103 mm a^{-1} that must occur across the PACI/NHON/AMUR plate boundary zone. Some proportion of this budget could also be accommodated within the northern Honshu block itself, not only on its boundaries as the elastic block model assumes. The influence on the measured GPS velocities from interseismic coupling on the block-bounding faults is shown in Figure 6.3b. This component of the GPS velocity field is assumed to be elastic (temporary) deformation, which will be recovered during slip on the block-bounding faults in future earthquakes. Removing this elastic component of the velocity field (due to interseismic coupling on the Japan Trench and Sea of Japan faults) from the GPS velocities allows us to estimate residual strains that may be due to active faulting within the northern Honshu region.

6.3.4 Residual upper plate strain in the Northern Honshu region

A variety of methods can be used to convert GPS site velocities to a map of regional strain. Perhaps the most widely used is one developed by John Haines and Bill Holt (e.g. Haines and Holt, 1993; Beavan and Haines, 2001). To employ their method, a grid (in latitude–longitude) is developed over the area of interest; for the purposes of the northern Honshu case study, a grid $\approx 5 \times 5$ km is used. Velocities are modeled as bicubic splines within each cell and the inversion matches the input velocities and minimizes the strain rates within each cell. A strain-rate variance parameter ($\frac{1}{v}$) in the terminology of Beavan and Haines (2001) is defined, with lower values of $\frac{1}{v}$ giving smoother solutions. This parameter is chosen such that the sum of squared residuals between the model and input velocities plus the sum of squares in matching the strain-rate constraints is approximately equal to the number of degrees of freedom in the GPS data set (i.e. twice the number of velocities). In other words, the reduced chi-squared statistic, χ_n^2 = (sum of squared residuals)/(degrees of freedom), is ≈ 1.

Figure 6.4 shows the shear and areal strain rates for the GPS velocity field without removing the elastic strain due to offshore fault coupling. These strain rates are quite high (up to -90 nanostrain a^{-1} for areal strain, and $>$ 120 nanostrain a^{-1} for the maximum shear strain), but this strain is largely elastic strain from interseismic coupling on the major faults (Japan Trench, Sea of Japan) and is unlikely to lead to permanent deformation in the same location where the strain is currently accumulating.

Figures 6.5a and 6.5b show the strain field (areal and shear components) after this elastic part of the GPS velocity field (Figure 6.3b) has been removed from the raw velocity field (Figure 6.2). The shear strain rate is greatly reduced (Figure 6.5b), and there is no coherent pattern of shear strain that might indicate a possible through-going fault zone. The lack of coherent shear strain in the residual strain map is consistent with the lack of evidence for active strike-slip faults within northern Honshu.

Areal strain is greatly reduced in Figure 6.5a compared to Figure 6.4a. However, a zone of elevated contractional strain ($\approx$ 20 nanostrain a^{-1}, $\chi_n^2 = 1.0$ case) persists along the

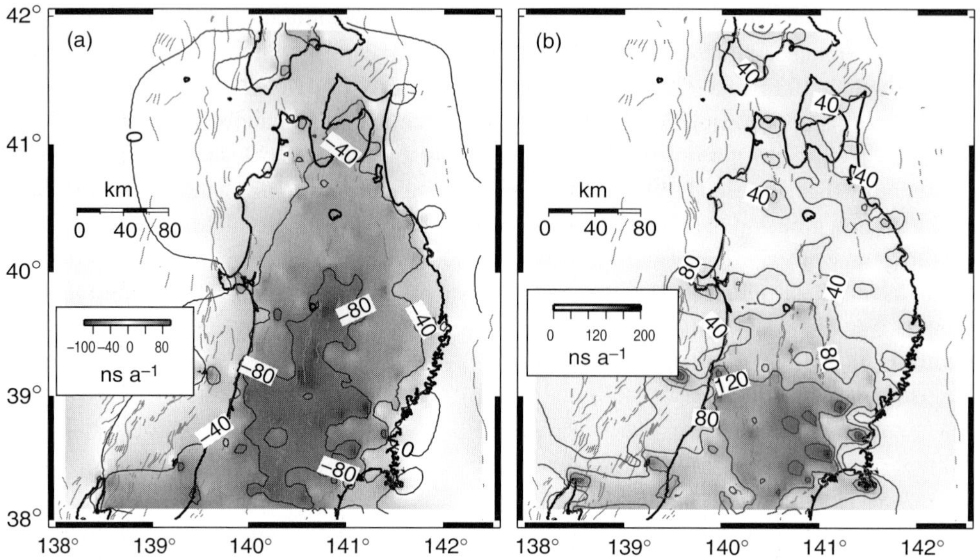

Fig. 6.4 (a) Areal and (b) shear strain rates from the raw GPS velocity field in nanostrain a^{-1} (Figure 6.3; no elastic strains are removed) estimated using the method of Haines and Holt (1993). See color plate section.

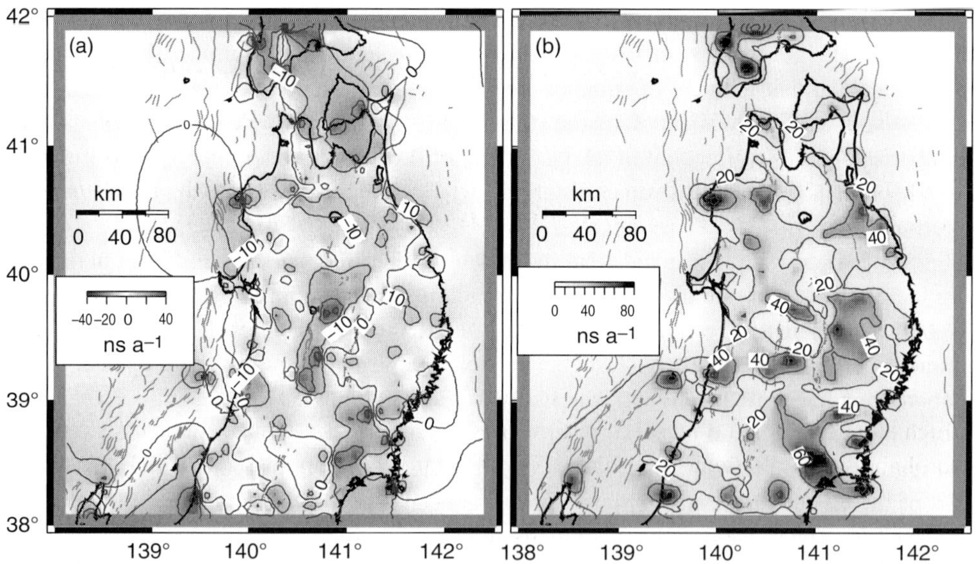

Fig. 6.5 (a) Areal and (b) shear strain rates are contoured in the residual GPS velocity field in northern Honshu, after the elastic component of the velocity field from coupling on block-bounding faults (e.g. Figure 6.3b) is removed from the raw velocity field (Figure 6.2). Strain rates shown are nanostrain a^{-1}. See color plate section.

central Backbone Range, coinciding with a zone of well-documented, active reverse faults thought to be accommodating a total horizontal slip $\approx$ 1–2 mm a^{-1} (AIST, 2008). Our residual strain-rate estimates suggest that a plate boundary convergence $\approx$ 1–3 mm a^{-1} is taken up within the Backbone Range. Moreover, the Senya fault (within the Backbone Range; Figure 6.2) ruptured in 1896, producing 3.5 m of reverse displacement in some locations (Matsuda *et al.*, 1980), providing independent evidence for ongoing contraction within the Backbone Range. Using a backslip approach without block rotations, Miura *et al.* (2004a) obtained a similar result of high contractional strain along the Backbone Range after removing the elastic strain due to subduction coupling (Japan Trench) from the GPS velocity field in northern Honshu. In addition to the Backbone Range contraction, elevated contraction is detected along the west coast of northern Honshu, possibly due to onshore active faulting associated with the Sea of Japan convergent zone. Due to our arguably inaccurate assumption that Sea of Japan convergence occurs on a single dipping fault offshore of the west coast of northern Honshu (Figure 6.3b), it is possible that the contraction in the GPS velocity field due to active faulting onshore near the west coast is underestimated. For example, much of the contraction on the west coast, possibly due to active faulting, has been interpreted as elastic strain due to coupling on the offshore Sea of Japan boundary in our block model (i.e. the models are non-unique). Estimating strain using a probabilistic approach could be used to incorporate this and other alternative tectonic models into our northern Honshu regional strain estimates. For a detailed discussion see Section 6.4 and Stirling *et al.* (Chapter 10, this volume).

6.4 A probabilistic approach to estimate tectonic hazards with GPS

Seismological and geological data are important inputs into probabilistic seismic hazard assessments (PSHA; e.g. Reiter, 1990; Senior Seismic Hazard Analysis Committee, 1997; Frankel *et al.*, 2002; Stirling *et al.*, 2002), which have been widely used to estimate the seismic shaking hazard at many locations around the world, including detailed PSHA work on some key nuclear facilities (e.g. Stepp *et al.*, 2001). Methods have been developed for PSHA that can embrace a variety of different techniques (and/or tectonic models) that might be used to convert seismological and geological data into a probability of "seismic hazard." Hazard estimates using a variety of techniques and data sets are combined, by weighting the results of different models, data and/or expert opinion in a logic tree environment, which is then sampled (thousands of times) by Monte Carlo sampling methods to provide a probability distribution of resulting seismic hazard estimates that collectively expresses the epistemic uncertainty.

Here, we demonstrate the initial steps of a method to assess the "rock deformation hazard" (i.e. likelihood of permanent rock deformation occurring at a given point on the Earth's surface in the future) by using GPS data to obtain contemporary surface strain rates. These steps combine McCaffrey's elastic block approach with the Haines/Holt method. To estimate the hazard probabilistically, alternative tectonic models (i.e. different block and/or fault configurations) need to be considered. An approach that incorporates all viable tectonic models and expert opinions similar to that used in PSHA methodologies may be the best way to incorporate all available knowledge about the potential tectonic hazard into a probabilistic rock deformation hazard assessment using GPS. Stirling *et al.* (Chapter 10, this volume) discuss the application of a PSHA-inspired method in northern Honshu to develop a multidisciplinary approach for estimating future rock deformation from strain rates derived from active fault, seismological and geodetic data.

Concluding remarks

Geodetic data, especially GPS, can provide a rich source of information regarding the likelihood of future, permanent rock deformation in regions where these data are available. However, various interpretations must be made of those data before they can provide information relevant to tectonic hazard assessment, due to a variety of transient signals that can influence GPS measurements in plate boundary zones. We have outlined an approach that extracts strain rates related to possible future rock deformation using the northern Honshu region of Japan. Our result of elevated contraction in the Backbone Range of central northern Honshu is consistent with geological and seismological evidence for active reverse faulting there. However, to better understand the uncertainty in these strain rate/rock deformation estimates, other viable tectonic models and techniques must be incorporated into the interpretation of the GPS velocities, using a probabilistic approach similar to the method outlined by Stirling *et al.* (Chapter 10, this volume). In addition to assessing the possible future hazards of permanent rock deformation for the purposes of nuclear facility siting, this method may also provide a practical means of incorporating GPS data into seismic hazard models.

Further reading

For an overview of GPS techniques and their application to crustal deformation studies, Dixon (1991) and Segall and Davis (1998) are good places to start. To read more about the elastic block method used here, McCaffrey (2002) provides a detailed description, and Haines and Holt (1993) and Beavan and Haines (2001) discuss the methodology we use to estimate strain rates from GPS velocities.

Acknowledgments

We thank Susan Ellis, Peter LaFemina and Nicola Litchfield for review comments that improved the manuscript. We also thank the Nuclear Waste Management Organization of Japan (NUMO) for their support of this work.

References

AIST (2008). Active fault database of Japan. National Institute of Advanced Industrial Science and Technology, http://riodb02.ibase.aist.go.jp/activefault/index_e.html.

Altamimi, Z., P. Sillard and C. Boucher (2002). ITRF2000: a new release of the International Terrestrial Reference Frame for earth science applications. *Journal of Geophysical Research*, **107**(B10), doi:10.1029/2001JB000561.

Beanland, S., G. H. Blick and D. J. Darby (1990). Normal faulting in a back-arc basin: geological and geodetic characteristics of the 1987 Edgecumbe earthquake, New Zealand. *Journal of Geophysical Research*, **95**, 4693–4707.

Beavan, J. and J. Haines (2001). Contemporary horizontal velocity and strain-rate fields of the Pacific–Australian plate boundary zone through New Zealand. *Journal of Geophysical Research*, **106**, 741–770.

Beavan, J., P. Tregoning, M. Bevis, T. Kato and C. Meertens (2002). The motion and rigidity of the Pacific plate and implications for plate boundary deformation. *Journal of Geophysical Research*, **107**, 2261, doi:10.1029/2001JB000282.

Bennett, R., A. M. Friedrich and K. P. Furlong (2004). Codependent histories of the San Andreas and San Jacinto fault zones from inversion of fault displacement rates. *Geology*, **32**(11), 961–964.

Beutler, G., H. Bock, E. Brockmann *et al.* (2001). Bernese GPS Software Ver. 4.2. U. Hugentobler, S. Schaer and P. Fridez (eds.). University of Berne: Astronomical Institute.

Bird, P. (2003). An updated digital model of plate boundaries. *Geochemistry, Geophysics, Geosystems*, **4**(3), doi:10.1029/2001GC000252.

Calais E., M. Vergnolle, V. San'kov *et al.* (2003). GPS measurements of crustal deformation in the Baikal–Mongolia area (1994–2002): implications for current kinematics of Asia. *Journal of Geophysical Research*, **108**(B10), doi:10.1029/2002JB002373.

Cox, A. and R. B. Hart (1986). *Plate Tectonics: How It Works*. Palo Alto, CA: Blackwell Science.

DeMets, C., R. G. Gordon, D. F. Argus and S. Stein (1994). Effect of recent revisions to the geomagnetic reversal time scale on estimates of current plate motions. *Geophysical Research Letters*, **21**, 2191–2194.

Dixon, T. H. (1991). An introduction to the Global Positioning System and some geological applications. *Reviews of Geophysics*, **29**(2), 249–276.

Dixon, T. H., A. Mao, M. Bursik *et al.* (1997). Continuous monitoring of surface deformation at Long Valley Caldera, California, with GPS. *Journal of Geophysical Research*, **102**, 12 017–12 034.

Dixon, T. H., E. Norabuena and L. Hotaling (2003). Paleoseismology and Global Positioning System: earthquake-cycle effects and geodetic versus geological fault slip rates in the Eastern California Shear Zone. *Geology*, **31**(1), 55–58.

Donnellan, A., B. H. Hager, R. W. King and T. A. Herring (1993). Geodetic measurement of deformation in the Ventura basin region, southern California. *Journal of Geophysical Research*, **98**, 21 727–21 739.

Dragert, H., K. Wang and T. James (2001). A silent slip event on the deeper Cascadia subduction interface. *Science*, **292**, 1525–1528.

Frankel, A. D., M. D. Petersen, C. S. Mueller *et al.* (2002). Documentation for the 2002 update of the national seismic hazard maps. US Geological Survey Open-File Report 02-420. Reston, VA: US Geological Survey.

GSI (2008). Crustal movement in Japan. Geographical Survey Institute, http://mekira.gsi.go.jp/.

Haines, A. J. and W. E. Holt (1993). A procedure for obtaining the complete horizontal motions within zone of distributed deformation from the inversion of strain rate data. *Journal of Geophysical Research*, **98**(B7), 12 057–12 082.

Harvard CMT (2008). Global CMT web page. Harvard Centroid-Moment-Tensor (CMT) Project, http://www.globalcmt.org

Hasegawa, A., S. Horiuchi and N. Umino (1994). Seismic structure of the northeastern Japan convergent margin: a synthesis. *Journal of Geophysical Research*, **99**(B11), 22 295–22 312.

Hauksson, E., L. M. Jones and K. Hutton (1995). The 1995 Northridge earthquake sequence in California: seismological and tectonic aspects. *Journal of Geophysical Research*, **100**(B7), 12 335–12 355.

Herring, T. A. (2001). GLOBK global Kalman filter VLBI and GPS analysis program, Ver. 5.03. Cambridge, MA: Massachusetts Institute of Technology.

Heki, K., S. Miyazaki and H. Tsuji (1997). Silent fault slip following an interplate thrust earthquake at the Japan Trench. *Nature*, **386**, 595–597.

Heki, K., S. Miyazaki, H. Takahashi *et al.* (1999). The Amurian plate motion and current plate kinematics in Eastern Asia. *Journal of Geophysical Research*, **104**, 29 147–29 155.

Hill, E. M. and G. Blewitt (2006). Testing for fault activity at Yucca Mountain, Nevada, using independent GPS results from the BARGEN network. *Geophysical Research Letters*, **33**, doi:10.1029/2006GL026140.

Hirose, H., K. Hirahara, F. Kimata, N. Fujii and S. Miyazaki (1999). A slow thrust slip event following the two 1996 Hyuganada earthquakes beneath the Bungo Channel, southwest Japan. *Geophysical Research Letters*, **26**(21), 3237–3240.

Hsu, Y-J., N. Bechor, P. Segall *et al.* (2002). Rapid afterslip following the 1999 Chi-Chi, Taiwan earthquake. *Geophysical Research Letters*, **29**(16), doi:10.1029/2002GL014967.

Kreemer, C., W. E. Holt, S. Goes and R. Govers (2000). Active deformation in eastern Indonesia and the Philippines from GPS and seismicity data. *Journal of Geophysical Research*, **105**(B1), 663–680.

LaFemina, P. C., T. H. Dixon, R. Malservisi *et al.* (2005). Geodetic GPS measurements in south Iceland: strain accumulation and partitioning in a propagating ridge system. *Journal of Geophysical Research*, **110**, doi:10.1029/2005JB003675.

Langbein, J. (2004). Noise in two-color electronic distance meter measurements revisited. *Journal of Geophysical Research*, **109**, B04406, doi:10.1029/2003JB002819.

Larson, K., A. Lowry, V. Kostoglodov *et al.* (2004). Crustal deformation measurements in Guerrero, Mexico. *Journal of Geophysical Research*, **109**(B4), B04409.

Mansinha, L. and D. E. Smylie (1971). The displacement fields of inclined faults. *Bulletin of the Seismological Society of America*, **61**(5), 1433–1440.

Matsuda, T., H. Yamazaki, T. Nakata and T. Imaizumi (1980). The surface faults associated with the Rikku earthquake of 1896. *Bulletin of the Earthquake Research Institute, University of Tokyo*, **55**, 795–855.

Mattioli, G. S., T. H. Dixon, F. Farina *et al.* (1998). GPS measurement of surface deformation around Soufriere Hills Volcano, Montserrat, from October 1995 to July 1996. *Geophysical Research Letters*, **25**, 3417–3420.

Mazzotti, S., X. Le Pichon, P. Henry and S. Miyazaki (2000). Full interseismic locking of the Nankai and Japan-west Kurile subduction zones: an analysis of uniform elastic strain accumulation in Japan constrained by permanent GPS. *Journal of Geophysical Research*, **105**(B6), 13 159–13 177.

McCaffrey, R. (1995). DEFNODE user's manual. Rensselaer Polytechnic Institute, Troy, NY, http://www.rpi.edu/~mccafr/defnode.

McCaffrey, R. (2002). Crustal block rotations and plate coupling, in plate boundary zones. In: Stein, S. and J. Freymueller (eds.) *AGU Geodynamics Series*, **30**, 100–122.

McCaffrey, R. (2005). Block kinematics of the Pacific–North America plate boundary in the southwestern United States from inversion of GPS, seismological and geologic data. *Journal of Geophysical Research*, **110**, doi:10.1029/2004JB003307.

McCaffrey, R., M. D. Long, C. Goldfinger *et al.* (2000). Rotation and plate locking at the southern Cascadia subduction zone. *Geophysical Research Letters*, **27**, 3117–3120.

McClusky, S., S. Balassanian, A. Barka *et al.* (2000). GPS constraints on plate kinematics and dynamics in the eastern Mediterranean and Caucasus. *Journal of Geophysical Research*, **105**, 5695–5719.

McClusky, S., S. C. Bjornstad, B. H. Hager *et al.* (2001). Present day kinematics of the eastern California shear zone from a geodetically constrained block model. *Geophysical Research Letters*, **28**, 3369–3372.

Meade, B. J. and B. H. Hager (2005). Block models of crustal motion in southern California constrained by GPS measurements. *Journal of Geophysical Research*, **110**, doi:10.1029/2004JB003209.

Miura, S., S. Ueki, T. Sato, K. Tachibana and H. Hamaguchi (2000). Crustal deformation associated with the 1998 seismo-volcanic crisis of Iwate Volcano, northeastern Japan, as observed by a dense GPS network. *Earth Planets Space*, **52**, 1003–1008.

Miura, S., T. Sato, A. Hasegawa, Y. Suwa, K. Tachibana and S. Yui (2004a). Strain concentration zone along the volcanic front derived by GPS observations in the NE Japan arc. *Earth Planets Space*, **56**, 1347–1355.

Miura, S., Y. Suwa, A. Hasegawa and T. Nishimura (2004b). The 2003 M8.0 Tokachi-Oki earthquake – how much has the great event paid back slip debts? *Geophysical Research Letters*, **31**, doi:10.1029/2003GL019021.

Morgan, W. J. (1968). Rises, trenches, great faults, and crustal blocks. *Journal of Geophysical Research*, **73**, 1959–1982.

NASA (2008). Crustal dynamics data information system (CDDIS). National Space and Aeronautics Administration, http://cddis.gsfc.nasa.gov/.

Nikolaidis, R. (2002). Observation of geodetic and seismic deformation with the Global Positioning System. Ph.D. thesis. University of California, San Diego.

Nishimura, T., T. Hirasawa, S. Miyazaki *et al.* (2004). Temporal change of interplate coupling in northeastern Japan during 1995–2002 estimated from continuous GPS observations. *Geophysical Journal International*, **157**, 901–916.

Okada, Y. (1985). Surface deformation due to shear and tensile faults in a half-space. *Bulletin of the Seismological Society of America*, **75**, 1135–1154.

Ozawa, S., M. Murakami, M. Kaidzu *et al.* (2002). Detection and monitoring of ongoing aseismic slip in the Tokai region, central Japan. *Science*, **298**, 1009–1012, doi: 10.1126/science.1076780.

Pollitz, F. (1997). Gravitational viscoelastic postseismic relaxation on a layered spherical Earth. *Journal of Geophysical Research*, **102**(B8), 17 921–17 941.

Prawirodirdjo, L. and Y. Bock (2004). Instantaneous global plate motion model from 12 years of continuous GPS observations. *Journal of Geophysical Research*, **109**, doi:10.1029/2003JB002944.

Reilinger, R., S. McClusky, P. Vernant *et al.* (2006). GPS constraints on continental deformation in the Africa–Arabia–Eurasia continental collision zone and implications for the dynamics of plate interactions. *Journal of Geophysical Research*, **111**, doi:10.1029/2005JB004051.

Reiter, L. (1990). *Earthquake Hazard Analysis*. New York, NY: Columbia University Press.

Rothacher, M. and L. Mervart (eds.) (1996). Documentation of the Bernese GPS Software Ver. 4.0. Bern, Switzerland: University of Bern.

Sagiya, T. and W. Thatcher (1999). Coseismic slip resolution along a plate boundary megathrust: the Nankai Trough, southwest Japan. *Journal of Geophysical Research*, **104**(B1), 1111–1129.

Sagiya, T., S. Miyazaki and T. Tada (2000). Continuous GPS array and present-day crustal deformation of Japan. *Pure and Applied Geophysics*, **161**, 2303–2322.

Savage, J. C. (1983). A dislocation model for strain accumulation and release at a subduction zone. *Journal of Geophysical Research*, **88**, 4984–4996.

Savage, J. C. and W. H. Prescott (1978). Asthenospheric readjustment and the earthquake cycle. *Journal of Geophysical Research*, **83**(B7), 3369–3376.

Savage, J. C. and J. L. Svarc (1997). Postseismic deformation associated with the 1992 Mw 7.3 Landers earthquake, southern California. *Journal of Geophysical Research*, **102**(B4), 7565–7578.

Segall, P. and J. L. Davis (1998). GPS applications for geodynamics and earthquake studies. *Annual Reviews of Earth and Planetary Science*, **25**, 301–336.

Segall, P., R. Bürgmann and M. Matthews (2000). Time-dependent triggered afterslip following the 1989 Loma Prieta earthquake. *Journal of Geophysical Research*, **105**(B3), 5615–5634.

Sella, G. F., T. H. Dixon and A. L. Mao (2002). REVEL: a model for recent plate velocities from space geodesy. *Journal of Geophysical Research*, **107**(B4), doi:10.1029/2000JB000033.

Senior Seismic Hazard Analysis Committee (1997). Recommendations for probabilistic seismic hazard analysis: guidance on uncertainty and use of experts. Report NUREG/CR-6372, UCRL-ID-122160. Washington, DC: National Academy Press.

Seno, T., T. Sakurai and S. Stein (1996). Can the Okhotsk plate be discriminated from the North American plate? *Journal of Geophysical Research*, **101**(B5), 11 305–11 315.

SIO (2008). Scipps orbit and permanent array center (SOPAC). Scripps Institute of Oceanography, http://sopac.ucsd.edu.

Stepp, J. C., I. Wong, J. Whitney *et al.* (2001). Probabilistic seismic hazard analyses for ground motions and fault displacement at Yucca Mountain, Nevada. *Earthquake Spectra*, **17**, 113–151.

Stevens, C. W., R. McCaffrey, Y. Bock *et al.* (2002). Evidence for block rotations and basal shear in the world's fastest slipping continental shear zone in NW New Guinea. In: Stein, S. and J. Freymueller (eds.) *Plate Boundary Zones*, AGU Geodynamics Series, **30**, 87–99.

Stirling, M. W., G. H. McVerry and K. R. Berryman (2002). A new seismic hazard model for New Zealand. *Bulletin of the Seismological Society of America*, **92**, 1878–1903.

Subaraya, C., M. Chilieh, L. Prawirodirdjo *et al.* (2006). Plate-boundary deformation associated with the great Sumatra–Andaman earthquake. *Nature*, **440**, 46–51, doi:10.1038/nature04522.

Suwa, Y., S. Miura, A. Hasegawa, T. Sato and K. Tachibana (2006). Interplate coupling beneath NE Japan inferred from three-dimensional displacement field. *Journal of Geophysical Research*, **111**, doi:10.1029/2004JB003203.

Takahashi, H., M. Kasahara, F. Kimata *et al.* (1999). Velocity field of around the Sea of Okhotsk and Sea of Japan regions determined from a new continuous GPS network data. *Geophysical Research Letters*, **26**(16), 2533–2536.

Thatcher, W. and J. B. Rundle (1984). A viscoelastic coupling model for the cyclic deformation due to periodically repeated earthquakes at subduction zones. *Journal of Geophysical Research*, **89**, 7631–7640.

Wald, D. J. and R. W. Graves (2001). Resolution analysis of finite fault source inversion using one- and three-dimensional Green's functions 2. Combining seismic and geodetic data. *Journal of Geophysical Research*, **106**, 8767–8788.
Wald, D. J., T. H. Heaton and K. W. Hudnut (1996). The slip history of the 1994 Northridge, California, earthquake determined from strong motion, teleseismic, GPS and levelling data. *Bulletin of the Seismological Society of America*, **86**, S49–70.
Wallace, K., G. Yin and R. Bilham (2004). Inescapable slow slip on the Altyn Tagh fault. *Geophysical Research Letters*, **31**, doi:10.1029/2004GL019724.
Wallace, L. M., J. Beavan, R. McCaffrey and D. Darby (2004). Subduction zone coupling and tectonic block rotations in the North Island, New Zealand. *Journal of Geophysical Research*, **109**(B12), doi:10.1029/2004JB003241.
Wallace, L. M., J. Beavan, R. McCaffrey, K. Berryman and P. Denys (2007). Balancing the plate motion budget in the South Island, New Zealand using GPS, geological and seismological data. *Geophysical Journal International*, **168**, doi:10.1111/j.1365–246X. 2006.03183.x.
Wei, D. and T. Seno (1998). Determination of the Amurian plate motion. In: Flower, J. F., S. Chung, C. Lo and T. Lee (eds.) *Mantle Dynamics and Plate Interactions in East Asia*, Geodynamic Series 27. Washington, DC: American Geophysical Union.
Wernicke, B., J. L. Davis, R. A. Bennett *et al.* (1998). Anomalous strain accumulation in the Yucca Mountain area, Nevada. *Science*, **279**, 2096–2100.
Wernicke, B., J. L. Davis, R. A. Bennett *et al.* (2004). Tectonic implications of a dense continuous GPS velocity field at Yucca Mountain, Nevada. *Journal of Geophysical Research*, **109**(12), doi:10.1029/2003JB002832.
Williams S. D. P., Y. Bock, P. Fang *et al.* (2004). Error analysis of continuous GPS position time series. *Journal of Geophysical Research*, **109**, doi:10.1029/2003JB002741.
Yamanaka Y. and M. Kikuchi (2004). Asperity map along the subduction zone in northeastern Japan inferred from regional seismic data. *Journal of Geophysical Research*, **109**, doi:10.1029/2003JB002683.
Zhang, J., Y. Bock, H. Johnson *et al.* (1997). Southern California Permanent GPS Geodetic Array: error analysis of daily position estimates and site velocities. *Journal of Geophysical Research*, **102**(B8), 18 035–18 056.
Zhu L. and L. A. Rivera (2002). A note on the dynamic and static displacements from a point source in multilayered media. *Geophysical Journal International*, **148**, 619–627.

7

Tectonic setting of volcanic centers in subduction zones: three-dimensional structure of mantle wedge and arc crust

Y. Tamura, J. Nakajima, S. Kodaira and A. Hasegawa

The most common eruptions observed by humans, and by far the most dangerous to human populations, are those from volcanoes above the world's subduction zones (Simkin and Siebert, 2000). Population growth and development of technology are also concentrated in areas such as the Pacific Rim, where subduction-zone volcanism is prevalent. Many new and proposed nuclear facilities are therefore located in regions of active subduction (Connor *et al.*, Chapter 3, this volume). Because nuclear facilities require low-risk sites, and because some nuclear facilities, such as high-level radioactive waste repositories, require very long performance periods, it is necessary to understand the nature of volcanism in subduction zones from a regional, plate tectonic perspective. This perspective will allow us to develop more robust hazard models for future volcanic activity on a variety of timescales, and to better assess assumptions made by these volcanic hazard models. The goal of this chapter is to provide state-of-the-art information about the geological processes operating on a regional scale in subduction zones. Subduction zones are locations where oceanic plates subduct into the mantle; they are characterized geomorphologically by deep ocean trenches and volcanic arcs or continental margins, seismically by landward-dipping deep seismic zones and magmatically by arcuate belts of volcanoes. Subduction and arc magmatism are fundamental processes in the evolution of the Earth, because they play crucial roles in the present-day differentiation of Earth's materials and are believed to be major sites of continental crust generation that have operated throughout geologic time (e.g. Taylor, 1967; Arculus, 1981; Gill, 1981; Eiler, 2003; Rudnick and Gao, 2003).

Processes of mantle melting and volcanic eruptions along subduction zones are often illustrated by the use of two-dimensional (2D) cross-section models of convergent margins (Figure 7.1). Initially, aqueous fluids released from the subducted oceanic sediments and crust rise into the mantle wedge, lowering the mantle solidus and stimulating magma generation and, ultimately, volcanism at the surface (e.g. McBirney, 1969; Tatsumi, 1989; Sisson and Layne, 1993; Schmidt and Poli, 1998). In addition, the descent of the plate stirs the mantle, bringing a flux of warmer mantle material from greater depth, thermally reinforcing the melt generation process (e.g. Furukawa, 1993; Kincaid and Sacks, 1997; Sisson and Bronto, 1998).

We review here the structure of the mantle wedge and arc crust beneath the northeastern Japan arc and the Izu–Bonin arc (Figure 7.2), respectively, and suggest that the

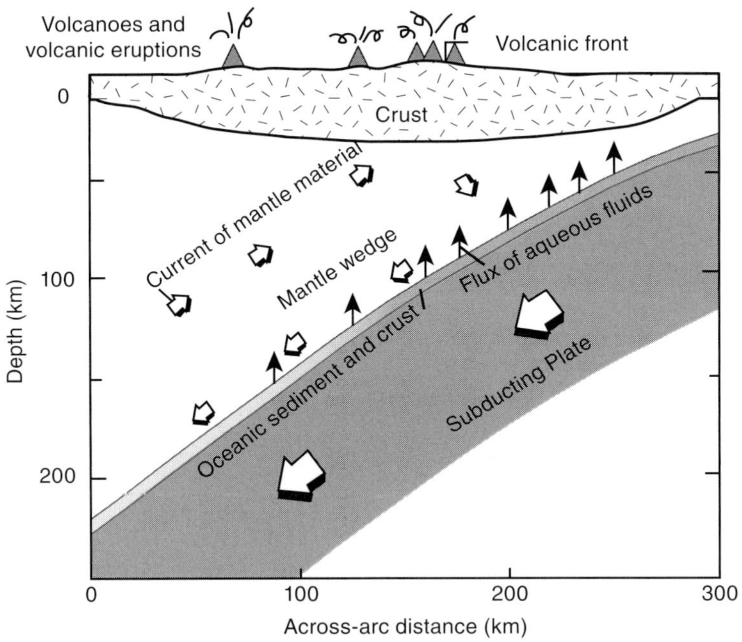

Fig. 7.1 Schematic diagram of a vertical cross-section of the crust and upper mantle of a subduction zone. Both fluxing of aqueous fluids form the subducting plate and upward current of warmer mantle material result in magma generation within the subduction zones.

third dimension, lying along the strike of the arc, is necessary to understand the actual production of magmas in subduction zones. These arcs are two of the best places in the world to understand the three-dimensional (3D) structure of the mantle wedge and arc crust. In this context, this 3D structure indicates that magma productivity is not uniform along a volcanic arc. In turn, this alters volcanic hazard rates.

The northeastern Japan arc lies above a typical subduction zone, where the Pacific plate descends beneath mainland Japan at a rate of 8–10 cm a^{-1}, and it is one of the most seismologically studied arcs. For example, Nakajima *et al.* (2001) studied 3D structure of P- and S-wave velocities V_p, V_s and V_p/V_s beneath northeastern Japan by using 169 712 P-wave and 103 993 S-wave arrival-time data from 4338 events recorded by 230 stations. The spacing between stations was 10–20 km in central northeastern Japan and 30–50 km in areas farther south and north.

The Izu–Bonin–Mariana (IBM) intra-oceanic arc, which extends for more than 2800 km from Sagami Bay in the north to Guam in the south, is one of the best oceanic arcs to investigate the evolution of arc crust. This is chiefly because the complicating effects of pre-existing continental crust have been minimal, since the Pacific plate began to subduct at ∼ 48 Ma along the IBM arc (Stern *et al.*, 2003). In an effort to investigate along-arc structural variations and their relation to arc volcanism, the Japan Agency for Marine–Earth Science

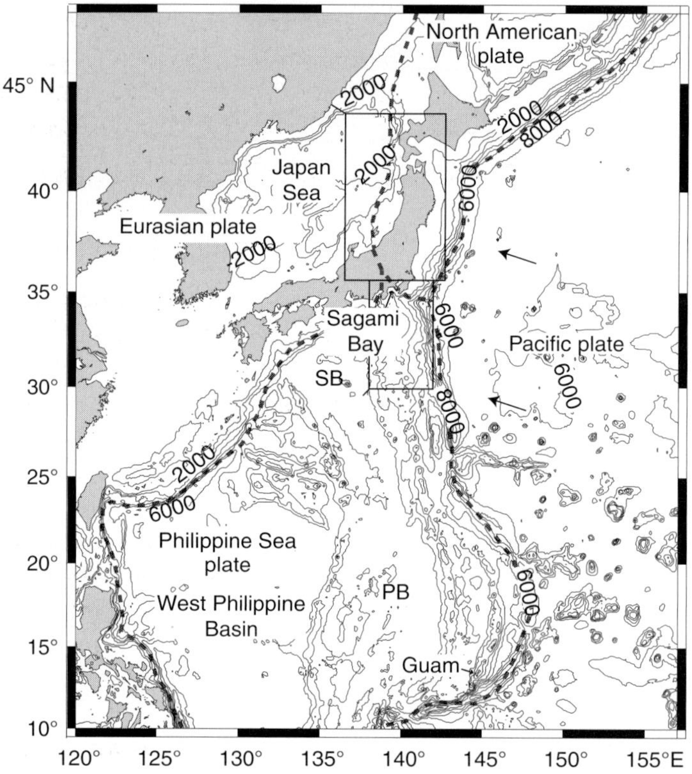

Fig. 7.2 Location map showing topography and plate configurations. The subduction direction of the Pacific plate is shown by arrows; (SB) Shikoku Basin; (PB) Parece Vela Basin. The northeastern Japan arc is built by subduction of the Pacific plate beneath the North American plate at the Japan trench. The Izu–Bonin–Mariana volcanic arc formed by subduction of the Pacific plate beneath the Philippine Sea plate, and is an excellent example of an intra-oceanic convergent margin.

and Technology (JAMSTEC) recently conducted an active-source seismic study along a 550 km profile of the volcanic front from Sagami Bay to Torishima volcano (Kodaira *et al.*, 2007). A linear array of densely deployed ($\approx$ 5 km spacing) ocean-bottom seismographs (OBSs) and a large air-gun array (eight Bolt long-life air guns with a total volume of 12 000 cubic inches, $\approx$ 197 L) were used to acquire the seismic data.

Information about 3D structures have come from independent studies of the mantle wedge (Nakajima *et al.*, 2001; Hasegawa and Nakajima, 2004) and arc crust (Kodaira *et al.*, 2007) in the northeastern Japan and Izu–Bonin arcs, respectively, and common periodic structural variations, having wavelengths of 80–100 km, can be observed in both areas. Thus we suggest here that the 3D thermal structure of mantle wedge has a direct link to the 3D structure of arc crust via production of arc magma within the mantle wedge. The "hot fingers" models (Tamura *et al.*, 2002) may play an important role in linking the 3D structures within the mantle wedge and overlying arc crust to volcanic eruptions at the surface.

7.1 "Hot fingers" in the mantle wedge beneath the northeastern Japan arc

There have been a number of petrological studies and discussions related to the genesis of arc magmas along the northeastern Japan arc (e.g. Kushiro, 1983; Tatsumi *et al.*, 1983; Sakuyama and Nesbitt, 1986; Kersting *et al.*, 1996; Shibata and Nakamura, 1997; Tamura, 2003; Kimura and Yoshida, 2006; Shuto *et al.*, 2006). Moreover, the accumulation of recent age data for volcanic rocks has clarified the temporal distribution of Quaternary volcanoes in Japan (Kondo *et al.*, 1998; Umeda *et al.*, 1999; Committee for Catalogue of Quaternary Volcanoes in Japan, 1999; Kondo *et al.*, 2004; Kondo, Chapter 12, this volume).

The volcanic front extends along the middle of the arc and is sub-parallel to the strike of the trench; many Quaternary volcanoes are distributed along the volcanic front and in back-arc areas in northeastern Japan (Sugimura, 1960). Kushiro (1983) and Schmidt and Poli (1998) suggest that the temperature within the mantle wedge exceeds the solidus along the leading edge of the convecting mantle within the wedge, which results in an abrupt and well-defined volcanic front. A fairly regular distribution of arc volcanoes along other convergent plate boundaries has been observed. For example, volcanoes of the Cascade Range in the United States, such as Mt. Hood, Mt. Jefferson and the Three Sisters, are separated by intervals of $\sim$ 80 km, and the volcanoes along the Aleutian arc and Alaska Peninsula are located at intervals of $\sim$ 70 km (Marsh and Carmichael, 1974). Such spatial distribution may provide insights into 3D dynamic processes in the underlying mantle wedge such as convection-controlled heterogeneous thermal structures and the non-uniform generation and migration of magmas (Marsh and Carmichael, 1974; Vogt, 1974; Marsh, 1979). On the other hand, d'Ars *et al.* (1995) examined the distribution of volcanoes along 16 active margins, containing 479 volcanic systems, and concluded that, in contrast to the previous suggestions, volcanoes are randomly distributed along active plate margins.

Apparently, individual Quaternary volcanoes along the northeastern Japan arc, as along many other arcs, exhibit no regular spacing. However, Tamura *et al.* (2002) suggested that Quaternary volcanoes in the northeastern Japan arc could be grouped into ten volcano clusters striking transverse to the arc (Figure 7.3); these have an average width of $\sim$ 50 km, and are separated by parallel gaps 30–75 km wide. Undulating basement highs, consisting of the ten topographic features, coincide with these groups of volcanoes. These workers suggested that this grouping may be related to locally developed hot regions within the mantle wedge that have the form of inclined, 50 km-wide fingers (hot fingers). This interpretation was based on the clustering of the volcanic centers, topographic profiles, low-velocity regions in the mantle wedge and local negative Bouguer gravity anomalies (Tamura *et al.*, 2002).

7.2 Tomographic evidence of mantle upwelling beneath northeastern Japan and its along-arc variation

Recent high-resolution seismic tomographic studies (Nakajima *et al.*, 2001; Hasegawa and Nakajima, 2004) have provided new insights into arc magmatism, and these have implications for along-arc variations of magma sources in the mantle.

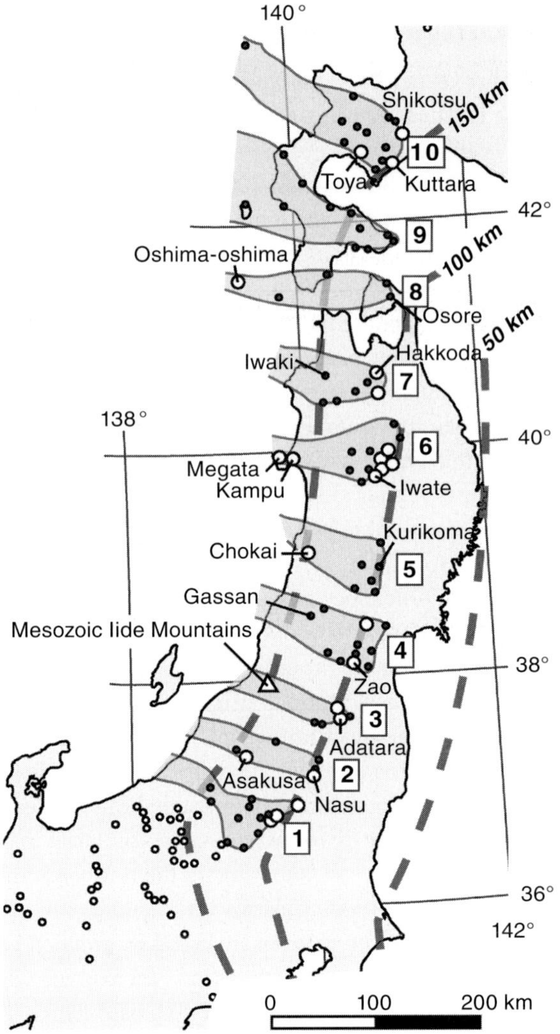

Fig. 7.3 Quaternary volcanoes in northeastern Japan (Committee for Catalogue of Quaternary Volcanoes in Japan, 1999) have been used to define ten volcanic groups (hot fingers) striking transverse to the volcanic front (Tamura *et al.*, 2002). Basalt-bearing volcanoes and basalt-free volcanoes on the fingers are shown by open circles and solid circles, respectively (Tamura, 2003). Mesozoic Iide Mountains, which are uplifted basement by possibly Quaternary magmas (Umeda *et al.*, 2007), are shown by a triangle. Dashed lines indicate depth to the subducted slab.

Figure 7.4 shows a cross-arc vertical cross-sections of S-wave velocity perturbations along the six transects shown in the insert map. A prominent inclined, low-velocity zone is enclosed in the mantle wedge, sub-parallel to the down-dip direction of the underlying Pacific slab. Interestingly, this low-velocity zone is seen in the cross-sections that contain active volcanoes (lines A, B, D, E and F) as well as those cross-sections without active

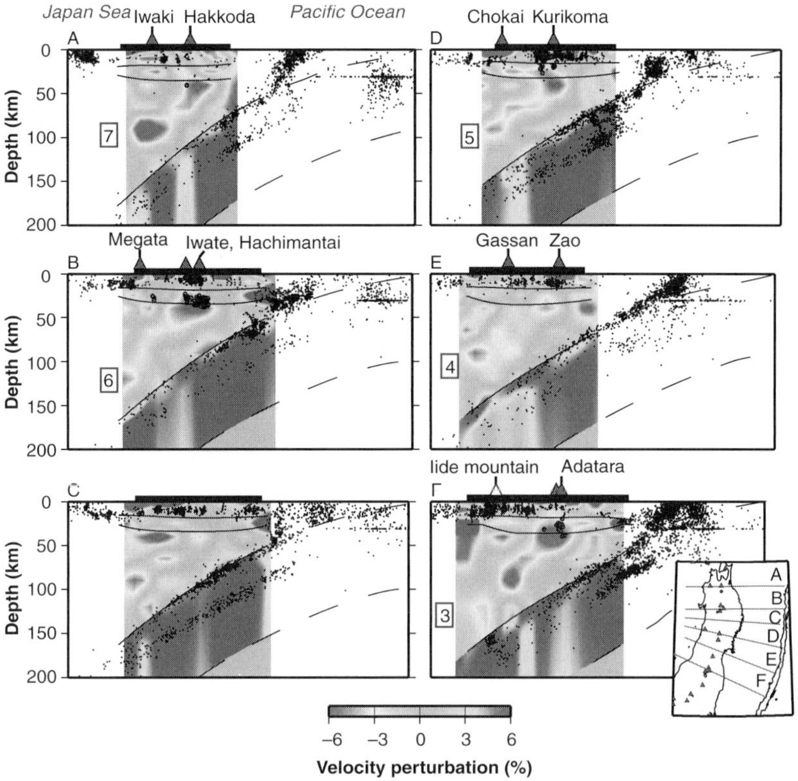

Fig. 7.4 Across-arc vertical cross-sections of S-wave velocity perturbations along lines in the inserted map of northeastern Japan (Nakajima *et al.*, 2001). The solid line and black triangles at the top represent land area and active volcanoes, respectively. The open triangle in F represents the Mesozoic Iide Mountains. Dots and circles denote earthquakes and deep, low-frequency microearthquakes, respectively. Copyright (2001) American Geophysical Union. Reproduced by permission of the American Geophysical Union. See color plate section.

volcanoes (line C). This suggests that the inclined low-velocity zone in the mantle wedge has the form of a single and extensive inclined sheet, with larger low-velocity perturbations imposed on this sheet, corresponding to areas of thermal anomalies.

Hasegawa and Nakajima (2004) conducted a tomographic inversion to image velocity structures in the mantle wedge beneath northeastern Japan. They estimated S-wave velocities within the mantle wedge portion alone by using the same data set from their previous study (Nakajima *et al.*, 2001), but with those of the nearby crust and slab being fixed to the values obtained during the previous study. Their results provide better spatial resolution ($\sim$ 10 km) in the mantle wedge in both the horizontal and vertical directions, except in the southernmost area of northeastern Japan. Figure 7.5 depicts S-wave velocity perturbations along the core of the inclined low-velocity zone in the mantle wedge, which correspond to velocity perturbations along the slowest S-wave velocity portion of

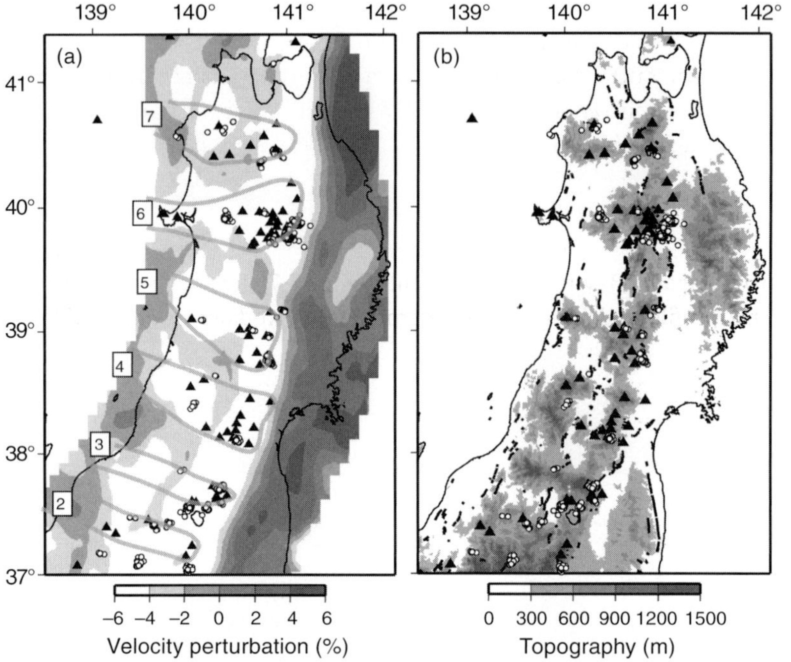

Fig. 7.5 (a) S-wave velocity perturbations along the inclined low-velocity zone in the mantle wedge of northeastern Japan. (b) Topography of northeastern Japan. Black triangles and white circles denote Quaternary volcanoes and deep, low-frequency microearthquakes, respectively (Hasegawa and Nakajima, 2004). Volcano groups are shown as in Figure 7.3. Copyright (2004) American Geophysical Union. Reproduced by permission of the American Geophysical Union.

the mantle wedge. Velocity reductions in the low-velocity zone vary in the along-arc direction, and have local maxima below areas where Quaternary volcanoes are clustered at the surface (Figure 7.5). In other words, the hot fingers proposed by Tamura *et al.* (2002) have extremely low S-wave velocities compared to the surrounding mantle wedge. Deep, low-frequency microearthquakes, which occur at Moho depths, are hence considered to be closely associated with magmatic activity (e.g. Hasegawa and Yamamoto, 1994), and these are also clustered above such localized low-velocity areas. Note that spatial resolution in the backarc side of southern northeastern Japan is low because of sparse distribution of stations; this causes an unclear image for the southernmost finger. Figure 7.5 also shows that volcanic groups and the elevated basement toward the back-arc side are distributed above the prominent low-velocity portions in the mantle wedge, which correspond to the hot mantle fingers.

7.3 High-resolution seismic image of the Izu intra-oceanic arc crust

Suyehiro *et al.* (1996) presented the first seismic profile crossing the entire Izu-Bonin arc. These workers observed a 4.5 km-thick layer with a P-wave velocity (V_p) of 6.0–6.4 km s^{-1}

in the middle of the crust beneath the volcanic front. Since such velocities correspond to the range of velocities found in the upper part of continental crust (Christensen and Mooney, 1995), Suyehiro *et al.* (1996) concluded that continental crust was created in mid-crustal regions. Rocks in the 6.0–6.4 km s^{-1} region have been interpreted to be tonalitic in composition (Taira *et al.*, 1998). Across- and along-arc data make it possible to obtain 3D structural information about arc crust.

Figure 7.6 shows the location of a 550 km wide-angle seismic profile extending along the arc front from Sagami Bay to Torishima, and Figure 7.7 shows the final seismic velocity model of this profile and its geologic interpretation (Kodaira *et al.*, 2007). The crustal structure modeled from seismic velocity and reflectivity imaging along the volcanic front of the Izu arc consists of five layers (Figure 7.7). The upper crust (V_p = 1.8–5.8 km s^{-1}, 5–7 km thick) was characterized by a steep velocity gradient and strong lateral structural variations

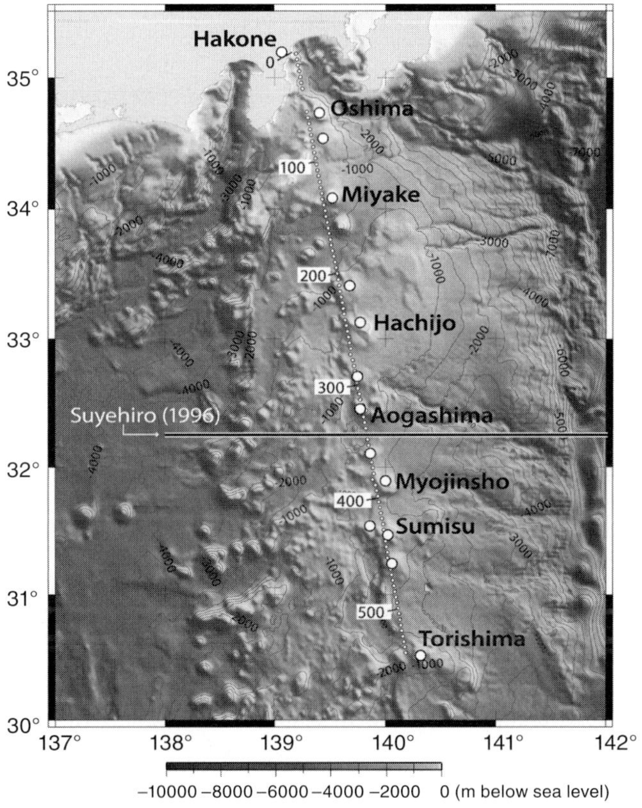

Fig. 7.6 Map showing the location of the wide-angle seismic profile along the Izu arc (Kodaira *et al.*, 2007). Small circles show ocean-bottom seismographs used in the study. Large circles indicate Quaternary volcanoes near the seismic profile. The location of the wide-angle seismic profile studied by Suyehiro *et al.* (1996), which cut across the Izu arc, is also shown. Numbers show distance (km) from the northern end of the profile.

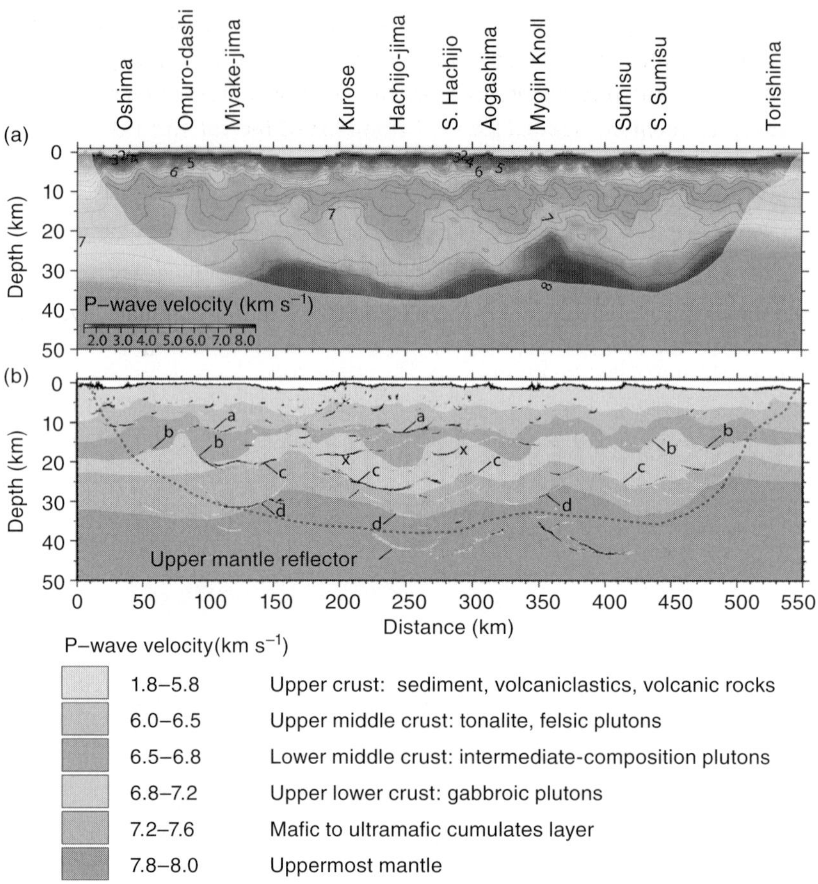

Fig. 7.7 (a) Seismic velocity model with the shaded area indicating the poorly resolved area identified by the checkerboard test. (b) Seismic reflectivity image and geological interpretation. Reflectors labeled **a–d** indicate the top of the lower part of the middle crust, the top of the upper part of the lower crust, the top of the lower part of the lower crust and the bottom of the lower part of the lower crust, respectively. Reflectors labeled **x** are interpreted as floating reflectors representing features such as laterally intruded sills and faults (after Kodaira *et al.*, 2007). Copyright (2007) American Geophysical Union. Reproduced by permission of the American Geophysical Union. See color plate section.

indicative of complex structures consisting of sediments, volcaniclastics and volcanic rocks. These workers further subdivided the middle crust layer (V_p = 6.0–6.8 km s^{-1}) into two layers (6.0–6.5 km s^{-1} in the upper part and 6.5–6.8 km s^{-1} in the lower part), interpreted to be made up of tonalites and intermediate-composition plutonic rocks, respectively. Seismic reflectors vaguely define the boundary between the upper and lower parts of this layer. The lower crust, which underlies the middle crust, was also subdivided into two layers (V_p = 6.8–7.2 km s^{-1} in the upper and V_p = 7.2–7.6 km s^{-1} in the lower). A strong, laterally continuous seismic reflector was imaged at the bottom of the upper layer, and another reflector defining the bottom of the lower crust was imaged in the northern and central parts of the profile. The thickness of the lower crust varies laterally from 12–18 km. Kodaira *et al.* (2007) interpreted the upper part of the lower crust to be made of gabbroic plutons; the high velocities modeled in the lower part of the lower crust were thought to indicate mafic to ultramafic cumulate rocks.

7.4 Crustal growth and arc volcanism

Figure 7.8a shows the average crustal P-wave velocity, excluding cumulate layer, and thickness of the middle crust (modified after Kodaira *et al.*, 2007). The upper crust was also

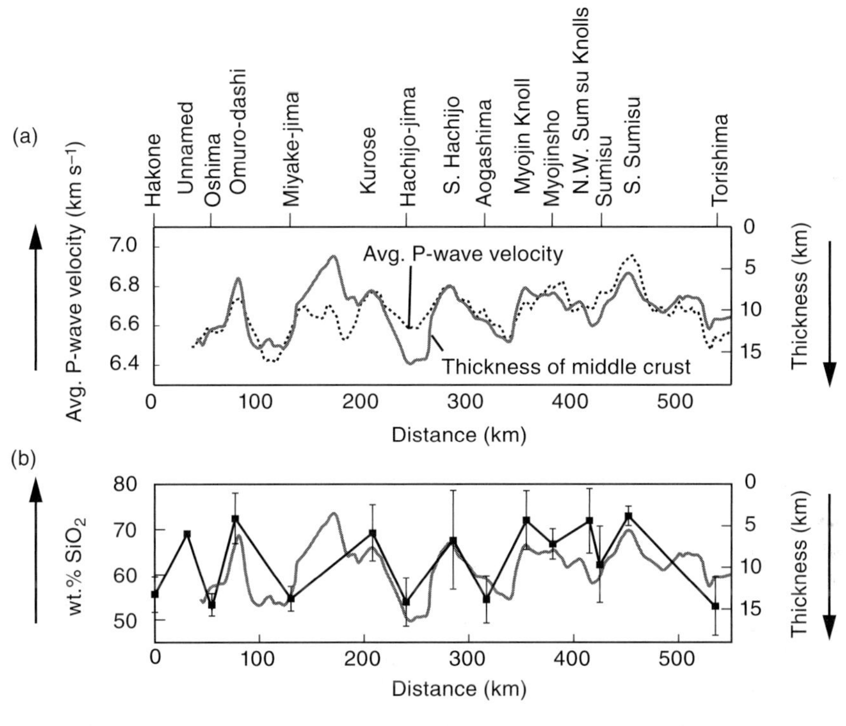

Fig. 7.8 (a) The average crustal P-wave velocity and thickness of middle-crust relationships shows that the thickness of the middle crust is negatively correlated with average seismic velocity; the dotted black line indicates the average crustal velocity excluding the lower part of the lower crust; the black line indicates the thickness of the middle crust. The average crustal velocity of typical continental crust is 6.45 km s^{-1} (Christensen and Mooney, 1995). The horizontal axis indicates distance from the northern end of the profile. (b) Average SiO_2 wt.% of volcanic rocks sampled and dredged from Quaternary volcanoes and thickness of middle-crust relationships. The observed bimodal distribution of SiO_2 contents correlates well with the thickness of the middle crust. Large volumes of continental crust, including the intermediate-composition middle crust, are generated predominantly below the basaltic volcanoes (figures modified after Kodaira *et al.*, 2007). Copyright (2007) American Geophysical Union. Reproduced by permission of the American Geophysical Union.

excluded from this profile, because upper-crustal velocities can be affected by several parameters other than crustal composition, such as variable fracture distribution and porosity (e.g. Kelemen and Holbrook, 1995). There is strong lateral variation in the thickness of the middle-crust layer (Figure 7.8a), and this variation relates to the location of volcanoes. The estimated average crustal velocities showed periodic lateral fluctuations of 6.4–6.9 km s^{-1}, with a constant wavelength of 80–100 km. It is clear from Figure 7.8a that the thickness of the middle crust is negatively correlated with average seismic velocity, as high average velocities are associated with thin middle crust.

Kodaira *et al.* (2007) investigated the possible relationship between the observed structural fluctuations and the chemistry of the Quaternary volcanoes that have been constructed at the surface (Figure 7.8b). The average SiO_2 wt.% calculated from samples of volcanic rocks shows a clear bimodal distribution (Figure 7.8b). Basalts and basaltic andesites ($<$ 55 wt. % SiO_2) are found at Oshima, Miyake-jima, Hachijo-jima, Aogashima and Torishima. Dacitic to rhyolitic rocks (66–74 wt. % SiO_2) are found at Omurodashi, Kurose, South Hachijo, Myojin knoll, Myojin-sho, northwestern Sumisu knolls and South Sumisu caldera. The observed bimodal distribution of SiO_2 contents correlates well with the thickness of the middle crust (Figure 7.8b). Thick mid-crustal sections (up to 13 km thick) occur beneath the large basaltic volcanoes (e.g. Miyake-jima, Hachijo-jima, Aogashima, Torishima), while the thin mid-crustal sections (3 km thick) lie beneath submarine rhyolitic calderas (e.g. Kurose, South Hachijo, Myojin knoll and South Sumisu) (Figure 7.8b). The volcanic rocks sampled in the Sumisu caldera have their own bimodal distribution (Tamura *et al.*, 2005), and the crustal structure beneath this volcano shows intermediate features between these two groups.

7.5 Significance of along-arc variations

The descent of the plate stirs the mantle, bringing a flux of warmer mantle material from greater depth. Moreover, aqueous fluids released from the subducted oceanic sediments and crust rise into the mantle wedge, lowering the mantle solidus and stimulating magma generation and, ultimately, volcanism at the surface.

7.5.1 Mantle convection

The subduction of an oceanic plate generates mechanically induced secondary convection in the overlying mantle wedge (e.g. McKenzie, 1969). Mantle materials near the slab are forced to flow downward by viscous coupling between the subducting plate and the overriding plate. This flow induces upward advection of mantle materials toward the corner of the mantle wedge (Figure 7.1). Velocity fields of upwelling flow in the mantle wedge are modeled to be sub-parallel to the down-dip direction of the subducting slab, and the high temperature of the upwelling flow causes an inclined low-velocity zone in the mantle wedge (e.g. Eberle *et al.*, 2002). Therefore, the inclined low-velocity zone obtained by the seismic tomography can be interpreted as mantle return flow (upwelling flow) (e.g. Hasegawa *et al.*, 1991; Hasegawa and Nakajima, 2004).

In addition to this large-scale convection in the mantle wedge, several papers by S. Honda (Honda *et al.*, 2002; Honda and Saito, 2003; Honda and Yoshida, 2005a, 2005b; Honda *et al.*, 2007) were stimulated by hot finger models; based on numerical simulations they propose the existence of small-scale convection under the volcanic arc having axes perpendicular to the arc. They concluded this convection influences the spatial and temporal evolution of arc volcanism in northeastern Japan and the Izu–Bonin arc.

A velocity image obtained by seismic tomography is a present-day snapshot and does not convey information about the direction of mantle return flow. Instead, shear-wave splitting measurements, using local S-waves, provide constraints on the orientation of mantle flow in subduction zones. Nakajima and Hasegawa (2004) and Nakajima *et al.* (2006) discovered in northeastern Japan a striking rotation of polarization direction of the leading shear wave from trench-parallel in areas closer to the trench to trench-normal in those far from the trench; they interpreted these relationships to be related to mantle flow. These workers concluded that the observed first directions correspond to the direction of mantle upwelling, indicating that the return flow is generated sub-parallel to the down-dip direction of the subducting Pacific slab.

7.5.2 Mantle Melting

Tamura *et al.* (2002) suggested that mantle melting and the production of magmas in northeastern Japan may be controlled by locally developed hot regions within the mantle wedge that have the form of inclined, 50 km-wide fingers. Although these hot fingers have been imaged by seismic tomography (Figure 7.5) (Hasegawa and Nakajima, 2004), the role they play in arc magma genesis is not fully understood. Do these hot mantle fingers melt to produce arc magmas, or, alternatively, do they play a role mainly as heat sources? Are arc magmas generated from mantle diapirs, which form in the lower part of the mantle wedge and rise through the overlying hot fingers in the mantle wedge (Tamura, 2003, Tamura *et al.*, 2005)? Gerya and co-workers (Gerya and Yuen, 2003, Gerya *et al.*, 2004, Gerya *et al.*, 2006) similarly suggest that seismic structures with strong positive and negative velocity anomalies in the mantle wedge above subduction zones are caused by the plume-like structures that evolve from the subducted plate.

When aqueous fluids are released by the dehydration reactions of serpentine and chlorite immediately above the slab at depths of 150–200 km (Iwamori, 1998), they may migrate buoyantly upward into the mantle wedge and meet the upwelling mantle flow at a depth of 100–150 km. The addition of these aqueous fluids to hot upwelling mantle flows would lower the solidus of peridotite. Consequently, partial melting could occur within the mantle upwelling. Nakajima *et al.* (2005) estimated quantitatively the melt fractions within the upwelling flow from P-wave and S-wave velocity reduction rates, as follows. They found a systematic change in melt-filled pore shapes with depth, suggesting the existence of 3–6 vol.% melts as grain boundary tubules at a depth of 90 km, 0.04–0.05 vol.% melts as thin cracks or dikes with an aspect ratio of $\sim$ 0.001 at a depth of 65 km, and 1–2 vol.% melts as cracks or dikes with an aspect ratio of 0.02–0.04 at a depth of 40 km. Degrees of mantle melting to produce arc magmas are estimated to be 10–20 wt.% (e.g. Tamura *et al.*, 2005);

thus, it is possible that hot upwelling of mantle flow (hot fingers) may not serve as magma sources, but instead as heat sources for rising diapirs (Tamura, 2003; Tamura *et al.*, 2005).

7.5.3 Arc crust

Some interesting geologic information concerning hot fingers was presented by Umeda *et al.* (2007). The rear-arc side of hot finger number 3 at 38.0° N is not capped by Quaternary volcanoes but by the Iide Mountains, consisting of uplifted Mesozoic sedimentary rocks and late Cretaceous to Paleogene granitic rocks (Figures 7.3 and 7.4). Interestingly, Umeda *et al.* (2007) presented new helium isotope data from high-density sampling in the Iide Mountains and surrounding areas and examined the origin of the heat source of associated hot springs of the Iide Mountains. They conclude that the anomaly beneath the Iide Mountains is due to newly ascending magmas in the present-day subduction system, and that the Iide Mountains probably have been uplifted as the result of repeated magma injections into the underlying crust (Umeda *et al.*, 2007). Thus, the hot fingers model predicted the existence of Quaternary magmas hidden beneath the Mesozoic sediments in the part of the northeastern Japan arc.

Kodaira *et al.* (2007) propose that large volumes of continental crust, including the intermediate-composition middle crust, are generated below the basaltic volcanoes, while the middle crust tends to be thinner beneath the dacitic to rhyolitic volcanoes (Figure 7.8). One possible interpretation is that hot fingers exist below basaltic volcanoes and produce thicker middle crust with a constant wavelength of 80–100 km. The origin of the dacitic to rhyolitic volcanoes is thought to involve remelting of the intermediate-composition middle crust (e.g. Tamura and Tatsumi, 2002). Tamura and Tatsumi (2002) analyzed rock samples from 17 Quaternary volcanoes of the Izu arc and concluded that the rhyolitic magmas may have been produced by dehydration–melting of calc-alkaline andesite in the upper to middle crust. They also suggested that a lateral influx of hot basaltic magma, probably from the adjacent basaltic magmatic systems, would have provided the heat source to remelt the middle crust. Although the distance between basaltic and rhyolitic volcanoes along the Izu arc is at least 30 km, the Miyake-jima eruption of 2000 demonstrated that lateral migration of a basalt dike over such a distance is possible (e.g. Geshi *et al.*, 2002; Nakada *et al.*, 2005; Uhira *et al.*, 2005). Remelting of the middle crust by basaltic magmas and production of rhyolitic magmas effectively transform middle crust to lower crust by altering composition. Thus, the thinning of the middle crust beneath rhyolitic volcanoes (Figure 7.8) may be caused by remelting of the middle crust. The resultant mafic restite may increase the volume of the lower crust.

Pearcy *et al.* (1990) showed that bulk compositions of the Tonsina–Nelchina segment of the Talkeetna island–arc and the island arc-related Canyon Mountain Complex are basaltic and magnesium-rich andesitic-to-basaltic, respectively. These arc compositions seem to be different from the Izu–Bonin arc, where along-arc variations in crustal evolution create important variations in composition. Thus the single sections studied by Pearcy *et al.* (1990) could not represent the whole arc system.

Concluding remarks

A remarkable characteristic of subduction zones is the along-arc periodic structural variations that occur with a wavelength of 80–100 km (Tamura *et al.*, 2002; Hasegawa and Nakajima, 2004; Kodaira *et al.*, 2007). Although information about 3D structures has come from independent studies of the mantle wedge and arc crust in the northeastern Japan and Izu–Bonin arcs, the common structural wavelength suggests a strong coupling between the mantle wedge, arc crust and the surface. Figure 7.9 is a schematic diagram showing a possible model of hot fingers in the mantle wedge of subduction zones. The hot fingers extend from $\sim$ 150 km below the backarc region towards the shallower mantle ($\sim$ 50 km) beneath the volcanic front and are characterized by along-arc wavelengths of 80–100 km. We suggest that, in addition to the dehydration of the subducting plate, 3D hot fingers strongly influence mantle melting and the production of magmas in subduction zones. The volcanic centers at the arc front and the rear-arc are thus located above the tips and the knuckles, respectively, of these fingers. In places, basaltic dikes from the large volcanic centers above the fingers extend laterally into adjacent non-volcanic areas, and melt the middle crust between fingers. This results in rhyolitic caldera volcanoes between fingers. The along-arc crustal structures, having the same periodic structural variations with the mantle wedge, suggest that fingers in the mantle wedge are stable for millions of years to produce such fundamental crustal variations.

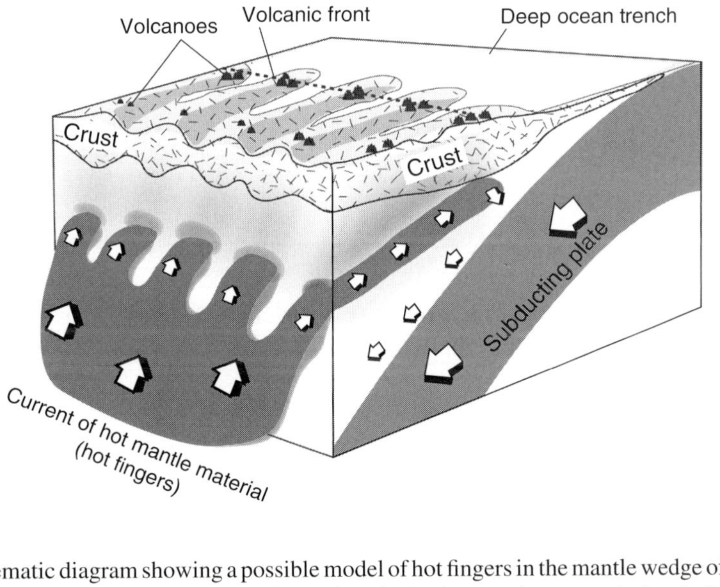

Fig. 7.9 Schematic diagram showing a possible model of hot fingers in the mantle wedge of subduction zones. The hot fingers extend from $\sim$ 150 km below the backarc region towards the shallower mantle ($\sim$ 50 km) beneath the volcanic front and are characterized by along-arc wavelengths of 80–100 km. The volcanic centers at the arc front and the rear-arc are located above the tips and the knuckles, respectively, of these fingers.

From the perspective of evaluation of long-term volcanic hazards, clarifying 3D structures of the mantle wedge and overlying arc crust along a single arc is essential. Hazard rates are non-uniform in subduction zones. Some areas close to persistent volcano clusters or behind-the-arc adjacent to volcanic clusters may be persistently more hazardous than regions between some volcano clusters. Similarly, our investigations suggest that the types of volcanism may vary along arcs in response to these 3D structures of the mantle and crust. As the types of volcanoes vary along arcs, so do volcanic hazards. Such geologic features of arcs should be considered in performing volcanic hazard assessment for nuclear facilities in these environments. Conversely, increased understanding of the geometry of subduction zones is required in order to develop robust hazard assessments of specific sites.

Thus, additional, higher-resolution geophysical and geochemical investigations have high potential to improve our understanding of hazard rates. In the near future, for example, JAMSTEC plans to conduct seismic surveys with the goal of producing high-resolution images of the mantle wedge and relation to crust. For example, densely covered along-arc and across-arc active source seismic profiles will make it possible to obtain 3D views of the structure of the crust and upper part of the mantle wedge. Passive seismic studies, which utilize natural earthquakes to determine seismic tomography, will provide additional 3D information about structural variations in the mantle wedge. Combined analysis and interpretation of results from both studies will provide new insights into arc volcanism and related crustal evolution processes.

Further reading

The following books are recommended as further reading both for subduction zones generally and seismic tomography, and volcano clustering in arcs: *Inside the Subduction Factory* (Eiler, 2003), *The State of the Planet: Frontiers and Challenges in Geophysics* (Sparks and Hawkesworth, 2004) and *Intra-Oceanic Subduction Systems: Tectonic and Magmatic Processes* (Larter and Leat 2003).

Acknowledgments

We greatly appreciate the review and comments by Richard S. Fiske, Charles B. Connor and Paul H. Wetmore. This work was supported by JSPS Grant-in-Aid for Scientific Research (B) (17340165) (YT).

References

Arculus, R. J. (1981). Island Arc, magmatism in relation to the evolution of the crust and mantle. *Tectonophysics*, **75**, 113–133.

Christensen, N. I. and W. D. Mooney (1995). Seismic velocity structure and composition of the continental crust: a global view. *Journal of Geophysical Research*, **100**, 9761–9788.

Committee for Catalogue of Quaternary Volcanoes in Japan (1999). *Catalogue of Quaternary Volcanoes in Japan*. Tokyo: Volcanological Society of Japan.

d'Ars, J. D, C. Jaupart and R. S. J. Sparks (1995). Distribution of volcanoes in active margins. *Journal of Geophysical Research*, **100**, 20 421–20 432.
Eberle, M. A., O. Grasset and C. Sotin (2002). A numerical study of the interaction between the mantle wedge, subducting slab, and overriding plate. *Physics of the Earth and Planetary Interiors*, **134**, 191–202.
Eiler, J. M. (ed.) (2003). *Inside the Subduction Factory*, Geophysical Monograph Series, 138. Washington, DC: American Geophysical Union.
Furukawa, Y. (1993). Magmatic processes under arc and formation of the volcanic front. *Journal of Geophysical Research*, **98**, 8309–8319.
Gerya, T. V. and D. A. Yuen (2003). Rayleigh–Taylor instabilities from hydration and melting propel "cold plumes" at subduction zones. *Earth and Planetary Science Letters*, **212**, 47–62.
Gerya, T. V., D. A. Yuen and E. O. D. Serve (2004). Dynamical causes for incipient magma chambers above slabs. *Geology*, **32**, 89–92.
Gerya, T. V., J. A. D. Connolly, D. A. Yuen, W. Gorczyk and A. M. Capel (2006). Seismic implications of mantle wedge plumes. *Physics of the Earth and Planetary Interiors*, **156**, 59–74.
Geshi, N., T. Shimano, T. Chiba and S. Nakada (2002). Caldera collapse during the 2000 eruption of Miyakejima volcano, Japan. *Bulletin of Volcanology*, **64**, 55–68.
Gill, J. (1981). *Orogenic Andesites and Plate Tectonics*. Berlin: Springer Verlag.
Hasegawa, A. and J. Nakajima (2004). Geophysical constraints on slab subduction and arc magmatism. In: Sparks, R. S. J. and C. J. Hawkesworth (eds.) *The State of the Planet: Frontiers and Challenges in Geophysics*, Geophysical Monograph Series, 150. Washington, DC: American Geophysical Union, 81–94.
Hasegawa, A. and A. Yamamoto (1994). Deep, low frequency microearthquakes in or around seismic low-velocity zones beneath active volcanoes in northeastern Japan. *Tectonophysics*, **223**, 233–252.
Hasegawa, A., D. Zhao, S. Hori, A. Yamamoto and S. Horiuchi (1991). Deep structure of the northeastern Japan arc and its relationship to seismic and volcanic activity. *Nature*, **352**, 683–689.
Honda, S. and M. Saito (2003). Small-scale convection under the back-arc occurring in the low viscosity wedge. *Earth and Planetary Science Letters*, **216**, 703–715.
Honda, S. and T. Yoshida (2005a). Application of the model of small-scale convection under the Island Arc, to the NE Honshu subduction zone. *Geochemistry, Geophysics, Geosystems*, **6**, Q01002, doi: 10.1029/2004GC000785.
Honda, S. and T. Yoshida (2005b). Effects of oblique subduction on the 3-D pattern of small-scale convection within the mantle wedge. *Geophysical Research Letters*, **32**, L13307, doi: 10.1029/2005GL023106.
Honda, S., M. Saito and T. Nakakuki (2002). Possible existence of small-scale convection under the back arc. *Geophysical Research Letters*, **29**, doi:10.1029/2002GL015853.
Honda, S., T. Yoshida and K. Aoike (2007). Spatial and temporal evolution of arc volcanism in the northeast Honshu and Izu–Bonin arcs: evidence of small-scale convection under the Island Arc? *The Island Arc*, **16**, 214–223.
Iwamori, H. (1998). Transportation of H_2O and melting in subduction zones. *Earth and Planetary Science Letters*, **160**, 65–80.
Kelemen, P. B. and W. S. Holbrook (1995). Origin of thick high-velocity igneous crust along the US East Coast Margin. *Journal of Geophysical Research*, **100**, 10 077–10 094.

Kersting, A. B., R. J. Arculus and D. A. Gust (1996). Lithospheric contributions to arc magmatism: isotope variations along strike in volcanoes of Honshu, Japan. *Science*, **272**, 1464–1468.

Kimura, J. I. and T. Yoshida (2006). Contributions of slab fluid, mantle wedge and crust to the origin of Quaternary lavas in the NE Japan arc. *Journal of Petrology*, **47**, 2185–2232.

Kincaid, C. and I. S. Sacks (1997). Thermal and dynamical evolution of the upper mantle in subduction zones. *Journal of Geophysical Research*, **102**, 12 295–12 315.

Kodaira, S., T. Sato, N. Takahashi *et al.* (2007). Seismological evidence for variable growth of crust along the Izu intraoceanic arc. *Journal of Geophysical Research*, **112**, doi:10.1029/2006JB004593.

Kondo, H., K. Kaneko and K. Tanaka (1998). Characterization of spatial and temporal distribution of volcanoes since 14 Ma in the northeast Japan arc. *Bulletin of Volcanological Society of Japan*, **43**, 173–180.

Kondo, H., K. Tanaka, Y. Mizuochi and A. Ninomiya (2004). Long-term changes in distribution and chemistry of middle Miocene to Quaternary volcanism in the Chokai–Kurikoma area across the northeast Japan arc. *The Island Arc*, **13**, 18–46.

Kushiro, I. (1983). On the lateral variations in chemical composition and volume of Quaternary volcanic rocks across Japanese arcs. *Journal of Volcanology and Geothermal Research*, **18**, 435–447.

Larter, R. D. and P. T. Leat (eds.) (2003). *Intra-Oceanic Subduction Systems: Tectonic and Magmatic Processes*, Special publication 219. London: Geological Society.

Marsh, B. D. (1979). Island Arc, development: some observations, experiments, and speculations. *Journal of Geology*, **87**, 687–713.

Marsh, B. D. and I. S. E. Carmichael (1974). Benioff zone magmatism. *Journal of Geophysical Research*, **79**, 1196–1206.

McBirney, A. R. (1969). Compositional variations in Cenozoic calc-alkaline suites of Central America. In: McBirney, A. R. (ed.) *Proceedings of the Andesite Conference*, 65. Portland, OR: Oregon Department of Geology and Mineral Industries, 185–189.

McKenzie, D. P. (1969). Speculations on the consequences and causes of plate motions. *Geophysical Journal of the Royal Astronomical Society*, **18**, 1–32.

Nakada, S., M. Nagai, T. Kaneko, A. Nozawa and K. Suzuki–Kamata (2005). Chronology and products of the 2000 eruption of Miyakejima Volcano, Japan. *Bulletin of Volcanology*, **67**, 205–218.

Nakajima, J. and A. Hasegawa (2004). Shear-wave polarization anisotropy and subduction-induced flow in the mantle wedge of northeastern Japan. *Earth and Planetary Science Letters*, **225**, 365–377.

Nakajima, J., T. Matsuzawa, A. Hasegawa and D. Zhao (2001). Three-dimensional structure of V_p, V_s, and V_p/V_s beneath northeastern Japan: implications for arc magmatism and fluids. *Journal of Geophysical Research*, **106**, 21 843–21 857.

Nakajima, J., Y. Takei and A. Hasegawa (2005). Quantitative analysis of the inclined low-velocity zone in the mantle wedge of northeastern Japan: a systematic change of melt-filled pore shapes with depth and its implications for melt migration. *Earth and Planetary Science Letters*, **234**, 59–70.

Nakajima, J., J. Shimizu, S. Hori and A. Hasegawa (2006). Shear-wave splitting beneath the southwestern Kurile arc and northeastern Japan arc: a new insight into mantle return flow. *Geophysical Research Letters*, **33**, L05305, doi:10.1029/2005GL025053.

Pearcy, L. G., S. M. DeBari and N. H. Sleep (1990). Mass balance calculations for two sections of Island Arc, crust and implications for the formation of continents. *Earth and Planetary Science Letters*, **96**, 427–442.

Rudnick, R. L. and S. Gao (2003). Composition of the continental crust. *Treatise of Geochemistry*, **3**, 1–64.

Sakuyama, M. and R. W. Nesbitt (1986). Geochemistry of the Quaternary volcanic rocks of the Northeast Japan arc. *Journal of Volcanology and Geothermal Research*, **29**, 413–450.

Schmidt, M. W. and S. Poli (1998). Experimentally based water budgets for dehydrating slabs and consequences for arc magma generation. *Earth and Planetary Science Letters*, **163**, 361–379.

Shibata, T. and E. Nakamura (1997). Across-arc variations of isotope and trace element compositions from Quaternary basaltic volcanic rocks in northeastern Japan: implications from interaction between subducted oceanic slab and mantle wedge. *Journal of Geophysical Research*, **102**, 8051–8064.

Shuto, K., H. Ishimoto, Y. Hirahara *et al.* (2006). Geochemical secular variation of magma source during Early to Middle Miocene time in the Niigata area, NE Japan: asthenospheric mantle upwelling during back-arc basin opening. *Lithos*, **86**, 1–33.

Simkin, T. and L. Siebert (2000). Earth's volcanoes and eruptions: an overview. In: Sigurdsson, H. (ed.) *Encyclopedia of Volcanoes*. London: Academic Press, 249–261.

Sisson, T. W. and S. Bronto (1998). Evidence for pressure-release melting beneath magmatic arcs from basalt at Galunggung, Indonesia. *Nature*, **391**, 883–886.

Sisson, T. W. and G. D. Layne (1993). H_2O in basaltic and basaltic andesite glass inclusions from four subduction-related volcanoes. *Earth and Planetary Science Letters*, **117**, 619–637.

Sparks, R. S. J. and C. J. Hawkesworth (eds.) (2004). *The State of the Planet: Frontiers and Challenges in Geophysics*, Geophysical Monograph Series, 150. Washington, DC: American Geophysical Union.

Stern, R. J., M. J. Fouch and S. L. Klemperer (2003). An overview of the Izu-Bonin-Mariana Subduction Factory. In: Eiler, J. (ed.) *Inside the Subduction Factory*, Geophysical Monograph Series, 138. Washington, DC: American Geophysical Union, 175–222.

Sugimura, A. (1960). Zonal arrangement of some geophysical and petrological features in Japan and its environs. *Journal of the Faculty of Science – Section 2: Geology, Mineralogy, Geography, Geophysics*. Tokyo: University of Tokyo, 133–153.

Suyehiro, K., N. Takahashi, Y. Ariie *et al.* (1996). Continental crust, crustal underplating and low-Q upper mantle beneath an oceanic Island Arc. *Science*, **272**, 390–392.

Taira, A., S. Saito, K. Aoike *et al.* (1998). Nature and growth rate of the northern Izu–Bonin (Ogasawara) arc crust and their implications for continental crust formation. *The Island Arc*, **7**, 395–407.

Tamura, Y. (2003). Some geochemical constraints on hot fingers in the mantle wedge: evidence from NE Japan. In: Larter, R. D. and P. T. Leat (eds.) *Intra-Oceanic Subduction Systems: Tectonic and Magmatic Processes*, Special publication 219. London: Geological Society, 221–237.

Tamura, Y. and Y. Tatsumi (2002). Remelting of an andesitic crust as a possible origin for rhyolitic magma in oceanic arcs: an example from the Izu–Bonin arc. *Journal of Petrology*, **43**, 1029–1047.

Tamura, Y., Y. Tatsumi, D. Zhao, Y. Kido and H. Shukuno (2002). Hot fingers in the mantle wedge: new insights into magma genesis in subduction zones. *Earth and Planetary Science Letters*, **197**, 105–116.

Tamura, Y., K. Tani, O. Ishizuka, Q. Chang, H. Shukuno and R. S. Fiske (2005). Are arc basalts dry, wet, or both? Evidence from the Sumisu Caldera Volcano, Izu–Bonin arc, Japan. *Journal of Petrology*, **46**, 1769–1803.

Tatsumi, Y. (1989). Migration of fluid phases and genesis of basalt magmas in subduction zones. *Journal of Geophysical Research*, **94**, 4697–4707.

Tatsumi, Y., M. Sakuyama, H. Fukuyama and I. Kushiro (1983). Generation of arc basalt magmas and thermal structure of the mantle wedge in subduction zones. *Journal of Geophysical Research*, **88**, 5815–5825.

Taylor, S. R. (1967). The origin and growth of continents. *Tectonophysics*, **4**, 17–34.

Uhira, K., T. Baba, H. Mori, H. Katayama and N. Harada (2005). Earthquake swarms preceding the 2000 eruption of Miyakejima volcano, Japan. *Bulletin of Volcanology*, **67**, 219–230.

Umeda, K., S. Hayashi, M. Ban *et al.* (1999). Sequence of the volcanism and tectonics during the last 2 Ma along the volcanic front in Tohoku district, NE Japan. *Bulletin of Volcanological Society of Japan*, **44**, 233–249.

Umeda, K., K. Asamori, A. Ninomiya, S. Kanazawa and T. Oikawa (2007). Multiple lines of evidence for crustal magma storage beneath the Mesozoic crystalline Iide Mountains, NE Japan. *Journal of Geophysical Research*, **112**, B05207, doi:10.1029/2006JB004590.

Vogt, P. R. (1974). Volcanic spacing, fractures, and thickness of the lithosphere. *Earth and Planetary Science Letters*, **21**, 235–252.

8

Conceptual model for small-volume alkali basalt petrogenesis: implications for volcanic hazards at the proposed Yucca Mountain nuclear waste repository

F. J. Spera and S. J. Fowler

Today, 31 countries operate $\sim$ 450 nuclear power reactors supplying electric power to $\sim$ 1 billion people, $\sim$ 15% of the world population. Nuclear reactors generate $\sim$ 17% of global electric power needs and a number of industrialized countries depend on nuclear power for at least half of their electricity. In addition, $\sim$ 30 nuclear power reactors are presently under construction worldwide (Macfarlane and Miller, 2007). A comprehensive summary of the principles, practices and prospects for nuclear energy may be found in Bodansky (1996). Concerns regarding energy resource availability, climate change, air quality and energy security imply a continuing demand for nuclear power in the world energy budget (Craig *et al.*, 2001). However, to date no country has solved the problem of long-term disposal or storage of nuclear waste. Without a long-term solution, the viability of nuclear energy as an increasingly significant contributor to power generation in the long-range future remains unclear. There is broad consensus that geologic disposal is the safest feasible long-term solution to high-level waste and spent-fuel disposal. Although a number of countries have ongoing geologic repository research programs, there is presently no operational geologic repository for spent fuel or high-level waste on Earth. In the United States, where spent nuclear fuel and high-level waste amounts to $\sim$ 50 000 metric tons, $\sim$ 15% of the world total, implementation has proven to be challenging both technically and politically. Nuclear waste is currently stored on-site at existing nuclear power stations and at several temporary storage facilities. Permanent geologic disposal, like the siting of a nuclear power plant, requires careful site selection. For geologic disposal, lithologies that can isolate radioactive waste from the surrounding environment and biosphere at geologic timescales $\sim 10^4$–10^6 a are a minimum requirement (Macfarlane and Ewing, 2006). Of particular importance in this regard are the nature, consequences and probabilities of volcanic hazards that can potentially compromise public, environmental and biospheric safety at long-term nuclear waste storage sites.

Yucca Mountain (YM) in Nevada, USA was identified in the early 1980s as a potential geologic repository for nuclear waste. Yucca Mountain is made up of silicic volcanic tuffs-rocks composed chiefly of pyroclastic flow and fall deposits. The proposed Yucca Mountain Repository (YMR) lies on the western boundary of the Nevada Test Site (NTS) within the Basin and Range geologic province (Thompson and Burke, 1974; Zoback *et al.*, 1981). This region is geologically active, with transtensional deformation manifested by faulting, related seismicity, high $^3He/^4He$ anomalies indicative of mantle degassing (Kennedy and van Soest, 2007) and volcanic activity. Intense study of the seismicity, seismic hazards, geohydrology, petrophysics, structural, tectonic and volcanic history of the YM region for over thirty years has provided the geologic foundation for the Total System Performance Assessment (TSPA) used by YM project geoscientists to make probabilistic forecasts of repository behavior. The TSPA considers all potential paths of radionuclides into the environment and defines the US Department of Energy's (DOE) understanding of expected repository performance if built and operated according to present plans. The TSPA forms an integral part of the license application that the DOE submitted to the Nuclear Regulatory Commission (NRC) in June 2008.

The consequences of magmatic disruption of the repository could be very significant in terms of the TSPA regarding radionuclide dispersion. An eruption beneath, into or through the repository could lead to wide dispersal of radionuclides via atmospheric, surficial (i.e. particulate sedimentation), fluvial and groundwater paths into the biosphere. Therefore it is important to evaluate the probability and consequences of potentially disruptive magmatic events. The consequences of disruption evidently depend on the characteristics of a magmatic event (e.g. eruptive style and volume, magma properties and dynamics, volatile content of magmas, etc.). A disruptive igneous event within the footprint of the repository has been estimated to have a rather low occurrence probability of 1 event in 70 Ma, with a 90% confidence interval ranging from 1 event in 20 Ma to 1 event in 180 Ma (Geomatrix Consultants, 1996). Independent of the DOE, the NRC has estimated volcanically-induced disruptive event probabilities in the range of 1 event in 10 Ma to 1 event in 100 Ma.

In this study we present a sketch of a conceptual model consistent with the eruption dynamics, petrology, and major and trace element geochemistry of small-volume alkali basaltic volcanism in the YM region. In particular, the generation, modification upon ascent, and eruption of alkali basaltic magma to form small-volume volcanic constructs of lava and tephra is considered in some detail. This style of eruption, in our opinion, is the type most relevant to possible future eruptions at or near YM. Quantitative evaluation of volcanic hazards including eruption forecasting ultimately requires integration of comprehensive quantitative models of petrogenesis (*sensu lato*) into the regional geological framework. The aim of this study is to present a birds-eye-view of small-eruptive volume alkali basalt petrogenesis relevant to evaluation of magmatic hazards at the proposed permanent nuclear waste repository at Yucca Mountain, Nevada, USA.

8.1 Volcanic hazard evaluation

Evaluation of volcanic hazards at YM is aided by understanding the fundamental nature of regional volcanism. Results of previous geologic investigations of YM and environs clearly indicate a proclivity for small-volume basalt eruptions, as opposed to, for example, large-volume, highly explosive silicic eruptions. However, evaluation of volcanic hazards at YM is complicated compared to traditional volcanic hazard studies that assess the possibility of future eruption at an existing volcano, because volcanism forecasting relevant to the

YMR involves an event at a specific location where volcanism has not occurred in the last 10 Ma. Volcanic hazard evaluation and prediction at YM therefore involves the coupled and difficult problems of predicting the timing, location, volume and eruptive style of possible eruptive events. Any single one of these issues can be complicated; together they represent a very challenging problem. In addition, high-resolution modeling of magmatic phenomena and eruption forecasting is itself complicated by the complex nature of dynamic processes in magmatic systems, which are non-linear at multiple spatial and temporal scales (Shaw, 1987), and the inherently stochastic distribution of heterogeneity at all scales in geologic media such as the upper crust beneath YM.

Information necessary for assessing the probability and consequences of a future disruptive magmatic event (volcanic eruption or intrusion by dike or sill at the repository depth) at YM includes the eruptive style(s) (e.g. pyroclastic flows or falls, lava flows, lahars, phreatomagmatic explosions, etc.), spatial and temporal distribution and number of previous events, as well as the distribution and geometric properties of the subsurface magma transport system (dikes, sills, hypabyssal intrusions, conduits). Assessing associated uncertainties is as important as defining mean values or average types of behavior. Constraints derive from, for example, careful study of analog volcanic provinces, including older, exhumed terrains, in the context of the regional geologic and geophysical setting, as well as application of dynamical, phase equilibria, and trace element models relevant to magma generation, transport, reaction, and eruption based on thermodynamics, fracture mechanics, fluid dynamics and geochemistry.

This chapter is organized as follows. In the next section, a brief summary of YM region volcanism is given. This provides a basis for inferring the most likely composition, volume, volatile content and eruptive style of volcanic events in the next ten thousand to one million years. Based on this determination, we consider stages in magma transport from source to surface. These stages include the phase equilibria and thermodynamics of partial melting; segregation; mobilization and ascent of magma through the lithosphere; and finally, near-surface flow of magma, driven mainly by volatile exsolution.

8.1.1 Volcanological and tectonic background

Over the past thirty years, many studies have addressed the age, geochemistry, petrology, volcanology and magma properties relevant to magmatic activity in the greater Yucca Mountain region. It is beyond the scope of this work to review this literature. A few recent studies that serve as links to earlier work include Smith *et al.* (1990), Fleck *et al.* (1996), Perry *et al.* (1998), Valentine *et al.* (1998), Perry and Youngs (2000) and Smith *et al.* (2002). The report of Detournay *et al.* (2003) summarizes the history of volcanism, describes magma thermodynamic and transport properties and addresses the likely characteristics of a future volcanic event at YM based on past volcanic activity, with a focus upon the past 5 Ma. The consequences of repository disruption by an igneous event are discussed in Detournay *et al.* (2003), Woods *et al.* (2002), Menand *et al.* (Chapter 17, this volume) and Valentine and Perry (Chapter 19, this volume). Crowe *et al.* (2006) present an overview of the volcanism problem, how it has been studied historically in the context of the YM project and its potential impact on an underground repository. The study of Fridrich *et al.* (1999) provides a thorough account of the tectonics, especially the evidence and timing (mainly Miocene) of extension in the YM region. Regional volcanism is summarized in Perry *et al.* (1998). Lathrop Wells, the youngest volcano in the region is described in detail by Perry and Straub (1996), Valentine and Perry (2007) and Valentine *et al.* (2006). We draw upon these studies in the summary below.

A regional map that highlights recent volcanism around YM is given in Figure 8.1. Yucca Mountain (*sensu stricto*) is composed of Miocene silicic volcanic rocks representing deposits from a series of large-volume eruptions associated with several large calderas north of the mountain. The most proximate one is the Timber Mountain caldera north of YM. Silicic pyroclastic eruptive activity began $\sim$ 15 Ma and ceased $\sim$ 8 Ma. Patterns of silicic ignimbrite-forming events in the Great Basin over the past 30 Ma indicate that further large-volume silicic pyroclastic flow and fall eruptions are not likely to recur in the YM region within the next few million years. Late Miocene–Quaternary basaltic volcanic activity succeeded Miocene silicic volcanism in the YM region. The basalts can be divided into two major episodes. Basalt of the silicic episode erupted during the waning stage of silicic volcanism ($>$ 8 Ma), whereas eruptions of post-caldera basalt began more recently and continue into the Quaternary. The post-caldera episode can be further subdivided into older post-caldera basalts outcropping north, northeast and southeast of YM (Figure 8.1), and younger post-caldera basalts, which crop out west, southwest and south of YM. The older post-caldera basalts range in age from $\sim$ 9–6 Ma. An apparent volcanic hiatus of about two and a half million years (from $\sim$ 7.2–4.7 Ma) separates the older and younger post-caldera basalts.

In order of decreasing age, the younger (Pliocene–Quaternary) basaltic volcanics (lava flows and tephra deposits) include the Pliocene basalts of Thirsty Mesa (4.7 Ma; volume, $V \sim 2.6\,\text{km}^3$) and southeast Crater Flat (3.7 Ma; $V \sim 0.6\,\text{km}^3$), the 3.1 Ma basaltic trachyandesites of Buckboard Mesa ($V \sim 0.8\,\text{km}^3$), the five Quaternary ($\sim$ 1 Ma) alkali basalt cones of Crater Flat of total volume $\sim 0.15\,\text{km}^3$, the 0.35 Ma basalt cones of Hidden Cone and Little Black Peak at Sleeping Butte ($V \sim 0.05\,\text{km}^3$), and the most recent eruption in the area, the 78 ka, $\sim 0.1\,\text{km}^3$ Lathrop Wells alkali basalt cone and lava field located $\sim$ 20 km south of the YMR footprint. Within the last million years, eruptions near YM have all been small volume; deposits include both tephra fallout and lava flows of alkali basalt (nepheline normative) composition.

In addition to exposed volcanic deposits, a number of buried basaltic lava flows or small intrusive bodies inferred from geomagnetic surveys lie beneath alluvial fan deposits in the shallow subsurface southwest (Crater Flat) and southeast (Jackass Flats) of YM (Figure 8.1). Drilling and geochronological, geochemical and petrographic examination of these bodies has been conducted at the following locations (refer to Figure 8.1): (i) the 11.3 Ma basalt of magnetic anomalies **Q** (**R** and **4**, probably related to **Q**) and **T** located northwest of the $\sim$ 1.1 Ma Red and Black cones in Crater Flat, east of the Bare Mountain fault at the western edge of the Amargosa trough, (ii) anomaly **A**, a 10.1 Ma basanite located due south of

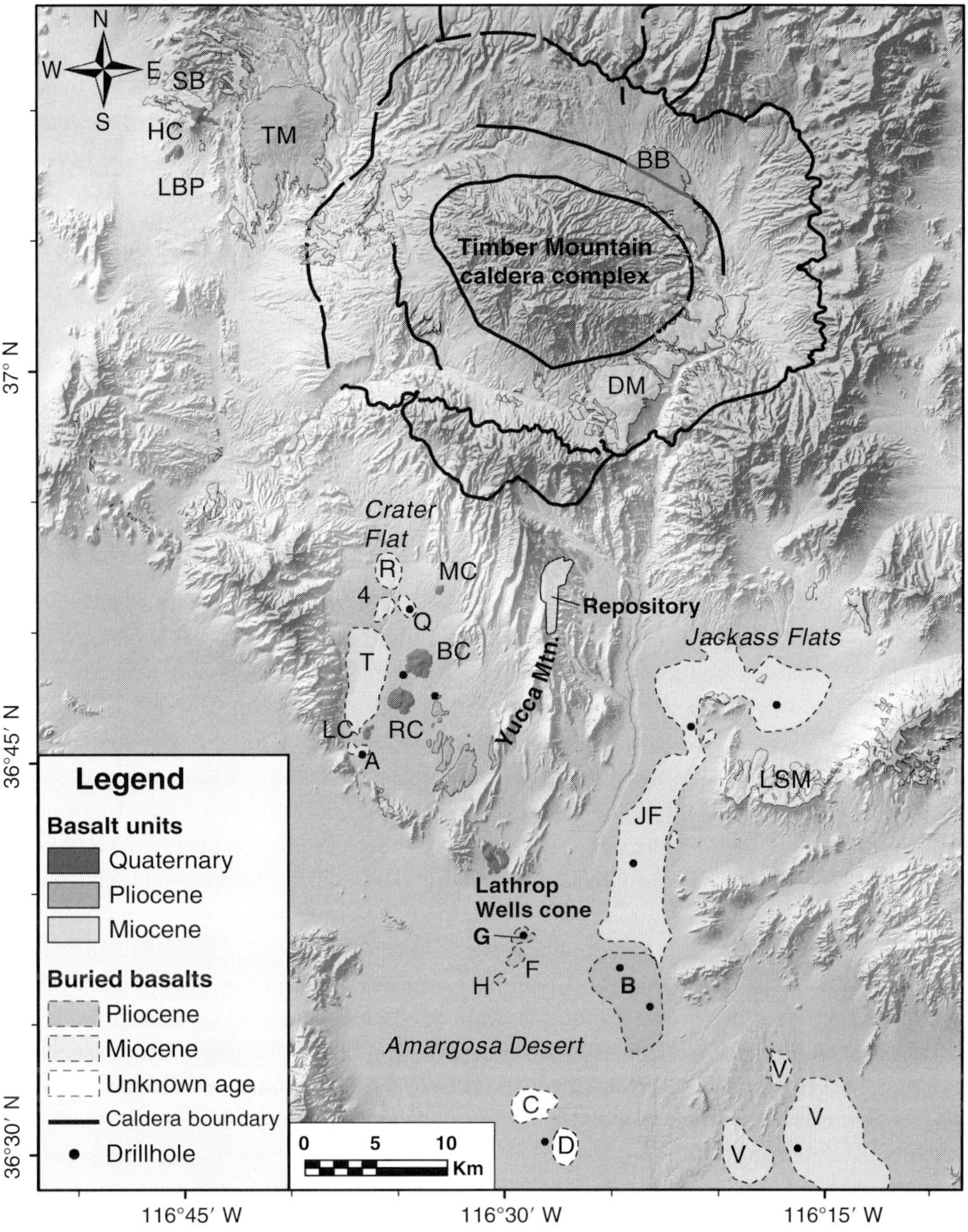

Fig. 8.1 Basaltic volcanoes in the YM region. Buried Pliocene and Miocene basalts (dashed outline) identified by magnetic anomalies, drilling and sample geochronology. Miocene caldera boundary is black line. Miocene volcanic rocks: Sleeping Buttes (SB), Dome Mountain (DM), Little Skull Mountain (LSM). Pliocene volcanic rocks: Thirsty Mesa (TM), Buckboard Mesa (BB). Quaternary volcanoes: (1.1 Ma): Makani (MC), Black Cone (BC), Red Cone (RC), Northeast and Southwest Little Cones (LC); (0.35 Ma): Little Black Peak (LBP), Hidden Cone (HC), 77 ka Lathrop Wells volcano. Magnetic anomalies: **A–D, F–H, Q, R, T, V, 4**. Map provided by Dr. Frank V. Perry of Los Alamos National Laboratory. See color plate section.

anomaly **T** and the Quaternary Little Cones, (iii) a large-area magnetic anomaly in Jackass Flats of age ~ 9.5 Ma mainly south and east of YM forming the eastern boundary of the Amargosa trough, and (iv) the buried basalts of anomalies **V** (9.5 Ma), **B** (3.8 Ma) and **G** (**F** and **H** probably related to **G**) dated at 3.8 Ma. Several additional magnetic anomalies, inferred to be buried volcanics based on their locations and magnetic signatures, have not been sampled (e.g. anomalies **C** and **D**). Although anomalies **C** and **D** have not been dated radiometrically, estimates based on depth of burial and alluvial fan sedimentation rates suggest an age 4.2–5.8 Ma.

Except for the basaltic andesite of Buckboard Mesa and the basalt of anomaly **B**, all younger post-caldera exposed basalts lie within a 10 km-wide northwest-trending zone west of YM in the Crater Flat–Amargosa trough corridor. This zone extends from the small-volume basalts of Sleeping Butte south through the Quaternary and Pliocene basalts of Crater Flat and continues south to Lathrop Wells and buried Pliocene basaltic deposits in the Amargosa Valley.

8.1.2 Expected event: composition, volume and style

Discrete events in the region in the last few million years are characterized by small-volume (0.01 km^3–1 km^3) alkali basaltic lava and tephra eruptions. Based on modern analogs, typical eruption durations are measured in terms of days, weeks or months and depend primarily on the total volume of magma erupted. Eruptive fluxes in the range 10^3–10^4 $kg\,m^{-2}\,s^{-1}$ lead to mass flows of order 10^5–10^7 $kg\,s^{-1}$ based on typical conduit cross-sectional areas observed in ancient analog systems (e.g. Detournay *et al.*, 2003). Eruptions at the high end of these ranges could produce Strombolian eruption plumes up to 10 km high; more typical eruptions would produce plume heights of order several kilometers. The volatile content, inferred from phase equilibria, the study of glass inclusions in phenocrysts and the order of low-pressure, near-surface crystallization of microphenocrysts, lies in the range 2.5−4 wt.%, with H_2O the dominant fluid species (Detournay *et al.*, 2003). The dissolved volatile content of rising magma plays a critical role in determining eruptive style. A dynamical transition occurs when the volume fraction of the fluid phase in magma exceeds the critical volume fraction (θ_{crit}) for magma fragmentation ~ 0.5−0.7. There is a rapid decrease in magma density and increase in magma (mixture) compressibility around this rheological transition. These magma property variations lead to rapid increases in magma eruptive velocity near the fragmentation depth. The depth at which magma fragmentation occurs depends upon the dissolved volatile content of magma and the dependence of volatile solubility on temperature, pressure and melt composition. It also depends on whether or not volatiles can leak from the magma into surrounding crustal rocks. These issues are quantified in Section 8.5.3. Although not discussed here, phreatomagmatic-style eruptions, in which magma encounters low-temperature, water-saturated permeable crust to generate streamrich violent eruptions, are possible in wetter climates than now observed at YM. Musgrove and Schrag (2006) have analyzed possible future climates, specifically wetter conditions associated with a higher water table, in the southern Great Basin. They noted that the most recent time in Earth history when CO_2 levels approached those anticipated in the next few hundred years was in the Eocene. Consideration of Eocene-like climate scenarios may therefore provide some lessons about possible climate changes due to increased (anthropogenic) CO_2. The design of the repository should account for the potential risk of significantly wetter conditions at YM that would enhance the chances for phreatomagmatic-style eruptions. In the remainder of this chapter, phreatomagmatic eruptions are not explicitly considered.

8.2 Magma generation and transport: a source to surface overview

Several topics related to the generation, segregation, ascent and eruption of basaltic magma, and the results in light of volcanic hazards at YM, are considered. We adopt as typical of future eruptive activity in the Yucca Mountain region the 78 ka, $\sim 0.1\,\mathrm{km}^3$ alkali basaltic lava and tephra eruption at Lathrop Wells, ≈ 20 km south of the proposed YMR. Although the generation and transport of magma is a continuum process, it is convenient to analyze successive stages from melt generation to eruption or shallow-level intrusion. The self-consistent, thermodynamically based pMELTS and MELTS phase-equilibria models of Ghiorso *et al.* (2002) are used to perform calculations that account for important sources of variability in liquid compositions and their physical properties. Extensive documentation of the phase-equilibria algorithms is presented elsewhere (e.g. Hirschmann *et al.*, 1998; Hirschmann *et al.*, 1999a, 1999b; Ghiorso *et al.*, 2002; Asimow and Longhi, 2004). Although no phase-equilibria model is perfect due to the thermodynamic complexities of multicomponent–multiphase silicate systems, the MELTS algorithm has been repeatedly shown to faithfully capture multicomponent phase relations in mafic–ultramafic systems. Petrogenesis from phase equilibria includes forward modeling of primary melt generation via partial melting of a peridotitic source and primary melt modification by fractional crystallization during upward ascent of primary melt. In particular, we compare melt compositions resulting from forward modeling of partial melting and fractional crystallization to the available volcanologic, geochemical and petrologic database for Lathrop Wells (Perry and Straub, 1996; Valentine *et al.*, 2006; Valentine and Perry, 2007). The phase-equilibria results link heat transfer between magma and surrounding lithosphere to the fracture mechanisms that drive magma ascent beneath YM, where the large-scale extensional stress environment of the Great Basin aids buoyancy-driven upward propagation of magma-filled crack networks. As magma ascends to the near surface, new dynamic processes become important. Once a magmatic mixture develops an appreciable volume fraction of fluid (bubbles), the behavior of the compressible dynamics of magma (melt plus fluid) becomes important. The dispersal of ash and lava from a small-volume alkali basalt volcano is very relevant to magmatic hazard analysis at YM.

The initial step leading to small-volume alkali basaltic eruption is partial melting of peridotite in the upper mantle. Factors that govern the composition of primary melt include the extent to which partial melt stays in chemical potential equilibrium with the crystalline residue, the composition of source peridotite (i.e. mineral compositions and abundances, water content and redox state) and the mean pressure of partial melting. The effects of each of these parameters are tested by performing approximately 80 partial melting simulations, systematically varying the governing parameters. Below, results for representative cases to illustrate parameter sensitivity are presented. The melting scenario that produces primary partial melt is analyzed in detail. This melt, upon further evolution by fractional crystallization, exhibits a composition corresponding most closely to Lathrop Wells basalt. Liquid ferric iron to ferrous iron ratios are based on the oxygen fugacity constraint along the QFM-1 buffer (i.e. one log base-ten unit below the Quartz–Fayalite–Magnetite buffer). A number of sensitivity tests are performed to explore the effects of varying oxygen fugacity. These effects are found to be minor compared to variations of other intensive parameters such as pressure or source fertility.

8.2.1 Source bulk compositions

Table 8.1 gives the anhydrous major element compositions of three end-member ultramafic compositions used in the phase-equilibria simulations. Composition **1B-33** (Bergman, 1982) corresponds to a depleted harzburgite xenolith from a 38 ± 10 ka lava flow within the Lunar Crater volcanic field, Nevada (Yogodzinski, 1996), located several hundred kilometers northeast of YM. Composition **1B-33** represents the most depleted of the compositions used in this study. Composition **PA-12** (Frey and Prinz, 1978) is a moderately fertile peridotite xenolith from the Pleistocene basaltic vent at Peridot Mesa in the San Carlos volcanic field, Arizona, USA. The most fertile peridotite composition used in this study, **KLB-1** (Hirose and Kushiro, 1993), is a garnet peridotite xenolith from Kilbourne Hole, New Mexico, USA (Padovani and Reid, 1989). The Kilbourne Hole maar, a phreatomagmatic explosion crater, formed when the Afton basalt intermingled with wet rift-fill sediments in the Camp Rice Formation of the Santa Fe Group. The age of the phreatomagmatic explosion and basalt eruption responsible for creating the Kilbourne Hole maar is 77 ka (Anthony and Poths, 1992). The compositions are peridotitic, but differ in terms of fertility. Fertility refers to the potential of a peridotite to generate basaltic liquid (melt) by partial melting. There is no strict definition of fertility; several measures of fertility are given in Table 8.2. In particular, the mass ratio of the phases amphibole + phlogopite + clinopyroxene + garnet relative to olivine + orthopyroxene at the solidus pressure; modal clinopyroxene; the mass ratio of FeO/MgO in the peridotite source; and relatively high values for Al_2O_3, TiO_2 and CaO all positively correlate with peridotite fertility. Fertility increases from **1B-33** (depleted) to **PA-12** (moderately fertile) to **KLB-1** (fertile). To each anhydrous bulk composition in Table 8.1, 0.2 wt.% H_2O has been added. At 1.5 GPa amphibole (Amp) and phlogopite (Phl) are present in the fertile compositions at the solidus. Garnet (Gt) and phlogopitic mica are present at 3.5 GPa.

8.2.2 Fractional or batch melting?

There are two end-member models relevant to partial melting. In batch partial melting, melt remains in chemical potential equilibrium with crystalline residue during progressive

Table 8.1. Anhydrous composition of peridotites

Wt.%	1B-33	PA-12	KLB-1
SiO_2	43.96	43.66	44.59
TiO_2	0.03	0.02	0.16
Al_2O_3	1.85	3.13	3.60
Cr_2O_3	na	0.30	0.31
FeO	7.48	7.98	8.12
MnO	0.15	0.15	0.12
MgO	45.42	42.89	39.32
CaO	1.02	1.41	3.45
Na_2O	0.05	0.31	0.30
K_2O	0.03	0.12	0.03
P_2O_5	na	0.02	na

Table 8.2. Mineralogy and fertility measures for peridotite starting compositions

	Mass fraction[1] Amp + Phl + Cpx + Gt/Ol + Sp + Opx				
Sample	1.5 GPa	3.5 GPa	Wt.% alkalies	Wt.% FeO/MgO	Wt.% MgO
1B-33	0.036	0.067	0.08	0.16	45.4
Lunar Crater	Ol[2],Opx,Gt,Cpx	Ol,Opx,Gt,Cpx,Phl			
PA-12	0.083	0.152	0.43	0.19	42.9
San Carlos	Ol,Opx,Cpx,Amp,Sp,Phl	Ol,Opx,Gt,Cpx,Phl,Sp			
KLB-1	0.190	0.325	0.33	0.21	39.3
Kilbourne Hole	Ol,Opx,Cpx,Amp,Sp,Phl	Ol,Opx,Cpx,Gt,Sp,Phl			

[1] Phase abundances are at solidus temperature at indicated pressure with 0.2 wt.% H_2O added to the anhydrous compositions of Table 8.1. The redox condition is QFM-1 in all cases.
[2] Order of phase listing is from most to least modal abundance at the prescribed pressure.

fusion until melt is segregated from source crystals at some fixed value of the extent of melting, either 5%, 10% or 15% melting (by mass) in this study. In fractional partial melting, melt is isolated from crystalline residue immediately and completely upon generation; no further reaction occurs between melt and residual crystals. In reality, the mode of partial melting in the Earth lies between these limits. We compared results of fractional and batch partial melting for all three starting compositions. Figure 8.2 shows MgO variation diagrams for the compositional sequence of partial melts generated by fractional and batch partial melting for starting composition **KLB-1** (fertile peridotite; Table 8.1) plus 0.2 wt.% H_2O.

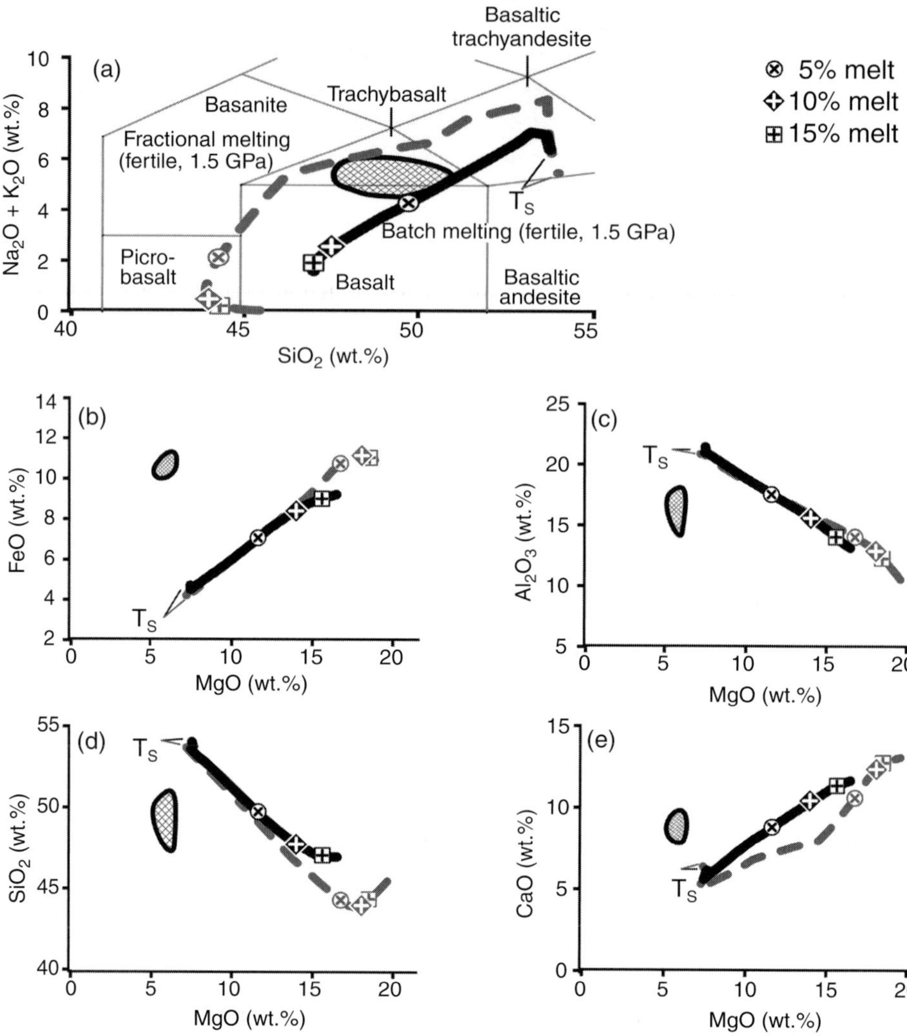

Fig. 8.2 Comparison of isobaric fractional (gray dashed line) and batch (solid black line) partial melting phase-equilibria calculations using the pMELTS algorithm for fertile peridotite bulk composition **KLB-1** (Hirose and Kushiro, 1993). Oxygen fugacity is fixed along the QFM-1 buffer. Increasing wt.% MgO and decreasing wt.% ($Na_2O + K_2O$) correspond to increasing magma temperature. The beginning of each pMELTS trend is labeled T_S. Cross marks within circle, diamond and square outlines, signify 5%, 10% and 15% melting, respectively. The termination of each pMELTS trend corresponds to a melt fraction of 20%. The shaded region represents the field of Lathrop Wells data (Perry and Straub, 1996; Valentine and Perry, 2007). The data field is not expected to coincide with the pMELTS trends due to the non-primary nature of Lathrop Wells basalt (see text for further details). Plots of major element variations versus MgO (a–e) document the trajectory of melt compositions during batch and fractional melting of fertile peridotite. The total alkalies–silica diagram is based on the classification scheme of Le Maitre *et al.* (1989).

Oxygen fugacity is constrained at QFM-1 and the pressure of isobaric melting is 1.5 GPa, corresponding to a depth ~ 50 km beneath YM. The solidus is marked T_s and points are labeled on the liquid curves representing 5%, 10% and 15% partial melting by mass. The melt composition at the thermodynamic solidus where the "trace" of melt is present, is, by definition, identical for batch and fractional melting. In Section 8.3 we present fractional crystallization calculation results indicating that batch partial melting better represents mantle partial melting than fractional partial melting. In short, fractional crystallization models based on melts derived from fractional partial melting produce liquids that deviate widely from observed compositions at Lathrop Wells, whereas batch partial melting generates liquids that, on undergoing fractional crystallization, coincide quite well with observed Lathrop Wells major and trace element data. Based on trace element arguments, it is generally agreed that the degree of melting (by mass) of a peridotite source to produce basalt is in the range 5–10%. All results reported on hereafter refer to batch partial melting with extent of melting between 0–15%.

8.2.3 Role of peridotite fertility

We investigate the importance of peridotite fertility via partial melting computations based on the three peridotites of Table 8.1; 0.2 wt.% H_2O has been added to each anhydrous bulk composition. Table 8.2 gives the solidus phase assemblage for each bulk composition, calculated by Gibbs energy minimization at 1.5 and 3.5 GPa. Small amounts of modal amphibole and phlogopite (hydrous phases) are present at the 1.5 GPa solidus for the more fertile compositions **PA-12** and **KLB-1**. Figure 8.3a–e depicts the composition of melts generated by batch partial melting at 1.5 and 3.5 GPa on MgO variation diagrams. The symbols along each curve indicate 5%, 10% and 15% melting by mass. The composition of melt at the solidus is marked T_s. The Lathrop Wells basalt compositional field, shown as a shaded field, is for reference only. Since Lathrop Wells basalt is not primary melt (its Mg # is far too low), calculated partial melt compositions are not expected to cross the Lathrop Wells field. However, because we expect that primary melt evolves to Lathrop Wells melt by fractional crystallization, the starting composition along the batch partial melting curve must lie at a location in composition space such that fractional crystallization of the primary melt drives the liquid composition into the Lathrop Wells field. We demonstrate in Section 8.3.1 that a fertile (**KLB-1**) or moderately fertile (**PA-12**) source undergoing partial melting at relatively low pressure (1.5–2 GPa) does indeed provide a composition, which upon subsequent fractional crystallization during ascent evolves into the Lathrop Wells compositional field. This is especially clear on the CaO and total alkalies MgO variation diagrams. The batch partial melting generated by 5%, 10% or even 15% partial melting of a depleted source cannot evolve by fractional crystallization into the Lathrop Wells field. In addition, batch partial melting at 3.5 GPa generates liquids that do not, upon subsequent crystal fractionation, evolve into the Lathrop Wells compositional field. We conclude that the mantle source for Lathrop Wells basalt was fertile or moderately fertile and that the mean pressure of partial melting was significantly closer to 1.5 GPa (~ 50 km) than 3.5 GPa (110 km). The effects of melting pressure are examined in more detail below.

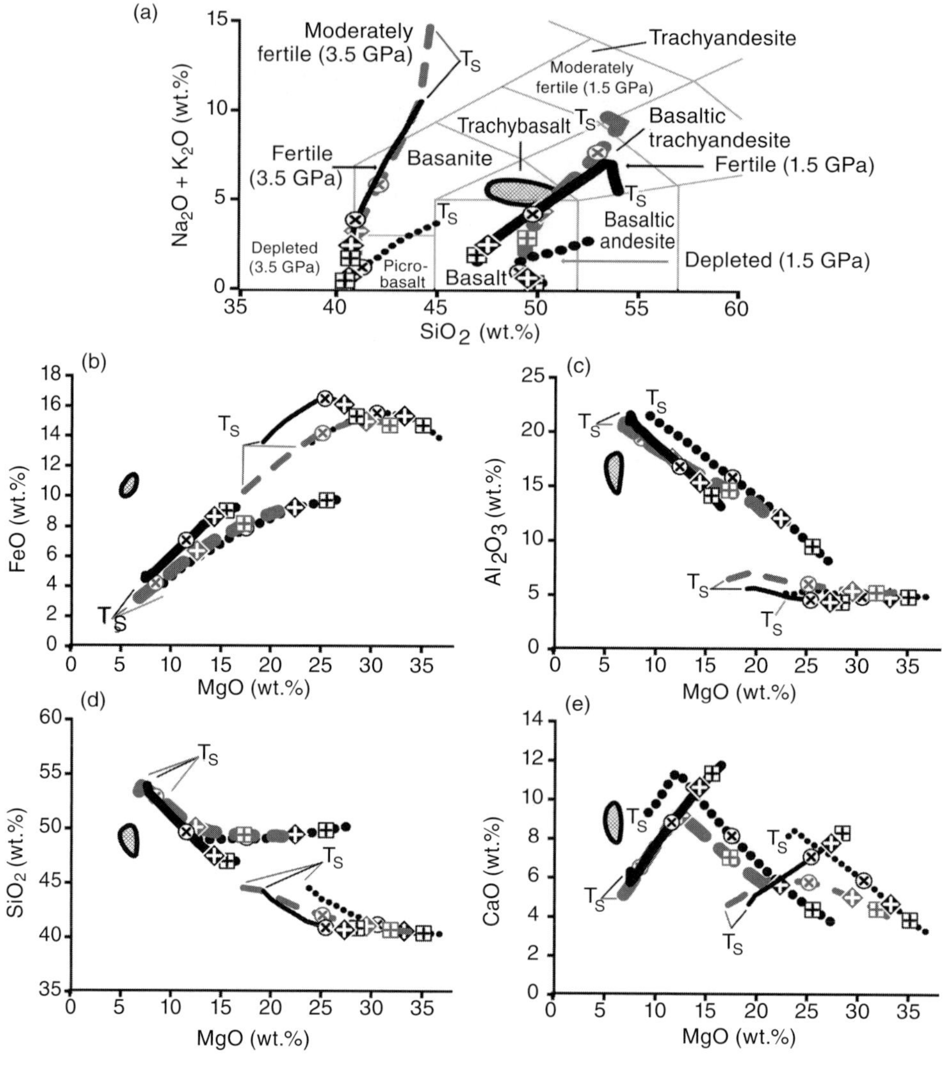

Fig. 8.3 Results of isobaric pMELTS batch partial melting calculations at 1.5 GPa (large symbols) and 3.5 GPa (small symbols) for depleted, moderately fertile and fertile peridotite. Compositions are given in Table 8.2; dotted line: depleted peridotite **1B-33** (Bergman, 1982); dashed gray line: moderately fertile peridotite **PA-12** (Frey and Prinz, 1978); solid black line: fertile peridotite **KLB-1** (Hirose and Kushiro, 1993). All other parameters and abbreviations are identical to Figure 8.2.

8.2.4 Role of pressure on composition of partial melting

The role of pressure in modifying the composition of peridotite batch partial melts has been studied in detail. The starting composition is fertile **KLB-1** plus 0.2 wt.% H_2O. The redox condition is fixed along the QFM-1 buffer. The MgO variation diagrams showing

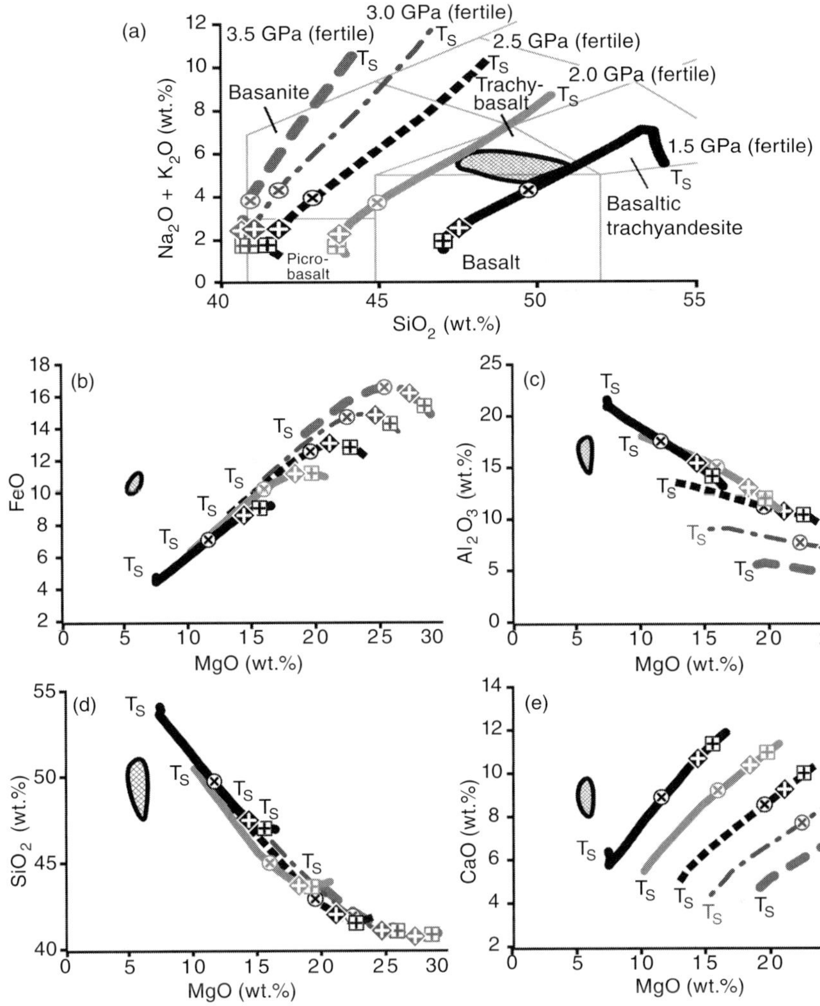

Fig. 8.4 pMELTS isobaric batch partial melting trends based on fertile peridotite **KLB-1** (Hirose and Kushiro, 1993) for varying pressures; solid black line: 1.5 GPa; solid gray line: 2.0 GPa; dotted black line: 2.5 GPa; dot-dashed line: 3.0 GPa; dashed gray line: 3.5 GPa. All other parameters and abbreviations are identical to Figure 8.2. Isobaric partial melting at pressures greater than 2 GPa generate melts that do not evolve into the Lathrop Wells field upon fractional crystallization. Melts generated by partial melting for pressures in the range of 1.5–2.0 GPa evolve into the Lathrop Wells field upon subsequent fractional crystallization.

the compositional sequence of batch partial melts starting at the solidus are shown in Figure 8.4 for pressures of 1.5, 2.0, 2.5, 3.0 and 3.5 GPa. These pressures span a depth range ∼ 50–110 km beneath YM. The effect of increasing pressure at fixed melt fraction (e.g. 5 wt.%) is to increase MgO and FeO and reduce the Al_2O_3 content of partial melt.

An important result of these calculations is that high-pressure partial melting can be ruled out. For example, the FeO vs. MgO plot (Figure 8.4b) shows that partial melts generated at pressures $\leq$ 2 GPa are less FeO rich than high-pressure melts at the same extent of melting. Because fractional crystallization of basaltic melt drives derivative liquids to higher FeO, it is clear that fractional crystallization of a high-pressure partial melt cannot drive a primary melt into the Lathrop Wells field. The conclusion is that the pressure of partial melting for generation of the primary Lathrop Wells basalt liquid is in the range 1.5–2 GPa ($\sim$ 52 – 67 km beneath Yucca Mountain).

8.2.5 Role of H_2O content in source peridotite

A series of hydrous melting calculations based on a variety of source H_2O contents was performed to explore the effect of H_2O on melting of fertile peridotite **KLB-1**. Figure 8.5 shows the calculated compositional paths for batch melting at 1.5 GPa, with oxygen fugacity at QFM-1. The plotted results correspond to four distinct source water contents: dry, 0.1 wt.% H_2O, 0.2 wt.% H_2O and 0.6 wt.% H_2O. The symbol boxes along the melt lines refer to 5%, 10% and 15% melting. The solidus is marked by T_S. An initial H_2O concentration of 0.6 wt.% exceeds the storage capacity of melt and hydrous minerals (amphibole and phlogopite) and leads to a hydrous fluid phase at the solidus, a somewhat unlikely condition in general given inferences of the average water content of typical mantle peridotites (Asimow and Langmuir, 2003; Carlson and Miller, 2003). For a fixed fraction of partial melt (e.g. 5%) the alkali–silica diagram (Figure 8.5a) shows that partial melts become increasingly silica-rich (from $\sim$ 46–53 wt.% SiO_2) with increasing initial H_2O content, although the total alkali content of the melt remains constant around 4 wt.%. The criterion that pMELTS liquid H_2O concentration predictions at the end of fractional crystallization must coincide with inferred pre-eruptive Lathrop Wells H_2O concentrations of 2.5–4 wt.% (Detournay *et al.*, 2003) leads to the conclusion that source water contents lie in the range 0.1–0.3 wt.%. We have chosen to present pMELTS results based on a representative starting H_2O concentration of 0.2 wt.%. Dry partial melts are too alumina and silica poor and too FeO rich to evolve by fractional crystallization into the Lathrop Wells field.

8.2.6 Summary of partial melting conditions

More than 100 phase-equilibria simulations were conducted using the pMELTS algorithm of Ghiorso *et al.* (2002) in which systematic variation of several critical factors (melting process, source fertility, pressure of melting and source H_2O content) was studied. Evaluation of the quality of results is based on the criterion that fractional crystallization of primary partial melts of a fertile peridotitic source should give rise to a predicted melt composition similar to that observed for Lathrop Wells basalt. The most plausible melting scenario is a 5–10% batch melting of a fertile peridotite (olivine, orthopyroxene, clinopyroxene, spinel, amphibole and phlogopite) with an initial water content of $\approx$ 0.2 wt.% H_2O along the QFM-1 buffer curve at 1.5–2 GPa ($\sim$ 50–70 km depth). At either 5% or 10% melting,

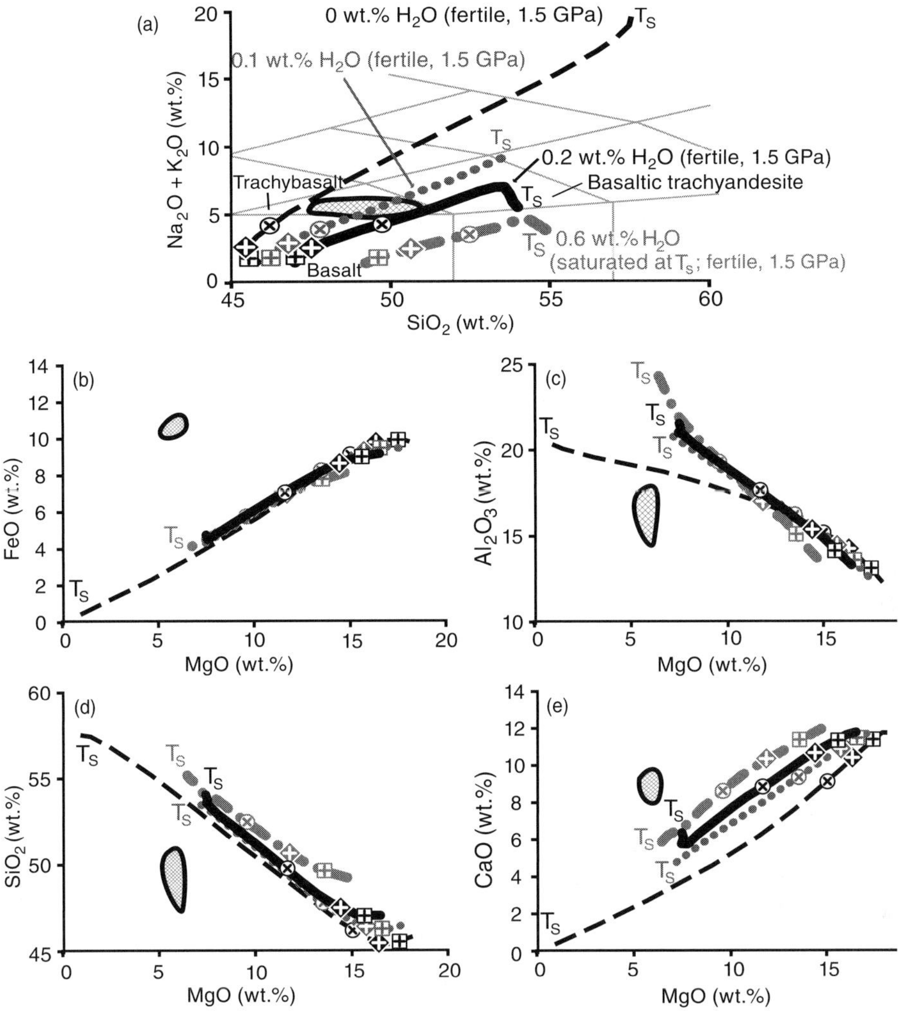

Fig. 8.5 Results of numerical experiments designed to examine batch partial melting of pMELTS liquid compositions with varying initial bulk water contents. The source composition is fertile peridotite **KLB-1** (Hirose and Kushiro, 1993); dashed black line: anhydrous; dotted gray line: 0.1 wt.% H_2O; solid black line: 0.2 wt.% H_2O; dashed gray line: 0.6 wt.% H_2O. All other parameters and abbreviations are identical to Figure 8.2. Hereafter, the 0.2 wt.% H_2O case is used as the reference case.

the residual source assemblage is olivine, orthopyroxene, clinopyroxene and spinel. The temperatures and dissolved H_2O contents of primary melt generated at 5% and 10% melting are 1280 °C and 1350 °C, and 3.3 wt.% H_2O and 1.74 wt.% H_2O, respectively at 1.5 GPa. The Mg # ($\equiv$ atomic Mg/(Mg + Fe^{+2})) of the 5% and 10% partial melts are 76 and 77, respectively. The mineral proportion diagram for batch partial melting of fertile peridotite

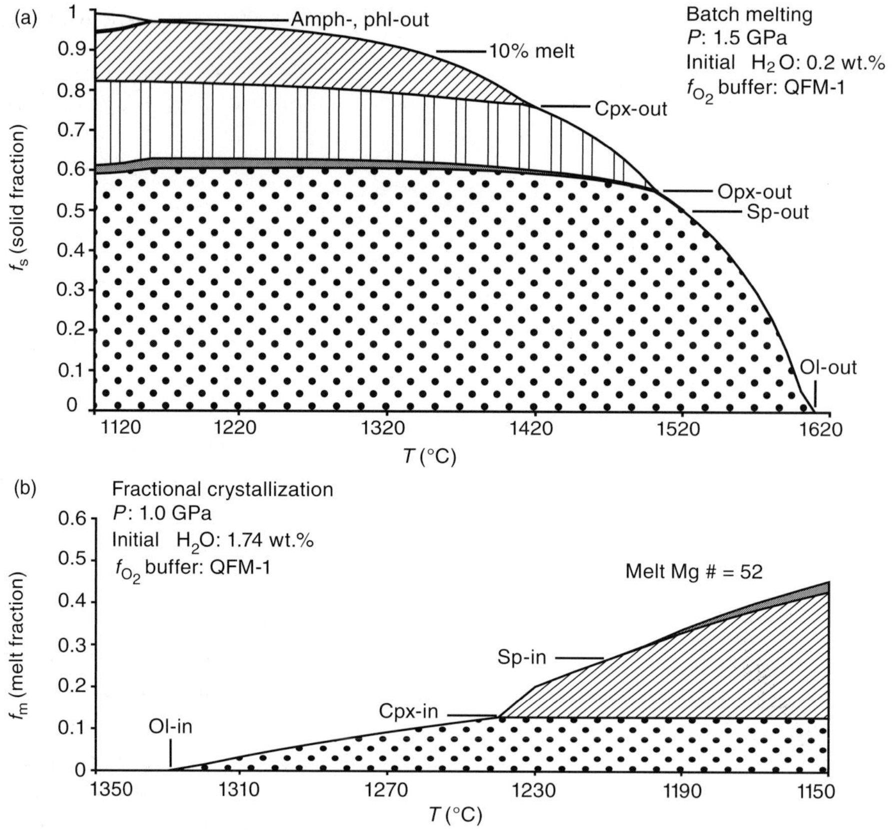

Fig. 8.6 Phase proportions as a function of temperature (T) based on (a) batch partial melting of fertile peridotite **KLB-1** (Hirose and Kushiro, 1993) plus 0.2 wt.% H_2O at 1.5 GPa; the 10% partial melt has a dissolved H_2O content of 1.74% at ~ 1310 °C at 1.5 GPa and is used as the starting composition for the fractional crystallization simulation; (b) mineral proportion diagram for isobaric (1 GPa) fractional crystallization starting with liquid generated at 10% batch melting (Figure 8.6a). Crystallizing phases are olivine (Ol), clinopyroxene (Cpx) and spinel (Sp), in the order given. Fractionation is stopped when Mg # = 52, the observed Mg # for Lathrop Wells basalt (abbr., amphibole (Amp), orthopyroxene (Opx), phlogopite (Phl)).

KLB-1 is shown in Figure 8.6a. The solidus at 1.5 GPa is 1120 °C; 10% batch melting is reached at ~ 1310 °C. The residual phase assemblage is olivine, orthopyroxene, spinel and clinopyroxene. All modal amphibole and phlogopite are consumed during partial melting within ≈ 30 °C of the solidus.

8.3 Fractional crystallization during ascent

The Mg # of Lathrop Wells basalt is 54 ± 2 (Perry and Straub, 1996). Given that primary liquid generated by partial melting of peridotitic sources has Mg # ~ 75–78, we conclude that Lathrop Wells basalt is not primary melt. Assuming that Lathrop Wells basaltic melt originated by peridotite partial melting followed by fractional crystallization during ascent, the mean pressure of fractional crystallization can be estimated based on phase-equilibria constraints. The primary melt products of peridotite batch partial melting are used as input compositions for calculating fractional crystallization compositional paths using the MELTS and pMELTS algorithms (Ghiorso and Sack, 1995; Ghiorso *et al.*, 2002). Fractional crystallization calculations are ended when the Mg # of the computed melt is 54 ± 2, the range observed for Lathrop Wells basalt. These calculations were performed where the source fertility, mean pressure (depth) of fractional crystallization and fraction of primary partial melt (and, therefore, primary melt composition) varied systematically. Numerous fractional crystallization calculations were based on primary melt fractions of 5%, 10% and 15%, but only the fractional crystallization results based on primary melt fractions of 10% are included here, as these coincide most closely with observed Lathrop Wells major element data (Figure 8.6b). In each fractional crystallization calculation, the initial water content corresponds to that of the primary melt product of batch partial melting. The quality of the results is judged by (i) comparison of calculated melt composition oxides SiO_2, Al_2O_3, FeO, CaO, K_2O and Na_2O with Lathrop Wells basalt at a computed melt Mg # of 54 and (ii) comparison of the computed H_2O content of derivative melt (Mg # = 54) with the inferred water content of Lathrop Wells eruptive basalts of 2.5–4 wt.% H_2O (Detournay *et al.*, 2003). In a later section, independent methods are used based on Lathrop Wells trace element data to test the validity of the phase-equilibria constraints. In particular, a mean pressure of fractional crystallization $\sim$ 1 GPa (depth $\sim$ 36 km) best fits observed trace element abundances. This depth lies in the shallow upper mantle beneath YM.

8.3.1 Mean pressure and implied phase relations

Magnesium oxide variation diagrams are shown in Figure 8.7 for isobaric fractional crystallization of primary melts formed by 10% partial melting of fertile peridotite **KLB-1** (Table 8.1) at 3.5 GPa and 1.5 GPa. A sequence of fractional crystallization calculations were performed at successively higher pressures with a maximum pressure of 3.5 GPa. Predicted major element concentrations do not compare well with those measured on Lathrop Wells basalts (Perry and Straub, 1996; Valentine and Perry, 2007) for fractional crystallization at pressures above about 2 GPa. Isobaric fractional crystallization paths at both 1 GPa and 0.5 GPa are shown. The major feature is that fractional crystallization of primary partial melts formed at high pressure generates evolved melts that are too rich in FeO, CaO and the alkalies, and too depleted in silica and alumina to match Lathrop Wells basalt. In contrast, partial melts generated at low pressure (1.5 GPa) evolve into the Lathrop Wells field during fractional crystallization at pressures in the range of 0.5–1.0 GPa. Liquids resulting from fractional crystallization at 1 GPa of primary melt generated at 1.5 GPa better coincide with Lathrop Wells compositions than do those resulting from lower pressure (0.5 GPa) fractional crystallization of the same primary melt. The fractionating phases at 1 GPa are olivine, clinopyroxene and spinel. At 0.5 GPa, the fractionating phases are

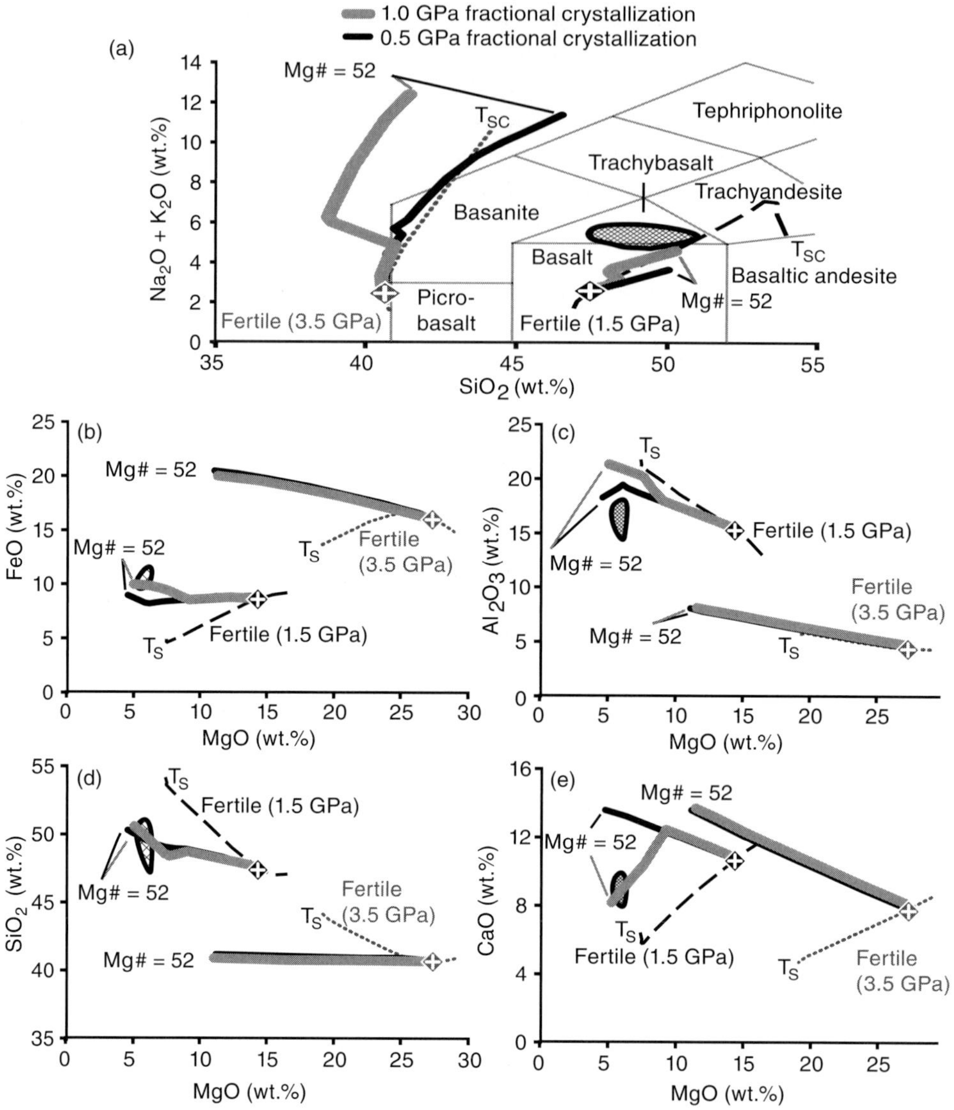

Fig. 8.7 Major element MgO variation diagrams showing Lathrop Wells data (shaded field; Perry and Straub, 1996) and pMELTS calculation results. Predicted trends represent evolution of a 10% **KLB-1** (Hirose and Kushiro, 1993) primary melt formed by batch partial melting at 1.5 GPa (dashed black line) and 3.5 GPa (dotted gray line) by isobaric, closed-system fractional crystallization at 1.0 GPa (solid gray line) and 0.5 GPa (solid black line). Thinner lines show the results of partial melting from Figure 8.4. A cross marks the beginning of each fractional crystallization trend. Decreasing wt.% MgO and increasing wt.% ($Na_2O + K_2O$) correspond to decreasing magma temperature. The termination of each fractional crystallization trend is labeled (Mg # = 52). See text for details. Note that for all oxides except alumina, the termination of the computed fractionation path lies in or very near the region of the Lathrop Wells composition for fractionation at 1 GPa of the 10% partial melt of fertile peridotite generated at 1.5 GPa.

olivine, plagioclase and trace clinopyroxene. The contrasting roles of plagioclase (low-pressure) versus clinopyroxene (high-pressure) fractionation account for the CaO–MgO variation diagram differences in Figure 8.7. In each case, the initial dissolved water content of the primary melt is 1.75 wt.% H_2O at the start of fractional crystallization. The dissolved water content of liquid at the conclusion of fractional crystallization (melt Mg # $\approx$ 52–54) is 3.2 wt.% H_2O at 1.0 GPa and 2.6 wt.% H_2O at 0.5 GPa, within the 2.5–4.0 wt.% H_2O range of Lathrop Wells basalt pre-eruptive H_2O contents estimated by Detournay *et al.* (2003). If this exercise is repeated using 5% partial melt as the starting liquid for otherwise identical conditions, the liquid line of descent departs significantly from the Lathrop Wells field at both 1 GPa and 0.5 GPa. This is because the crystallizing phases of the 5% partial melt are dominated by olivine and amphibole. Amphibole is not found as a phenocryst in Lathrup Wells basalts.

8.3.2 Summary of solidification conditions

The composition of the Lathrop Wells basalts has been used to define an approximate set of compositional and geophysical parameters relevant to partial melting and fractional crystallization based on phase-equilibria modeling. The representative model suggests batch partial melting at $\sim$ 50−70 km of fertile peridotite containing $\approx$ 0.2 wt.% of H_2O, with redox conditions near the QFM-1 buffer. The extent of melting is in the range 7−10%, perhaps closer to 10% by mass. Magma ascends and undergoes fractional crystallization at a mean pressure of $\approx$ 1 GPa (depth equivalent to 35−40 km) in the upper mantle. Because the extent of partial melting is about 10% and about half the primary melt freezes at depth, the volume of contributing source zone mantle can be estimated. The eruptive volume (dense rock equivalent of lava plus tephra) of Lathrop Wells is approximately 0.12 km^3. Allowing for the effects of distal ash dispersal and the effects of solidification during ascent ($\sim$ 50% by mass), the total volume of melt generated by partial melting in the source is liberally estimated to be $\sim$ 0.5 km^3. If the extent of partial melting was about 10 wt.%, then the mantle melting volume was $\sim$ 5 km^3. This suggests a small "melting" footprint compared to, for instance, the map distance between the southernmost and northernmost basaltic cones of the Quaternary in Crater Flats (Figure 8.1) of $\sim$ 15 km. For example, if the melting region is assumed spherical, then the radius of the melting region is $\sim$ 1 km.

8.4 Trace element geochemistry: a test of the phase equilibria model

It is important to test independently the conclusions of phase-equilibria modeling. One test is provided by comparing trace element concentrations in Lathrop Wells basalts with predictions based on trace element modeling using the phase proportions and compositions taken from the phase equilibria model. Once the trace element abundances in the bulk source and mineral-melt partition coefficients are fixed, then the consequences of partial melting and fractional crystallization can be forward modeled by numerical solution to the equations for batch partial melting and subsequent fractional crystallization. The trace element models

Table 8.3. Initial trace element concentrations (ppm) used for trace element modeling of combined batch partial melting of **KLB-1** fertile peridotite and subsequent fractional crystallization

Element	(ppm)	Element	(ppm)	Element	(ppm)
Rb	1.90	Nd	2.67	Lu	0.043
Ba	33	Sm	0.47	Co	112
Th	0.71	Zr	21	Cr	2690
Nb	4.8	Hf	0.27	Ni	2160
Ta	0.4	Eu	0.16	Sc	12.2
La	2.60	Tb	0.07	V	56
Ce	6.29	Y	4.4	Zn	65
Sr	49	Yb	0.26		

Results presented in Figure 8.8; trace element concentrations from McDonough (1990).

are based upon the phase assemblages computed from MELTS self-consistently (Spera *et al.*, 2007). Source trace element abundances used in the calculations are collected in Table 8.3. The values of the mineral-melt trace element partition coefficients are given in Table 8.4. These data together with phase assemblages from the phase-equilibria modeling were used to compute the trace element contents of partial melts by batch melting followed by fractional crystallization. The absolute trace element abundance pattern for Lathrop Wells basalts is portrayed as a band in Figure 8.8. The starting melt composition is the point marked T_l. The cross marks the spot where the melt has a Mg # of 52. Recall that Lathrop Wells basalts have Mg # in the range 52–56. Twenty-three trace element absolute abundances have been calculated including representatives from the transition metals (e.g. Ni, Co, Cr, V), rare earth elements (La through Lu), large-ion lithophile elements (e.g. Rb, Ba, Sr, Pb) and the high-field-strength elements (e.g. Nb, Ta, Zr, Hf) among others. Overall the computed results agree well with abundances observed for Lathrop Wells basalts (Perry and Straub, 1996; Valentine and Perry, 2007). We conclude that trace element modeling based upon the major element phase equilibria results are mutually consistent.

8.5 Ascent of magma: overview

One can distinguish several dynamical regimes relevant to the segregation and upward transport of magma associated with small-volume alkali basalt lava/tephra eruptions of the Lathrop Wells type. In the source region at depths of 50–60 km, percolative flow (i.e. Darcy flow) of melt in response to small pressure gradients represents the first stage in the segregation of melt from residual crystals into networks of contiguous volumetric domains. If the melt generation rate exceeds the rate at which ductile flow can accommodate the increased volume associated with melting, magma pressure will rise above the prevailing

Table 8.4. Solid-melt partition coefficients (K_{sm}) used in partial melting and fractional crystallization trace element modeling

	K_{sm}					
Element	Phl	Cpx	Amp	Ol	Opx	Sp
Rb	1.9	0.001	0.02	0.0002	0.0006	0.15
Ba	1.1	0.00011	0.1	0.0000022	0.0000036	0.028
Th	0.12	0.04	0.11	0.03	0.013	0.1
Nb	0.088	0.05	0.8	0.11	0.15	0.4
Ta	0.56	0.261	0.62	0.17	0.11	2.0
La	0.028	0.002	0.045	0.0004	0.0003	0.0029
Ce	0.03	0.017	0.09	0.0001	0.0008	0.01
Sr	0.08	0.04	0.022	0.00001	0.000511	0.077
Nd	0.0255	0.065	0.2	0.0003	0.0050	0.01
Sm	0.03	0.09	0.033	0.00014	0.0023	0.0072
Zr	2.5	0.001	0.15	0.0035	0.02	0.02
Hf	0.146	0.004	0.19	0.001	0.0063	0.14
Eu	0.03	0.09	0.3	0.001	0.0033	0.01
Tb	0.7	0.28	0.32	0.0015	0.019	0.01
Y	0.018	0.467	0.4	0.009	0.18	0.0039
Yb	0.033	0.48	0.46	0.014	0.11	1.5
Lu	0.0494	0.28	0.4	0.018	0.17	0.32
Co	23.0	1.2	13.0	2.0	3.0	8.0
Cr	5.0	5.0	30.0	1.5	10.0	11.0
Ni	1.3	1.5	1.0	6.0	1.1	29.0
Sc	8.3	2.0	2.18	0.08	1.5	0.67
V	0.5	1.81	1.49	0.06	0.6	0.5
Zn	7.0	0.49	0.69	0.86	0.41	2.6

Results shown in Figure 8.8. Data from GERM (2008), Henderson (1982), Best and Christiansen (2001). Phl (trioctahedral phyllosilicate), Cpx (clinopyroxene), Amp (amphibole), Ol (olivine), Opx (orthopyroxene), Sp (spinel).

mean normal stress, $\sigma_n(\equiv (\sigma_1 + \sigma_2 + \sigma_3)/3)$ and crack propagation will be favored. The minimum principal stress, σ_3, is horizontal to sub-horizontal for regional transtensional stress states, and therefore vertical to sub-vertical magma-pressured cracks could open. Melt slightly above the mean normal pressure can then segregate from crystalline residue and begin ascent under the combined forces of buoyancy and small ($\sim$ few MPa) principal stress differences. Swarms of melt-filled cracks propagating upwards transport melt at rates depending on factors such as the fracture width, fracture resistance, the driving pressure gradient and melt viscosity. It is during this stage that fractional crystallization, probably along crack margins, occurs as melt fractures rise through cooler lithospheric mantle. Rates of ascent in this stage for alkali basalts are in the range cm s^{-1} to m s^{-1} (Spera, 1986).

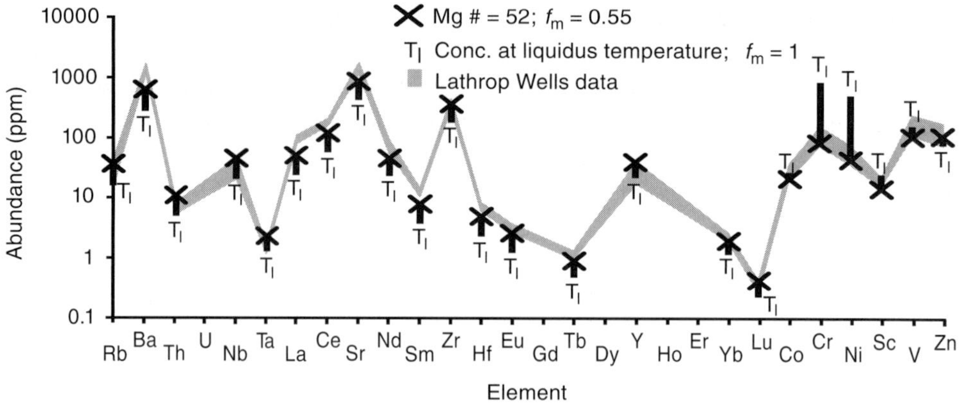

Fig. 8.8 Comparison of Lathrop Wells trace element data (absolute abundances; Perry and Straub, 1996) and calculation results based on 10% isobaric batch partial melt of **KLB-1** (Hirose and Kushiro, 1993) followed by closed-system fractional crystallization isobarically at 1 GPa terminating at a liquid Mg # = 52; symbol T_l gives the concentration of the trace element at the beginning of isobaric fractionation. The large **X** gives the trace element concentration at the end of crystal fractionation when the Mg # = 52, the low end of the small range (52–56) for Lathrop Wells. See text for details.

Alkali basalts moving at the high end of this range commonly transport ultramafic xenoliths from mantle depth to the surface. Because mantle-derived xenoliths have not been found at Lathrop Wells, mean ascent rates at mantle depths were probably in the cm s^{-1} to 0.1 m s^{-1} range. At shallow depth a transition from incompressible to compressible flow occurs due to the exsolution of volatiles triggered by magma decompression. The flow regime changes from a single phase melt flow to one with an increasing larger volume fraction of bubbles. Finally, when the volume fraction of bubbles exceeds the fragmentation limit, the flow becomes significantly compressible and accelerates upon further decompression. The incompressible to compressible flow transition is relevant to the analysis of magmatic hazards for the proposed repository.

8.5.1 Incompressible flow and thermal regime

The transport of melt in the incompressible regime has been discussed by many authors including Shaw (1980), Atkinson (1987), Pollard (1987), Spence and Turcotte (1990), Lister (1990, 1991), Lister and Kerr (1991); Rubin (1995), Takada (1994), and Heimpel and Olson (1994). Here, we draw on these to present a scale analysis specifically applicable to the eruption at Lathrop Wells. In general, two source conditions pertinent to magma transport serve as reasonable idealizations: (i) a constant volume source condition in which the magma source is rapidly depleted compared to the time required for magma transport through the crack network, and (ii) a constant discharge (volumetric rate) condition, applicable provided flow into the magma fracture plexus proceeds long after the fracture has opened. In the case of the relatively small-volume ($\sim 10^8$ m^3) Lathrop Wells eruption, condition (i) seems more applicable since the volume of a hypothetical single fracture of strike length 1 km and width of several meters extending to the source depth of $\sim$ 50 km is an appreciable fraction of the entire Lathrop Wells eruptive volume. Hence buoyant fractures with constant melt volume that close at the tail are considered appropriate to magma transport feeding the Lathrop Wells eruptive fissure-vent system.

There are two limiting models for the control of the propagation speed of a magma-filled crack. If magma viscosity is the limiting factor, then although the fracture mechanics determines the crack-tip shape, the effect of the crack-tip fracture resistance on the crack propagation velocity is negligible. Models based on this concept give rise to fractures with a slightly bulging head that tapers to a narrow conduit. The crack propagation velocity is equivalent to the Poiseuille velocity of magma in the crack. In this case, the relationship between the average melt ascent velocity, $\bar{v}$, fracture width, *h*, melt viscosity, η and driving pressure gradient, $\Delta p/\ell$ is

$$\bar{v} = \frac{h^2 \Delta p}{12\eta\ell} \tag{8.1}$$

If it is assumed that melt buoyancy drives the flow ($\Delta p/\ell \sim \Delta\rho g$), then adopting the scale parameters $h = 1$ m, $\eta = 500$ Pa s and $\Delta\rho = 200$ kg m^{-3}, a typical crack propagation rate, equivalent to the magma ascent speed for this model is $\bar{v} = 0.3$ m s^{-1}. In contrast, for buoyancy-driven crack propagation where the fracture resistance of the solid is important, the fracture velocity, v_F depends on the buoyancy of the melt and the elastic properties of the surrounding lithosphere. In particular, Heimpel and Olson (1994) found that when the melt volume in the buoyancy-driven fracture is sufficient for K_I, the mode I (tensile) crack-tip stress intensity factor (Atkinson, 1987), to exceed K_{Ic}, the mode I critical fracture toughness, unstable dynamic crack-tip propagation results. In typical dynamic fracture applications, the load is applied externally such as by external torsion, and the fracture propagation velocity rapidly approaches the elastic wave velocity of the solid, v_E, of order several km s^{-1}. However, for constant melt volume buoyancy-driven fracture propagation (the condition relevant to magma transport), the load on the crack tip is not externally defined but instead due to the body force associated with melt buoyancy within the propagating crack of height, ℓ. Therefore, for any volume of melt in the crack, the loading configuration reaches a time-averaged steady state in which the crack-tip propagation velocity is matched by the crack-tail closure velocity. The crack-tip propagation velocity in this case is

$$v_F = c\frac{v_E K_I^2}{2\ell\sigma_y^2}\left|\left(1 - \frac{K_I^2}{K_{Ic}^2}\right)\right| \tag{8.2}$$

where the mode I, stress factor K_I is

$$K_I \sim \frac{4}{\pi}\Delta p\sqrt{\frac{\pi\ell}{2}} \tag{8.3}$$

In these expressions, σ_y represents the extensional yield strength of host rock, v_E is the elastic wave propagation velocity of the lithosphere ($\sim 4\,\mathrm{km\,s^{-1}}$), Δp is the difference between the magma pressure (approximately lithostatic) and the crack-normal principal stress and c is a constant of order unity in SI units. Adopting representative scale parameters for basaltic magma transport through the upper mantle lithosphere ($K_{Ic} = 300\,\mathrm{MPa\,m^{1/2}}$, $v_E = 4\,\mathrm{km\,s^{-1}}$, $\sigma_y = 0.3 - 1\,\mathrm{GPa}$, $\ell = 1\,\mathrm{km}$, $\Delta p = 1 - 10\,\mathrm{MPa}$), the fracture propagation speed, $v_F \sim 0.1\,\mathrm{m\,s^{-1}}$. One needs to remember that the stress intensity factor and fracture toughness can vary considerably in geologic media due to heterogeneity, the geometry of the crack and "process zone" effects (Lawn and Wilshaw, 1975; Atkinson and Meredith, 1987; Kostrov and Das, 1988; Rubin, 1995). It appears that magma fracture propagation speeds in the range $\mathrm{cm\,s^{-1}}$ to $\mathrm{m\,s^{-1}}$ are appropriate for alkali basalt ascent through the lithosphere. That is, the two physical models controlling ascent velocity, one limited by melt viscosity and the other by fracture propagation resistance, provide roughly comparable ascent speed estimates. The absence of ultramafic xenoliths in the Lathrop Wells eruptive products is consistent with ascent rates near the lower end of the range, perhaps ~ 0.01–$0.1\,\mathrm{m\,s^{-1}}$. Adopting a mean ascent rate of $0.05\,\mathrm{m\,s^{-1}}$ gives a melt travel time of about two weeks from a depth of 60 km. During this period roughly half of the initial volume of magma crystallizes at depth to generate small pods or dike selvages of wehrlite (olivine + clinopyroxene cumulates). Interestingly, such ultramafic xenoliths are common at other Quaternary alkali basalt localities in the western Great Basin. The eruptive volume of Lathrop Wells is $0.12\,\mathrm{km^3}$. Based on phase-equilibria modeling, about half of the primary melt volume crystallizes at depth. This implies transport of heat between ascending magma and the surrounding cooler lithosphere. The amount of heat transfer is estimated according to

$$Q_{\mathrm{loss}} = \rho_m V_{\mathrm{magma}} \left(C_p (T_{\mathrm{liquidus}} - T_{\mathrm{solidus}}) + \Delta h_{\mathrm{cry.}} \right) \tag{8.4}$$

where ρ_m, V_{magma}, C_p and $\Delta h_{\mathrm{cry.}}$ represent the melt density, total volume of magma generated, isobaric specific heat capacity of melt and specific enthalpy of crystallization, respectively. Adopting the parameters $\rho_m = 2800\,\mathrm{kg\,m^{-3}}$, $V_{\mathrm{magma}} = 0.12\,\mathrm{km^3}$, $C_p = 1300\,\mathrm{J\,kg^{-1}\,K}$, $\Delta h_{\mathrm{cry.}} = 375\,\mathrm{kJ\,kg^{-1}}$ and a liquidus to solidus temperature interval of 400 K, the total heat loss is 3×10^{17} J or about 125 kJ per kg of primary melt. It is reasonable to ask if this inferred heat loss based on the phase-equilibria and fracture mechanical picture is consistent with the implied requirements for heat transport. Recall that in order to match the Mg # of the Lathrop Wells basalt, roughly half of the primary melt generated by partial melting should crystallize within the lithosphere during ascent. Is this inference consistent with elementary heat transfer theory? The model involves a swarm (Shaw, 1980; Takada, 1994) of magma-filled cracks (pods) migrating upwards with concomitant heat loss to surrounding cooler lithosphere. To perform the heat transfer scale analysis, we use an estimate of the volume of a single melt-filled crack (V_c) based on fracture mechanics to determine the total number of magma-filled cracks (n_c) needed to match the Lathrop Wells eruptive volume accounting for partial solidification based on equilibria constraints. We then use results based on non-isothermal laminar flow of magma in a propagating crack to estimate the duration of magma flow such that the required heat could be extracted. We then compare the calculated duration to the one estimated from magma ascent fracture mechanics.

The volume of a single melt-filled crack is $V_c = \ell w h$ and hence the total number (n_c) of propagating magma-filled cracks associated with the Lathrop Wells eruption is $n_c = V_T/V_c$ where ℓ is the crack height, w is the crack length (parallel to strike), h is the crack width and V_T represents the total volume of magma that freezes during ascent (roughly equal to the volume of the Lathrup Wells eruption based on phase equilibria calculations). The heat loss ($\dot{Q}$) for flow in a single vertically propagating crack is

$$\dot{Q} = \frac{4kNu\ell w(T_m - T_w)}{h} \tag{8.5}$$

where Nu is the non-dimensional Nusselt number, equal to 7.54 (Bird *et al.*, 1987) for planar crack flow, k is the melt thermal conductivity (0.3 W m^{-1} K) and the temperature difference is the mean difference between magma and surrounding lithosphere. The total heat loss (Q_T) for an ensemble (swarm) of such propagating melt-filled cracks is

$$Q_T = \frac{V_T}{V_c}\dot{Q}t_{event} \tag{8.6}$$

where t_{event} represents the duration of the swarm migration based on the phase equilibration requirement that sufficient heat is removed from ascending magma to crystallize about half its starting mass (primary melt). Using typical parameters ($\ell = w = 1$ km, $h = 2$ m, $T = 500$ K), $t_{event} \sim$ one month is estimated. This agrees with estimates based on fluid and fracture mechanics of ascent from 60 km at a mean ascent rate in the range 0.01–0.05 m s^{-1} (2 months to 2 weeks). The number of melt-filled cracks that rise through the lithosphere to feed the Lathrop Wells eruption is $\sim$ 60 in this model.

8.5.2 Compressible flow and explosive eruption regime

Magma ascent rates are approximately constant along the lithospheric ascent path because magma behaves approximately as an incompressible fluid before volatile-saturation is attained. However, because the solubility of volatile species (e.g. H_2O, CO_2, SO_2) depends strongly on pressure, rising melt eventually becomes saturated with volatiles at some pressure, $p = p_S$, and a discrete fluid phase forms. Volatile-saturation and continued exsolution and growth of fluid bubbles initiate the regime of bubbly multiphase flow. The pressure (translated to depth with knowledge of the local stress field) at which fluid saturation occurs depends on the composition and abundance of volatiles and the dependence of volatile solubility on temperature, pressure and melt composition. At some pressure less than p_S the volume fraction of fluid in magma may exceed the critical threshold for magma fragmentation. This pressure is identified as the fragmentation pressure, $p = p_F$. In this regime, magma is fluid-dominated volumetrically and blobs of melt (pyroclasts) are carried upwards in a rapidly expanding and accelerating fluid phase. The depth at which fragmentation takes place depends on the same parameters that control the volatile-saturation pressure and additionally on whether exsolved volatiles can leak from the magmatic mixture into the surrounding host rhyolitic tuff. The end-member scenarios are: (i) open system behavior in which volatiles are immediately and completely expelled from rising magma upon exsolution from the melt, or (ii) closed system behavior in which volatiles exsolved from the melt remain within the magmatic mixture. The dynamics of magma in the compressible regime is clearly relevant to volcanic hazard analysis and is discussed briefly below.

8.5.3 Thermodynamic volatile model

The thermodynamic mixed volatile (H_2O–CO_2) model of Papale (1999) is used to estimate the composition and abundance of dissolved and exsolved volatiles in melt of Lathrop Wells composition as a function of pressure, temperature and volatile bulk composition. Closed system behavior is assumed; the depth estimate for magma fragmentation is therefore a maximal value. Given the total abundance of volatiles and the bulk-system mass ratio of H_2O to CO_2, the mass fraction of H_2O and CO_2 in co-existing volatile-saturated melt and fluid can be computed. Results exhibiting the properties of melt, fluid and magma are portrayed in Figure 8.9 for basaltic magma of Lathrop Wells composition at 1150 °C with 4 wt.% total volatiles (H_2O + CO_2) and a bulk fluid mass ratio H_2O/CO_2 of 20:1. Figures 8.9a–b show the magma, fluid and melt density and viscosity as a function of pressure. Although the melt viscosity and density increase upon decreasing pressure, the viscosity (not shown) and density of the magma decrease due to the increasing volume fraction of fluid in magma upon decompression. Especially note the very rapid decrease in magma density as magma rises close to the surface. In Figure 8.9c, the variation of mass fraction and volume fraction of fluid in magma is given as a function of pressure. The critical volume fraction at fragmentation, $\theta_{crit} = 0.7$, is marked by the vertical line. This critical rheological limit is attained at a magma pressure $\sim$ 17 MPa for the assumed parameters listed on the figure. Finally, in Figure 8.9d, the fragmentation pressure is plotted versus the total initial dissolved volatile abundance. The fragmentation pressure varies from 4 MPa to 17 MPa at 1150 °C as total volatile concentrations increase from 1 wt.% to 4 wt.%.

Conversion of the saturation pressure, p_S and the fragmentation pressure, p_F to depth beneath YM requires knowledge of the state of stress with depth. Yucca Mountain lies within a region undergoing active east-southeast–west-northwest extension (Zoback, 1989). The state of stress around YM has been studied using hydraulic fracturing stress measurements, breakout and drilling-induced fractures, earthquake focal mechanisms and fault-slip orientations (Carr, 1974; Rogers *et al.*, 1983; Springer *et al.*, 1984; Stock *et al.*, 1985; Warren and Smith, 1985; Harmsen and Rogers, 1986; Frizzell and Zoback, 1987; SNL, 1997). A review of these data is given in Stock and Healy (1988). Stress measurements at $<$ 1.5 km depth give the ratio of horizontal to vertical stress $S_h/S_v \sim 0.35$–0.7. If it is assumed that S_v is lithostatic ($S_v = \rho_{tuff}\, gz$, ρ is the density of tuff, and z is depth), then p_S and p_F can be

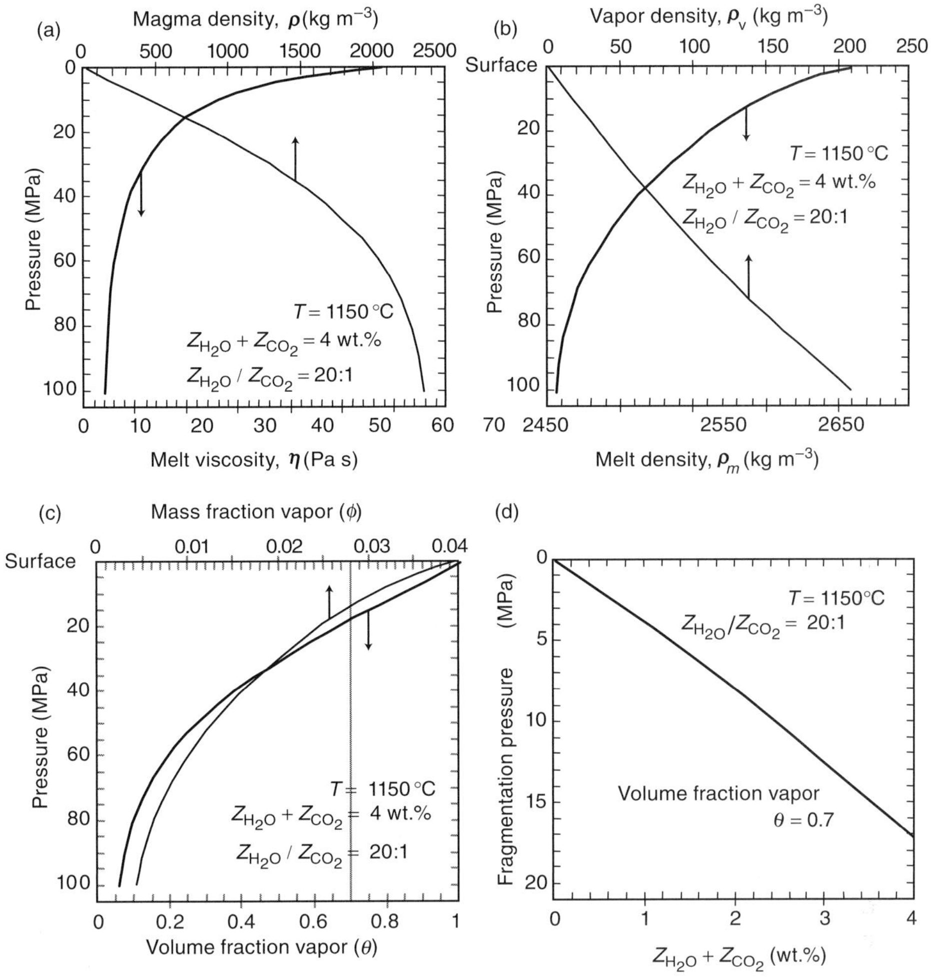

Fig. 8.9 Variation of magma properties for Lathrop Wells basaltic magma as function of pressure: (a) melt viscosity and magma density vs. pressure; (b) melt density and fluid (vapor) density vs. pressure; (c) mass fraction and volume fraction of exsolved supercritical fluid vs. pressure; (d) fragmentation pressure of Lathrop Wells basalt magma as a function of total volatile content; the fragmentation pressure is defined as the pressure at which the volume fraction of fluid equals 0.7. See text for details.

converted to depth. The dynamic constraint on magma pressure (p_{m}) for crack propagation is that $p_{\mathrm{m}} \geq \sigma_3$. Consistent with the state of stress at YM, we assume $S_{\mathrm{h}} = \sigma_3$ and $S_{\mathrm{v}} = \sigma_1$ and that $S_{\mathrm{h}} = \frac{2}{3} S_{\mathrm{v}}$. Accordingly, the relationship between depth beneath YM and magma pressure is approximately

$$z = \frac{3 p_{\mathrm{m}}}{2 g \rho_{\mathrm{tuff}}}$$

Magma fragmentation pressures of 4 MPa and 17 MPa correspond to depths $\sim$ 270 m and $\sim$ 1300 m for total volatile contents of 1 wt.% and 4 wt.%, respectively, assuming closed-system behavior (no leakage of volatiles from magma to tuff host rock).

Finally, we can obtain an estimate of the increase in magma ascent velocity in the compressible flow regime. The theory is based on a model in which volatile-saturated magma, treated as a compressible, viscous, homogeneous pseudofluid, flows upwards in a vertical conduit of constant cross-sectional area (Spera, 1984). Magma is assumed to flow through the conduit at constant flux $\dot{M} = \rho_0 u_0$. The initial magma density and velocity are taken from the incompressible regime with $\rho_0 = 2700\,\mathrm{kg\,m^{-3}}$ and $u_0 = 1\,\mathrm{m\,s^{-1}}$. Scaling of the conservation of mass expression then leads to an expression for the increase in the magma velocity

$$\delta u \sim \frac{\dot{M}}{\rho^2}\delta\rho$$

between the depth of volatile saturation and the surface. With $\dot{M} = 2.7 \times 10^3\,\mathrm{kg\,m^{-2}\,s}$, $\delta\rho \approx 2000\,\mathrm{kg\,m^{-3}}$ and $\rho \approx 250\,\mathrm{kg\,m^{-3}}$ (Figure 8.9a), $\delta u \sim 90\,\mathrm{m\,s^{-1}}$. Evidently, magma accelerates rapidly as the surface is approached. These magma exit velocities are consistent with eruption plumes on the order of several kilometers in height. Shallow crustal xenoliths of silicic tuff are common in the cinder cone at Lathrop Wells, consistent with the discharge rates inferred above (Doubik and Hill, 1999).

8.5.4 Compressible regime summary

In summary, the presence of a few percent by mass of dissolved volatiles in basaltic melts will ensure attainment of volatile-saturation at depths on the order of a few kilometers and magma fragmentation at depths on the order of one kilometer. If volatiles can leak out of rising magma by expulsion into surrounding country rock, the fragmentation depth will move upwards to shallower depths. The extent of upward migration of the fragmentation depth depends on the rate of fluid expulsion relative to the rate of volatile exsolution. To first order, the rate of fluid exsolution depends on the ascent rate of rising magma. This can be modeled with knowledge of the permeability of the country rock and information on the physical and thermodynamic state of rising magma; this calculation is not performed here. We have shown in Section 8.3.1 that dissolved water contents consistent with those inferred for Lathrop Wells are easily achieved provided a small amount of H_2O (e.g. 0.2 wt.%) is present in the source peridotite. The ultimate source of these fluids may be related to subduction along the western margin of the North American plate during the Mesozoic and Early Cenozoic although this is speculative. The saturation pressure (p_S) and fragmentation pressure (p_F) can be converted to depth once the state of stress in a region is specified. Beneath YM, the minimum principal stress is horizontal and the measured state of stress implies that magma fragmentation takes place at depths in the range 0.5–1.5 km depending on volatile contents assuming closed-system behavior. If volatiles do indeed escape from rising magma, then the depth at which magma fragmentation occurs will be shallower.

An eruption with a conduit anywhere within the YMR footprint would be expected to mobilize some fraction of radioactive waste; that fraction is beyond the scope of this study to estimate (see Menand *et al.*, Chapter 17, this volume; Lejuene *et al.*, Chapter 18, this volume) although we can say that the larger the eruptive volume and the higher the initial volatile-content of the magma the greater the amount of radioactive materials would be dispersed. Eruption plume heights scale with the one-fourth root of the mass discharge, $\dot{M} A_c$, where A_c is the cross-sectional area of the volcanic conduit, typically in the range 10^2-10^4 m^2 and $\dot{M}$ is the mass flux defined in Section 8.5.3. Once in the atmosphere, ash dispersal depends very strongly on the structure of winds aloft, the particle size distribution of entrained particles and other parameters (see Volentik *et al.*, Chapter 9, this volume).

Concluding remarks

This study shows that it is possible to obtain a sketch of the thermodynamics and mechanics of magma generation, partial crystallization, ascent and eruption for small-volume alkali basaltic volcanoes. These volcanoes are typically present in continental extensional provinces such as the Great Basin of western North America where YM lies. The melting of a fertile peridotitic source containing a small amount of H_2O at depths in the range $\sim 50-60$ km followed by significant fractional crystallization (about 50% by mass) during upward ascent is consistent with the petrology, geochemistry and inferred eruptive history of Lathrop Wells volcano, a 78 ka volcanic construct of tephra and lava ~ 20 km south of the proposed nuclear waste repository. Ascent rates at depths greater than a few kilometers are in the range $0.01-0.1$ m s^{-1} and are limited either by the fracture toughness of the surrounding mantle or by the shear viscosity of the melt. Ascent rates are roughly constant as magma rises through the upper mantle and lower crust. At shallow depths (a few kilometers) magma contains about 3 wt.% H_2O and becomes volatile-saturated. As magma continues to ascend and decompress the volume fraction of exsolved supercritical fluid phase, dominated by H_2O component but also undoubtedly containing small amounts of CO_2 and other constituents (e.g. SO_2, H_2S, CO), rapidly increases approaching and then exceeding the fragmentation limit, provided fluid does not leak out of magma into the surrounding volcanic tuff. Around this critical rheological transition from melt-dominated to fluid-dominated magma limit, the density and viscosity of magma rapidly declines. This leads to significant magma acceleration from initial velocity of ~ 1 m s^{-1} to final velocity (at the vent) of $\sim 100-200$ m s^{-1}. For total magma volatile contents around 3 wt.%, the fragmentation limit is reached at ~ 1 km depth, well below the depth of the proposed repository at YM of 300 m depth. If volatiles can escape rising magma, fragmentation depths move upwards to shallower depths.

An analytical survey of the consequences of repository disruption by magmatic processes has been presented elsewhere (Detournay *et al.*, 2003; Menand *et al.*, Chapter 17, this volume; Lejuene *et al.*, Chapter 18, this volume; Valentine and Perry, Chapter 19, this volume) although many important issues remain to be more fully explored quantitatively. Although such an analysis is beyond the scope of the present study, it is clear that quantitative conceptual models of alkali basalt petrogenesis of the type outlined in this study form the basis of informed volcanic hazard analysis.

Further reading

Recommended reading on this subject includes articles by Shaw (1980), Spera (2000) and Ghiorso *et al.* (2002), and the section by Walker (1993) entitled Basaltic-volcano systems, in the book *Magmatic Processes and Plate Tectonics*, edited by Prichard *et al.*

Acknowledgments

We gratefully acknowledge the efforts of the Geographic Information Systems group at Los Alamos National Laboratory (LANL).

References

Anthony, E. and J. Poths (1992). ^{3}He surface exposure dating and its implications for magma evolution in the Potrillo volcanic field, Rio Grande Rift, New Mexico, USA. *Geochimica et Cosmochimica Acta*, **56**(11), 4105–4108.

Asimow, P. D. and C. H. Langmuir (2003). The importance of water to oceanic mantle melting regimes, *Nature*, **421**, 815.

Asimow, P. D. and J. Longhi (2004). The significance of multiple saturation points in the context of polybaric near-fractional melting. *Journal of Petrology*, **45**, 2349–2367.

Atkinson, B. K. (1987). Introduction to fracture mechanics and its geophysical applications. In: Atkinson, B. K. (ed.) *Fracture Mechanics of Rock*. London: Academic Press, 1–27.

Atkinson, B. K. and P. G. Meredith (1987). The theory of subcritical crack growth with application to minerals and rocks. In: Atkinson, B. K. (ed.) *Fracture Mechanics of Rock*. London: Academic Press, 111–166.

Bergman, S. C. (1982). Petrogenetic aspects of the alkali basaltic lavas and included megacrysts and nodules from the Lunar Crater Volcanic Field, Nevada, USA. Ph.D Thesis, Princeton University, USA.

Best, M. G. and E. H. Christiansen (2001). *Igneous Petrology*. Oxford: Blackwell.

Bird, R. B., R. C. Armstrong and O. Hassager (1987). *Dynamics of Polymeric Liquids*. New York: Wiley-Interscience.

Bodansky, D. (1996). *Nuclear Energy: Principles, Practices and Prospects*. Woodbury, NY: American Institute of Physics.

Carlson, R. and J. Miller (2003). Mantle wedge water contents estimated from seismic velocities in partially serpentinized peridotites. EGS–AGU–EUG Joint Assembly. Nice, France, April 6–11, #7463.

Carr, W. J. (1974). Summary of tectonic and structural evidence for stress orientation at the Nevada Test Site, *USGS Open-File Report* 74-176. Denver, CO: US Geological Survey.

Craig, J. R., D. J. Vaughan and B. J. Skinner (2001). *Resources of the Earth*. New Jersey: Prentice Hall.

Crowe, B. M., G. A. Valentine, F. V. Perry and P. K. Black (2006). Volcanism: the continuing saga. In: Macfarlane, A. M. and R. C. Ewing (eds.) *Uncertainty*

Underground: Yucca Mountain and the Nations High-Level Nuclear Waste. Cambridge, MA: MIT Press, 131–148.
Detournay, E., L. G. Mastin, A. Pearson, A. M. Rubin and F. J. Spera (2003). Final Report of the Igneous Consequences Peer Review Panel. Las Vegas, NV: Bechtel SAIC Company LLC Report.
Doubik, P. and B. E. Hill (1999). Magmatic and hydromagmatic conduit development during the 1975 Tolbachik eruption, Kamchatka, with implications for hazards assessment at Yucca Mountain, NV. *Journal of Volcanological and Geothermal Research*, **91**, 43–64.
Fleck, R. J., B. D. Turrin, D. A. Sawyer *et al.* (1996). Age and character of basaltic rocks of the Yucca Mountain region, southern Nevada. *Journal of Geophysical Research*, **101**, 8205–8227.
Frey, F. A. and M. Prinz (1978). Ultramafic inclusions from San Carlos, Arizona: petrologic and geochemical data bearing on their petrogenesis. *Earth and Planetary Science Letters*, **38**, 129–176.
Fridrich, C. J., J. W. Witney, M. R. Hudson and B. M. Crowe (1999). Space–time patterns of late Cenozoic extension, vertical axis rotation, and volcanism in the Crater Flat basin, southwest Nevada. In: Wright, L. A. and B. W. Troxel (eds.) *Cenozoic Basins of Death Valley Region*, Geological Society of America Special paper 333, 197–212.
Frizzell, V. A. and M. L. Zoback (1987). Stress orientation determined from fault slip data in the Hampel Wash area, Nevada, and its relation to contemporary regional stress field. *Tectonics*, **6**(2), 89–98.
Geomatrix Consultants (1996). Probabilistic volcanic hazard analysis for Yucca Mountain, Nevada, Report BA0000000-1717-220-00082. San Francisco, CA: Geomatrix Consultants.
GERM (2008). Partition Coefficient (Kd) Database. Geochemical Earth Reference Model, http://earthref.org/GERM/index.html.
Ghiorso, M. S. and R. O. Sack (1995). Chemical mass transfer in magmatic processes; IV, A revised and internally consistent thermodynamic model for the interpolation and extrapolation of liquid–solid equilibria in magmatic systems at elevated temperatures and pressures. *Contributions to Mineralogy and Petrology*, **119**(2–3), 197–212.
Ghiorso, M. S., M. M. Hirschmann, P. W. Reiners and V. C. Kress (2002). The pMELTS: a revision of MELTS for improved calculation of phase relations and major element partitioning related to partial melting of the mantle to 3 GPa. *G-Cubed–Geochemistry, Geophysics, Geosystems*, **3**(5), doi:10.1029/2001GC000217.
Harmsen, S. C. and A. M. Rogers (1986). Inferences about the local stress field from focal mechanisms: applications to earthquakes in the southern Great Basin of Nevada. *Bulletin of the Seismological Society of America*, **76**, 1560–1572.
Heimpel, M. and P. Olson (1994). Buoyancy-driven fracture and magma transport through the lithosphere: models and experiments. In: Ryan, M. P. (ed.) *Magmatic Systems.* New York, NY: Academic Press, 223–240.
Henderson, P. (1982). *Inorganic Geochemistry*. Oxford: Pergamon Press.
Hirose, K. and I. Kushiro (1993). Partial melting of dry peridotites at high pressures: determination of compositions of melts segregated from peridotite using aggregates of diamond. *Earth and Planetary Science Letters*, **114**, 477–489.
Hirschmann, M. M., M. S. Ghiorso, L. E. Wasylenki, P. D. Asimow and E. M. Stolper (1998). Calculation of peridotite partial melting from thermodynamic models of minerals and melts. I. Review of methods and comparison to experiments. *Journal of Petrology*, **39**, 1091–1115.

Hirschmann, M. M., P. D. Asimow, M. S. Ghiorso and E. M. Stolper (1999a). Calculation of peridotite partial melting from thermodynamic models of minerals and melts. III. Controls on isobaric melt production and the effect of water on melt production. *Journal of Petrology*, **40**, 831–851.

Hirschmann, M. M., M. S. Ghiorso and E. M. Stolper (1999b). Calculation of peridotite partial melting from thermodynamic models of minerals and melts. II. Isobaric variations in melts near the solidus and owing to variable source composition. *Journal of Petrology*, **40**, 297–313.

Kennedy, B. M. and M. C. van Soest (2007). Flow of mantle fluids through the ductile lower crust: helium isotope trends. *Science*, **318**, 1433–1436.

Kostrov, B. V. and S. Das (1988). *Principles of Earthquake Source Mechanics*. Cambridge: Cambridge University Press.

Lawn, B. R. and T. R. Wilshaw (1975). *Fracture of Brittle Solids*. Cambridge: Cambridge University Press.

Le Maitre, R. W., P. Bateman, A. Dudek *et al.* (1989). *A Classification of Igneous Rocks and Glossary of Terms*. Oxford: Blackwell.

Lister, J. R. (1990). Buoyancy-driven fluid fracture: the effects of material toughness and of low-viscosity precursors. *Journal of Fluid Mechanics*, **210**, 263–280.

Lister, J. R. (1991). Steady solution for feeder dikes in a density stratified lithosphere. *Earth and Planetary Science Letters*, **107**, 233–242.

Lister, J. R. and R. C. Kerr (1991). Fluid-mechanical models of crack propagation and their application to magma transport in dykes. *Journal of Geophysical Research*, **96**, 10 049–10 077.

Macfarlane, A. M. and R. C. Ewing (2006). Introduction. In: Macfarlane, A. M. and R. C. Ewing (eds.) *Uncertainty Underground: Yucca Mountain and the Nations High-Level Nuclear Waste*. Cambridge, MA: MIT Press, 1–26.

Macfarlane, A. M. and M. Miller (2007). Nuclear energy and uranium resources. *Elements*, **3**(3), 185–192.

McDonough, W. F. (1990). Constraints on the composition of the continental lithospheric mantle. *Earth and Planetary Science Letters*, **101**, 1–18.

Musgrove, M. and D. P. Schrag (2006). Climate history at Yucca Mountain: lessons learned from Earth History. In: Macfarlane, A. M. and R. C. Ewing (eds.) *Uncertainty Underground: Yucca Mountain and the Nations High-Level Nuclear Waste*. Cambridge, MA: MIT Press, 149–162.

Padovani, E. R. and M. R. Reid (1989). Field guide to Kilbourne Hole maar, Dona Ana County, New Mexico. In: Chapin, C. E. and J. Zidek (eds.) *Field Excursions to Volcanic Terranes in the Western United States, Volume I. Southern Rocky Mountain Region*, Memoir 46. New Mexico Bureau of Mines and Mineral Resources, 174–185.

Papale, P. (1999). Modeling of the solubility of a two-component $H_2O + CO_2$ fluid in silicate liquids. *American Mineralogist*, **84**, 477–492.

Perry, F. V. and K. T. Straub (1996). Geochemistry of the Lathrop Wells volcanic center, Report LA-13113-MS. Los Alamos, NM: Los Alamos National Laboratory.

Perry, F. V. and B. Youngs (2004). Characterize framework for igneous activity at Yucca Mountain, Nevada, US DOE Report ANL-MGR-GS-000001(02). Las Vegas, NV: Office of Civilian Radioactive Waste Management.

Perry, F. V., B. M. Crowe, G. A. Valentine and L. M. Bowker (1998). Volcanism studies: final report for the Yucca Mountain project, Report LA-13478-MS. Los Alamos, NM: Los Alamos National Laboratory.

Pollard, D. D. (1987). Elementary fracture mechanics applied to the structural interpretation of dykes. In: Halls, H. C. and W. F. Fahrig (eds.) *Mafic Dyke Swarms*, Special Paper 34. Toronto, Ontario: Geological Association of Canada, 5–24.

Rogers, A. M., S. S. Harmsen, W. J. Carr and W. Spence (1983). Southern Great Basin seismological data report for 1981 and preliminary data analysis, USGS Open-File Report 83-699. Denver, CO: US Geological Survey.

Rubin, A. M. (1995). Propagation of magma-filled cracks. *Annual Reviews in Earth and Planetary Science*, **23**, 287–336.

Shaw, H. R. (1980). The fracture mechanism of magma transport from the mantle to the surface. In: Hargraves, R. B. (ed.) *Physics of Magmatic Processes*. Princeton, NJ: Princeton University Press, 201–264.

Shaw, H. R. (1987). Uniqueness of volcanic systems. In: Decker, R. W., T. L. Wright and P. H. Stauffer (eds.) *Volcanism in Hawaii*, US Geological Survey Professional Paper 1350(2), 1357–1394.

Smith, E. I., D. L. Feuerbach, T. R. Naumann and J. E. Faulds (1990). The area of most recent volcanism near Yucca Mountain, Nevada: implications for volcanic risk assessment. In: *High level Radioactive Waste Management 1990*, vol. 1. New York, NY: American Society of Civil Engineers, 81–90.

Smith, E. I., D. L. Keenan and T. Plank (2002). Episodic volcanism and hot mantle: implications for volcanic hazard studies at the proposed nuclear waste repository at Yucca Mountain, Nevada. *GSA Today*, **12**, 4–12.

SNL (1997). Hydraulic fracturing stress measurements in test hole ESF-AOD-HDFR#1, thermal test facility, Exploratory Studies Facility at Yucca Mountain, WA-0065. Albuquerque, NM: Sandia National Laboratories, TIC: 237818, TDIF: 305878, DTN: SNF37100195002.001.

Spence, D. A. and D. L. Turcotte (1990). Buoyancy-driven magma fracture: a mechanism for ascent through the lithosphere and the emplacement of diamonds. *Journal of Geophysical Research*, **95**, 5133–5139.

Spera, F. J. (1984). Carbon dioxide in petrogenesis: III. Role of volatiles in the ascent of alkaline magma with special reference to xenolith-bearing mafic lavas. *Contributions to Mineralogy and Petrology*, **88**, 217–232.

Spera, F. J. (1986). Fluid dynamics of ascending magma and mantle metasomatic fluids. In: Menzies, M. and C. Hawkesworth (eds.) *Mantle Metasomatism*. London: Academic Press, 241–259.

Spera, F. J. (2000). Physical properties of magma. In: Sigurdsson, H. (ed.) *Encyclopedia of Volcanoes*. New York: Academic Press, 171–190.

Spera, F. J., W. A. Bohrson, C. B. Till and M. S. Ghiorso (2007). Partitioning of trace elements among coexisting crystals, melt and supercritical fluid during isobaric crystallization and melting. *American Mineralogist*, **92**(10–11), 1568–1586.

Springer, J. E., R. K. Thorpe and H. L. McKague (1984). Borehole elongation and its relation to tectonic stress at the Nevada Test Site, Report UCRL-53528. Livermore, CA: Lawrence Livermore National Laboratory, OSTI ID 6592868.

Stock, J. M. and J. H. Healy (1988). Stress field at Yucca Mountain, Nevada. In: Carr, M.D. and J. C. Yount (eds.) *Geologic and Hydrologic Investigation of a Potential Nuclear Waste Disposal Site at Yucca Mountain, Southern Nevada*, US Geological Survey Bulletin 1790, 87–94.

Stock, J. M., J. H. Healy, S. H. Hickman and M. D. Zoback (1985). Hydraulic fracturing stress measurements at Yucca Mountain, Nevada, and relationship to the regional stress field. *Journal of Geophysical Research*, **90**(B10), 8691–8706.

Takada, A. (1994). Accumulation of magma in space and time by crack interaction. In: Ryan, M. P. (ed.) *Magmatic Systems*. New York, NY: Academic Press, 241–257.

Thompson, G. A. and D. B. Burke (1974). Regional geophysics of the Basin and Range Province. *Annual Review of Earth and Planetary Science*, **2**, 213–238.

Valentine, G. A. and F. V. Perry (2007). Tectonically controlled, time-predictable basaltic volcanism from a lithospheric mantle source (central Basin and Range Province, USA). *Earth and Planetary Science Letters*, **261**, 201–216.

Valentine, G. A., G. Wolde-Gabriel, N. D. Rosenberg *et al.* (1998). Physical processes of magmatism and effects on the potential repository: synthesis of technical work through fiscal year 1995. In: Perry, F. V., B. M. Crowe, G. A. Valentine and L. M. Bowker (eds.) Volcanism Studies: Final Report for the Yucca Mountain Project, Los Alamos National Laboratory Report LA-13478. Los Alamos, NM: Los Alamos National Laboratory.

Valentine G. A., F. V. Perry, D. J. Krier *et al.* (2006). Small volume basaltic volcanoes: eruptive products and processes, and posteruptive geomorphic evolution in Crater Flat (Pleistocene), southern Nevada. *Geological Society of America, Bulletin*, **118**, 1313–1330.

Walker, G. P. L. (1993). Basaltic-volcano systems. In: Prichard, H. M. *et al.* (eds.) *Magmatic Processes and Plate Tectonics*, Special Publication 76. London: Geological Society, 3–39.

Warren, W. E. and C. W. Smith (1985). In situ stress estimates from hydraulic fracturing and direct observation of crack orientation. *Journal of Geophysical Research*, **90**, 6829–6839.

Woods, A. W., S. Sparks, O. Bokhove *et al.* (2002). Modeling magma-drift interaction at the proposed high-level radioactive waste repository at Yucca Mountain, Nevada, USA. *Geophysical Research Letters*, **29**, 1–4.

Yogodzinski, G. M., T. R. Naumann, E. I. Smith and T. K. Bradshaw (1996). Evolution of a mafic volcanic field in the central Great Basin, south central Nevada. *Journal of Geophysical Research*, **101**, 17 425–17 445.

Zoback, M. L. (1989). State of stress and modern deformation of the northern Basin and Range Province. *Journal of Geophysical Research*, **94**(B6), 7105–7128.

Zoback, M. L., R. E. Anderson and G. A. Thompson (1981). Cenozoic evolution of the state of stress and style of tectonism of the Basin and Range province of Western United States. *Royal Society of London Philosophical Transactions – Series A*, **300**, 407–434.

9

Aspects of volcanic hazard assessment for the Bataan nuclear power plant, Luzon Peninsula, Philippines

A. C. M. Volentik, C. B. Connor, L. J. Connor and C. Bonadonna

How would the eruption of a volcano affect a nearby nuclear power plant (NPP)? Specifically, would the products of a volcanic eruption impact the operation of an NPP located near an erupting volcano? The answer to this question begins with an assessment of the geological phenomena that result from volcanic eruptions. These phenomena are diverse, and include tephra fallout, pyroclastic flows and lahars, among others (Connor *et al.*, Chapter 3, this volume). The effects of these phenomena depend on a host of factors, such as the proximity of the volcano to the NPP, the size and character of the eruption, wind direction and topography around the volcano.

The complexity and uncertainty associated with these phenomena suggest that their potential impacts be assessed probabilistically. One important aspect of probabilistic assessment involves forecasting the timing of eruptions. When will the next eruption occur? Or, phrased another way, how much time must elapse before a volcano no longer has a credible potential for future eruptions? This question is not easily resolved, as volcanoes may go thousands of years, or even tens of thousands of years without erupting. A second aspect of volcanic hazard assessment is estimation of the effects of volcanic eruptions, once they occur. Which areas might be inundated by lahars, or experience tephra fallout? As eruption magnitudes and their effects vary widely, this question must also be answered probabilistically. Admittedly, assessment of the timing and consequences of potential eruptions is a daunting task, requiring site-specific data, a refined understanding of volcanic processes and computational tools to actually estimate probabilities.

In the face of these complexities, a systematic approach is warranted. Hill *et al.* (Chapter 25, this volume) recommend guidelines for volcanic hazard assessment for surface nuclear facilities that provide a systematic approach. In this chapter these recommended guidelines are applied to a specific NPP site located in the Philippines. Our goal is to illustrate key points of the application of the recommended guidelines to volcanic hazard assessment for surface nuclear facilities.

We illustrate aspects of volcanic hazard assessment using the Bataan nuclear power plant (BNPP) site, located on Napot Point on the west coast of the Bataan Peninsula, western Luzon Peninsula, Philippines, at 14°38′ N, 120°19′ E, or, in UTM Zone 51 N coordinates, 210 500 E, 1 619 000 N (Figure 9.1). This NPP was sited and constructed during the late 1970s and early 1980s, using then current designs for a pressurized water reactor. Although

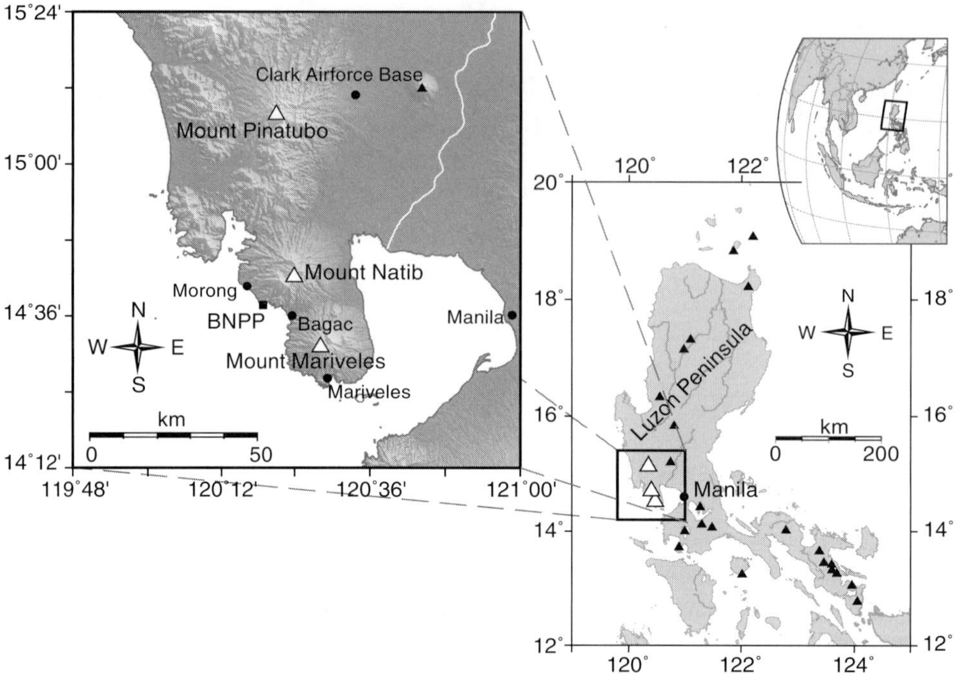

Fig. 9.1 Location map showing the Bataan Peninsula, forming the southern part of the Luzon Peninsula within the Philippines archipelago. Black triangles indicate active volcanoes. White triangles indicate active volcanoes closest to the BNPP. The Bataan Peninsula is formed from two volcanoes, Mt. Natib to the north and Mt. Mariveles to the south. The location of the BNPP is marked with a black square (digital elevation model data from CGIAR-CSI, 2004).

some nuclear fuel was delivered, the reactor never operated. The project was quite controversial at the time of siting and construction. In the United States, for example, questions arose about whether hazard assessments at the site were partly the responsibility of the US Nuclear Regulatory Commission, because US companies exported technology used to construct the BNPP (D'Amato and Engel, 1988). The US Nuclear Regulatory Commission ultimately decided that it had no legal role in reviewing the hazard assessments for the BNPP. Nevertheless, concerns about the siting assessment for the BNPP remained. The Union of Concerned Scientists cited the proximity of the site to the potentially active Mt. Natib volcano as a major source of concern (D'Amato and Engel, 1988). The conclusions of volcanic hazard assessments performed by a US consulting company (EBASCO, 1977, 1979) on behalf of the Philippine Atomic Energy Commission were questioned by US scientists (Newhall, 1979), experts from the International Atomic Energy Agency (IAEA, 1978) and oversight panels in the Philippines.

It is not our intent to review, or recreate, this controversy. Rather, data gathered during the site investigation and after the site investigation are used to assess hazards, within the guidelines outlined by Hill *et al.* (Chapter 25, this volume), as an illustration of the application of these guidelines. This assessment, some 30 a after construction of the NPP, utilizes modern methods for numerical modeling of volcanic phenomena, particularly with regard to assessment of tephra fallout hazards and susceptibility of the site to pyroclastic flows and lahars. As mentioned previously, this assessment stops short of a comprehensive volcanic hazard assessment. In this regard, one criticism of the original hazard assessment was the lack of adequate geologic mapping of Mt. Natib volcano (IAEA, 1978; Newhall, 1979). To our knowledge, such comprehensive mapping has not yet been undertaken for Mt. Natib volcano and hence a comprehensive hazard assessment, fully meeting the recommendations described by Hill *et al.* (Chapter 25, this volume), is not possible at this time.

9.1 Volcanological setting

The BNPP site is located within a Quaternary volcanic province known as the Bataan Lineament (Wolfe and Self, 1983), formed by the eastward subduction of the South China Sea floor along the Manila trench off the west coast of Luzon Peninsula. The Bataan Lineament is 320 km long and comprises at least 27 volcanoes, including Mt. Natib (Figure 9.1). The summit of Mt. Natib volcano is located about 15 km northeast of the BNPP. Mt. Pinatubo and Mt. Mariveles volcanoes, which may be relevant to the hazard assessment, lie about 57 km north and 22 km southeast of the site, respectively (Figure 9.1).

Mt. Natib and Mt. Mariveles are both Quaternary composite volcanoes and together form the dominant topographic features of the Bataan Peninsula. These volcanoes have not erupted historically. Geologic mapping and radiometric dating of Mt. Natib deposits indicate that this volcano has produced violent explosive eruptions during the last several hundred thousand years. These eruptions produced tephra fallout, pyroclastic density currents (pyroclastic flows) and secondary lahars (EBASCO, 1977; Newhall, 1979; Wolfe and Self, 1983; Siebert and Simkin, 2007). The Napot Point tuff, as described by Newhall (1979), is a pyroclastic flow deposit that resulted from such eruptions. The Napot Point tuff and lahar deposits are located within the BNPP site area. Details of past eruptions are difficult to decipher, however, because both Mt. Natib and Mt. Mariveles volcanoes have poorly defined stratigraphies. Geologic mapping (Newhall, 1979; Ruaya and Panem, 1991) is challenging on the peninsula, in part due to poor exposures in this tropical environment and in part due to complexities of volcanic stratigraphy in an arc terrain.

The volcanic hazard assessment made by EBASCO (1977) preceded the dramatic volcanic eruptions of Mt. Pinatubo in 1991. After dormancy of about 540 a, Mt. Pinatubo reawakened in April 1991. The volcanic activity culminated in an explosive Plinian eruption on June 15, 1991, which produced a strong vertical plume and several pyroclastic density currents (Newhall and Punongbayan, 1996). This eruption is important for hazard assessment at the BNPP in two respects. First, the eruption directly affected the site, depositing $\sim$ 6 cm of tephra in the site area. Second, the Mt. Pinatubo eruption provides an analog for potential future eruptions of Mt. Natib and Mt. Mariveles, at which no historical eruptions have occurred and for which geologic mapping is incomplete.

The 1991 Mt. Pinatubo eruption column reached a maximum height of 35–40 km (Koyaguchi and Tokuno, 1993; Koyaguchi, 1996; Holasek *et al.*, 1996; Koyaguchi, 1996; Paladio-Melosantos *et al.*, 1996; Rosi *et al.*, 2001). Tephra fallout occurred throughout the entire eruption and deposited several layers of pumiceous lapilli and ash, with two distinct tephra layers associated with the climatic eruption (Koyaguchi and Tokuno, 1993; Koyaguchi, 1996; Paladio-Melosantos *et al.*, 1996; Koyaguchi and Ohno, 2001). Voluminous pyroclastic flows were generated during the eruption that traveled as much as 12–16 km radially from the vent and were able to overcome topographic ridges as high as 400 m in proximal areas (Scott *et al.*, 1996). These pyroclastic deposits were remobilized by heavy rainfalls, triggering large lahars around Mt. Pinatubo shortly after and more than six years following the eruption (Rodolfo *et al.*, 1996; Wolfe and Hoblitt, 1996; Daag, 2003; van Westen and Daag, 2005).

Although Mt. Natib and Mt. Mariveles seem to have erupted volcanic products slightly more mafic than Mt. Pinatubo (Defant *et al.*, 1991; Newhall *et al.*, 1996), these volcanoes are similar in several respects. Summit calderas truncate the three edifices. Mt. Natib caldera is 6 × 7 km, Mt. Mariveles caldera is 4 km in diameter and Mt. Pinatubo caldera is the smallest, only 2.5 km in diameter. The caldera of Mt. Pinatubo formed as a result of one collapse event following the 1991 eruption. It is unclear if the calderas on Mt. Natib and Mt. Mariveles were similarly formed following only one explosive eruption or by incremental collapses through time associated with multiple eruptions. At least for Mt. Natib, pyroclastic flows seem to have traveled more than 10 km from the caldera, to a point where they reached the sea, and secondary lahars associated with these pyroclastic flows were generated and traveled in main drainages. No tephra-fall deposits have been reported or mapped for eruptions from Mt. Natib or Mt. Mariveles (EBASCO, 1977, 1979; Newhall, 1979). However, this does not imply that extensive tephra fall did not accompany past volcanic eruptions from these volcanoes. For comparison, the tephra deposit from the 1991 eruption of Mt. Pinatubo has been almost completely eroded away (Daag, 2003; Newhall, 2007, personal communication).

Today, volcanic activity at Mt. Natib is manifest by thermal springs that are located within the summit caldera (Ruaya and Panem, 1991), suggesting the presence of a hydrothermal system within the volcano. Mt. Natib and Mt. Mariveles are not currently monitored, so nothing more is known about the current state of activity at these volcanoes.

Much of what is known about the history of eruptions at Mt. Natib volcano is based on radiometric age determinations from samples collected by EBASCO (1977), Wolfe and Self (1983) and the Philippine Institute of Volcanology and Seismology (PHIVOLCS). Without adequate stratigraphic constraints, these age determinations can only provide a snapshot of volcanic activity, rather than information about the stratigraphic sequence or variations in the rate of volcanic activity. The consulting company EBASCO (1977) concluded that Mt. Natib volcano was active between 0.069–1.6 Ma, based on a series of 27 K/Ar dates on lavas and pyroclastic flows exposed on the flanks of the volcano. A total of 31 K/Ar dates reported by Wolfe and Self (1983) suggest a range of activity from 0.54–3.9 Ma. In addition, fission track ages on seven pumice samples range from 20–59 ka, but EBASCO (1977, 1979) suggested these may be underestimates of eruption age due to potential uranium migration in these samples. In 1999, PHIVOLCS made an uncalibrated ^{14}C age determination of 27 ± 0.63 ka on charcoal within a young pyroclastic flow deposit on the eastern flank of Mt. Natib (C. Newhall, 1999, written communication). More recently, Cabato *et al.* (2005) found evidence to support an even younger explosive eruption on the western flanks of Mt. Natib. Based on a high-resolution seismic study of the Subic Bay, Cabato *et al.* (2005) proposed an age of 11.3–18 ka for a potential pyroclastic deposit interlayered with sediments in the eastern Subic Bay. This pyroclastic deposit is thought to originate from an explosive eruption in the northwestern area of the breached Mt. Natib caldera. These age determinations are much younger than the youngest age of 69 ka reported by EBASCO (1977) in their hazard assessment of Mt. Natib, highlighting that reconnaissance mapping and dating may have missed these younger units on the western side. The occurrence of young pyroclastic flows on the western flanks of the volcano emphasizes the potential for volcanic hazards around Mt. Natib on both the western and eastern flanks (see EBASCO, 1977, 1979). Unfortunately, there are no available ages for the Napot Point tuff, which crops out in the BNPP site vicinity. A reconnaissance suite of radiometric dates have been reported for Mt. Mariveles volcano. Wolfe and Self (1983) reported a range of activity from 0.19–4.1 Ma, based on 20 samples. However, the most recent eruption at Mt. Mariveles may be as young as 2050 BC (uncalibrated ^{14}C date, Siebert and Simkin, 2007).

9.2 Assessment of volcano capability

For the closest volcanoes to the BNPP site, the frequency and timing of past volcanic events are incompletely understood and thus highly uncertain. The concept of a capable volcano (Hill *et al.*, Chapter 25, this volume) is used to assess the potential for Mt. Natib and Mt. Mariveles volcanoes to produce hazardous phenomena that may reach the BNPP. Following its 1991 eruption, Mt. Pinatubo is clearly a capable volcano, as ~ 6 cm of tephra fell on the BNPP during that eruption. As described in detail in Chapter 25, a capable volcano is one for which both (i) a future eruption or related volcanic event is credible; and (ii) such an event has the potential to produce phenomena that may affect a site. If Mt. Natib or Mt. Mariveles volcanoes are capable, a detailed, site-specific volcanic hazard assessment is warranted that considers the likelihood of occurrence and associated uncertainties for volcanic phenomena that may reach a site.

One step in determining a volcano's capability is to evaluate its potential for future eruptions. Activity documented during the Holocene (i.e. within the last 10 ka) is one criterion used to determine that a volcano appears capable of future volcanic eruptions (e.g. Hill *et al.*, Chapter 25, this volume). There is no definitive evidence that Mt. Natib or Mt. Mariveles volcanoes have erupted during the Holocene. Nevertheless, determining Holocene eruptive activity is difficult, especially in the tropical environment of these volcanoes and where mapping is incomplete. In such cases evidence of current volcanic activity includes ongoing volcanic unrest, or the presence of an active hydrothermal system and related phenomena. As Mt. Natib and Mt. Mariveles volcanoes are not monitored, there is no information available regarding current unrest, such as the occurrence of volcano-tectonic earthquakes or ground deformation. However, the presence of thermal springs within Mt. Natib's caldera (Ruaya and Panem, 1991) is indicative of an active hydrothermal system. Thus, future eruptions of Mt. Natib are possible (cf. Siebert and Simkin, 2007). Therefore, an analysis should be made to assess the possibility of volcanic phenomena reaching the site of the BNPP, given a potential eruption of Mt. Natib volcano. Hydrothermal activity has not been reported at Mt. Mariveles volcano, but, as previously noted, one ^{14}C date suggests that Holocene activity has occurred. Thus Mt. Mariveles volcano may also be a capable volcano, based on the timing of past eruptions.

The probability of future eruptions of Mt. Natib is highly uncertain, given the incomplete record of radiometric age determinations and lack of detailed stratigraphic control on important geologic units. Based on this incomplete record, EBASCO (1977) estimated the probability of a future volcanic eruption of Mt. Natib volcano to be $\sim 3 \times 10^{-5}\,a^{-1}$. The global record of repose intervals preceding large Plinian eruptions (i.e. volcano explosivity index (VEI) 6–7) of long-dormant volcanoes (Siebert and Simkin, 2007) provides one means of evaluating this probability. Connor *et al.* (2006) found that repose intervals preceding VEI 6–7 eruptions of long-dormant volcanoes follow a log-logistic probability distribution. Applying this probability model and using a repose interval of 14.65 ka, based on the date of the youngest known pyroclastic flow (11.3–18 ka) on Mt. Natib (Cabato *et al.*, 2005), the probability of a VEI 6–7 eruption of Mt. Natib is $\approx 1 \times 10^{-4} - 2 \times 10^{-4}\,a^{-1}$, with 95% confidence, which is almost one order of magnitude greater than the EBASCO (1977) result. Such probabilities appear sufficient to consider future eruptions as credible events, and indicate that a hazard analysis for these eruptions appears warranted (Hill *et al.*, Chapter 25, this volume). Following the same approach but using a repose interval of 4.05 ka, based on the youngest date on volcanic products from Mt. Mariveles, the probability of a VEI 6–7 eruption of Mt. Mariveles is $\sim 3.5 \times 10^{-4}$–$6 \times 10^{-4}\,a^{-1}$, with 95% confidence.

Had we applied this probabilistic method in 1990, before the eruption of Mt. Pinatubo, the probability of a VEI 6–7 eruption of Mt. Pinatubo would have been $\sim 0.6 \times 10^{-3} - 1 \times 10^{-3}\,a^{-1}$, using a repose interval of 0.54 ka based on its most recent known explosive eruption prior to 1991. Repose intervals between eruptive episodes for the most recent eruptions of Mt. Pinatubo (Newhall *et al.*, 1996), not including the 1991 eruption, are chronologically: 2.5 ka, 2.5 ka, 3.5 ka, 8.4 ka and 17.6 ka (Newhall *et al.*, 1996; Siebert and Simkin, 2007). Assuming that the timing of eruptions is described by a Poisson process (Connor *et al.*, Chapter 3, this volume), the interval estimate of repose for Mt. Pinatubo is 3.4–21.2 ka, with 95% confidence. This interval corresponds to a probability of an eruption of Mt. Pinatubo to be 0.5×10^{-4}–$3 \times 10^{-4}\,a^{-1}$. A bootstrap with replacement procedure (Efron and Tibshirani, 1991) yields an interval estimate of 3.8–9.9 ka, with 95% confidence, corresponding to a probability of an eruption of Mt. Pinatubo of 1×10^{-4}–$2.6 \times 10^{-4}\,a^{-1}$. Note that the period between eruptions of Mt. Pinatubo becomes shorter with time. This may reflect nonstationarity in repose intervals between eruptions, or may simply reflect sampling bias, due to difficulty in distinguishing eruptive units in the older stratigraphic record. Regardless, it appears that the probability of eruptions of Mt. Pinatubo, prior to the 1991 eruptions, were $\sim$ one to two orders of magnitude greater than probabilities of large explosive eruptions currently estimated for Mt. Natib. Our interpretation of this comparison is that, although the probability of an explosive eruption of Mt. Natib appears to be much lower than the probability of eruptions of Mt. Pinatubo, explosive eruptions of Mt. Natib are credible.

For Mt. Natib volcano, evidence of an active hydrothermal system and probability estimates both indicate that the volcano has a credible potential for future eruptions. The potential for volcanic phenomena to impact the BNPP site should be estimated. This step is accomplished by estimating screening-distance values for volcanic phenomena, such as tephra fallout, lahars and pyroclastic flows, which have the highest potential to affect the BNPP site adversely. These screening-distance values consider the potential for volcanic phenomena to reach the BNPP site, using conservative assumptions about the magnitudes of volcanic eruptions and simplified numerical models of potential hazards. The remainder of this chapter describes numerical and probabilistic techniques to estimate screening-distance values for these phenomena. These values, in turn, are used to determine the capability of Mt. Natib, Mt. Mariveles and Mt. Pinatubo volcanoes to affect the BNPP site.

9.3 Estimating screening-distance values

Geological maps of the Mt. Natib volcanic deposits indicate that pyroclastic density currents and secondary lahar (volcanic mudflow) deposits occur in the BNPP site area (EBASCO, 1977; Newhall, 1979). The presence of these deposits is clear evidence that the BNPP site is located within a screening distance for these phenomena. A screening distance is defined as the distance from a volcano that a specific volcanic phenomena, such as pyroclastic flows, may plausibly reach (Hill *et al.*, Chapter 25, this volume). Screening distance depends on a number of factors, such as topography of the volcano, possible magnitudes of future eruptions and the types of volcanic phenomena involved. There is uncertainty in the estimate of screening distances, but it is often practical to determine if a specific site is beyond, or within, a screening distance for specific phenomena originating from a volcano using scoping calculations. Such scoping calculations are used here to assess volcanic hazards at the Bataan site.

The site may be exposed to both proximal and distal effects of volcanic eruptions from Mt. Natib (Connor *et al.*, Chapter 3, this volume). In addition, the site may be exposed to far-field effects from Mt. Pinatubo and Mt. Mariveles volcanoes, particularly from tephra fallout. We can refine our understanding of potential hazards using a variety of numerical methods to consider the magnitudes of potential volcanic eruptions that produce phenomena that could impact the site. Analysis is limited to products of explosive eruptions, including tephra fallout, pyroclastic flows and lahars. Other phenomena, such as new vent formation and lava flows are not considered in this chapter, and require additional analyses in order to assess their potential hazards to the BNPP site.

9.3.1 Hazards from tephra fallout

Tephra fallout creates loads on engineered structures and may disrupt ventilation, electrical and cooling systems at nuclear power plants. Thick tephra accumulation might render a site temporarily inoperable, and very rapid erosion of tephra deposits may generate potentially damaging lahars. The TEPHRA2 computer program (Bonadonna *et al.*, 2005; Bonadonna, 2006; Connor and Connor, 2006; Connor *et al.*, 2008) is used to estimate potential accumulations of tephra fallout at the BNPP site and on the flanks of Mt. Natib volcano above the BNPP site with a series of deterministic and probabilistic analyses. The numerical simulation of tephra accumulation is based on the advection–diffusion equation (Suzuki, 1983; Armienti *et al.*, 1988; Connor *et al.*, 2001), which is expressed by a simplified mass-conservation equation:

$$\frac{\partial C_j}{\partial t} + w_x \frac{\partial C_j}{\partial x} + w_y \frac{\partial C_j}{\partial y} - v_{l,j} \frac{\partial C_j}{\partial z} = K \frac{\partial^2 C_j}{\partial x^2} + K \frac{\partial^2 C_j}{\partial y^2} + \Phi \qquad (9.1)$$

where x, y and z are spatial coordinates expressed in meters; C_j is the mass concentration of particles (kg m^{-3}) of a given particle size class, j; w_x and w_y are the x and y components of the wind velocity (m s^{-1}); K is a horizontal diffusion coefficient for tephra in the atmosphere (m^2 s^{-1}); $v_{l,j}$ is the terminal settling velocity (m s^{-1}) for particles of size class, j, as these particles fall through a level in the atmosphere, l; Φ is the change in particle concentration at the source with time, t (kg m^{-3} s^{-1}). The algorithm implemented in TEPHRA2 assumes negligible vertical wind velocity and diffusion, and assumes a constant and isotropic horizontal diffusion coefficient ($K = K_x = K_y$). The terminal settling velocity, v, is calculated for each particle size, j, at each atmospheric level, l, as a function of the particle's Reynolds number, which varies with atmospheric density. Wind velocity is allowed to vary as a function of height in the atmosphere, but it is assumed to be constant within a specific atmospheric level.

Tephra fallout hazard studies are most concerned with mass accumulation at specific locations. The TEPHRA2 program calculates tephra accumulation, M (kg m^{-2}), at each location, (x, y):

$$M(x, y) = \sum_{l=0}^{H_{\max}} \sum_{j=d_{\min}}^{d_{\max}} m_{l,j}(x, y) \qquad (9.2)$$

where $m_{l,j}(x, y)$ is the mass fraction of the particle size, j, released from atmospheric level, l, accumulated at location (x, y); $H_{\max}$ is the maximum height of the erupting column, and $d_{\min}$ and $d_{\max}$ are, respectively, the minimum and maximum particle diameters. Thus, the distribution of tephra mass following an eruption depends on both the distribution of mass in the eruption column and the distribution of mass by grain size. The algorithm implemented in TEPHRA2 assumes that mass is uniformly distributed in the eruption column, or can be specified to be uniformly distributed in some fraction of the uppermost column, to be consistent with observations of strong volcanic plumes. Grain-size distribution is assumed to be log-normal, and is deduced from comparison with studies of well-preserved deposits.

Deterministic analysis

Hazard assessments rely on probabilistic methods to forecast accurately the potential occurrence of disruptive phenomena. As part of this modeling process, deterministic assessments can be useful for estimating potential tephra accumulation resulting from eruptions of a specific size during specific meteorological conditions. For example, one can estimate tephra fallout at the BNPP resulting from a large-volume, explosive eruption of Mt. Natib, when the wind is blowing from the volcano toward the site. These analyses are accomplished by completely specifying the eruption and meteorological parameters, and calculating an isomass map based on these parameters using TEPHRA2. Such bounding calculations are particularly useful for estimating screening distances and may provide useful supplementary material for the interpretation of probabilistic assessments.

Five deterministic scenarios are presented based on the following eruption parameters: location of the vent, column height, total erupted mass and grain-size distribution. Meteorological parameters include wind direction and speed, as a function of elevation in the atmosphere. Eruptions span VEI 3–7 with associated maximum column heights ($H_{\max}$) of 8, 14, 25, 35 and 45 km, respectively (Newhall and Self, 1982). A small VEI 3 eruption is represented by an 8 km-high erupting column. A 14 km-high column represents the approximate boundary between VEI 3 and VEI 4 eruptions; this column height is also the lower limit of sub-Plinian eruptions according to Pyle (1989). As EBASCO (1977, 1979) proposed an analysis of tephra hazard at the BNPP site based on analogy with the 1912 Katmai eruption, 25 km was selected to match the maximum column height of that VEI 5 eruption (Fierstein *et al.*, 1997). The 35 km column height reflects the 1991 eruption (VEI 6) of Mt. Pinatubo, and 45 km represents an upper limit scenario based on the historical eruption (VEI 7) of Tambora in 1815 (Sigurdsson and Carey, 1989).

Eruption duration, T, and maximum column height, $H_{\max}$, are used to calculate the total erupted mass for each scenario, assuming steady-state conditions. Koyaguchi and Tokuno (1993) proposed an eruption duration of 5 hr for the 1991 climactic eruption of Mt. Pinatubo, based on the expansion of the umbrella cloud in the stratosphere. This is, perhaps, an overestimate of eruption duration, as the umbrella cloud may have continued to expand in the stratosphere after cessation of appreciable mass discharge from the vent. Tahira *et al.* (1996) proposed an eruption duration of 3.5 hr, based on infrasonic and acoustic waves generated by the explosive eruption. According to Paladio-Melosantos *et al.* (1996), the peak activity of the eruption was sustained for about 3 hr, followed by waning activity for 6 hr. For these deterministic scenarios, we chose an eruption duration of 3 hr, which maximizes flow rate for specific column heights.

For steady eruptions, the mass discharge rate of an eruption is empirically related to the column height (Sparks *et al.*, 1997):

$$H_{\max} = 1.67Q^{0.259} \tag{9.3}$$

where Q is the magma discharge rate ($m^3\,s^{-1}$). From the density of the deposit (ρ_{dep}) and the duration of the eruption (T), the magma discharge rate (Q) is:

$$Q = \frac{M_o}{T\rho_{\text{dep}}} \tag{9.4}$$

Table 9.1. Eruption column height and total mass inputs for deterministic tephra models are based on analog eruptions and volcano explosivity index (VEI)

Parameter	VEI 3	VEI 4	VEI 5	VEI 6	VEI 7
Column height (km)	8	12	25	35	45
Total mass (kg)	5.7×10^9	5.3×10^{10}	5.4×10^{11}	2.1×10^{12}	5.7×10^{12}

where M_o is the total mass of the deposit in kilograms. The bulk density of the deposit, ρ_{dep} (kg m^{-3}), is assumed to be 1000 kg m^{-3}, corresponding well with the average density of the 1991 phenocryst-rich dacitic pumice of Mt. Pinatubo (977 kg m^{-3} proposed by Pallister *et al.*, 1996). This value is also in good agreement with the range 500–1500 kg m^{-3}, the bulk density of known Plinian deposits (Sparks *et al.*, 1997). Total mass is related to eruption column height and eruption duration by:

$$M_o = T\rho_{dep}\left(\frac{H_{max}}{1.67}\right)^4 \tag{9.5}$$

Thus, assuming maximum eruption column heights, total eruption duration and deposit density, total eruption mass is calculated for each scenario. Table 9.1 lists the column heights and mass used by each scenario to estimate tephra accumulation at the BNPP site. For comparison, the VEI 6 scenario essentially matches source parameters that are described in the literature for the 1991 climactic Plinian eruption of Mt. Pinatubo (Koyaguchi and Tokuno, 1993; Holasek *et al.*, 1996; Koyaguchi, 1996; Paladio-Melosantos *et al.*, 1996; Rosi *et al.*, 2001).

Tephra dispersion also depends on the size distribution of particles (grain-size distribution) erupted from the volcano. Particle (clast) size distributions can be characterized in terms of several parameters (Inman, 1952): minimum and maximum volcanic clast diameter; median clast diameter (Md_ϕ); graphic standard deviation, or sorting (σ_ϕ); and the graphical skewness, a measure of the asymmetry of the grain-size distribution. Complete and reliable total grain-size distribution data for explosive volcanic eruptions are difficult to establish from field data and are rarely reported (Bonadonna and Houghton, 2005). Problems in determining the total grain-size distribution for an eruption stem from difficulty in sampling all facies of the deposit. Much of the tephra of the 1991 eruption of Mt. Pinatubo fell into the South China Sea (Wiesner *et al.*, 1995), so direct estimation of the total grain-size distribution is not practical. Nevertheless, based on a numerical model, Koyaguchi and Ohno (2001) proposed a grain-size distribution of class II fragments (pyroclasts that accumulate at medial distances from the vent) at the top of the eruption column for two depositional layers of the 1991 Plinian eruption of Mt. Pinatubo. Although not optimal, as proximal and distal pyroclasts are missing in Koyaguchi and Ohno's (2001) model, the grain-size data from those two tephra layers can be combined (Figure 9.2) to obtain a total grain-size

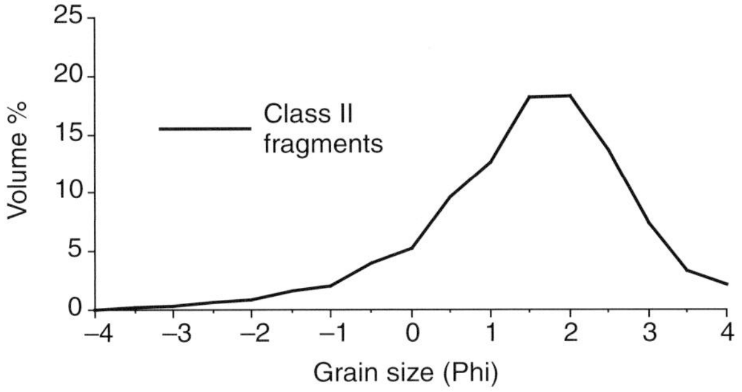

Fig. 9.2 The total grain-size distribution used in the tephra models is derived from analysis of the climactic Plinian eruption of Mt. Pinatubo in June 1991 (Class II fragments). Modified from Koyaguchi and Ohno (2001).

distribution for the tephra fallout of the 1991 Plinian eruption. The median diameter of volcanic clasts is 1.35 ϕ ($\phi = -\log_2(d)$), d being the particle diameter in millimeters. The sorting of the deposit is $\sigma_\phi = 1.16\phi$, representing good sorting for tephra deposits (Fisher and Schmincke, 1984; Cas and Wright, 1987). These values are used for all simulations in the deterministic analysis of tephra fallout, although the actual total grain-size distribution of the 1991 eruption of Mt. Pinatubo may be finer and less sorted. The graphical skewness of the total grain-size distribution is assumed to be zero.

Tephra accumulation at a site is strongly dependent on wind speed and direction during the timespan of eruption. Two different wind estimates are used for each scenario. One estimate uses wind velocities averaged for the year 2006 based on reanalysis data from the National Center for Environmental Prediction Reanalysis project (Kalnay *et al.*, 1996). The reanalysis data consists of wind-velocity estimates at 17 pressure levels which are are linearly interpolated to 30 heights from 1–30 km (above 30 km, wind conditions are assumed to be constant) at 1 km intervals (Figures 9.3a and b). The second estimate represents an upper limit, whereby the wind is assumed to blow toward the BNPP with a speed, at each level, similar to the average wind speeds from the reanalysis data for 2006. Wind conditions, very similar to this upper-limit estimate, occurred in 2006 $\sim$ 3% of the time for Mt. Pinatubo, $\sim$ 9% of the time for Mt. Natib and $\sim$ 11% of the time for Mt. Mariveles (Figure 9.3c).

Results of the different deterministic scenarios are given in Table 9.2. Estimated potential accumulation at the BNPP site varies from trace amounts to 3.6 m for a VEI 7 eruption at Mt. Natib with wind blowing toward the site. These thicknesses correspond to a range in dry tephra load of about 0.01 kg m^{-2} to 3600 kg m^{-2}. Rainfall saturates tephra deposits and may double these estimated loads (Blong, 1984).

Isomass maps are shown in Figure 9.4. Note that the scenario presented in Figure 9.4a for Mt. Pinatubo shows many similarities with the isopach maps proposed for the 1991 eruption in the literature (Koyaguchi, 1996; Paladio-Melosantos *et al.*, 1996), although the

Table 9.2. Tephra fallout thickness (cm) at the BNPP site for each eruption scenario in the deterministic analysis

Volcano	Wind field	VEI 3	VEI 4	VEI 5	VEI 6	VEI 7
Natib	Wind 2006[1]	1.0	6.7	39	100	180
	Max wind[2]	1.6	12	74	190	360
Mariveles	Wind 2006	0.001	0.01	0.03	0.1	0.7
	Max wind	0.4	5.3	36	98	200
Pinatubo	Wind 2006	0.005	0.3	4.7	13	30
	Max wind	0.01	0.5	8.0	26	58

[1] Average wind velocity in 2006.

[2] Average wind speed in 2006, but wind blows toward site.

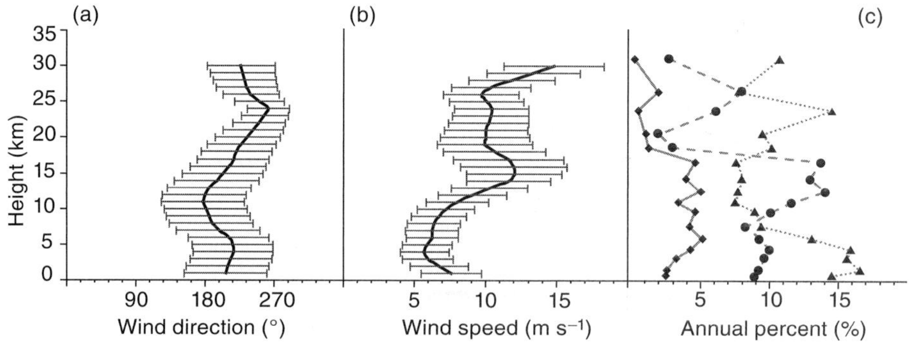

Fig. 9.3 Hazards associated with tephra fallout are strongly dependent on meteorological conditions. Here, a compilation of reanalysis data for the BNPP site illustrates the average (dark line) and one standard deviation (horizontal bars) of wind conditions in 2006, for (a) the direction toward which the wind is blowing and (b) wind speed ($\mathrm{m\,s^{-1}}$), as a function of height above sea level. These data are used as input parameters for TEPHRA2 to estimate tephra accumulation in the site region. (c) Tephra deposition at the BNPP site is maximum when the wind blows from the volcano toward the site. The percentage of the time the wind blew toward the site ($\pm 15°$) from Mt. Natib (circles), Mt. Pinatubo (diamonds) and Mt. Mariveles (triangles) is graphed as a function of height above sea level.

latter are more circular and dispersed toward the west-southwest (Koyaguchi and Tokuno, 1993; Wiesner *et al.*, 1995; Koyaguchi, 1996; Paladio-Melosantos *et al.*, 1996). This discrepancy is due to wind direction. Average wind directions at stratospheric altitudes for 2006 were mainly toward the southwest. Stratospheric winds were more toward the west during the actual eruption. The model predicts a tephra fallout thickness at the site of $\approx$ 13 cm, roughly double the observed accumulation during the eruption of Mt. Pinatubo, owing to the difference in wind direction. Furthermore, although the effect of the passage of typhoon Yunya during the eruption had little effect on the settling of high-Reynolds-number

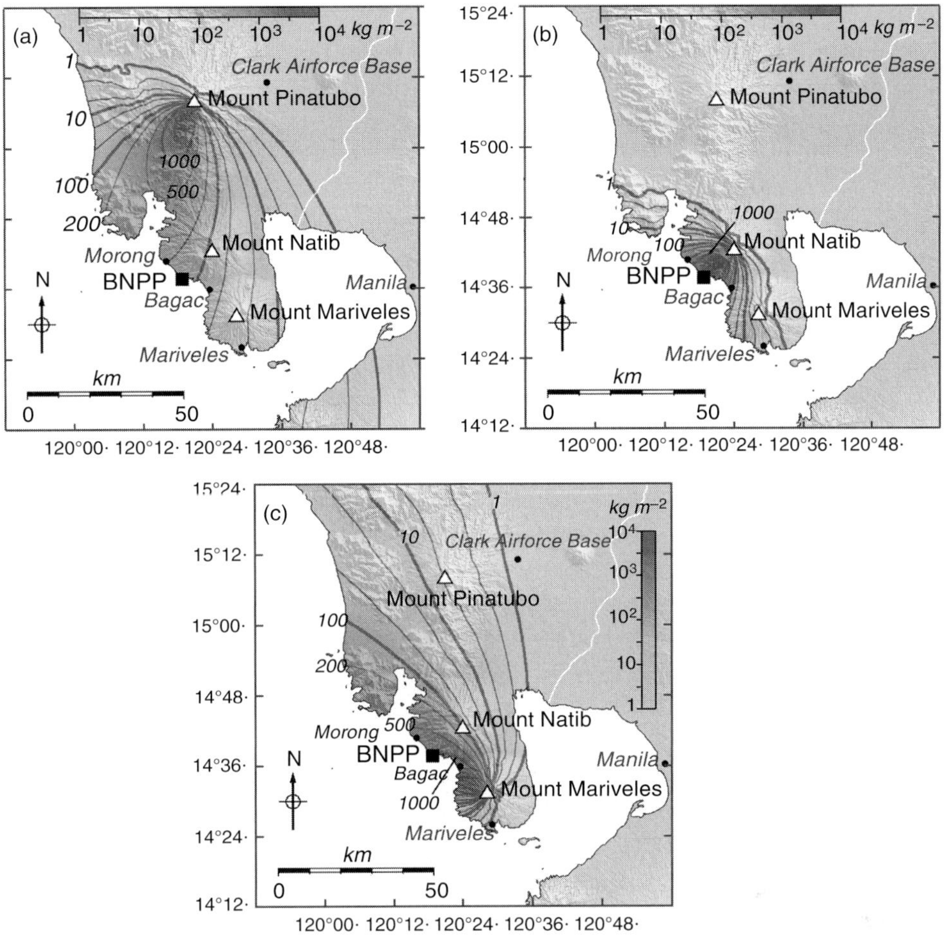

Fig. 9.4 Explosive eruptions of (a) Mt. Pinatubo, (b) Mt. Natib and (c) Mt. Mariveles volcanoes may result in substantial accumulation of tephra at the BNPP site. These examples, based on output from TEPHRA2, show isomass maps for eruptions of various magnitudes and wind conditions. In (a) a VEI 6 eruption of Mt. Pinatubo during average wind conditions for 2006 results in an isomass map that is very similar to the actual tephra distribution following the 1991 eruption of Mt. Pinatubo. In this example, tephra accumulation at the site is $\sim 100\,\mathrm{kg\,m^{-2}}$, a mass load sufficient to cause damage to some structures, and to adversely affect electrical and water filtration systems. In contrast, a much smaller magnitude eruption, VEI 4, from Mt. Natib would potentially result in much larger tephra accumulation at the site, $>1000\,\mathrm{kg\,m^{-2}}$ (b). In this simulation wind is assumed to blow from Mt. Natib toward the site at average speed as a function of elevation for the region. Similarly, the model suggests that a VEI 5 eruption of Mt. Mariveles would result in $>1000\,\mathrm{kg\,m^{-2}}$ tephra accumulation at the site, if winds blew from the volcano toward the site. Contours are in mass of tephra accumulation per unit area ($\mathrm{kg\,m^{-2}}$, dry, where $\mathrm{kg\,m^{-2}}$ is roughly equivalent to 10 cm tephra thickness).

particles (Rosi *et al.*, 2001), typhoon Yunya may have been responsible for more spherical dispersion of low- to intermediate-Reynolds-number clasts, resulting in the more spherical isopach maps proposed in the literature (Koyaguchi, 1996; Paladio-Melosantos *et al.*, 1996). Comparison of the eruption and the simulations suggest that had the wind blown toward the site on June 15, 1991, the BNPP might have experienced tephra fallout as thick as $\sim$ 25 cm (Table 9.2).

Although average wind conditions for 2006 closely mimic the shape of the tephra deposit of the 1991 eruption of Mt. Pinatubo (Figure 9.4a), these conditions poorly estimate extreme events. As an example, a VEI 6 from Mt. Mariveles will deposit only 1 mm of tephra with the average wind conditions, while with the wind blowing toward the site, the tephra thickness could reach 1 m (Table 9.2). Figure 9.3c shows that this scenario (wind blowing toward the site) occurs $\sim$ 11% of the time for Mt. Mariveles. The average wind conditions happen to deposit tephra away from the BNPP, but a large fraction of individual wind fields actually blows closer to the BNPP area.

These isomass maps also point to the possibility that secondary phenomena resulting from tephra fallout could potentially affect the site area. Although tephra accumulations at the BNPP site are not significant (e.g. not exceeding 10 cm) for many scenarios (e.g. < VEI 6), lower explosivity eruptions on Mt. Natib may result in significant tephra accumulations up-slope from the site. Such deposits may be sufficient to remobilize and form lahars that could possibly affect the site area.

Probabilistic analysis

Probabilistic methods more thoroughly assess the effects of random variation in eruption parameters and meteorological conditions on estimates of tephra accumulation. Our probabilistic analysis uses TEPHRA2 to calculate the distribution of tephra accumulation at the BNPP site for potentially explosive eruptions of Mt. Natib, Mt. Mariveles and Mt. Pinatubo volcanoes. For each of the three source volcanoes, a Monte Carlo analysis is completed, each consisting of 1000 simulations. Eruption column height is randomly sampled from a log-uniform distribution of range 14–40 km. A log-uniform distribution is used because this truncates possible values at the lower limit of credible column heights for small explosive eruptions. As noted previously, this minimum column height (14 km) represents the approximate boundary between VEI 3–4. The upper bound of the range also has practical significance. Although higher columns may be possible, the properties of the atmosphere at these altitudes are such that higher columns would have little additional impact on the dispersion of tephra particles. The use of a logarithmic function reflects the higher frequency of lower-altitude volcanic plumes (Simkin and Siebert, 1994).

The total erupted mass of tephra is calculated using (9.5) from eruption duration and column height. Duration is randomly sampled from a uniform distribution of range 1–9 hr. This range is consistent with eruption durations reported for VEI 4–6 eruptions, and is consistent with the eruption duration for the 1991 Plinian eruption of Mt. Pinatubo (Koyaguchi and Tokuno, 1993; Paladio-Melosantos *et al.*, 1996; Tahira *et al.*, 1996). No correlation is assumed between eruption column height and eruption duration for the purpose of estimating

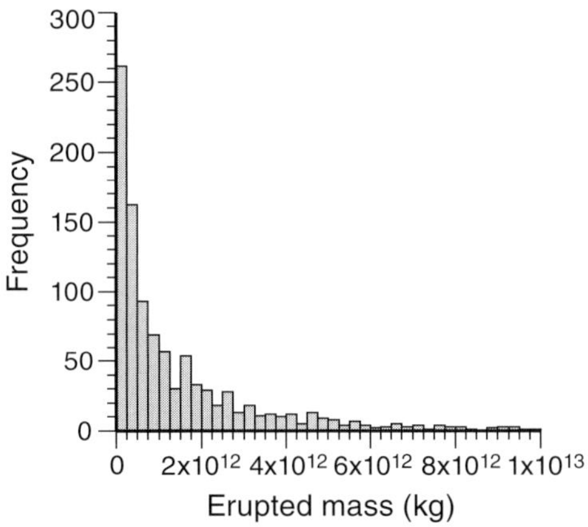

Fig. 9.5 Log-normally distributed values for total amount of tephra erupted (kg). The probabilistic assessment of tephra fallout randomly chooses total eruption mass values from this log normal distribution. The range of possible values is initially calculated from a range of probable eruption column heights and eruption durations.

total eruption mass. The resulting distribution of total eruption mass is log-normally distributed (Figure 9.5), emphasizing the higher probability of smaller eruptions (Simkin and Siebert, 1994).

Large variations in grain-size distribution appear possible for different types of Plinian eruptions from similar volcanoes. The 1991 Plinian eruption of Mt. Pinatubo has estimates of $Md_\phi = 1.35$, and $\sigma_\phi = 1.16$, for class II fragments at the top of the volcanic eruption column (Koyaguchi and Ohno, 2001). Typically, Md_ϕ estimates for entire tephra-fallout deposits might range from $-1.0\,\phi$–$4.0\,\phi$, or even smaller (i.e. $Md_\phi = 4.4$–4.8 for the 1980 eruption of Mount St. Helens; Carey and Sigurdsson, 1982; Durant *et al.*, submitted). The variation in the sorting of a tephra-fall deposit (σ_ϕ) may be smaller, $2\,\phi$–$3\,\phi$ for Plinian eruptions. Given this uncertainty, Md_ϕ and σ_ϕ are sampled from uniform distributions with ranges of $-1\,\phi$ to $5\,\phi$, and $2\,\phi$–$3\,\phi$, respectively. No correlation is assumed between these grain-size distribution parameters and column height or eruption mass.

Reanalysis data are again used to describe the variation in wind velocity with height; a set of 1460 wind profiles (acquired four times daily during 2006, Kalnay *et al.*, 1996) are randomly sampled. Although the eruption duration may be longer than 6 hr, only one randomly selected profile per simulation is used.

The results of this probabilistic analysis indicate that tephra accumulation at the BNPP site from possible eruptions of Mt. Natib and Mt. Mariveles would likely exceed tephra accumulation from possible eruptions of Mt. Pinatubo, by approximately one order of magnitude (Figure 9.6a). For comparison, the EBASCO (1979) hazard curve (Figure 9.6b) indicates that the probability of exceeding 1 m tephra accumulation at the site, given an explosive eruption of Mt. Natib, is $\approx P\{\text{accumulation} > 1\text{ m} \mid \text{explosive eruption}\} = 55\%$. However, using TEPHRA2, all calculated hazard curves, given an explosive eruption of

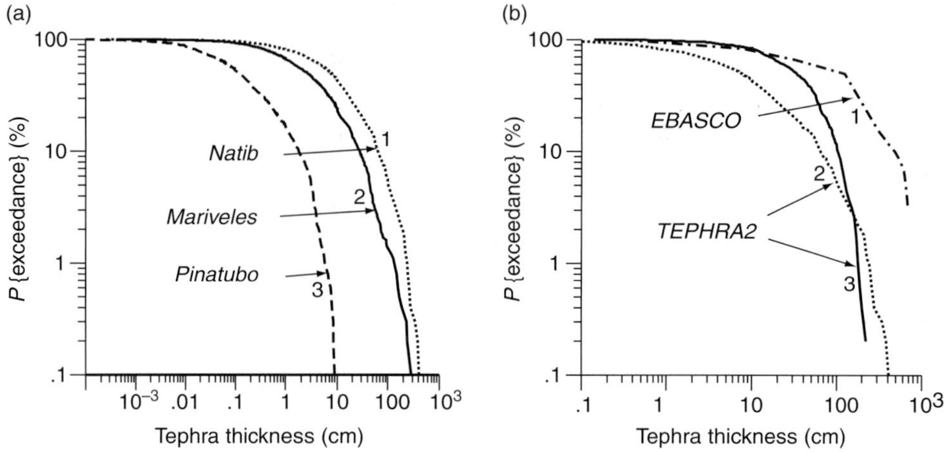

Fig. 9.6 Hazard curves show the conditional probability of exceeding different thicknesses of tephra at the location of the BNPP, given a volcanic eruption. Graph (a) compares tephra thicknesses modeled for Natib (1), Mariveles (2) and Pinatubo (3). The curves were generated from TEPHRA2 output, based on 1000 simulations using wind values randomly selected from reanalysis data for 2006 and eruption parameters randomly selected from a range of explosive eruption conditions. This graph indicates that given an eruption, tephra accumulation at the BNPP from eruptions of Mt. Natib and Mt. Mariveles are similar, and would likely exceed tephra accumulations associated with a Mt. Pinatubo eruption by one order of magnitude. Graph (b) compares the EBASCO hazard curve, based on the 1912 eruption of Mt. Katmai in Alaska (1), with two hazard curves generated by 1000 simulations using TEPHRA2: curve 2 is identical to curve 1 in graph (a), curve 3 is based on eruption parameters similar to the 1991 eruption of Mt. Pinatubo and a random wind field, also based on 2006 reanalysis data. This graph indicates that using the 1912 Katmai eruption as an analog for tephra accumulation overestimates the hazard at the site by approximately one order of magnitude. These curves could be weighted by the probability of occurrence as part of a comprehensive hazard assessment.

Mt. Natib, yield lower probabilities for exceeding 1 m of tephra accumulation at the BNPP site. It appears that the EBASCO (1979) assessment using Mt. Katmai analog eruption data may overestimate the tephra-fall hazard at the site compared to the results of numerical simulation.

The results of probabilistic analyses can also be represented as probability maps. These maps show the probability of exceeding a given threshold of tephra accumulation over an area of interest. Thresholds of tephra accumulation can be chosen to reflect potential damage to buildings (i.e. tephra load leading to partial or complete roof collapse) in the area, potential accumulation that might lead to lahar formation, or reflect design factors for NPP structures. A value of 10 cm reflects the onset of roof collapse, 20 cm reflects widespread roof collapse, especially if the tephra layer is saturated with water (Spence *et al.*, 1996). The probability of tephra accumulation exceeding 100 kg m^{-2} (dry accumulation, $\approx$ 10 cm in thickness) in the region around the BNPP is shown in Figure 9.7 and is $\approx$ 55% near the BNPP site. Of course, roofs of nuclear facilities may be designed to withstand higher loads than typical

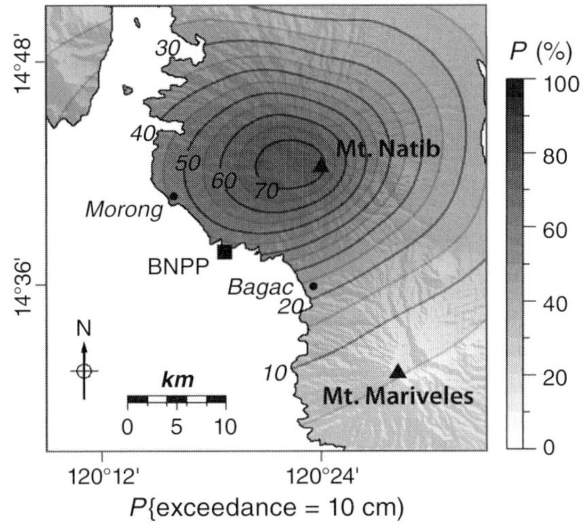

Fig. 9.7 The map contours the probability of tephra accumulation exceeding 10 cm ($\sim$100 kg m^{-2}), given an explosive eruption of Mt. Natib. Note that these simulations indicate that tephra accumulation is most likely on the western and southwestern flanks of Mt. Natib, suggesting these areas are potential sources for lahars following explosive volcanic activity. Map contoured in 5% intervals.

buildings and houses. Nevertheless, the probability map indicates that widespread damage to community infrastructure in the region of the NPP is likely in the event of an eruption of significant intensity (e.g. $\geq$ VEI 4). Such conditions are important to consider in site suitability assessment and design (Hill *et al.*, Chapter 25, this volume).

The probability map also indicates that the central part of Mt. Natib, and the western and southwestern flanks of the volcano are the most likely areas to be subjected to tephra fallout. This indicates that in the event of an explosive eruption, lahars would likely occur, potentially over widespread areas, on this flank of the volcano. These lahars would impact community infrastructure and possibly directly impact the BNPP site area.

In summary, it appears that the BNPP site is located within the screening distance for tephra fallout from both Mt. Natib and Mt. Mariveles. A screening threshold of approximately 10 cm tephra accumulation is used, based on a damage threshold commonly observed for residential and commercial buildings. A different screening threshold may be desirable for nuclear facilities, which may have greater resiliency for roof loads, but may be more sensitive to particulates in water and electrical systems. In a comprehensive hazard analysis, the design of the facility should be evaluated for the range of tephra loads such as those described in our analysis (cf. Hill *et al.*, Chapter 25, this volume).

9.3.2 Lahar source regions

Several types of volcanic phenomena, such as pyroclastic flows and lahars, can be strongly influenced by topography. Estimation of screening-distance values for these volcanic phenomena should account for major topographic features of the volcano and of the site. On Mt. Natib volcano, for instance, a steep caldera wall on the west side of the volcano may prevent many types of flows that originate in the caldera from reaching the site. Instead, such flows might drain from the caldera through a break in its northwestern rim (EBASCO, 1977). It is therefore somewhat counter-intuitive that pyroclastic flow deposits and lahars are mapped at the BNPP site. Either these units were deposited when the volcano had a very different form, prior to the formation of the summit caldera, or these phenomena can reach the site area despite the topographic barrier provided by the western caldera rim. The purpose of screening-distance-value calculations is to determine whether such flows could reasonably develop in the future, considering current or likely future conditions.

On Mt. Pinatubo volcano, lahars were generated as a result of the accumulation of pyroclastic material (e.g. tephra and pyroclastic flow deposits) on steep slopes and as a result of tropical rainfalls that worked to erode these deposits rapidly (Daag, 2003). For potential lahars from Mt. Natib, experience on Mt. Pinatubo volcano (Newhall and Punongbayan, 1996) suggests that complex scenarios accompany explosive volcanic eruptions, and these scenarios might cause lahar source regions to develop outside the caldera, including high on the west flank of the volcano. The exceedance-probability map for tephra accumulation in the site region makes it clear that if explosive activity were to occur on Mt. Natib or a nearby volcano, conditions for lahar formation may develop (Figure 9.7). The steep slopes on the west and south flanks of Mt. Natib could serve as source regions for lahars that would descend river valleys lower on flanks of the volcano.

Several different models can be used to assess the potential hazard posed by lahars to the BNPP site. For example, a statistical model, such as LAHARZ (Iverson *et al.*, 1998) might be used to assess potential flow-paths. Daag (2003) proposed a water runoff model for lahars following the 1991 eruption of Mt. Pinatubo to predict lahar runout and magnitude, utilizing a cell-based distributed model coupled with a high resolution digital elevation model (DEM). Daag's (2003) model is catchment-scale and uses physical laws of flow dynamics to describe lahars on Mt. Pinatubo. Although these models help delineate potential areas of lahar inundation, they require high-resolution (i.e. ideally < 10 m grid) digital elevation data that are currently not available for the Mt. Natib region. Here, we focus on the coupled nature of tephra fallout and lahar generation by considering two empirical models. The first model is based on the potential for gravitationally induced failure of the tephra deposit on the volcano slopes (Iverson, 2000), thus triggering lahars. The second model is based on the increase in water and sediment runoff as tephra accumulates (Daag, 2003; Yamakoshi *et al.*, 2005).

Iverson's (2000) model assumes that slope failure is described by a Coulomb failure criterion expressed as a yield condition:

$$|\tau| = c + \sigma_{\mathrm{n}} \tan \beta \tag{9.6}$$

where τ is shear stress, c is tephra cohesion, σ_{n} is normal stress (perpendicular to the slope) and β is the angle of internal friction. This Coulomb failure model can be expressed as a ratio of resisting and driving forces, known as the Factor of Safety (FS):

$$FS = \frac{\text{Resisting force}}{\text{Driving force}} = \frac{c + \sigma_n \tan\beta}{|\tau|} \tag{9.7}$$

where $\tau = -Zy_t \sin\alpha$, $\sigma_n = Zy_t \cos\alpha$, Z is the layer thickness (here derived from estimates of potential tephra accumulation), y_t is the total unit weight of the deposit per unit area and α is the slope of the slip surface (pre-tephra deposition topography). It follows that:

$$FS = \frac{c}{Zy_t \sin\alpha} + \left(1 - \frac{y_w}{y_t}\right)\frac{\tan\beta}{\tan\alpha} \tag{9.8}$$

where y_w is the unit weight of water added to the deposit by rainfall. Slope failure occurs when $FS < 1$. Lahars will be generated primarily on steep slopes after deposition of tephra units that become saturated by water infiltration, greatly reducing the shear strength of these tephra layers (Daag, 2003). This slope-failure model for lahar generation is thus coupled to a tephra fallout model, and, assuming deposit saturation by infiltrating water, areas of likely slope failure ($FS < 1$) can be inferred (Figure 9.8a).

The actual slope-failure process depends on the nature of the slope geology (i.e. infiltration, strength characteristics) underlying the potential tephra deposit and the cover of vegetation, grain-size properties of the tephra fallout and infiltration rates both into the tephra deposit and into the underlying units. These factors are complex, spatially variable and not explicitly addressed by the Iverson (2000) model, but should be considered in interpreting model results. Given these caveats, a Factor of Safety map (Figure 9.8a) indicates zones of potential lahar generation, from where those flows may follow main drainages and inundate areas lower on the flanks of the volcano, an idea completely consistent with the screening-distance calculation.

The potential lahar source region (Figure 9.8a) covers an area of about 12 km^2. The corresponding total volume of tephra deposit occurring on these steep slopes ($FS < 1$) is 1.7×10^7 m^3. Iverson *et al.* (1998) developed an empirical relationship between lahar volume and the planimetric area inundated by lahars, based on their analysis of numerous lahar deposits: $B = 200V^{2/3}$, where B (m^2) is the area inundated by a lahar of volume V (m^3). Using this relationship, the planimetric area inundated by one or more lahars resulting from slope failure is about 13 km^2, a moderate lahar event, comparable to the one that occurred at Nevado del Ruiz volcano in 1985 (see Connor *et al.*, Chapter 3, this volume). Therefore, this model suggests that areas on the southwest flank of the volcano would potentially be impacted by lahars following explosive volcanic activity.

Alternatively, lahars can be triggered when even thin, fine-grained tephra layers accumulate, because these layers may impede infiltration and increase surface runoff (Yamakoshi *et al.*, 2005). In a study of lahar generation following recent eruptions of Miyakejima volcano, Yamakoshi *et al.* (2005) found a positive, non-linear correlation between tephra thickness and decreased infiltration, resulting in increased surface runoff that may trigger lahars with very low sediment load by volume. Such flows are called

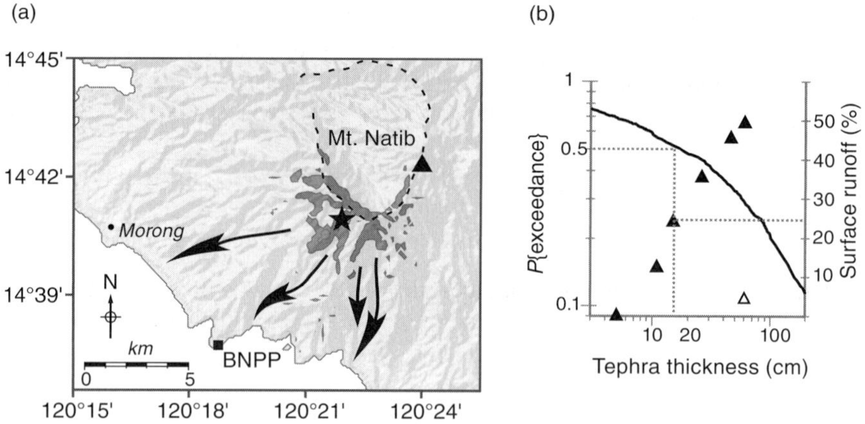

Fig. 9.8 (a) Potential lahar source regions (dark shaded areas) resulting from a hypothetical VEI 5 eruption from Mt. Natib (wind blowing toward the site), identified as those areas where the Factor of Safety, $FS \leq 1$. Arrows highlight main drainages on the south-southwest part of Mt. Natib where lahars have the potential to occur and affect the NPP site region. The black star indicates the location for the hazard curve shown in (b). (b) Exceedance probability, based on the VEI 5 eruption used in (a), and surface runoff % (100 × water and sediment runoff divided by the amount of rainfall) plotted as a function of tephra thickness. This plot indicates that lahar potential increases with higher tephra accumulation and higher runoff. Surface runoff vs. tephra thickness values (solid triangles) are for fine-grained tephra on Miyakejima volcano (Japan), modified after Yamakoshi *et al.* (2005). Runoff is diminished for coarse-grained deposits (open triangle) on Miyakejima volcano. For example, given an explosive eruption (VEI 5) of Mt. Natib, the TEPHRA2 model indicates that the probability of tephra exceeding 17 cm is 50%. Empirical observations on Miyakejima volcano suggest that for this thickness of tephra, ~ 25% of rainfall and sediment by volume will runoff into drainages, forming hyperconcentrated flows.

hyperconcentrated flows. Essentially, as the tephra deposit gets thicker, water is less likely to infiltrate the whole tephra deposit, thus increasing potential surface runoff. Daag (2003) observed, from rainfall simulations on tephra deposits from Mt. Pinatubo, that tephra fall-out deposits have low infiltration capacities due to the abundance of fine particles, thus increasing water runoff. In applying this alternative model, a hazard curve for tephra accumulation in a specific area up-slope of the site can be evaluated in terms of lahars triggered by increased runoff of surface water and tephra (Figure 9.8b).

The hazards due to hyperconcentrated flows are different from those associated with slope failure. Rather than a single voluminous debris flow, the increased surface runoff results in persistent hyperconcentrated flows and floods, which may continue to affect the area at the base of the volcano for years following the eruption. Individual events, however, are of much smaller volume. For example, in response to 9 mm rainfall and using the empirical relationship shown in Figure 9.8b, the tephra deposit would cause an increase in runoff of approximately 4×10^5 m^3. As with larger-volume debris flows, these persistent hyperconcentrated flows would adversely affect the region on the southwest flank of the volcano.

It appears that tephra dispersal can potentially result in flow-paths for lahars toward the site vicinity, following the major drainages on Mt. Natib southern flanks (Figure 9.8a), despite near-vent topographic barriers presented by the caldera wall. Because tephra deposition would likely be widespread on the southwest flank of the volcano following an explosive eruption, the nature of sedimentation on this flank of the volcano would change, resulting in lahars, hyperconcentrated flows and/or water floods. This coupled nature of volcanic phenomena is quite important to consider in estimation of screening-distance values.

Cumulatively, these analyses indicate that the BNPP site is within screening distance for lahars, given their potential volume and the areas potentially inundated by them. Of course, the geographic position of the BNPP site on Napot Point may protect the site from inundation by lahars. Comprehensive analysis of lahar flow-paths with a high-resolution DEM may verify this possibility. Regardless, lahars would significantly disrupt roads and communities in the site region, a factor important to determination in site suitability.

9.3.3 Hazards from pyroclastic density currents

Topographic barriers also may prevent future pyroclastic density currents from reaching the BNPP site. We consider a basic but widely used model (i.e. the energy-cone model; Sheridan, 1979) for the potential runout of pyroclastic density currents for the purposes of evaluating this potential hazard; and to determine if the site lies within or beyond a screening-distance value that is representative for such highly mobile flows. The energy-cone model was first proposed by Sheridan (1979; also see Connor *et al.*, Chapter 3, this volume). Essentially this model uses the height, H, from which pyroclastic density currents originate, directly related to the potential energy of the flows, to estimate their runout, L, the horizontal distance the flows are likely to travel from their source. The ratio, H/L depends on the mobility of the pyroclastic density current. Examples in the literature commonly range from $H/L = 0.2$ for small flows, to $H/L < 0.01$ for large-volume, highly mobile pyroclastic density currents.

For pyroclastic density currents originating from dome-building eruptions (e.g. the ongoing eruption of Soufrière Hills volcano, Montserrat) and from low-volume explosive eruptions (< VEI 5), our analysis shows that the caldera wall will likely act as a topographic barrier for pyroclastic flows traveling toward the site from a central vent eruption of Mt. Natib. Such flows would have insufficient potential energy to overcome the 300–500 m-high topographic barrier of the caldera wall and likely would be channelized toward the northwest, possibly exiting the caldera through a gap in the caldera wall (Figure 9.9a). Therefore, assuming low-energy explosive eruptions occur within the existing caldera, the site seems to be outside the screening distance for pyroclastic density currents released from comparatively low-lying sources within the caldera.

In the case of an explosive eruption of ~ VEI 5 or greater, or an eruption occurring from a new vent located on the southern flanks of the volcano, the energy-cone model suggests that pyroclastic density currents may reach the site. Flows associated with such eruptions often

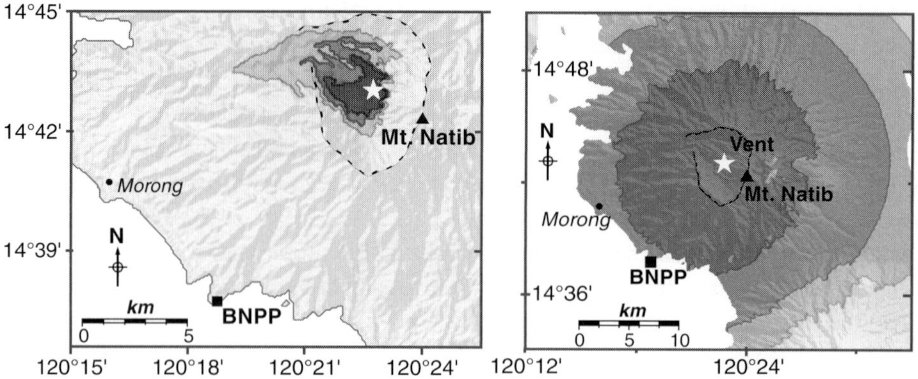

Fig. 9.9 (a) Three potential pyroclastic flow runouts from the caldera floor of Mt. Natib, estimated using the energy-cone model. The three gray-shaded regions represent possible areas inundated by pyroclastic flows originating from the collapse of a 100 m-high dome. The different shaded regions represent areas inundated by pyroclastic flows of increasing potential energy, represented by the ratio of dome height to runout length: $H/L = 0.2$ (darkest gray area), $H/L = 0.15$ (medium gray area), $H/L = 0.1$ (light gray area). Uncertainty in the appropriate value of H/L results in uncertainty in the total runout of the flow. In all of these cases, the pyroclastic flows do not overtop the caldera wall, and thus flow away from the BNPP site. (b) In contrast, higher release heights (e.g. 1000 m above the caldera floor) associated with eruption-column collapse and higher intensity eruptions result in inundation of the BNPP site. Shaded areas show inundation by pyroclastic density currents for $H/L = 0.15$ (closest to the vent, darkest shading), $H/L = 0.1$ and $H/L = 0.075$ (farthest from the vent, lightest shading).

generate pyroclastic density currents as a result of collapse of the eruption column or by boiling-over of a particularly dense eruption column. In such circumstances, the potential energy of the flow may be sufficient to overcome topographic barriers approximately 500 m high, such as the caldera wall. Based on the energy-cone model, a potential column collapse assumed to initiate at the top of the gas-thrust region (Connor *et al.*, Chapter 3, this volume), would need to originate at no more than 1 km elevation above the caldera floor. Once overcoming the southwest wall of the caldera, the topographic slope is such that the flow extends beyond the site area for $H/L < 0.15$ (Figure 9.9b). Based on this simplified analysis, it appears that the BNPP site is located within the screening-distance value of pyroclastic density currents for eruptions VEI 5 or greater. As is the case for tephra fallout, such flows could also generate voluminous lahars, which may have the potential to affect the site.

Numerical models of pyroclastic density currents (e.g. Todesco *et al.*, 2002) might greatly improve this assessment and could be considered as part of a comprehensive hazard analysis for pyroclastic density currents. The energy-cone calculation strongly suggests such an assessment would be useful for understanding a range of pyroclastic density current hazards for the BNPP site. In addition, a complete analysis of hazards also should consider the potential for new vent formation on the flanks of Mt. Natib. As observed at other composite volcanoes, such vents might also be a source of pyroclastic density currents that may create additional hazards at the site.

Concluding remarks

This analysis is intended to illustrate several important steps in a volcanic hazard assessment for nuclear facilities. For the BNPP site, this means using available data and available numerical methods to assess the capability of nearby volcanoes to erupt in the future and to produce potentially hazardous phenomena at the site. Do Mt. Natib, Mt. Mariveles and Mt. Pinatubo have a credible potential for future eruptions? Future eruptions appear highly likely from Mt. Pinatubo, considering its last explosive eruption in 1991 and many other eruptions in the Holocene. Several lines of evidence indicate that future eruptions from Mt. Natib and Mt. Mariveles are credible, including the existence of an active hydrothermal system within Mt. Natib volcano, the presence of little-eroded volcanic features (e.g. caldera depressions truncating both Mt. Natib and Mt. Mariveles) and probabilistic assessment based on the estimated repose since the last-dated eruptive event. The probability estimate is highly uncertain, due to uncertainties in ages of past eruptions and possibly underestimates of recurrence rate due to poor preservation of smaller eruptions in the geologic record. This uncertainty supports a conservative approach to hazard assessment that assumes future eruptions are possible from Mt. Pinatubo, Mt. Natib and Mt. Mariveles.

We have made such a preliminary assessment for a subset of potential volcanic phenomena utilizing a screening-distance-value approach. This analysis, made using relatively simple and widely available numerical techniques, indicates that the BNPP site has the potential to be affected by phenomena such as tephra fallout, lahars and pyroclastic density currents in the event of a future eruption. Cumulatively, these analyses indicate that Mt. Natib and Mt. Mariveles and, for tephra-fall hazards, Mt. Pinatubo, are capable volcanoes, following the definition provided by Hill *et al.* (Chapter 25, this volume).

Identification of Mt. Natib and Mt. Mariveles as capable volcanoes indicates that a more comprehensive volcanic hazard assessment appears warranted for the BNPP site. Goals of such a comprehensive assessment would include analysis of the current state of volcanic unrest at both Mt. Natib and Mt. Mariveles through implementation of monitoring techniques, such as a seismic network on the volcano, deformation monitoring and perhaps seismic tomography to ascertain the origin and extent of the hydrothermal system within the volcano. Similarly, geochemical analyses might be extremely useful to delineate a magmatic component in thermal springs located in the caldera of Mt. Natib. A second goal of the comprehensive assessment would be to map these two volcanoes in sufficient detail to develop a more complete understanding of each volcano's stratigraphy, and to place radiometric dates in stratigraphic context. An integrated program including new radiometric dates and paleomagnetic analysis appears critical to developing a suitable record of past volcanic activity. Such a geologic program would be necessary in order to assess more fully the probability of future volcanism, the timing of most recent volcanism and the important characteristics of past eruptions. Finally, more detailed analysis of volcanic hazards could make full use of a variety of numerical models that might further elucidate site hazards, particularly of lahar and pyroclastic flow phenomena. All of these numerical models of surface flows require use of a high-resolution (preferably < 10 m resolution) digital elevation model, which was not available to the authors at the time of this analysis. Acquisition of such a model would be an important step in the volcanic hazard assessment. Regardless of the details involved, the analyses presented herein demonstrate that several capable volcanoes exist within the area of the BNPP site. Based on recommendations in, for example, Hill *et al.* (Chapter 25, this volume), a comprehensive analysis appears necessary to support discussions or decisions regarding the suitability of the BNPP site.

This case study also illustrates some general factors to consider in volcanic hazard assessments of nuclear facilities. The timing and recurrence rate of volcanism at Mt. Natib and Mt. Mariveles are a major source of uncertainty in estimates of the likelihood of future activity. It is unlikely that the probability of future eruptions from Mt. Natib and Mt. Mariveles could be narrowed much below one order of magnitude by additional analyses, unless the global stratigraphic and chronological framework of these volcanoes were improved or very young volcanic deposits were identified. This is a common situation where nuclear facilities are considered in volcanically active regions. Screening based on the probability of occurrence of volcanic eruptions may have large uncertainties, giving a weak basis for decision making.

Screening distances are an effective method of assessing the potential for various phenomena to impact a site. Numerical techniques can have an important role to play in estimating these screening distances. For example, EBASCO (1977) did not have methods to simulate tephra fallout at the BNPP site numerically. Instead, a large eruption from a presumably analogous volcano was used. This resulted in a possible overestimate of potential tephra fallout hazards at the site. In contrast, probabilistic models yield hazard curves for the BNPP site that reasonably reproduce observed deposits from the 1991 Mt. Pinatubo. The great advantage of these models is that various scenarios of activity can be evaluated, providing a more robust perspective on the parameters that contribute to the potential hazards at the site.

Similarly, at the time of siting of the BNPP, it was argued that the topography of the Mt. Natib summit caldera protects the site from potential pyroclastic flows. This effect appears supportable for relatively small eruptions ($<$ VEI 5), but our analysis suggests pyroclastic flows from $>$ VEI 5 eruptions may reach the site. Furthermore, the coupled nature of volcanic phenomena (e.g. the potential of lahars resulting from tephra fallout) warrants further consideration. Fortunately, volcanology now possesses many of the tools required to make such assessments at an appropriate level of detail.

Further reading

Articles in the volume *Statistics in Volcanology* (Mader *et al.*, 2006) provide an overview of the literature on the timing of volcanic eruptions and forecasting activity at long-dormant volcanoes. The TEPHRA2 code is freely available on the Worldwide Web (Connor *et al.*,

2008). Modeling of volcanic phenomena is evolving rapidly. Models of volcanic eruptions and eruption phenomena are widely discussed in the *Bulletin of Volcanology* and *Journal of Volcanology and Geothermal Research*. See *Fire and Mud* (Newhall and Punongbayan, 1996) for comprehensive discussion of the eruptions of Mt. Pinatubo.

Acknowledgments

This manuscript was improved by the comments of Neil Chapman, Britt Hill and an anonymous reviewer. ACMV was supported by a grant from the University of South Florida.

References

Armienti, P., G. Macedonio and M. T. Pareschi (1988). A numerical model for simulation of tephra transport and deposition: applications to May 18, 1980, Mount-St-Helens eruption. *Journal of Geophysical Research*, **93**, 6463–6476.

Blong, R. J. (1984). *Volcanic Hazards: A Sourcebook on the Effects of Eruptions*. Sydney: Academic Press.

Bonadonna, C. (2006). Probabilistic modelling of tephra dispersion. In: Mader, H. M., S. G. Cole, C. B. Connor and L. J. Connor (eds.) *Statistics in Volcanology*, Special Publications of IAVCEI 1. London: Geological Society, 243–259.

Bonadonna, C. and B. F. Houghton (2005). Total grain-size distribution and volume of tephra-fall deposits. *Bulletin of Volcanology*, **67**, 441–456.

Bonadonna, C., C. B. Connor, B. F. Houghton *et al.* (2005). Probabilistic modeling of tephra dispersal: hazard assessment of a multiphase rhyolitic eruption at Tarawera, New Zealand. *Journal of Geophysical Research*, **110**, doi:10.1029/2003JB002896.

Cabato, M. E. J. A., K. S. Rodolfo and F. P. Siringan (2005). History of sedimentary infilling and faulting in Subic Bay, Philippines revealed in high-resolution seismic reflection profiles. *Journal of Asian Earth Sciences*, **25**, 849–858.

Carey, S. and H. Sigurdsson (1982). Influence of particle aggregation on deposition of distal tephra from the May 18, 1980, eruption of Mount St. Helens volcano. *Journal of Geophysical Research*, **87**, 7061–7072.

Cas, R. A. F. and J. V. Wright (1987). *Volcanic Successions, Modern and Ancient: A Geological Approach to Processes, Products, and Successions*. Berlin: Springer Verlag.

CGIAR-CSI (2004). Void-filled seamless SRTM data V1, International Centre for Tropical Agriculture (CIAT). CGIAR–Consortium for Spatial Information, SRTM 90 m Database, http://srtm.csi.cgiar.org.

Connor, L. J. and C. B. Connor (2006). Inversion is the key to dispersion: understanding eruption dynamics by inverting tephra fallout. In: Mader, H. M., S. G. Cole, C. B. Connor and L. J. Connor (eds.) *Statistics in Volcanology*, Special Publications of IAVCEI 1. London: Geological Society, 231–242.

Connor, C. B., B. E. Hill, B. Winfrey, N. M. Franklin and P. C. La Femina (2001). Estimation of volcanic hazards from tephra fallout. *Natural Hazards Review*, **2**, 33–42.

Connor, C. B., A. R. McBirney and C. Furlan (2006). What is the probability of explosive eruption at a long-dormant volcano? In: Mader, H. M., S. G. Cole, C. B. Connor and L. J. Connor (eds.) *Statistics in Volcanology*, Special Publications of IAVCEI 1. London: Geological Society, 39–46.

Connor, L. J., C. B. Connor and C. Bonadonna (2008). Forecasting tephra dispersion using TEPHRA2. http://www.cas.usf.edu/~cconnor/vg@usf/tephra.html.
Daag, A. S. (2003). Modelling the erosion of pyroclastic flow deposits and the occurrences of lahars at Mt. Pinatubo, Philippines. Unpublished Ph.D thesis, ITC Dissertation number 104. Utrecht: University of Utrecht.
D'Amato, A. and K. Engel (1988). State responsibility for the exportation of nuclear power technology. *Virginia Law Review*, **74**, 1011–1066.
Defant, M. J., R. C. Maury, E. M. Ripley, M. D. Feigenson and D. Jacques (1991). An example of island-arc petrogenesis: geochemistry and petrology of the southern Luzon Arc, Philippines. *Journal of Petrology*, **32**, 455–500.
Durant, A. J., W. I. Rose, A. M. Sarna-Wojcicki, S. Carey and A. C. M. Volentik (2008). Hydrometeor-enhanced tephra sedimentation: constraints from the 18 May 1980 eruption of Mount St. Helens (USA). *Journal of Geophysical Research*, **114**, doi:10.1029/2008JB005756.
EBASCO (1977). Preliminary safety analysis report, Philippine Nuclear Power Plant #1. Philippine Atomic Energy Commission Open-File Report and response to questions. Manila: Philippine Atomic Energy Commission.
EBASCO (1979). Evidence substantiating the incredibility of volcanism on the west flank of Mt. Natib, and the assessment of volcanic hazards at Napot Point. Response to Philippine Atomic Energy Commission question #3. Manila: Philippine Atomic Energy Commission.
Efron, B. and R. Tibshirani (1991). Statistical data analysis in the computer age. *Science*, **253**, 390–395.
Fierstein, J., B. F. Houghton, C. J. N. Wilson and W. Hildreth (1997). Complexities of Plinian fall deposition at vent: an example from the 1912 Novarupta eruption (Alaksa). *Journal of Volcanology and Geothermal Research*, **76**, 215–227.
Fisher, R. V. and H.-U. Schmincke (1984). *Pyroclastic Rocks*. Berlin: Springer Verlag.
Holasek, R. E., S. Self and A. W. Woods (1996). Satellite observations and interpretation of the 1991 Mount Pinatubo eruption plumes. *Journal of Geophysical Research*, **101**, 27 635–27 655.
Inman, D. L. (1952). Measures for describing the size distribution of sediments. *Journal of Sedimentary Petrology*, **22**, 125–145.
IAEA (1978). Report of the IAEA Safety Mission to the Philippines Nuclear Power Plant No.1. Vienna: International Atomic Energy Agency.
Iverson, R. M. (2000). Landslide triggering by rain infiltration. *Water Resources Research*, **36**, 1897–1910.
Iverson, R. M., S. P. Schilling and J. W. Vallance (1998). Objective delineation of lahar-inundation hazard zones. *Geological Society of America Bulletin*, **110**, 972–984.
Kalnay, E., M. Kanamitsu, R. Kistler *et al.* (1996). The NCEP/NCAR 40-year reanalysis project. *Bulletin of the American Meteorological Society*, **77**, 437–471.
Koyaguchi, T. (1996). Volume estimation of tephra-fall deposits from the June 15, 1991, eruption of Mount Pinatubo by theoretical and geological methods. In: Newhall, C. G. and R. S. Punongbayan (eds.) *Fire and Mud*. Seattle: University of Washington and Quezon City–PHIVOLCS, 583–600.
Koyaguchi, T. and M. Ohno (2001). Reconstruction of eruption column dynamics on the basis of grain size of tephra fall deposits: 2. Application to the Pinatubo 1991 eruption. *Journal of Geophysical Research*, **106**, 6513–6534.
Koyaguchi, T. and M. Tokuno (1993). Origin of the giant eruption cloud of Pinatubo, June 15, 1991. *Journal of Volcanology and Geothermal Research*, **55**, 85–96.

Mader, H. M., S. G. Cole, C. B. Connor and L. J. Connor (eds.) (2006). *Statistics in Volcanology*, Special Publications of IAVCEI 1. London: Geological Society of London.

Newhall, C. G. (1979). Review of volcanologic discussions in the PSAR and related documents, Philippine Nuclear Power Plant #1. Washington DC: US Nuclear Regulatory Commission.

Newhall, C. G. and R. S. Punongbayan (1996). *Fire and Mud*. Seattle: University of Washington and Quezon City–PHIVOLCS.

Newhall, C. G. and S. Self (1982). The volcanic explosivity index (VEI): an estimate of explosive magnitude for historical volcanism. *Journal of Geophysical Research*, **87**, 1231–1238.

Newhall, C. G., A. S. Daag, F. G. Delfin Jr. *et al.* (1996). Eruptive history of Mount Pinatubo. In: Newhall, C. G. and R. S. Punongbayan (eds.) *Fire and Mud*. Seattle: University of Washington and Quezon City–PHIVOLCS, 165–195.

Paladio-Melosantos, M. L. O., R. U. Solidum, W. E. Scott *et al.* (1996). Tephra falls of the 1991 eruptions of Mount Pinatubo. In: Newhall, C. G. and R. S. Punongbayan (eds.) *Fire and Mud*. Seattle: University of Washington and Quezon City–PHIVOLCS, 513–535.

Pallister, J. S., R. P. Hoblitt, G. P. Meeker, R. J. Knight and D. F. Siems (1996). Magma mixing at Mount Pinatubo: petrographic and chemical evidence from the 1991 deposits. In: Newhall, C. G. and R. S. Punongbayan (eds.) *Fire and Mud*. Seattle: University of Washington and Quezon City–PHIVOLCS, 687–731.

Pyle, D. M. (1989). The thickness, volume and grainsize of tephra fall deposits. *Bulletin of Volcanology*, **51**, 1–15.

Rodolfo, K. S., J. V. Umbal, R. A. Alonso *et al.* (1996). Two years of lahars on the western flanks of Mount Pinatubo: initiation, flow processes, deposits and attendant geomorphic and hydraulic changes. In: Newhall, C. G. and R. S. Punongbayan (eds.) *Fire and Mud*. Seattle: University of Washington and Quezon City–PHIVOLCS, 989–1013.

Rosi, M., M. L. Paladio-Melosantos, A. Di Muro, R. Leoni and T. Bacolcol (2001). Fall vs. flow activity during the 1991 climactic eruption of Pinatubo Volcano (Philippines). *Bulletin of Volcanology*, **62**, 549–566.

Ruaya, J. R. and C. C. Panem (1991). Mt. Natib, Philippines: a geochemical model of a caldera-hosted geothermal system. *Journal of Volcanology and Geothermal Research*, **45**, 255–265.

Scott, W. E., R. P. Hoblitt, R. C. Torres *et al.* (1996). Pyroclastic flows of the June 15, 1991, climactic eruption of Mount Pinatubo. In: Newhall, C. G. and R. S. Punongbayan (eds.) *Fire and Mud*. Seattle: University of Washington and Quezon City–PHIVOLCS, 545–570.

Sheridan, M. F. (1979). Emplacement of pyroclastic flows: a review. In: Chapin, C. E. and W. E. Elston (eds.) *Ash-flow Tuffs*, Geological Society of America Special Paper 180, 125–136.

Siebert, L. and T. Simkin (2007). Volcanoes of the world: an illustrated catalog of Holocene volcanoes and their eruptions. Smithsonian Institution, Global Volcanism Program Digital Information Series, GVP-3, http://www.volcano.si.edu/world/.

Sigurdsson, H. and S. Carey (1989). Plinian and co-ignimbrite tephra fall from the 1815 eruption of Tambora volcano. *Bulletin of Volcanology*, **51**, 243–270.

Simkin, T. and L. Siebert (1994). *Volcanoes of the World*. Tucson, AZ: Geoscience Press.

Sparks, R. S. J., M. I. Bursik, S. N. Carey *et al.* (1997). *Volcanic Plumes*. New York, NY: John Wiley & Sons.

Spence, R. J. S., A. Pomonis, P. J. Baxter *et al.* (1996). Building damage caused by the Mount Pinatubo eruption of June 15, 1991. In: Newhall, C. G. and R. S. Punongbayan (eds.) *Fire and Mud*. Seattle: University of Washington and Quezon City–PHIVOLCS, 1055–1061.

Suzuki, T. (1983). A theoretical model for dispersion of tephra. In: Shimozuru, D. and I. Yokoyama (eds.) *Arc Volcanism; Physics and Tectonics*. Tokyo: Terra Scientific Publishing, 95–113.

Tahira, M., M. Nomura, Y. Sawada and K. Kamo (1996). Infrasonic and acoustic-gravity waves generated by the Mount Pinatubo eruption of June 15, 1991. In: Newhall, C. G. and R. S. Punongbayan (eds.) *Fire and Mud*. Seattle: University of Washington and Quezon City–PHIVOLCS, 601–613.

Todesco, M., A. Neri, T. Esposti Ongaro *et al.* (2002). Pyroclastic flow hazard assessment at Vesuvius (Italy) by using numerical modeling. I. Large-scale dynamics. *Bulletin of Volcanology*, **64**, 155–177.

van Westen, C. J. and A. S. Daag (2005). Analysing the relation between rainfall characteristics and lahar activity at Mount Pinatubo, Philippines. *Earth Surface Processes and Landforms*, **30**, 1663–1674.

Wiesner, M. G., Y. Wang and L. Zheng (1995). Fallout of volcanic ash to the deep South China Sea induced by the 1991 eruption of Mount Pinatubo (Philippines). *Geology*, **23**, 885–888.

Wolfe, E. W. and R. P. Hoblitt (1996). Overview of the eruptions. In: Newhall, C. G. and R. S. Punongbayan (eds.) *Fire and Mud*. Seattle: University of Washington and Quezon City–PHIVOLCS, 3–20.

Wolfe, J. A. and S. Self (1983). Structural lineaments and Neogene volcanism in southwestern Luzon. In: Hayes, D. E. (ed.) *The Tectonic and Geological Evolution of Southeast Asian Seas and Islands: Part 2*, American Geophysical Union Monograph 27, 157–172.

Yamakoshi, T., Y. Doi and N. Osanai (2005). Post-eruption hydrology and sediment discharge at the Miyakejima volcano, Japan. *Zeitschrift für Geomorphologie*, **140**, 55–72.

10

Multi-disciplinary probabilistic tectonic hazard analysis

M. W. Stirling, K. R. Berryman, L. M. Wallace, N. J. Litchfield, J. Beavan and W. D. Smith

Previous chapters have described how plate tectonics is the driving mechanism for tectonic hazards at the surface of the Earth. This chapter provides an overview of the methodologies being used today to quantify these hazards, describes some of the long-standing methodology-related issues that are being addressed as a result of the engineering demands of the nuclear industry, and presents a case study of evolving probabilistic methods to evaluate tectonic hazards for high-level radioactive waste disposal in Japan.

Traditionally, seismic catalogs and active fault data sets have been the basis for earthquake shaking (seismic) hazard assessment, and to a lesser extent surface rupture (fault displacement or rock deformation) hazard. These data sets are frequently criticized for being "patchy," in that there is considerable uncertainty regarding the completeness of the active fault databases, and the questionable long-term representativeness of seismicity catalogs. Not surprisingly, the advent of global positioning system (GPS) techniques as a means for quantifying the contemporary rates of active tectonic processes has been seen as a much needed augmentation of the other data sets for tectonic hazard analysis. Such data sets are very likely to become major components of tectonic hazard analysis, once a better understanding of the relevance of GPS-derived crustal strain rates to tectonic hazards has been achieved (Wallace *et al.*, Chapter 6, this volume). In this context, a new method for the melding of GPS data with the traditional data sets of tectonic hazard analysis (seismological and geologically based surface deformation/active fault data) is presented in this chapter. The methodology involves reducing the geological, seismological and GPS data sets to a parameter common to all three that can be used to quantify tectonic hazards, this being the rate of strain in the crust. Crustal strain rate gives the rate at which the crust deforms in response to tectonic forcing, and as such it can be a proxy for the potential for hazardous tectonic events such as earthquakes, fault displacement (sudden release of elastic strain in an earthquake, or steady-state creep in some restricted cases) and uplift and tilting. Development of crustal strain rate models therefore provides a basis for the quantification of a variety of tectonic hazards. We provide an overview of the developmental methodology, using the Tohoku region of Honshu, Japan, as a case study.

10.1 Probabilistic seismic hazard analysis

Cloos (Chapter 2, this volume) has provided the background to the plate tectonics theory, and how plate tectonic processes are expressed in the Earth by the occurrence of features such as active faults and earthquakes. The tectonic hazards of earthquake shaking and abrupt fault displacement are usually characterized by way of a probabilistic methodology. Probabilistic hazard analyses allow the consideration of multi-valued or continuous events and models, in contrast to deterministic analyses which are based on discrete events or models. The methodology of probabilistic seismic hazard analysis (PSHA) was originally developed by Cornell (1968), and the methods have now become a standard for combining information on earthquake occurrence, radiation of seismic energy and the influence of site geology on the strength of earthquake shaking. Deterministic methods were used extensively for engineering design prior to PSHA, but the need to consider hazards from all sources and for a range of probabilities or return periods led to the use of probabilistic methods. Today most hazard analyses combine the two techniques to suit the needs of a wide variety of end-users.

Some of the most obvious products of PSHA are the national seismic hazard maps developed for countries like the United States and New Zealand (Frankel *et al.*, 2002; Stirling *et al.*, 2002). Detailed PSHAs have also been carried out for nuclear facilities, such as the proposed Yucca Mountain high-level nuclear waste repository in Nevada (Stepp *et al.*, 2001). Probabilistic methodology characterizes hazards from a combination of all potential events that may impact a site; it is usually the case that more events can produce moderate levels of impact than the few events closest to the site producing the most severe of impacts. Major facilities such as nuclear power plants are typically designed to operate throughout the occurrence of frequent events of moderate severity (e.g. peak ground accelerations (PGAs) of 0.2 g or less that might be expected at return periods of tens of years), and also to be safely shut down in the event of a rare, potentially damaging event.

A four-step methodology describes PSHA (Cornell 1968; Figure 10.1). Step 1 is to locate all known earthquake sources in the area, usually within 100–200 km of the site. The two classes of earthquake sources considered are large earthquakes, assumed to be represented by faults, and moderate to large background earthquakes, represented by historical seismicity patterns. Fault sources are usually characterized from geological data, and provide information on the location, size and recurrence intervals (number of earthquakes through time) of the largest earthquakes (typically of magnitude $M \geq 6.5$, and with recurrence intervals of $10^2 - 10^5\ a^{-1}$, depending on the tectonic setting). Since most faults have not ruptured during the timespan of written historical records, estimates of their size and recurrence intervals are made by way of empirical methods. Empirical scaling relationships based on historical earthquake data use fault rupture dimensions such as length, width and displacement to estimate magnitudes that would be expected to accompany rupture of the faults. The scaling relationships are regressions of worldwide sets of historical earthquake data, and there are numerous examples of these relationships in the literature (e.g. Aki and Richards, 1980; Wells and Coppersmith, 1994; Hanks and Bakun 2002; Berryman and Villamor, 2004). The recurrence intervals assigned to these faults are either based on interpretation

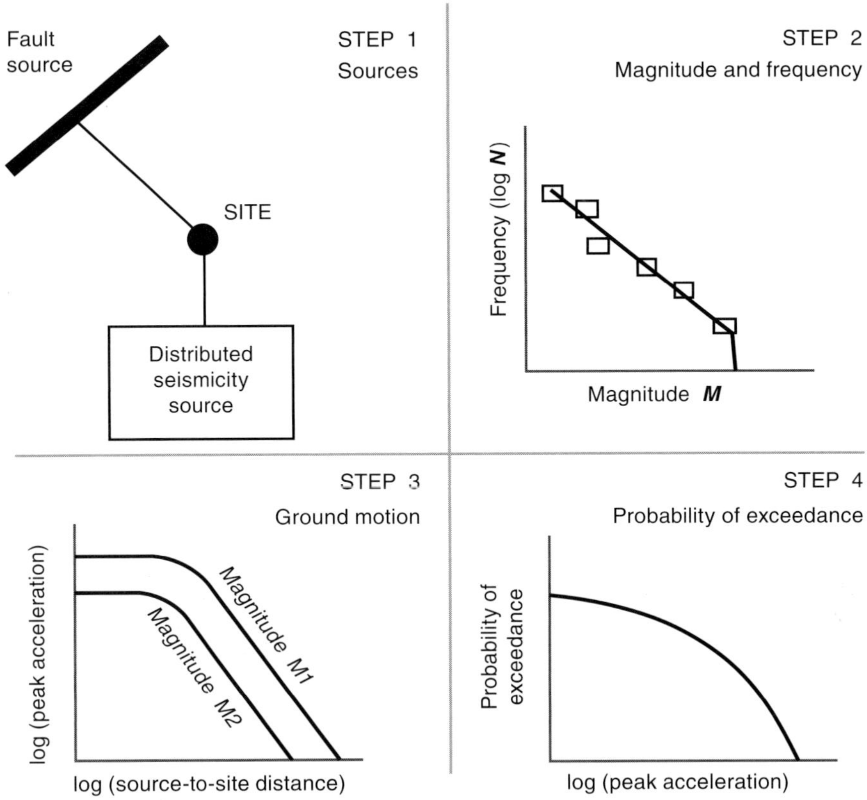

Fig. 10.1 The four steps of probabilistic seismic hazard analysis (PSHA).

of geological data taken from detailed study of prehistoric earthquake activity (paleoseismology) or are estimated through consideration of parameters like slip rate and magnitude (e.g. Berryman and Villamor, 2004).

Background earthquakes are less than or equal in size to the large fault-derived events, and can occur both on and away from the faults. They are well illustrated by maps of seismicity. These events are considered in a PSHA for two reasons. First, a large percentage of earthquakes in the historical record have not occurred directly on mapped faults. These earthquakes are presumably due to interseismic strain accumulation in areas between the major faults or are due to displacements on unmapped or blind faults. Second, earthquakes of $M \leq 6.5$ generally do not produce surface ruptures that contribute to the measurable (geological) displacement of the ground surface across the faults (e.g. Wesnousky, 1986). The rupture widths of these earthquakes are less than the width of the fault plane, hence the lack of surface expression. Background earthquakes are usually characterized from historical seismicity data, though some source characterization has occasionally come from geodetic data, mainly through providing crustal strain rates that are converted to seismic moment rates and then to earthquake rates (e.g. Working Group of California Earthquake Probabilities, 1995). The earthquakes tend to be in the range of $M \leq 7$, and occur more frequently than the larger events according to the Gutenberg–Richter relationship (see below). In PSHA, background earthquakes are generally only used to model the magnitude–frequency for $M > 5$, given that smaller events generally do not produce significant damage (e.g. Reiter, 1990).

Step 2 is to determine the magnitude and frequency of earthquakes that have been produced by these sources in the past. The log-linear Gutenberg–Richter relationship (Gutenberg and Richter, 1944) typically describes the total size distribution of earthquakes in a region, and is shown schematically in Step 2 of Figure 10.1. This relationship shows that the logarithm of the frequency of earthquake occurrence is a decreasing function of magnitude, and as such is of considerable value as it allows the prediction of the frequency of larger events from the observed frequency of smaller events. Gutenberg and Richter found that this type of distribution of seismicity applies to large areas, and it has also been shown to describe the earthquakes that occur along a given fault zone that are smaller than the maximum magnitude ($M_{\max}$) expected on a given fault (e.g. Stirling *et al.*, 1996). The equation of the Gutenberg–Richter relationship is:

$$\log N = a - bM \tag{10.1}$$

where N is the number of events per year of magnitude greater than or equal to M, and a and b are empirical constants. The b value is typically ~ 1 for most regions of the world. Gutenberg–Richter distributions typically represent the seismicity inside regions (area sources, such as the one shown schematically in Step 1, Figure 10.1).

Step 3 is to calculate the ground-motion levels (e.g. PGA) that would be produced by the earthquakes in Step 2, and Step 4 is to estimate the frequency or probability of these ground-motion levels from all sources (usually expressed as a probability or frequency of exceedance in a unit time period, e.g. one year). The PSHA can be expressed mathematically as:

$$E(z) = \sum_{i=0}^{n} \alpha_i \int_{M_o}^{M_u} \int_0^{\infty} f_i(M) f_i(r) P\{Z > z|M, r\} \mathrm{d}r \mathrm{d}M \tag{10.2}$$

in which $E(z)$ is the expected number of exceedances of ground-motion level z during a specified time period t, α_i is the mean rate of occurrence of earthquakes between lower and upper bound magnitudes (M_o and M_u) being considered in the ith source, $f_i(M)$ is the frequency of occurrence for magnitude M (i.e. derived from the recurrence relationship) for source i, $f_i(r)$ is the probability density distribution of epicentral (or source) distance between the various locations within source i and the site for which the hazard is being estimated, and $P\{Z > z|M, r\}$ is the probability that a given earthquake of magnitude M and epicentral distance r will exceed ground-motion level z (Reiter, 1990). An attenuation relationship is used to predict a median ground-motion level, and then the standard deviation for the attenuation model is used with the median to predict $P\{Z > z|M, r\}$. The fundamental output of Step 4 is a hazard curve, which gives the frequency or probability of exceedance for a suite of ground-motion levels (e.g. PGA) from all sources. The hazard curve is thus a monotonically decreasing function of probability with respect to ground motion, but this is accentuated by the fact that more earthquake sources in a region can contribute to the lower levels of earthquake shaking relative to the limited number of close-by sources that can produce the strongest motions. The strongest levels of ground motion are defined by the uppermost extremes of $P\{Z > z|M, r\}$ in (10.1), which can either be truncated at n standard deviations above the median (3 standard deviations are used for the national maps of the USA and New Zealand), or left unbounded, as in the case of the Yucca Mountain PSH model (Stepp *et al.*, 2001).

The actual likelihood or probability calculation in Step 4 in PSHA is usually estimated by assumption of a Poisson process, in which probability is assumed to be time-independent (constant over time). In the context of earthquake hazards the following equation provides the probability (p) of a given severity of hazard (e.g. ground-motion level) according to a Poisson process:

$$p = 1 - e^{-rt} \tag{10.3}$$

in which r is the rate of equaling or exceeding a given severity of hazard (from Step 2 for displacement hazard, or Step 3 for seismic hazard, i.e. (10.2)) in time t. Exceptions to the exclusive use of a Poisson process occur when there is sufficient information on the recurrence interval and elapsed time since the last event on a fault for a conditional probability model to be developed. In such cases a fault "almost ready to go" would be given a higher conditional probability of earthquake occurrence than a fault that had recently produced an earthquake and was thought to be in the early stages of the earthquake cycle.

10.2 Probabilistic fault-displacement hazard analysis

Despite being developed to model earthquake shaking hazard, the PSHA methodology has also been successfully adapted to probabilistic fault-displacement hazard analysis (PFDHA; see Youngs *et al.*, 2003); and to hazards such as tsunami, landslide and volcanic eruptions (e.g. Stirling and Wilson, 2002; Power *et al.*, 2007). In the case of PFDHA, abrupt fault displacement replaces seismic shaking as the hazard in question, so it is the potential activity of fault sources directly beneath the site that is of concern. The probabilistic model is developed to quantify the likely magnitude and frequency of tectonic displacements at the site in question. Such assessment requires information on active or potentially active faults in the immediate vicinity of the site, and as such PFDHAs typically require detailed field investigations in and around the site. In the context of the four steps of PSHA, Step 2 is often greatly simplified for PFDHA in the case where a single fault source occurs at the site. Step 3 is unnecessary, and Step 4 may be greatly simplified by there typically being a very limited number of sources known or suspected beneath the site.

10.3 Treatment of uncertainty: logic trees and expert elicitation

Fundamental to the development of a probabilistic hazard model is the identification and treatment of uncertainty in the model parameters and resulting hazard. Uncertainty can be separated into two categories. *Aleatory variability* is randomness that is inherent in many Earth system processes, such as the differences between ground motions for earthquakes of identical magnitude. *Epistemic uncertainty* is knowledge uncertainty, such as the differences in earthquake magnitude estimated from two published equations. Deciding the degree of epistemic uncertainty and aleatory variability that should be assigned to a PSH model is frequently the topic of considerable debate (e.g. Bommer *et al.*, 2004).

The development of a probabilistic hazard model for a major facility such as a nuclear reactor or hydroelectric dam usually involves the formation of a panel of experts with relevant knowledge and experience to facilitate full treatment of uncertainty, and the incorporation of a full range of viable alternative models. This process is often referred to as expert elicitation (Coppersmith *et al.*, Chapter 26, this volume). An example would involve scientists with expertise in the faults and seismicity of a specific region. All of the parameters, along with the quantification of aleatory variability and epistemic uncertainty in those parameters are combined into a probabilistic model by way of a logic tree. A logic tree expresses the choices of each parameter diagrammatically as branches on a tree (Figure 10.2), and can be exceedingly complex when all parameters are fully identified and quantified. The logic tree developed in the PSHA for Yucca Mountain possesses millions of limbs (Stepp *et al.*, 2001). Weights are assigned to each choice of a given parameter (each branch of the logic tree) by the expert panel, such that the weights sum to one for all choices of that parameter. A distribution of hazard curves (Step 4) can be constructed by calculating every possible path (or a large sample of paths by way of Monte Carlo analysis; e.g. Figure 10.3) through the logic tree, and then combining them on one graph. Alternatively, a probability distribution can be defined for a single value of hazard.

In general, a distribution of hazard estimates allows the statistical uncertainty in hazards to be quantified for the site, such that hazard estimates can be expressed as the mean or

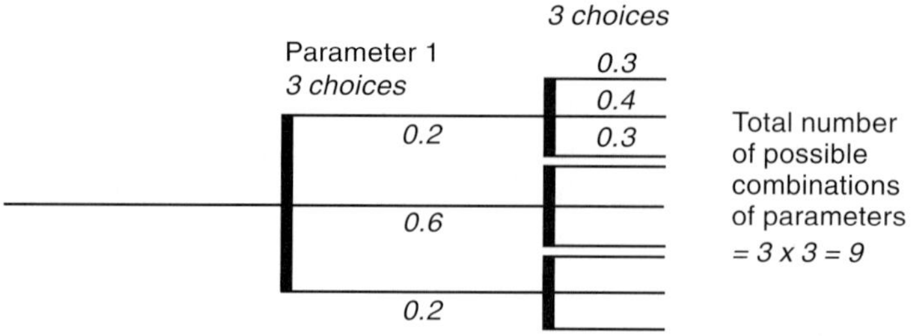

Fig. 10.2 In this schematic example of a logic tree, the numbers in italics represent the weights assigned to each branch of the logic tree. These weights sum to one for each set of choices for a given parameter.

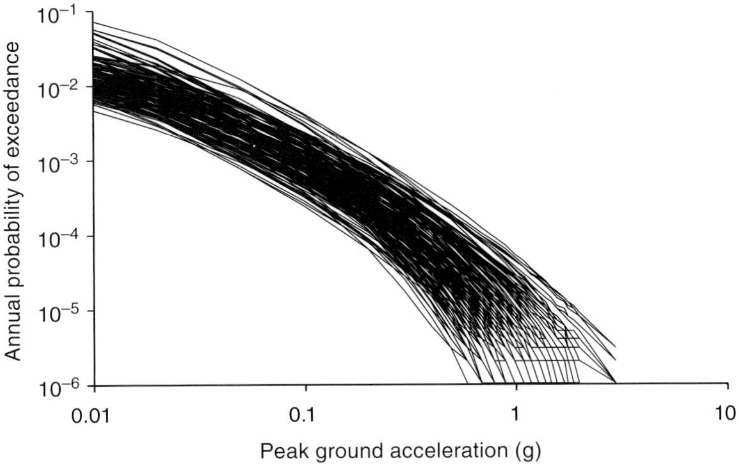

Fig. 10.3 A suite of hazard curves produced by a large number of samples of a logic tree, and the associated percentiles of hazard for that distribution of hazard curves, indicates overall uncertainty in this analysis. This example is for a research nuclear facility site in Australia (Stirling and Berryman, 2003).

the xth percentile of a distribution. The distribution of hazard curves in Figure 10.3 was developed for a research nuclear reactor facility in Australia (Stirling and Berryman, 2003). In this case, the design level of hazard was taken from the suite of hazard curves based on the relevant return period and percentile level of hazard for the facility. While not the sole means of expressing uncertainty in probabilistic models, the example in Figure 10.3 is useful for showing how uncertainty has been quantified in previous studies, and how large the resulting uncertainty in hazard can be.

The potentially serious consequences of damage to major facilities such as nuclear facilities and hydroelectric dams are such that they are typically designed to consider ground motions with return periods $\geq 10^4$ a. The challenge is therefore to define a probabilistic hazard model that can provide realistic estimates of hazard for such long return periods, especially when the input data are often only representative of return periods $< 10^4$ a. The methodology of PSHA (10.2) produces ground-motion estimates that increase as a function of return period, or equivalently increases as the annual frequency of exceedance decreases (e.g. Figure 10.3). In the case of untruncated or unbounded ground-motion distributions (in which the probability distributions are not limited to a particular number of standard deviations) this behavior inevitably leads to very strong ground motions (e.g. PGA > 1 g) at very long return periods (i.e. $\geq 10^4$ a, and sometimes 10^3 a in regions characterized by rapid tectonic processes, such as California and New Zealand). However, (10.2) embodies the ergodic assumption, which, in the context of seismic hazard, implies that the full distribution of ground motions derived from spatially diverse strong motion data sets (i.e. spatially diverse earthquake source, path and site combinations) will be of the same order as the ground-motion distribution at a single site over time. In the context of seismic hazard the ergodic assumption implies that the full distribution of ground motions derived from spatially diverse strong-motion data sets (i.e. spatially diverse earthquake source, path and site combinations) will be of the same order as the ground-motion distribution at a single site over time. Intuitively, application of this assumption to a single site would seem to be overly conservative, given that the single site will only be influenced by a limited number of source, path and site combinations. Not surprisingly then, an important area of current research is currently aimed at testing the ergodic assumption to determine the true natural upper limits of ground motions. The motivation for this research is that some PSHAs are now having to consider return periods of 10^6 a or more (e.g. Yucca Mountain), and the resulting ground motions turn out to be considerably stronger than the strongest levels that have been measured anywhere in the world (e.g. Bommer *et al.*, 2004). Geological criteria such as ancient delicate landform features and rock-masses are being used to provide preliminary upper bounds on these ground motions, and the challenge is to determine if these upper bounds are sufficient for engineering applications (e.g. Brune 1999; Stirling and Anooshehpoor, 2006; Purvance *et al.*, 2008, in press). Studies thus far, combined with examination of recently available high-quality seismicity data sets, are strongly suggestive of non-ergodicity at the site-specific level (e.g. Purvance *et al.*, 2008, in press).

Each of the hazard curves in Figure 10.3 includes the effects of aleatory variability, and their positions on the *y*-axis have been raised as a result. The epistemic uncertainty controls the spread, or difference between highest and lowest values, on the *y*-axis. This difference is typically a factor of 10 for most studies (e.g. Figure 10.3), which represents a considerable range of hazard. Design earthquakes are usually chosen to represent this range of possible hazard. For instance, points on the hazard curve distribution such as the median, mean or 84th percentile of ground motions for the 10^4 a return period are often used as a basis for defining the relevant design earthquake (e.g. Stirling and Berryman, 2003). In this example, a probabilistic seismic hazard model is examined to determine the earthquake magnitudes and source-to-site distances that contribute most to the median, mean or 84th percentile ground motions for the 10^4 a return period. The 84th percentile is one standard deviation above the median, in which both median and standard deviation are in natural log units of ground motion (i.e. components of a log-normal distribution, which is typically assumed to describe ground-motion distributions). These dominant magnitudes and distances are then used as a basis to search strong-motion databases to find the time histories from actual historical earthquakes that have similar magnitudes and distances. The time histories are then used directly for detailed seismic performance analysis in the engineering design.

10.4 Case example: Tohoku region, northern Japan

Probabilistic hazard assessment has been the standard for the quantification of seismic and, to a lesser extent, other tectonic hazards for the nuclear industry for > 20 a, and there is now a standard set of guidelines for development of a PSHA for a nuclear facility (SSHAC,

1997). In recent years we have focused considerable efforts into developing a probabilistic methodology for the quantification of tectonic hazards (e.g. rock deformation hazard) at prospective nuclear waste repository sites in Japan. The following sections describe our evolving methodology and preliminary results in this developmental area of research.

Our fundamental approach to developing a methodology for tectonic hazard analysis is to investigate multiple lines of evidence for rock deformation using a common measure: the rate of strain in the crust of the Earth in the vicinity of a prospective repository site. Strain is a dimensionless parameter that describes deformation, and strain rate is the change in length of a section of material relative to the original length per unit time (e.g. $10^{-9}\ a^{-1}$). The strain rate could be used to estimate, through probabilistic methods and assumptions regarding the relationship between strain rate and earthquake occurrence, the expected amount of tectonic displacement, or other tectonic hazards that could directly impact a repository over a range of time periods.

Strain rates can be quantified in many ways, and the emphasis of our work thus far has been to develop a methodology that may later be applied to compare the viability of alternative volunteer repository sites in Japan. The approach has been to utilize available data sets to develop several independent strain-rate models. Focusing on the Tohoku region of northern Honshu as an initial case study region, we have developed preliminary strain rate per unit area models from two data sets routinely used in seismic hazard analysis (geologically based surface deformation/active fault data and historical seismicity data), and a data set that has not yet seen significant application in hazard analysis (the geodetic record of contemporary crustal deformation). In Japan, these three data sets tend to be representative of time periods ~1–125 ka, 10–1000 a and 10–100 a, respectively, and each has relative strengths and weaknesses. The geological data provide estimates of strain for geological time periods, but are thought to provide incomplete information with respect to the low slip-rate faults (Y. Ota *et al.*, personal communication). The geodetic data have excellent spatial coverage but represent a very short time period of observation, and are very sensitive to modeling parameters (e.g. the degree of interseismic coupling on the subduction zone of the Japan trench; see Wallace *et al.*, Chapter 6, this volume). On the other hand, a major strength of geodetic data is that they can help to identify "hidden" active faults that have not yet been recognized by seismological or geological studies (e.g. Donnellan *et al.*, 1993; Stevens *et al.*, 2002). The historical seismicity model utilizes the much longer ~ 400 a historical record of Japan, but there are considerable uncertainties associated with the recurrence parameters calculated for this model. The following section provides a brief description of the data sets and methods used to derive strain rates from three data sets.

10.4.1 Determining strain rates from geological, GPS and seismicity data

Geologically derived surface strain rates are calculated from two main data sets: active faults and surface uplift. Onshore active fault data are obtained primarily from AIST (2008) and offshore data from a summary map published by Ohtake *et al.* (2002). The Active Fault Research Centre database (AIST, 2008) is restricted to the 100 most active faults in Japan,

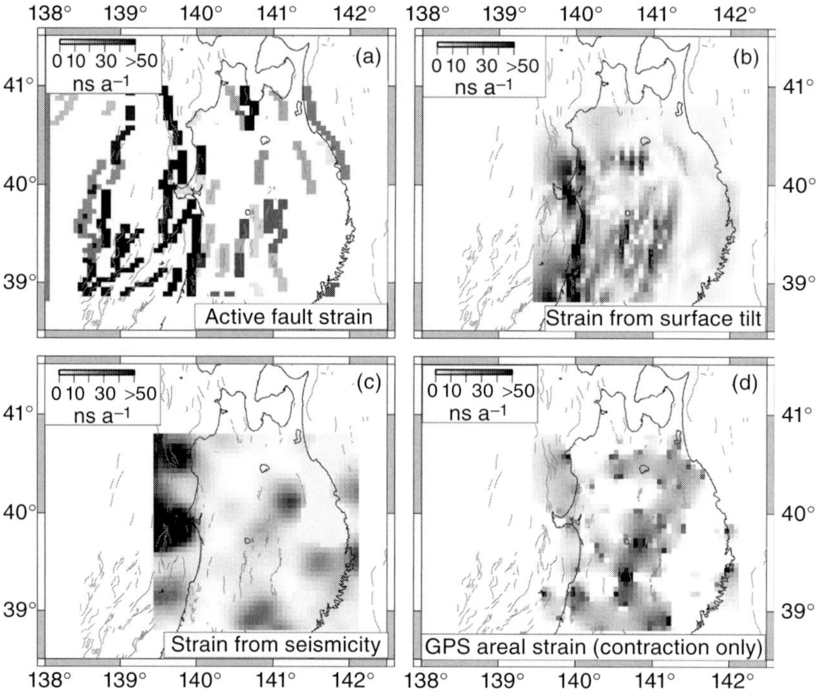

Fig. 10.4 Maps of the Tohoku region indicate strain rates derived from geologically based surface deformation data, including (a) active fault strain and (b) strain derived from surface tilt. Strain rates are also derived from seismicity data (c), and GPS data (d).

but a small number of Class *C* faults (those with slip rates of 0.01–0.1 mm a^{-1}; Y. Ota, personal communication, 2005) are also included in this database. In addition, a number of unmapped offshore faults are inferred to account for uplift of marine terraces along the Pacific coast. Strain rates are calculated from active faulting data by converting the vertical slip rate to horizontal (west–east) shortening (mm a^{-1}), divided by fault width (km), noting that 10^{-6} strain rate = 1 mm km^{-1} a^{-1}. Thus the main sources of uncertainty captured in the logic tree are the uncertainty in vertical slip rate, fault dip and fault width. The dip and width values, along with the logic tree branch weightings, are assigned by a panel of Japanese geology experts. Figure 10.4a shows an example of an active fault strain map for the highest-weighted branch of the active fault logic tree (dip 50°, width 10 km, mean vertical slip rate). Active fault strains are calculated for only those grid squares that contain active faults, which results in very high strain rates in those squares, no strain outside of them and a blocky appearance.

Surface uplift data are obtained from the surface uplift contour map for the onshore Tohoku region since 125 ka (Tajikara, 2005) from pairs of river terraces and marine terraces. Surface uplift can be converted to strain rates in two ways. The first is to assume that all surface uplift occurs on active faults, in which case the strain-rate calculation is the same as outlined above. The second is by assuming that the regional tilting is caused by ongoing strain, and approximating the strain rate by dividing the tilt (mm km^{-1}) by the age of the tilted surface (125 ka), noting again that 10^{-6} strain rate = 1 mm km^{-1} a^{-1}. Given that the active fault strain rates above are calculated in a west–east direction, a west–east tilt direction is given the highest weighting. Thus the main uncertainties captured by the branches of the logic tree are the tilt (a function of the uplift measurements) and the age. In addition, the tilt can be calculated by fitting a surface to the uplift contour map, and thus the tension on that surface (or the tightness of the fit to the structure contour map) is also varied in the logic tree; this acts as a smoothing function. Figure 10.4b shows the tilting strain map of the highest-weighted branch of the tilting logic tree (west–east tilt, tension of 0.15, 125 ka age). Comparison of Figure 10.4a and 10.4b shows good correspondence (note the tilt strain map is calculated from onshore data only), suggesting that the active faults are a major driver of the surface-uplift in the Tohoku region. Thus, to calculate a combined surface uplift strain rate, the contribution of active faults versus regional uplift data needs to be weighted by an expert panel.

Velocities of GPS sites throughout Japan are derived from a combination of SINEX (Solution INdependent EXchange format) files provided by the Geographical Survey Institute (GSI, 2008). A detailed description of the GPS data set, associated analysis and its interpretation is provided in Wallace *et al.* (Chapter 6, this volume). Maps of crustal strain rates can be generated from the GPS velocity field (e.g. Beavan and Haines, 2001), although a variety of short-term, earthquake-cycle related processes must be accounted for before such crustal strain-rate measurements can be useful for tectonic hazard estimations (Wallace *et al.*, Chapter 6, this volume). For example, in northern Japan a large part of the deformation that GPS techniques measure is elastic (recoverable) strain related to interseismic coupling on the offshore subduction thrust, and is not of concern with regards to the tectonic hazard of on-land faults; thus, this signal must be removed before estimating the "residual" strain potentially due to faulting in the on-land portion of northern Japan.

We have developed a logic tree in consultation with a panel of Japanese GPS experts that encompasses some of the alternative models and methods that should be considered in the estimation of crustal strains from GPS velocities in northern Honshu. For example, some of the alternative models in the logic tree involve varying the maximum depth of interseismic coupling; along-strike smoothing of coupling on the Japan trench subduction thrust; and the dip of the fault representing the tectonic block boundary in the Japan Sea (this influences the upper-plate strain estimates we obtain for the west coast). We also consider different methods of mapping the strain rates from GPS techniques (e.g. Beavan and Haines, 2001; Miura *et al.*, 2004), as well as using results from published studies where the elastic strain from subduction-zone coupling in northern Japan has been removed (e.g. Nishimura *et al.*, 2004). Figure 10.4d shows an areal strain map derived from the highest-weighted branch of the GPS logic tree. Of particular note in Figure 10.4d is the band of elevated contractional strain (up to 10–20 ns a^{-1}) following the north–south trending mountains ("Backbone Range") of central Tohoku.

The source of the seismicity data used to develop seismic strain rates in this study is the Japanese Meteorological Agency (JMA) network catalog. Japanese earthquakes have been routinely recorded since 1926 at a detection threshold magnitude of around M 4.5 (e.g. Stirling *et al.*, 1996). That detection threshold has improved with time and progressive development of the JMA network. The catalog is also supplemented by a complete record of major historical earthquakes (M $\geq$ 6.9) for the period 1581–1925 (Wesnousky *et al.*, 1982).

Our methodology to quantify and map seismic strain rates is first to develop a seismicity model in the same overall way that background seismicity models are developed for PSHAs. Two different methods are used to develop the seismicity model, each providing a set of gridded point sources that have a set of earthquake parameters assigned to them, and with the parameters described by the Gutenberg–Richter relationship. The first is the traditional method of defining large seismicity zones from seismo-tectonic considerations, assigning seismicity parameters to each zone, and then uniformly distributing the parameters across the zone. The second method is to allow the seismicity parameters to vary within the zones according to the spatial distribution of seismicity within the zone. The latter method was developed by Frankel (1995) to characterize the probabilistic seismic hazards from background earthquakes, and has been successfully adapted to develop probabilistic seismic hazard models in other regions (e.g. Stirling *et al.*, 2002). Use of the two alternative methods for treatment of the seismicity data represents important epistemic uncertainties in the seismic strain model. The traditional method allows for the possibility that the current seismicity patterns (varying over distances of tens of kilometers) do not necessarily represent the long-term parent distribution of seismicity, and instead relies on the broader seismo-tectonic regions as the guiding definition. The spatially varying method in contrast assumes that the current seismicity patterns are a reasonable representation of long-term seismicity. Frankel (1995) initially made such arguments based on a spatial correlation between small, frequently occurring events and less-frequently occurring M $\geq$ 5 earthquakes. These alternative modeling techniques, along with alternative regionalization (area source zonation) schemes, use of declustered versus raw seismicity catalogs and the uncertainties in the seismicity parameters are the basis for definition of logic trees for the seismic strain-rate model.

The seismicity parameters (a-value and b-value of the Gutenberg–Richter relationship; (10.1)), are calculated by the maximum-likelihood method of Weichert (1980), which allows the use of different magnitude completeness levels for various time periods to calculate parameter b. The $M_{\max}$ is based on the maximum magnitudes assigned to the various parts of the Tohoku region by the Research Group for Active Faults of Japan (1991), which is based on a combination of historical earthquake information and consideration of the likely maximum magnitudes derived from the fault database. The final step in the seismic strain model is to convert the seismicity rates into equivalent strain rates. This is achieved by the method of Kostrov (1974), which converts the equivalent seismic-moment rate from each of the Gutenberg–Richter-distributed earthquakes into strain rate through assumption of a crustal volume. Our convention is to use $5 \times 5 \times 20$ km crustal volumes, 20 km being the average depth to the base of crustal seismicity, based on interpretation of seismicity cross-sections. The map in Figure 10.4c shows an example of the spatial distribution of seismic strain rate.

10.4.2 Strain-rate comparisons

Much of our work has thus far gone into separately developing the three strain-rate models. An important focus of ongoing work is how best to integrate the three measures of strain, given the various strengths and weaknesses of the three data sets, along with the underlying assumptions and uncertainties regarding what the individual data sets represent. The GPS data set (here, from the interseismic period) reflects the accumulation of elastic strain in the crust (Wallace *et al.*, Chapter 6, this volume), whereas the seismicity and geological data reflect the release of elastic strain across the region over time. Simply combining these strain-rate signatures by way of a simple weighted averaging procedure would therefore be inappropriate, and at present it is more informative to compare the three strain signatures independently. Not surprisingly, comparison of the strain-rate maps (Figure 10.4) reveal considerable differences in magnitudes of strain rate among different locations. A certain degree of correlation is evident from the strain-rate maps from the geological, seismological and geodetic data sets, particularly in terms of the crudely north–south trending zone of high strain rate that roughly coincides with the central mountains of Tohoku. Higher strain-rate zones are also evident for all three data sets in the west, and lower strain rate in the east.

The strain-rate maps allow visual comparison of the results produced from the three data sets, though reliance has to be placed on choosing a representative output (sample) from the logic trees to plot gridded values on a map (with the exception of the GPS-derived strain map which is a weighted average of strain rates from full sampling of the logic tree). Site-specific comparisons allow much more complete comparisons to be made (Figure 10.5). The graphs of strain rates are for three sample sites from different tectonic settings in the Tohoku region, and show the total distribution of strain rates calculated from every possible limb of the logic tree (108 to 126 limbs for the geologically based surface-deformation tree, and 148 limbs for the GPS tree), or by calculating from a large sample (1000) of logic-tree limbs in the case of the seismological data set. Sampling for the latter is achieved by way of Monte Carlo methods, whereby a large number of random parameter samples are made from the logic tree according to the weights assigned to each branch of the logic tree. Two graphs are shown for each site in Figure 10.5. The left-hand graphs are histograms that show the probability that strain rate will be equal to a given suite of values at the site (i.e. discrete probability). These histograms are all compiled with a bin size of 1 ns a^{-1}. The right-hand graphs show the probability of exceedance for the same suite of strain rates (i.e. cumulative probability) and are analogous to hazard curves if strain rate is considered a proxy for tectonic hazard (e.g. Figures 10.1 and 10.2). In Figure 10.5, the geologically based strain-rate estimates are derived solely from tilt deformation, and do not include any strain rates calculated from active fault data. The familiar form of the cumulative graphs (i.e. analogous to hazard curves in PSHA) therefore allow for easy comparison of the strain-rate probabilities for the different data sets.

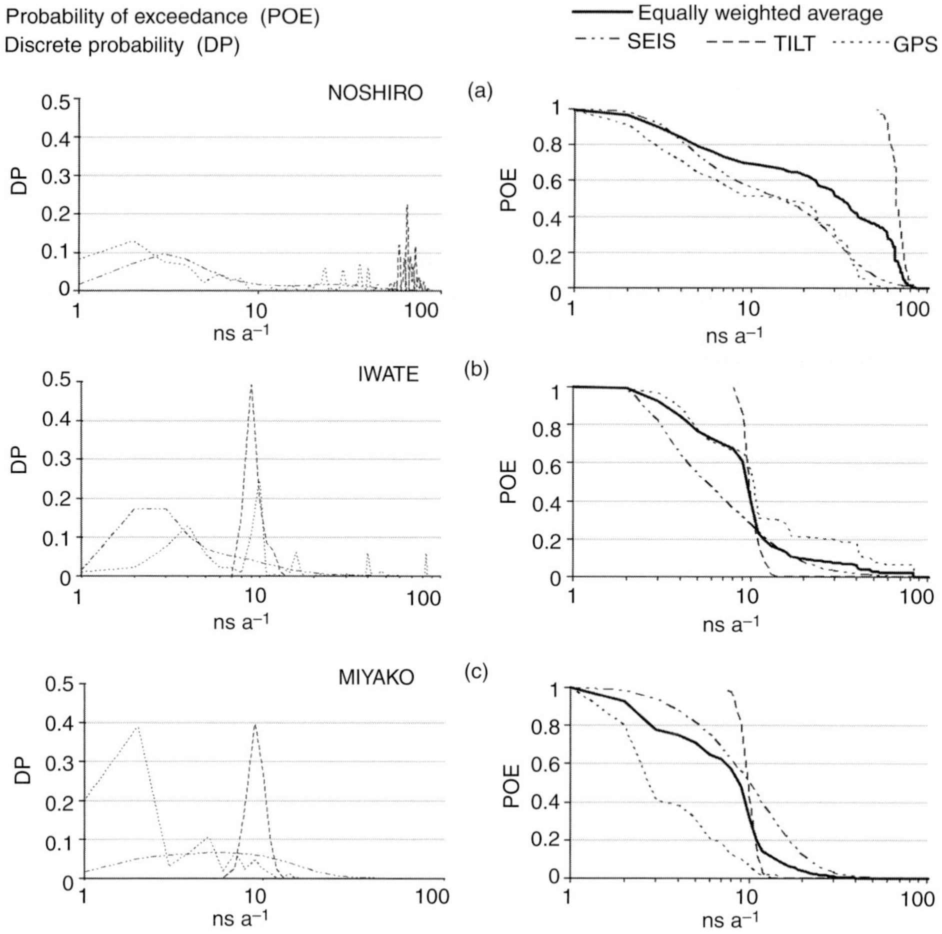

Fig. 10.5 Comparisons are shown of geologically (tilt deformation, TILT), GPS- and seismically derived (seis) strain rates for three example sites: (a) Noshiro (western Tohoku); (b) Iwate (central Tohoku); and (c) Miyako (eastern Tohoku). For each, the left graph shows a histogram of discrete strain-rate probabilities (i.e. probability = a given strain rate), and the right graph shows the equivalent probability of exceedance for those strain-rates. The solid lines shows the equally weighted average.

Comparison of the three graphs shows there to be considerable differences in strain rate across the Tohoku region; but much more noticeable are the large uncertainties in site-specific strain rate within each graph due to data set. This is particularly the case for the Noshiro site (Figure 10.5a), where the probability of exceedance for a given strain rate is considerably higher for geologically derived strain rate than for the other data sets. Close proximity to an active fault could be the cause of high tilt-induced strain rates, and non-stationarity of strain rates over timespans greater than those of the seismicity and GPS data sets could also be a likely physical explanation for these differences. To explain, the seismicity catalog spans about 400 a but this record is too short to have recorded the large earthquakes implied by the presence of active faults (recurrence intervals of 10^3–10^4 a). Epistemic uncertainties associated with modeling methodologies for the data sets is also a valid explanation for the observed differences. A good example is that the geologically derived strain rates may be strongly influenced by the nearby presence of active faults (the likely case for Noshiro), whereas the seismologically and GPS-derived strain rates are distributed more widely by way of various smoothing techniques. The GPS-derived strain rates are also inherently smoothed owing to the spatial distribution of GPS sites and the fact that the source of the strain-accumulation signal is often at significant depth in the Earth. Even the different smoothing techniques represent epistemic uncertainties, and the effects are possibly revealed by the Miyako site graph (Figure 10.5c), which shows considerable differences between the seismically and GPS-derived strain rates. Specifically, a spatially varying strain-rate model is used for the GPS-derived strain rates whereas a uniformly distributed strain-rate model is most widely used for the seismological model. Strain rates appear comparable for the Iwate site across all three data sets (Figure 10.5b). Not surprisingly, our present efforts are being focused on resolution of the differences between the results for the sites showing the large between-data set differences.

10.4.3 Strain rate to hazard

Tectonic hazards can potentially be modeled using strain-rate estimates. The strain-rate graphs (Figure 10.5) give a probability distribution of possible strain rates for each of the sample sites, and are a direct product of the wide range of realizations of strain rate that are possible from the logic trees. Since strain rate is a direct manifestation of tectonic forces, it follows that the higher the strain rate, the greater the tectonic hazard in terms of frequency or likelihood of occurrence.

Using fault displacement as an example of a relevant hazard for the siting of a repository, the strain-rate graphs could be used to develop equivalent site-specific graphs of displacement hazard. The exceedance probability curves (Figure 10.5, right-hand graphs) could, for instance, be converted to show the probability of exceedance for a suite of coseismic displacements (Figure 10.6).

We are confident that probabilistic strain-rate estimates will provide a means to quantify the likelihood that a range of tectonic hazards will adversely impact a repository facility in the future. Once the relevant hazards and threshold values are defined in terms of magnitude and frequency, then it will be straightforward to work backwards to determine the limiting strain rate and associated probability of exceedance for the relevant threshold levels of hazard. In this respect strain-rate-based "tools" such as maps and graphs may become the basis for early appraisal of candidate sites for repositories. Given the orders-of-magnitude differences in strain rate across the Tohoku region, there may be some sites where the strain rate may turn out to be of small consequence to a proposed facility over the timespan of interest. As an example, preliminary probabilistic fault displacement hazard calculations indicate

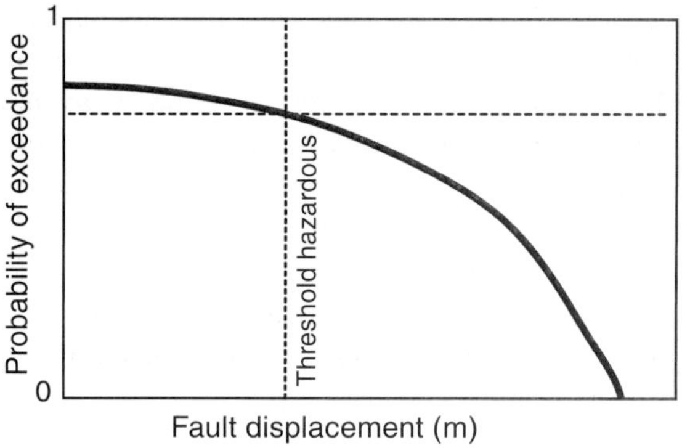

Fig. 10.6 Schematic diagram illustrating one possible way to express probabilistic fault displacement hazard. The graph shows the probability of exceedance for "hazardous" displacement (i.e. displacements beyond a threshold value that would preclude the siting of a repository).

that the strain rates of around 10 ns a^{-1} in eastern Tohoku translate to recurrence intervals of 10^5 a or more for a "hazardous" displacement of 0.5 m or more for a 5×5 km crustal area (we choose 0.5 m as an arbitrary, illustrative threshold value of hazardous displacement). In western Tohoku the strain rates are more typically in the range of 10–100 ns a^{-1}, and these translate to recurrence intervals of 10^4–10^5 a for hazardous displacements. The PFDH is therefore likely to vary by an order of magnitude across the Tohoku region, which is a direct consequence of the differences in strain rate across the region.

Concluding remarks

We have provided examples of how tectonic hazards are quantified probabilistically, and in particular, how these methods are being adapted to provide a basis for the characterization of tectonic hazards at potential nuclear waste repository sites in Japan. Probabilistic methodology is the essential tool for quantifying tectonic hazards, as it considers all hazardous sources in a region, allows uncertainty to be fully quantified and provides hazard estimates for a spectrum of return periods. However, the considerable challenge of providing hazard estimates for the long return periods relevant to nuclear facilities cannot be overstated. High-level waste repositories need to be designed to accommodate natural hazards that will occur over time periods of 10^4–10^5 a, yet our data sets collectively only sample 10^4 a and are incomplete. In addition, data quality and quantity are variable in space and time, and the standard methods of probabilistic hazard analysis are very basic and generalized. These methods simplistically base future hazards on what has happened or not happened in the past, so only limited allowance is made for totally unanticipated hazardous events in previously inactive areas or low-activity areas. Repository design can bypass these issues through conservatism of design, but this can be a costly and crude approach, especially when having to consider very long time periods like the 10^6 a performance period being considered at present for the Yucca Mountain site. However, realization of these issues is also the first step towards addressing them.

Our early approach to developing a hazard analysis methodology for Japan is being undertaken from a multi-disciplinary perspective. Understanding the similarities and differences between independent hazard measures derived from probabilistic geologically, GPS and seismicity-based data sets is a robust and defensible approach to hazard analysis. The continued accumulation of high-quality data sets, improving methods of physical modeling of tectonic hazards (e.g. earthquake source modeling) and the current research revealing the limited validity of the ergodic assumption in probabilistic hazard analyses can only contribute positively to the reliability of future hazard estimates and engineering design.

Further reading

Reiter (1990) gives a comprehensive but down-to-earth overview of seismic hazard analysis, including good examples of the application of probabilistic seismic hazard analysis to the nuclear industry.

Acknowledgments

This work has been funded by Nuclear Waste Management Organization of Japan (NUMO) for a project to develop a methodology for long-term tectonic hazard assessment. The authors thank Japanese and international geoscientists in the International Tectonics Meeting (ITM) and Takashi Kumamoto for useful discussions and earthquake catalog data.

References

AIST (2008). Active fault database of Japan. National Institute of Advanced Industrial Science and Technology, http://riodb02.ibase.aist.go.jp/activefault/index.html.

Aki, K. and P. G. Richards (1980). *Quantitative Seismology: Theory and Methods*. San Francisco, CA: W. H. Freeman.

Beavan, J. and J. Haines (2001). Contemporary horizontal velocity and strain-rate fields of the Pacific–Australian plate boundary zone through New Zealand. *Journal of Geophysical Research*, **106**, 741–770.

Berryman, K. and P. Villamor (2004). Surface rupture of the Poulter fault in the 1929 March 9 Arthur's Pass earthquake, and redefinition of the Kakapo fault, New Zealand. *New Zealand Journal of Geology and Geophysics*, **47**, 341–351.

Bommer, J., N. Abrahamson, F. Strasser *et al.* (2004). The challenge of defining upper bounds on earthquake ground motions. *Seismological Research Letters*, **75**, 82–95.

Brune, J. N. (1999). Precarious rocks along the Mojave section of the San Andreas Fault, California: constraints on ground motion from great earthquakes. *Seismological Research Letters*, **79**, 29–33.

Cornell, C. A. (1968). Engineering seismic risk analysis. *Bulletin of the Seismological Society of America*, **58**, 1583–1606.

Donnellan, A., B. H. Hager, R. W. King and T. A. Herring (1993). Geodetic measurement of deformation in the Ventura basin region, southern California. *Journal of Geophysical Research*, **98**, 21 727–21 739.

Frankel, A. (1995). Mapping seismic hazard in the central and eastern United States. *Seismological Research Letters*, **66**, 8–21.

Frankel, A. D., M. D. Petersen, C. S. Mueller *et al.* (2002). Documentation for the 2002 Update of the National Seismic Hazard Maps, USGS Open-File Report 02-420. Denver, CO: US Geological Survey.

GSI (2008). Crustal movement in Japan. Geographical Survey Institute, http://mekira.gsi.go.jp/.

Gutenberg, B. and C. F. Richter (1944). Frequency of earthquakes in California. *Bulletin of the Seismological Society of America*, **34**, 185–188.

Hanks, T. C. and W. H. Bakun (2002). A bilinear source-scaling model for M-log A observations of continental earthquakes. *Bulletin of the Seismological Society of America*, **92**, 1841–1846.

Kostrov, B. V. (1974). Seismic moment and energy of earthquakes and seismic flow of rock. *Izvestiya, Academy of Sciences, USSR Physics, Solid Earth*, **1**, 23–40.

Miura, S., T. Sato, A. Hasegawa *et al.* (2004). Strain concentration zone along the volcanic front derived by GPS observations in the NE Japan arc. *Earth Planets Space*, **56**, 1347–1355.

Nishimura, T., T. Hirasawa, S. Miyazaki *et al.* (2004). Temporal change of interplate coupling in northeastern Japan during 1995–2002 estimated from continuous GPS observations. *Geophysical Journal International*, **157**, 901–916.

Ohtake, M., A. Taira and Y. Ota (2002). *Active Faults and Seismo-Tectonics of the Eastern Margin of the Japan Sea*. Tokyo: University of Tokyo Press.

Power, W., G. Downes and M. Stirling (2007). Estimation of tsunami hazard in New Zealand due to South American earthquakes. *Pure and Applied Geophysics*, **164**, 547–564.

Purvance, M. D., R. Anooshehpoor and J. N. Brune (2008). Overturning of freestanding blocks exposed to earthquake excitations 2: shake table validation. *Earthquake Engineering and Structural Dynamics*, **37**, doi:10.1002/eqe.789.

Purvance, M. D., J. N. Brune, J. G. Anderson and N. A. Abrahamson (in press). Consistency of precariously balanced rocks with vector valued probabilistic seismic hazard estimates. *Bulletin of the Seismological Society of America*.

Reiter, L. (1990). *Earthquake Hazard Analysis: Issues and Insights*. New York, NY: Columbia University Press.

Research Group for Active Faults of Japan (1991). *Active Faults in Japan: Sheet Maps and Inventories, Revised Edition*. Tokyo: University of Tokyo Press.

SSHAC (1997). Recommendations for probabilistic seismic hazard analysis: guidance on uncertainty and use of experts, Techinal Report NUREG/CR-6372(1), UCRL-ID-122160. Washington, DC: Senior Seismic Hazard Analysis Committee, US Nuclear Regulatory Commission.

Stepp, J. C., I. Wong, J. Whitney *et al.* (2001). Probabilistic seismic hazard analyses for ground motions and fault displacement at Yucca Mountain, Nevada. *Earthquake Spectra*, **17**, 113–151.

Stevens, C. W., R. McCaffrey, Y. Bock *et al.* (2002). Evidence for block rotations and basal shear in the world's fastest slipping continental shear zone in NW New Guinea. In: Stein, S. and J. Freymueller (eds.) *Plate Boundary Zones*, Geodynamics Series 30. Washinton, DC: American Geophysical Union, 87–99.

Stirling, M. W. and R. Anooshehpoor (2006). Constraints on probabilistic seismic-hazard models from unstable landform features in New Zealand. *Bulletin of the Seismological Society of America*, **96**, 404–414.

Stirling, M. W. and K. R. Berryman (2003). Earthquake ground motion and fault hazard studies at the Lucas Heights Research Reactor Facility, Sydney, Australia: evolution of methods and changes in results. Presented at: International Symposium on Seismic Evaluation of Existing Nuclear Facilities, August 25–29. Vienna, Austria: International Atomic Energy Agency.

Stirling, M. W. and C. J. N. Wilson (2002). Development of a volcanic hazard model for New Zealand: first approaches from the methods of probabilistic seismic hazard analysis. *Bulletin of the New Zealand Society of Earthquake Engineering*, **35**, 266–277.

Stirling, M. W., S. G. Wesnousky and K. Shimazaki (1996). Fault trace complexity, cumulative slip, and the shape of the magnitude–frequency distribution for strike-slip faults: a global survey. *Geophysical Journal International*, **124**, 833–868.

Stirling, M. W., G. H. McVerry and K. R. Berryman (2002). A new seismic hazard model for New Zealand. *Bulletin of the Seismological Society of America*, **92**, 1878–1903.

Tajikara, M. (2005). Vertical displacement during the last 120 000 years in central part of northeast Japan, estimated from TT value and FS value. In: Koike, K. *et al.* (eds.) *Landforms of Japan*, vol. 3, Tohoku district. Tokyo: University of Tokyo Press.

Weichert, D. H. (1980). Estimation of the earthquake recurrence parameters for unequal observation periods for different magnitudes. *Bulletin of the Seismological Society of America*, **70**, 1337–1346.

Wells, D. L. and K. J. Coppersmith (1994). New empirical relationships among magnitude, rupture length, rupture width, rupture area, and surface displacement. *Bulletin of the Seismological Society of America*, **84**, 974–1002.

Wesnousky, S. G. (1986). Earthquakes, quaternary faults and seismic hazard in California. *Journal of Geophysical Research*, **91**(B12), 12 587–12 632.

Wesnousky, S. G., C. H. Scholz and K. Shimazaki (1982). Deformation of an island arc: rates of moment-release and crustal shortening in intraplate Japan determined from seismicity and Quaternary fault data. *Journal of Geophysical Research*, **87**, 6829–6852.

Working Group of California Earthquake Probabilities (1995). Seismic hazards in Southern California: probable earthquakes, 1994–2024. *Bulletin of the Seismological Society of America*, **85**(2), 379–439.

Youngs, R., W. Arabasz, R. Anderson *et al.* (2003). A methodology for probabilistic fault displacement hazard analysis (PFDHA). *Earthquake Spectra*, **19**, 191–219.

11

Tsunami hazard assessment

W. Power and G. Downes

No natural disaster in modern history has impacted a broader region, or impacted more lives in more diverse communities, than the 2004 Indian Ocean tsunami. For many, this disaster redefined the scale of conceivable impacts of natural disasters. The scales of natural processes that contributed to this disaster were equally tremendous, including the magnitude of the tsunamigenic earthquake, the largest on Earth since the Chilean earthquake of 1960; the volume of water displaced during the earthquake and the speed with which the resulting wave could traverse an entire ocean basin; and the force and extent of the wave runup where the tsunami reached coastal areas. Yet on geological timescales, such events are common.

We must prepare for such natural events. In the context of nuclear facilities, preparation involves understanding, or forecasting, the impact of potential tsunami on coastal sites (McKinley and Alexander, Chapter 22, this volume). This includes developing an understanding of the sources of tsunami, the propagation of tsunami waves, and their wave height, wavelength and runup. This chapter reviews tsunami processes, and introduces the concept of probabilistic tsunami hazard assessment, parallel to other types of hazard analysis already common for nuclear facilities.

11.1 What is a tsunami?

Tsunami comes from the Japanese word meaning "harbor wave." In Japanese, the word tsunami is the same in both the singular and plural form. In English *tsunamis* is often used for the plural, although both *tsunami* and *tsunamis* are correct. A tsunami is a series of waves that are rapidly generated when a large volume of water (e.g. a lake, the sea, the ocean) is vertically displaced by an impulse disturbance such as an explosion, earthquake, volcanic eruption, landslide or meteorite impact. Tsunami can inundate coastlines violently and unexpectedly, causing devastating damage to buildings, infrastructure and the environment, and also causing injuries and loss of life. High-profile, infrequent, catastrophic events, such as the 2004 Indian Ocean tsunami, often cause people to lose sight of the fact that moderate tsunami are also a significant problem and even small tsunami, that do not inundate beyond the shoreline, can threaten life and cause significant disruption to coastal environments.

Tsunami are gravity waves, meaning that the force of gravity acts to restore displaced water back towards equilibrium. This process causes a tsunami to propagate away from the source in a way similar to other water waves. There are, however, several significant differences between tsunami and other sea waves: (i) tsunami usually involve the whole ocean depth and not just the top few tens of meters; (ii) for beach or sea waves, the period (time between successive crests) varies from a few seconds to about a minute, while for tsunami, the period varies from several minutes to a few hours; (iii) the wavelength (distance between successive crests) for beach or sea waves is on the order of tens to hundreds of meters, while for tsunami, wavelengths can be several kilometers to > 400 km (in deep water). Hence, a huge volume of water can be involved in a tsunami, especially when it is generated in some of the deepest parts of the ocean, near ocean trenches created by subduction zones.

Very large initial disturbances are required to generate tsunami that are still damaging at great distances from their source. Great earthquakes ($M \geq 8$), particularly on a subduction interface, are the principal causes of major ocean-wide tsunami. The fault that ruptured in the 1960 Chilean M 9.4–9.5 earthquake was ~ 1000 km long by 120 km wide, with an average slip of ~ 20 m (Plafker, 1972; Barrientos and Ward, 1990). This earthquake caused a tsunami that reached a maximum height of 25 m above ambient sea-level locally, > 10 m in Hawaii, > 6 m in Japan and ~ 4 m in New Zealand, with thousands of deaths recorded in Chile and hundreds of deaths recorded in Japan and Hawaii. Other sources that cause ocean-wide impact, such as volcanic flank collapse or bolide impact, are very rare.

To be large and damaging at nearby shores, tsunami do not need very large disturbances and can have a variety of source types. In 2006, the M 7.7 earthquake that occurred 200 km off the southern coast of Java caused > 500 fatalities and injuries and temporarily displaced $> 50\,000$ people. This tsunami significantly affected only 300 km of coastline, with inundation depths at the shoreline in the 7–8 m range over tens of kilometers (Reese *et al.,* 2007). Submarine and coastal landslides are also capable of producing tsunami with large wave heights (e.g. 1929 Grand Banks, Canada, submarine landslide) and in some cases extreme wave heights (e.g. the 1958 Lituya Bay subaerial landslide), but often these affect only a limited extent of coastline.

Because most tsunami waves have wavelengths that are much longer than the depths of water in which they travel, they can generally be treated as "shallow-water waves," although there are situations in which shallow-water wave theory is inadequate to describe tsunami completely (Ward, 1980, 2001a). The velocity, c, of shallow-water waves is given by:

$$c = \sqrt{gh} \tag{11.1}$$

where g is gravitational acceleration (m s^{-2}) and h is water depth (m). The high velocities of tsunami (e.g. ~ 200 m s^{-1} in water 4000 m deep) combined with large volumes of water mean that tsunami can have huge momentum. Furthermore, the simple dependence of the velocity of a tsunami on the water depth has important consequences for tsunami propagation behavior and characteristics in deep ocean and as the tsunami approaches the shore.

In deep water, tsunami can travel > 700 km hr^{-1} while the wave height (i.e. vertical crest-to-trough height) is $\lesssim 1-2$ m, producing only a gentle rise and fall of the sea surface over a wavelength of hundreds of kilometers. This is not noticeable by ships, nor able to be seen by aircraft, although satellites with sea-surface elevation technology can detect large tsunami in the deep ocean (Gower, 2005).

When tsunami waves move into shallow waters, their speed decreases rapidly while their height increases. The front of each wave slows down and the back of each wave, which is moving faster, catches up to the front, piling the water up higher. A tsunami that is only half a meter high in the open ocean can shoal to become as much as 10 m high when it reaches shore. Furthermore, a tsunami's long period and wavelength mean that when tsunami reach the shore, they do not withdraw in a few seconds like a normal beach wave, but continue to flood inland for many minutes. Overland flow velocities of 10–75 km hr^{-1} have been measured (Matsutomi *et al.*, 2006; Choowong *et al.*, 2008).

The propagation path of a tsunami is governed by the bathymetry over which it propagates. Since speed is controlled by the water depth, the tsunami is subject to familiar wave phenomena such as reflection, refraction and waveguiding. Refraction bends the wavefronts toward areas of shallower bathymetry. Where an underwater ridge is present some energy may become trapped by repeated refraction from the edges of the ridge, leading to a waveguiding effect. This effect was a very noticeable feature of the 2004 Sumatran tsunami (Titov *et al.*, 2005). The effect of refraction and scattering by bathymetric features has been studied extensively by Mofjeld *et al.* (2000).

In situations when it is not valid to use the shallow-water wave theory – for example, when the wavelength is not substantially larger than the depth, as may occur for a small source in deep water – there is a discrepancy between the phase and group velocity of tsunami. Both of these are less than that given by the shallow-water formula, with the consequence that dispersion occurs, leading to a longer train of waves with a lower peak height (Ward, 2001a). This is one reason why most tsunami generated by landslides or volcanoes, which tend to have comparatively small source areas, often do not have as severe an impact on distant shores as do earthquake-generated tsunami of similar, near-source height. Such smaller sources also tend to produce waves that spread out like point sources, unlike the directed energy of earthquake-generated tsunami that is perpendicular to the strike of the source fault.

11.2 What causes a tsunami?

Any event that suddenly displaces a large volume of water over a sufficiently large area is capable of generating a tsunami. These events include: (i) large submarine or coastal earthquakes, where significant uplift or subsidence of the seafloor or coast occurs; (ii) underwater landslides and large landslides from coastal or lakeside cliffs, which may be triggered by an earthquake or volcanic activity; (iii) volcanic activity such as underwater explosions, eruptions or caldera collapse, pyroclastic flows and atmospheric pressure waves; and (iv) meteorite (bolide) splashdown or an atmospheric air-burst over the ocean. In order to generate a tsunami efficiently, the width of the displaced water surface must be greater than the water depth. Wiegel (1970) describes, in detail, how the source parameters determine the form of the resulting waves.

Man-made events such as nuclear explosions are capable of generating tsunami-like waves, as are extreme weather events (for example, storm surges and "meteorological tsunami"; Pelinovsky *et al.*, 2001). Such sources generally lie at the extreme short- and long-period ends of the tsunami spectrum, respectively. It has also been suggested that catastrophic destabilization of gas hydrates may cause slopes to become destabilized, potentially causing tsunami-generating underwater landslides (Pecher *et al.*, 2005). We concentrate on the accepted geological sources.

11.2.1 Earthquakes as sources of tsunami

Earthquakes are the most common and the most easily recognized sources of tsunami; in many ways, they are also the most easily modeled and understood. The main determinant of tsunami generation is the distribution of seafloor vertical deformation, but temporal rupture characteristics are also important. Factors, such as non-uniform distribution of slip, rupture duration and velocity of rupture can significantly affect the tsunamigenic potential of an earthquake, especially near its source (Geist, 1999; Geist *et al.*, 2007).

Advances in seismographic and geodetic techniques, including tsunami inversion, as well as geological field data, now provide the means to estimate the key rupture parameters of past events. To some extent, application of these techniques also provides constraints on the nature of future events. For example, geodetically determined slip-deficit can place constraints on prospective slip-distributions (Cloos, Chapter 2, this volume). Whereas past tsunami source models were based on the assumption of instantaneous uniform slip along a planar fault in a uniform half-space, they can now be modeled more realistically.

The 2004 Sumatra–Andaman earthquake, detailed in the following section, provides an example of a large-magnitude earthquake that generated tsunami. Many lessons can be learned from its occurrence and effects (Bilek *et al.*, 2007). The general mechanism of tsunami generation by subduction-zone earthquakes is shown in Figure 11.1.

Another recent event, the 2006 Java M 7.7 earthquake (Mori, 2007), belongs to a different class of subduction-zone earthquakes, called "tsunami earthquakes." This type of earthquake, first identified by Kanamori (1972), has several identifying characteristics (Bilek and Lay, 2002). These include the capacity to generate greater than expected tsunami for

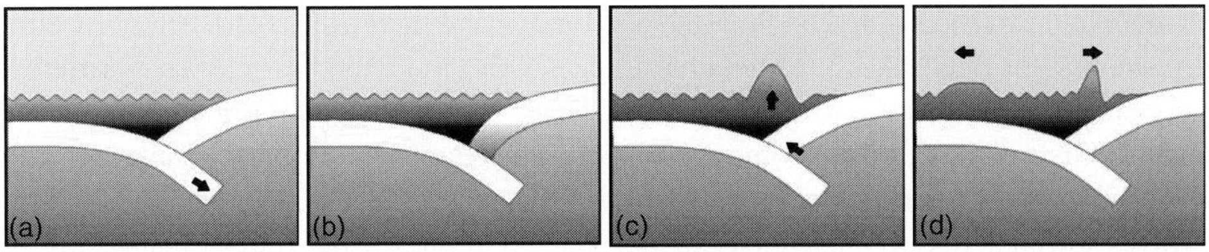

Fig. 11.1 Stages in tsunami generation by a subduction-zone earthquake: (a) subduction of lower plate, (b) strain accumulation at locked interface, (c) sudden strain release and surface deformation during earthquake, (d) tsunami propagation.

the magnitude of the earthquake, long rupture durations for the magnitude of the earthquake, location near to the trench and low-angle thrust mechanism consistent with being on the very shallow part of plate interface, slow rupture velocity, high-energy release at low frequencies and possibly an association with subducted seamounts.

The 2004 Sumatra–Andaman Islands earthquake

The most destructive tsunami of modern times occurred on December 26, 2004 in the Indian Ocean. The cause of the tsunami, an earthquake with an estimated magnitude of Mw 9.3 (e.g. Stein and Okal, 2007), took place on the subduction-zone interface between the India plate and the Burma microplate that borders the larger Sunda plate. This earthquake is the first large earthquake to have occurred since the advent of modern space-based geodesy and broadband seismology, and has provided the opportunity to compare and integrate estimates of spatial and temporal rupture characteristics derived from seismological, geodetic, geological and tsunami data. Much has been published individually in scientific journals and as collections of papers in special issues (e.g. *Bulletin of the Seismological Society of America* in January 2007). During the earthquake, rupture occurred along a 1300–1600 km segment of the plate boundary (Ishii *et al.*, 2005; Lay *et al.*, 2005). Slip estimates (estimated from geodetic, geological, tsunami and seismological data) averaged $\sim$ 9 m, over the entire length of the rupture, and averaged 15 m over the southern segment (Ammon *et al.*, 2005; Banerjee *et al.*, 2007; Geist *et al.*, 2007). Maximum slip of up to 25–30 m occurred off northern Sumatra (Fujii and Satake, 2007; Piatanesi and Lorito, 2007). The rupturing process initiated near Sumatra and took $\sim$ 10 minutes (Lambotte *et al.*, 2007) to travel from south to north. Rupture velocities were calculated to be $\approx 0.7-2.5\ \mathrm{km\,s^{-1}}$, with the slow slip occurring on a timescale $\approx$ 40–50 minutes longer than the fast slip (Lay *et al.*, 2005; Seno and Herata, 2007). The seabed was deformed over an area of $\geq 100\,000\ \mathrm{km}^2$ and this dislocation initiated the tsunami.

Within a few minutes of the earthquake, a large tsunami reached the nearest shores of Sumatra and the Andaman Islands (Jaffe *et al.*, 2006; Stein and Okal, 2007). At some locations along the Sumatra coast wave runup measured $>$ 30 m above sea level (Borrero, 2005). The greatest loss of life occurred in the city of Banda Aceh, at the northern tip of Sumatra, after a wave reaching $\sim$ 10 m above sea level struck the coast. This wave caused complete devastation of the town areas closest to the coast, and traveled as far as 4 km inland (Borero, 2005; Umitsu *et al.*, 2007). Flooding of rivers extended 8.5 km upstream, and inundation from Lho-nga on the west coast may have traveled $>$ 6 km before meeting the wave that hit Banda Aceh. Although estimates vary, $\sim$ 170 000 people were presumed dead (including persons classified as missing) in Aceh Province alone (Doocy *et al.*, 2007), mostly from drowning or from being crushed or impaled by debris. Many people also died from disease in the aftermath of the tsunami.

Waves spread across the Indian Ocean and reached the coasts of Sri Lanka and Thailand after 90–120 minutes. As a result of the Thailand coast being a popular area with tourists, there were a large number of videos and photographs taken of the tsunami, providing an unparalleled source of scientific information regarding near-shore and onshore tsunami behavior. On the Thailand and Sri Lanka coasts, maximum runup heights were typically in the range of 2–10 m (Liu *et al.*, 2005; Tsuji *et al.*, 2006). The tsunami continued to spread, causing casualties on the African coast, and extended beyond the Indian Ocean to be detected on tide gauges worldwide (Titov *et al.*, 2005). In total > 220 000 people were known to be killed, > 43 000 were still listed as missing one year after the event, 400 000 homes were destroyed, 1.4 million people lost their livelihood and >$10 billion in damage was estimated by the United Nations. Most of the damage was a consequence of the tsunami rather than the earthquake itself.

Of critical concern for nuclear facilities, the large magnitude of the 2004 Sumatra–Andaman earthquake has raised questions about the validity of basing assumptions about the maximum magnitude earthquake on any particular subduction zone in the historical record (McCaffrey, 2008). Recent paleotsunami research in Japan (Nanayama *et al.*, 2001) and in southern Chile (Cisternas, 2003; Salgado, 2003) has shown that, along some subduction margins, the largest events may occur at very long return times, although there may be smaller events (some up to M > 8) in the interim. In addition, the length and duration of earthquake rupture has implications for the usual assumption of instantaneous dislocation and tsunami generation along the entire fault length, made in many tsunami models. Geist *et al.* (2007) showed that the triggering of the tsunami successively along the fault as rupture proceeded created a different form of tsunami than would have resulted from instantaneous rupture.

11.2.2 Landslides as sources of tsunami

Landslides causing tsunami may originate either above or below the water line. Subaerial landslides (starting above the water line, and potentially including ice falls) often produce dramatic effects locally, but no historical examples have been large enough to spread a destructive tsunami ocean-wide. The largest historically recorded tsunami runup produced by a subaerial landslide occurred in Lituya Bay, Alaska, on July 9, 1958. About 4×10^7 m^3 of rock, loosened by movement on, or by shaking from, a nearby fault, fell suddenly into the head of the bay. The resulting waves surged to > 500 m on nearby slopes before propagating at speeds up to 200 km hr^{-1} down the bay at heights of 20–50 m (Miller, 1960) and crossing the bar at the bay mouth. Many, but not all, subaerial landslides are initiated by earthquakes. Some occur spontaneously from oversteepened coastline. For example, the 1999 landslide in the Marquesas Islands (Okal *et al.*, 2002) and the December 2002 landslides on Stromboli (Tinti *et al.*, 2005) each initiated localized tsunami that caused injuries to three people and damaged villages on the island of Stromboli (maximum runup equaled 11 m) and on Panarea Island, 20 km away. Extremely large subaerial landslides, effectively sector collapses, may occur where volcanoes become oversteepened and induce landsliding beneath the water. Subaerial landslides, proposed to have occurred in the Canary Islands, might be large enough to produce ocean-wide effects (Ward and Day, 2001). The Nuuanu landslide, caused by the collapse of Koolau volcano on the Hawaiian island of Oahu and initially partially subaerial, was one of the largest submarine landslides on Earth (Moore *et al.*, 1989; Bryant, 2001).

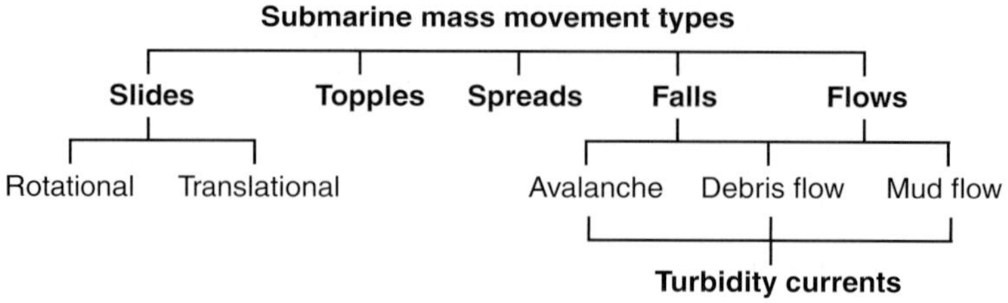

Fig. 11.2 Classification of submarine mass movements adapted from subaerial classification proposed by the ISSMGE Technical Committee on Landslides (TC-11), from Locat and Lee (2000). The nature of the landslide is categorized by basic type of mass movement, rather than its underlying cause.

Submarine landslides can also generate tsunami. Landslides that are triggered by earthquakes are the cause of much debate, since it is often difficult to disentangle the landslide contribution to a tsunami from the direct effects (i.e. seafloor deformation) of the earthquake. For example, it is suggested that the tsunami following the 1946 Aleutian earthquake was, to a large extent, the result of a submarine landslide (Okal *et al.*, 2003; Fryer *et al.*, 2004), although the degree to which landsliding contributed to the far-field tsunami is still unclear.

Submarine landslides occur on a wide variety of scales and via a variety of mechanisms (Figure 11.2; Moore, 1978; Locat and Lee, 2000; Masson *et al.*, 2006). Many of these submarine mechanisms are the same as those that occur on land, but some are peculiar to the marine environment. These include turbidity currents, in which sediment is maintained in suspension by fluid turbulence (Masson *et al.*, 2006) and very long runout slides, > 100 km in some cases. The 8 ka Storegga slide off the coast of Norway, for example, extends 160 km over the seafloor, while deposits from the 1929 Grand Banks, Canada, landslide, transported by a turbidity current, can be found 1000 km from the slide initiation area (Locat and Lee, 2000). Table 11.1 lists key parameters for a selection of major submarine landslides.

As with subaerial landslides, strong earthquake shaking is a key trigger and, based on historical evidence, it is generally recognized that the majority of large submarine landslides are triggered by earthquakes (Masson *et al.*, 1996). Susceptible terrain for both subaerial and submarine landslides includes oversteepened slopes, deltas and some geological conditions, such as weak layers that are predisposed to slip (Masson *et al.*, 1996). Modern techniques for imaging bathymetry data make it possible to identify the evidence of both historical and pre-historical landslides (e.g. Lamarche *et al.*, 2003).

Key parameters in determining the size of tsunami generated by a submarine landslide include the volume of material displaced, the depth of water in which it takes place, the speed of mass movement and the nature of the landslide process (slide, debris flow, turbidity current, etc.), among others. Numerical modeling of tsunami initiation by submarine landslides has been reported by Pelinovsky and Poplavsky (1996) and Grilli and Watts (2005), among others. Wave-tank experiments are useful for constructing empirical models and validating numerical models of landslide processes (e.g. Watts, 1998; Grilli and Watts, 2005).

Table 11.1. Selected major slides, debris flows and turbidity currents

Name or location	Date	Type	Area (km^2)	Volume (km^3)	Maximum slope (°)
Nuuanu, Hawaii	$\gtrsim$ 1.5 Ma	Slide/avalanche	23 000	5000	∼ 5.0
Storegga, North Sea	∼ 8 ka	Debris flow	95 000	2400–3200	∼ 1.4
Canary Islands	∼ 15 ka	Debris flow	40 000	400	∼ 1.0
Grand Banks	1929	Turbidite[1]	160 000	200	∼ 5.0
Ruatoria, New Zealand	∼ 170 ka	Debris flow	8 000	3000	∼ 10

[1] Turbidite: geological formation has its origin in turbidity current deposits.
Data from Moore *et al.* (1989), Piper *et al.* (1999), Collot *et al.* (2001), Haflidason *et al.* (2004), Fine *et al.* (2005) and Masson *et al.* (2006).

The Grand Banks tsunami of 1929

The Grand Banks tsunami of 1929 was caused by a submarine landslide triggered by a M 7.1 earthquake. The earthquake appears to have initiated several slides, which coalesced into debris flows that grew as they collected more material while traveling downslope. Eventually, these debris flows joined together and formed a single turbidity current containing ∼ 185–200 km^3 of material (Heezen and Ewing, 1952; Piper *et al.*, 1999; Bryant, 2001; Fine *et al.*, 2005). Several telephone cables were severed in sequence over a period of 12 hr as the landslide cut across them, implying a slide velocity of 15–20 $m\,s^{-1}$ (Heezen and Ewing, 1952; Masson *et al.*, 1996). The tsunami, initiated by the landslide, reached the southern coast of Newfoundland ≈ 2.5 hr later, claiming 29 lives, the largest death toll following any earthquake in Canada (Whelan, 1994). The waves were ∼ 3–8 m high (Fine *et al.*, 2005) and destroyed houses and boats around the Burin Peninsula.

11.2.3 Volcanic eruptions as sources of tsunami

Volcanic eruptions can act as sources of tsunami, but infrequency of events, few nearby survivors and lack of scientific measurements raise questions about the primary mechanisms involved. Undoubtedly, several mechanisms can occur together. For example, an earthquake, the collapse of a coastal lava-built platform and a volcanic eruption occurred very close in time to the 1868 tsunami on the island of Hawaii (Wood, 1914). Latter (1981) identified ten potential means by which a volcano could initiate a tsunami: (i) deformation caused by earthquakes accompanying eruptions, (ii) the action of submarine explosions, in particular the collapse of water domes caused by explosions, (iii) displacement of water by pyroclastic flows, (iv) collapse or subsidence of calderas, and subsequent infilling by water, (v) landslides of cold rock, caused by shaking or gravitational collapse, impacting into the sea, (vi) the effect of base surges and accompanying shock waves, (vii) avalanches of hot rock impacting into the sea, (viii) lahars entering the sea, (ix) forcing of the sea surface

Table 11.2. Historical tsunami induced by volcanic eruptions

Name, location	Date	Mechanism	Maximum runup (m)
Krakatau	1883	Submarine explosion, pyroclastic flow, caldera collapse	42
Stromboli, Italy	Many	Hot rock avalanches	10
Unzen, Japan	1792	Cold rock landslide	9
Ritter Island, Papua New Guinea	1888	Caldera collapse	15
New Hebrides	1878	Volcanic earthquake	17
Matavanu, Samoa	1906	Lava	3.6

Data based on Latter (1981), Bryant (2001) and Tinti *et al.* (2005).

due to air waves caused by explosions, and (x) lava avalanching into water. Some of these mechanisms have been studied in detail by Nomanbhoy and Satake (1995), Ward and Day (2001) and Watts and Waythomas (2003).

Notable tsunami-causing volcanic events (Table 11.2) include the eruption of Krakatau in 1883, the eruption of Santorini in the eastern Mediterranean in ∼ 1470 BC (Bryant, 2001; Lockridge, 1988), which is believed to have played an important role in the collapse of the Minoan civilization, and the eruption of Mt. Unzen in southern Japan in 1792, which led to ∼ 15 000 fatalities from the ensuing tsunami (Lockridge, 1990). Although such events are rare, their impact is certainly large and they cannot be ignored when considering tsunami hazards for nuclear facilities.

The Krakatau eruption of 1883

The Krakatau eruption of August 27, 1883 was one of the largest eruptions in recorded history, producing sound waves that could be heard 4000 km across the Indian Ocean. Most of the 34 000 casualties were caused by the ensuing tsunami, which destroyed hundreds of villages and thousands of boats around the Sunda Strait (Verbeek, 1884; Bryant, 2001). The events of that day are well documented, but there were few scientific instruments available at the time to make quantitative measurements. Consequently, the details of how the eruption caused a tsunami are still debated (Self and Rampino, 1981; Nomanbhoy and Satake, 1995). What is known is that a sequence of volcanic activities, starting in May 1883, culminated in four main explosions on August 27. Of these explosions, the third and largest explosion is believed to have caused the main tsunami, based on the arrival time at Batavia (Verbeek, 1884; Latter 1981). Smaller tsunami were caused prior to and after the main event. Three main candidates have emerged for the tsunami-causing mechanism: (i) collapse of northern Krakatau Island into the sea, (ii) a pyroclastic flow entering the water, and (iii) a submarine explosion caused by water entering the magma chamber. These different hypotheses are examined in detail by Nomanbhoy and Satake (1995).

The tsunami waves are estimated to have been $\sim$ 15 m high within the Sunda Strait, with a maximum runup of 42 m, and maximum inland penetration of 5 km in low-lying areas. Waves were observed far beyond the islands of Java and Sumatra. A tsunami reached the northwestern cape of Western Australia four hours after the explosion and traveled up to 1 km inland; nine hours after the eruption, 300 riverboats were sunk by tsunami waves in Calcutta. Sea-level variations were observed on tide gauges worldwide, although controversy still surrounds the question of whether these waves were tsunami waves propagated through the ocean or whether the waves were related to atmospheric pressure waves that accompanied the eruption (Choi *et al.*, 2003).

11.2.4 Bolide impacts as sources of tsunami

Comets and asteroids occasionally pass close to Earth and sometimes collisions ensue. Comets normally orbit outside the known planets but, occasionally, gravitational interactions will send them into orbits that cross Earth's orbit. Asteroids mostly orbit between Mars and Jupiter but are occasionally sent into orbits that cross Earth's orbit. A direct collision between a comet and a planet is an immense event, as was witnessed in 1994 when the disintegrating pieces of the comet Shoemaker–Levy 9 entered the atmosphere of Jupiter. Probably the biggest well-recognized impact on Earth in the last few hundred million years is the impact that occurred at Chicxulub, in Mexico. This event has been implicated in causing the extinction of the dinosaurs $\sim$ 65 Ma, although this idea and the precise date of the event are still debated. The global effects from such an event today would be so devastating that the tsunami hazard would be only one of many interrelated hazards. Furthermore, such an event is probably too rare to be considered in most probabilistic tsunami hazard assessments. However, the more common and less devastating impacts of smaller asteroids and cometary fragments may need to be considered, depending on level of hazard assessment for specific nuclear facilities.

The 2.15 Ma Eltanin impact

During the late Pliocene era, $\sim$ 2.15 Ma, an asteroid of estimated 1–4 km dimensions collided with Earth, impacting in the Bellinghausen Sea, which lies in the eastern Pacific sector of the Southern Ocean (Gersonde *et al.*, 1997). This event, known as the Eltanin impact, is unusual as it is the only known example of a substantial bolide impacting in the deep ocean, even though the deep ocean covers 60% of Earth's surface. For comparison, there are $\approx$ 140 known impacts on other parts of Earth's surface. The tsunami waves caused by the Eltanin impact have been estimated to be $\sim$ 25–80 m high along the west coast of South America and between 5–40 m around the wider Pacific rim (Mader, 1998).

11.3 Historical tsunami databases

The National Geophysical Data Center (NGDC) on-line global tsunami database (NGDC, 2008) contains over 2300 tsunami events ranging from 2000 BC–2007. Of these events,

1126 are considered probable or definite tsunami and 902 have occurred since AD 1800. Of these 902 tsunami events, nearly 75% were sourced in the Pacific (excluding Indonesia and Malaysia), 733 were caused by an earthquake alone, 31 were caused by a landslide, 47 were caused by an earthquake and associated landslide, 38 were caused by a volcanic eruption alone (i.e. there was no recognized accompanying earthquake or landslide), and the remaining have multiple other causes, or the cause is unknown. Linked to the events database is a database of $> 10\,000$ runup observations. Both have display options and can be searched using combinations of fields including source region, runup region/country, validity, source mechanism, etc. The Novosibirsk Tsunami Laboratory (NTL , 2008) maintains several on-line databases which cover the globe, but have been divided into various regional databases (e.g. the Pacific, Mediterranean, Atlantic). These two global databases, the NTL and the NGDC, have many common events but they differ in structure and detail.

Many tsunami databases are compiled at a national level. Japan has, perhaps, the most detailed record of any Pacific country, extending back > 1.3 ka. New Zealand also has a particularly comprehensive database, but it only spans $\lesssim 170$ a, since the time of significant European settlement. Stories and legends relating to tsunami originating from pre-historical times may also be useful in understanding present-day risk (Atwater *et al.*, 2005; McFadgen, 2008).

In the past it has been necessary to draw conclusions based on the long-term probability of tsunami occurrence and impact from historical databases. When attempting to do so, it is important to consider the level of completeness of these databases, not only in number of events but also in correct identification of causes and the extent and level of impact. To be featured in the database, a tsunami (or at least its consequences) must: (i) be observed, (ii) be recorded in writing, (iii) have its written record survive into the modern era and finally, (iv) come to the attention of the database compiler. For example, a large number of tsunami take place in the Pacific Ocean where a written culture did not exist in many of the surrounding lands until the last few hundred years. Therefore it is not surprising that the majority of database entries are from the last two centuries. It is also not surprising that the majority of recorded tsunami sources are earthquakes, as these sources tend to be the easiest to identify. There are, for example, probably many cases where submarine landslides have been triggered by earthquakes but have not been recognized. A more important consideration is whether historical databases span a long enough time period to reflect the full range of events that might occur, even for return periods as short as 500 a (a typical return period of interest for many hazard and risk assessments).

11.4 Paleotsunami data

Paleotsunami are tsunami that have occurred within the geological past, particularly prior to the written record of historical events. Evidence for their occurrence comes from sediments and debris that have been deposited (tsunami deposits). It is only within the last decade that techniques have been developed to identify key characteristics of tsunami deposits, thereby enabling the recording of tsunami much further back in time than the historical and instrumental record. Tsunami deposits, in addition to providing evidence for the occurrence of past tsunami, can also provide information about tsunami sources, frequency and magnitude.

Source

The aspect and length of coast over which a tsunami deposit is found can provide information about the direction and offshore distance of the tsunami source (helping to determine whether it was a local, regional or distant event). The type of source can sometimes be inferred from co-existence of the tsunami deposit with physical evidence of deformation (e.g. subsidence and liquefaction features would imply an earthquake source). Correlation of the deposit with a known tsunami-causing event can be used to infer a source, in cases where high-resolution age control is available.

Frequency

It is possible to estimate recurrence intervals for paleotsunami when a long geological record of tsunami deposits exists. This type of information is particularly important where no large tsunami have occurred in historical times, but when large events are represented in the geological record frequently enough to suggest future risk.

Magnitude

Sedimentary deposits are usually evidence of moderate to large paleotsunami, since small tsunami are unlikely to leave obvious evidence in the geological record. The physical extent of tsunami deposits along and across coastal topography, and the height above sea level that deposits reach, provide estimates for tsunami inundation distance and runup height.

Although paleotsunami data sets have a unique contribution to make to tsunami hazard assessment, there are some major limitations that must be taken into account. Paleotsunami data sets will always be incomplete because: (i) many paleotsunami are not preserved in the geological record, because not all tsunami leave a recognizable deposit, and not all deposits are preserved for long periods of time; (ii) many paleotsunami cannot be identified, because not all deposits contain unique tsunami signatures, deposition is patchy, so evidence may be missing from a particular site, and storm-surge deposits may be misinterpreted as tsunami deposits.

Paleotsunami research is in its infancy, so there are relatively few researchers working in this field. There is, as yet, a lack of coverage of many key sites and little detail at many of the sites that have been studied. Paleotsunami research is time-consuming, so the focus of many studies has been on the initial identification of tsunami deposits. Additional work that is crucial for the assessment of tsunami source, frequency and magnitude, such as detailed mapping of the extent of the deposit, high-resolution age control and investigation of multiple events at any one site, has yet to be carried out in many cases.

11.5 Numerical modeling of tsunami

Numerical modeling of tsunami serves a dual purpose: it allows us to estimate the effects of events that have yet to happen; and it enables us to evaluate our understanding of past tsunami. The process of numerical tsunami modeling can be considered in three main stages: *source modeling*, in which the generation of the tsunami, either by earthquake, landslide, volcanic eruption or bolide impact, is simulated; *propagation modeling*, in which the dispersal of the tsunami waves in the ocean, sea or lake is simulated; and *impact modeling*, in which the consequences of tsunami impact on specific coastal settings are assessed. Impact modeling is usually considered as two parts: *inundation modeling*, in which water flow over dry land is simulated; and *loss modeling*, in which the damage caused by the tsunami is estimated.

The modeling process is usually performed using specially designed computer programs. The latest three-dimensional tsunami models can include source, propagation and inundation stages, overcoming the difficulty of interfacing different models, particularly for boundary conditions near the shore, which is the most dynamic and complex phase of tsunami development. Ideally, loss modeling would be integrated with inundation modeling, but this is not yet standard practice. Tsunami damage is not always a direct consequence of inundation; for example, the recession which sometimes precedes a tsunami or occurs between waves may threaten power-station water intakes.

Tsunami source models are well developed for earthquakes, where surface deformation can be estimated by assuming that the earthquake represents a finite dislocation within an elastic body (Okada, 1985). These techniques have been tested against data from numerous real events and generally demonstrate reasonable agreement, although the December 26, 2004 Sumatra–Andaman Islands earthquake has highlighted some areas for improvement (Lay *et al.*, 2005). Both landslides and volcanic eruptions tend to have greater variability in the mechanisms by which they initiate tsunami, and the physics of those mechanisms is, in some cases, only partly understood. Consequently, while modeling of past events can be undertaken and specific scenarios for future events can be investigated, it is harder to develop general insights that would support a predictive ability.

Propagation modeling, in which the processes by which tsunami waves spread out from the source are simulated, is well understood in terms of the underlying physics, although uncertainty in some parameters remain. This area of modeling is now at a stage where many useful insights can be gained (e.g. Figure 11.3).

Impact modeling is an area in which numerical modeling is at a preliminary stage, not only because of its inherent complexity and site-specificity, but also because resources and data availability are limited. There are many different processes taking place as tsunami approach and impact a coastal area, each of which may be well understood in isolation, but effective modeling of the combined processes remains challenging. Developing high-resolution models can capture these processes, but are time-consuming and require high-capacity computing. Useful insights for more empirical impact modeling can, however, be gained from studying the impacts of real tsunami.

11.5.1 General insights

Earthquake-generated tsunami typically propagate in such a way that most of the wave energy is directed perpendicular to the fault on which the earthquake occurred, and the initial

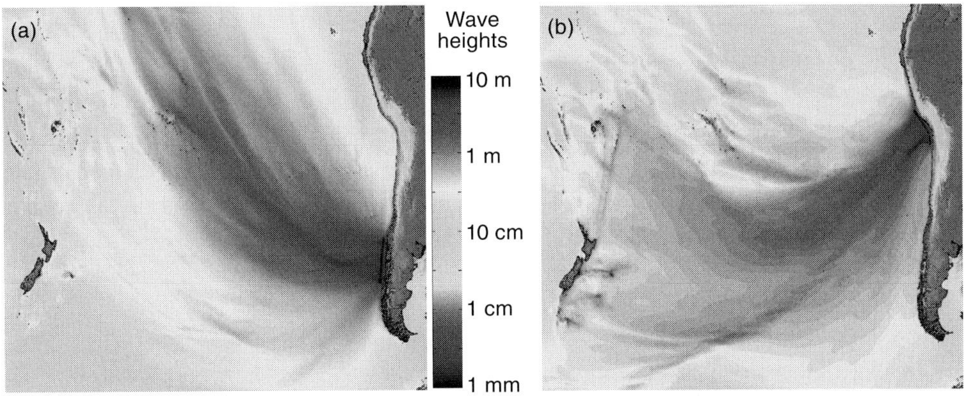

Fig. 11.3 Tsunami modeling scenarios using the MOST modeling programs (Titov and Gonzalez, 1997). This two-scenario comparison of a South American tsunami affecting New Zealand illustrates the effect that directivity of the source can have on distant locations. The source for model (a) is the 1960 Chilean M 9.5 earthquake; the source for model (b) is an equivalent M 9.5 earthquake originating in the area of southern Peru/northern Chile. The models estimate maximum wave heights offshore (> 25 m deep). Wave heights may increase by several times close to the coast. To estimate wave heights at the shore, higher resolution "nested grid" models can be used (modeling by W. Power and V. Titov). See color plate section.

wave is separated into two components traveling in opposite directions. Landslide sources can be highly directional, sending a fairly concentrated tsunami "beam" perpendicular to the slope, in the direction of the landslide movement (Ward, 2001b; Walters *et al.*, 2006). Many volcano sources can also be highly directional, but more typically radiate waves in a circular pattern.

Where the dimensions of the tsunami source are small, less than a few tens of kilometers in the case of ocean sources, the resulting waves may be subject to significant dispersion, in which the different frequencies present in the tsunami propagate at different speeds. This leads to stretching out of the tsunami wave train and, consequently, generally lower amplitudes. This is one reason why landslides and volcanic eruptions are usually not a tsunami hazard at large distances.

Tsunami waves tend to become concentrated above undersea ridges because of refraction. In this situation the ridge can act as a "waveguide," leading to enhanced tsunami heights at locations where these ridges focus towards the shore (Koshimura *et al.*, 2001).

Bays and inlets around the coast have specific natural frequencies, determined by the time it takes for water to pass into and out of the bay (e.g. Walters and Goff, 2003). If the natural frequency of a bay matches that of the tsunami, then amplification will occur. This can often explain variations in tsunami height, which may at first appear random, along a given section of coastline. Identifying the natural frequencies of coastal bays and comparing them with characteristic frequencies for tsunami is a useful step towards identifying those areas most at risk.

11.5.2 Problems and limitations

In many areas of the world there is very limited historical data on, for example, wave period, the number of waves and their variability along a coast during tsunami, which can be used to calibrate (or validate) models. A critical input to propagation models is the bathymetry of the seafloor. This is because the speed, and ultimately the direction, of the tsunami are controlled by the depth of water. Consequently the model results are only as good as the bathymetry data allow. Good bathymetry data exist, but the processes of combining different sources of bathymetry and processing it into the required form is one of the most labor-intensive aspects of tsunami modeling.

Many propagation models assume that coastlines behave as perfect reflectors of tsunami waves; this omits the natural dissipation of tsunami energy which occurs when they run up against the shore (Dunbar *et al.*, 1989), leading to a gradual reduction of the accuracy of the model. This is a particular problem for modeling the effect of tsunami from distant sources, as incoming waves may arrive over the course of several hours and interact with earlier waves, especially in locations where tsunami waves may become "trapped" within bays and inlets.

Impact modeling requires very detailed data on coastal bathymetry (which may vary significantly with time in some areas), the topography of the areas being considered (ideally with a vertical resolution of less than 0.25 m) and the distribution of different types of land cover, including man-made structures (variable on an even shorter timescale). Few parts of the world have regularly updated databases with appropriate resolution. Ideally, the near-shore bathymetry and on-land topography and surface structures would be obtained as a seamless digital elevation data set, to enable simulations using the full power of high-resolution hydraulic modeling software.

Source characterization represents a problem for tsunami modeling. Where models are used for real-time forecasting, it is usually only possible to determine very basic information on the characteristics of the source in the time available. This problem also applies to modeling of past tsunami, because there may be little source information available. This is particularly true for local-source tsunami, because the waves are often strongly influenced by the details of the source (e.g. the distribution of fault-slip in an earthquake). However, deep-water wave buoys may be useful in forecasting the potential effects of distant tsunami, as they record the local characteristics of the resulting wave train without requiring explicit source characterization.

The modeling techniques outlined here are useful for estimating what may happen in a particular scenario; yet a comprehensive estimate of tsunami hazards also requires an understanding of the likelihood of particular scenarios occurring.

11.6 Tsunami hazard and risk assessment

As with other natural physical hazards, there are two standard approaches to tsunami hazard and risk analysis. The older deterministic method involves the development of scenarios for particular events. This approach is useful where there are only a small number of possible events that may affect any one location, and in this case, it is possible to model all of the relevant scenarios.

A potential weakness of the deterministic approach is that the likelihood of the scenario event is often not fully taken into account. Therefore, assessing hazards in a scenario-based approach, often keyed to the return period of the damaging event, may be a difficult task. In the case of tsunami, there are often a large number of possible tsunami scenarios; here it is often more useful to take a probabilistic approach.

A probabilistic approach (Figure 11.4) that incorporates the magnitude and frequency distribution for each of the tsunami-generating phenomena could be the goal for hazard and risk assessment at critical facilities (Yanagisawa *et al.*, 2007). Because of source uncertainties and modeling complexity, methods for such assessments have only recently been developed (Rikitake and Aida, 1988; Burbidge *et al.*, 2006; Annaka *et al.*, 2007; Power *et al.*, 2007), but they generally build on long-standing approaches taken for earthquake ground motion (Stirling *et al.*, Chapter 10, this volume).

Overall, probabilistic tsunami hazard assessment (PTHA) for nuclear facilities depends on a variety of regional- and local-scale characteristics. At the regional scale, these include

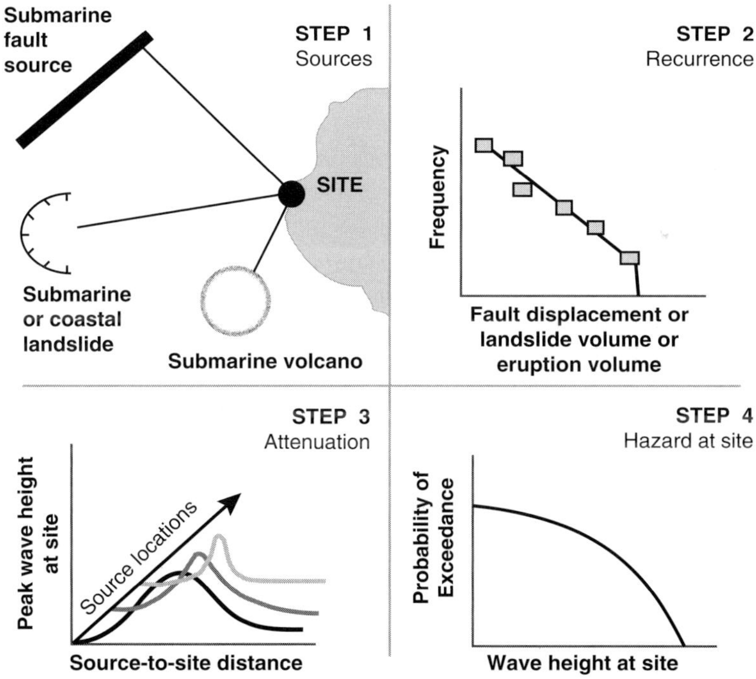

Fig. 11.4 The general four-step procedure of probabilistic tsunami hazard assessment (PTHA), compare with the four-step procedure outlined by Stirling *et al.* (Chapter 10, this volume) for tectonic hazards.

potential tsunami-generating sources (size and frequency of earthquakes, landslides, volcanoes) and the nature of global/regional wave propagation from the sources. Local-scale characteristics include coastal features that influence the form of the tsunami; specific vulnerabilities to, for example, wave impact force, backwash, flooding, saltwater contamination, sea-level draw down, etc.; and defensive measures (either passive or active, implemented in response to a tsunami warning) that may be implemented to mitigate risk. Elements in the development of PTHA for nuclear facilities are described in the following. The basic treatment of probabilities is similar to that in other types of hazard analyses (e.g. Volentik *et al.*, Chapter 9, this volume; Stirling *et al.*, Chapter 10, this volume; Valentine and Perry, Chapter 19, this volume).

11.6.1 Tsunami sources

Earthquakes may be the most frequent generator of damaging tsunami, but not necessarily the source causing the largest runups. Their tsunami-generating potential is essentially capped by the amount of slip that can occur during a single rupture on a given fault plane, and this, in turn, sets an upper limit on the size of seafloor deformation and subsequent tsunami. Extrapolating relationships developed by Wells and Coppersmith (1994) between earthquake magnitude, M, fault length, L, and average fault slip, u, suggests that the slip scales approximately as

$$u \approx 2.5 \times 10^{-5} L \tag{11.2}$$

The largest earthquake-induced tsunami occur on subduction-zone plate boundaries, and it is not always clear where subduction zones terminate or where heterogeneities in the subducting plate produce barriers to earthquake rupture. Assuming an upper limit to the length of rupture on subduction zones of 2000 km, (11.2) would suggest a maximum possible average slip of $\sim$ 50 m, with seafloor deformation equal to up to about half of this. Such events are likely to be rather rare; the largest rupture and greatest seismic moment of any earthquake in instrumental history was the 1960 Chilean Mw 9.5 earthquake, which had a rupture length of $\sim$ 1000 km, a width $>$ 60 km and an estimated average slip $\sim$ 20–40 m (Plafker and Savage, 1970; Plafker, 1972). Paleotsunami studies in the source area of the 1960 Chilean earthquake suggest that four such events have occurred in this region within the last 800–970 a (Salgado *et al.*, 2003). These larger events are interspersed with smaller earthquakes and tsunami which are still large enough to be recorded historically, but too small to leave an obvious geological record.

Other processes such as landslides, bolide impacts and volcanic eruptions have the potential to produce tsunami with larger runup heights than earthquakes. Very large landslides [e.g. those found around the islands of the Hawaiian ridge (Lipman *et al.*, 1988; Garcia and Hull, 1994; Moore *et al.*, 1995)] or volcanic eruptions on the scale of the Krakatau 1883 eruption, may, at least locally, result in tsunami at $\geq$ 5–10 times larger than those created by earthquakes. Bolide impacts are reasonably well understood and their potential to produce tsunami can possibly exceed those from any other source. The tsunami caused by the largest of such processes have been called "mega-tsunami;" fortunately, however, these are very rare events. Nevertheless, these phenomena could be of concern for critical facilities located on the coast where disruptive events with return periods $\sim 10^4$–10^5 a need to be considered.

Earthquake magnitude–frequency

The basis for determining the magnitude–frequency distribution for tsunamigenic earthquakes is essentially the same as determining the magnitude–frequency distribution for seismic hazards (Stirling *et al.*, Chapter 10, this volume; Coppersmith *et al.*, Chapter 26, this volume). The principal difference is that earthquakes incapable of directly producing significant tsunami can be ignored (i.e. those inland, or submarine earthquakes with maximum Mw $\lesssim 6.5$ (Ward, 2001a). Note, however, that this does not include consideration of earthquakes as a trigger for other processes that then cause tsunami. Additionally, for a site-specific analysis, it may be possible to ignore earthquake sources for which the maximum magnitude event is still insufficient to produce a damaging wave at the site. The historical record and use of empirical relationships are useful in making this determination.

Typically, earthquake sources are described using either a Gutenberg–Richter distribution or in terms of characteristic earthquakes. The frequency of events with magnitude M or above ($N(M)$) is given by:

$$\begin{aligned} \log N(M) &= a - b \log M, &\quad \text{for } M < M_{\max} \\ N(M) &= 0, &\quad \text{for } M > M_{\max} \end{aligned} \tag{11.3}$$

where, for a Gutenberg–Richter distribution, the key parameters are the maximum magnitude ($M_{\max}$) and the empirical parameters, termed the a- and b- values. For a characteristic earthquake the key parameters are the magnitude, the variability in magnitude (standard deviation) and the return time. It also is important to estimate the uncertainty in all model parameters (e.g. a- and b- values). These uncertainties can be estimated, for example, with a logic-tree approach (Stirling *et al.*, Chapter 10, this volume).

Landslide magnitude–frequency

The determination of magnitude–frequency distributions for landslides is an active, but as yet immature, field of research. Volume of material and velocity are the most obvious parameters determining tsunami size, but other factors such as the depth of initiation, the slope gradient and the mode of transport (e.g. slide, debris flow, turbidity current) are likely to be important in defining accurate source terms. Grilli and Watts (2005) have studied the influence of various landslide parameters on tsunami generation, using numerical models supported by wave-tank experimental data. A survey of landslide source regions where the ages and properties of landslides are well recorded (bearing in mind the possibility that older landslides may be hidden beneath younger ones) would enable estimation of the probability

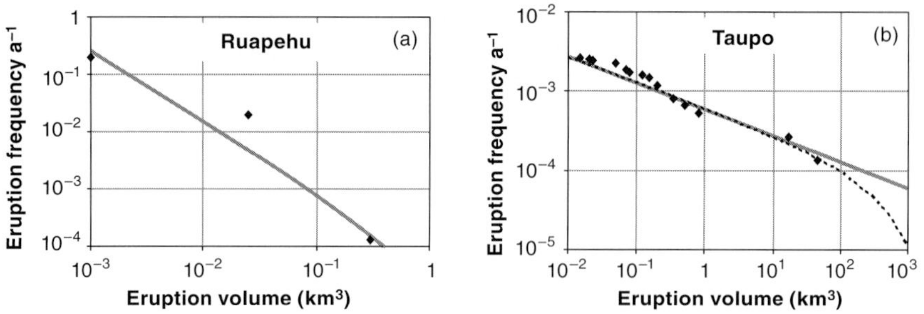

Fig. 11.5 Eruption frequency is related to eruption volume for (a) Ruapehu and (b) Taupo volcanoes (Stirling and Wilson, 2002; Hurst and Smith, 2004). Power-law models (gray lines) explain this relationship reasonably well, but may roll off (dashed line), or truncate, for the largest eruption volumes.

distribution for generating landslides of given properties within regions of similar geology. In the simplistic case, where only volume is considered important, it is likely that volume distribution will, like earthquakes, follow a truncated power-law (e.g. Gutenburg–Richter) relationship. The generalization of such an approach to a wider set of source parameters is likely to become quite complicated (Ward, 2001b).

Volcanic-eruption magnitude–frequency

Similarly, the magnitude and frequency characteristics of eruptions from individual New Zealand volcanoes (Stirling and Wilson, 2002) and eruptions globally (e.g. Simkin and Siebert, 1994) exhibit a truncated power-law relationship (Figure 11.5). The relationship between eruption volume and tsunami size is not a straightforward one, since the tsunami generation depends on many factors such as the time over which the eruption occurs. However, at present, eruption volume is the only parameter for which magnitude–frequency curves have been routinely produced.

Ideally, the magnitude–frequency relationship would be obtained in terms of other parameters that are considered more directly correlated to tsunami size, such as the volcano explosivity index or eruption intensity (Newhall and Self, 1982; see Connor *et al.*, Chapter 3, this volume). In the meantime we must work under the assumption that, for any one volcano eruption, volume and tsunami size are meaningfully correlated.

Bolide-impact magnitude-frequency

The flux of small near-Earth objects colliding with the Earth also follows a power-law distribution (Brown *et al.*, 2002) with, however, no clear size cut-off. The frequency, N, of impact of objects with diameters $> D$ is given by:

$$\log N(D) = 1.57 - 2.70 \log D \qquad (11.4)$$

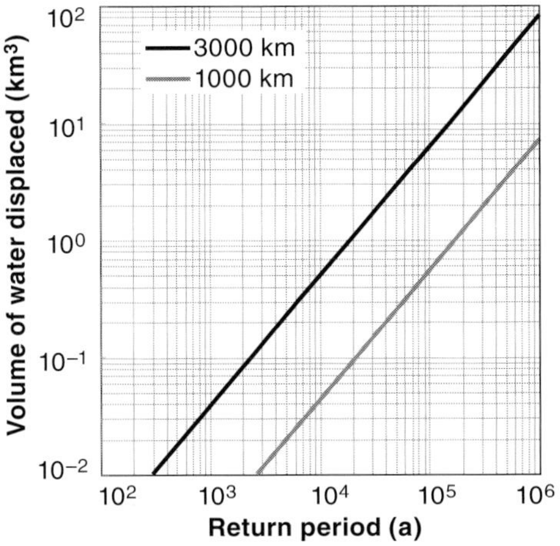

Fig. 11.6 Estimated probability (as return period) of volume of water displaced by a bolide hitting the ocean within 3000 km (black line) and 1000 km (gray line) of a coastal location (surrounded by water) for various return periods (Berryman, 2005).

which can also be expressed in terms of total energy released in the impact, E (in kilotons of TNT, where one kiloton of TNT equivalent is 4.185×10^{12} Joules):

$$\log N(E) = 0.57 - 0.90 \log E \tag{11.5}$$

and, compared to other tsunami sources, this power law is relatively well defined. The flux is more-or-less uniformly distributed over Earth's surface so that the impact frequency is directly proportional to the area being considered. This can be used to calculate probability of displaced volume as a function of area of ocean considered, as indicated in Figure 11.6.

Within a probability horizon of return periods of a few thousand years, bolide-impact tsunami do not feature as a significant risk; they are lost in the background uncertainty in defining other large and more probable events. However, at longer event horizons, bolide tsunami are the largest that can hit extensive areas of coast. There is, however, a bolide size ($\sim$ 1 km in diameter) above which tsunami are dwarfed by other environmental disruptions caused by the impact (Hills and Goda, 1993). Such large events are well known from the geological record and have estimated return periods on the scale of hundreds of thousands of years, comparable to the performance period of geological repositories.

11.6.2 Estimating wave heights from tsunami sources

A procedure is needed for estimating the wave height at the site of interest for a source of specified location and source parameters (magnitude, volume, etc.). Ideally, this can be accomplished by explicit numerical modeling of the interaction of the source with the ocean and of the subsequent wave propagation (Rikitake and Aida, 1988; Power *et al.*,

2007). However, as this is often a computationally demanding process, empirical and semi-empirical methods are also sometimes used (Berryman, 2005) and this simplified procedure is described in the following.

Earthquakes

Based on a compilation of historical, largely Pacific Ocean, data, Abe (1979) proposed an equation for estimating the wave height, H, of a tsunami at a distant shore due to an earthquake of magnitude Mw:

$$H = 10^{(Mw - B_{ij})} \tag{11.6}$$

where B_{ij} is a parameter that varies for each combination of shore, i, and earthquake source, j. The parameter B_{ij} can be determined using either historical data (the preferred method if sufficient data are available), numerical modeling or a combination of both (Berryman, 2005). When using (11.6) for estimating wave height, H, it is important to be aware of the type of data used to determine B_{ij}. Abe (1979) used tide-gauge amplitudes where possible, and adjusted runup data where necessary; Berryman (2005) used inundation heights at the coast. The interpretation of H must match that of the data used for the estimation of B_{ij}. The data that Abe (1979) based this equation on show considerable scatter, so the relationship has significant uncertainty.

For local-source tsunami, the equivalent Abe relationship is given by:

$$H = 10^{(Mw - \log R - 5.55 + C)} \tag{11.7}$$

where H is the wave height at a local coast in meters (strictly speaking it is the peak-to-trough amplitude as measured by a tide gauge, as this is what Abe (1981, 1995) used to determine the constant in the equation), R is the source-to-site distance in kilometers and C is an adjustment factor that depends on the tectonic environment of the earthquake source. The best available values of C are derived from Japanese data and have possible values of 0.0 and 0.2, depending upon factors such as the crustal rigidity (Abe, 1985). This empirical equation estimates the tsunami height based only on earthquake magnitude and distance, and thus takes no explicit account of the effects of bathymetry or source orientation. It is important to account for this uncertainty in the wave-height estimate.

Landslides

An empirical or semi-empirical procedure for estimating wave heights from landslide sources has not yet been established. By analogy to the procedure for distant-source earthquakes we suggest that the form of such an equation, for the simplified case where only the volume is specified, may be:

$$H = aV^b \tag{11.8}$$

where H is the wave height at the coast, caused by a landslide of volume, V, in a particular source region, which is here assumed to lie within a relatively narrow range of distance from the site. Explicitly incorporating distance into this equation is difficult, because the attenuation of wave height with distance is strongly dependent on dispersion, which is itself dependent on the source dimensions (Ward, 2001a, 2001b). The constants a and b are determined by explicit modeling of a set of example landslides of varying volume from within the source region to the site of interest, and plotting of the resulting wave heights against the source volume on a log–log plot. It is here assumed that within any specific source region a particular mode of slope failure (Figure 11.2) is dominant. Hence, other statistical models may need to be explored for specific site assessments.

Volcanic eruptions

Similarly, (11.8) might be used to estimate wave heights from sub-sea volcano eruption sources, based on eruption volume, V, at a specific volcano. As previously discussed, wave height will not correlate with volume of volcanic eruptions that are subaerial, or partly subaerial, or when eruptions are dominantly effusive. Consequently, there is a great deal of uncertainty in relating wave height to the volume of volcanic eruptions, and the types of volcanic activity expected at specific volcanoes must be considered carefully.

Bolide impacts

Relationships describing the expected maximum wave height at a particular site caused by a bolide impact of specified radius at a given distance have been studied in detail by Ward and Asphaug (2000). Their formula (which are complex and not included here) also incorporate the depth of water into which the bolide falls and the influence of the ray path between source and site.

11.6.3 Coastal impact models

Estimating tsunami impact is challenging for several reasons. Most importantly, the number, heights and wavelength of future tsunami will be highly variable depending on source, propagation and shoaling effects. The effect of this variability on probabilistic models is enhanced by uncertainty related to the very complicated flow of water across rough surface topography. Features such as dunes and coastal vegetation, buildings, topographic irregularities and rivers all significantly affect impacts. In the future, numerical models incorporating some of these complexities will provide more detailed scenario descriptions for particular "reference" tsunami. When considering facilities that are constructed with the intention that they last for long periods of time, consideration must also be given to the possible changes that may occur to the surrounding environment over the timeframe of concern (e.g. McKinley and Alexander, Chapter 22, this volume).

Extent of flooding and sea-level drop

Geographic information system–based modeling has been used to provide a relatively simple model of tsunami inundation that has been used as part of the process for estimating aggregate losses at population centers (Berryman, 2005). For studies of the impact on critical infrastructure a more complete model of the inundation physics is needed, often using a model such as the ANUGA finite-volume model (Nielsen and Gray, 2005). Such models can also be used to estimate the extent of sea-level retreat prior to inundation, which may be important in some circumstances. Usually, such models assume that the tsunami inundates an unchanging environment; however, it is expected that future models will also incorporate the impact of tsunami on structures (see the following section) and on the movement of entrained objects, and the influence these processes have on subsequent water flow.

Tsunami forces and building strength

The forces exerted by a tsunami depend on the depth and velocity of the water in it and entrained debris. The velocity is highly variable depending on whether the tsunami is acting like a rapidly rising tide or is surge-like in behavior. When the tsunami is tide-like the velocities are likely to be low, typically $1\ \mathrm{m\,s^{-1}}$, and most of the initial damage will result from buoyant and hydrostatic forces and the effects of flooding. Higher velocities and greater damage often occur during the subsequent withdrawal of the water (Camfield, 1980).

When the tsunami takes the form of a surge, the current velocities are proportional to the square root of the inundation depth (Camfield, 1980):

$$V = 2\sqrt{gD} \tag{11.9}$$

where V is the inundation velocity, D is the inundation depth and g is gravitational acceleration. Velocities in surging flows are usually much higher than $1\ \mathrm{m\,s^{-1}}$ and damage arising from surge and drag forces is much greater than that due to buoyant and hydrostatic forces. Expected velocities are given in Table 11.3, along with surge and drag forces estimated using formula presented in Camfield (1980). Field surveys of tsunami-damaged areas indicate that relatively weak houses typical of coastal villages of the Philippines and Indonesia are likely to be pushed off their foundations by about 1 m depth of water (Imamura *et al.*, 1995; Tsuji *et al.*, 1995a, 1995b), whereas relatively well-built Japanese houses that are bolted to concrete foundations require 1.5 to 2 m depth (Shimamoto *et al.*, 1995). Sensitive nuclear structures are built to much higher standards but, nevertheless, for tsunami inundation depths significantly > 1 m, the risk of physical damage, particularly from the drag force of larger waves, needs to be carefully considered.

Other impacts

Tsunami damage and casualties are dependent on the characteristics and duration of the tsunami impact. The tsunami may arrive as a smoothly rising and falling tide-like fluctuation in water levels, or as a bore or breaking wave. The damage is usually caused in one of three ways: (i) impacts of swiftly flowing torrent, (ii) fires and chemical contamination, and (iii) inundation and saltwater contamination. Where coastal margins are inundated, impacts of swiftly flowing torrent, or traveling bores, on vessels in navigable waterways, canal

Table 11.3. Estimated surge and drag forces for tsunami waves impacting on flat walls

Inundation depth (m)	Water velocity ($m\,s^{-1}$)	Surge force ($kN\,m^{-1}$)	Drag force ($kN\,m^{-1}$)
0.2	2.8	30	4
0.5	4.4	70	10
1.0	6.3	140	40
2.0	8.9	270	160
5.0	14.0	720	1000
10.0	20.0	1600	4000
20.0	28.0	4200	16 000

In each case, the wall is assumed to be higher than depth of the wave. Forces are expressed as $kN\,m^{-1}$-length-of-wall perpendicular to the direction of flow of the wave.

estates and marinas, and on buildings, infrastructure and people, can be very damaging. Torrents (inundating and receding) and bores can also cause substantial erosion both of the coast and the seafloor. They can destroy coastal defenses, scour roads and railways, land and associated vegetation. The receding flows, or "out-rush," when a large tsunami wave recedes are often the main cause of drowning, as people are swept out to sea. Fire may occur when fuel installations are floated or breached by debris, or when home heaters are overturned. Breached fuel tanks, and broken or flooded sewerage pipes or works can cause contamination. Homes and many businesses contain many harmful chemicals that can be spilled. Inundation and saltwater contamination by the ponding of potentially large volumes of seawater will cause medium- to long-term damage to buildings, electronics, fittings and to farmland. Impacts specific to nuclear facilities are described by McKinley and Alexander (Chapter 22, this volume), but, in addition, safe operation and design of nuclear facilities should consider these potential impacts of tsunami on communities and community infrastructure.

Concluding remarks

Tsunami are waves generated by the displacement of water in large water bodies; they can travel large distances and still be highly destructive. The most frequent cause of tsunami is earthquakes. However, other causes such as landslides and volcanoes can potentially cause tsunami with larger runup heights than earthquakes during infrequent extreme events. The historical record of tsunami goes back many hundreds of years but is not sufficiently complete to be a reliable guide to future events. Paleotsunami studies can partially address this deficiency, but in many locations a record of past events is not adequately preserved in the environment.

An understanding of the likely sources, combined with sophisticated hydraulic computer models, enables scenarios of potential tsunami to be investigated. These models can incorporate the initiation, propagation and impact of the hypothetical tsunami. Ideally, the scenarios are incorporated into a probabilistic framework similar to that used for seismic hazards, although much can be learned from individual scenarios, especially in situations where a small number of sources dominates the risk.

Tsunami have the potential to cause very serious consequences for nuclear power plants due to inundation, debris impact or loss of cooling during withdrawal of the sea. This is true whether they originate far away, most often as a consequence of a major earthquake, or locally, where they may also be caused by processes such as cliff failure or submarine landslide. Tsunami hazard and risk assessment needs to become a standard procedure for development of critical facilities in coastal environments.

References

Abe, K. (1979). Size of great earthquakes of 1837–1974 inferred from tsunami data. *Journal of Geophysical Research*, **84**, 1561–1568.

Abe, K. (1981). Physical size of tsunamigenic earthquakes of the northwestern Pacific. *Physics of the Earth and Planetary Interiors*, **27**, 194–205.

Abe, K. (1985). Quantification of major earthquake tsunamis of the Japan Sea. *Physics of the Earth and Planetary Interiors*, **38**, 214–223.

Abe, K. (1995). Modeling of the runup heights of the Hokkaido–Nansei–Oki tsunami of 12 July 1993. *Pure and Applied Geophysics*, **144**, 735–745.

Ammon, C. J., C. Ji, H. K. Thio *et al.* (2005). Rupture process of the 2004 Sumatra–Andaman earthquake. *Science*, **308**, 1133–1139.

Annaka, T., K. Satake, T. Sakakiyama, K. Yanagisawa and N. Shuto (2007). Logic-tree approach for probabilistic tsunami hazard analysis and its applications to the Japanese coasts. *Pure and Applied Geophysics*, **164**, 577–592.

Atwater, B. F., S. Musumi-Rokkaku, K. Satake *et al.* (2005). *The Orphan Tsunami of 1700*. Seattle, WA: University of Washington Press.

Banerjee, P., F. Pollitz, B. Nagarajan and R. Burgmann (2007). Coseismic slip distributions of the 26 December 2004 Sumatra–Andaman and 28 March 2005 Nias earthquakes from GPS static offsets. *Bulletin of the Seismological Society of America*, **97**, 86–102.

Barrientos, S. E. and S. N. Ward (1990). The 1960 Chile earthquake; inversion for slip distribution from surface deformation. *Geophysical Journal International*, **103**, 589–598.

Berryman, K. R. (2005). Review of tsunami hazard and risk in New Zealand, Client Report 2005/104. Lower Hutt, New Zealand: Institute of Geological and Nuclear Sciences.

Bilek, S. L. and T. Lay (2002). Tsunami earthquakes possibly widespread manifestations of frictional conditional stability. *Geophysical Research Letters*, **29**, 1673.

Bilek, S. L., K. Satake and K. Sieh (2007). Introduction to the special issue on the 2004 Sumatra–Andaman earthquake and the Indian Ocean tsunami. *Bulletin of the Seismological Society of America*, **97**, 1–5.

Borrero, J. C. (2005). Field survey of northern Sumatra and Banda Aceh, Indonesia after the tsunami and earthquake of 26 December 2004. *Seismological Research Letters*, **76**, 312–320.

Brown, P., R. E. Spalding, D. O. ReVelle, E. Tagliaferri and S. P. Worden (2002). The flux of small near-Earth objects colliding with the Earth. *Nature*, **420**, 294–296.

Bryant, E. (2001). *Tsunami: The Underrated Hazard.* Cambridge: Cambridge University Press.

Burbidge, D. R., P. R. Cummins and H. K. Thio (2006). A probabilistic tsunami assessment for western Australia and the south coast of Java. *Eos, Transactions American Geophysical Union*, **87**(52), Fall Meeting Supplement, Abstract T31G-07.

Camfield, F. E. (1980). Tsunami Engineering, Special Report CERC-SR-6. Vicksburg, MS: Coastal Engineering Research Center.

Choi, B. H., E. Pelinovsky, K. O. Kim and J. S. Lee (2003). Simulation of the trans-oceanic tsunami propagation due to the 1883 Krakatau volcanic eruption. *Natural Hazards and Earth System Sciences*, **3**, 321–322.

Choowong, M., N. Murakoshi, K. I. Hisada *et al.* (2008). 2004 Indian Ocean tsunami inflow and outflow at Phuket, Thailand. *Marine Geology*, **248**(3-4), 179–192.

Cisternas, M., B. F. Atwater, G. Machuca *et al.* (2003). Buried soils, tree rings, and old maps suggest that the 1960 Chile earthquake was larger than its predecessors of 1837 and 1737. *Abstracts with Programs: Geological Society of America*, **35**, 478.

Collot, J. -Y., K. Lewis, G. Lamarche and S. Lalleman (2001). The giant Ruatoria debris avalanche on the northern Hikurangi margin, New Zealand; results of oblique seamount subduction. *Journal of Geophysical Research*, **106**, 19.

Doocy, S., Y. Gorokhovich, G. Burnham, D. Balk and C. Robinson (2007). Tsunami mortality estimates and vulnerability mapping in Aceh, Indonesia. *American Journal of Public Health*, **97**, 146–151.

Dunbar, D., P. H. LeBlond and T. S. Murty (1989). Maximum tsunami amplitudes and associated currents on the coast of British Columbia. *Science of Tsunami Hazards*, **7**, 3–44.

Fine, I. V., A. B. Rabinovich, B. D. Bornhold, R. E. Thomson and E. A. Kulikov (2005). The Grand Banks landslide-generated tsunami of November 18, 1929: preliminary analysis and numerical modeling. *Marine Geology*, **215**, 45–57.

Fryer, G. J., P. Watts and L. F. Pratson (2004). Source of the great tsunami of 1 April 1946; a landslide in the upper Aleutian forearc. *Marine Geology*, **203**, 201–218.

Fujii, Y. and K. Satake (2007). Tsunami source of the 2004 Sumatra–Andaman earthquake inferred from tide gauge and satellite data. *Bulletin of the Seismological Society of America*, **97**, S192–S207.

Garcia, M. O. and D. M. Hull (1994). Turbidites from giant Hawaiian landslides; results from Ocean Drilling Program Site 842. *Geology*, **22**, 159.

Geist, E. L. (1999). Local tsunamis and earthquake source parameters. *Advances in Geophysics*, **39**, 117–209.

Geist, E. L., V. V. Titov, D. Arcas, F. F. Pollitz and S. L. Bilek (2007). Implications of the 26 December 2004 Sumatra–Andaman earthquake on tsunami forecast and assessment models for great subduction-zone earthquakes. *Bulletin of the Seismological Society of America*, **97**, S249–S270.

Gersonde, R., F. T. Kyte, U. Bleil *et al.* (1997). Geological record and reconstruction of the late Pliocene impact of the Eltanin asteroid in the Southern Ocean. *Nature*, **390**, 357–363.

Gower, J. (2005). Jason 1 detects the 26 December 2004 tsunami. *Eos, Transactions, American Geophysical Union*, **86**, 37–38.

Grilli, S. T. and P. Watts (2005). Tsunami generation by submarine mass failure; I. Modeling, experimental validation, and sensitivity analyses. *Journal of Waterway, Port, Coastal and Ocean Engineering*, **131**, 283–297.
Haflidason, H., H. P. Sejrup, A. Nygrard *et al.* (2004). The storegga slide: architecture, geometry and slide development. *Marine Geology*, **213**, 201–234.
Heezen, B. C. and W. M. Ewing (1952). Turbidity currents and submarine slumps, and the 1929 Grand Banks [Newfoundland] earthquake. *American Journal of Science*, **250**, 849–873.
Hills, J. G. and M. P. Goda (1993). The fragmentation of small asteroids in the atmosphere. *Astronomical Journal*, **105**, 1114–1144.
Hurst, T. and W. Smith (2004). A Monte Carlo methodology for modeling ashfall hazards. *Journal of Volcanology and Geothermal Research*, **138**, 393–403.
Imamura, F., C. E. Synolakis, E. Gica *et al.* (1995). Field survey of the 1994 Mindoro Island, Philippines tsunami. *Pure and Applied Geophysics*, **144**, 875–890.
Ishii, M., P. M. Shearer, H. Houston and J. E. Vidale (2005). Extent, duration and speed of the 2004 Sumatra–Andaman earthquake imaged by the Hi–Net array. *Nature*, **435**, 933–936.
Jaffe, B. E., J. C. Borrero, G. S. Prasetya *et al.* (2006). Northwest Sumatra and offshore islands field survey after the December 2004 Indian Ocean tsunami. *Earthquake Spectra*, **22**, S105–S135.
Kanamori, H. (1972). Mechanism of tsunami earthquakes. *Physics of the Earth and Planetary Interiors*, **6**, 346–359.
Koshimura, S. -I., F. Imamura and N. Shuto (2001). Characteristics of tsunamis propagating over oceanic ridges; numerical simulation of the 1996 Irian Jaya earthquake tsunami. *Natural Hazards*, **24**, 213–229.
Lamarche, G., P. Barnes, K. Lewis, I. Wright and J. Y. Collot (2003). Submarine landslide hazards on the New Zealand continental margin. In: Cochran, U. A. (ed.) *Program and Abstracts, The International Workshop: Tsunamis in the South Pacific: Research Towards Preparedness and Mitigation, September 25–27, Wellington, New Zealand*, Institute of Geological and Nuclear Sciences, Information Series 58. Lower Hutt, New Zealand: Institute of Geological and Nuclear Sceinces, 21.
Lambotte, S., L. Rivera and J. Hinderer (2007). Constraining the overall kinematics of the 2004 Sumatra and the 2005 Nias earthquakes using the Earth's gravest free oscillations. *Bulletin of the Seismological Society of America*, **97**, S128–S138.
Latter, J. H. (1981). Tsunamis of volcanic origin; summary of causes, with particular reference to Krakatoa, 1883. *Bulletin Volcanologique*, **44**, 467–490.
Lay, T., H. Kanamori, C. J. Ammon *et al.* (2005). The great Sumatra–Andaman earthquake of 26 December 2004. *Science*, **308**, 1127–1133.
Lipman, P. W., W. R. Normark, J. G. Moore, J. B. Wilson and C. E. Gutmacher (1988). The giant submarine Alika debris slide, Mauna Loa, Hawaii. *Journal of Geophysical Research*, **93**, 4279–4299.
Liu, P. L. F., P. Lynett, H. Fernando *et al.* (2005). Observations by the International Tsunami Survey Team in Sri Lanka. *Science*, **308**, 1595.
Locat, J. and H. J. Lee (2000). Submarine landslides: advances and challenges. In: Bromhead, E., N. Dixon and M. -L. Ibsen, M. (eds.) *Landslides: In Research, Theory and Practice: Proceedings of the 8th International Symposium on Landslides Held in Cardiff on 26–30 June*. London: Thomas Telford, 1–30.
Lockridge, P. A. (1988). Volcanoes generate devastating waves. *Earthquakes and Volcanoes*, **20**, 190–195.

Lockridge, P. A. (1990). Nonseismic phenomena in the generation and augmentation of tsunamis. *Natural Hazards*, **3**, 403–412.

Mader, C. L. (1998). Modelling the Eltanin asteroid tsunami. Science of tsunami hazards. *International Journal of the Tsunami Society*, **16**, 17–20.

Masson, D. G., N. H. Kenyon and P. P. E. Waever (1996). Slides, debris flows, and turbidity currents. In: Summerhayes, C. P. and S. A. Thorpe (eds.) *Oceanography: An Illustrated Guide*. London: Manson Publishing, 136–151.

Masson, D. G., C. B. Harbitz, R. B. Wynn, G. Pedersen and F. Lovholt (2006). Submarine landslides; processes, triggers and hazard prediction. *Philosophical Transactions of the Royal Society A: Mathematical, Physical and Engineering Sciences*, **364**, 2009–2039.

Matsutomi, H., T. Sakakiyama, S. Nugroho and M. Matsuyama (2006). Aspects of inundated flow due to the 2004 Indian Ocean tsunami. *Coastal Engineering Journal*, **48**, 167–195.

McCaffrey, R. (2008). Global frequency of magnitude 9 earthquakes. *Geology*, **36**, 263–266.

McFadgen, B. G. (2008). *Hostile Shores: Catastrophic Events in Prehistoric New Zealand and Their Impacts on Maori Coastal Communities*. Auckland, NZ: Auckland University Press.

Miller, D. J. (1960). The Alaska earthquake of July 10, 1958: giant wave in Lituya Bay. *Bulletin of the Seismological Society of America*, **50**(2), S253–S266.

Mofjeld, H. O., V. V. Titov, F. I. Gonzalez and J. C. Newman (2000). Analytical theory of tsunami wave scattering in the open ocean with application to the North Pacific, NOAA Technical Memorandum OAR PMEL-116. Seattle, WA: Pacific Marine Environmental Laboratory.

Moore, D. G. (1978). Submarine slides. In: Voight, B. (ed.) *Rockslides and Avalanches*, Developments in geotechnical engineering, 14A–B. Amsterdam, the Netherlands: Elsevier Scientific, 563–266.

Moore, J. G., D. A. Clague, R. T. Holcomb *et al.* (1989). Prodigious submarine landslides on the Hawaiian Ridge. *Journal of Geophysical Research*, **94**, 17 465–17 484.

Moore, J. G., W. B. Bryan, M. H. Beeson *et al.* (1995). Giant blocks in the South Kona Landslide, Hawaii. *Geology*, **23**, 125–128.

Mori, J., W. D. Mooney, S. Afnimar *et al.* (2007). The 17 July 2006 tsunami earthquake in west Java, Indonesia. *Seismological Research Letters*, **78**, 201–207.

Nanayama, F., A. Makino, K. Satake *et al.* (2001). Twenty tsunami event deposits in the past 9000 years along the Kuril subduction zone identified in Lake Harutori-ko, Kushiro City, eastern Hokkaido, Japan. *Annual Report on Active Fault and Paleoearthquake Research*, **1**, 233–294.

Newhall, C. G and S. Self (1982). The volcanic explosivity index (VEI): an estimate of explosive magnitude for historical volcanism. *Journal of Geophysical Research*, **87**, 1231–1238.

NGDC (2008). Tsunami data at NGDC. National Geophysical Data Center, http://www.ngdc.noaa.gov/hazard/tsu_db.shtml.

Nielsen, O. and D. Gray (2005). Hydrodynamic inundation modeling for disaster risk management. In: *Hydrometerological Applications of Weather and Climate Modelling*. Extended abstracts of presentations at the 17th annual BMRC Modelling Workshop, October 3–6, BMRC Research Report No. 111, 63–66.

Nomanbhoy, N. and K. Satake (1995). Generation mechanism of tsunamis from the 1883 Krakatau eruption. *Geophysical Research Letters*, **22**, 509–512.

NTL (2008). Historical Tsunami Databases for the World Ocean (HTDB/WLD). Novosibirsk Tsunami Laboratory, http://tsun.sscc.ru/On_line_Cat.htm.

Okada, Y. (1985). Surface deformation due to shear and tensile faults in a half-space. *Bulletin of the Seismological Society of America*, **75**, 1135–1154.

Okal, E. A., G. J. Fryer, J. C. Borrero and C. Ruscher (2002). The landslide and local tsunami of 13 September 1999 on Fatu Hiva (Marquesas Islands; French Polynesia). *Bulletin de la Societe Geologique de France*, **173**, 359–367.

Okal, E. A., G. Plafker, C. E. Synolakis and J. C. Borrero (2003). Near-field survey of the 1946 Aleutian tsunami on Unimak and Sanak Islands. *Bulletin of the Seismological Society of America*, **93**, 1226–1234.

Pecher, I. A., S.A. Henrys, S. Ellis, S.M. Chiswell and N. Kukowski (2005). Erosion of the seafloor at the top of the gas hydrate stability zone on the Hikurangi margin, New Zealand. *Geophysical Research Letters*, **32**, L24603, doi:10.1029/2005GL024687.

Pelinovsky, E. and A. Poplavsky (1996). Simplified model of tsunami generation by submarine landslides. *Physics and Chemistry of the Earth*, **21**, 13–17.

Pelinovsky, E. N., T. Talipova, A. Kurkin and C. Kharif (2001). Nonlinear mechanism of tsunami wave generation by atmospheric disturbances. *Natural Hazards and Earth System Sciences*, **1**, 243–250.

Piatanesi, A. and S. Lorito (2007). Rupture process of the 2004 Sumatra–Andaman earthquake from tsunami wave form inversion. *Bulletin of the Seismological Society of America*, **97**, s223–s231.

Piper, D. J. W., P. Cochonat and M. L. Morrison (1999). The sequence of events around the epicenter of the 1929 Grand Banks earthquake; initiation of debris flows and turbidity current inferred from sidescan sonar. *Sedimentology*, **46**, 79–97.

Plafker, G. (1972). Alaskan earthquake of 1964 and Chilean earthquake of 1960; implications for arc tectonics. *Journal of Geophysical Research*, **77**, 901–925.

Plafker, G. and J. C. Savage (1970). Mechanism of the Chilean earthquakes of May 21 and 22, 1960, Alaskan earthquake of 1964 and Chilean earthquake of 1960; implications for arc tectonics. *Geological Society of America Bulletin*, **81**, 1001–1030.

Power, W., G. Downes and M. Stirling (2007). Estimation of tsunami hazard in New Zealand due to South American earthquakes. *Pure and Applied Geophysics*, **164**, 547–564.

Reese, S., W. J. Cousins, W. L. Power *et al.* (2007). Tsunami vulnerability of buildings and people in South Java: field observations after the July 2006 Java tsunami. *Natural Hazards and Earth System Sciences*, **7**(5), 573–589.

Rikitake, T. and I. Aida (1988). Tsunami hazard probability in Japan. *Bulletin of the Seismological Society of America*, **78**, 1268–1278.

Salgado, I., A. Eipert, B. F. Atwater *et al.* (2003). Recurrence of giant earthquakes inferred from tsunami sand sheets and subsided soils in south-central Chile. *Abstracts with Programs, Geological Society of America*, **35**(6), 584.

Self, S. and M. R. Rampino (1981). The 1883 eruption of Krakatau. *Nature*, **294**, 699–704.

Seno, T. and K. Hirata (2007). Did the 2004 Sumatra–Andaman earthquake involve a component of tsunami earthquakes? *Bulletin of the Seismological Society of America*, **97**, S296–S306.

Shimamoto, T., A. Tsutsumi, E. Kawamoto, M. Miyawaki and H. Sato (1995). Field survey report on tsunami disasters caused by the 1993 southwest Hokkaido earthquake. *Pure and Applied Geophysics*, **144**, 665–691.

Simkin, T. and L. Siebert (1994). *Volcanoes of the World*, 2nd edn. Tucson, AZ: Geoscience Press.

Stein, S. and E. A. Okal (2007). Ultralong period seismic study of the December 2004 Indian Ocean earthquake and implications for regional tectonics and the subduction process. *Bulletin of the Seismological Society of America*, **97**, S279–S295.
Stirling, M. W. and C. J. N. Wilson (2002). Development of a volcanic hazard model for New Zealand; first approaches from the methods of probabilistic seismic hazard analysis. *Bulletin of the New Zealand National Society for Earthquake Engineering*, **35**, 266–277.
Tinti, S., A. Manucci, G. Pagnoni, A. Armigliato and F. Zaniboni (2005). The 30 December 2002 landslide-induced tsunamis in Stromboli: sequence of the events reconstructed from the eyewitness accounts. *Natural Hazards and Earth Systems Science*, **5**, 763–775.
Titov, V. V. and F. I. Gonzalez (1997). Implementation and testing of the method of splitting tsunami (MOST) model, NOAA Technical Memorandum, ERL PMEL-112. Seattle, WA: NOAA/Pacific Marine Environmental Laboratory.
Titov, V. V., A. B. Rabinovich, H. O. Mofjeld, R. E. Thomson and F. I. Gonzalez (2005). The global reach of the 26 December 2004 Sumatra tsunami. *Science*, **309**, 2045–2048.
Tsuji, Y., H. Matsutomi, F. Imamura *et al.* (1995a). Damage to coastal villages due to the 1992 Flores Island earthquake tsunami. *Pure and Applied Geophysics*, **144**, 481–524.
Tsuji, Y., F. Imamura, H. Matsutomi *et al.* (1995b). Field survey of the east Java earthquake and tsunami of June 3, 1994. *Pure and Applied Geophysics*, **144**, 839–854.
Tsuji, Y., Y. Namegaya, H. Matsumoto *et al.* (2006). The 2004 Indian tsunami in Thailand; surveyed runup heights and tide gauge records. *Earth, Planets and Space*, **58**, 223–232.
Umitsu, M., C. Tanavud and B. Patanakanog (2007). Effects of landforms on tsunami flow in the plains of Banda Aceh, Indonesia, and Nam Khem, Thailand. *Marine Geology*, **242**(1–3), 141–153.
Verbeek, R. D. M. (1884). The Krakatoa eruption. *Nature*, **30**, 10–15.
Walters, R. A. and J. Goff (2003). Assessing tsunami hazard along the New Zealand coast. Science of tsunami hazards. *International journal of the Tsunami Society*, **21**(3), 137–153.
Walters, R. A., P. Barnes, K. Lewis, J. R. Goff and J. Fleming (2006). Locally generated tsunami along the Kaikoura coastal margin. Part 1, Submarine landslides. *New Zealand Journal of Marine and Freshwater Research*, **40**(1), 17–28.
Ward, S. N. (1980). Relationships of tsunami generation and an earthquake source. *Journal of Physics of the Earth*, **28**, 441–474.
Ward, S. N. (2001a). Tsunamis. In: Meyers, R. A. (ed.) *The Encyclopedia of Physical Science and Technology*. San Diego, CA: Academic Press.
Ward, S. N. (2001b). Landslide tsunami. *Journal of Geophysical Research*, **106**, 11 201–11 215.
Ward, S. N. and E. Asphaug (2000). Asteroid impact tsunami: a probabilistic hazard assessment. *Icarus*, **145**, 64–78.
Ward, S. N. and S. Day (2001). Cumbre Vieja Volcano; potential collapse and tsunami at La Palma, Canary Islands. *Geophysical Research Letters*, **28**, 3397–3400.
Watts, P. (1998). Wavemaker curves for tsunamis generated by underwater landslides. *Journal of Waterway, Port, Coastal and Ocean Engineering*, **124**, 127–137.
Watts, P. and C. F. Waythomas (2003). Theoretical analysis of tsunami generation by pyroclastic flows. *Journal of Geophysical Research*, **108**, 2563–2584.

Wells, D. L. and K. J. Coppersmith (1994). New empirical relationships among magnitude, rupture length, rupture width, rupture area, and surface displacement. *Bulletin of the Seismological Society of America*, **84**, 974–1002.
Whelan, M. (1994). The night the sea smashed Lord's Cove. *Canadian Geographic*, **114**, 70–73.
Wiegel, R. L. (1970). Tsunamis. In: Wiegel, R. L. (ed.) *Earthquake Engineering*. Englewood Cliffs, NJ: Prentice-Hall, 253–306.
Wood, H. O. (1914). On the earthquakes of 1868 in Hawaii. *Bulletin of the Seismological Society of America*, **4**, 169–203.
Yanagisawa, K., F. Imamura, T. Sakakiyama *et al.* (2007). Tsunami assessment for risk management at nuclear power facilities in Japan. *Pure and Applied Geophysics*, **164**, 565–576.

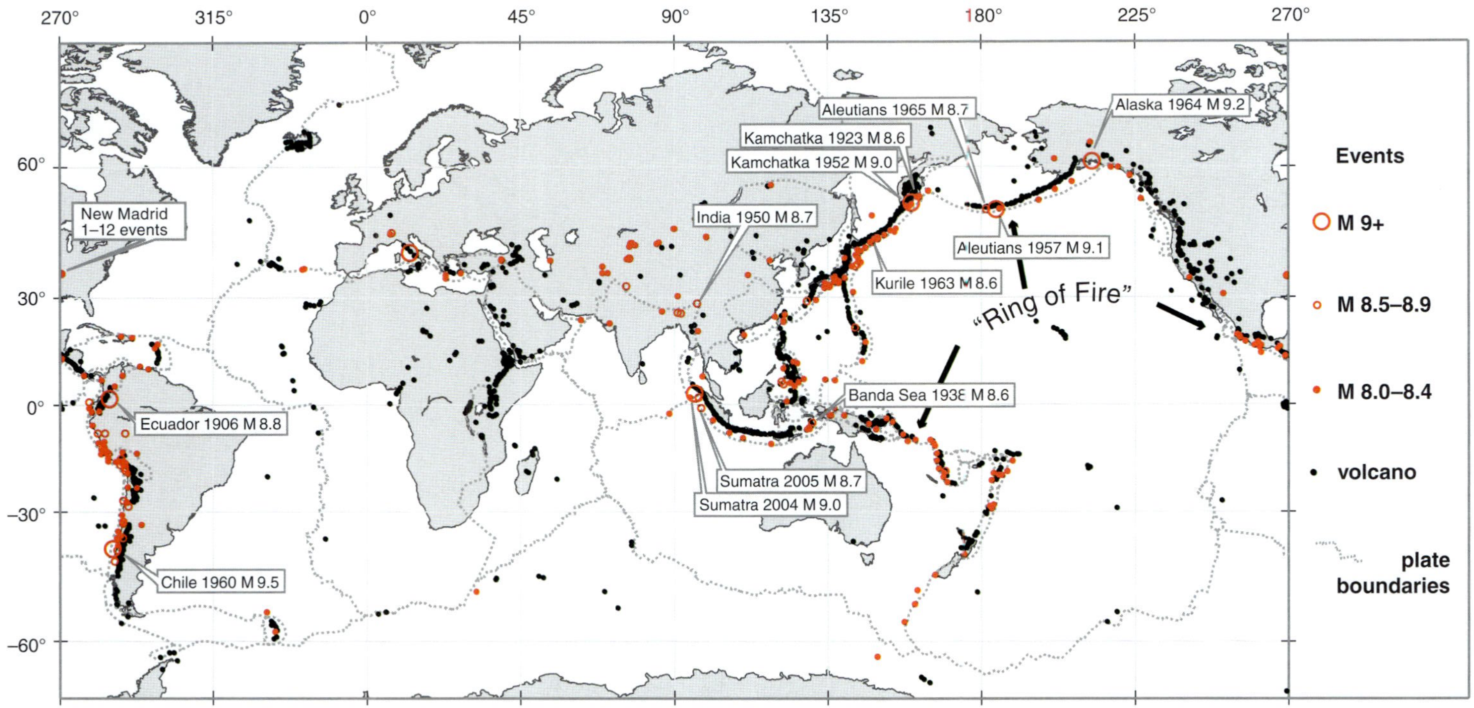

Fig. 2.1 Map showing the locations of all large (M > 8) earthquakes since AD 1900. Most of the volcanic activity and large earthquakes occur around the margins of the Pacific Ocean basin. This concentration of activity is commonly referred to as the "Ring of Fire."

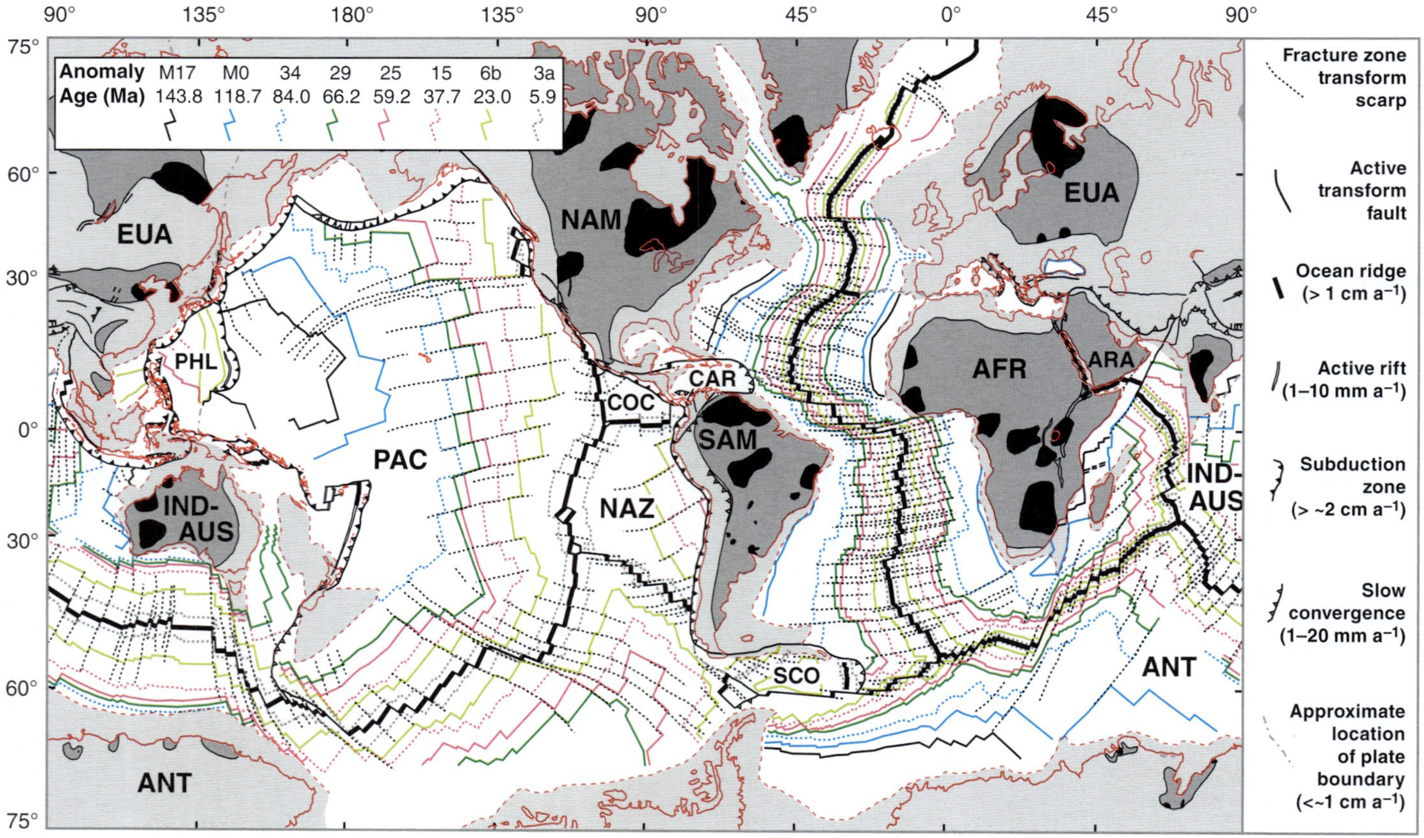

Fig. 2.3 Magnetic anomalies delineate age of ocean floor. Continental crust (light gray), Precambrian crust (dark gray), Archean crust (black). Major plates: North America (NAM), South America (SAM), Nazca (NAZ), Eurasia (EUR), India-Australian (IND-AUS), Antarctic (ANT), Africa (AFR). Minor plates: Philippine (PHL), Cocos (COC), Caribbean (CAR), Scotia (SCO), Arabia (ARA).

Fig. 3.7 Volcanoes have widely varying morphologies, reflecting the processes that lead to their construction. (a) Mauna Kea, a shield volcano in Hawaii, is constructed predominantly from effusion of basaltic lava flows (photo by A. Leonard–Hintz). (b) San Cristobal, a composite volcano in Nicaragua, consists of interbedded lavas and tephra fallout deposits (photo by C. Connor). (c) Quilotoa Caldera is 1.3 km in diameter and formed during an explosive eruption 0.8 ka (photo by A. Volentik). (d) Monogenetic cinder cones in the Yerevan basin, Armenia formed along a major strike-slip fault in the basin. The large Ararat composite volcano appears in the background (photo by C. Connor).

Fig. 3.8 A basaltic feeder dike and sill exposure in the San Rafael subvolcanic field, Utah. This sill formed at a depth of $\sim$ 800 m, based on paleotopographic reconstruction and stratigraphy (photo by M. Díez).

Fig. 3.12 (b) The Shoshone Ignimbrite in California is part of a thick (300 m) sequence of pyroclastic flows and tephra fallout deposits that inundated the region $\sim$ 20 Ma. The glassy, welded interior of one flow creates the prominent black middle layer (photo, courtesy of Sean Callihan).

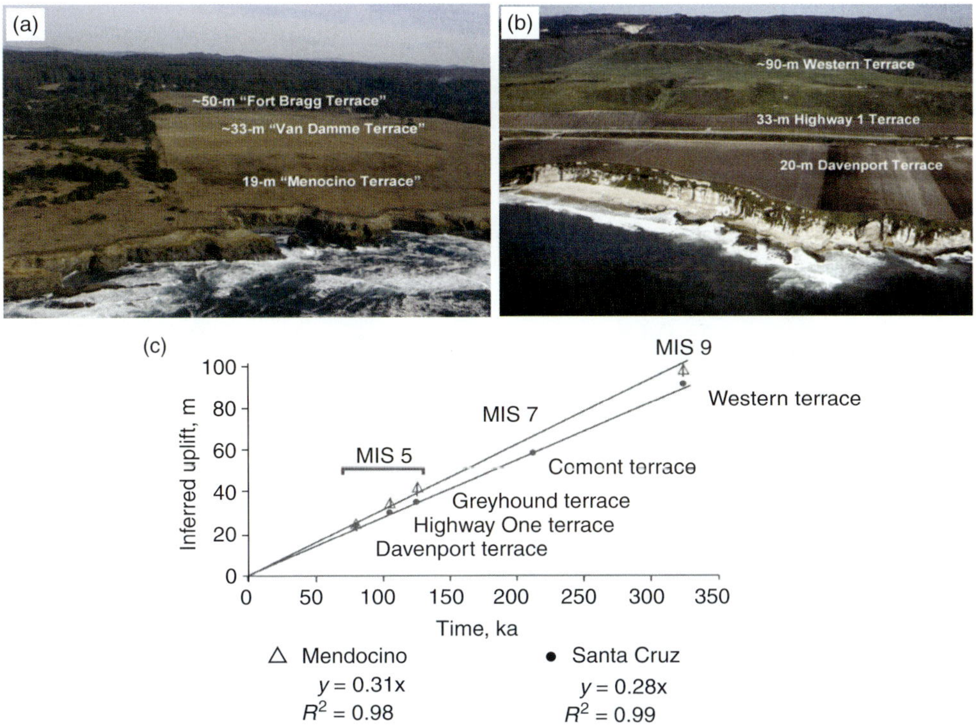

Fig. 4.4 (a) Air photo of the northern California coast south of the town of Mendocino (area of Van Damme State Park), taken November 14, 2002 (Adelman and Adelman, 2002; terrace surveying and mapping by Merritts and students, 1999–2006, unpublished data). Altitude estimates are ±1 m, but uncertainty of inner-edge estimate is ±2 m for the 2nd and 3rd terraces because of sedimentary cover on the platforms. (b) Air photo of the northern California coast near Laguna Creek, Santa Cruz taken in 1972 (Adelman and Adelman, 2002; terrace altitudes are from Bradley and Griggs, 1976). (c) Inferred uplift-rate diagram compares terrace inner-edge altitudes from the locales shown in (a) and (b).

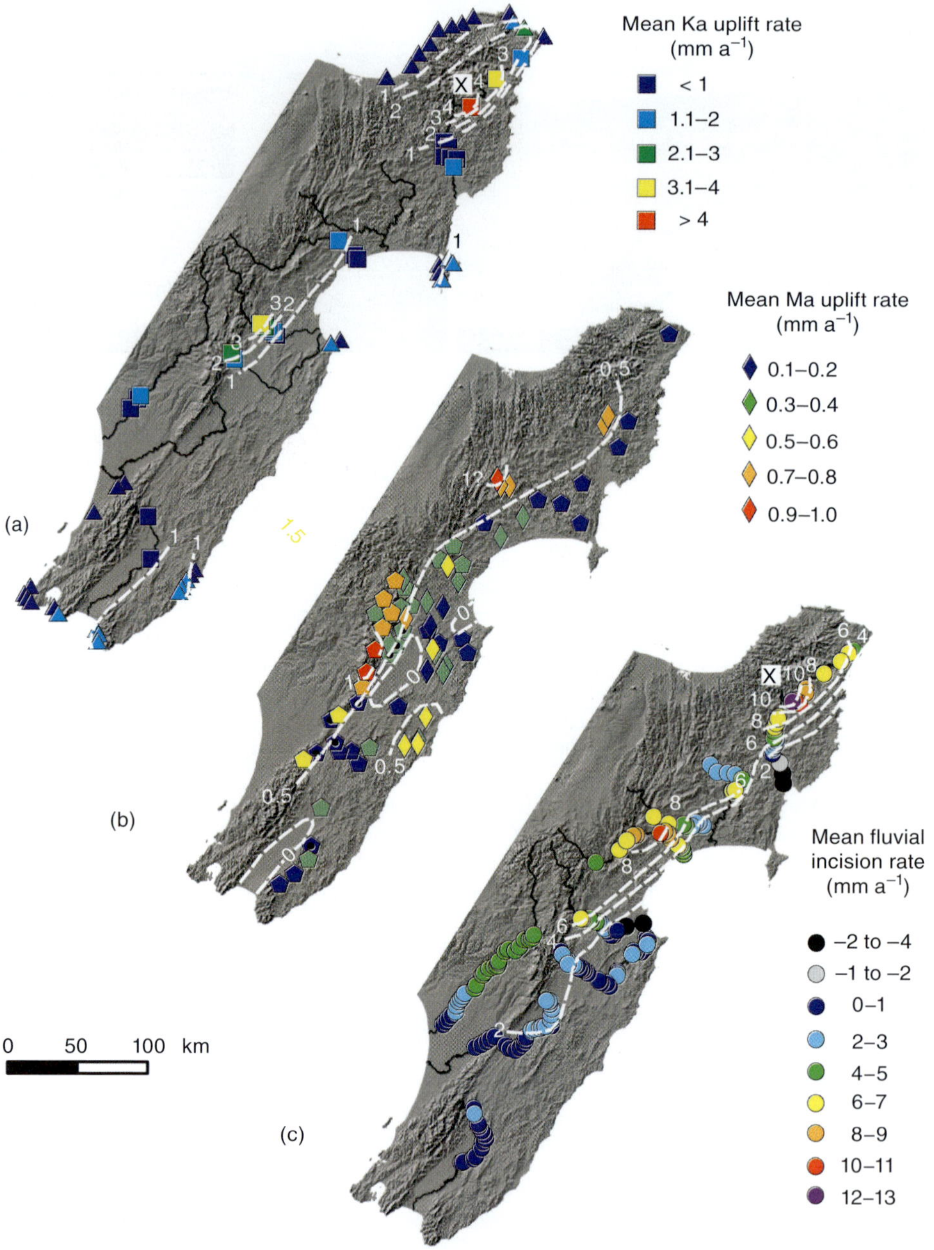

Fig. 4.7 Uplift and fluvial incision rate maps for the Hikurangi margin, New Zealand. Map (a) shows late Quaternary (ka) mean uplift rates derived from the altitude difference between pairs of fluvial fill terraces (~ 55 ka and ~ 18 ka) (squares) and the present-day altitude of the MIS 5e (125 ka) marine terrace (triangles). Map (b) shows million-year (Ma) mean uplift rates derived from the present-day altitude of geologic markers (pentagons) and mudstones for which maximum burial depth has been calculated from porosity (diamonds). Map (c) shows mean fluvial incision rates derived from the altitude difference between T1 (~ 18 ka) and the present-day river bed (Litchfield and Berryman, 2006; Litchfield *et al.*, 2007).

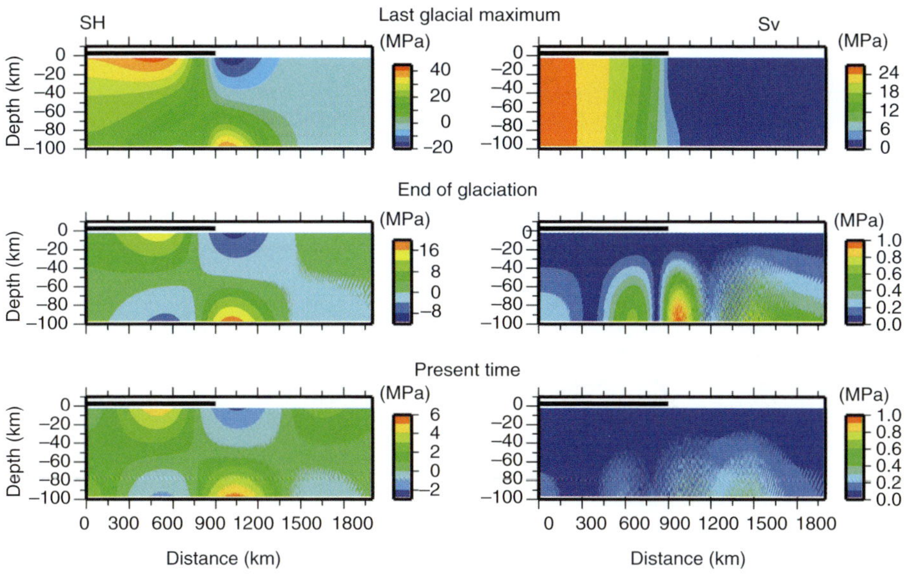

Fig. 5.2 Glacial stresses induced in a generic model of the Earth. Maximum horizontal (left) and vertical (right) stresses at three different times: the last glacial maximum, the end of glaciation and present. The Earth model is a simple 2D model with a 100 km elastic lithosphere overlying a viscoelastic half-space subject to a generic elliptic cross-section ice load with 900 km lateral extent and 25 MPa central pressure at the last glacial maximum. Adapted after Lund (2005).

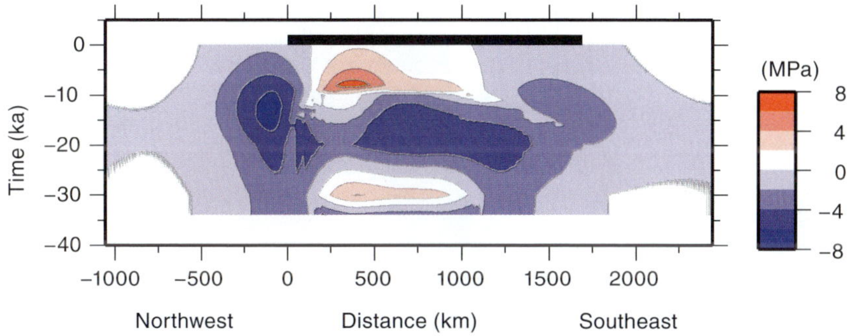

Fig. 5.3 Stability of optimally oriented faults at different times along a 2D profile across northern Fennoscandia at the latitude of the large glacially induced faults. Zero along the profile corresponds to the westernmost extent of the ice sheet, off the coast of Lofoten, Norway. To the east, the profile extends well into Russia. Positive, red colors, indicate unstable faults and negative, blue colors indicate stable faults. The black bar across the top shows the maximum lateral extent of the ice along the profile. See the text for information about the Earth and ice models.

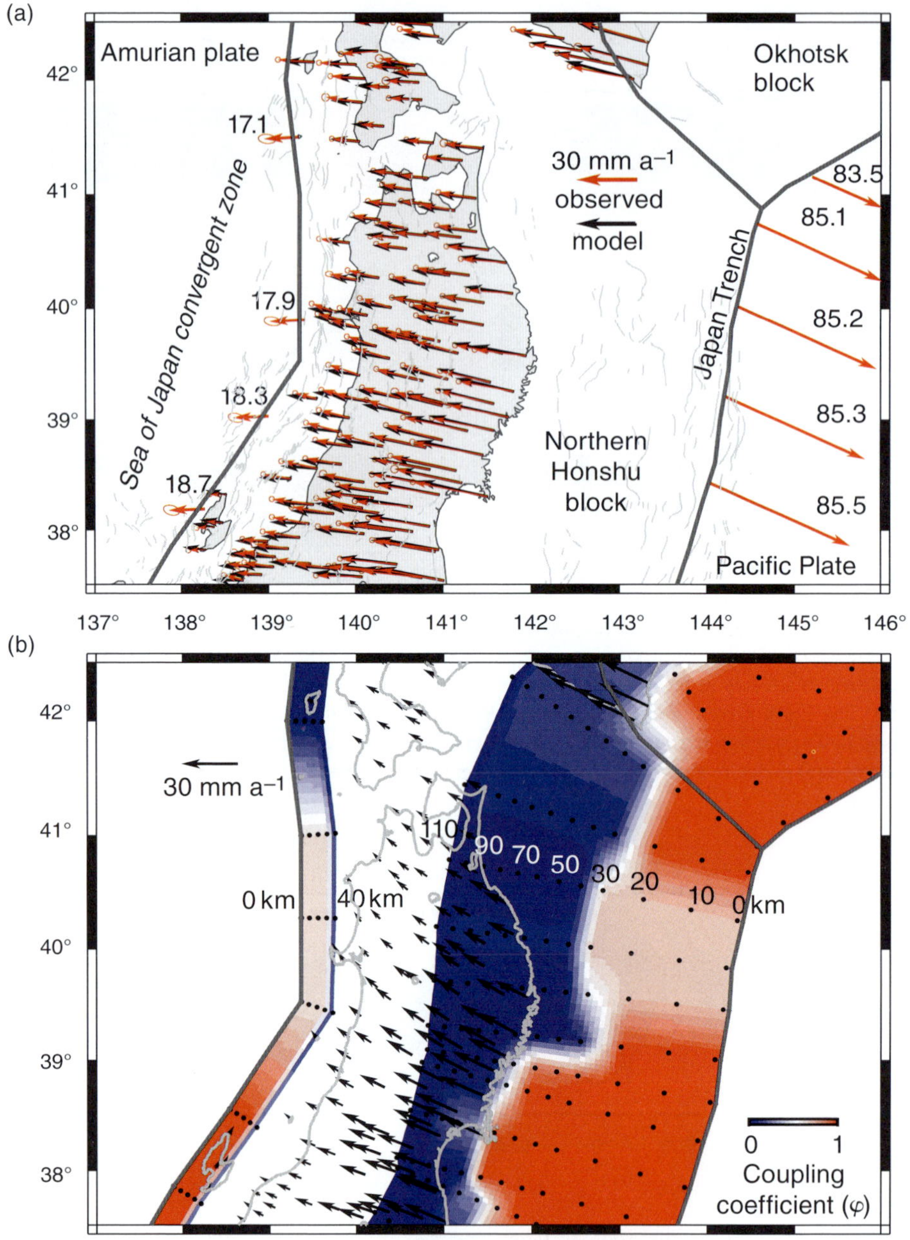

Fig. 6.3 (a) Block boundaries used in elastic block modeling are shown as heavy gray lines. Red or gray arrows represent GPS velocities. Corresponding black arrows show the best-fitting model. Gray arrows (and values) along the boundaries show Northern Honshu block motion relative to Amurian and Pacific plates (mm a^{-1}). (b) Coupling coefficients (ϕ) were estimated for each node (block dot) along the Japan Trench and Sea of Japan fault boundaries (values indicate depth of the fault nodes in km). Degree of coupling between nodes was calculated by bilinear interpolation (blue-to-red color scale). Black vectors show influence of interseismic coupling on the GPS velocity field. This influence is due to elastic (temporary) processes and is removed from the GPS velocity field to isolate long-term upper-plate strain rate in the Tohoku region (Figure 6.5).

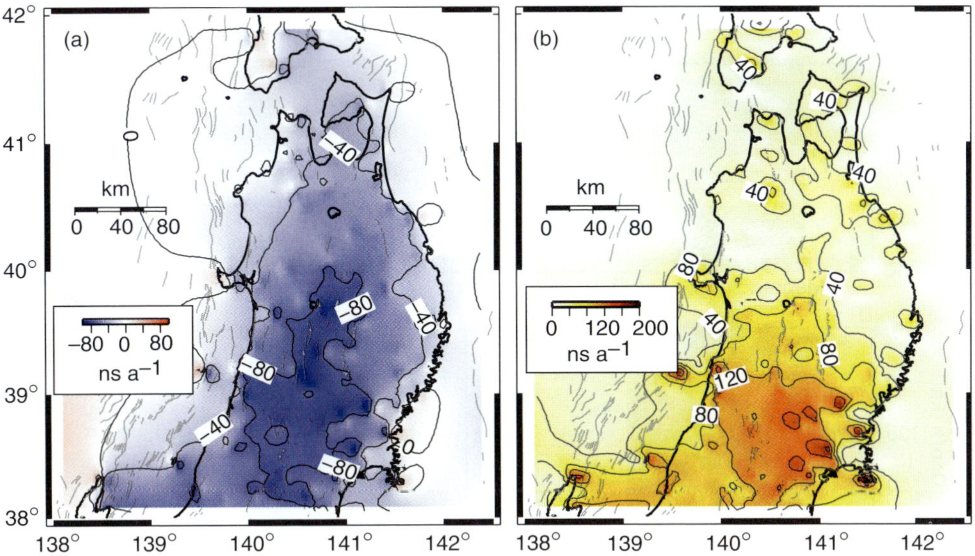

Fig. 6.4 (a) Areal and (b) shear strain rates from the raw GPS velocity field in nanostrain a^{-1} (Figure 6.3; no elastic strains are removed) estimated using the method of Haines and Holt (1993).

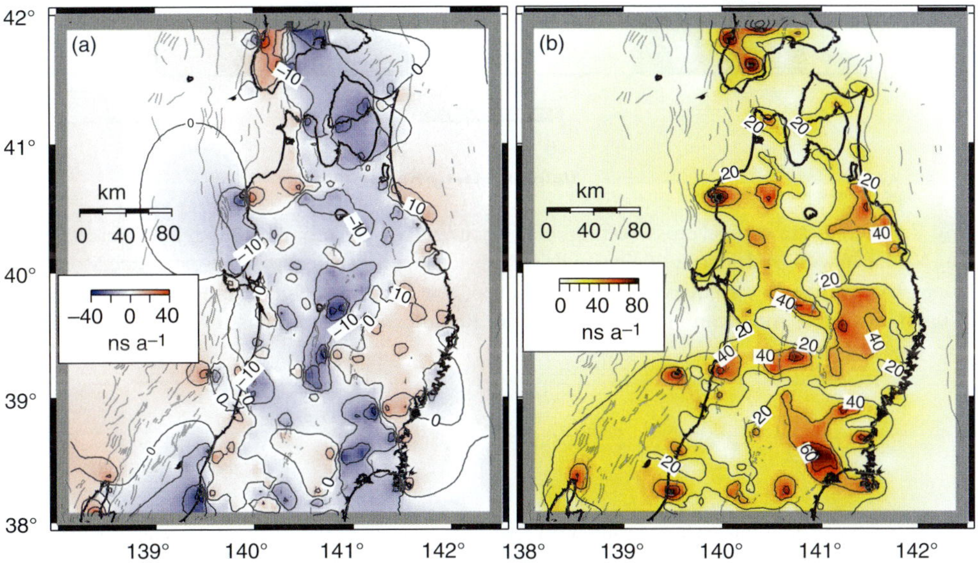

Fig. 6.5 (a) Areal and (b) shear strain rates are contoured in the residual GPS velocity field in Northern Honshu, after the elastic component of the velocity field from coupling on block-bounding faults (e.g. Figure 6.3b) is removed from the raw velocity field (Figure 6.2). Strain rates shown are nanostrain a^{-1}.

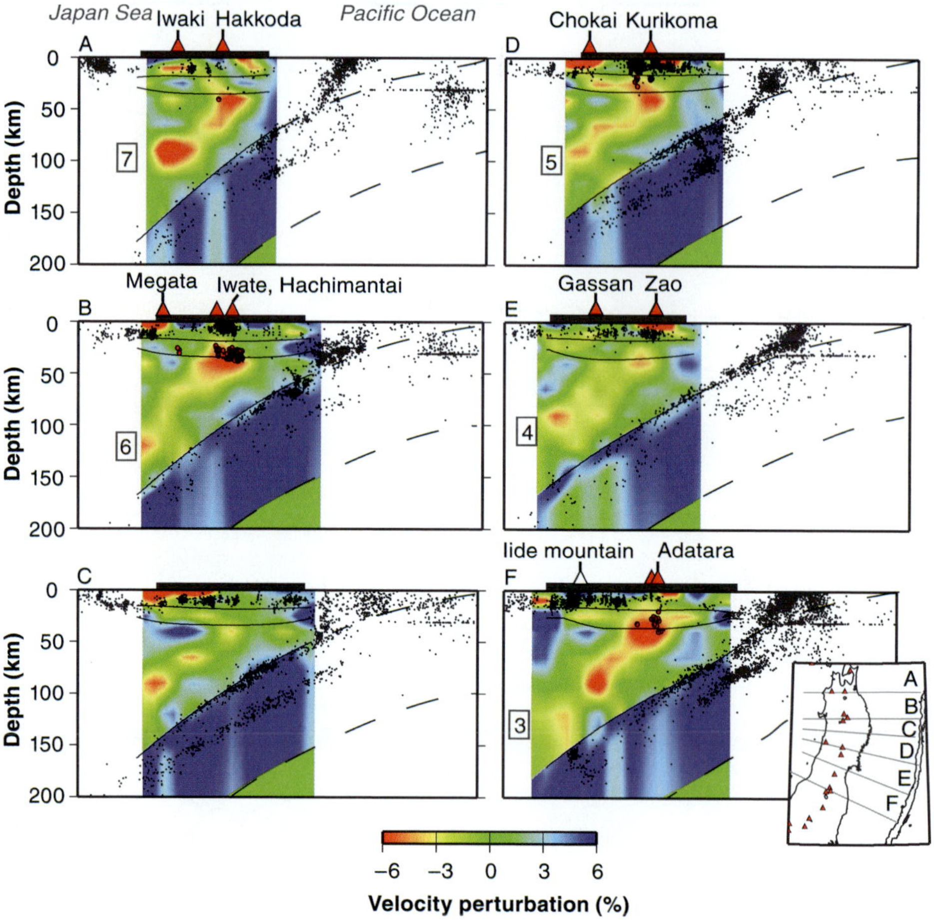

Fig. 7.4 Across-arc vertical cross-sections of S-wave velocity perturbations along lines (insert) crossing NE Japan with land area (solid line) and active volcanoes (red triangles), Mesozoic Iide Mountains (open triangle) in F, earthquakes (dots), and deep, low-frequency microearthquakes (red circles). Copyright (2004) American Geophysical Union. Reproduced with permission from the American Geophysical Union.

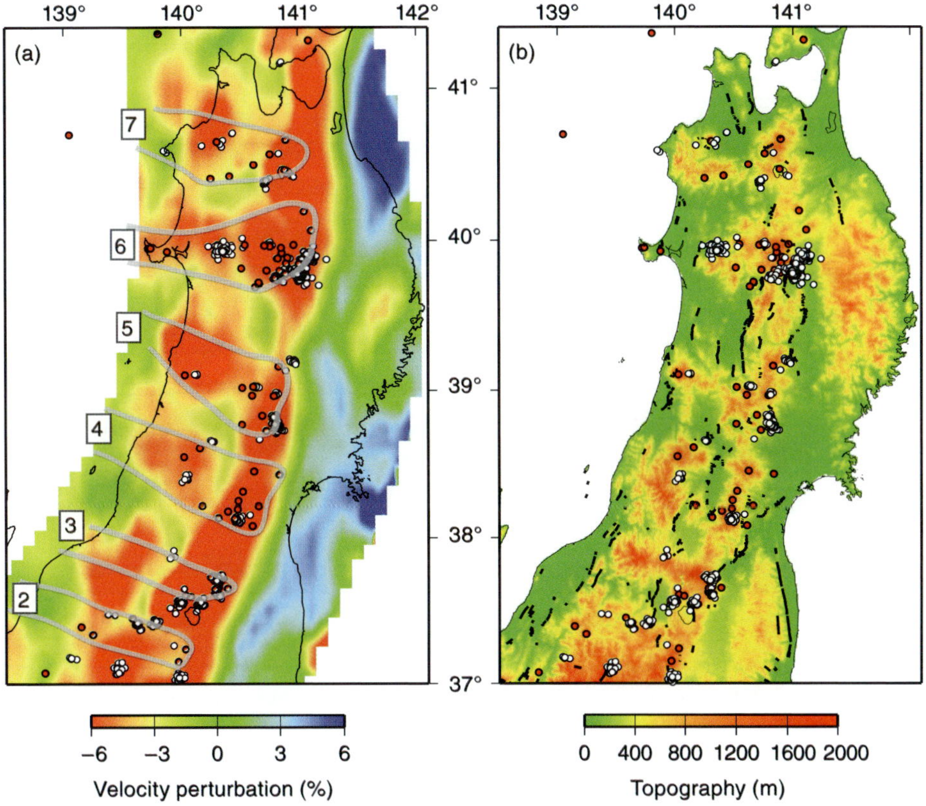

Fig. 7.5 (a) Map view of S-wave velocity perturbations along the inclined low-velocity zone and (b) topography with Quaternary volcanoes (red circles) and deep, low-frequency microearthquakes (white circles). Copyright (2004) American Geophysical Union. Reproduced with permission from the American Geophysical Union.

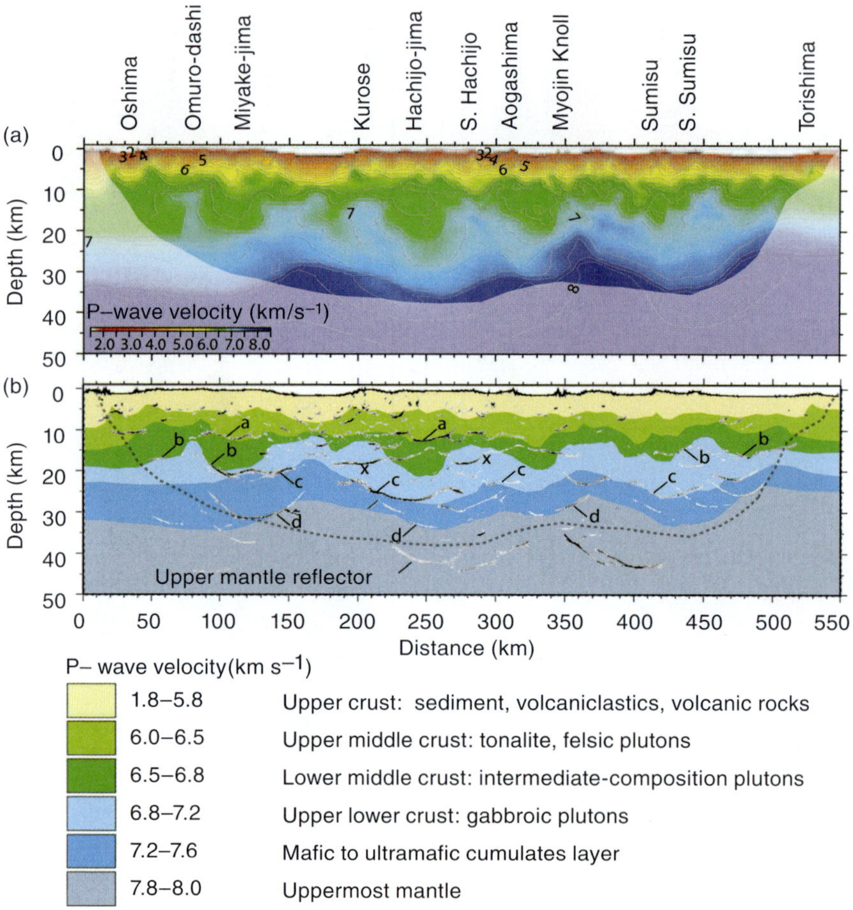

Fig. 7.7 (a) Seismic velocity model with shaded area indicating the poorly resolved area identified by the checkerboard test. (b) Seismic reflectivity image and geological interpretation. Reflectors labeled a–d indicate the top of the lower part of the middle crust, the top of the upper part of the lower crust, the top of the lower part of the lower crust and the bottom of the lower part of the lower crust, respectively. Reflector labeled x is interpreted as floating reflectors representing features such as laterally intruded sills and faults (after Kodaira *et al.*, 2007).

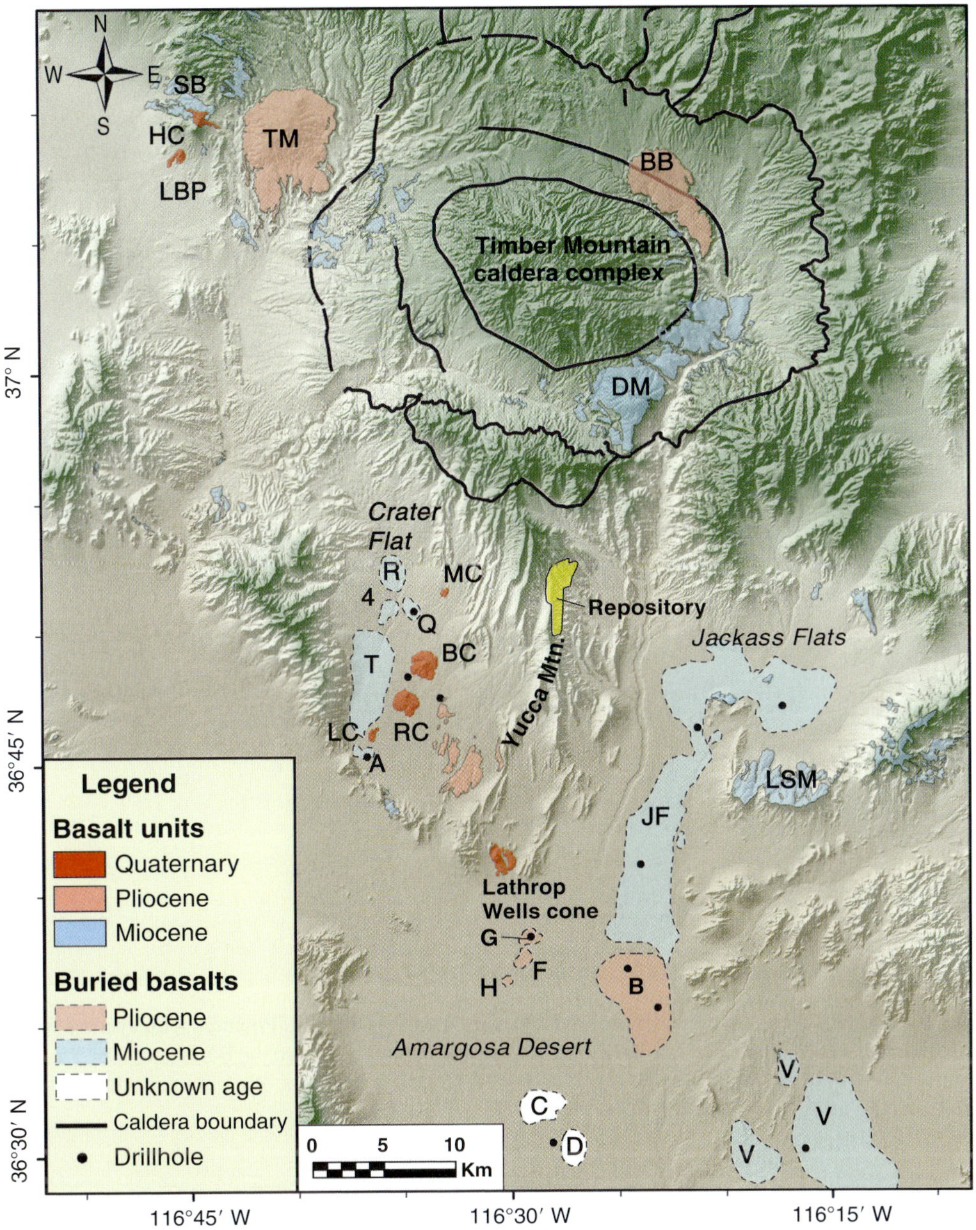

Fig. 8.1 Shaded relief map of Pliocene and Quaternary basaltic volcanoes in the YM region. Buried Pliocene and Miocene basalts are identified by drilling and sample geochronology; geomagnetics are outlined. Miocene caldera boundary is black line. Miocene volcanic rocks (lava and tephra): Sleeping Buttes (SB), Dome Mountain (DM), Little Skull Mountain (LSM). Pliocene volcanic rocks: Thirsty Mesa (TM), Buckboard Mesa (BB). Quaternary volcanoes (1.1 Ma): Makani (MC), Black Cone (BC), Red Cone (RC), NE and SW Little Cones (LC); (0.35 Ma): Little Black Peak (LBP), Hidden Cone (HC). 77 ka Lathrop Wells volcano (LW). Magnetic anomalies: **Q, R, 4, T, A, V, B, G, F, H, C, D**. Map provided by Dr. Frank V. Perry of LANL.

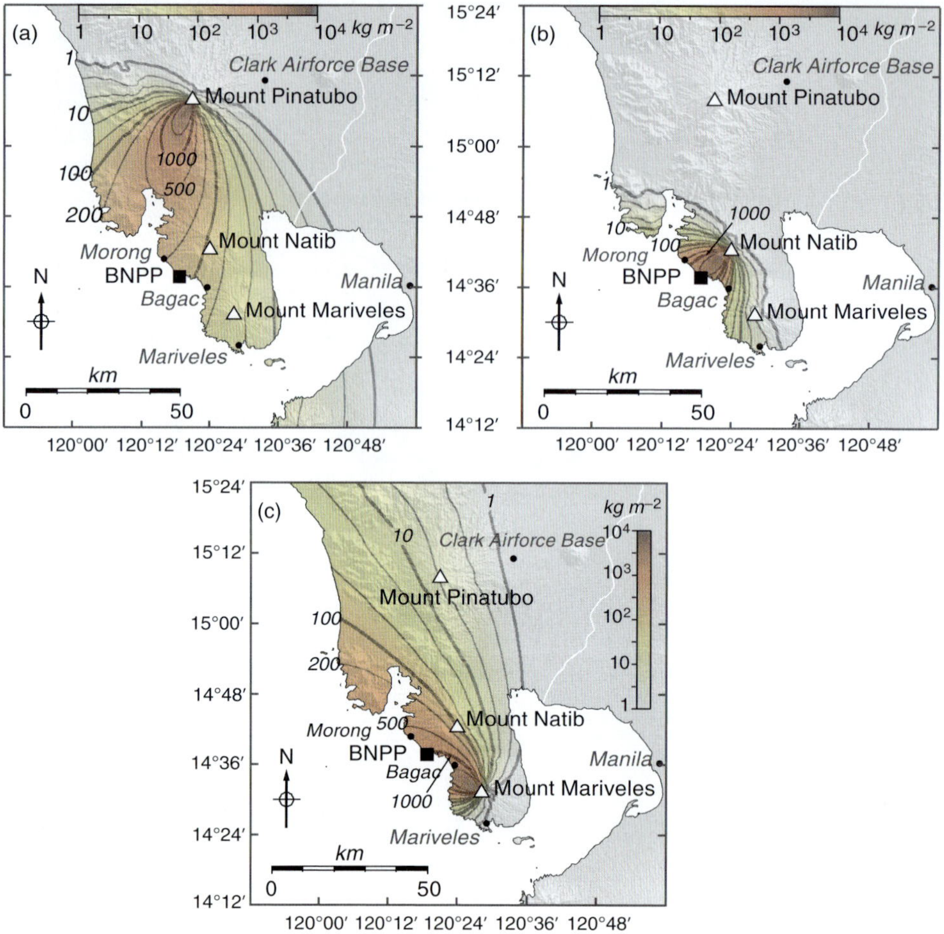

Fig. 9.4 Isomass maps for three deterministic scenarios: (a) VEI 6 at Mt. Pinatubo with average wind conditions for 2006; (b) VEI 4 at Mt. Natib with the wind blowing toward the BNPP; (c) VEI 5 at Mt. Mariveles with the wind blowing toward the BNPP. Contours are mass of tephra accumulation per unit area (kg m^{-2}, dry, where 100 kg m^{-2} is $\sim$ 10 cm tephra thickness).

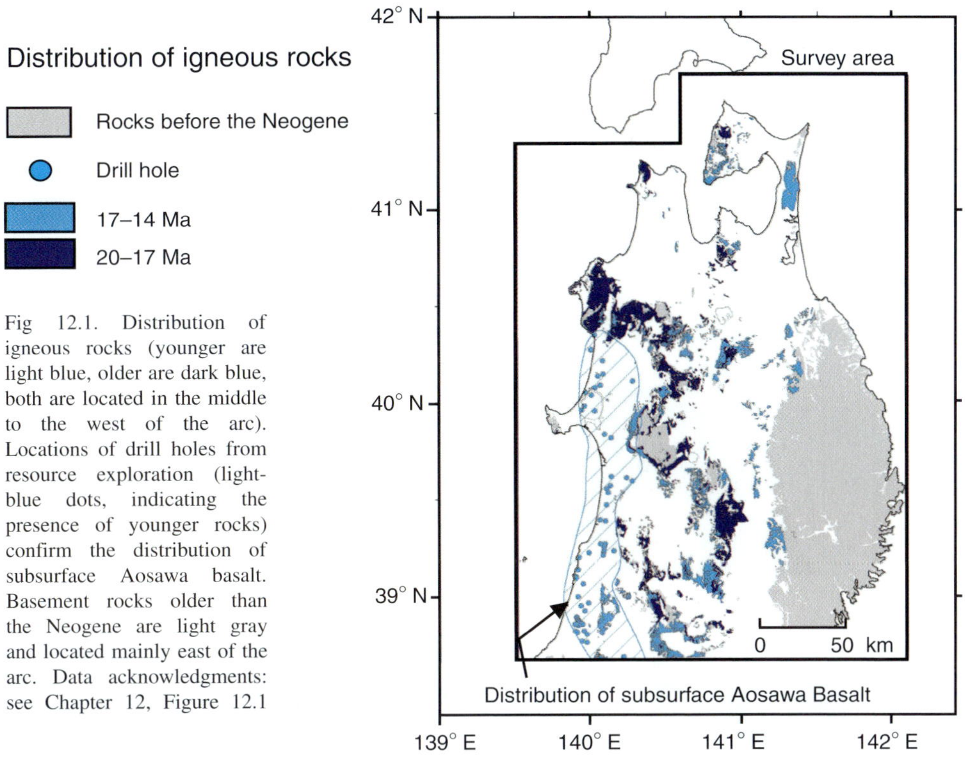

Fig 12.1. Distribution of igneous rocks (younger are light blue, older are dark blue, both are located in the middle to the west of the arc). Locations of drill holes from resource exploration (light-blue dots, indicating the presence of younger rocks) confirm the distribution of subsurface Aosawa basalt. Basement rocks older than the Neogene are light gray and located mainly east of the arc. Data acknowledgments: see Chapter 12, Figure 12.1

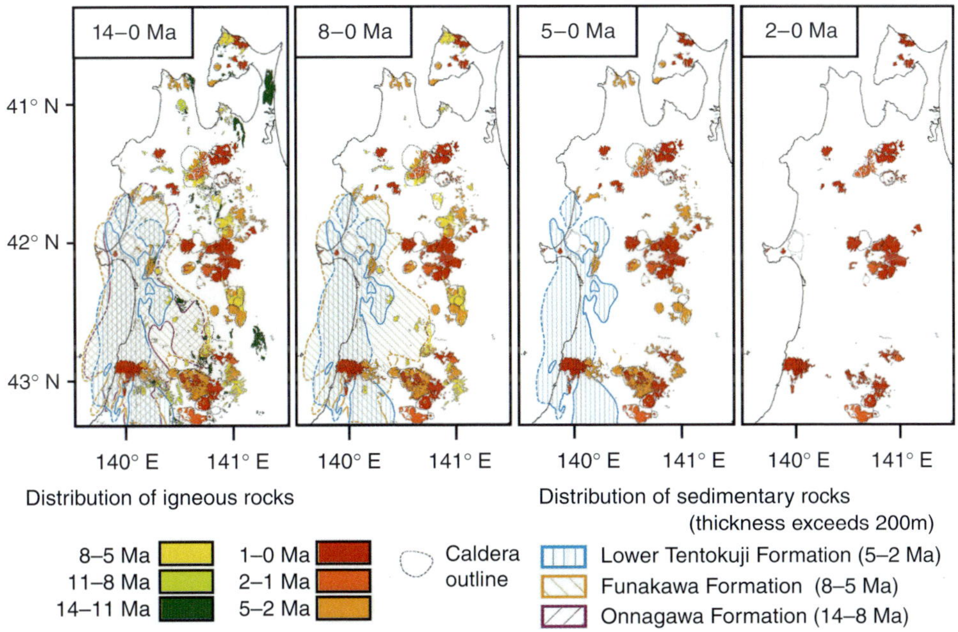

Fig. 12.2 Cumulative distribution of volcanism for four time periods. Six distributions of igneous rocks are differentiated by color (note map key for colors). Three distributions of sedimentary rocks within the backarc (thicknesses > 200 m) are defined by hatched areas with vertical or oblique lines. Data acknowledgments: see Chapter 12, Figure 12.2.

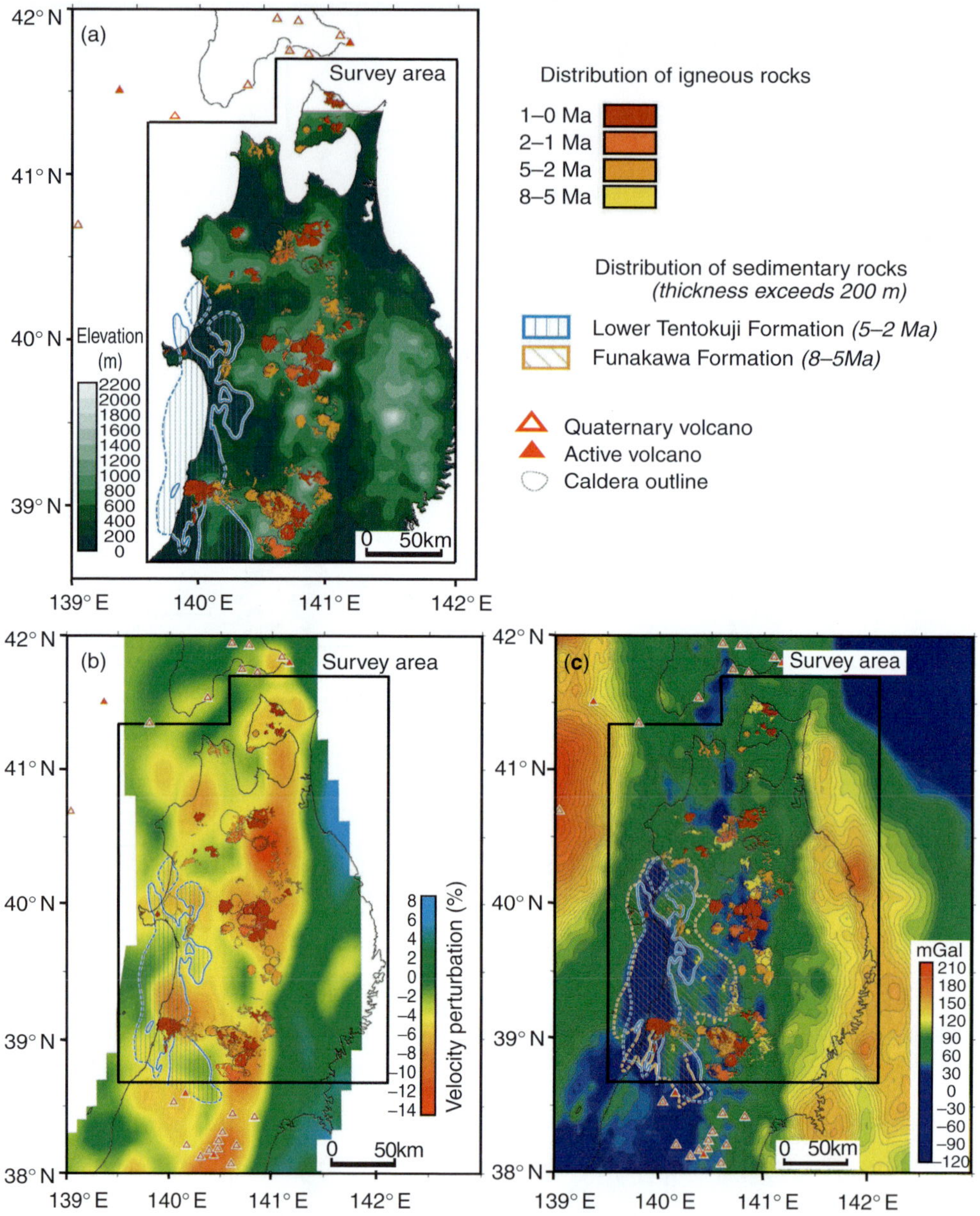

Fig. 12.3 Comparison of the distribution of volcanism with (a) summit level topography, (b) low-velocity anomalies within the mantle wedge (S-wave velocity perturbations along the core of the inclined low-velocity zone), and (c) Bouguer gravity anomalies. Igneous rocks are divided into four periods and differentiated by color. Two distributions of sedimentary rocks within the backarc (thickness > 200 m) are defined by hatched areas with vertical or oblique lines. Data acknowledgments: caldera outlines (Yoshida *et al.*, 1999); Quaternary volcanoes (Committee for the Catalogue of Quaternary volcanoes in Japan, 1999; Japan Meteorological Agency, 2003); gravity data (Komazawa, 2004); sedimentary rocks (Japan Natural Gas Association and Japan Offshore Petroleum Development Association, 1992); seismic data (Hasegawa and Nakajima, 2004).

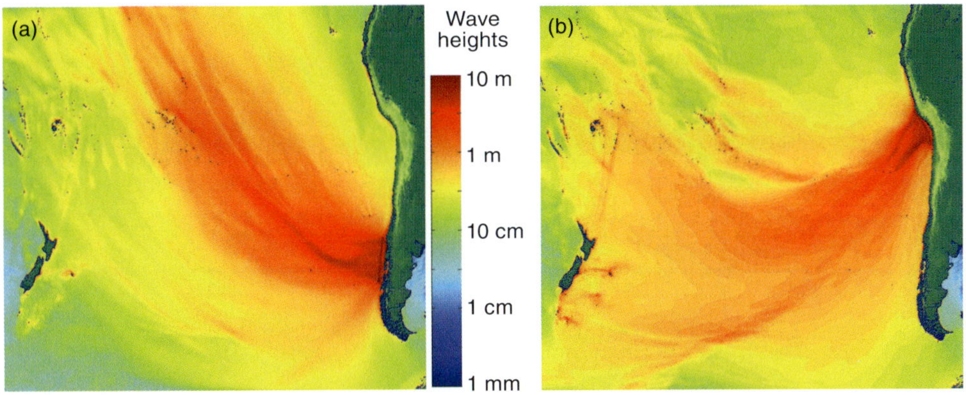

Fig. 11.3 Tsunami modeling scenarios using the MOST modeling programs (Titov and Gonzalez, 1997). This two-scenario comparison of a South American tsunami affecting New Zealand illustrates the effect that directivity of the source can have on distant locations. The source for model (a) is the 1960 Chilean M 9.5 earthquake; the source for model (b) is an equivalent M 9.5 earthquake originating in the area of southern Peru/northern Chile. The models estimate maximum wave heights offshore (> 25 m deep). Wave heights may increase by several times close to the coast.

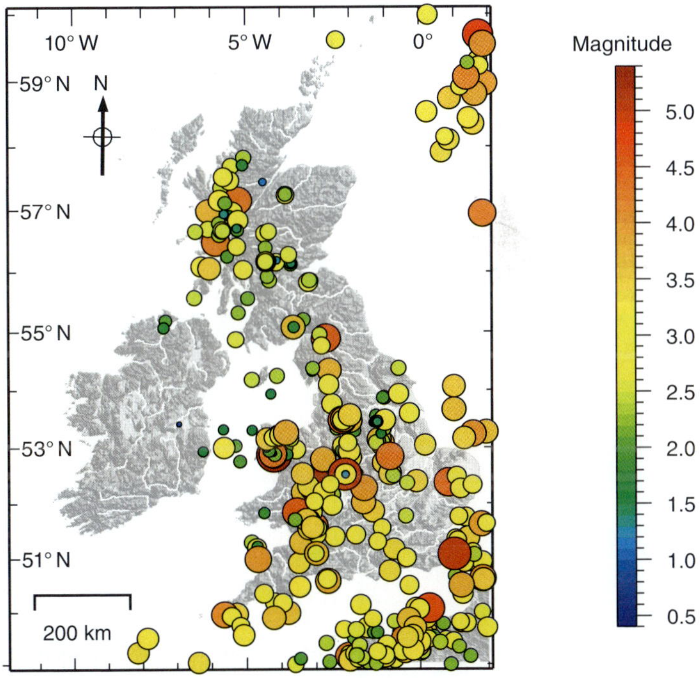

Fig. 14.2 Earthquake epicenters are plotted as shaded circles that increase in diameter, proportional to earthquake magnitude. Earthquake catalog data are from the NEIC/USGS database, January 1973–January 2007. These earthquake epicenters, together with a subset found in the ISC catalog, serve as a basis for spatial intensity estimation. Visually, the number of earthquake epicenters decreases west of 7° W.

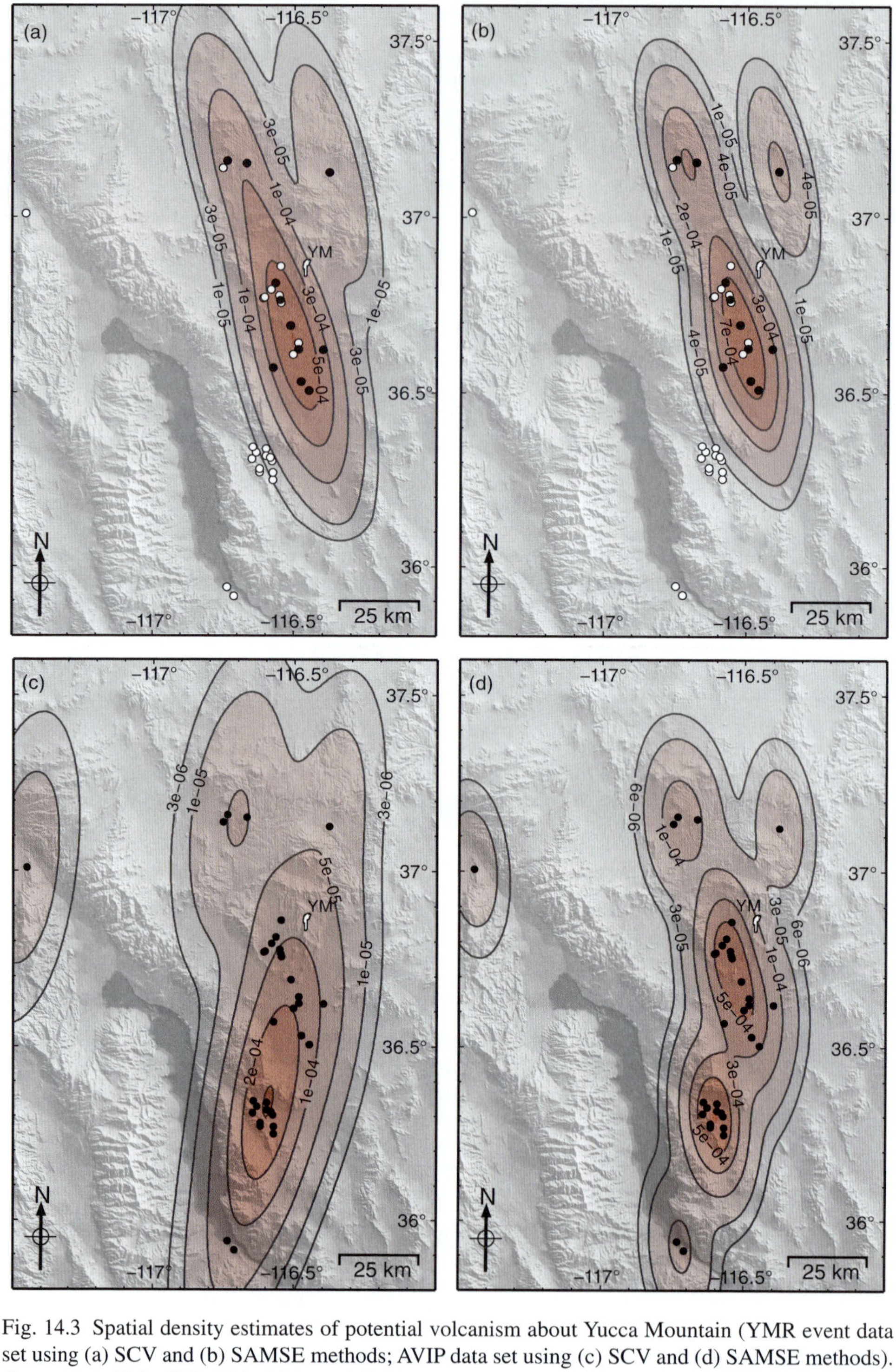

Fig. 14.3 Spatial density estimates of potential volcanism about Yucca Mountain (YMR event data set using (a) SCV and (b) SAMSE methods; AVIP data set using (c) SCV and (d) SAMSE methods). Contours at 25th, 50th, 75th, 95th, and 99th percentiles (e.g. on map (a) locations within the 1e-04 contour (i.e. 75th quartile) have spatial density $> 1 \times 10^{-4}\,\mathrm{km}^{-2}$; given a volcanic event, there is a 75% chance it will occur within this quartile).

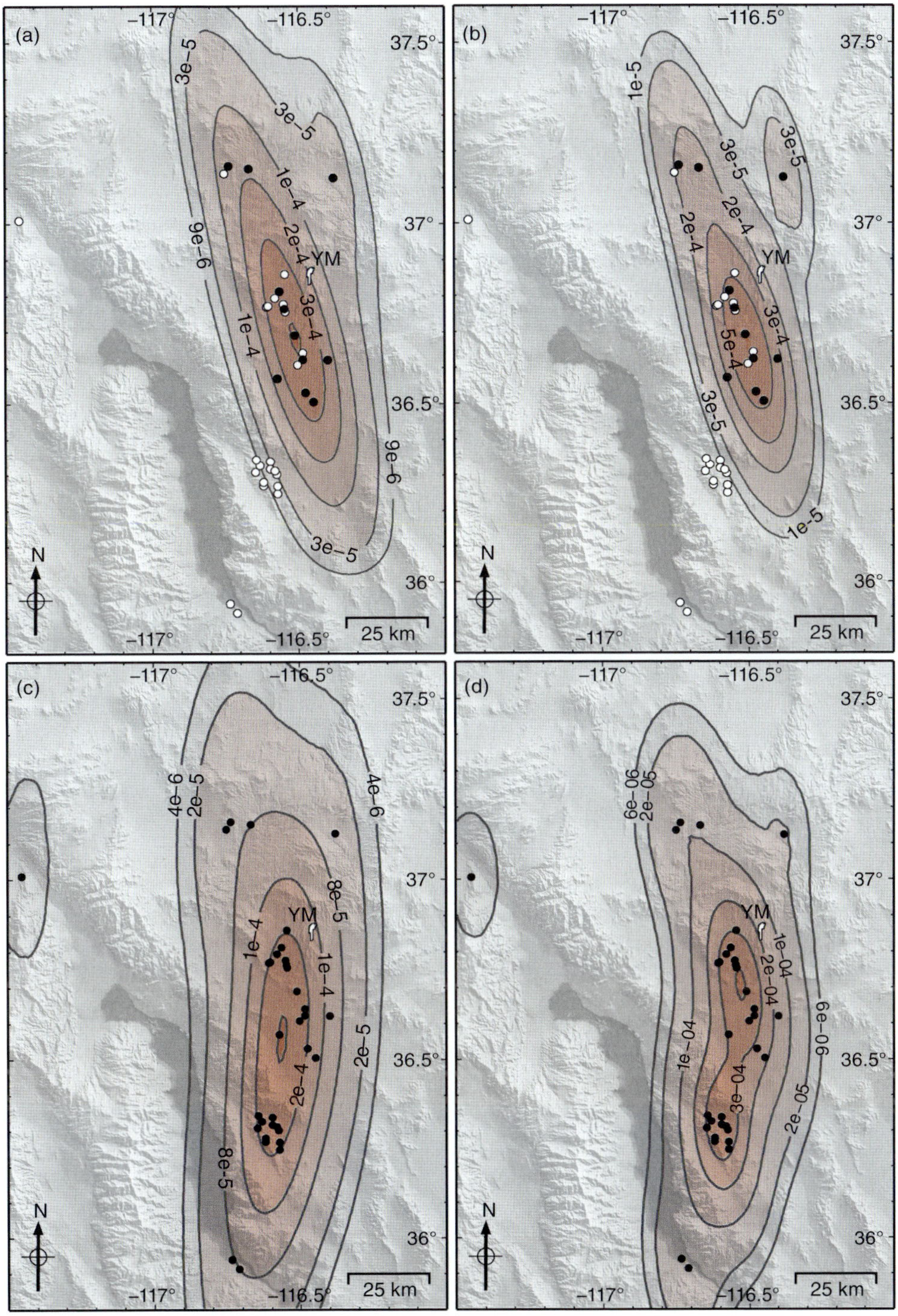

Fig. 14.7 Median spatial density at each grid location (1 km grid spacing) based on 1200 simulations: YMR event dataset with (a) SCV and (b) SAMSE bandwidth, AVIP event data set with (c) SCV and (d) SAMSE bandwidth. Contours are 1st, 25th, 50th, 75th, 95th, and 99th percentiles of median spatial density.

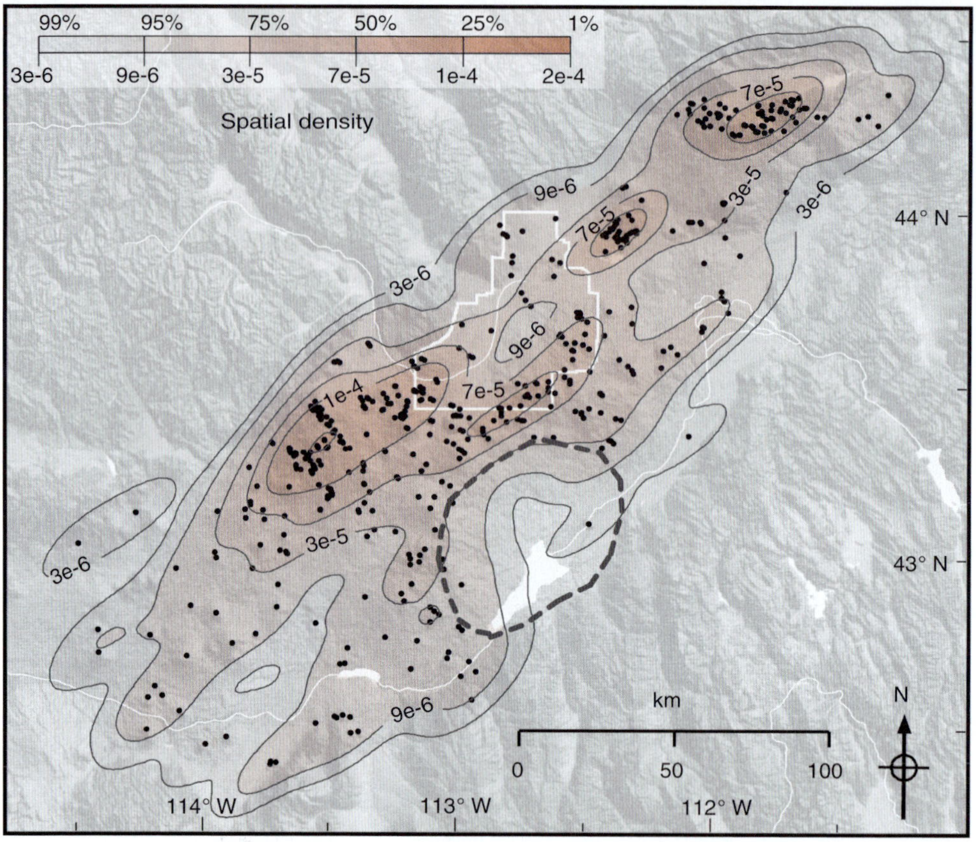

Fig. 16.6 This contour plot of the spatial density estimation of exposed vent locations within the ESRP is based on a grid that was calculated using a Gaussian kernel function and an optimal 2×2 smoothing bandwidth matrix. The contours indicate 25%, 50%, 75%, 95% and 99% of the total spatial density (e.g. 50% of the total spatial density falls within the 7×10^{-5} contour). Black dots represent the locations of exposed vents. The INL boundary is outlined in white. Notice the centrally located hole of lower spatial density within the center of the INL due to the absence of exposed vents. The dark gray (dashed) circle represents the approximate location of Taber caldera (Pierce and Morgan, 1992).

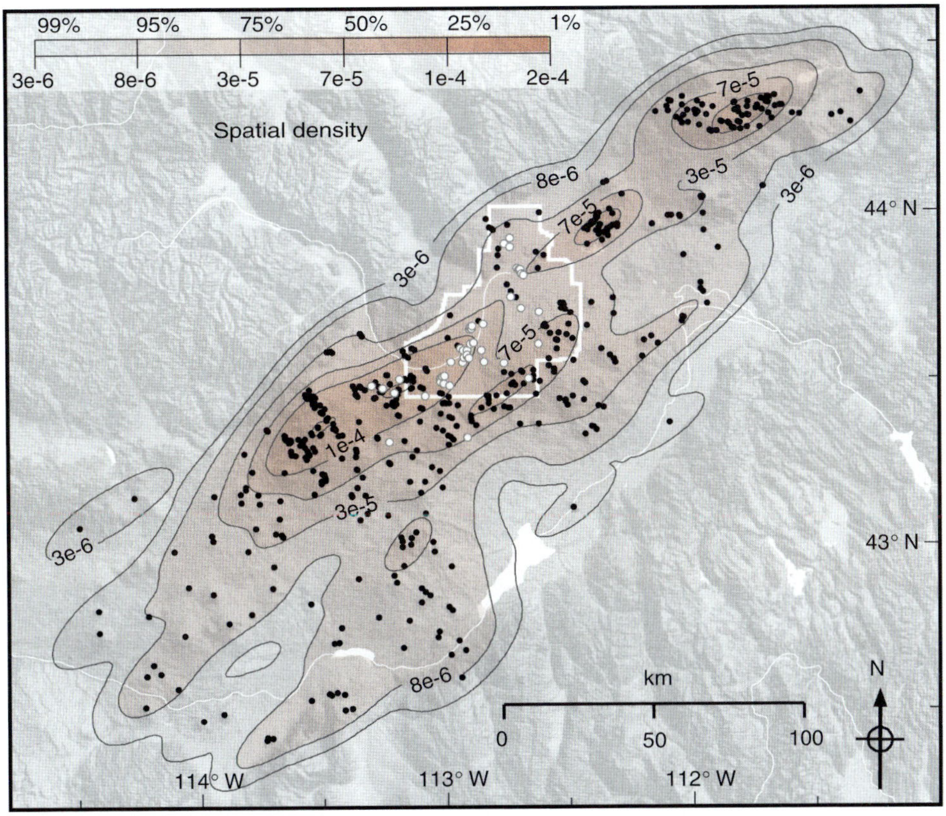

Fig. 16.7 This contour plot of spatial density was based on the population of exposed (black dots) and sub-surface (inferred) vent locations (white circles) within the ESRP. This map was calculated and contoured as described in Figure 16.5. Notice the change in contours within the center of the INL (white lined boundary). The hole of lower spatial density, apparent in Figure 16.5, is no longer evident due to the inclusion of sub-surface (inferred) vent locations in the spatial density calculation.

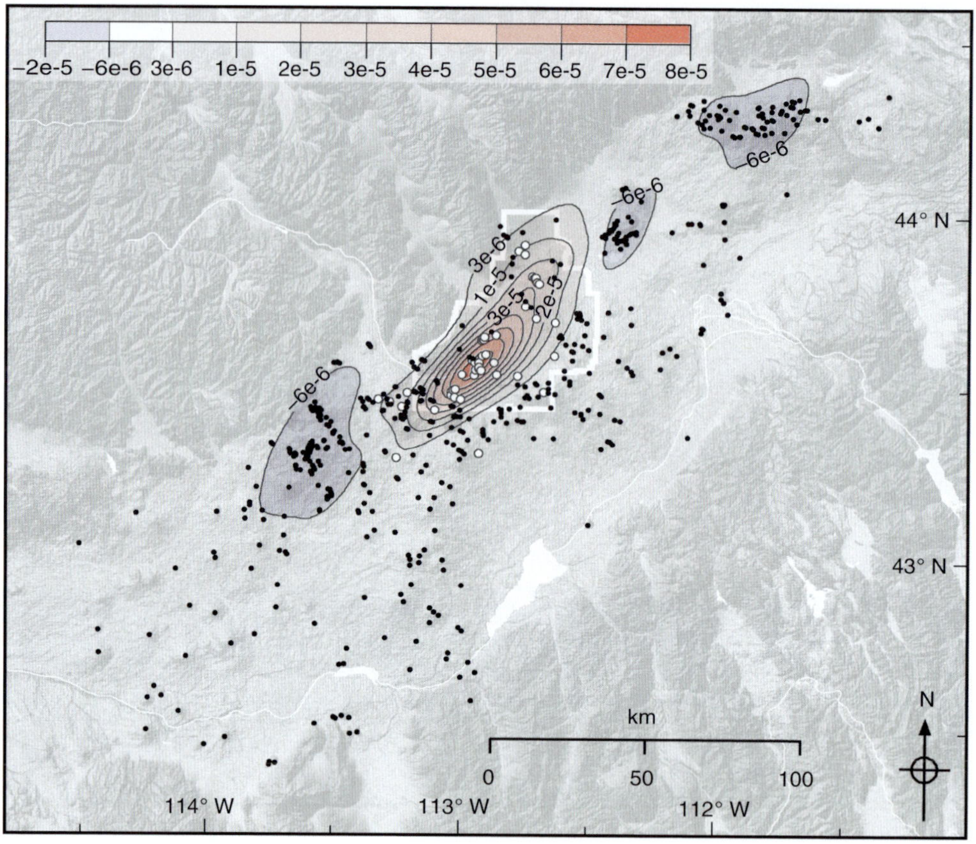

Fig. 16.8 This map shows the difference between the spatial density map in Figure 16.5 and the spatial density map in Figure 16.6. The difference is contoured at a $1 \times 10^{-5}\,\mathrm{km}^2$ interval. The centrally located positive anomaly is caused by an increase in spatial density due to the inclusion of sub-surface vents. The smaller negative anomalies to either side of the "high" are a relative effect due to the centrally located positive anomaly.

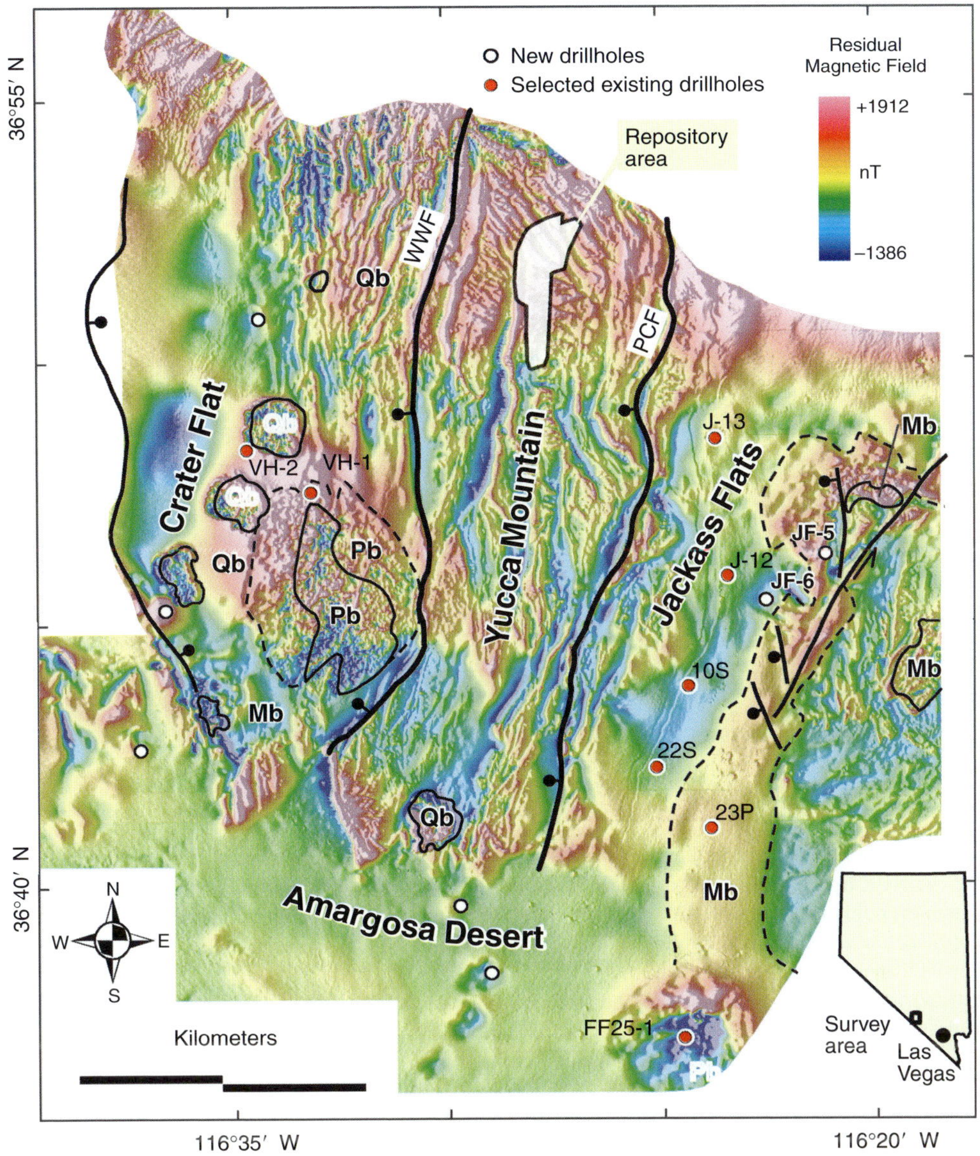

Fig. 19.4 Aeromagnetic map of Yucca Mountain and surrounding basins. Windy Wash fault (WWF) and Paintbrush Canyon fault (PCF) in the central part of the survey area define the approximate boundaries between uplifted Miocene tuffs of the Yucca Mountain range block and the Crater Flat and Jackass Flats basins. The Bare Mountain fault defines the western edge of the Crater Flat Basin. Single-character alphanumeric labels indicate anomalies suspected of representing buried basalt prior to drilling, which shows only buried basalts that were confirmed by post-survey drilling). Solid lines enclose outcrops of Quaternary (Qb), Pliocene (Pb) and Miocene (Mb) basalt. The four Quaternary basalts in Crater Flat are 1.1 Ma scoria cone volcanoes. The Quaternary basalt south of Yucca Mountain is the ~ 77 ka Lathrop Wells volcano. Dashed lines enclose areas of inferred buried basalt associated with outcrops of Pliocene and Miocene lava flows. Red circles indicate selected previous drillholes, and white circles indicate drillholes from the recent drilling program. The bar and ball symbols on faults denote the downthrown side.

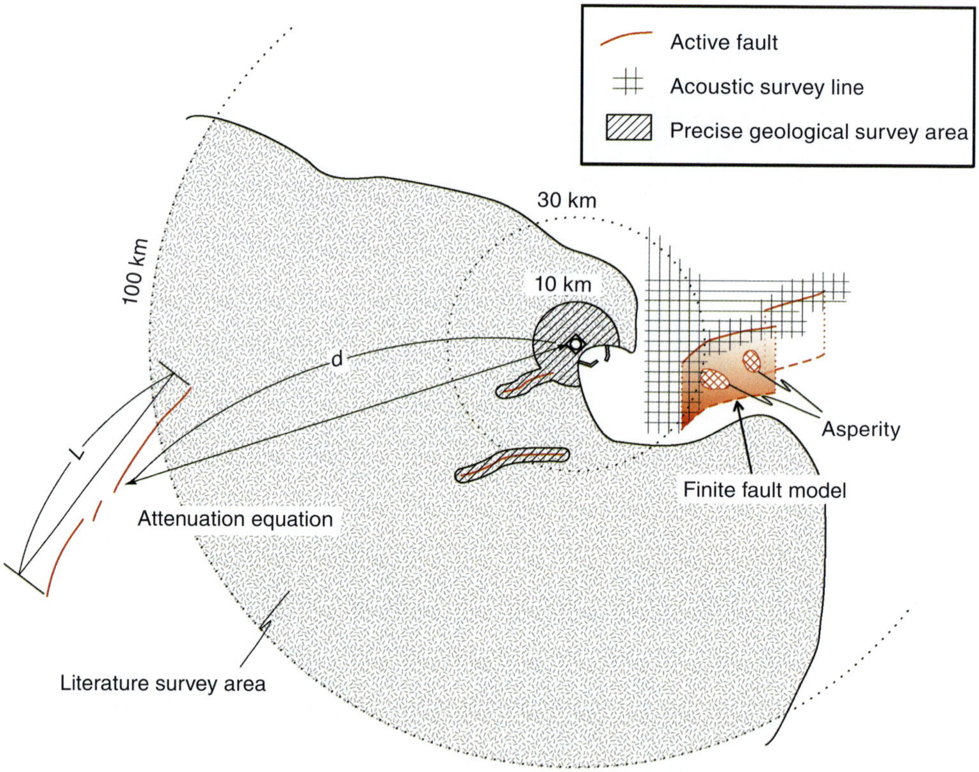

Fig. 21.3 Schematic image of an active fault investigation of a NPP in Japan. The area within 10 km of the NPP, both onshore and offshore, is subject to detailed investigations by field geological and geomorphological study, including geophysical exploration and trenching. In order to not overlook some active faults, the area within 30 km is further investigated by literature study and supplemented by additional investigations. The strong-motion on NPPs is calculated using two methods. First, by strong-motion evaluation based on the attenuation equation using the length, L, of the fault and the distance, d, between the fault center and the NPP. Second, by strong-motion evaluation based on finite fault models using parameters such as fault dimension and asperity.

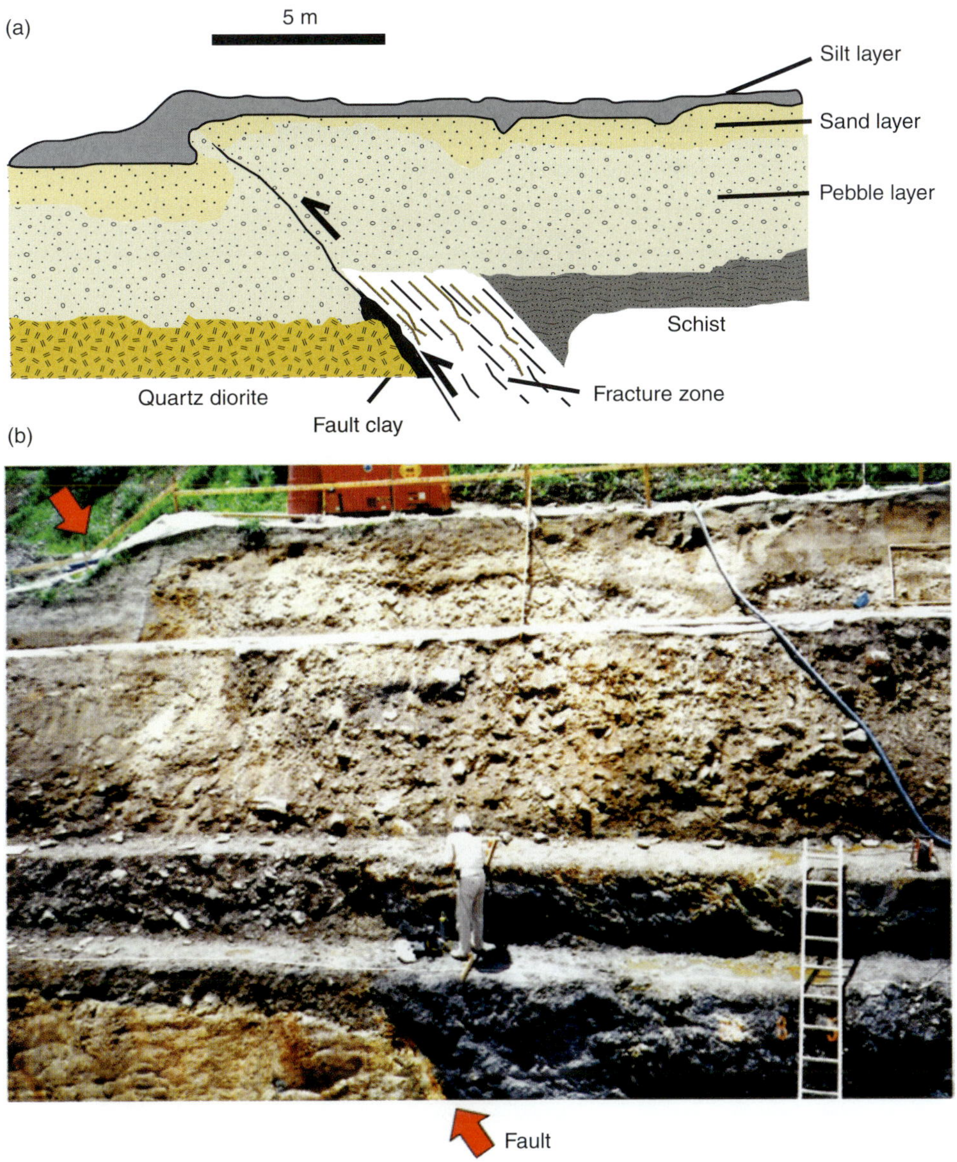

Fig. 21.4 Schematic geologic section (a) and photo (b) of a trench across the Fukouzu fault, Aichi Prefecture, Japan. This fault moved during the Mikawa earthquake, 1945. The earthquake fault continues from basement rocks of quartz porphyry and schist into Quaternary sediments. The crush zone is 4 m in width in the basement rocks. The vertical slip at the uppermost basement rocks is 1.3 m. The pebbles along the fault plane were rotated by the fault displacement so that the long-axes of pebbles are parallel to the fault. Borings adjacent to the trench indicate that the fault dip lessens with depth (Sone and Ueta, 1990) (Photo: D. Inoue).

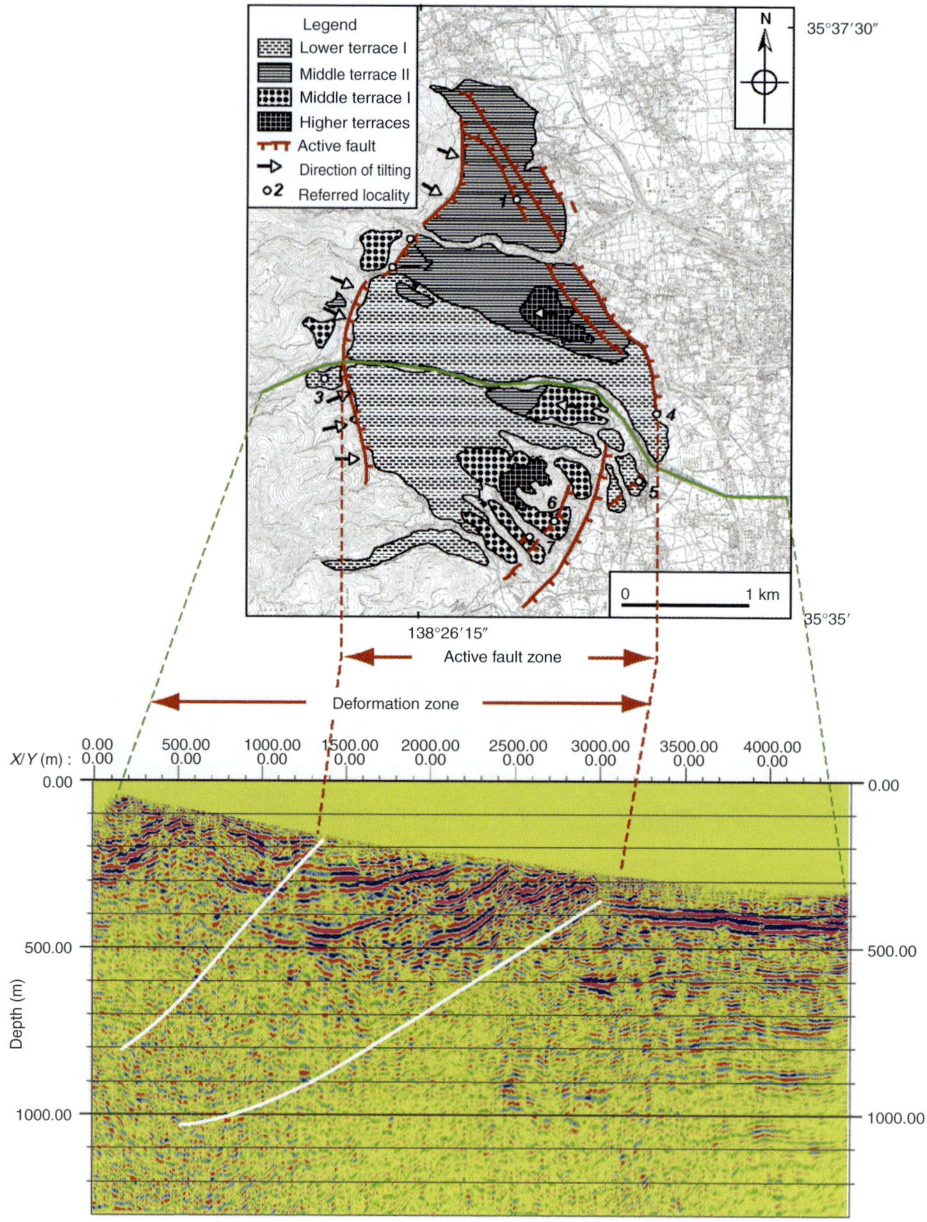

Fig. 21.5 An example of detailed investigations of a thrust fault, the Ichinose fault. This fault is part of the Itoigawa-Shizuoka tectonic line in Japan. The geological map of active faults and tilted marine terraces is used to define the active fault zone between the two active faults. The seismic profile was collected along the gray line shown on the geologic map (horizontal scale shows meters from of the west end of the profile) and shows that deformation of the basement layer extends beyond the western margin of the active fault zone. Additional trenching may indicate whether this part of the fault zone is active or not. The original active fault map is from Miura *et al.* (2002), the seismic profile is from JSCE (2004).

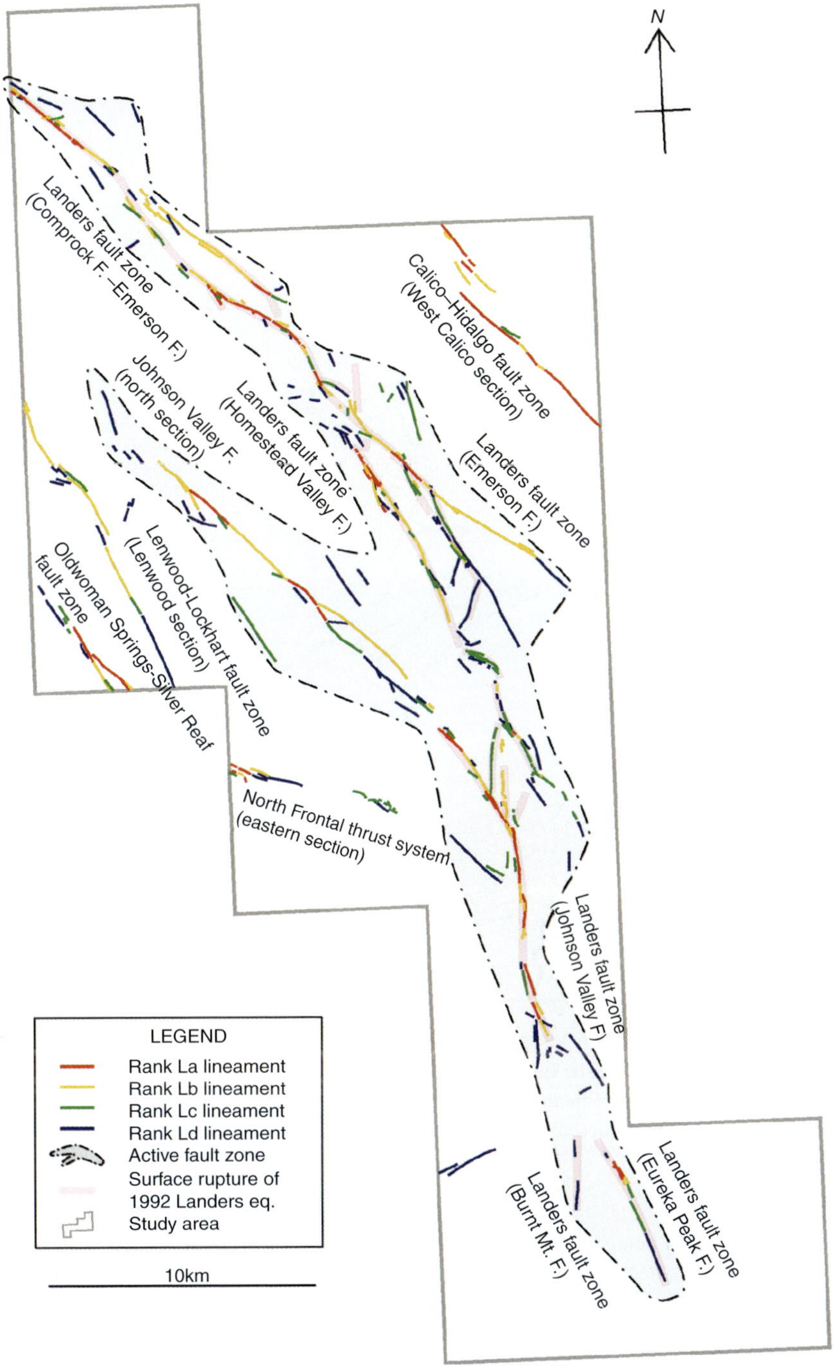

Fig. 21.6 An active strike slip fault, the Landers fault system in California (Sieh *et al.*, 1993). The active fault zone is defined as a region between the eastern and western edge of the lineaments. The active fault zone widens where the lineaments merge. Ranks of lineaments are defined by Inoue (2002).

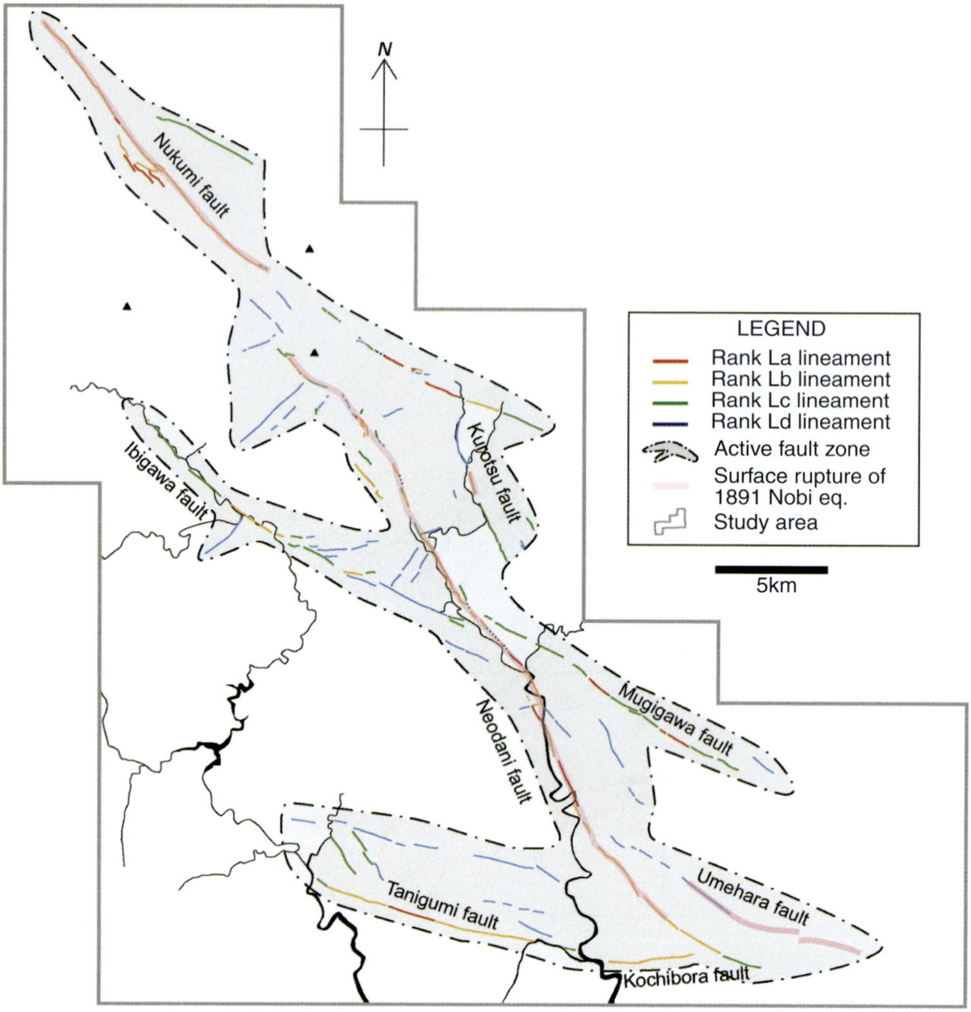

Fig. 21.7 Example of an active strike-slip fault, the Neodani fault system in Japan, which experienced a Mj = 8 earthquake in 1891. The definition of active fault zone is the same as in Figure 21.5. The active fault zone widens where lineaments merge and at the end of lineaments. The lineaments should be checked in the field, as the active fault zone could be smaller than shown.

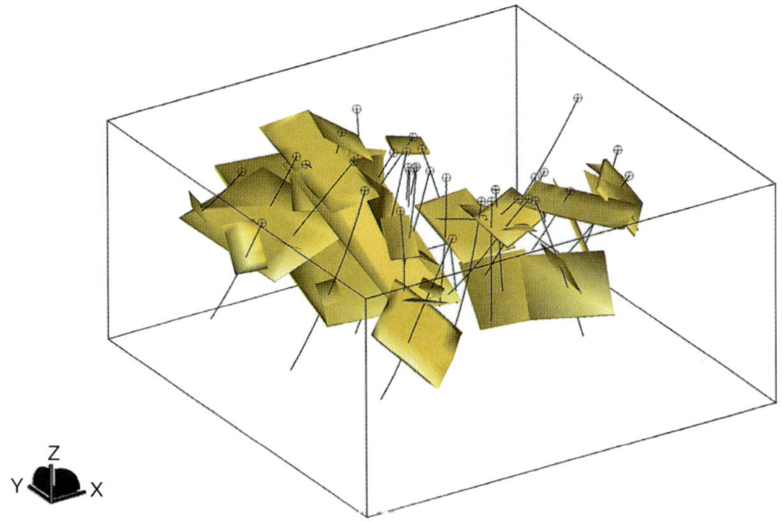

Fig. 23.9 Fault zones at Olkiluoto in one of the fault groups, borehole traces are also shown. Copyright (2006) Posiva Oy. Reproduced by permission of Posiva Oy.

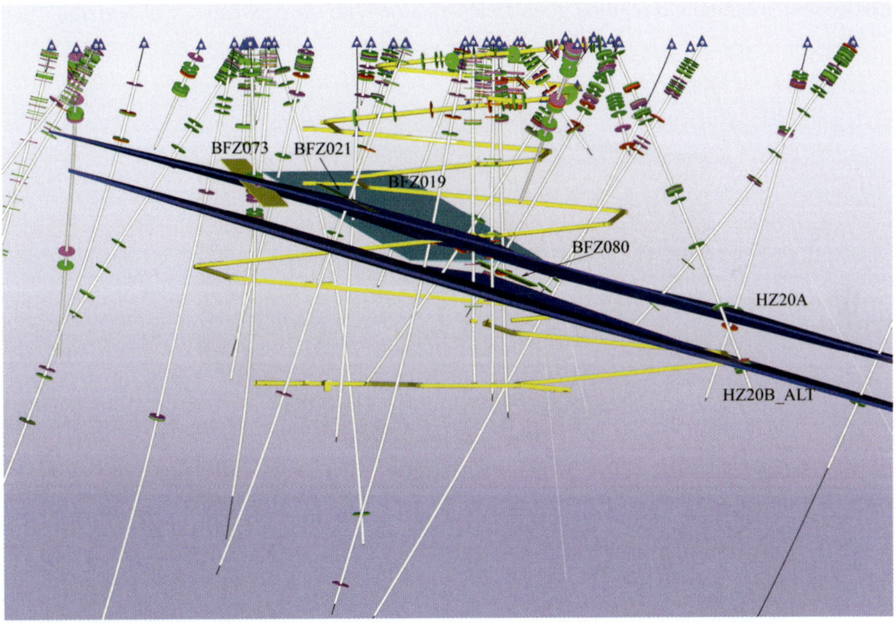

Fig. 23.10 Hydrogeological zones HZ20A (upper blue) and HZ20B_ALT (lower blue) are shown together with nearby deformation or brittle fracture zones (BFZs) with approximately the same orientation. Boreholes (essentially vertical) are used to characterize these zones. Measured transmissivities are shown as discs along the boreholes (classified as: $T > 10^{-5}\,\mathrm{m^2\,s^{-1}}$, red; $10^{-5} > T > 10^{-6}\,\mathrm{m^2\,s^{-1}}$, purple; $10^{-6} > T > 10^{-7}\,\mathrm{m^2\,s^{-1}}$, green). The discs, which indicate the locations of the data, are unoriented. The triangles indicate surface locations of the drillholes. The tunnel (essentially horizontal) is also shown. View is from the SW. Copyright (2006) Posiva Oy. Reproduced by permission of Posiva Oy.

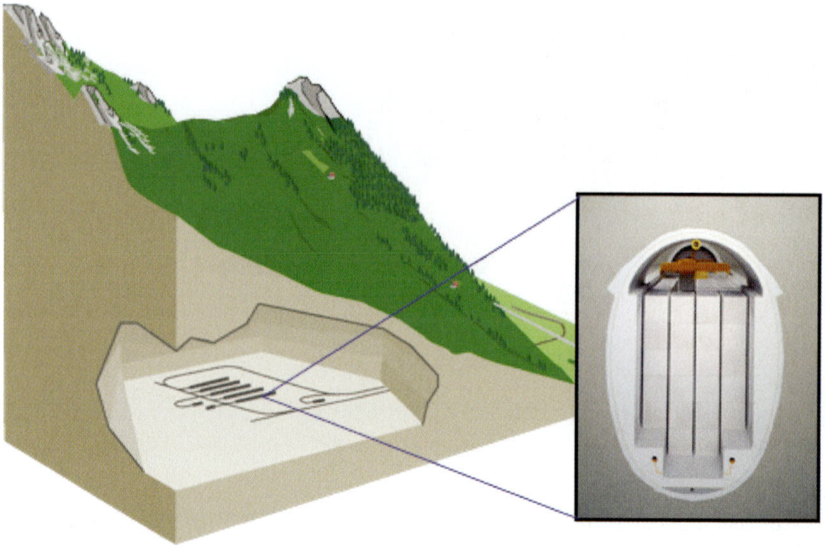

Fig. 24.2 Conceptual layout for a repository at Wellenberg. Disposal caverns for packaged L/ILW are located under a mountain, providing geological isolation with the convenience of horizontal access.

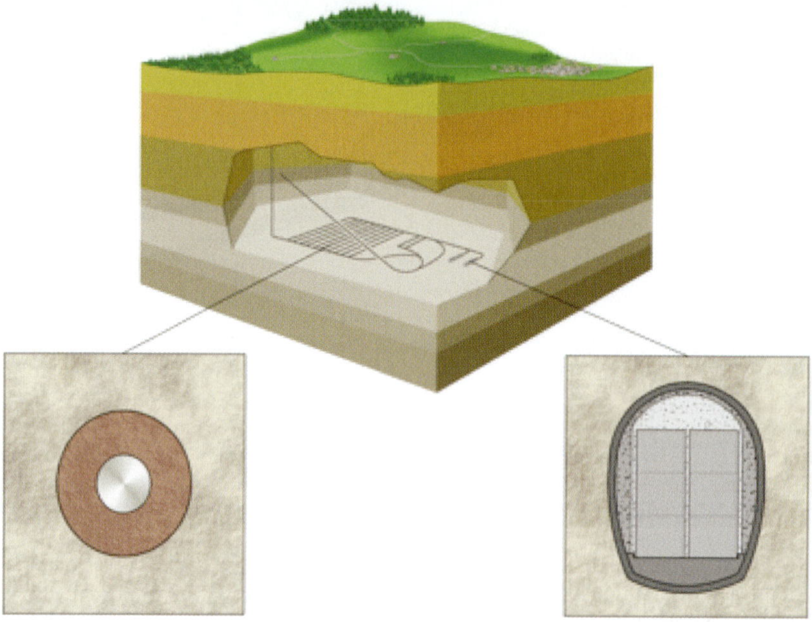

Fig. 24.3 Concept for co-disposal of HLW, SF and long-lived ILW in Opalinus Clay in northern Switzerland. HLW and SF within massive steel overpacks are emplaced axially in small diameter tunnels while ILW is emplaced in larger caverns in a separate part of the repository.

Fig. 24.4 (a) A marked redox front at the Osamu Utsumi uranium mine, Brazil. (b) On a small scale it can be seen that uranium is mobilized on the oxidizing (left) side of the front and concentrated as black nodules on the reduced (right) side.

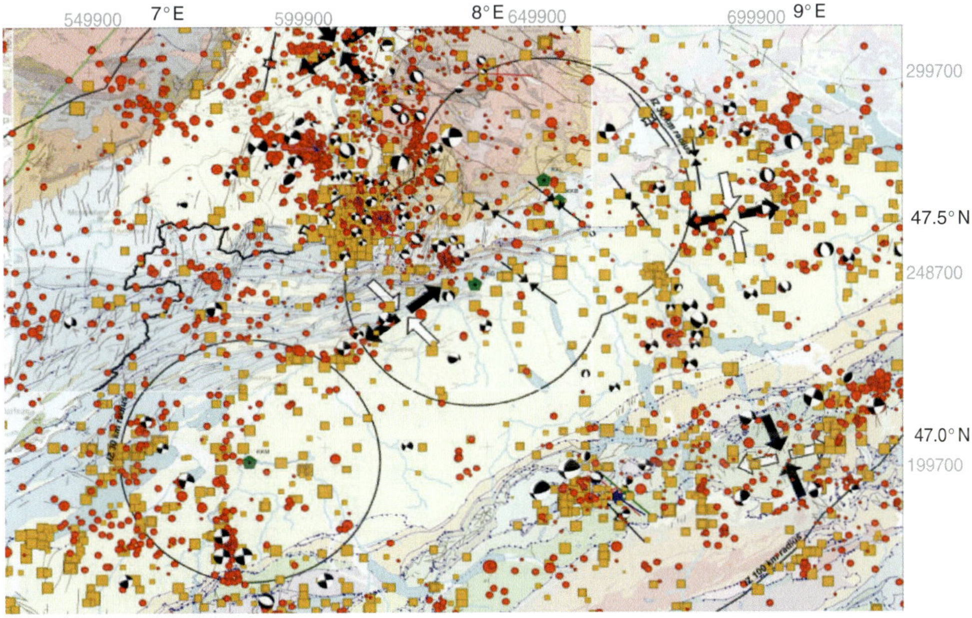

Fig. 26.1 Example of a map developed for the PEGASOS seismic source characterization experts. The GIS layers include seismicity (instrumental shown by red circles; historical by yellow boxes), regional stress orientations (large arrows), focal mechanisms and regional geology. The four Swiss nuclear power plant sites are marked as green pentagons. Circular black lines indicate 25 km distance ranges from the sites (NAGRA, 2004).

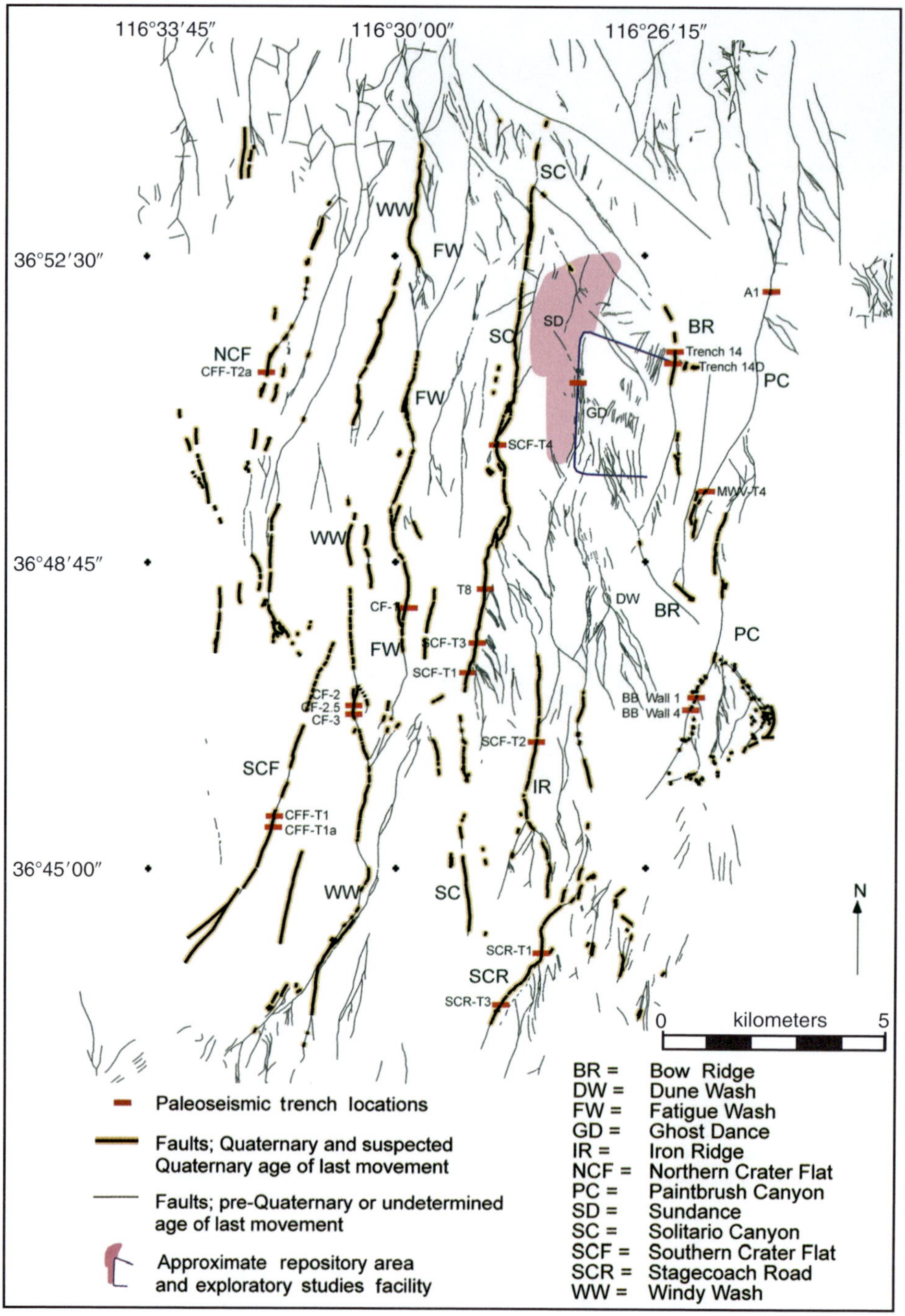

Fig. 26.2 Example of a map provided to the experts of the Yucca Mountain probabilistic seismic hazard analysis. Map shows faults and paleoseismic trench locations in the Yucca Mountain region. Fault map is modified from Simonds *et al.* (1995) and Pezzopane *et al.* (1996).

12

Regional-scale volcanology in support of site-specific investigations

H. Kondo

A fundamental policy for selecting sites for high-level radioactive waste (HLW) geological disposal is to exclude regions of future volcanism to avoid direct damage to the repository due to magmatic intrusion or volcanic eruption. For this purpose, it is important to estimate the potential position, timing, lateral extent and style of future volcanic activity. This estimation is done based on the characteristics of past volcanism and related phenomena that have occurred near the site. Two scales of evaluation should be considered in the context of a deterministic igneous intrusion and volcanic hazard assessment, as will be described in this chapter. These are: (i) evaluation at the scale of an individual volcano, in terms of the potential of migration of magma to the repository from existing volcanoes and their magma plumbing systems; and (ii) evaluation at the scale of a volcano cluster, in terms of the potential for generation of a new volcano in the site area during the performance period of the repository.

In order to make these two evaluations, we should endeavor to understand the geologic processes that give rise to volcanism and to establish plausible geological models for volcanism in specific site regions using available data. These data include spatio-temporal patterns in volcanism, and their relationship to topographical, geological and geophysical data. With regard to geological models, recent research has revealed much useful information about the clustered distribution of volcanism in volcanic arcs, and related geologic processes responsible for the formation of new volcanoes. These processes persist for at least several millions of years in the subduction zone of the Japanese Islands (Kondo *et al.*, 1998, 2004; Tamura *et al.*, 2002; Hasegawa and Nakajima, 2004), resulting in the formation and persistence of volcano clusters within the arc. The occurrence of these processes means that the probability of future volcanism varies along and across volcanic arcs.

In this chapter, a basic framework is presented for an empirical approach to the evaluation of volcanism in the Japanese arc. Special emphasis is given to evaluation of regions of new volcano formation, for the purposes of site evaluation for HLW repositories in such environments. This empirical approach is based on results of a case study in Tohoku (the northeastern district of Honshu), but provides a basic example for the evaluation of future volcanism in a variety of subduction-zone environments.

12.1 Framework for the evaluation of future volcanism in the Japanese Islands

Volcano distribution and the basic processes of magmatic generation and ascent strongly reflect subduction conditions, essentially controlled by the movement modes (e.g. movement rate, direction, inclination) of the oceanic plate in a subduction zone. The duration of certain conditions of plate movement, particularly that of the subducting oceanic plate, is one of the essential factors to consider during evaluation of the potential of future volcanism near a site. The basic framework of the plate system around the Japanese Islands has been unchanged since the cessation of the opening of its backarc basins (e.g. Japan Sea, Kurile Basin, Shikoku Basin) $\sim$ 15 Ma (Jolivet *et al.*, 1994). Detailed investigations of changes in the movements of oceanic plates suggest that their current configuration and rates of motion have persisted since 2.5 Ma for the Pacific plate and since $\sim$ 1.5 Ma for the Philippine Sea plate (Pollitz, 1986; Kamata and Kodama, 1999).

A policy statement by the Advisory Committee on Nuclear Fuel Cycle Backend Policy (1997) stated that it is possible to evaluate the extent and pattern of natural phenomena for the next one hundred thousand years based on analysis and interpretation of the regularity and continuity of the phenomena during the past several hundred thousand years. This view is supported by information about the stability of the plate tectonic regime, and also by the continuity of stress conditions in and around the Japanese Islands. In accordance with this view, a period of around one hundred thousand years has been adopted as the projected timescale for evaluating future geological events for HLW geological disposal (JNC, 1999; JSCE, 2001; NUMO, 2004), although a timescale for repository safety assessment has not yet been stipulated in Japan.

The lifetime of a volcano in a volcanic arc such as the Japanese Islands is generally several hundreds of thousands of years (Kaneoka and Ida, 1997). Using the next one hundred thousand years as the projected timescale for evaluating future geological events for HLW geological disposal, both the possibility of formation of a new volcano and associated intrusive systems in the site area, and the possibility of migration of magma to the repository from existing volcanoes in the site area, should be considered. This is because the timescale of repository performance has long been compared to the longevity of individual volcanoes. For the purpose of evaluating new volcano development and potential changes in the position of magma activity in a region, information about the spatio-temporal variation of volcanism for the last 10^5–10^7 a is essential, as this is the timescale on which regional magmatic activity changes in response to plate tectonic processes.

Consequently, the evaluation of volcanism during site selection for HLW geological disposal demands the investigation of the past volcanism and related phenomena since at least 2 Ma. Investigation of the past volcanism for a longer period, which is dependent on the continuity of the present subduction conditions around the volcanic arc, also provides useful information for evaluating regions of future volcanism.

Information about the characteristics and growth history of volcanoes around the site area on the scale of individual volcanoes is required for evaluation of the possibility of magmatic disruption of a HLW repository. Similarly, information about the spatio-temporal variation of volcanism among volcanoes or volcanic units on the scale of an entire volcano cluster is required for evaluation of the possible generation of a new volcano in the site area.

12.2 Evaluation on the scale of an individual volcano

A key objective of evaluating volcanism on the scale of an individual volcano is to elucidate the lateral extent of magma activity, and the factors controlling the migration of magma during the growth history of each existing volcano. The Nuclear Waste Management Organization of Japan (NUMO, 2002) established an exclusion area within a 15 km radius from the center of each Quaternary volcano as a nationwide evaluation factor (a primary screening criterion) of volcanism for selecting sites. Sites located > 15 km from Quaternary volcanoes may be considered as preliminary investigation areas, in the context of a stepwise approach for HLW site selection (Chapman *et al.*, Chapter 1, this volume). This criterion is based on results of case studies of the area affected by direct magmatic activity at Quaternary volcanoes, which is roughly estimated through the distribution of parasitic volcanoes, craters and dikes in the Japanese Islands using available data. Following this initial screening, detailed surveys of volcanoes and development of models of the magma plumbing systems of individual volcanoes may be necessary to verify the potential of magma migration at a specific repository site.

Evaluation of the potential for magma intrusion into a repository from existing volcanic systems can be evaluated in part using information obtained from published literature and on-site surveys. This information can be used to assess: (i) the growth history of each Quaternary volcano (in particular, temporal changes in the position, style and volume of its magmatic activity), (ii) the structure of the magma plumbing system beneath the volcanic edifice, (iii) the evolution of magma (e.g. crystallization, differentiation, assimilation, magma mixing), and (iv) the lateral extent of magma migration in each volcanic system. This assessment requires a broad range of data obtained through geological and petrological methods, and geophysical and geodetic methods. Geological and petrological data include, for example, the distribution and arrangement of craters and dikes, the history of intrusion and eruption of magma, and the volume and chemistry of intrusions and eruptive products. Geophysical and geodetic data augment such data sets, and are used to assess the position and scale of magma reservoirs and conduits.

Much of the structure of magma plumbing systems is controlled by regional and local stress fields. The geological structure of the basement rock also possibly influences the ascent and lateral intrusion of magma. Furthermore, regional and local stress fields influence the stability of volcanic vents, the style of eruptions, the direction of extension of dikes and the arrangement of craters (Nakamura, 1977; Takahashi, 1997). Differences in the physical and chemical properties of magma, such as mafic or silicic compositions, can influence the length and frequency of radial dikes derived from the same position of the central conduit of a composite volcano (Perry *et al.*, 2001). Moreover, temporal changes in the eruptive rate of volcanic products during the lifetime of a magma plumbing system possibly result in varying conditions of magma migration. This interpretation is supported by changes in the lateral extent of crater distribution over the lifetime of individual volcanoes (Ishizuka, 1999). All of these factors provide useful information to evaluate the lateral migration of magma to the repository (e.g. by establishing an offset distance) from each existing volcano near a site.

12.3 Evaluation on the scale of a volcano cluster

Evaluation of the possible formation of a new volcano in the site region, independent of an existing shallow magmatic system, must consider two factors: (i) estimation of the places where new volcanoes will form, and (ii) estimation of the scale and characteristics of this new volcanism. Volcanism is not uniformly distributed within volcanic arcs. Consequently, when assessing the possibility of new volcano formation it is especially important to focus on the duration of volcano clusters and interpretation of the geologic processes causing such phenomena within a volcanic arc, in the context of persistent subduction conditions. Regions of future volcanism can be estimated by extrapolation based on plausible geological models that consider trends or regularity in volcanic activity.

For the estimation of the scale and characteristics of new volcanism, it is important to consider that the plate tectonic regime and regional and local stress conditions can control the spatial structure and the eruptive style of a specific magma plumbing system. Thus, the stability of the plate tectonic regime and the continuity of the stress conditions in and around the Japanese Islands for the past several hundreds of thousands of years enable us to estimate the rough scale and characteristics of future volcanism by analogy with Quaternary volcanism in the region. Information necessary for the estimation of the scale and characteristics of future volcanism is basically equivalent to that for the evaluation of volcanism on the scale of an individual volcano, described in Section 12.2.

In the Japanese Islands, the ocean-trench-side distribution of volcanoes of various ages is defined as the volcanic front. The generation of magma in a subduction zone in general is controlled by both decompression melting processes within the mantle wedge associated with mantle convection induced by oceanic plate subduction, and by dehydration of the subducting slab and the affected hydrous layer of the mantle wedge (Tatsumi, 1995; Iwamori, 1998; Takahashi, 2000; Fujii, 2002). Sudden changes in both the number of volcanoes and the volume of volcanic products occur passing through the volcanic front. Such sudden changes strongly suggest that the volcanic front is a boundary line dividing two regions, and that the conditions of magma generation and ascent arise rapidly at or behind the front in the subduction zone, although the exact causes of the volcanic front remain an open question. In one rare circumstance in Japan, the volcanic front is understood as a boundary line between two regions: one is a region of alkali-basalt monogenetic volcanism that is not related to the above-mentioned processes characteristic of subduction; the other is a region of no volcanism where the subducting oceanic plate (the Philippine Sea plate) has been covering its magma source in the mantle wedge (Chugoku District, western Honshu;

Uto, 1995; Kimura *et al.*, 2003, 2005). In any case, the rate of migration of the volcanic front since the late Miocene can be estimated to be several kilometers per million years or less. In other words, the volcanic front is stable on the order of several millions of years (e.g. Uto, 1995; Yoshida *et al.*, 1995). Therefore, it is extremely unlikely that new igneous activity will commence on the ocean trench side of the present volcanic front during the next hundred thousand years or so (JSCE, 2001). Thus, current plate tectonic models for the Tohoku region strongly suggest that potential HLW repository sites located trenchward of the volcanic front are essentially free from volcanic hazard.

Of course the situation behind the volcanic front is more complicated. Quaternary volcanism is unevenly distributed on the backarc side of the volcanic front in the Japanese Islands (JNC, 1999). In the Tohoku region, volcano clusters occur in east–west-trending "branches" at intervals of 50–100 km. Volcanism is rare between these branches in the backarc. This regular spacing of magmatism has developed under constant subduction conditions since the cessation of the backarc opening, ~ 14 Ma (Kondo *et al.*, 1998, 2004). Apparently, the evolution of the thermal structure within the mantle wedge since the cessation of the backarc opening has resulted in a tendency for volcanoes to cluster during the late Miocene to Quaternary (Kondo *et al.*, 2004). These volcano clusters correlate with other phenomena, such as topographic highs (including elevated basement), local negative Bouguer gravity anomalies (indicative of thickening of the crust) and low-velocity anomalies in seismic tomography images (interpreted as hot regions) of the mantle wedge (Tamura *et al.*, 2002, Chapter 7, this volume; Hasegawa and Nakajima, 2004). Based on this model of arc volcanism, regularly spaced volcano clusters reflect convection-controlled, heterogeneous thermal structures within the mantle wedge along the northeast Japan arc.

The observed distribution of volcanoes in clusters, together with geophysical correlations, suggests that geologic mapping can reveal much about the persistence of clusters, and hence the potential of future volcanism in areas within and between clusters. For example, if volcanoes are restricted to basement highs with coincident mantle seismic tomographic anomalies, then this information should be used to forecast the future distribution of volcanism.

12.4 Volcano clusters in arcs: the Tohoku case

On the backarc side of the volcanic front in the northeast Japan arc, a geological model for evaluating regions of future volcanism should consider at least three key phenomena (Tamura *et al.*, 2002; Hasegawa and Nakajima, 2004; Kondo *et al.*, 1998, 2004): (i) hot mantle regions capable of supplying magma to the crust are distributed in specific regions within the mantle wedge with east–west-trending patterns arranged along the arc; (ii) volcanism occurring within specific regions in the form of volcano clusters, accompanied by its uplifted basements and crustal thickening, has been caused by repeated injection or underplating of magmas derived from the hot regions in the mantle; and (iii) geologic mapping indicates there is a tendency for the distribution of volcanism to become localized and concentrated into clusters since 14 Ma. These phenomena appear to have been caused by the presence of convection-controlled heterogeneous thermal structures and their evolution within the mantle wedge along the arc under persistent, long-term subduction conditions.

In order to assess these factors, a set of maps was prepared that show: (i) the cumulative distribution of volcanism during several time intervals from 14 Ma to the present, (ii) the distribution of summit-level topography, (iii) the distribution of Bouguer gravity anomalies, and (iv) S-wave velocity perturbations along the inclined low-velocity zone within the mantle wedge. In the following, the geophysical interpretation of these maps and the meaning of correlation among these disparate observations is provided in the context of volcanic hazard assessments for HLW repositories.

Cumulative distribution of volcanism

The distribution of middle Miocene to Quaternary igneous rocks was investigated in an area bounded by 139°30'−142°00'E, and 38°40'−41°40'N. This area includes typical Quaternary volcano clusters and gaps in the northern part of Tohoku. This area also includes the Chokai–Kurikoma area (Kondo *et al.*, 2004) in its southern part, where one of the east−west-trending volcano clusters is located. The igneous rocks formed within the region since 20 Ma were mapped to elucidate the differences in the distribution of igneous rocks before and after the cessation of backarc opening ~ 14 Ma. The distribution of igneous rocks was determined mainly using comprehensive geological publications, such as geological maps at 1:200 000 scale from the Geological Survey of Japan, and the Japan Institute of Construction Engineering. In addition, data obtained from deep drilling for resource exploration provided useful information about the distribution of sedimentary strata and buried igneous rocks in the backarc region (Tsuchiya, 1988; Japan Natural Gas Association and Japan Offshore Petroleum Development Association, 1992).

Some interpretation of these data is required in order to differentiate areas of actual magma ascent from surface areas inundated by surface volcanic activity. For this purpose, the locations of "volcanic central facies" (Ohguchi *et al.*, 1989; Kondo *et al.*, 1998, 2004) were extracted from these map compilations. Such facies are defined by intrusive rocks, lavas, coarse-grained pyroclastic rocks and thick piles of welded tuffs, all of which are considered to be emplaced or deposited in the vicinity of volcanic vents. In contrast, fine-grained pyroclastic rocks (tuffs and lapilli tuffs) were not considered to represent volcanic central facies, as they may be deposited far from volcanic vents. The location and geological age range for each volcanic unit in the area was determined on the basis of the distribution of igneous rocks, the stratigraphic correlation between igneous rocks and the overlying and underlying sedimentary strata, and reliable radiometric age determinations. For this purpose, the reliability of the available radiometric age determinations obtained from previous studies was examined, considering such factors as the degree of alteration of the rock samples, the consistency with stratigraphic correlation data, non-radiogenic ^{40}Ar content in the case of K–Ar data, and similar information. Stratigraphic correlation data obtained from microfossil biochronology were effective for the determination of geological age range of each volcanic unit (Kitamura, 1986), even if no reliable radiometric age data were available.

In the northeast Japan arc, there is a tendency for the distribution of volcanism to become localized and concentrated into more specific areas since the cessation of the backarc opening ~ 14 Ma. Changes in the distribution of volcanism with time can be characterized for periods of 3 Ma or 6 Ma duration, based on results of previous studies (Kondo *et al.*, 1998, 2004). Time ranges for cumulative distribution maps of volcanism were fixed at: 14–0 Ma, 8–0 Ma, 5–0 Ma, and 2–0 Ma. The distribution of igneous rocks in the period of 20–14 Ma was also compiled for comparison of differences in the distribution of igneous rocks before and after the cessation of backarc opening ~ 14 Ma ago.

After the termination of the opening of the Japan Sea backarc basin, sedimentary basins widely developed in the backarc region, indicating thermal subsidence due to lithospheric cooling (Sato, 1992; 1994). In these basins, argillaceous marine sediments were mainly deposited after 14 Ma before a crustal shortening stage (3.5 Ma–present). The total thickness of the sediments in these basins amounts to ~ 2000 m. In order to clarify the contrastive relationship between regions of volcanism and sedimentary basins, particularly in the backarc region, the distribution of sedimentary rocks whose thickness exceeds 200 m, as an outline of the distribution of sedimentary basins, was also drawn on each map. This distribution of sedimentary rocks was extracted from isopach maps, based on data obtained from deep drilling for resource exploration (Japan Natural Gas Association and Japan Offshore Petroleum Development Association, 1992).

Distribution of summit-level topography

Quaternary volcano clusters are closely correlated with topographic highs, including elevated basement along-strike profiles on both the volcanic front and the back-arc sides of the northeast Japan arc (Tamura *et al.*, 2002). Moreover, the distribution patterns of mountainous regions composed of both a north–south-trending range and east–west-trending mountainous regions can be correlated with the distribution patterns of low-velocity anomalies within the mantle wedge on the backarc side of the volcanic front (Hasegawa and Nakajima, 2004; Hasegawa *et al.*, 2005). There is a complementary distribution between clustering of Quaternary volcanoes and north–south-trending active faults along the Backbone Range of northeast Japan in general. These active faults slipped repeatedly throughout the late Quaternary, and possibly longer. Furthermore, these faults have controlled the formation of north–south-trending mountains during crustal shortening (Sato, 1992, 1994). However, the distribution of east–west-trending mountainous regions is remarkably discrepant to that of north–south-trending faults. The formation of these east–west-trending topographically elevated regions is not explained by north–south-trending fault movements. Rather, it has been suggested that the formation of east–west-trending mountainous regions can be explained by warping (Imaizumi, 1999). Consequently, the elevated basement associated with volcano clusters is explained as a result of long-term injection, or underplating, of magma in the lower crust. Based on this model, the elevation of the mountainous regions on the backarc side of the volcanic front is a rough relative index of regions experiencing long-term uplift related to magmatism.

For the purpose of examining the correlation between the cumulative distribution of volcanism and regions of uplift related to volcanism, a contour map was constructed showing the distribution of summit level topography over a 10 km mesh, using data from the Geographical Survey Institute of Japan (2001). It is noted that the region of the Kitakami Mountains on the ocean trench side of the volcanic front is not considered here, because it is clear that the Kitakami Mountains have formed by warping due to tectonic movements, and are unrelated to volcanism (Koike, 2005).

Distribution of Bouguer gravity anomalies

It has been pointed out that there is some cyclic occurrence of local negative anomalies in the Bouguer gravity profile along the coastline of the Japan Sea. The gravity troughs in the profile can be correlated with the across-arc topographic highs and regions of volcano clustering, possibly associated with thickening of the crust (Tamura *et al.*, 2002).

For the purpose of examining the correlation between the cumulative distribution of volcanism and regions of local negative Bouguer gravity anomalies, a contour map was created showing the distribution of Bouguer gravity anomalies, using data from the Geological Survey of Japan, National Institute of Advanced Industrial Science and Technology (Komazawa, 2004).

S-Wave velocity perturbations

Seismic tomography studies in the northeast Japan arc (Hasegawa and Nakajima, 2004; Hasegawa *et al.*, 2005) have revealed the presence of inclined seismic low-velocity and high-attenuation zones. These zones appear to delineate a single inclined sheet shallower than ~ 150 km within the mantle wedge, sub-parallel to the subducting slab. These anomalies probably correspond to the upwelling-flow portion of subduction-induced convection. Within this upwelling flow, it is inferred that temperatures are higher than the wet solidus of peridotite, and melt inclusions with volume fractions of 0.1 to several percent exist (Hasegawa *et al.*, 2005). There is striking along-arc variation of the inclined low-velocity zone in the mantle. Very low-velocity regions periodically occur along the strike of the arc, corresponding to clusters of Quaternary volcanoes and topographic highs at the surface.

For the purpose of examining the correlation between the cumulative distribution of volcanism and regions of low-velocity anomalies within the mantle wedge, a map showing the projection of the distribution of S-wave velocity perturbations along the core of the inclined low-velocity zone within the mantle wedge was prepared. These data were obtained from a specialized tomographic inversion of S-wave velocity structure of the mantle wedge beneath northeast Japan using the same data set of Nakajima *et al.* (2001) (Hasegawa and Nakajima, 2004). These high-resolution images reveal perturbations along the inclined low-velocity zone, obtained by taking velocity perturbations along the slowest S-wave velocity portion of the mantle wedge.

Results of correlations

Comparing all of these data sets, there is a striking contrast between the distributions of the 20–14 Ma and the 14–0 Ma igneous rocks: volcanism was widespread and had a

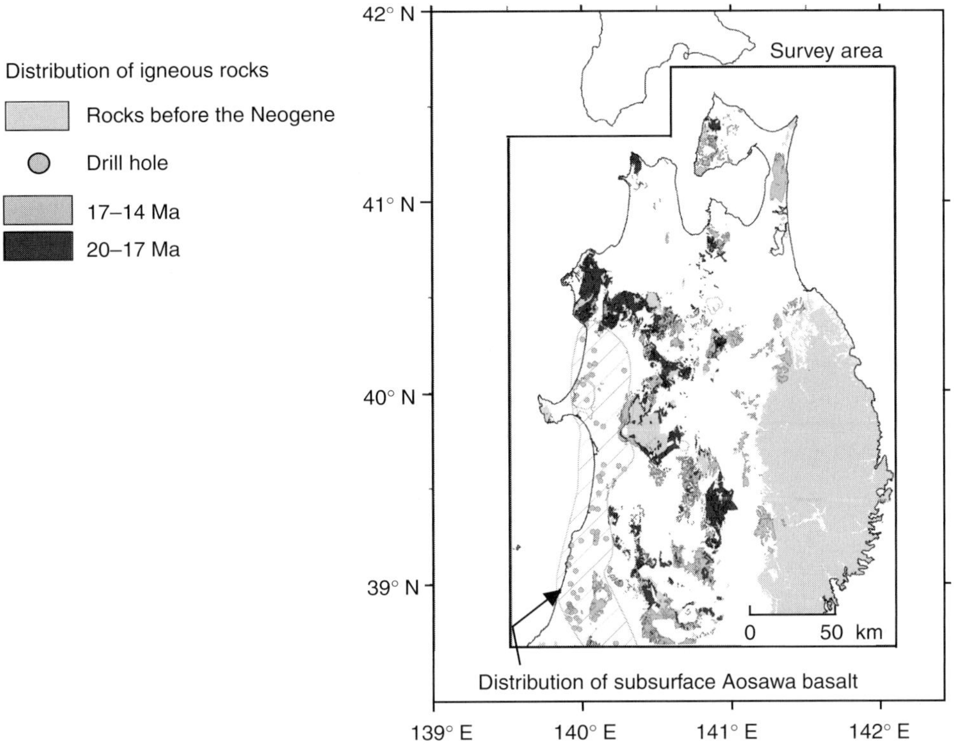

Fig. 12.1 Distribution of igneous rocks 20–14 Ma (younger rocks are light colored, older rocks are dark colored, both located in the middle to the west of the arc). The locations of drill holes (light-colored dots, indicating the presence of younger rocks) from resource exploration confirm the distribution of subsurface Aosawa basalt. Basement rocks older than the Neogene are colored light gray and located mainly east of the arc. Data acknowledgments: distribution of igneous rocks (Tsushima, 1964; Fujioka *et al.*, 1976, 1977; Hata *et al.*, 1972, 1984; Ozawa and Suda, 1978, 1980; Ikebe *et al.*, 1979; Ozawa *et al.*, 1981, 1982, 1983, 1984, 1985, 1986, 1988a, 1988b, 1993; Tsuchiya, 1981; Tsuchiya *et al.*, 1984; Yoshida *et al.*, 1984; Editorial Committee for Civil Engineering Geological Maps of Tohoku District, 1988; Kamata *et al.*, 1991; Japan Natural Gas Association and Japan Offshore Petroleum Development Association, 1992; Nakano and Tsuchiya, 1992); extent of subsurface Aosawa basalt (Tsuchiya, 1988). See color plate section.

north–south trend in the 20–14 Ma period (the last period of opening of the Japan Sea), whereas regions of volcanism have become more localized and concentrated since 14 Ma (after the end of backarc opening) (Kondo *et al.*, 1998, 2004; Figures 12.1 and 12.2). Since 14 Ma, sedimentary basins developed in a wide area of the backarc, indicating regional subsidence, possibly due to cooling within the mantle wedge. At the same time, volcanism developed in specific regions characterized by both a "trunk," the north–south-trending volcanic front, and east–west-trending "branches" in the backarc. The comparison between cumulative distribution maps of volcanism for different periods of time reveals that the shorter the period, the more conspicuous the characteristics of the distribution of volcano

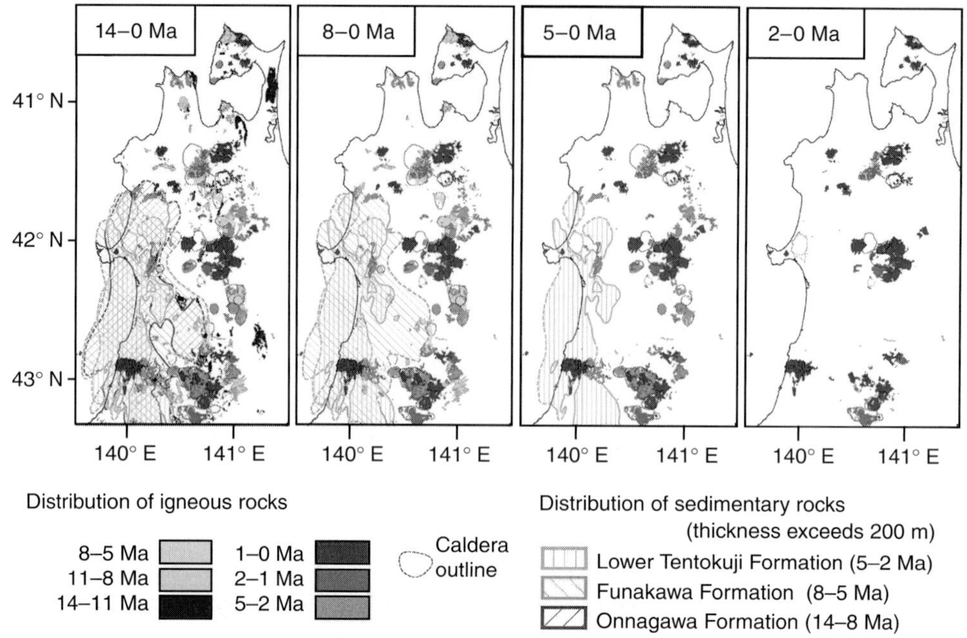

Fig. 12.2 Cumulative distribution of volcanism for four time periods. Six distributions of igneous rocks are differentiated by color (note map key for colors). Three distributions of sedimentary rocks within the backarc (thicknesses > 200 m) are defined by hatched areas with vertical or oblique lines. Data acknowledgments: igneous rocks (see Figure 12.1); caldera outlines (Yoshida *et al.*, 1999); sedimentary rock locations (Japan Natural Gas Association and Japan Offshore Petroleum Development Association, 1992). See color plate section.

clusters become, and volcano clusters in the 2–0 Ma period are the most conspicuous (Figure 12.2). However, the cumulative distribution pattern of volcanism in the 2–0 Ma period is not the most consistent with the patterns of topography, geological and geophysical data. In connection with this observation, it should be considered that the presence of ascending magmas was estimated through geochemical and geophysical methods beneath the Iide Mountains, $\approx$ 100 km south of this survey area, although Quaternary volcanism has not occurred there (Umeda *et al.*, 2007). This mountainous region is situated above one of the "hot fingers" of the mantle wedge based on interrelationships between topographic profiles, seismic tomography and Bouguer gravity anomalies (Tamura *et al.*, 2002).

Overall, in the survey area the cumulative distribution pattern of volcanism in the 5–0 Ma period is well correlated with the distribution pattern of regions of uplift related to volcanism based on the contour map, and with hot regions indicated by low-velocity anomalies on the map showing the distribution of S-wave velocity perturbations (Figures 12.3a and 12.3b). The distribution of volcanism since 5 Ma is closely related to mountainous regions that include east–west-trending warping, whereas it is discrepant to the distribution of sedimentary basins in the backarc regions in the same period (Figure 12.3a). The top

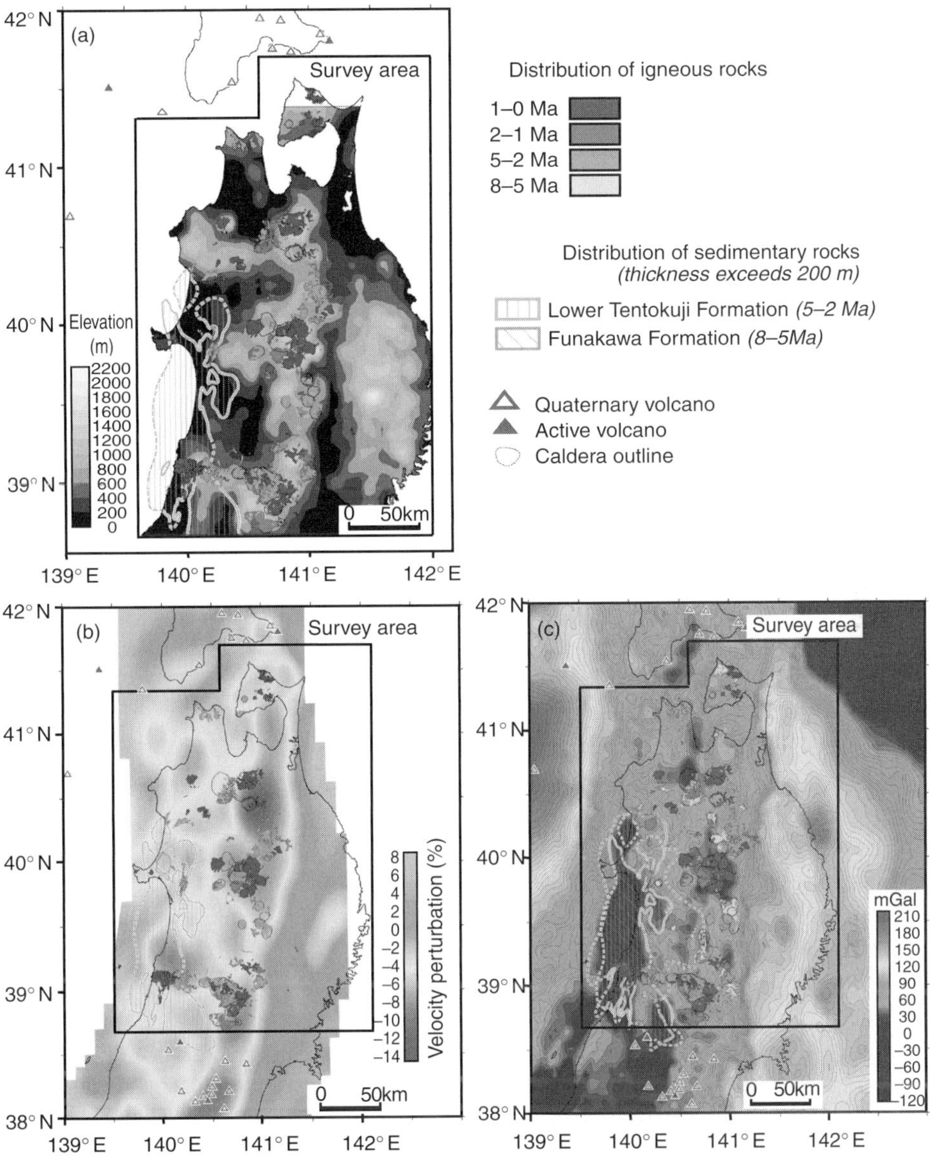

Fig. 12.3 Comparison of the distribution of volcanism with (a) summit-level topography, (b) low-velocity anomalies within the mantle wedge (S-wave velocity perturbations along the core of the inclined low-velocity zone), and (c) Bouguer gravity anomalies. Igneous rocks are divided into four periods and differentiated by color. Two distributions of sedimentary rocks within the backarc (thicknesses > 200 m) are defined by hatched areas with vertical or oblique lines. Data acknowledgments: caldera outlines inside the survey area (Yoshida *et al.*, 1999); Quaternary volcanoes and active volcanoes outside the survey area (Committee for the Catalogue of Quaternary Volcanoes in Japan, 1999; Japan Meteorological Agency, 2003); gravity data (Komazawa, 2004); sedimentary rocks (Japan Natural Gas Association and Japan Offshore Petroleum Development Association, 1992); seismic data (Hasegawa and Nakajima, 2004). See color plate section.

of the elevated basement of post-5 Ma volcanic products is situated > 1000 m above sea level in each east–west-trending mountainous region in the backarc region within the survey area. On the other hand, sedimentary basins have developed in regions between these warping mountains since 14 Ma: the thickness of sedimentary rocks in basins decreases toward regions of warping mountains, based on drilling data. These facts suggest differential movement between the locally uplifted mountains related to volcanism and the regional subsidence of sedimentary basins since 14 Ma. However, the coincidence of distribution patterns between volcanism since 5 Ma and the related phenomena observable at the present time indicates that the timeframe of the continuity of these phenomena is on order of several millions of years.

Figure 12.3c is a map showing the relation between Bouguer gravity anomalies and the distribution of volcanism and sedimentary basins since 8 Ma. In the backarc region, the wide distribution of negative Bouguer gravity anomalies can be explained by the presence of sedimentary basins, where thick piles of sediments accumulated for more than several millions of years. In the volcanic front region, some areas of local negative Bouguer gravity

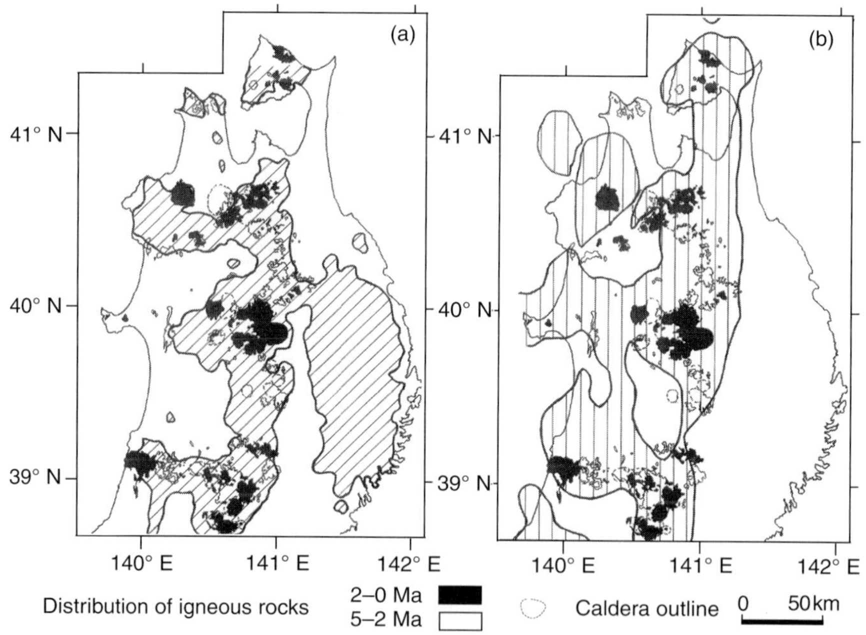

Fig. 12.4 Map (a) is an overlay of the distribution of igneous rocks since 5 Ma (older rocks are light-colored; younger rocks are dark-colored, see map key) on to the range of mountainous regions with elevations > 500 m (hatched area with oblique lines). Data are based on Figure 12.3a. Map (b) is an overlay of the distribution of igneous rocks since 5 Ma on to the range of S-wave low-velocity anomalies along the core of the inclined low-velocity zone within the mantle wedge, where the velocity perturbation is < −4% (hatched area with vertical lines). Data are based on Figure 12.3b. Data acknowledgments: see Figure 12.3.

anomalies roughly correspond to volcano clusters associated with local depressions since 8 Ma. This interrelationship in the volcanic-front region will possibly become clearer by extracting short wavelength components through high-pass filtering analysis of the gravity anomalies. Through the result of correlation, it is concluded that Bouguer gravity anomalies cannot simply indicate the undulation of crustal thickness, which is assumed to be caused by repeated injection or underplating of magmas.

In order to specify possible regions of future volcanism, a threshold should be fixed appropriately, based on results of correlation between these key phenomena. As the values of indexes of the related phenomena such as the elevation of the mountainous regions and the velocity perturbations within the mantle wedge can change continuously, plausible bases are needed for explaining the meaning of the thresholds as a result of the correlation of volcano distributions and related phenomena. Within the scope of the whole survey area, tentative thresholds can be fixed qualitatively at $\sim$ 500 m above sea level in terms of uplift of mountains related to volcanism, and at $\sim -4\%$ of velocity perturbation in the inclined low-velocity zone in terms of hot regions within the mantle wedge (Figures 12.4a and 12.4b). More detailed investigation focusing on the region around the site area is required in order to determine optimal thresholds through the synthesis of the key phenomena and also to interpret the meaning of discrepancies between volcano distributions and topographic and seismic tomographic anomalies.

Concluding remarks

Through this examination for a typical region of Tohoku in the northeast Japan arc, the following key information for the evaluation of regions of future volcanism can be identified: (i) volcano clusters become conspicuous in east–west-trending “branch-like” patterns during the last several millions of years; (ii) volcanism since 5 Ma is associated with mountainous regions, including east–west-trending warping mountains, suggesting that uplift is a result of repeated injection or underplating of magmas in the crust; and (iii) uneven distribution patterns of low-velocity anomalies in the inclined low-velocity zone suggest hot regions within the mantle wedge, also coincident with the distribution of volcanism since 5 Ma. These observations are the basis for development of an empirical model of volcanism in the region.

From a more general viewpoint, a flowchart showing the basic procedure for evaluating regions of future volcanism is developed (Figure 12.5). At the beginning of evaluation, the position of the volcanic front and the extent of its migration during the last several million years near the site area is investigated. Detailed investigations are required if the site area is located on the backarc side of the volcanic front, or near the volcanic front even on the ocean-trench side. Information about temporal changes in tectonics, in particular the continuity of subduction conditions, is useful for the determination of the appropriate time range of the evaluation. Within the time period of constant subduction conditions, investigation of spatio-temporal patterns of volcanism around the site provides fundamental information for evaluating future volcanism. The investigation of crustal structure and movement indicating

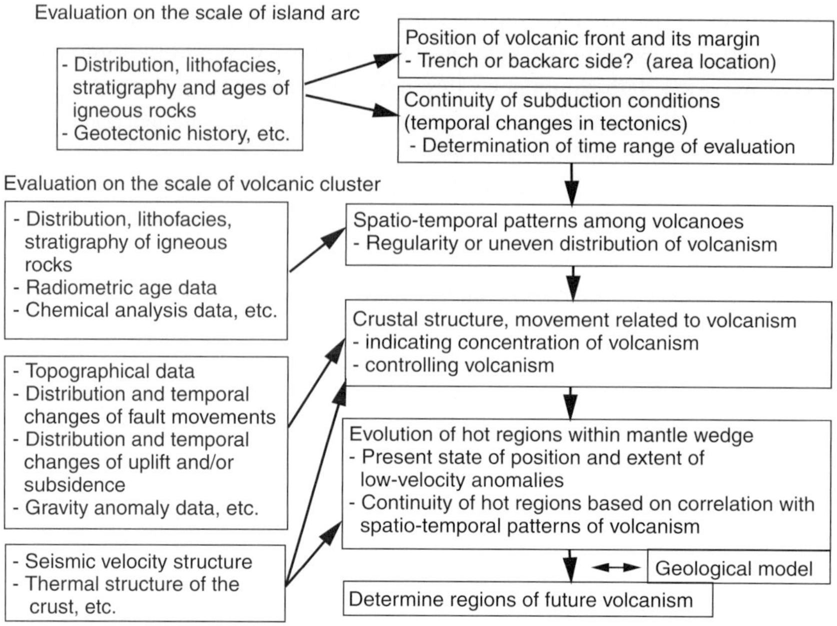

Fig. 12.5 Flowchart outlines the basic procedure for evaluating regions of future volcanism.

(or controlling) the concentration of volcanism possibly provides information useful for specifying regions of future volcanism (e.g. regional uplift due to continuous injection or underplating of magma in the crust as was supposed in the Tohoku case). Of course, different tectonic models for evaluating future volcanism may be appropriate in different tectonic settings (e.g. Nakada *et al.*, 1997). Investigations of correlations with other data sets, such as seismic tomography and gravity, have proven to be invaluable in assessing areas of future potential volcanism in Tohoku. Such data sets will likely be useful in other subduction zones as well. Regions of future volcanism are determined through the above-mentioned investigations in consideration of plausible geological models, by establishment of conservative thresholds.

Further reading

Japan Nuclear Cycle Development Institute, which was succeeded by Japan Atomic Energy Agency, published the second progress report (JNC, 1999) that demonstrates the technical reliability of geological disposal of HLW in Japan and provides technical bases for the selection of candidate disposal sites. This report also reviews overall results of research and development until then, including evaluation of igneous activity with special consideration of the characteristics of tectonic settings of the Japanese Islands. The Nuclear Waste Management Organization of Japan, NUMO (2002) established siting factors for selecting preliminary investigation areas (including an exclusion criterion for evaluation of volcanism) for only the first-stage criteria in the context of a stepwise approach for HLW site selection in Japan. A later publication by the Nuclear Waste Management Organization of Japan, (NUMO, 2004) provides scientific and technical bases for siting factors, and describes the related key issue of geological stability of the Japanese Islands.

Acknowledgments

The author thanks N. A. Chapman, R. S. J. Sparks and C. B. Connor for their constructive comments in the International Tectonics Meeting (ITM) project meetings for evaluation of volcanism and rock deformation. The author is much indebted to A. Hasegawa and J. Nakajima for permitting him to use their seismic tomography data. The author is also much indebted to M. Komazawa for providing his gravity anomaly data in the area of the sea around Japan for this examination. Reviews and helpful suggestions by C. B. Connor and L. J. Connor are much appreciated. This study was financially supported by the Nuclear Waste Management Organization of Japan.

References

Advisory Committee on Nuclear Fuel Cycle Backend Policy (1997). Guidelines on research and development relating to geological disposal of high-level radioactive waste in Japan. Tokyo: Atomic Energy Commission of Japan (in Japanese).

Committee for the Catalogue of Quaternary Volcanoes in Japan (eds.) (1999). Catalogue of Quaternary Volcanoes in Japan (CD-ROM). Tokyo: Volcanological Society of Japan.

Editorial Committee for Civil Engineering Geological Maps of Tohoku District (eds.) (1988). Civil engineering geological maps of Tohoku district at 1:200 000 scale. Tokyo: Japan Institute of Construction Engineering (in Japanese).

Fujii, T. (2002). How to create the Earth's crust. In: Kawakatsu, H. (ed.) *Earth Dynamics and Tomography. New Development of Earth Science*, vol. 1, Tokyo: Asakura Publishing (in Japanese), 73–95.

Fujioka, K., A. Ozawa and Y. Ikebe (1976). Geology of the Ugo-Wada District, with geological sheet map at 1:50 000. Tsukuba: Geological Survey of Japan (in Japanese with English abstract).

Fujioka, K., A. Ozawa, T. Takayasu and Y. Ikebe (1977). Geology of the Akita District, with geological sheet map at 1:50 000. Tsukuba: Geological Survey of Japan (in Japanese with English abstract).

Geographical Survey Institute of Japan (2001). Digital elevation map 50 m grid (CD-ROM). Tsukuba: Geographical Survey Institute of Japan.

Hasegawa, A. and J. Nakajima (2004). Geophysical constraints on slab subduction and arc magmatism. In: Sparks, R. S. J. and C. J. Hawkesworth (eds.) *The State of the Planet: Frontiers and Challenges in Geophysics*, Geophysical Monograph Series, 150, Washington, DC: American Geophysical Union, 81–94.

Hasegawa, A., J. Nakajima, N. Umino and S. Miura (2005). Deep structure of the northeastern Japan arc and its implications for crustal deformation and shallow seismic activity. *Tectonophysics*, **403**, 59–75.

Hata, M., K. Tsushima, Y. Suda and Y. Ono (1972). Geological map of Japan 1:200 000, Hakodate and Shiriyazaki. Tsukuba: Geological Survey of Japan.

Hata, M., F. Uemura and T. Hiroshima (1984). Geological map of Japan 1:200 000, Hakodate and Oshima–O-shima. Tsukuba: Geological Survey of Japan.

Ikebe Y., A. Ozawa and H. Inoue (1979). Geology of the Sakata District, with geological sheet map at 1:50 000. Tsukuba: Geological Survey of Japan (in Japanese with English abstract).

Imaizumi, T. (1999). Topographical relief in the Tohoku region from the viewpoint of the distribution of active faults – several questions. *Earth Monthly*, extra issue, **27**, 113–117 (in Japanese).

Ishizuka, Y. (1999). Eruptive history of Rishiri volcano, northern Hokkaido, Japan. *Bulletin of the Volcanological Society of Japan*, **44**, 23–40 (in Japanese with English abstract).

Iwamori, H. (1998). Transportation of H_2O and melting in subduction zones. *Earth and Planetary Science Letters*, **160**, 65–80.

Japan Meteorological Agency (2003). Selection of active volcanoes and classification (rank division) of their volcanic activities by the Coordinating Committee for Prediction of Volcanic Eruptions. Press release, January 21. Tokyo: Japan Meteorological Agency (in Japanese).

Japan Natural Gas Association and Japan Offshore Petroleum Development Association (1992). *Oil and Natural Gas Resources in Japan*, 2nd edn. (in Japanese).

JNC (1999). Project to establish the scientific and technical basis for high-level radioactive waste disposal in Japan: second progress report on research and development for the geological disposal of high-level radioactive waste in Japan, JNC TN1400 99-020 (Japanese version), JNC TN1410 2000-001 (English version). Tokai: Japan Nuclear Cycle Development Institute (currently Japan Atomic Energy Agency).

Jolivet, L., K. Tamaki and M. Fournier (1994). Japan Sea, opening history and mechanism: a synthesis. *Journal of Geophysical Research*, **99**(B11), 22 237–22 259.

JSCE (2001). Geological disposal and the geological environment: geological factors to be considered in the selection of preliminary investigation areas for high-level radioactive waste disposal. Sub-Committee on the Underground Environment, Civil Engineering Committee of the Nuclear Power Facilities. Tokyo: Japan Society of Civil Engineers (in Japanese).

Kamata, H. and K. Kodama (1999). Volcanic history and tectonics of the southwest Japan arc. *The Island Arc*, **8**, 393–403.

Kamata, K., M. Hata, K. Kubo and T. Sakamoto (1991). Geological map of Japan 1:200 000, Hachinohe. Tsukuba: Geological Survey of Japan.

Kaneoka, I. and Y. Ida (eds.) (1997). *Volcanoes and Magma*. Tokyo: University of Tokyo Press (in Japanese).

Kimura, J., T. Kunikiyo, I. Osaka *et al.* (2003). Late Cenozoic volcanic activity in the Chugoku area, southwest Japan arc during back-arc basin opening and reinitiation of subduction. *The Island Arc*, **12**, 22–45.

Kimura, J., R. J. Stern and T. Yoshida (2005). Reinitiation of subduction and magmatic responses in SW Japan during Neogene time. *Geological Society of America Bulletin*, **117**, 969–986.

Kitamura, N. (ed.) (1986). *Cenozoic Arc Terrane of Northeast Honshu, Japan*, 3 volumes. Sendai: Hobundo (in Japanese).

Koike, K. (2005). Macroscopic topography and the classification of landform in the Tohuku region. In: Koike, K., T. Tamura, K. Chinzei and T. Miyagi (eds.) *Geomorphology of Tohuku Region. Regional Geomorphology of the Japanese Islands*, vol. 3, Tokyo: University of Tokyo Press (in Japanese), 22–26.

Komazawa, M. (2004). Gravity Grid Database of Japan. Gravity CD-ROM of Japan, 2, Digital Geoscience Map Series, P-2. Tsukuba: Geological Survey of Japan, National Institute of Advanced Industrial Science and Technology.

Kondo, H., K. Kaneko and K. Tanaka (1998). Characterization of spatial and temporal distribution of volcanoes since 14 Ma in the northeast Japan arc. *Bulletin of the Volcanological Society of Japan*, **43**, 173–180.

Kondo, H., K. Tanaka, Y. Mizuochi and A. Ninomiya (2004). Long-term changes in distribution and chemistry of middle Miocene to Quaternary volcanism in the Chokai–Kurikoma area across the northeast Japan arc. *The Island Arc*, **13**, 18–46.

Nakada, M., T. Yanagi and S. Maeda (1997). Lower crustal erosion induced by mantle diapiric upwelling: constraints from sedimentary basin formation followed by voluminous basalt volcanism in northwest Kyushu, Japan. *Earth and Planetary Science Letters*, **146**, 415–429.

Nakajima, J., T. Matsuzawa, A. Hasegawa and D. Zhao (2001). Three-dimensional structure of VP, VS, and VP/VS beneath northeastern Japan: implications for arc magmatism and fluids. *Journal of Geophysical Research*, **106**, 21 843–21 857.

Nakamura, K. (1977). Volcanoes as possible indicators of tectonic stress orientation – principle and proposal. *Journal of Volcanology and Geothermal Research*, **2**, 1–16.

Nakano, S. and N. Tsuchiya (1992). Geology of the Chokaisan and Fukura District, with geological sheet map at 1:50 000. Tsukuba: Geological Survey of Japan (in Japanese with English abstract).

NUMO (2002). Siting factors for the selection of preliminary investigation areas. Information package of open solicitation for candidate sites for safe disposal of high-level radioactive waste. Tokyo: Nuclear Waste Management Organization of Japan.

NUMO (2004). Evaluating site suitability for a HLW repository: scientific background and practical application of NUMO's siting factors. NUMO-TR-04-02 (Japanese version), NUMO-TR-04-04 (English version). Tokyo: Nuclear Waste Management Organization of Japan.

Ohguchi T., T. Yoshida and K. Okami (1989). Historical change of the Neogene and Quaternary volcanic field in the northeast Honshu arc, Japan. *Memoirs of the Geological Society of Japan*, **32**, 431–455 (in Japanese with English abstract).

Ozawa, A. and Y. Suda (1978). Geological map of Japan 1:200 000, Hirosaki and Fukaura. Tsukuba: Geological Survey of Japan.

Ozawa, A. and Y. Suda (1980). Geological map of Japan 1:200 000, Akita and Oga. Tsukuba: Geological Survey of Japan.

Ozawa, A., H. Kano, T. Maruyama *et al.* (1981). Geology of the Taiheizan District, with geological sheet map at 1:50 000. Tsukuba: Geological Survey of Japan (in Japanese with English abstract).

Ozawa, A., Y. Ikebe, Y. Arakawa *et al.* (1982). Geology of the Kisakata District, with Geological sheet map at 1:50 000. Tsukuba: Geological Survey of Japan (in Japanese with English abstract).

Ozawa, A., N. Tsuchiya and K. Sumi (1983). Geology of the Nakahama District, with geological sheet map at 1:50 000. Tsukuba: Geological Survey of Japan (in Japanese with English abstract).

Ozawa, A., Y. Ikebe, J. Hirayama, Y. Awata and T. Takayasu (1984). Geology of the Noshiro District, with Geological sheet map at 1:50 000. Tsukuba: geological Survey of Japan (in Japanese with English abstract).

Ozawa, A., A. Kujiraoka and Y. Awata (1985). Geology of the Ugo-Hamada District, with geological sheet map at 1:50 000. Tsukuba: Geological Survey of Japan (in Japanese with English abstract).

Ozawa, A., T. Katahira and N. Tsuchiya (1986). Geology of the Kiyokawa District, with geological sheet map at 1:50 000. Tsukuba: Geological Survey of Japan (in Japanese with English abstract).

Ozawa, A., T. Hiroshima, M. Komazawa and Y. Suda (1988a). Geological map of Japan 1:200 000, Shinjo and Sakata. Tsukuba: Geological Survey of Japan.

Ozawa, A., T. Katahira, S. Nakano, N. Tsuchiya and Y. Awata (1988b). Geology of the Yashima District, with geological sheet map at 1:50 000. Tsukuba: Geological Survey of Japan (in Japanese with English abstract).

Ozawa, A., K. Mimura and T. Hiroshima (1993). Geological map of Japan 1:200 000, Aomori (2nd edn.). Tsukuba: Geological Survey of Japan.

Perry, F. V., G. A. Valentine, E. K. Desmarais and G. WoldeGabriel (2001). Probabilistic assessment of volcanic hazard to radioactive waste repositories in Japan: intersection by a dike from a nearby composite volcano. *Geology*, **29**, 255–258.

Pollitz, F. R. (1986). Pliocene change in Pacific plate motion. *Nature*, **320**, 738–741.

Sato, H. (1992). Late Cenozoic tectonic evolution of the central part of northern Honshu, Japan. *Bulletin of the Geological Survey of Japan*, **43**, 119–139 (in Japanese with English abstract).

Sato, H. (1994). The relationship between late Cenozoic tectonic events and stress field and basin development in northeast Japan. *Journal of Geophysical Research*, **99**(B11), 22 261–22 274.

Takahashi, M. (1997). The structure of high level crustal magmatic plumbing systems beneath the Quaternary arc volcanoes in Japanese Islands. *Bulletin of the Volcanological Society of Japan*, **42**, S175–S187 (in Japanese with English abstract).

Takahashi, M. (2000). *Island Arc, Magma and Tectonics*. Tokyo: University of Tokyo Press (in Japanese).

Tamura, Y., Y. Tatsumi, D. Zhao, Y. Kido and H. Shukuno (2002). Hot fingers in the mantle wedge: new insights into magma genesis in subduction zones. *Earth and Planetary Science Letters*, **197**, 105–116.

Tatsumi, Y. (1995). *Subduction Zone Magmatism – A Contribution to Whole Mantle Dynamics*. Tokyo: University of Tokyo Press (in Japanese).

Tsuchiya, N. (1981). Geology of the Osawa District, with geological sheet map at 1:50 000. Tsukuba: Geological Survey of Japan (in Japanese with English abstract).

Tsuchiya, N. (1988). Distribution and chemical composition of the Middle Miocene basaltic rocks in Akita–Yamagata oil fields of northeastern Japan. *Journal of the Geological Society of Japan*, **94**, 591–608 (in Japanese with English abstract).

Tsuchiya, N., A. Ozawa and Y. Ikebe (1984). Geology of the Tsuruoka District, with geological sheet map at 1:50 000. Tsukuba: Geological Survey of Japan (in Japanese with English abstract).

Tsushima, K. (1964). Geological map of Japan 1:200 000, Noheji. Tsukuba: Geological Survey of Japan.

Umeda, K., K. Asamori, A. Ninomiya, S. Kanazawa and T. Oikawa (2007). Multiple lines of evidence for crustal magma storage beneath the Mesozoic crystalline Iide Mountains, northeast Japan. *Journal of Geophysical Research*, **112**, B05207.

Uto, K. (1995). Volcanoes and age determination: now and future of K–Ar and $^{40}Ar/^{39}Ar$ dating. *Bulletin of the Volcanological Society of Japan*, **40**, S27–S46 (in Japanese with English abstract).

Yoshida, T., A. Ozawa, M. Katada and J. Nakai (1984). Geological Map of Japan 1:200 000, Morioka. Tsukuba: Geological Survey of Japan.

Yoshida, T., T. Ohguchi and T. Abe (1995). Structure and evolution of source area of Cenozoic volcanic rocks in northeast Honshu arc. *Memoirs of the Geological Society of Japan*, **44**, 263–308 (in Japanese with English abstract).

Yoshida, T., K. Aizawa, Y. Nagahashi *et al.* (1999). Geological history and formation of late Cenozoic calderas in the stage of island arc volcanism in northeast Honshu arc. *Earth Monthly*, extra issue, **27**, 123–129 (in Japanese).

13

Exploring long-term hazards using a Quaternary volcano database

S. H. Mahony, R. S. J. Sparks, L. J. Connor and C. B. Connor

Statistics play an increasing part in the role of forecasting the probabilities of where and when future hazards such as volcanic events may occur. Spatio-temporal models allow forecasts of where volcanoes are most likely to form over designated timescales in specific regions. Owing to the long performance periods required of geological repositories for radioactive wastes (see Chapman *et al.*, Chapter 1, this volume), this type of forecasting is especially relevant in repository siting projects in regions that are clearly prone to volcanism. This chapter looks at a key component of forecasting the spatio-temporal likelihood of volcanism in Japan where a site for a geological repository for high-level waste is being sought, using an open volunteer process that potentially opens up large parts of the country for consideration. The constraints on siting a repository are that it should have an extremely low probability of being directly impacted by future magma intrusion in the next 10 ka, with the probability of such events over a 100 ka timescale (possibly out to 1 Ma) being an important aspect of the safety assessment of potential sites.

The context of the work reported in this chapter is a case study of the Tohoku region of northeast Japan. Our aim is to develop a generic methodology to assist with the definition of spatio-temporal probabilities of future volcanism and associated magmatic disruption of a repository. The choice of this area for a case study is illustrative. There is no implication that the Nuclear Waste Management Organization of Japan (NUMO), which is responsible for the repository development program, is actively considering sites in this region.

It is unlikely that any nuclear facility would be located in an area with a high probability of future volcanism during its operational lifetime. However, geological repositories need to provide isolation for tens of thousands of years and, over such timescales, the possibility of future volcanism in areas that either have no previous record or are close to young volcanoes needs to be evaluated. This is true of any volcanically active country or region of the world. This study considers the location of future volcanism. We use the Gaussian kernel density estimation method (Connor and Hill, 1995) to generate spatial probability distributions of volcano locations in order to assess site-specific hazards of future volcanism. The method is based on the analysis of data related to the spatio-temporal distributions of volcanoes. The results of such studies, however, depend on how three main issues are dealt with, namely: how to define the meaning, in terms of past style of activity, of the over-simplified "dots on the map" used to depict volcanoes in typical databases; how to treat the problem that volcanoes have finite lifetimes with the density of dots potentially being a function of the time interval chosen for analysis; and how to account for the wide range of volcano sizes and related hazardous phenomena which may have different hazard implications.

The first issue is defining a mapped volcano for the purpose of investigating the spatial and temporal density of volcanism. For this work, our main concern is with discrete volcanic events that may occur in the future. We define an event as an episode of volcanism that forms a distinct center, which can be characterized by its spatial position on a map. The volcanic event has a lifetime and past events may also be recognized as extinct, dormant or active. The identification of individual past volcanic events to be depicted as a "dot on a map" requires the interpretation of geological data. Volcano databases may make over-simplified representations of spatially and temporally complex edifices and structures as a single, centralized "dot" or location. However, some volcanic edifices that are closely located in space may be considered as either different stages of the same volcano or alternatively as independent volcanoes. Alternative identifications of volcanic events may result from alternative interpretations of the same geological data and thus result in different maps of volcano location. Analysis of these alternative maps will result in different probability distributions and, consequently, variations in estimates of hazard, here defined as the probability of a new volcano occurring at a specific location over a specified period of time. An approach to this problem is to use alternative data sets to assess the robustness and related uncertainties in hazard. All alternatives are considered valid unless compelling scientific reasons can be advanced to eliminate them as untenable. The Japanese database includes data relating to named volcanoes and volcanic edifices related to them. Each edifice is considered here to be a sub-part of a volcano, so spatially it could be suggested that edifices may not be independent events when compared to volcanoes. However, when time becomes a factor they could be treated as independent events.

The second issue relates to time. Volcanoes are born and become extinct. As a consequence, a map of volcano locations for analysis depends first on the time interval chosen, and second on whether it is possible to identify a volcano as either extinct or dormant. Indeed, robust criteria for designating a volcano as dormant or extinct have not been established. The quality and quantity of age and geological data are highly variable but critical to a credible hazard assessment. Age determinations of younger volcanoes are more likely to give a better representation of their true age since more recent material allows for more accurate dating. It is necessary to consider the average lifetime of a volcano, the timescale over which the volcanic hazard is to be assessed and the errors in the volcanic-event age data. Analysis of age data can establish if the spatial density of volcanoes is stationary or non-stationary.

The third issue to consider is the geological variability of volcanic events and their influence on hazard. The style and size of a volcanic event will affect the area of influence within which volcanic hazards will be expected. Even within a high-quality database, as is the case for Japan, volcanic events may vary from a small monogenetic cinder cone to a large caldera volcano with a long and complex history. There is a wide variety of hazardous volcanic phenomena, which may have different implications for the hazards and associated risk at a repository. Different kinds of volcanoes may thus vary in their hazard potential.

Here we consider alternative volcano definitions for volcanoes in the Tohoku region of northeast Japan. Our source of data is the Catalog of Quaternary Volcanoes in Japan (Committee for Catalog of Quaternary Volcanoes in Japan, 1999), which provides data on volcanoes that have been active during the last 2.5 Ma (note: because the database extends to 2.5 Ma it is by some definitions a Plio-Quaternary database; here we use Quaternary to include the last 2.5 Ma). Where any such database is applied for a specific purpose, it requires interpretation. This database contains both temporal and spatial information, makes variable assumptions on how an individual volcano is defined and located, and may include biases caused, for example, by artifacts of geological preservation potential.

The internal inconsistencies, which include different definitions of volcanoes and interpretations of volcanic landscapes, have led us to form alternative data sets. We investigate alternative volcano data sets to generate two-dimensional (2D) probability surfaces that are compared at 14 sample locations, selected to represent varying tectonic and volcanic environments within the case-study region. Alternative data sets would be expected to generate a range of probability estimates for future magmatic disruption, with a different consequent hazard potential for each group of volcano types considered. Hazard is expressed here as the probability of new volcano formation occurring in a specified location in the next 1 Ma, with "location" being defined as a 5 $\times$ 5 km area within a grid covering the case-study area. It is assumed that if a volcano does form in the specified area then the chance of disruption of a hypothetical geological repository at that location by a magmatic intrusion or related igneous phenomenon is high. However, we do not investigate either the probability of disruption or the nature of the disruption (see Menand *et al.*, Chapter 17, this volume; Valentine and Perry, Chapter 19, this volume).

The spatial distribution of Japanese Quaternary volcanoes forms the basis of empirical and probabilistic models (Kondo, Chapter 12, this volume; Jaquet and Lantuéjoul, Chapter 15, this volume). We show how different definitions of a volcano produce a range of probabilities for a new volcano to form at a given location. Assessing the likelihood of future volcanoes appearing in areas where there has been no volcanism during the Quaternary is a key issue in identifying suitable sites for a geological repository. This study provides an approach to estimating uncertainties in hazard and developing confidence in low hazard estimated at specific sites. Variations in calculated hazard, resulting from using different data sets, constitutes an aspect of epistemic uncertainty. If the wide range of alternative data sets produces hazard maps and probability values that are similar to a reference data set, then it could be concluded that alternative volcano definitions do not need to be investigated. This might also suggest that data set completeness is not a major cause for concern and that we could be confident in forecasting the likelihood of future volcanoes forming, despite lacking a complete knowledge of all existing volcanoes, and notwithstanding the difficulties and ambiguities of defining volcanoes. In addition, if the hazard at a site is very low for all alternative data sets, then this approach can give confidence in the hazard assessment and provide evidence that the site has promise for a nuclear facility from the perspective of future volcanism.

13.1 Volcanism in the Tohoku arc

The Tohoku volcanic arc is a mature double volcanic arc with a backarc marginal sea (Figure 13.1a; Tamura *et al.*, Chapter 7, this volume). Volcanism in this region can be characterized by several distinct evolutionary stages from 13 Ma to the present (Kondo *et al.*, 1998; Kondo, Chapter 12, this volume). The last two stages in the Quaternary involved a regional caldera-forming episode (8–1.7 Ma) and development of the current volcanic arc (1.7–0 Ma; Yoshida *et al.*, 1995). Umeda *et al.* (1999) used eruptive volumes to suggest that from 2.0 to 1.2 Ma there were large-scale silicic eruptions related to calderas, with the development of stratovolcanoes along the volcanic front from 1.2 to 0.5 Ma. There is a suggestion that this change in style of volcanism was related to a change to a compressive stress regime (Umeda *et al.*, 1999). Kondo *et al.* (1998) identified localization of volcanic centers during the last 0.5 Ma and weak north–south trends of volcanism. Tamura *et al.* (2002) divided this area into ten clusters of volcanoes along and behind the arc.

The Sengan volcano cluster, a prominent area of volcanism in Tohoku (Figure 13.1b), is used to illustrate some of the problems of defining volcanoes and interpreting the results of probabilistic hazard assessments. The volcanoes in this region are mostly polygenetic stratovolcanoes forming along a north–south trend, parallel to the arc. There are two monogenetic volcanoes towards the west, behind the volcanic front.

13.2 Nature of the database

Hazard analyses were made in the Tohoku region using data from the Catalog of Quaternary Volcanoes of Japan. This database is comprehensive, with information on location, age, volumes, rock types, dominant eruptive styles, morphology and miscellaneous other features, such as craters, intrusions, hot springs and fissures. Hone *et al.* (2007) discussed problems with dividing volcanoes in this database into constituent "edifices" and some minor inconsistencies of definition of a volcano. The categorization of volcanic features into volcanoes, and edifices belonging to a volcano, were based on the interpretation of independent local experts; in particular, a volcano was defined as a single coherent magmatic system (S. Hayashi and K. Umeda, personal communication) with edifices being part of that system. This process of human interpretation of volcanic landforms as objects for classification potentially introduces subjectivity and inconsistency, due to its qualitative nature and the involvement of many committee members (> 50) in the compilation of the database. Differences and inconsistencies regarding the exact location of the volcanoes and edifices in the database have been recognized in 7 of the 59 volcanoes in the Tohoku case-study area (Hone *et al.*, 2007). Otherwise the Japanese volcano database is systematic and consistent.

There are distinct "types" of volcanic edifices: steep- and shallow-sided stratovolcanoes, lava cones, lava flows, lava domes, maar deposits, collapse caldera, pyroclastic flow plateaus, volcanic necks and alluvial fans consisting of debris- and mud-flow deposits. However, the edifices in the database are not subdivided into different types; they are simply

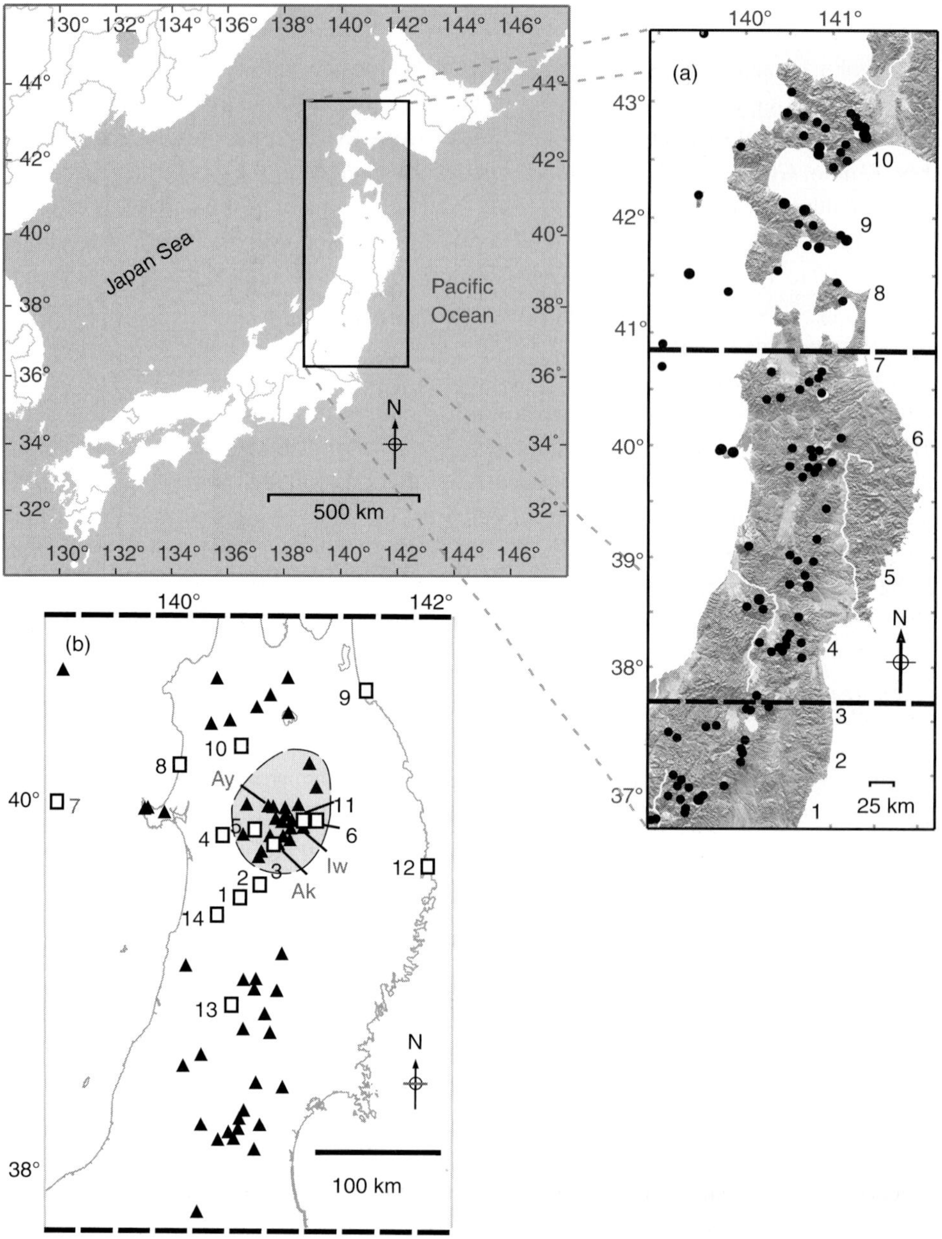

Fig. 13.1 (a) Map of the Tohoku region of Japan, showing the locations of the Quaternary volcanoes and identifying the ten volcano clusters described by Tamura *et al.* (2002). (b) We focus on volcanoes (triangles) in the Tohoku region. The Sengan cluster (gray area) has the highest volcano density of any of the ten clusters, with three young active volcanoes: Iwate (Iw), Akita-Yakeyama (Ay), Akita-Komagatake (Ak). The 14 test locations are squares labeled with numbers.

Table 13.1. Alternative data sets and their associated probability estimates calculation using test location 3.

ID	Description	N	BW	$\hat{\lambda}_t$	$^1P_{mean}$	$^2P_{log-mean}$	$^3P_{max}$	$^4P_{range}$
Volcanoes								
DS 1	All	59	22	29.5	0.046	0.032	0.170	0.160
DS 3	< 0.25 Ma	25	NA	100.0	0.120	0.074	0.630	0.600
DS 4	< 1 Ma	44	28	44.0	0.052	0.038	0.220	0.210
DS 7a	Explosive	24	33	12.0	0.016	0.010	0.088	0.088
DS 7b	Extrusive	35	12	17.5	0.030	0.021	0.086	0.086
DS 7	Expl. + eff.	59	NA	NA	0.046	0.031	0.170	0.170
DS 8a	Spacing (< 3 km)	57	22	28.5	0.044	0.031	0.170	0.160
DS 8b	Spacing (< 5 km)	53	23	26.5	0.037	0.025	0.170	0.160
DS 8c	Spacing (< 7 km)	44	22	22.0	0.019	0.014	0.061	0.061
DS 8d	Spacing (< 10 km)	39	21	19.5	0.011	0.009	0.022	0.022
Edifices								
DS 2	All	129	18	64.5	0.120	0.079	0.370	0.370
DS 5	< 0.25 Ma	52	NA	208	0.220	0.160	0.730	0.730
DS 6	< 1 Ma	101	20	101	0.160	0.110	0.480	0.480

ID: alternative data sets derived from database. (Committee for Catalog of Quaternary Volcanoes in Japan, 1999).
BW: estimated threshold bandwidth in kilometers.
N: number of volcanoes in alternative data set.
$\hat{\lambda}_t$: estimated recurrence rate (volcanoes Ma^{-1}).
[1]Arithmetic mean, [2]log of mean, [3]maximum and [4]range of probabilities.

classed as volcanoes; each volcano consists of one or more edifices. For our basic derivative data set, we define an edifice simply as "a place where magma reached the surface." Not all the edifice types in the database satisfy this definition (e.g. debris avalanches) so edifices were removed from our derivative data set. In the case of lava flows, the source vent represents the location that is included in our analyses. This type of censoring only affected a few data points, namely one volcano (which comprised only one edifice) and eight edifices from different volcanoes.

Clearly, there are many ambiguities and uncertainties in volcano databases with respect to applications to long-term probabilistic hazard assessment. We cannot be certain that every volcano is recorded in the database, due to preservation and under-recording problems. Different classifications of volcanoes can lead to different numbers of events, and hence different hazard rates. One solution to these problems is to investigate how alternative data sets derived from the database influence the hazard (Hill *et al.*, Chapter 25, this volume). This approach can be regarded as the use of multiple alternative hypotheses to assess hazard. The principle idea is that any hazard assessment should include viable alternative views and interpretations of data as part of the evaluation of uncertainty and robustness. In our development of alternative data sets, information from the database regarding the ages of eruptions, erupted volume for each volcano or edifice, location and eruption style was considered. A total of 13 alternative data sets were derived according to age, dominant eruptive style and distance to the nearest volcano (Table 13.1). These 13 data sets, described in more detail in the following, are used to investigate the dependence of estimated hazard on volcano definition.

13.3 Test locations

The organisation NUMO has defined a 15 km exclusion zone around each Quaternary volcano listed in the database within which a geological repository cannot be sited (NUMO, 2004). Fourteen test locations were selected within the study area (Figure 13.1b) where estimates are made of the probability of new volcanic formation. Test location 3 is within the Sengan cluster and is where the highest hazard is expected. Location 3 is used as a benchmark to assess the reduction in hazard in areas outside NUMO's exclusion zone. Other locations are selected for their particular tectonic setting. For example, locations 1 and 2 are south of the Sengan cluster, in the backarc, and in a gap region between clusters, far from Quaternary volcanoes. Location 6 is on the boundary of a 15 km exclusion zone and in the forearc, an area where volcanic hazard is expected to be extremely low. Comparison of probability estimates at these 14 locations provides a sense of the variation in hazard estimates across the region, with tectonic setting and choice of data set.

13.4 Volcano definition and alternative data sets

Defining a volcano or a volcanic event is not straightforward, but is a fundamental concern in the creation or use of any volcano database. When using these data in hazard assessments, it is particularly important to use volcano definitions that are robust and consistent. Within the database, volcanoes are commonly defined by their morphological descriptors, such as stratovolcano, maar, lava dome, caldera or cinder cone. Young volcanoes are easily mapped and cataloged; for example, the volcano location is defined as the summit or main crater. However, difficulties arise when cataloging older, eroded volcanic systems. In these cases, a more detailed study of near-vent breccias or radial dikes is required to infer the appropriate location for the volcano (Connor and Hill, 1995). In the following discussion "spatial intensity" is taken as the number of volcanoes per unit area, "spatial density" is the normalized intensity appropriate for probability estimates. Stationary is a term used to describe a series of "events" (here, volcano formation) occurring at a constant average rate through time. Temporal clustering implies a non-stationary series of volcano formations, where "episodes" or trends in volcano formation are observed. Spatio-temporal recurrence rate is a term used to describe the number of volcanic events occurring per unit area per unit time. Here the hazard is defined as the probability of a new volcano forming in a 25 km^2 area during a fixed period of time, which is chosen as 1 Ma.

A major problem with using a volcano database to generate probabilistic hazard maps is that a province of volcanoes represents a dynamic system with both spatial and temporal attributes. In principle each polygenetic volcano will have a finite life with a birth age and an extinction age. If the time period is short (e.g. the Holocene), compared to the typical lifetime of a volcano, then the spatial intensity may be adequately described as stationary, with no temporal clustering. However, for periods that are either comparable or longer than typical volcano lifetimes, the spatial intensity of volcanoes will tend to increase as the time window increases and older volcanoes are added to the map; even if the system is, in fact, stationary. The Quaternary period is a convenient but arbitrary time window. It is, however, longer than the lifetime of many polygenetic volcanoes, so temporal variations in the rate of formation of new volcanoes may influence the spatial intensity. A further complication is that volcano lifetimes may be highly variable: a monogenetic volcano may be formed in a single eruption lasting a few weeks, while a complex polygenetic stratovolcano may have a lifetime of over one million years. Older volcanoes may also be eroded or buried and potentially under-represented in a data set. The Quaternary catalog is likely to include volcanoes that are, in fact, dormant or even extinct. These extinct volcanoes may or may not provide relevant information about the potential distribution of volcanoes forming in the future. These factors will complicate any hazard analysis undertaken.

The Sengan cluster is used to illustrate the problems of defining volcano distributions in space and time. Using the database, the Sengan volcano cluster can be defined as having 50 edifices and 20 volcanoes (Figure 13.1b). The cluster, however, becomes poorly defined and less convincing as a cluster if only the three volcanoes that are < 0.25 Ma in age and defined as "active" (i.e. Iwate, Akita-Yakeyama and Akita-Komagatake volcanoes) are selected (Figure 13.1b). The present-day Sengan cluster is the product of several stages of evolution (Kondo, Chapter 12, this volume). Early, large-scale caldera-forming eruptions led to widespread silicic ignimbrites at 1–2 Ma. This volcanism was followed by a period (1–0.5 Ma) of dominantly andesite volcanism with clustered domes and lavas. Finally the volcanism became increasingly mafic with construction of the basaltic to andesitic stratovolcanoes of Iwate, Akita-Yakeyama and Akita-Komagatake. During this compositional evolution, volcanism focused increasingly towards the edge of the Sengan cluster.

The rate of volcano formation apparently decreases back in time in the Tohoku area (Figure 13.2). There are several possible explanations for this apparent change in rate. One is that older volcano edifices are not preserved and may be lumped together, while younger volcanoes are easier to identify and distinguish from one another. Another explanation is that, as illustrated for the Sengan cluster, a systematic evolution of volcanic systems is a feature of the whole northern Tohoku region. Large-scale ignimbrites of similar ages (1–2 Ma) are observed in volcano clusters to the north (Lake Towada) and to the south (Narugo/Oni-kobe/Mukaimachi) of Sengan. Following ignimbrite formation is a temporal compositional evolution to andesitic stratovolcanoes, followed by late Holocene volcanoes with volumetrically significant basalt products. Thus the data shown in Figure 13.2 could reflect the fact that there were smaller numbers of larger volcanoes in the early Quaternary.

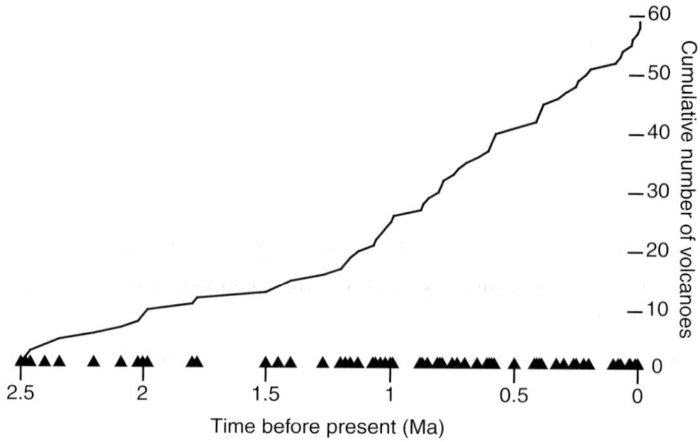

Fig. 13.2 Cumulative frequency of the formation of new volcanoes in the Tohoku region during the last 2.5 Ma. Triangles indicate the time of formation of individual volcanoes.

There may also be biases in the age data: (i) the oldest rocks of a volcano are the most likely to have been buried or eroded; (ii) some of the volcanoes only have one age date and (iii) the oldest age recorded in the Japanese database is likely a minimum. Changes in rate probably reflect all of these dating biases, and raise questions about the robustness of assuming a stationary system and the biases associated with the under-recording of events. A key issue is what time window should be used in the hazard analysis.

Analyses of volcano spatial intensity over a specific time period allows estimation of the spatio-temporal recurrence rate (Connor and Hill, 1995). This parameter can be used to make probabilistic forecasts (Martin *et al.*, 2004). There are, however, uncertainties when deciding what time period to choose for the analysis. Uncertainty in the analysis increases as the time period decreases because fewer data become available for use in the analysis. In contrast, although increasing the time period reduces the uncertainties related to the number of data used, it also increases uncertainties related to the stationarity of the system, and increases the chance that long-extinct volcanoes will be included in the analysis. These uncertainties are not easy to quantify, as they depend on interpretation of the geological record. Here, several alternative data sets are used to explore issues of temporal variations in spatial density and hazard. The two primary data sets (DS) used are 59 Quaternary volcanoes taken directly from the database without any modification (DS 1), and 129 volcanic edifices (DS 2) (Table 13.1). These data sets contain all of the reported Quaternary volcanoes and are the benchmarks used for comparisons with the alternative data sets. Four alternative data sets include age-dependent groupings of volcanoes and edifices: volcanoes formed < 0.25 Ma (DS 3), volcanoes formed < 1 Ma (DS 4), volcanic edifices formed < 0.25 Ma (DS 5) and volcanic edifices formed < 01 Ma (DS 6).

Another problem for hazard analyses is that, within the database, volcanoes vary greatly in their scale and range of hazardous volcanic phenomena. For example, a large caldera dominated by silicic explosive volcanism poses different hazards than a stratovolcano dominated by basaltic or andesitic lava eruptions. Such variations are not taken into account by representing volcanoes as simple "dots on a map." Therefore, it may be useful to distinguish between different kinds of volcanoes and undertake independent hazard analyses using each type. Volcanoes have many different attributes, such as chemical composition, volume, types of activity and morphology, so there are potentially many different ways of splitting a large group of volcanoes into different smaller subgroups. A systematic approach to classification is the use of cladistics (Hone *et al.*, 2007). Cladistics is known in biology for its use in creating the evolutionary "tree of life." Cladistics is a method for analyzing evolutionary relationships between groups, or clusters, focusing on evolutionary history or common ancestry rather than on morphology or chronology. The cladistical approach used by Hone *et al.* (2007) compares all of the diverse features of volcanoes in the database (in cladistics these features are called taxons) and groups the volcanoes, parsimoniously, into distinct groupings, called clades. This type of analysis reduces human bias in deciding how volcanoes should be categorized and uses all the different parameters that characterize each volcano. It is beyond the scope of this study to assess whether different clades pose different hazards. Nevertheless, one way to explore estimated hazard from two subgroups of volcanoes is by splitting the larger Tohoku data set into two smaller subsets, dominantly explosive volcanoes (DS 7a) and dominantly extrusive volcanoes (DS 7b), and compare hazards resulting from the use of these two subsets (Table 13.1; Figure 13.3).

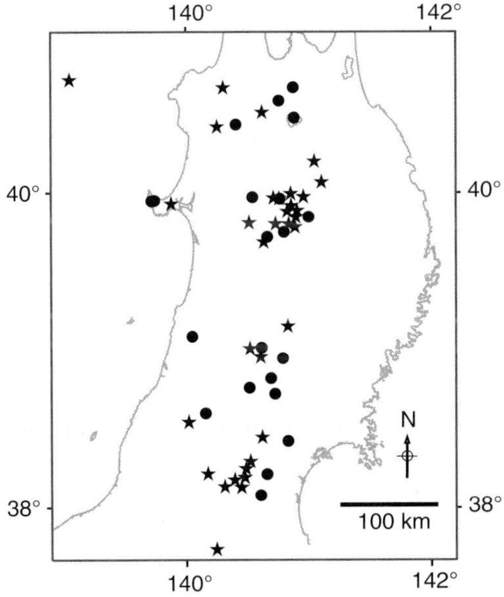

Fig. 13.3 Map showing the locations of explosive volcanoes (DS 7a) as stars, and the locations of extrusive volcanoes (DS 7b) as circles, within the Tohoku region.

Alternative volcano definitions can also lead to quite different spatial distributions. During fieldwork in the Sengan volcano cluster, it became clear that defining events is non-trivial since edifices may represent a variety of different geological phenomena. For example, Hachimantai volcano in the Sengan cluster is classified as an active volcano (i.e. has magma eruptions). However, its last magmatic product is a Pleistocene lava ($\sim$ 0.6 Ma; Oba and Umeda, 1999), indicating that the volcano has long been dormant and is probably even extinct. Reported Holocene activity consisted of the formation of phreatic explosion craters associated with landslide-related fissures. These features may reflect phenomena related to the Sengan geothermal field, but are unrelated to either magmatic unrest or the older Hachimantai volcano. Since Hachimantai is classified as active within the database, it has as much statistical weight in probabilistic analyses as a currently active volcano, such as Iwate. In the database, Iwate is noted as "Iwate and Amihari", where Amihari is the old volcano and Iwate is the active stratovolcano, and both are are assumed to have the same magma source. However, although Iwate is physically a distinct peak, Amihari and several other volcanic peaks form a volcanic complex, where the volcanic edifices have grown over one another. The separately named volcanic edifices might be treated either as different volcanoes or as overlapping parts of the same volcano of a different age. Similarly, other geological information is treated very differently in the database. Within the Sengan cluster, many of the volcanoes of the 0.5–1 Ma period are andesite domes and lava complexes, with the mapped location of each "volcano" identified as the location of highest elevation. In contrast, Kuju volcano in Kyushu consists of over 20 different overlapping dome complexes that cover a similar area to Sengan. However, Kuju is classified as a single volcano within the database. The extent of the Sengan cluster would be much diminished if all the andesite domes and lavas of 1–0.5 Ma were classified as a single volcano.

Several alternative data sets (DS 8a–d) combine volcanoes whose nearest neighbors are within a specified distance. In this study, 3, 5, 7 and 10 km near-neighbor distances are used, respectively. Each volcanic event in these data sets has a single location based on the mean location of its near-neighbor groups. For example, DS 8a groups volcanoes with near neighbors $<$ 3 km apart as single volcanic events. Using DS 8a–d, we assess how hazard estimates vary across the map region (with spatial-density contour maps), and in more detail (by calculating the probability of a new volcano forming) at the 14 test locations.

13.5 Calculating spatial density and hazard

The analysis of the alternative data sets uses two techniques: namely a forward modeling technique; and the calculation of the probability of a new volcano forming at each of the 14 test locations, each of which has an area of 5 $\times$ 5 km, within the next 1 Ma.

13.5.1 Technique 1: forward modeling

Martin *et al.* (2004) used Bayesian inference to combine several data sets of geophysical data and then tested the Gaussian and Cauchy kernels on the one agglomerative data set.

Connor and Hill (1995) modeled the distribution and timing of basaltic volcanism using three non-homogeneous methods: spatio-temporal nearest neighbor, kernel and nearest-neighbor kernel, on data from the Yucca Mountain region of Nevada. These previous studies assumed that the original data are complete and correct, allowing a range of statistical models to be tested to see which gives the best estimate of the observed spatial intensity of volcanoes. Here we construe that our original data set may not be complete and correct, so we use one single analysis technique to allow comparison between the alternative data sets to explore how different data sets (and so how incomplete data sets) might affect hazard analyses.

Kernel density methods have been used by many workers to estimate spatial density of volcanoes (e.g. Connor and Hill, 1995; Lutz and Gutmann, 1995; Martin *et al.*, 2004; Weller *et al.*, 2006; Connor and Connor, Chapter 14, this volume). The Gaussian kernel is used in this study because it is a straightforward statistical method of quantifying volcano distribution as a probabilistic surface. The Gaussian distribution also arises in problems of heat and mass transfer, which are similar diffusion processes to those expected in volcanic systems. It has the properties of smoothing density declining radially from a maximum, which in this case centers on each volcano, with tails that approach zero asymptotically (Weller *et al.*, 2006). The sum of these individual functions around each volcano forms a 2D spatial density surface, which in this case was used to produce maps showing the spatial density for each alternative data set.

The rate of change in spatial density with distance from volcanoes included in the data set depends on kernel bandwidth, a distance equal to one-half of the standard deviation of the Gaussian kernel function. Kernel bandwidth selection is a major issue in spatial density estimation for natural hazards. A small bandwidth implies strong clustering with a high probability that future volcanoes will form close to existing ones. A large bandwidth means that spatial density is more uniform over a larger area. These different bandwidths imply very different hazards for specific sites. Thus, bandwidth selection is a major area of research with techniques ranging from subjective bandwidth selection, as described by Wand and Jones (1995), to more sophisticated bandwidth selector algorithms, such as those used by Connor and Connor (Chapter 14, this volume).

Our approach is to perform analyses using volcanoes older than 0.25 Ma in each data set to evaluate how well the distribution of younger volcanoes (formed < 0.25 Ma) is predicted. First, a spatial density map is made using volcanoes > 0.25 Ma and a constant bandwidth of 15 km. A 15 km bandwidth was used because this distance was selected by NUMO as the radius of the exclusion-zone around each volcano. Thus the approach can also help evaluate the choice of 15 km as an exclusion zone criterion. Second, a plot is made of the volcanoes < 0.25 Ma on these spatial density maps. An arbitrary spatial density contour was selected at $1 \times 10^{-10}\,\text{km}^{-2}$ and young volcanoes that fall outside this contour are deemed to be poorly predicted. Third, the bandwidth is adjusted and spatial density recalculated using the old volcanoes until all young volcanoes fall within the $1 \times 10^{-10}\,\text{km}^{-2}$ contour. Hereafter, this adjusted bandwidth is simply referred to as the "threshold bandwidth." This approach gives a systematic way to select a bandwidth for each data set, and also allows comparison of the alternative data sets. The value of $1 \times 10^{-10}\,\text{km}^{-2}$ is chosen, as probabilities of

$1 \times 10^{-8}\,\text{km}^{-2}$ to $1 \times 10^{-6}\,\text{km}^{-2}$ are considered high in previous studies of volcanic hazard assessments for nuclear facilities (e.g. Martin *et al.*, 2004; Weller *et al.*, 2006). The choice of a constant threshold also allows us to observe how strongly the volcanoes are clustered as measured by the "minimum" bandwidth required to enclose all the younger volcanoes within the threshold contour. The threshold contour value of $1 \times 10^{-10}\,\text{km}^{-2}$ can be further considered a posteriori. Variation in the threshold bandwidth among the alternative data sets is a guide to the relative degree of confidence in the estimated spatial density, and ultimately hazard.

13.5.2 Technique 2: location-specific probability estimation

It is also useful to calculate the probability of volcano formation at specific sites using the alternative data sets and a range of kernel bandwidths. Volcanic hazard (the probability of a future volcano in a particular place over a specified period of time) is calculated from four parameters. Two of these are determined by the nature of the nuclear facility: the area of the facility (A) and performance period of the facility (Δt). The third parameter is the spatial density, ($\hat{\lambda}_s$), estimated from volcano distribution using the Gaussian kernel density estimation techniques described in the previous section. The fourth parameter is the temporal recurrence rate of volcano formation ($\hat{\lambda}_t$), which was estimated for each alternative data set using a maximum-likelihood technique. Treating volcanism as a Poisson process, homogeneous in time and non-homogeneous in space (Jaquet and Lantuéjoul, Chapter 15,

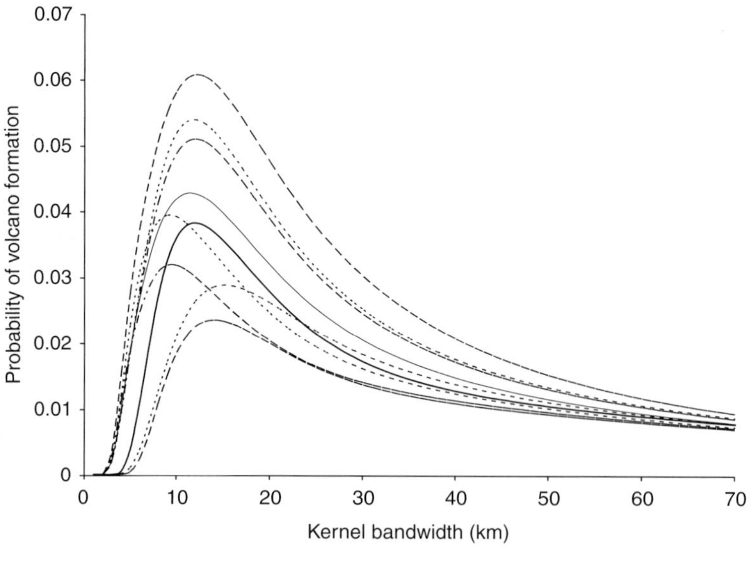

Fig. 13.4 This plot demonstrates that each data set (individual curves) tested at a single location over a range of bandwidths 1–70 km will produce a maximum peak in probability at a certain bandwidth value. This example is based on calculations using location 5 (Figure 13.1b).

this volume), the probability of N new volcanoes forming during $\Delta t = 1$ Ma within an area $A = 25\,\text{km}^2$ is:

$$P\{N \geq 1|\Delta t, A\} = 1 - \exp\left[-\hat{\lambda}_s A \hat{\lambda}_t \Delta t\right] \tag{13.1}$$

Probability is calculated for each of the 14 test locations and each alternative data set, using bandwidths at 1 km intervals over the range 1–70 km. This method generates a large number of realizations for each location, and the resulting distribution of probability values can then be analyzed. Probability estimates among the 14 locations are compared using the range (P_{range}), the arithmetic mean (P_{mean}), the logarithmic mean ($P_{\text{log-mean}}$) and maximum (P_{max}) probability values. The arithmetic mean was preferred over the logarithmic mean, as this approach weights the mean value towards higher probability values. In hazards assessments the precautionary principle favors statistical analyses that are weighted towards the higher probability realizations. As bandwidth increases using a specific alternative data set the probability values go through a maximum (Figure 13.4). Based on the precautionary principle we used this maximum value to define the maximum hazard. A sample of the results for the alternative data sets is presented in Table 13.1 for location 3, which generally produces the highest probabilities relative to the other localities.

13.6 Effects of alternative data sets on hazard estimates

There is a marked variation in the tendency of volcanoes or edifices in different alternative data sets to cluster. The threshold bandwidth for the complete set of volcanoes (DS 1) is 22 km (Figure 13.5a). In contrast, explosive volcanoes (DS 7a, Figure 13.5b) have the largest threshold bandwidth (33 km), forming a broad spatial density map. These explosive volcanoes exhibit only a weak tendency to cluster. Extrusive volcanoes have a much shorter threshold bandwidth (12 km), suggesting a much stronger tendency to cluster. Note that although the large estimated bandwidth for explosive volcanoes might be attributed to the small sample size of this data set ($N = 24$), the short bandwidth found for extrusive volcanoes cannot be attributed to the size of the extrusive volcano data set ($N = 35$). The higher degree of clustering observed for extrusive volcanoes appears to be an important feature of the distribution of this type of volcano, and suggests that, in this case, volcano type influences spatial density.

Changes in threshold bandwidth, and hence clustering, are less marked when comparing volcanoes (DS 1) and edifices (DS 2), or when comparing subsets using volcanoes of different ages (e.g. compare DS 2 and DS 6 in Table 13.1). Instead, marked differences in probability at specific sites result from differences in temporal recurrence rates among these smaller subsets. For data sets of edifices, the larger numbers of events translates to higher probabilities, a clear indication that event definition affects hazard estimates. As suggested by Figure 13.2, the timespan of considered events also influences estimated hazard. Essentially, hazard increases as the timespan of considered events decreases, as exemplified by

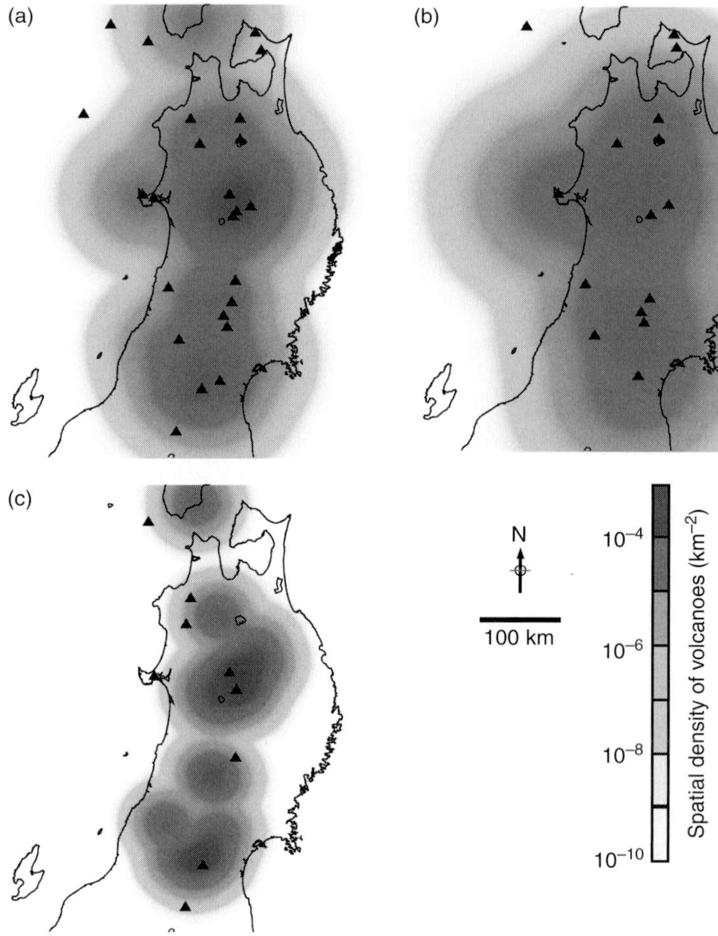

Fig. 13.5 Forward modeling results for (a) the original complete data set (DS 1) with a bandwidth of 22 km, (b) explosive volcanoes (DS 7a) with a bandwidth of 33 km and (c) extrusive volcanoes (DS 7b) with a bandwidth of 12 km. Volcanoes $>$ 0.25 Ma were used to estimate spatial density, and volcanoes $<$ 0.25 Ma (triangles) were used to assess model fit and adjust the threshold bandwidth for each data set.

the results for location 3 (Table 13.1). This change in hazard with time is attributed to under-recording of older volcanic events and, possibly, a regional change in rates of volcanism prior to 1 Ma.

Similarly, grouping volcanoes that are located in close proximity to one another, which may represent single volcanic events, does not appear to change spatial density significantly. Estimated threshold bandwidths for DS 8a–d (Table 13.1) are essentially identical and also identical to the threshold bandwidth estimated using all volcanoes (DS 1). Threshold bandwidths are also similar, or only slightly smaller, for the edifice data sets (DS 2,

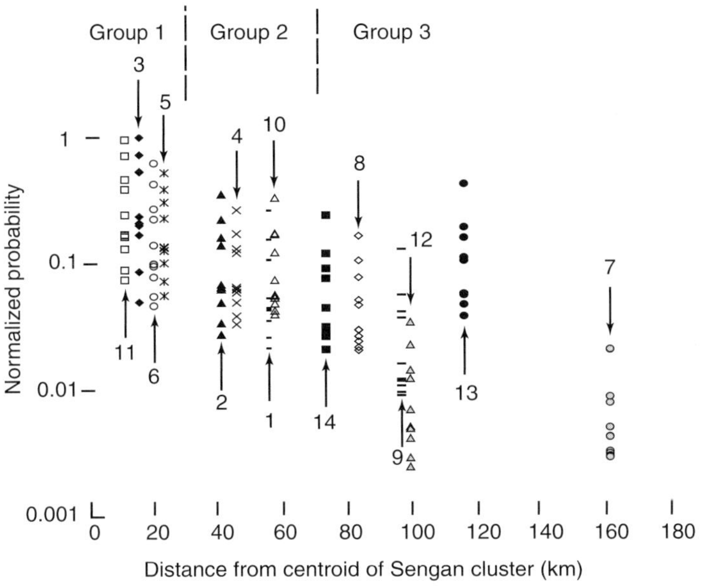

Fig. 13.6 Relative probabilities calculated for each site, from all the alternative data sets, each over a range of bandwidths with integer values from 1–70 km. The calculations are shown versus the distance of the test location from the center of the Sengan cluster. The probabilities are normalized relative to the maximum probability, found at location 3. Each site has several symbols of the same type attached to it, each of these symbols represents the results from analysis of an alternative data set. Groups 1–3 are defined based on changes in probability with distance from the center of the Segan cluster.

DS 6). Apparently, distinguishing these alternative data sets has less significance to spatial density estimates than does distinguishing volcanoes by type of activity. But again, as with the contrast between edifice and volcano data sets, as volcanoes are grouped the number of events (N) declines and probability decreases.

The 14 test locations can be roughly divided into three groups, based on their probability ranges (due to varying kernel bandwidth) and distance from the center of the Sengan cluster (Figure 13.6). The variation in probability is most easily visualized relative to the maximum probability found for location 3 (located within the Sengan cluster). The test locations < 30 km from the Sengan cluster show the highest hazard, decreasing with distance from the volcano cluster (i.e. the calculated interquartile range, median, maximum, logarithmic mean and arithmetic mean are all highest for these localities). The hazards at this distance are only slightly less (55–85%) than the hazard at location 3. Locations that are 30–70 km from the center of the Sengan cluster all have similar probabilities and these do not change significantly with distance from the Sengan cluster (Figure 13.6). The hazard at these probabilities is about 14–30% of the hazard at location 3. The third group comprises those localities > 70 km from the center of the Sengan cluster, with hazards $< 1-10\%$ of the hazard at location 3. However, location 13 is an exception, with elevated hazard

$\sim$ 110 km from the center of the Sengan cluster (30% compared to location 3). This locality is near cluster 4 (Figure 13.1) along the volcanic arc, and increased probabilities at this offset distance reflect the wavelength of volcano clustering in this part of the arc.

Recall that in technique 1, a threshold bandwidth is chosen such that all young volcanoes fall in areas of spatial density $\geq 1 \times 10^{-10}\,\text{km}^{-2}$, while in technique 2 the maximum hazard is identified for 1 km bandwidth intervals between 1–70 km. These maximum probabilities identified using technique 2 at specific locations (Figure 13.6) approximate equivalent spatial densities. These range from about $1 \times 10^{-6.25}\,\text{km}^{-2}$ at location 3 to $1 \times 10^{-10.8}\,\text{km}^{-2}$ at location 7. At 30 km distances from the center of the Sengan cluster the spatial density is $\sim 1 \times 10^{-8}\,\text{km}^{-2}$. This supports the view that $1 \times 10^{-10}\,\text{km}^{-2}$ is a conservative threshold.

A graphical summary of the probability distributions calculated for all data sets at locations 1, 2, 3 and 6 is presented in Figure 13.7a–d. Note that the four plots have similar forms. For example, the volcano edifice data sets (DS 5–6) give the highest probabilities for all of these localities. Given the very different positions of all these localities, with locations 3 and 6 near or in the Sengan cluster and locations 1 and 2 in the backarc (Figure 13.7), overall it appears that event definition has the strongest affect on recurrence rate, and a secondary effect on spatial density for this particular region.

Nevertheless, the influence of event definition on spatial density is apparent in specific cases. Location 3 is 15 km from the center of the Sengan cluster on the volcanic front

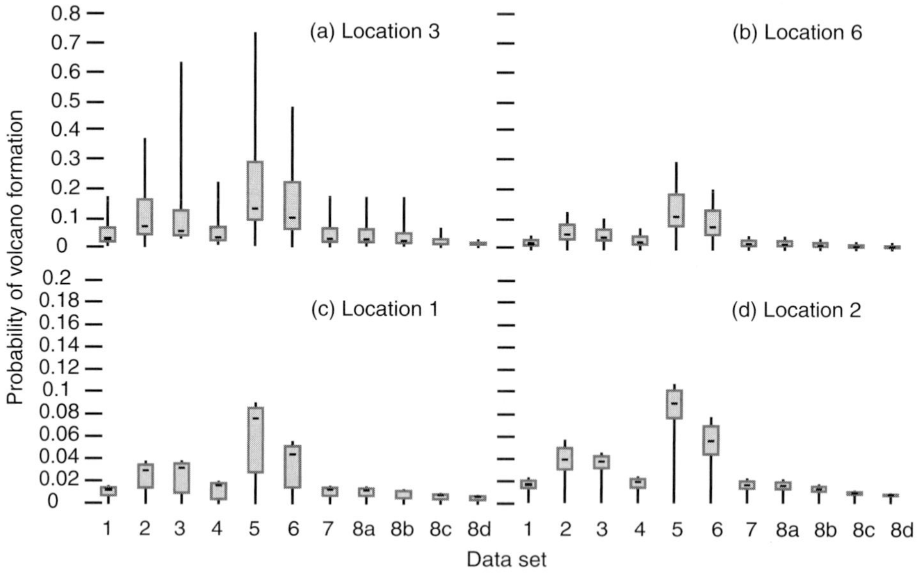

Fig. 13.7 Box and whisker plots of the variation in probability of new volcano formation estimated at four locations (a–d), calculated for each data set (Table 13.1) using an integer range of bandwidths of 1–70 km. Shown are the median (horizontal bar), interquartile range (box) and full probability range (vertical line).

and has the highest hazard for nearly all of the alternative volcano data sets. However, location 3 does not have the highest probability for DS 8c–d, or DS 4 (see Figure 13.7). The data sets DS 8c–d group together near-neighbor volcanoes and therefore have fewer volcanoes within the Sengan area (where location 3 is sited) than the other alternative data sets. Thus, a slight decrease in the hazard at location 3 for DS8c–d is explicable, but again, is indicative of the importance of volcano event definition in hazard analysis. This result suggests that it is unreasonable simply to group different volcanoes into single events based solely on their proximity.

Probability variation among these localities is also illustrative in the context of volcano-tectonic regimes. Location 6 is 15 km east of the main Sengan volcanic cluster center (Figure 13.1b), in a lowland area on the edge of the high plateau defined by the volcanic front. This locality is only 5 km further from the center of the Sengan cluster than location 3, but the hazard is reduced by about one third (Figure 13.6). Given that location 6 is in the forearc, however, relatively lower probabilities might be expected, as the potential for volcanism decreases to essentially zero across the volcanic front. This result therefore suggests limitations in this simple model resulting from use of a symmetrical kernel density function, as discussed by Connor and Connor (Chapter 14, this volume) and Jaquet and Lantuéjoul (Chapter 15, this volume).

Concluding remarks

The siting of nuclear facilities in a region characterized by active volcanism is challenging. Such facilities must be sited in areas of very low hazard, where it is highly unlikely that new volcanoes will form during the performance period. As part of site characterization in such regions, the probability of future volcanic events is estimated using the depiction of volcanoes as a stochastic point process, literally dots on maps. A volcano, however, is a complex manifestation of the flux of magma through the crust and it is not straightforward to define or interpret what is really meant by each dot. Our analysis of Tohoku region volcanism suggests that using national catalogs of volcanic activity for such hazards assessments is not ideal. Such catalogs represent tremendous effort, but are not constructed with application to hazards assessment of nuclear facilities in mind. Rather, databases must be constructed that reflect the complexity of volcanic systems, the need to consider alternative data sets and multiple working hypotheses, and the need to present hazard assessments in a probabilistic framework. As analysis of volcanism in the Tohoku region indicates, of primary concern is assessment of the independence of volcanic events, and hence the estimated temporal recurrence rate of volcanism. Alternative volcano definitions result in significant variation in probability estimates at specific sites (e.g. Table 13.1). Volcanologists developing national catalogs do not necessarily consider assessment of the independence of volcanic events to be a primary objective. Therefore, site-specific hazard assessments should take full advantage of national or region catalogs of volcanic activity, but these catalogs should be considered a starting point for developing a database dedicated to the hazard assessment. Here we have taken the Japanese database on Quaternary volcanoes as a starting point to generate several alternative data sets. From a hazard perspective each data set is potentially viable, unless proven otherwise, and so the results of analysis of all alternatives gives a measure of epistemic uncertainty.

Volcanoes are born and then become extinct. Patterns of volcanic activity may shift with time. Thus the spatio-temporal point process that describes volcanic activity in a particular region is not necessarily stationary. In the Tohoku region, we have found evidence (Figure 13.2) that volcanism is temporally non-stationary, although this result is uncertain due to potential recording bias, a problem common to most geological data sets. Because of non-stationarity, significant probability variation occurs depending on the age span of the alternative data set used. It is not necessarily possible to resolve uncertainties associated with non-stationarity because of the comparatively low resolution of the geologic record, even in areas where volcanologists have developed an extensive understanding of the geological record and made numerous radiometric age determinations, such as in Tohoku. This reinforces the argument that multiple alternative hypotheses should be used to take account of the uncertainties and ambiguities related to temporal and spatial patterns of volcanic activity. Such alternative hypotheses include use of alternative data sets, but should also include alternative statistical models.

Perhaps the most dramatic result of our analysis is that different types of volcanoes can exhibit different spatio-temporal patterns. In the Tohoku region, effusive volcanoes are more tightly clustered and apparently form more frequently in the recent geologic past. In contrast, explosive volcanoes are more widely dispersed. These different spatio-temporal patterns imply differences in the generation and ascent of magmas that produce these types of volcanic activity. Furthermore, these types of volcanoes potentially have very different impacts on nuclear facilities, so as hazard analyses progress to include the consequences of volcanic eruptions, these different spatio-temporal patterns may have a significant impact on risk estimates. This result has significant implications for how volcano databases for hazard assessment of nuclear facilities should be constructed and used. Traditional volcano classification has centered on morphology rather than process. Cinder cones and shield volcanoes, for example, have very distinctive morphologies, but both are associated with lateral dike injection and may have episodes of violent Strombolian activity. Alternative classification schemes, such as those based on quantitative cladistics (Hone *et al.*, 2007) or process-oriented classifications, are much more relevant to volcanic hazard assessments for nuclear facilities.

Acknowledgments

We would like to acknowledge the Nuclear Waste Organization of Japan for their continued scientific and technical interest and support, and MCM for administration of that support. Our Tohoku research has greatly benefited from discussions with numerous colleagues, including: K. Berryman, N. A. Chapman, J. Goto, T. Hasenaka, O. Jaquet, K. Kitayama, H. Kawamura, H. Kondo, S. Nakada, Y. Tamura, H. Tsuchi and L. Wallace. Contents of this chapter do not necessarily reflect the views or technical position of the Nuclear Waste Organization of Japan.

References

Committee for Catalog of Quaternary Volcanoes in Japan (1999). Catalog of Quaternary Volcanoes in Japan (CD-ROM). Tokyo: Volcanological Society of Japan.

Connor, C. B. and B. E. Hill (1995). Three nonhomogeneous Poisson models for the probability of basaltic volcanism: application to the Yucca Mountain region, Nevada. *Journal of Geophysical Research*, **100**, 10 107–10 125.

Hone, D. W. E., S. Mahony, R. S. J. Sparks and K. Martin (2007). Cladistics analysis applied to the classification of volcanoes. *Bulletin of Volcanology*, **70**, 203–220.

Kondo, H., K. Kaneko and K. Tanaka (1998). Characterization of spatial and temporal distribution of volcanoes since 14 Ma in the northeast Japan arc. *Bulletin of the Volcanological Society of Japan*, **43**, 173–180.

Lutz, T. M. and J. T. Gutmann (1995). An improved method of determining alignments of point-like features and its implications for the Picante volcanic field, Mexico. *Journal of Geophysical Research*, **100**, 17 659–17 670.

Martin, A. J., K. Umeda, C. B. Connor *et al.* (2004). Modeling long-term volcanic hazards through Bayesian inference: an example from the Tohoku volcanic arc, Japan. *Journal of Geophysical Research*, **109**, B10208, doi:10.1029/2004JB003201.

NUMO (2004). Evaluating site suitability for a HLW repository site, scientific background and practical application of NUMO's siting factors. Nuclear Waste Management Organization of Japan Report NUMO-TR-04-04. Tokyo: Nuclear Waste Management Organization of Japan.

Oba, T. and K. Umeda (1999). Geology of Hachimantai volcanic field and temporal, spatial variation of the magma compositions. *Journal of Mineralogy, Petrology and Economic Geology*, **94**, 187–202.

Tamura, Y., Y. Tatsumi, D. Zhao, Y. Kido and H. Shukuno (2002). Hot fingers in the mantle wedge: new insights into magma genesis in subduction zones. *Earth and Planetary Science Letters*, **197**, 105–116.

Umeda, K., S. Hayashi, M. Ban *et al.* (1999). Sequence of volcanism and tectonics during the last 2.0 million years along the volcanic front in Tohoku district, NE Japan. *Bulletin of the Volcanological Society of Japan*, **44**, 233–249.

Wand, M. P. and M. C. Jones (1995). *Kernel Smoothing*. Boca Raton, FL: Chapman and Hall.

Weller, J. N., A. J. Martin, C. B. Connor, L. J. Connor and A. Karakhanian (2006). Modelling the spatial distribution of volcanoes: an example from Armenia. In: Mader, H. M., S. G. Coles, C. B. Connor and L. J. Connor (eds.) *Statistics in Volcanology*, Special Publications of IAVCEI, 1. London: Geological Society, 77–87.

Yoshida, T., T. Oguchi and T. Abe (1995). Structure and evolution of source area of the Cenozoic volcanic rocks in northeast Honshu arc, Japan. *Memoirs of the Geological Society of Japan*, **44**, 263–308.

14

Estimating spatial density with kernel methods

C. B. Connor and L. J. Connor

Hazard assessments are invariably a blend of expert interpretations of geophysical events and statistical descriptions of these events. Analyses of the recurrence rate and magnitude of events, their spatial density and their potential effects are essential components of hazard assessment for nuclear facilities. This chapter explores a robust approach to estimating spatial density using kernel methods and describes new methods of quantifying the uncertainty in these estimations using statistical techniques. Some of the spatial density estimation methods presented in this chapter have been used since the mid 1990s. In addition, new tools are emerging that offer improved understanding of spatial density estimates and their application in hazard assessments. For example, algorithms have been developed for numerical optimization of estimates of spatial density. Smoothed bootstrap techniques provide a mechanism for assessing uncertainty in spatial density, especially where information on past events is sparse. Methods in parallel processing have revolutionized the way we explore models of spatial density, in ways that were not practical even a decade ago. These developments are exceedingly encouraging. Although purely quantitative descriptions of spatial density, by themselves, are unlikely to ever be sufficient for assessment of hazard and risk, these quantitative estimations combined with expert judgment provide a powerful tool for improving these assessments. Thus, recent developments in quantitative density estimation will have a significant impact on the quality of geologic hazard assessments for nuclear facilities.

Hazard assessments need to rely on robust regional spatial density estimates of the likely occurrence of future hazardous events, such as the location of new volcanic vents or the epicenters of large earthquakes. In volcanology, the distribution of older volcanoes or volcanic vents provides some of the clearest information about the probable location of future volcanoes and volcanic vents (e.g. Crowe *et al.*, 1983; Connor and Hill, 1995; Lutz and Gutmann, 1995; Jaquet and Carniel, 2006; Jaquet *et al.*, 2008). In seismology, historical earthquake catalogs provide information about the probable locations of future large earthquakes (e.g. Musson and Winter, 1996; Stirling and Wesnousky, 1998; Stock and Smith, 2002). In the siting of nuclear facilities, however, estimating the spatial density of such hazardous events has often proved contentious. Why the controversy? The underlying geologic processes controlling the distribution of these events are complex and incompletely understood. The frequency of such potentially catastrophic events is low, especially within many regions considered for nuclear facilities, so data used in these analyses are often sparse. The selection of specific statistical models to estimate spatial density is often subjective. These factors result in uncertainty.

Non-parametric methods, such as kernel density estimation, are efficient and objective estimators of spatial density. In this chapter the basis for kernel density estimation is described and these methods are applied to two data sets: (i) volcano distribution in the Yucca Mountain region, USA; and (ii) the distribution of earthquakes in the United Kingdom (UK). We include an introduction to techniques for estimating the smoothing bandwidth, a key feature of analyses that use kernel estimation methods. Because the spatial density is estimated from a limited number of previous events, there is uncertainty when utilizing these estimates to forecast future activity. We describe a Monte Carlo method for quantifying the uncertainty, which can be as significant to the hazard assessment as the estimation of the spatial density itself.

14.1 What is spatial density?

In the context of volcanic and seismic hazard assessments for nuclear facilities, the reason for estimating spatial density is to determine possible locations of future geophysical events; or to estimate the probability of an event occurring at a specific location, given that such events occur within the site region. Unfortunately, there is some ambiguity in the literature regarding the use of the terms density and intensity. In the geosciences, variation in the number of events per unit area (say the number of volcanic vents or earthquake epicenters) is described using the term density. For example, one might report the density of earthquake epicenters in a region as the number of epicenters per 1000 km^2. Intensity, in geoscience contexts, often refers to the magnitudes of these events. The intensity of ground shaking due to an earthquake can be described with the European macroseimic scale. The intensity of a volcanic eruption can be characterized in terms of its total mass of eruptive products or related indices (Pyle, 2000).

Density and intensity are defined differently in spatial statistics. In this context, spatial intensity refers to the expected number of events per unit area defined at a point, $\mathbf{s}$, a matrix containing the x and y coordinates of the location of the point (Diggle, 1985; Diggle and Marron, 1988; Gatrell *et al.*, 1996). Suppose there exists a set of events (e.g. earthquake epicenters or volcano locations) that occur within a given region, R. These events can be designated as $\mathbf{x}_n (n = 1, 2, ..., N) \in R$ where N is the total number of events, each consisting of the spatial location, x and y of the event (possibly given in Easting and Northing coordinates, or, latitude and longitude). One way we can create a model of spatial intensity from these events is to imagine they are realizations of a random variable, $\mathbf{X}$, a function that describes the set of all possible realizations. For example, $\mathbf{X}$ might be the distribution of potential earthquakes or the distribution of potential volcanoes, from which a set of observed realizations (e.g. those found in the earthquake catalog or on a geologic map) are drawn. The spatial intensity is formally written as (Gatrell *et al.*, 1996)

$$\lambda(\mathbf{s}) = \lim_{\mathrm{d}\mathbf{s} \to \mathbf{0}} \left\{ \frac{E(\mathbf{X}(\mathrm{d}\mathbf{s}))}{\mathrm{d}\mathbf{s}} \right\} \tag{14.1}$$

where $E(\mathbf{X}(\mathbf{ds}))$ is the expected number of events that fall within a small area d$\mathbf{s}$ about the point $\mathbf{s}$ (hence, if the location, $\mathbf{s}$, is given as Easting and Northing with units of meters, then the units of $\lambda(\mathbf{s})$ are m^{-2}). At first glance it appears that the statistical definition of *intensity* is equivalent to the term *density* as commonly used in the geosciences. This is not quite true. The geological processes that result in a given event distribution are incompletely known. We can think of these geological processes as giving rise to a stochastic point process that describes the relationship between the set of events and the geological processes that led to their formation. As the stochastic point process is incompletely known, the true value of the local spatial intensity, $\lambda(\mathbf{s})$, is also unknown. That is, the observed distribution of events is only one realization of the underlying process that gives rise to these events. Our goals are to find an estimate of the spatial intensity, $\hat{\lambda}(\mathbf{s})$, that approximates the true but unknown value of spatial intensity, $\lambda(\mathbf{s})$, and to understand the uncertainty in this estimate.

In hazard assessments there is a further requirement that this information be used to forecast the spatial distribution of possible future events. Often we consider spatial intensity in terms of the probable location of some future event, given that one occurs within our region of interest. This conditional probability can be estimated by:

$$\hat{f}(\mathbf{s}) = \frac{\hat{\lambda}(\mathbf{s})\mathrm{d}(\mathbf{s})}{\int_R \hat{\lambda}(\mathbf{s})\mathrm{d}(\mathbf{s})}. \tag{14.2}$$

Integrating $\hat{f}(\mathbf{s})$ across the region of interest, R, gives unity, if R is sufficiently large. Since all values of $\hat{f}(\mathbf{s})$ within this region are greater than or equal to zero, this makes $\hat{f}(\mathbf{s})$ a probability density function and this function may be used in probabilistic hazard models. The term $\hat{f}(\mathbf{s})$ is referred to as one estimate of the spatial density, and one can consider the spatial density per unit area in terms of conditional probability (e.g. given a volcanic event in the region, what is the probability that the event will occur within some small area about the point $\mathbf{s}$?). In addition, care is required in the selection of the region R, as external events located close to the border may have a non-negligible contribution to spatial density. A practical approach is to select R to be quite large compared to the region of specific interest (e.g. the site). This chapter will refer to spatial intensity and spatial density within the context of spatial statistics.

14.2 Assumptions behind spatial density estimates

How does one develop a best estimate of spatial density? In the real world, there is only one realization of an underlying geologic process, the observed distribution of past events. Unfortunately, geology is not conducive to repeating the experiment in a natural system. For a given region there is just one earthquake catalog, or one geologic map of volcano distribution. Presumably, if there existed a complete geophysical model for these events, we would use this information to better forecast the locations of future events. For example, if we knew the distribution of melt in the asthenosphere and lithosphere, and if we knew the state of the lithosphere through which the magma rises, we might have a better sense of where volcanoes are most likely to form next. Currently, we lack such a complete geophysical perspective. Some data sets give an idea of where partial melting of the mantle might occur, for example seismic tomographic models of "slowness" in the lithosphere and asthenosphere (e.g. Zhao, 2001). Other data, such as variations in gravity across a region (Connor *et al.*, 2000; Parsons *et al.*, 2006), show some correlation with the existing distribution of volcanoes in some circumstances, but the mechanisms relating gravity anomalies to the origin of magmas are not completely understood. As a result, these types of data have been used to support estimates of spatial density (e.g. Connor *et al.*, 2000; Martin *et al.*, 2004; Jaquet and Lantuéjoul, Chapter 15, this volume), but no model has yet been proposed that does not rely principally on the spatial distribution of past events. Similarly, in seismology there have been attempts to create blended hazard maps based on a variety of geophysical criteria, but these methods generally rely on the earthquake catalog (e.g. Ward, 1994).

The reliance on the distribution of past events implies that these realizations are representations of some underlying random variable, $\mathbf{X}$, that will govern the distribution of potential events in the future. This assumption immediately raises a fundamental question. Which are the past events that should be used to develop the spatial intensity estimate, $\hat{\lambda}(\mathbf{s})$ and density, $\hat{f}(\mathbf{s})$? Event data sets used to estimate the spatial density of future events need to be consistent with several features of geological processes.

First, any spatial intensity function for a geologic process must change with time. On timescales of tens of millions of years, plate boundaries change, volcanic arcs wax, wane and migrate, and major fault systems reorganize. In very long-term probabilistic hazard assessments for high-level waste repositories, which may have 10^6 a performance periods, these factors have to be considered in weighing the validity of using specific data in developing spatial intensity models. For processes like volcanism, where a geologic record of past events usually persists for tens of millions of years, consideration needs to be given to which events best represent the distribution of future volcanism. For example, the distribution of Miocene volcanoes in a given area might be much less relevant than the distribution of Pliocene and Quaternary volcanoes. Thus, in order to develop an estimate of the spatial intensity, a model of the geological evolution of the system is required. This geological model is used to justify the inclusion of some geological features in the event data set, and the exclusion of others.

Second, it is necessary to assess the completeness of the geologic record. In seismology, it is particularly clear that short earthquake catalogs carry the risk of biasing estimates of spatial intensity (Sirling *et al.*, Chapter 10, this volume). That is, the record of earthquakes in a given region collected on a short timescale might give an incomplete picture of the unknown distribution of potential earthquakes, $\lambda(\mathbf{s})$. Even volcanic events might be missed in initial geological investigations, as volcanic vents might be buried in sediment or otherwise obscured (e.g. Connor *et al.*, 1997; Wetmore *et al.*, Chapter 16, this volume).

Third, geological events, even when they are all identified, may be so rare as to present an incomplete picture of the underlying process. Consider an earthquake as a single event, $\mathbf{x}_n$, one realization of the random variable, $\mathbf{X}$. If, for example, $\mathbf{X}$ can be characterized by a uniform random distribution, then it is likely that the observed set of realizations will have a spatially random distribution within the region of interest, R. However, the underlying density usually has additional structure, causing independent realizations to cluster. For example, earthquake epicenters tend to cluster along plate boundaries and volcanoes cluster above zones of partial melting in the mantle. For random variables with a great deal of statistical structure, such as many modes in spatial intensity, a great number of events might be required to identify the statistical structure of the random variable.

Fourth, it is critical to ascertain which geological features are actually independent events. The true statistical structure of the random variable, $\mathbf{X}$, might be obscured if some events included in the event data set are not independent. For example, great earthquakes are followed by aftershocks. An earthquake aftershock, however, is not a random sample of the random variable "*spatial distribution of great earthquakes*," because these aftershocks are not realizations of this particular random variable. Rather, they are independent realizations of another random variable, say "*spatial distribution of aftershocks about a great earthquake*." So, the distribution of aftershocks does not necessarily give the best sense of the spatial intensity of great earthquakes, although these two random variables are correlated.

Similarly, volcanoes are complex geological structures. The spatial distribution of polygenetic volcanoes (Connor *et al.*, Chapter 3, this volume) reflects processes of magma generation and rise through the crust. The distribution of small vents (sometimes referred to as parasitic or adventive cones) does not necessarily reflect the distribution of polygenetic volcanoes, so a spatial intensity estimate that includes all vents as events would not correctly model the underlying random variable. Furthermore, in monogenetic volcanic fields alignments of volcanic cones develop in response to single magmatic events, episodes of magma rise through the shallow crust. This is because single igneous dikes ascending through the crust might form segments and rotate within the shallow crust, each segment feeding a separate vent and each building a volcanic cone. If the goal of analysis is to forecast the distribution of future magmatic events, each of which might produce more than one monogenetic volcano, geological data must be gathered; and volcanoes formed by the same magmatic event must be somehow grouped as single events.

Independence of events is not necessarily easy to determine. Rather than simply counting earthquakes in a catalog or volcanoes on a geological map, one must make a geological assessment of the independence of these data. For volcanoes, this is generally accomplished through detailed analyses of radiometric age determinations, stratigraphic correlations and related geological data. Often, even detailed analyses do not resolve whether or not specific features should be grouped as single events or treated as separate, independent events.

Consequently, a major task in preparing a spatial intensity estimate for a nuclear facility is defining the data set of events to be used. Certainly a major expense in site characterization is data gathering to support interpretation of geological features as events. Mahony *et al.* (Chapter 13, this volume) and Stirling *et al.* (Chapter 10, this volume) provide examples of such analyses. Hazard assessments often consider alternative event data sets and account for the effect of these varying data sets on spatial density estimates. This strategy will be employed in the following examples.

14.3 Two data sets

Two data sets are used to demonstrate spatial density estimation. One data set is used to assess the volcanic hazard at a proposed high-level radioactive waste repository in the USA. The other data set is used to estimate the spatial density of earthquakes in the UK, a factor in seismic hazard assessment for UK nuclear power plants (NPPs). Our goal is to illustrate modern techniques in kernel estimation, and not to produce hazard assessments of these areas. Indeed, these data sets have been integral to hazard assessments for nuclear facilities in both areas, and spatial intensity estimates form only a part, albeit an important one, of those hazard assessments (see Valentine and Perry, Chapter 19, this volume, and references therein; and Musson, 1997, and references therein).

Yucca Mountain, the proposed site of the first high-level radioactive waste repository in the USA, is located in the Basin and Range, a broad zone of active tectonism. One hazard issue associated with the site is the potential for basaltic eruptions to occur within the repository, and to transport radionuclides into the biosphere. This transport is thought to occur either through the direct eruption of radionuclides via volcanic eruption columns or through the destruction of geologic and engineered barriers, with subsequent accelerated transport of radionuclides within groundwater. The immediate site area has not experienced volcanic activity since about 11 Ma, when a basaltic dike intruded faults on the west side of Yucca Mountain (Smith *et al.*, 1997). Nevertheless, monogenetic volcanism has persisted, primarily to the west and south of Yucca Mountain, with $\approx$ 34 volcanic vents, mostly scoria cones, forming within the last 5 Ma (Figure 14.1). The most recent of these events ($\sim$ 0.08 Ma) created the $\sim$ 200 m-high Lathrop Wells cinder cone, located $\sim$ 20 km from the repository site (Valentine and Perry, Chapter 19, this volume), and Ubehebe maar, located within Death Valley, which erupted during the Holocene, $\sim$ 6 ka (Figure 14.1). Given this distribution of events, one task in the volcanic hazard assessment is to estimate the spatial density of potential future volcanism.

Extensive radiometric dating indicates that some volcanoes, especially those in Crater Flat, should be grouped into single events that include numerous eruptions over a comparatively short period of time. In addition, many hazard assessments (e.g. Crowe *et al.*, 1983; Connor *et al.*, 2000) have included the volcanoes of the Amargosa Desert and nearby areas (Figure 14.1). Based on the isotope geochemistry of basalts, Smith *et al.* (1997) have argued that the area of volcanism should encompass Death Valley and the Greenwater Range. This larger area is termed the Amargosa Valley isotopic province (AVIP). Volcanism in this area appears to be dominated by lithospheric melts, as opposed to asthenoshperic melts that predominate the volcanic fields to the north and south of the region (Yogodzinski *et al.*, 1996; Smith *et al.*, 1997; Valentine and Perry, 2007). Do different event definitions and different regions of interest affect the spatial density estimate for potential monogenetic volcanism in the region about Yucca Mountain? We explore this with two event data sets derived from the geologic data: the YMR (excludes Death Valley and the Greenwater Range) and AVIP (includes Death Valley and the Greenwater Range) (Figure 14.1).

The second example consists of earthquake epicenters that occurred within and near the UK (latitude 49° N–60° N, longitude 12° W–2° E), between January 1973 and January

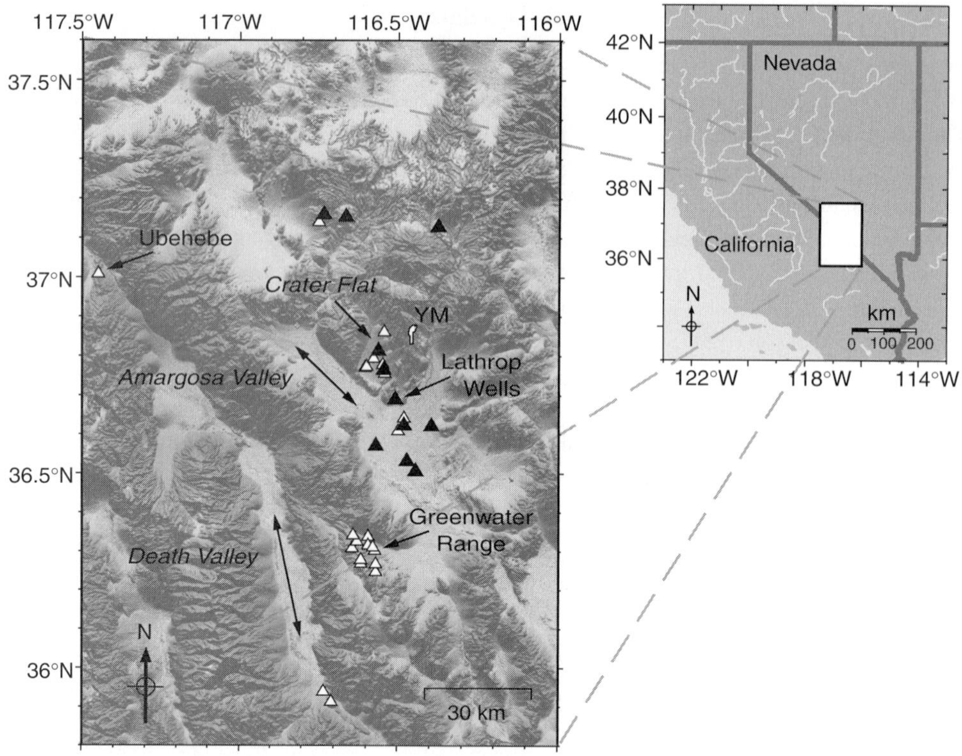

Fig. 14.1 Locations of volcanoes about the proposed Yucca Mountain (YM) high-level radioactive waste repository (solid black triangles – YMR event data set, white and black triangles – AVIP data set). These two data sets serve as the basis for spatial intensity estimation. The shaded-relief digital elevation model is based on Shuttle Radar Topography Mission data (Jarvis *et al.*, 2006). The map datum is WGS84.

2007. The UK experiences low seismic hazard associated with intraplate seismicity (e.g. Coppersmith and Youngs, 1986; Musson and Winter, 1996). Nevertheless, seismic hazard in the UK exists and has been the topic of numerous studies, particularly with regard to the siting and performance of NPPs (see Musson, 1997, and references therein). Two earthquake catalogs are used here. The first is from the NEIC/USGS catalog (NEIC/USGS, 2007), January 1973–January 2007, and includes a total of 445 earthquake epicenters (Figure 14.2). Although this catalog is a complete record of UK seismicity during the time interval, the catalog includes aftershocks. The second catalog includes larger-magnitude earthquakes located by the British Geological Survey only, gathered from the International Seismological Centre catalog (ISC, 2007). Large magnitude earthquakes are rare in the UK, only 23 of the earthquakes in the ISC catalog have magnitudes $ML > 3.0$. Spatial intensity estimates based on these data may provide some insight into the extent of clustering and relationships between earthquake spatial intensity and seismic hazard zonation models,

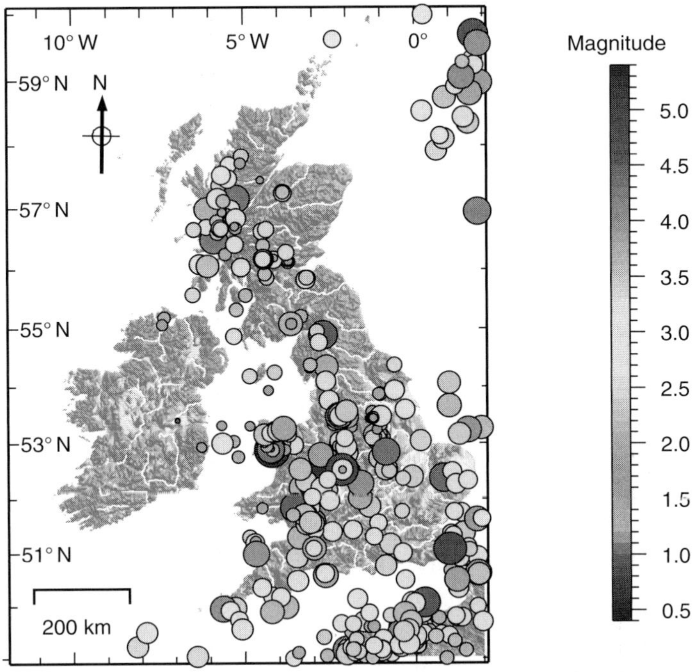

Fig. 14.2 Earthquake epicenters are plotted as shaded circles that increase in diameter, proportional to earthquake magnitude. Earthquake catalog data are from the NEIC/USGS database, January 1973–January 2007. These earthquake epicenters, together with a subset found in the ISC catalog, serve as a basis for spatial intensity estimation. Visually, the number of earthquake epicenters decreases west of 7° W. Topographic data, shown as a shaded-relief digital elevation model, is based on 1 km GLOBE data (GLOBE Task Team, 1999); map datum WGS84. See color plate section.

commonly used to assess seismic hazard at nuclear facilities. We compare spatial density estimates, made using these two catalogs, with special reference to 17 nuclear facilities in the UK.

14.4 Estimating spatial intensity with kernel methods

Spatial intensity models based on the distribution of past volcanic or seismic events might be parametric or non-parametric. Parametric models involve fitting a distribution, usually one from a common set of distributions (e.g. uniform random or bivariate Gaussian) to the distribution of events throughout a region or within zones (i.e. subsets of the region of interest). This estimate yields a set of parameters (e.g. mean location of the volcanic field in Northing and Easting coordinates, variance in Northing and Easting coordinates or rotation). Uncertainty in the distribution fit, and uncertainty in parameter estimates of spatial intensity, can be calculated using maximum likelihood estimation. A significant drawback of these parametric methods is that they assume a priori that the distribution of volcanoes is explained by the parametric distribution, for example that volcano distribution is reasonably described as a bivariate Gaussian density. This is not necessarily the case. In fact, it has been shown repeatedly that volcanoes cluster within volcanic fields. Such clustering may be completely smoothed by simple parametric models. To our knowledge, parametric models have not been used to model earthquake intensity, although completely spatially random models within zones are parametric in the most general sense (Beauvall *et al.*, 2006).

A non-parametric approach for estimating the spatial intensity involves kernel density estimation (Silverman, 1978; Diggle, 1985; Silverman, 1986; Wand and Jones, 1995). With this technique, the observed event locations are used to estimate the spatial intensity at any point in the region using a kernel function. For example,

$$\hat{\lambda}(\mathbf{s}) = \frac{1}{2\pi h^2} \sum_{i=1}^{N} \exp\left[-\frac{1}{2}\left(\frac{d_i}{h}\right)^2\right] \tag{14.3}$$

is a two-dimensional radially symmetric kernel function where the spatial intensity decreases with distance from events based on a bivariate Gaussian function. The local spatial intensity estimate, $\hat{\lambda}(\mathbf{s})$, depends on its distance, d_i, to each event location, and the smoothing bandwidth, h. The rate of change in spatial intensity with distance from events depends on the size of the bandwidth, which, in the case of a Gaussian kernel function, is equivalent to the standard deviation of the kernel. In this example, the kernel is radially symmetric, that is, h is constant in all directions. Nearly all kernel estimators used in geological hazard assessments have been of this type (e.g. Connor and Hill, 1995; Condit and Connor, 1996; Woo, 1996; Stock and Smith, 2002). The bandwidth is selected using some criterion, often visual smoothness of the resulting spatial intensity plots, and the spatial intensity function is calculated using this bandwidth. Alternatively, an adaptive kernel function can be used, in which the spatial intensity varies as a function of event spatial intensity (e.g. Stock and Smith, 2002; Weller *et al.*, 2006). These adaptive kernel functions are also radially symmetric.

Equation (14.3) is a simplification of the more general case, whereby the amount of smoothing by the bandwidth, h, varies in magnitude depending on direction. A two-dimensional elliptical kernel with a direction-varying bandwidth is given by (Wand and Jones, 1995),

$$\hat{\lambda}(\mathbf{s}) = \frac{1}{2\pi\sqrt{|\mathbf{H}|}} \sum_{i=1}^{N} \exp\left[-\frac{1}{2}\mathbf{b}^{\mathbf{T}}\mathbf{b}\right] \tag{14.4}$$

where,

$$\mathbf{b} = \mathbf{H}^{-1/2}\mathbf{d} \tag{14.5}$$

The bandwidth, $\mathbf{H}$, is a 2×2 element matrix that is positive and definite (important because the matrix must have a square root), $|\mathbf{H}|$ is the determinant of this matrix and $\mathbf{H}^{-1/2}$ is the inverse of its square root. The parameter $\mathbf{d}$ is a 1×2 distance matrix (i.e. the x-distance and y-distance from $\mathbf{s}$ to an event), $\mathbf{b}$ is the cross product of $\mathbf{d}$ and $\mathbf{H}^{-1/2}$, and $\mathbf{b}^{\mathbf{T}}$ is its transform. The resulting spatial intensity at each point location, $\mathbf{s}$, is usually distributed on a grid which has total extent that defines the region, R.

One difficulty with elliptical kernels is that all elements of the bandwidth matrix must be estimated. Several methods have been developed for estimating an optimal bandwidth matrix based on the locations of the event data (Wand and Jones, 1995), most recently summarized in the statistics literature by Duong (2007). Here we utilize two techniques, a modified asymptotic mean integrated squared error (AMISE) method, developed by Duong and Hazelton (2003), called the SAMSE pilot bandwidth selector, and the smoothed cross-validation (SCV) method of Hall *et al.* (1992), to estimate optimally the smoothing bandwidth for our Gaussian kernel function. These bandwidth estimators are found in the freely-available *R* statistical package (Hornik, 2007; Duong, 2007).

Using the YMR volcanic event data set, the SAMSE selector yields the following optimal bandwidth matrix and corresponding square-root matrix:

$$\mathbf{H} = \begin{bmatrix} 29.3 & -45.1 \\ -45.1 & 213.1 \end{bmatrix} \quad \sqrt{\mathbf{H}} = \begin{bmatrix} 4.9 & -2.3 \\ -2.3 & 14.4 \end{bmatrix} \tag{14.6}$$

The upper-left and lower-right diagonal elements represent smoothing in the east–west and north–south directions, respectively. The $\sqrt{\mathbf{H}}$ indicates an east–west smoothing distance of $\approx (2 \times 4.9) = 9.8$ km and a north–south smoothing distance of $\approx (2 \times 14.4) = 28.8$ km. The negative, off-diagonal elements represent a counter-clockwise rotation of this kernel. Given the values shown, the matrix forms an elliptical kernel, trending $\approx 25°$ to the north-northwest.

The SCV bandwidth selector yields the following optimal bandwidth and square-root matrices:

$$\mathbf{H} = \begin{bmatrix} 57.4 & -105.4 \\ -105.4 & 440.8 \end{bmatrix} \quad \sqrt{\mathbf{H}} = \begin{bmatrix} 6.5 & -3.9 \\ -3.9 & 20.6 \end{bmatrix} \tag{14.7}$$

In this case, the $\sqrt{\mathbf{H}}$ indicates an east–west smoothing distance of 6.5 km and a north–south smoothing distance of 20.6 km, with the bandwidth again trending toward the north-northwest. Both estimators generate bandwidths that are quite similar in shape and size.

For comparison, shown below are the optimal bandwidth matrices and their corresponding square roots generated for the AVIP data set. The SAMSE method generates

$$\mathbf{H} = \begin{bmatrix} 27.4 & -10.9 \\ -10.9 & 165.1 \end{bmatrix} \quad \sqrt{\mathbf{H}} = \begin{bmatrix} 5.2 & -0.6 \\ -0.6 & 12.8 \end{bmatrix} \tag{14.8}$$

and the SCV method generates,

$$\mathbf{H} = \begin{bmatrix} 36.9 & 15.6 \\ 15.6 & 432.9 \end{bmatrix} \quad \sqrt{\mathbf{H}} = \begin{bmatrix} 6.1 & 0.6 \\ 0.6 & 20.5 \end{bmatrix} \tag{14.9}$$

Again, both methods generate similar optimal bandwidths.

The spatial density estimates using the above optimally generated bandwidths are shown in Figure 14.3. Using the YMR data set, the estimated regional spatial density maps (SCV bandwidth, Figure 14.3a, and SAMSE bandwidth, Figure 14.3b) show predominant north-northwest density trends, roughly parallel to the topographic expression of the Death Valley fault system and adjacent regional tectonic structures. The most prominent difference between the two maps is the tendency for the SAMSE map to create clusters, whereas the SCV map is smoother. As clustering is a predominant feature of cinder-cone volcanism, and a feature of the YMR in particular (Connor and Hill, 1995), the SAMSE map appears to do a better job of representing this feature of the spatial density.

The kernel bandwidths change when events in Death Valley and the Greenwater Range are included (the AVIP data set), and hence, the map pattern of spatial density changes (Figure 14.3c–d). Unfortunately, radiometric age determinations are few in the Greenwater Range, so it is not possible to group vents in this area into events. Consequently, on these two maps all known Pliocene–Quaternary vents are used to estimate spatial density. The trend of the map changes to the north-northeast when these additional vents are included, clearly following the regional distribution of volcanism described by Yogodzinski *et al.* (1996) and colleagues. Also of note is Ubehebe crater, the westernmost vent on the map (Figure 14.1). This vent is a spatial outlier, and its occurrence is not well predicted by these spatial density models. As before, the SAMSE model appears to honor the clustering of events much better than the SCV kernel, which appears to over-smooth the map.

Spatial density estimates for UK seismicity (bandwidths chosen using the SAMSE method) are shown in Figure 14.4. Earthquakes are not uniform across the region, and this is reflected in both spatial density maps. Specifically, very few earthquakes occur in the western portion of the region, and there appears to be a slight tendency for earthquakes to cluster in the southern and central portions of England. The NEIC catalog of 445 earthquakes (Figure 14.4a) creates more-concentrated clusters, a result of aftershocks in the data set, whereas the 26 larger-magnitude earthquakes of the ISC catalog create a much smoother map.

The bandwidth matrices used in calculating the UK spatial density estimates of seismicity are

$$\mathbf{H} = \begin{bmatrix} 996 & -473 \\ -473 & 2687 \end{bmatrix} \quad \sqrt{\mathbf{H}} = \begin{bmatrix} 31.0 & -5.7 \\ -5.7 & 51.5 \end{bmatrix} \tag{14.10}$$

for the NEIC data set including aftershocks, and

$$\mathbf{H} = \begin{bmatrix} 5847 & 508 \\ 508 & 21528 \end{bmatrix} \quad \sqrt{\mathbf{H}} = \begin{bmatrix} 76.4 & 2.3 \\ 2.3 & 146.7 \end{bmatrix} \tag{14.11}$$

for the SIC data set of > 3 ML earthquakes. Note that in comparison to the Yucca Mountain volcano-event data sets, bandwidth estimates for UK seismicity are much larger and have

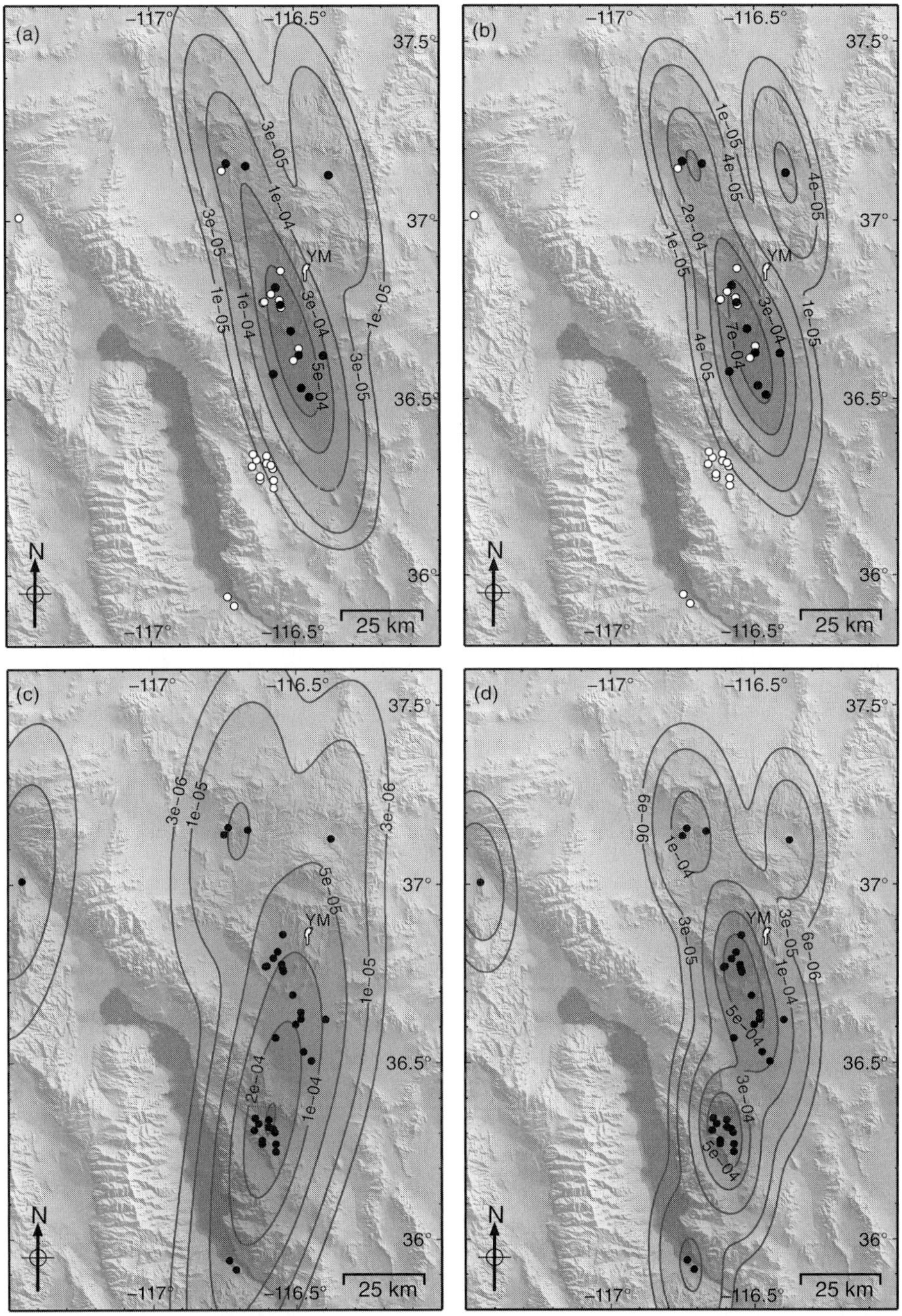

Fig. 14.3 Spatial density estimates of potential volcanism about Yucca Mountain (YMR event data set using (a) SCV and (b) SAMSE methods; AVIP data set using (c) SCV and (d) SAMSE methods). Contours at 25th, 50th, 75th, 95th and 99th percentiles (e.g. on map (a) locations within the 1e-04 contour (i.e. 75th quartile) have spatial density $> 1 \times 10^{-4}$ km^{-2}; given a volcanic event, there is a 75% chance it will occur within this quartile). See color plate section.

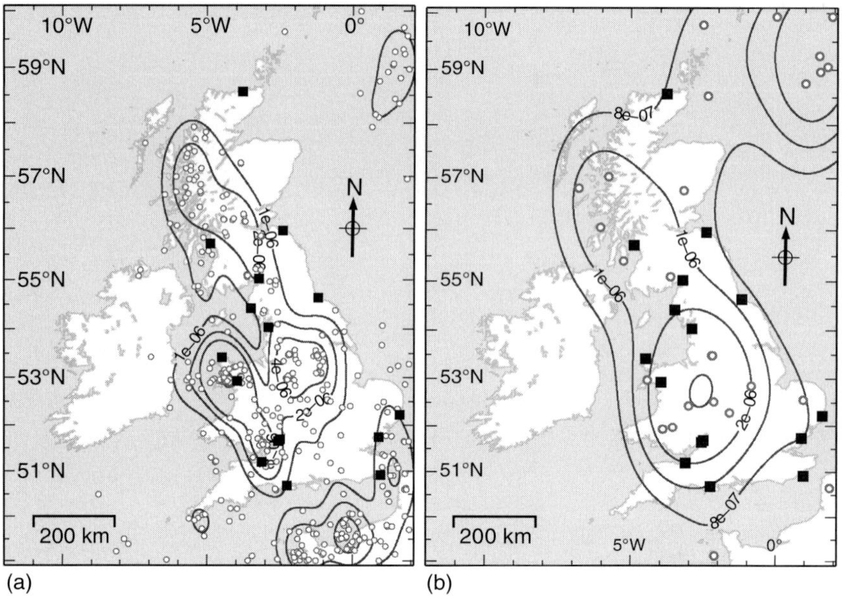

Fig. 14.4 Map (a) shows the regional spatial density estimate based on all earthquakes, including aftershocks, from January 1973–January 2007, based on the NEIC catalog. Map (b) includes only those earthquake events of magnitude > 3 ML, based on the ISC catalog. Both spatial density maps use bandwidths chosen by the SAMSE method. The maps are contoured at the 1st, 25th, 50th and 75th spatial density percentiles. As an example, using map (b), locations within the 1e-06 (75th percentile) contour have a spatial density greater than 1×10^{-6} per $100\,\text{km}^2$, or $1 \times 10^{-8}\,\text{km}^{-2}$. The smoother density map (b) may better reflect the true potential distribution of larger earthquakes since aftershocks are not included in this estimate. The black squares represent the locations of nuclear facilities.

lower eccentricity. The bandwidth for > 3 ML magnitude earthquakes is also much larger than the bandwidth estimated from the NEIC catalog, supporting the idea that clustering in the NEIC catalog is related to aftershock sequences, as opposed to statistical structure in the potential distribution of large earthquakes.

We are unaware of any previous applications using kernel functions with elliptical bandwidths for geologic hazard assessments; clearly these functions offer potential for improved estimates of spatial intensity, as geological events are often associated with linear trends, such as volcanic zones fault zones or similar tectonic structures.

14.5 Uncertainty in spatial density estimates

Uncertainty exists in the estimates of spatial density. This uncertainty stems from: (i) ambiguity in event data sets used to develop kernel estimates, (ii) application of the kernel density function, (iii) uncertainty in the bandwidth estimate used in the kernel density estimation, and (iv) few event data, a common problem in hazard assessment for nuclear facilities.

Each of these in considered in the following, with particular emphasis on the treatment of uncertainty arising from sparse data.

14.5.1 Event definition

Event definition affects the total number of events used to estimate density, and may introduce bias in density estimates. In the YMR volcano example, volcanoes that may have formed during single episodes of activity are grouped as single events. This tends to diminish the weight associated with events in the center of the distribution in this particular case, as these events all formed multiple volcanoes (Figure 14.1). Furthermore, geophysical investigations led to the recognition of four subsurface events in the Amargosa Valley (Figure 14.1). These events are associated with Pliocene ($\sim$ 3.8 Ma) volcanoes and volcano alignments. Without extensive aeromagnetic and ground-magnetic surveys, and ultimately drilling, these events would not have been included in the density estimate and the estimate would have been biased as a result. See Wetmore *et al.* (Chapter 16, this volume) for an additional example of a spatial density estimation of volcanism that is changed by including additional evidence of volcanism from subsurface investigations.

In the UK seismic example, the entire earthquake catalog was used to estimate the regional spatial density. This catalog includes low-magnitude earthquakes ($<$ 3 ML). It is uncertain whether or not these earthquakes should be included in the density estimate for larger, potentially hazardous earthquakes, or which earthquakes should be eliminated from the data set. The ISC data set for larger-magnitude earthquakes makes a particularly smooth map.

14.5.2 Kernel functions

Spatial density estimates made using kernel functions, as opposed to hazard zonation models (e.g. Crowe *et al.*, 1983; Musson, 1997) or parametric models, are explicitly data driven. A basic advantage of this approach is that any spatial density estimate will be consistent with the known data. Equations (14.3) and (14.4) are bivariate Gaussian kernels. Numerous authors have shown that the use of other kernels, such as the Epanechnikov kernel (Connor and Hill, 1995) or the Cauchy kernel (Martin *et al.*, 2004) has little impact on the final density estimate (e.g. Silverman, 1986; Wand and Jones, 1995). In hazard assessment, kernel functions with infinite tails (e.g. Gaussian) are preferred, as the probability is positive and real everywhere, albeit very small at locations far from past events. A potential disadvantage of these kernel functions is that they are not inherently sensitive to geological boundaries. One might hope that a complete understanding of the geology would result in a modification of the density estimate derived from a mathematical function. In fact, Connor *et al.* (2000) and Martin *et al.* (2004) discuss various methods of weighting density estimates in light of geological or geophysical information, in a manner similar to Ward (1994). A difficulty with such weighting is the subjectivity involved in recasting geological observations as density functions. Furthermore, geological insight is not always consistent with event distributions. For example, Musson (1997) noted, with some frustration, that epicenter distribution in the

UK does not correlate well or consistently with known major geologic structures that would be expected from a geophysical perspective to host earthquakes. Thus, even if a zonation model is used in hazard assessments, it makes sense to calculate the kernel-based density estimates and to consider any discrepancies in these models explicitly.

14.5.3 Kernel bandwidth

Bandwidth selection is a key feature of kernel density estimation, and is particularly relevant to seismic and volcanic hazard studies (e.g. Connor *et al.*, 2000; Molina *et al.*, 2001; Stock and Smith, 2002; Abrahamson, 2006; Jaquet *et al.*, 2008). Bandwidths that are narrow focus density near past events. Conversely, a large bandwidth may over-smooth the density estimate, resulting in unreasonably low density estimates near clusters of past events, and overestimate density far from past events. This dependence on bandwidth can create ambiguity in the interpretation of spatial density if bandwidths are arbitrarily selected (Abrahamson, 2006).

Bivariate bandwidth selectors like the SCV and SAMSE methods appear to be very promising, because although they are mathematically complex, they find optimal bandwidths using the actual data locations, removing subjectivity from the process. The bandwidth selectors used in this chapter provide global estimates of density, in the sense that one bandwidth or bandwidth matrix is used to describe variation across the entire region. An alternative method is to use adaptive kernel estimates, in which case the bandwidth changes with event density (Stock and Smith, 2002; Weller *et al.*, 2006). These adaptive bandwidths are calculated assuming radially symmetric kernel functions. Future research will likely involve developing bandwidth selectors that are adaptive across the map region.

14.5.4 Sparse event data

Often in hazard assessment for nuclear facilities there is a "problem" that there are few data available from which to forecast future events. That is, nuclear facilities are not likely to be planned where events are so frequent that the geological hazards are completely obvious. Instead, hazard analysis is most often required were few geologically hazardous events have occurred in the past (e.g. Figures 14.3 and 14.4). This is paradoxical because, by definition, uncertainty in hazard assessments must be comparatively high in these regions. If a spatial density is estimated using thousands of earthquakes or hundreds of volcanoes, we can assume that the true density is well represented by this model. Conversely, if the spatial intensity estimate is based on a handful of events, we might expect high uncertainty in the estimate. For example, the discovery of a single additional volcano, buried in sediment, might alter the shape of the estimated regional spatial density.

In order to assess the impact of uncertainty in $\hat{f}(\mathbf{s})$ due to sparse data, we modified a smoothed bootstrap method proposed by Press *et al.* (1992) based on the bootstrap methods of Efron and Tibishrani (1991). If we believe that the given spatial density estimates from

Figures 14.3 and 14.4 reflect the true density, then we should be able to draw a new set of N events from the original density estimate and, using the same chosen bandwidth matrix, $\mathbf{H}$, derive a new spatial density estimate based on this new set of events. This is a bootstrap procedure because we are resampling the original event distribution to derive a new event distribution. The bootstrap is smoothed because the N new events are sampled from the density function, rather than from a subset of the original event locations. If N is large, there will be very little difference between the new and the original spatial density estimates. However, if N is small, the new spatial density estimate might have significant differences from the original. Although the model has not changed, the locations of event data have changed. The differences between the new spatial density estimate and the original give a sense of the uncertainty resulting from the size, N, of the event data set.

This resampling of the estimated density suggests a Monte Carlo procedure. In our case, each of the six original spatial density surfaces are resampled to derive 1200 new event distributions for each original estimation. A total of 1200 spatial density estimations are then made at specific locations ($\mathbf{s}$) until a confidence interval for each $\hat{f}(\mathbf{s})$ emerges. In each of the 1200 simulations, the original spatial density is resampled for N new events and a new spatial density estimate is made. For the Yucca Mountain area, the density estimates are calculated at the Yucca Mountain site (YM on the maps). For the UK area, the density estimates are made at each of 17 NPP sites.

For the Yucca Mountain site, the spatial density estimate does not vary much with bandwidth selector or event data set (Figure 14.5). All analyses yield spatial densities around $1 \times 10^{-4}\,\text{km}^{-2}$. The median values produced by the smoothed bootstrap procedure yield slightly higher spatial densities, around $2 \times 10^{-4}\,\text{km}^{-2}$, because the Yucca Mountain site is relatively close to a small cluster of vents in Crater Flat. The kernel bandwidths are comparatively large and each simulation has few events, so resampling tends to increase the spatial density at the site compared to the original estimate. The uncertainty in spatial density determined using the smoothed bootstrap method is consistently less for the AVIP event data set compared to the YMR event data set because more events are used in the AVIP dataset. Uncertainty is not markedly different when using the SAMSE-selected or the SCV-selected bandwidths. For the density of volcanic events in the Yucca Mountain region, the event data set selection has more impact on the uncertainty than the choice of bandwidth selector.

The spatial density of earthquakes in the UK is also highly dependent on the data set used in the analysis. Using the NEIC data set (Figure 14.6a), most NPPs occur within areas of spatial density of earthquakes $\approx 2 \times 10^{-8}\,\text{km}^{-2}$ (i.e. 2×10^{-6} per $100\,\text{km}^2$). Two NPP sites, Trawsfynydd and Wylfa, appear to be located in areas of slightly higher seismic hazard. When only larger-magnitude earthquakes (> 3 ML) are used in the analysis, these anomalies completely disappear and the spatial density of earthquakes at NPP sites is remarkably uniform throughout the UK (Figure 14.6b).

The Monte Carlo procedure can be run for an entire grid by calculating (1200 times) the spatial density at each grid point across the map region. This process gives a sense of the uncertainty in density or spatial intensity estimates across the entire map region, but also

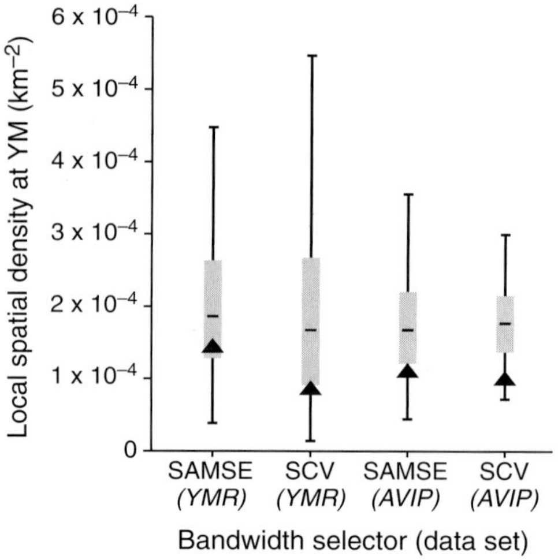

Fig. 14.5 Box plots illustrating uncertainties in local spatial density estimates at Yucca Mountain (YM) are obtained from a smoothed bootstrap resampling of four spatial density estimates of volcanism within the region. A total of 1200 estimates were made for each of the four combinations of bandwidths (SAMSE, SCV estimators) and event data set (YMR, AVIP). The quartile boundaries for each 1200 sample set are enclosed in gray with the median (50%) value indicated by the black horizontal bar. Black triangles represent the spatial density at the YM site generated using the original data sets. Uncertainties decrease in the AVIP analyses because of the increased number of events, but spatial density estimates and median estimates are consistent, at $\approx 1 \times 10^{-4}$ km^{-2} and 2×10^{-4} km^{-2}, respectively. Note that the median local spatial density estimates are always larger than the original estimates at YM, because the site is relatively close to a cluster of volcanoes and the optimal bandwidths are relatively large compared to this distance.

requires a great deal of computing power and time. We have accomplished this analysis using a system of parallel computers to speed computations.

One way to represent uncertainty across the entire map region is to plot the median spatial densities derived from the Monte Carlo procedure. Such maps no longer illustrate probability density functions, as the integral of the median values across the map region is no longer necessarily unity. One can either normalize this map, or simply plot values that are not normalized. For the Yucca Mountain area, contours of the median values from 1200 simulations at each grid point (not normalized) show that clusters tend to smooth (Figure 14.7). That is, we can be reasonably confident that some features of the spatial density are robust, such as the clustering of volcanoes west of Yucca Mountain and in the Greenwater Range, but less confident of the clusters of fewer volcanoes in southeast Death Valley or in the northern part of the map area.

Concluding remarks

The intent of this chapter has been to illustrate methods for estimating spatial density using kernel functions and optimal bandwidth selectors. Spatial density maps created in this way

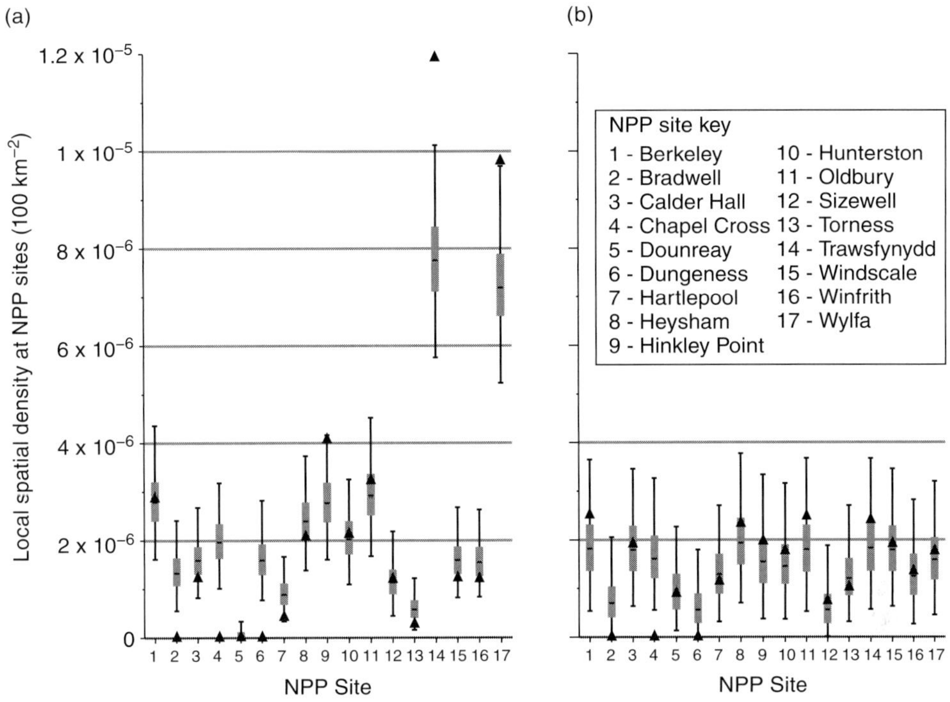

Fig. 14.6 (a) Box plots show the uncertainty in spatial density at 17 NPP sites in the UK, using the NEIC catalog for all earthquakes (a), and the ISC catalog for earthquakes of magnitude > 3 ML (b). The quartile boundaries for each 1200 sample set are enclosed in gray with the median (50%) value indicated by the black horizontal bar. Black triangles represent the original spatial density at each NPP site generated using the original data sets. Spatial densities are remarkably consistent at these 17 sites, with the exception of sites 14 and 17 (a). Removing smaller earthquakes, mostly aftershocks, from the analysis reduces the uncertainty in local spatial density at these two sites (b).

are smooth and differentiable. One advantage of this approach, over say hazard zonation maps, is that the basis of these maps is quantitative and they are entirely reproducible. They do not depend on expert judgment, but only on the appropriate use of the algorithm. Of course, geological boundaries do exist in reality and spatial density can vary quite rapidly across these boundaries (e.g. Connor *et al.*, 2000; Martin *et al.*, 2004). In the absence of many events, such boundaries will likely be missed in the spatial density estimate, and expert judgment becomes necessary. Even in these circumstances, the statistical estimates of spatial intensity retain their utility. Large departures from these estimates in a hazard assessment would have to be very strongly justified.

It is worth comparing the results of the Yucca Mountain volcanic hazard assessment and the UK seismic hazard assessment from the perspective of spatial density estimation, despite their completely different tectonic settings and geological meanings. All of the Yucca Mountain analyses point to complex structure in the spatial density. Orders of magnitude variation in spatial density occur over a narrow region (~ 30 km wide) through the central

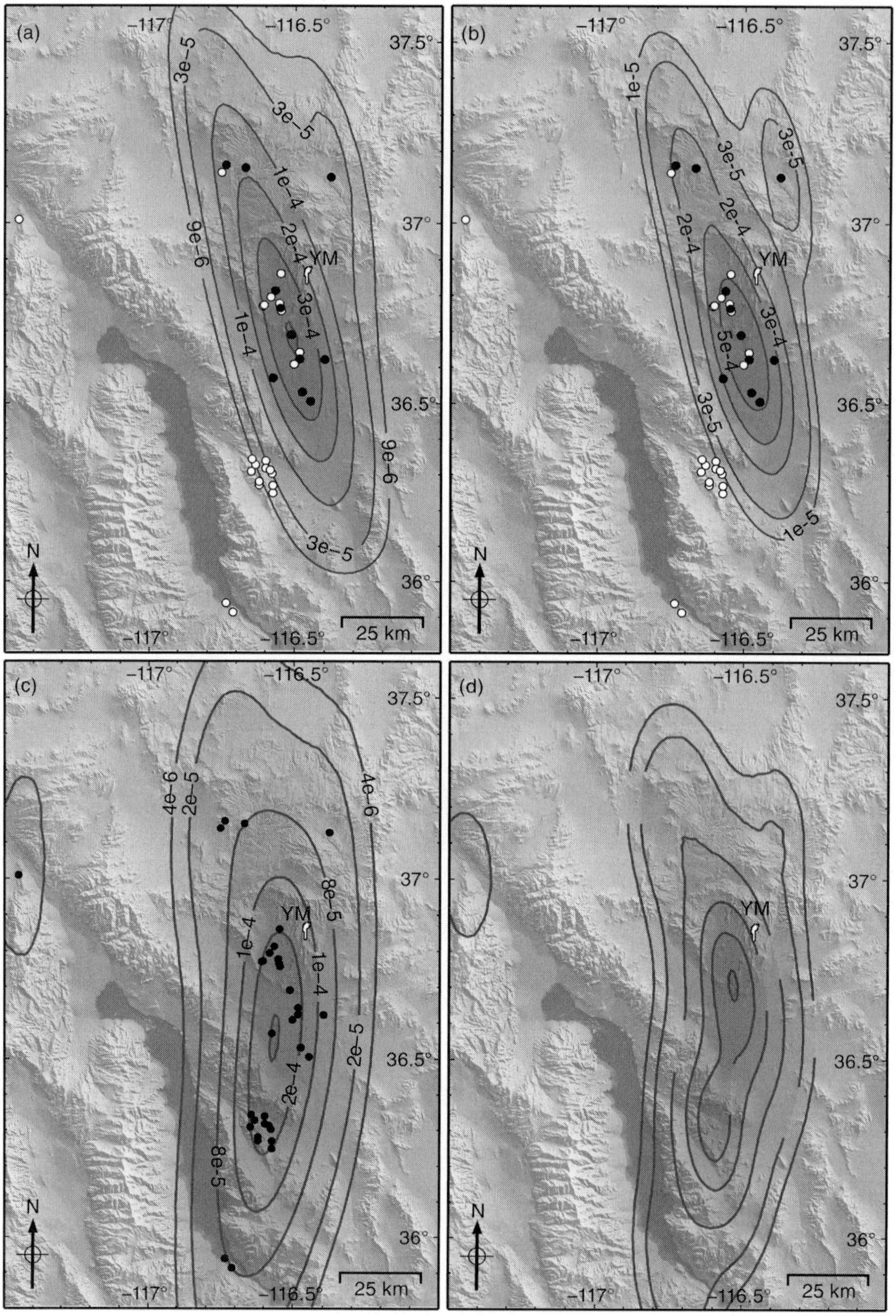

Fig. 14.7 Median spatial density at each grid location (1 km grid spacing) based on 1200 simulations: YMR event data set with (a) SCV and (b) SAMSE bandwidth, AVIP event data set with (c) SCV and (d) SAMSE bandwidth. Contours are 1st, 25th, 50th, 75th, 95th and 99th percentiles of median spatial density. See color plate section.

part of the volcanic zone, and these spatial density values fall off to near zero outside of this zone. In addition, the SAMSE bandwidth selector suggests that significant modes (clusters) occur within the larger, roughly north–south trending zone (Figure 14.3). In contrast, the spatial density of UK earthquakes (> 3 ML) is comparatively featureless. Other than a notable decrease in seismicity in the western portion of the map, seismicity is widely distributed. As a result, there is very little variation in the estimate of spatial density of earthquakes about the 17 NPP sites. Thus, the bandwidth selectors appear to be robust, in the sense that comparatively small and asymmetrical bandwidths were found from the YMR and AVIP event data sets, where significant statistical structure exists, and large and comparatively symmetrical bandwidths were selected for UK seismicity, which appears to be devoid of complex statistical structure.

In the analyses presented here for Yucca Mountain volcanism and UK earthquakes, use of alternative event data sets caused the most significant changes in spatial density models. For example, the greatest change in uncertainty in the Yucca Mountain analysis depended on whether Death Valley and Greenwater volcanism was included or not. The effects of the bandwidth selector were secondary. This follows from the fact that in these examples, and very often in geological hazard assessment, there are relatively few events in total. Thus, in practice, a great deal of care is required in assessing events. This assessment often requires substantial geological or geophysical investigations of the data, and development of geological models of events in the region. Often, it makes sense to carry several data sets through the analysis. We point out that the number of events also affects estimates of the temporal recurrence rate of events, so in a full probabilistic assessment these various event data sets must be used consistently to achieve an unbiased probability estimate.

Spatial density estimation continues to be an area of active research, both in computational statistics and geological hazard assessment. Future research directions that will likely be directly applicable to hazard assessments for nuclear facilities include improvements in automatic bandwidth selection, the use of adaptive kernel functions with automatic bandwidth selectors, and improvements for including geological data directly in spatial density estimates. On this last point, we have not yet been able to incorporate process-level geophysical models of volcanism or seismicity directly in spatial density estimates. Perhaps the greatest improvements in hazard assessments will rely on development of these geophysical models.

Further reading

Texts by Silverman (1986) and Wand and Jones (1995) are excellent starting points for learning about density estimation, especially using kernel methods.

Acknowledgments

Volcanological data discussed in this chapter were largely compiled by Britt Hill, Eugene Smith, Frank Perry and colleagues. We gratefully acknowledge their efforts, and those of the

US Geological Survey and the British Geological Survey in collection of earthquake data. Figures in this chapter were created using the Generic Mapping Tools software of Wessel and Smith (1991). We gratefully acknowledge thoughtful reviews by Neil Chapman, Gordon Woo, Olivier Jaquet and Christian Lantuéjoul, which improved this manuscript.

References

Abrahamson, N. (2006). Seismic hazard assessment: problems with current practice and future developments. Presented at the 1st European Conference on Earthquake Engineering and Seismology, Geneva, Switzerland, September 3–8, keynote address.

Beauvall, C., S. Hainzl and F. Scherbaum (2006). The impact of the spatial uniform distribution of seismicity on probabilistic seismic-hazard estimation. *Bulletin of the Seismological Society of America*, **96**, 2465–2471, doi:10.1785/0120060073.

Condit, C. D. and C. B. Connor (1996). Recurrence rate of basaltic volcanism in volcanic fields: an example from the Springerville volcanic field, AZ, USA. *Geological Society of America Bulletin*, **108**, 1225–1241.

Connor, C. B. and B. E. Hill 1995. Three nonhomogeneous Poisson models for the probability of basaltic volcanism: application to the Yucca Mountain region. *Journal of Geophysical Research*, **100**, 10 107–10 125.

Connor, C. B., S. Magsino and J. Stamatakos *et al.* (1997). Magnetic surveys help reassess volcanic hazards at Yucca Mountain. *Eos, Transactions of the American Geophysical Union*, **78**(7), 7377–7378.

Connor, C. B., J. Stamatakos, D. Ferrill *et al.* (2000). Volcanic hazards at the proposed Yucca Mountain, Nevada, high-level radioactive waste repository. *Journal of Geophysical Research*, **105**, 417–432.

Coppersmith, K. J. and R. R. Youngs (1986). Capturing uncertainty in probabilistic seismic hazard assessments within intraplate tectonic environments. *Proceedings of the Third U.S. National Conference on Earthquake Engineering, Charleston*, **1**, 301–312.

Crowe, B. M., D. T. Vaniman and W. J. Carr (1983). Status of volcanic hazard studies for the Nevada nuclear waste storage investigations, Lab Report LA-9325-MS. Los Alamos National Lab, NM.

Diggle, P. (1985). A kernel method for smoothing point process data. *Applied Statistics*, **34**, 138–147.

Diggle, P. and J. S. Marron (1988). Equivalence of smoothing parameter selectors in density and intensity estimation. *Journal of the American Statistical Association*, **83**, 793–800.

Duong, T. (2007). Kernel density estimation and kernel discriminant analysis for multivariate data in R. *Journal of Statistical Software*, **21**(7), 1–16.

Duong, T. and M. L. Hazaelton (2003). Plug-in bandwidth selectors for bivariate kernel density estimation. *Journal of Nonparametric Statistics*, **15**, 17–30.

Efron, B. and R. Tibishrani (1991). Statistical data analysis in the computer age. *Science*, **253**, 390–394.

Gatrell, A. C., T. C. Bailey, P. J. Diggle and B. S. Rowlingson (1996). Spatial point pattern analysis and its application in geographical epidemiology. *Transactions Institute of British Geographers*, **21**(1), 256–274.

GLOBE Task Team and others (Hastings, D. A., P. K. Dunbar, G. M. Elphingstone *et al.*, (eds.) (1999). *The Global Land One-kilometer Base Elevation (GLOBE) Digital*

Elevation Model, Version 1.0. Boulder, CO: National Oceanic and Atmospheric Administration, National Geophysical Data Center. Digital database at http://www.ngdc.noaa.gov/mgg/topo/globe.html and available on CD-ROMs.

Hall, P., J. S. Marron and B. U. Park (1992). Smoothed cross-validation. *Probability Theory and Related Fields*, **92**, 1–20.

Hornik, K. (2007). The R-FAQ, http://www.r-project.org/, ISBN 3-900051-08-9.

ISC (2007). On-line Bulletin, http://www.isc.ac.uk, International Seismology Centre, Thatcham, United Kingdom.

Jaquet, O. and R. Carniel (2006). Estimation of volcanic hazards using geostatistical methods. In: Mader, H. M. *et al.* (eds.), *Statistics in Volcanology*, Special Publications of IAVCEI 1. London: Geological Society, 77–87.

Jaquet, O., C. B. Connor and L. Connor (2008). Long-term volcanic hazard assessments for nuclear facilities. *Nuclear Technology*, **163**, 180–189.

Jarvis A., H. I. Reuter, A. Nelson and E. Guevara (2006). Hole-filled seamless SRTM data V3, International Centre for Tropical Agriculture (CIAT), available from http://srtm.csi.cgiar.org.

Lutz, T. M. and J. T. Gutmann (1995). An improved method for determining alignments of point-like features and its implications for the Pinacate volcanic field, Sonora, Mexico. *Journal of Geophysical Research*, **100**, 17 659–17 670.

Martin, A. J., K. Umeda, C. B. Connor *et al.* (2004). Modeling long-term volcanic hazards through Bayesian inference: example from the Tohoku volcanic arc, Japan. *Journal of Geophysical Research*, **109**, B10208, doi:10.1029/2004JB003201.

Molina, S., C. D. Lindholm and H. Bungum (2001). Probabilistic seismic hazard analysis: zoning free versus zoning methodology, *Bollettino di Geofisica Teorica et Applicata* **42**(1–2), 19–39.

Musson, R. M. W. (1997). Seismic hazard studies in the UK: source specification problems of intraplate seismicity. *Natural Hazards*, **15**, 105–119.

Musson, R. M. W. and P. W. Winter (1996). Seismic hazard maps of the UK. *Natural Hazards*, **14**(2–3), 141–154.

NEIC/USGS (2007). USGS National Earthquake Information Centre, US Geological Survey, On-line EQ database, http://neic.usgs.gov/neis/epic/epic.html.

Parsons, T., G. A. Thompson and A. H. Cogbill (2006). Earthquake and volcano clustering via stress transfer at Yucca Mountain, Nevada. *Geology*, **34**, 785–788, doi:10.1130/G22636.1.

Press, W. H., B. P. Flannery, S. A. Teukolsky and W. T. Vetterling (1992). *Numerical Recipes in C*, 2nd edn. Cambridge: Cambridge University Press.

Pyle, D. M. (2000). Sizes of volcanic eruptions. In: Sigurdsson, H. *et al.* (eds.), *Encyclopedia of Volcanoes*. New York, NY: Academic Press, 263–271.

Silverman, B. W. (1978). Choosing the window width when estimating a density. *Biometrika*, **65**, 1–11.

Silverman, B. W. (1986). *Density Estimation for Statistics and Data Analysis*. Monographs on Statistics and Applied Probability 26. London: Chapman and Hall.

Smith, E. I., S. Morikawa and A. Sanchez (1997). Volcanism studies related to the probabilistic volcanic hazard at Yucca Mountain for the period 1986–1996. Las Vegas, NV: University of Nevada. See http://www.state.nv.us/nucwaste/yucca/volcan01.htm.

Stirling, M. and S. G. Wesnousky (1998). Comparison of recent probabilistic seismic hazard maps for southern california. *Bulletin of the Seismological Society of America*, **88**, 855–861.

Stock, C. and E. G. C. Smith (2002). Comparison of seismicity models generated by different kernel estimations. *Bulletin of the Seismological Society of America*, **92**, 913–922.

Valentine, G. A. and F. V. Perry (2007). Tectonically controlled, time-predictable basaltic volcanism from a lithospheric mantle source (central Basin and Range province, USA). *Earth and Planetary Science Letters*, **261**, 201–216.

Wand, M. P. and M. C. Jones (1995). *Kernel Smoothing*. Monographs on Statistics and Applied Probability 60. London: Chapman & Hall.

Ward, S. N. (1994). A multidisciplinary approach to seismic hazard in southern California. *Bulletin of the Seismological Society of America*, **84**, 1293–1309.

Weller, J. N., A. W. Martin, C. B. Connor, L. Connor and A. Karakhanian (2006). Modelling the spatial distribution of volcanoes: an example from Armenia. In: Mader, H. M. *et al.* (eds.), *Statistics in Volcanology*, Special Publications of IAVCEI 1. London: Geological Society, 77–87.

Wessel, P. and W. H. F. Smith (1991). Free software helps map and display data. *Eos, Transactions of the American Geophysical Union*, **72**, 441.

Woo, G. (1996). Kernel estimation methods for seismic hazard area source modelling. *Bulletin of the Seismological Society of America*, **86**, 353–362.

Yogodzinski, G. M., T. R. Naumann, E. I. Smith and T. R. Bradshaw (1996). Evolution of a mafic volcanic field in the central Great Basin, south central Nevada. *Journal of Geophysical Research*, **101**, 17 425–17 445.

Zhao, D. (2001). Seismological structure of subduction zones and its implications for arc magmatism and dynamics. *Physics of the Earth and Planetary Interiors*, **127**, 197–214.

15

Cox process models for the estimation of long-term volcanic hazard

O. Jaquet, C. Lantuéjoul and J. Goto

Long-term volcanic hazard is gaining relevance due to increasing societal demands on timescales of hundreds to hundreds of thousands of years as regards the siting for critical facilities (Chapman *et al.*, Chapter 1, this volume). Volcanic hazard represents the probability of occurrence of a potentially damaging volcanic event within a specific period of time in a given region (UNDRO, 1979). For sites near volcanically active regions, long-term volcanic hazards often constitute the dominant source of uncertainty as input for risk assessments. Uncertainty is mainly related to imperfect knowledge of non-linear volcanic processes, to space–time variability of the distribution and intensity for volcanic events and to a limited amount of monitoring information. For these reasons the estimation of volcanic hazard is based on a probabilistic formalism (Sparks, 2003; Sparks and Aspinall, 2004).

Stochastic models have been developed in connection with the proposed high-level radioactive waste (HLW) repository at the Yucca Mountain site (Nevada, USA) in the vicinity of a Quaternary volcanic field to assess the potential for a repository disruption due to basaltic volcanism (Perry *et al.*, 2000). The simplest approach for the estimation of volcanic hazard used a homogeneous Poisson model (Crowe *et al.*, 1982) under the assumption of complete spatial and temporal randomness of the events. A non-homogeneous Weibull–Poisson model applied by Ho (1991) at Yucca Mountain made it possible to estimate the recurrence rate of new volcanoes being formed as a function of time. As shown by Connor and Hill (1994), the hazard estimates provided using this model are very sensitive to the total time under consideration. Indeed, Bebbington and Lai (1996) have demonstrated that the non-homogeneous and non-stationary model of Ho (1991) was an unsatisfactory tool for describing the time behavior of volcanic activity. These authors concluded that since data sets are small and likely incomplete, in the absence of a simple systematic trend, a stationary model is a better choice as it imposes less assumptions on the volcano behavior. Ho and Smith (1998) proposed a non-homogeneous space–time Poisson process for the identification and quantification of volcanic phenomena distributed through space and evolving in time. With the help of this model, they could estimate the probability of at least one volcanic disruption of a repository at the Yucca Mountain site during the next 10 ka. In order to account for the spatial distribution in basaltic volcanic fields, Connor and Hill (1995) presented nearest-neighbor non-homogeneous Poisson models that allowed the estimation of spatial intensity and the associated probability of volcanic eruption occurring in the Yucca Mountain region at a local scale. The application of these models requires using specific kernel methods for the estimation of their parameters (Connor and Connor, Chapter 14, this volume). The non-homogeneous Poisson models were extended by Martin *et al.* (2004) in a Bayesian framework to integrate additional geological and geophysical information. They were used to estimate future locations of volcanoes for siting purposes for a nuclear waste repository in Japan. All these stochastic models are of Poissonian or Markovian type and volcanic eruptions are considered as point processes that can occur either in the time or in the space–time domain. They do not describe space–time structured (non-Markovian) behavior as is likely to be observed in volcanic regions. In addition, no uncertainty is associated with the potential of volcanism, as within the framework of such models it is considered to be deterministic.

Recent efforts to develop a geological HLW repository in Japan (Apted *et al.*, 2004) have motivated the development of specific stochastic models for improving uncertainty characterization. The theoretical basis and concepts of these models are given and then an illustration is provided using a subset of a database of Quaternary volcanoes of Japan, in this case the Quaternary Tohoku volcanic arc in northern Honshu. We emphasize that this case study is only used as an example; no region in Japan is yet considered specifically as a HLW candidate site.

15.1 Model development

The characterization of distribution and occurrence of volcanic events needs to be performed for the purpose of long-term volcanic hazard assessment. From a conceptual point of view, the following pieces of evidence and hypotheses have to be incorporated within our modeling perspective: (i) the spatial distribution of Quaternary volcanic events tends to be clustered, (ii) the spatial distribution of these events is statistically correlated to the geophysical signature of crustal and mantle structures underneath, and (iii) future volcanic events are more likely located in zones of past activity and above zones of partial melting within the mantle.

15.1.1 Potential of volcanism

In order to integrate geological and geophysical information, we introduce the notion of *potential of volcanism* which represents the propensity of a given region to be affected by volcanic events. The potential of volcanism, being unknown, is considered as randomly structured within the context of the stochastic model. While the structured part represents the current geological and geophysical knowledge, the random part describes the uncertainty associated with this information (Figure 15.1). The notion of potential of volcanism can be extended to the time domain where it describes the tendency of a period to be affected by volcanic events.

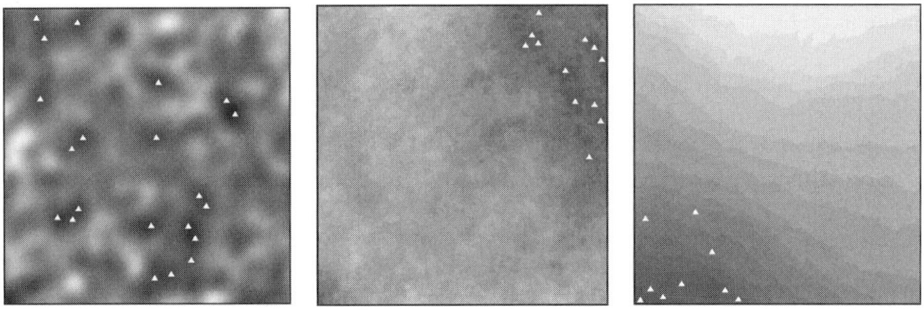

Fig. 15.1 Three regions are shown with different maps for the potential of volcanism. Volcanic events (triangles) are likely to be located in zones where the potential exhibits the darkest shade.

15.1.2 Stochastic model

In the stochastic model, the volcanic events that occurred during the Quaternary period are considered as a particular realization of a (random) point process. Similarly, the potential of volcanism during this period is interpreted as a particular realization of a (positive) random intensity function Z. Partitioning the volcanic region of interest into small domains $\{a_i, i \in I\}$, the potential of domain a_i is denoted by Z_i. In what follows, the two properties of the Cox process (Cox, 1955; Lantuéjoul, 2002) are assumed to be satisfied: (1) the number N_i of volcanic events that occurred in the domain a_i is Poisson distributed with random mean Z_i; (2) given the potentials $Z_i, Z_j, \ldots, Z_k$ of the domains $a_i, a_j, \ldots, a_k$ the numbers of volcanic events $N_i, N_j, \ldots, N_k$ are mutually independent. It is pointed out that property (2) means that the numbers of volcanic events in pairwise disjoint domains are only conditionally independent. They are not independent because the potential conveys its own structure on the Cox process.

Property (1) can be formulated as:

$$P\{N_i = n\} = E\left(e^{-Z_i} \frac{Z_i^n}{n!}\right), \quad n = 0, 1, 2, \ldots \tag{15.1}$$

which implies that N_i is not Poisson distributed unless Z_i is deterministic. Note, however, that N_i and Z_i possess the same mean, and even the same conditional mean:

$$E(N_i|Z) = Z_i \tag{15.2}$$

In other words, given the potential of all domains of interest, the mean number of events occurring in domain a_i is precisely the potential of that domain. Among all factors that contribute to the potential, the location of zones of partial melting at depth is experimentally available from seismic tomographic studies. This geophysical information is also interpreted as a particular realization of a random function $S = \{S_i, i \in I\}$. Owing to the integration of geophysics to the potential, the distribution of the number of volcanic events is not directly dependent on the geophysical information when the potential is known:

$$D(N_i|Z,S) = D(N_i|Z) \tag{15.3}$$

A flexible way to model the dependence relationships between the potential and the geophysical information is to write that their respective transforms in Gaussian space are bigaussian. More precisely, it is assumed that Z and S can be written as:

$$Z_i = \phi_Z\left(Y_i^Z\right), \; S_i = \phi_S\left(Y_i^S\right) \tag{15.4}$$

where ϕ_Z and ϕ_S are two monotonic increasing functions (i.e. Gaussian anamorphosis functions; cf. Section 15.2.2), and where Y^Z and Y^S are two standardized Gaussian random functions. Moreover, Y^Z and Y^S are related by the regression formula:

$$Y_i^Z = \rho Y_i^S + \sqrt{1-\rho^2}\, Y_i^R, \; i \in I \tag{15.5}$$

which involves the correlation coefficient ρ and a third standardized (residual) Gaussian random function Y^R independent of Y^S.

For the estimation of volcanic hazard, the proposed stochastic model (Cox process) is characterized by a multivariate potential of volcanism, since this potential presents dependencies with past volcanic activity as well as with geophysical data.

15.1.3 Conditional simulation

The estimation of volcanic hazard is performed by simulating the distribution of volcanic events likely to occur during a certain period of time in the future within the region of interest. In addition, the simulation has to deliver volcanic events that are more likely to be located in zones of past activity. Therefore, the simulation requires to be conditioned to all available data corresponding to the location of past volcanic events. The idea is to simulate the potential of volcanism conditioned on the number of volcanic events and on the geophysical information known for the analyzed region. The conditioning is achieved only on the numbers of past volcanic events, n_i, and the geophysical data, s_i, known in each domain, a_i. In particular, the exact location of each volcanic event is not taken into account. The proposed algorithm for the conditional simulation of the potential is based on the Gibbs sampler (Geman and Geman, 1984). This is an iterative algorithm that comprises the following steps:

(1) generate $Y_i^R \sim \text{Gaussian}(0,1)$ for each $i \in I$;
(2) select an index i at random;
(3) generate $y_0 \sim D(Y_i^R|Y_j^R = y_j^R, j \neq i)$;
(4) compute $y_i^S = \phi_S^{-1}(s_i)$ and $z_0 = \phi_Z\left(\rho y_i^S + \sqrt{1-\rho^2}\, y_0\right)$;
(5) generate $n_0 \sim \text{Poisson}(z_0)$;
(6) if $n_0 = n_i$ then put $y_i^R = y_0$ and $z_i = z_0$;
(7) go to (2).

Step 1 is used to initialize the Gaussian residuals Y_i^R. This can be done either by generating them separately, as written in the algorithm, or jointly, by resorting to the standard techniques used for simulating Gaussian random functions, such as Choleski decomposition (Wilkinson, 1965), circulant embedding (Dietrich and Newsam, 1997) or turning bands method (Lantuéjoul, 2002). The usual procedure for Step 3 is to sample y_0 from a Gaussian distribution. Its mean is the simple kriging estimate of Y_i^R starting from all y_j^R except y_i^R. Its variance is the corresponding kriging variance (Wackernagel, 2003). The correlation between the potential of volcanism and the geophysics is introduced at Step 4. In its design, the algorithm runs forever. In practice, it is stopped when each potential z_i has been effectively updated more than several 100 times.

At this stage, only the potential $\{z_i, i \in I\}$ of the past volcanic events has been generated. This potential is representative only of the period of time t_p from which all data originate. If the objective is to simulate the volcanic events that will occur during the future period of time t_f, then the future potential $\{z_i^\mathrm{f}, i \in I\}$ is required. Provided that the potential varies very slowly through time, there is no inconvenience to assume that the past and future potentials are proportional. This leads to the following algorithm to simulate the future volcanic events, n_i^f:

(1) compute $z_i^\mathrm{f} = z_i(t_\mathrm{f}/t_\mathrm{p})$ for each $i \in I$;
(2) generate $n_i^\mathrm{f} \sim \mathrm{Poisson}(z_i^\mathrm{f})$ for each $i \in I$.

The conditional simulation algorithm allows the estimation of volcanic hazard for each domain of the region of interest during the period of time considered (Figure 15.2). A Monte Carlo approach is performed using several thousand simulations in order to derive

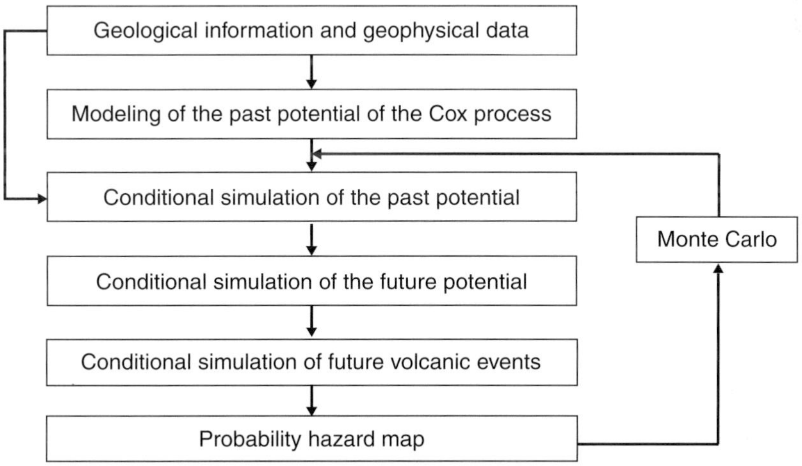

Fig. 15.2 This flow chart displays the stochastic modeling steps for the estimation of volcanic hazard.

stable probability estimates:

$$P\{N_i^{\mathrm{f}} \geq 1\} \approx \frac{1}{K_{\mathrm{sim}}} \sum_{k=1}^{K_{\mathrm{sim}}} 1_k(n_i^{\mathrm{f}} \geq 1) \tag{15.6}$$

where K_{sim} is the total number of simulations and

$$1_k(n_i^{\mathrm{f}} \geq 1)$$

equals 1 when the kth simulation assigns the domain a_i one or more volcanic events, and 0 otherwise.

15.2 Case study of Tohoku region

The issue of concern is the estimation of the probability for the formation of new polygenetic volcanoes over a proposed performance period of 100 000 a. All Quaternary polygenetic volcanoes (those active within the last 1.8 Ma) of the Tohoku region are considered as representing potentially active volcanic areas for the future. The Tohoku volcanic arc is located in northern Honshu, Japan and consists of more than 100 volcanic edifices erupted during the Quaternary (Figure 15.3). A volcanic edifice corresponds to the volcanic event considered for the Tohoku case study; i.e. a polygenetic volcano is composed of one or more edifices. Polygenetic volcanoes are characterized by a geomorphology and geological structures that are created by many episodes of eruptive activity likely to affect broad areas. The formation of new polygenetic volcanoes can take place at locations up to tens of kilometers away from sites of previous eruptions. Some of the volcanoes of the Tohoku region remain active today. We illustrate our methodology using a subset of a database of Quaternary volcanoes of Japan (cf. Mahony *et al.*, Chapter 13, this volume).

15.2.1 Geological and geophysical concept

According to Tamura *et al.* (2002, Chapter 7, this volume), magma production may be controlled by locally developed hot regions within the mantle wedge beneath the Tohoku volcanic arc having the shape of inclined fingers with an average width of 50 km. These structures of the mantle wedge were revealed by various seismic tomography studies (Zhao *et al.*, 1992, 1994, 2000; Hasegawa and Nakajima, 2004). They correspond to inclined seismic low-velocity zones located at depths shallower than about 150 km with an orientation sub-parallel to the down-dip direction of the slab. Such low-velocity zones are likely related to the upwelling-flow portion of the convection mechanically induced by slab subduction (Hasegawa and Nakajima, 2004). This upwelling flow of mantle materials encountering aqueous fluids is expected to produce partial melting in the low-velocity zones (Figure 15.4). The spatial distribution of volcanic events from the Tohoku volcanic arc is characterized by clusters which are located above these zones of partial melting. The

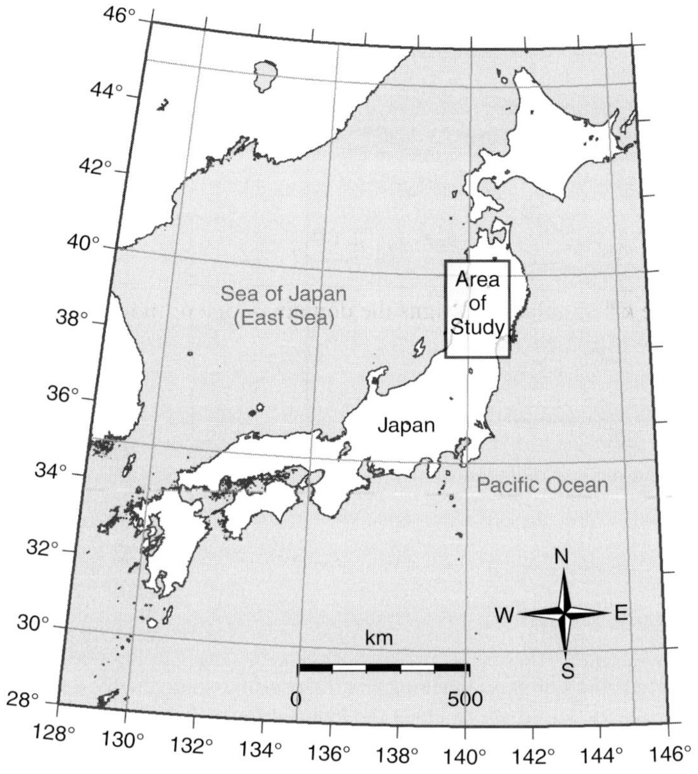

Fig. 15.3 The study region is situated in northern Honshu, Japan.

spacing of volcano clusters is governed by temperature variations along the Tohoku arc, ascent velocities and/or melt content in the upwelling flow of the mantle wedge underneath (Hasegawa and Nakajima, 2004; Hasegawa *et al.*, 2005).

Other geophysical evidence for zones of preferential magma generation beneath the Tohoku volcanic arc include low gravity anomalies and high topographic anomalies that both correlate with volcano clusters, even when Quaternary volcano products are not included in the analysis (Tamura *et al.*, 2002).

During the Quaternary and for the next tens of thousands of years (up to *c.* 0.1 Ma), the distribution of the volcanic events in the Tohoku region is assumed to be controlled by the space–time variability of the magma generation potential of the mantle wedge overlaying the subducting slab (cf. Kondo, Chapter 12, this volume).

The potential of volcanism for the Tohoku region is believed to assimilate geological data related to Quaternary volcanic events as well as seismic tomographic data indicating zones of partial melting at depth. The latter consists of S-wave velocity perturbations along the inclined low-velocity zone in the mantle wedge of northeast Japan (Hasegawa and Nakajima, 2004). Due to the generality of the developed model, other kinds of geophysical

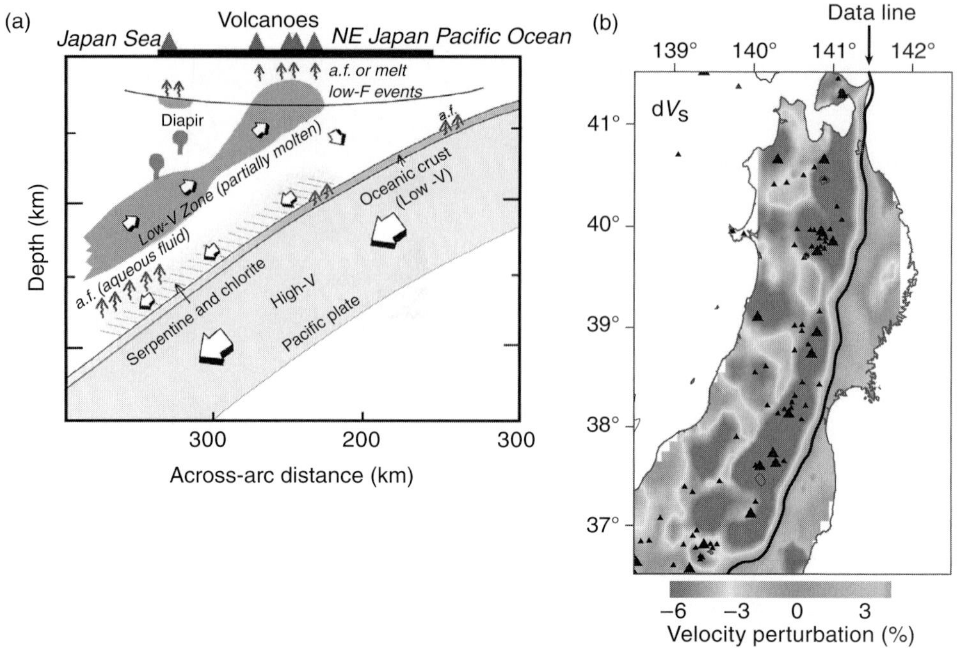

Fig. 15.4 (a) Inferred zones of partial melting and their seismic tomography signature, expressed in terms of S-wave velocity perturbations along the inclined low-velocity zone in the northeast Japan mantel wedge, are correlated with the location of volcanoes (triangles) after Hasegawa and Nakajima (2004). (b) Only the seismic velocity perturbations $\mathrm{d}V_S \leq 0$ located west of the data line were considered for the case study.

data could be applied, if the presence of a statistical correlation can be revealed between volcanic events and the geophysical data.

15.2.2 Estimation of model parameters

The statistical inference of the potential requires the following parameters to be estimated: the potential and the geophysical anamorphoses, ϕ_Z and ϕ_S; the correlation coefficient ρ; and the variograms of the three Gaussian random functions Y^Z, Y^S and Y^R. Formally, the potential anamorphosis is defined by (Chilès and Delfiner, 1999):

$$\phi_Z(y) = F_Z^{-1} \circ G(y)$$

where F_Z is the cumulative distribution function of the potential, and G is that of a standard Gaussian variable. The estimation of F_Z is facilitated by the existence of a bijective correspondence between the distributions of N_i and of Z_i. In the present case, it turns out that a good agreement exists between the experimental distribution of the number of volcanic

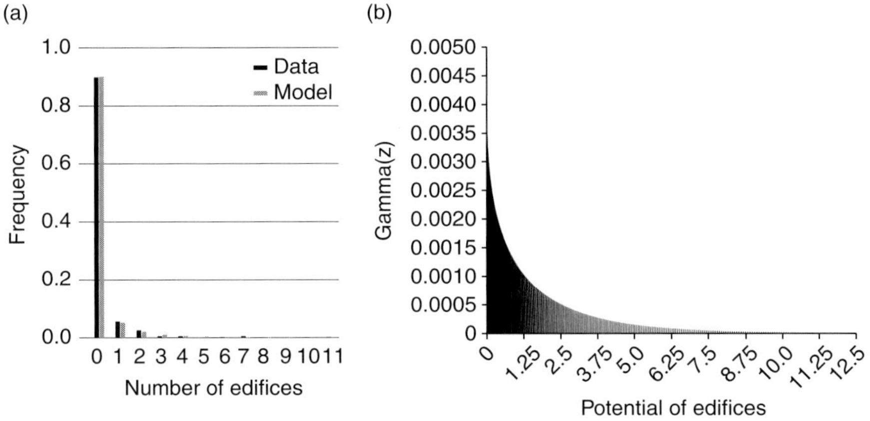

Fig. 15.5 (a) This experimental histogram shows the number of volcanic events (i.e. volcanic edifices) modeled with a negative binomial distribution and (b) the corresponding gamma distribution for the potential of volcanism.

events, and a negative binomial distribution (Figure 15.5):

$$P\{N_i = n\} = \frac{\Gamma(\alpha + n)}{\Gamma(\alpha) n!}(1-p)^\alpha p^n, \; n = 0, 1, 2, \ldots \tag{15.7}$$

with parameter $\alpha = 0.08$ and proportion $p = 0.73$. This automatically ensures that the potential can have its distribution modeled by a gamma distribution (Lantuéjoul, 2002):

$$f(z) = \frac{b^\alpha}{\Gamma(\alpha)} e^{-bz} z^{\alpha-1} \tag{15.8}$$

with the same parameter α and with the scale factor $b = (1-p)/p = 0.37$.

Similarly, the geophysical anamorphosis is defined as:

$$\phi_S(y) = F_S^{-1} \circ G(y)$$

Its estimation is simpler, because F_S is directly available from the data (Figure 15.4). A dichotomic approach is used to compute the explicit values of $\phi_Z(y)$ and $\phi_S(y)$ for y ranging from -5 to $+5$ with steps of 0.01. The other values are retrieved by linear interpolation.

The next stage is the estimation of the correlation ρ. The starting point is the regression $E(N_i|S_i)$ of the number of volcanic events on the seismic tomographic data which is experimentally available. Because the potential encompasses the seismic tomographic data and due to the conditional independence (15.3), this regression coincides with that of the potential on the seismic tomographic data:

$$E(N_i|S_i) = E(Z_i|S_i) \tag{15.9}$$

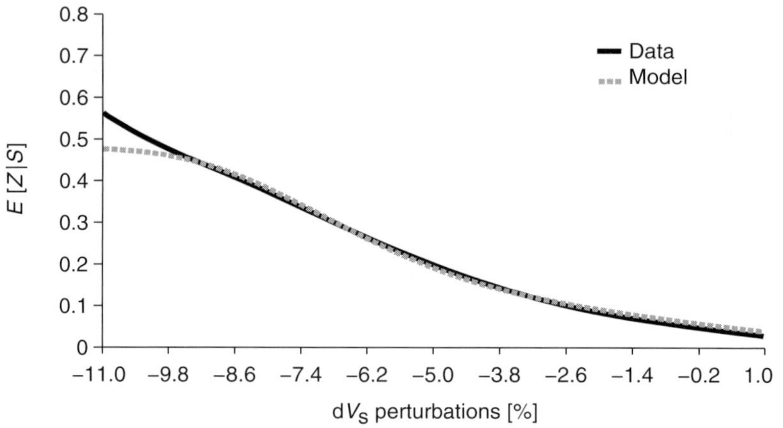

Fig. 15.6 The differences between experimental and modeled conditional mean in the data space are shown, based on the best-fit model. The graphical fit was obtained using value of $\rho = -0.3$ for the coefficient of correlation associated to the conditional mean modeled in the Gaussian space.

On the other hand, the regression of the right-hand side of (15.9) is analytically tractable after the estimation of the potential anamorphosis:

$$E(Z_i|S_i) = E(\phi_Z(Y_i^Z)|Y_i^S) = \int_{-\infty}^{+\infty} \phi_Z\left(\rho Y_i^S + \sqrt{1-\rho^2}\, y\right) g(y)\mathrm{d}y \qquad (15.10)$$

where g stands for the standard Gaussian probability density function. Accordingly, successive guesses are made to find the ρ value that best fits the empirical regression. A satisfactory agreement is obtained for the value $\rho = -0.3$ (Figure 15.6).

The final stage requires the characterization of the spatial behavior for the three Gaussian random functions. It can be specified by their variogram (Chilès and Delfiner, 1999). The variogram allows the detection of spatial correlation and the analysis of randomly structured behaviors. For example, if such a correlation exists for the number of volcanic events, then the closer two domains are, the more similar their numbers of volcanic events; and, vice versa, the further away they are from each other, the more independent their number of volcanic events. The computation of the variogram enables the estimation of the spatial scale at which correlation occurs for the number of volcanic events. This correlation scale can be considered as an estimate of the scale at which clustering of volcanic events occurs for the Tohoku region. This correlation scale, corresponding to the distance at which the variogram stabilizes, was estimated at 30 km using the variogram calculated along the direction 300° N (Figure 15.7). Since no obvious anisotropy was detected, this direction was selected for the graphical fitting of an isotropic spherical variogram model, a function of conditionally negative type (Chilès and Delfiner, 1999). Using such a variogram model implies that the behavior of the number of volcanic events is assumed stationary up to a scale of several hundred kilometers.

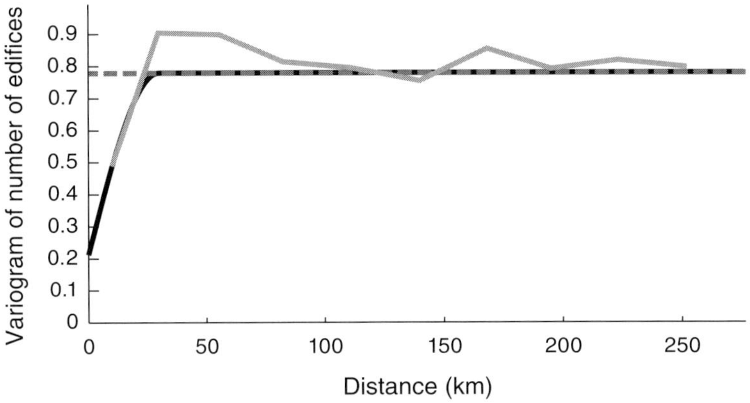

Fig. 15.7 This experimental variogram, calculated along the azimuth 300°, for the number of volcanic events fitted with a spherical variogram model (smooth curve), has a correlation scale of 30 km and a discontinuity at the origin (nugget effect) representing the random variability below the sampling scale (i.e. the 10 × 10 km domain). The horizontal line corresponds to the experimental variance of the data.

Starting from (15.5), the following relationship is obtained for the variogram of Y^Z:

$$Var\left(Y_i^Z - Y_j^Z\right) = \rho^2 Var\left(Y_i^S - Y_j^S\right) + (1-\rho^2) Var\left(Y_i^R - Y_j^R\right) \quad (15.11)$$

only the variograms of Y^S and Y^R have to be inferred. The variogram of Y^S is obtained by computing the experimental variogram of the $\phi_S^{-1}(s_i)$ and modeling it using a spherical variogram.

For the variogram of Y^R, the procedure rests on the fact that the variogram of the potential is experimentally available as a function of the variogram and the mean of the number of volcanic events (Brown *et al.*, 2005):

$$Var\left(Z_i - Z_j\right) = Var\left(N_i - N_j\right) - 2E\left(N_i\right),\ i \neq j \quad (15.12)$$

and also numerically computable from the formula (Chilès and Delfiner, 1999):

$$Var\left(Z_i - Z_j\right) = \int_{-\infty}^{+\infty}\int_{-\infty}^{+\infty} \left(\phi_Z(u) - \phi_Z(v)\right)^2 g_r(u,v)\mathrm{d}u\mathrm{d}v \quad (15.13)$$

where

$$r = 1 - Var\left(Z_i - Z_j\right)/2$$

is the correlation between Y_i^Z and Y_j^Z, and g_r is the standard bigaussian probability density function associated with r. The following algorithm allows the estimation of the variogram

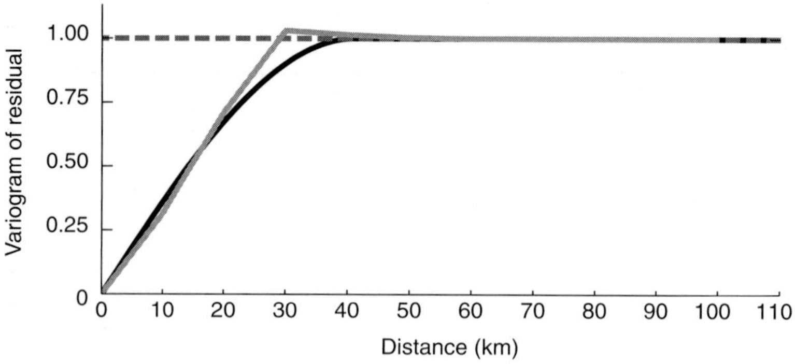

Fig. 15.8 This experimental variogram, calculated for Gaussian residual fitted with a spherical variogram model (smooth curve), has a correlation scale of 41 km. The horizontal line corresponds to the variance of the Gaussian data.

of Y^R at different lags:

(1) make a guess of the value $\gamma = Var\left(Y_i^R - Y_j^R\right)$;
(2) derive from γ a value for r;
(3) compute a value for $Var\left(Z_i - Z_j\right)$ using (15.13);
(4) compare this result with that given by (15.12);
(5) if these values are not comparable then go to (1);
(6) stop.

The resulting (discrete) variogram is also fitted using a spherical variogram model (Figure 15.8) for the purpose of simulation.

15.2.3 Estimation of volcanic hazard

The stochastic model described in Section 15.1 was applied for the estimation of volcanic hazard. With this model, conditional simulations of the number of future volcanic events were carried out over a period of 0.1 Ma for the Tohoku region (Figure 15.9). A Monte Carlo approach was performed using 10 000 simulations in order to obtain stable probability estimates. The likelihood of future volcanic events was displayed in the form of a hazard map for the period of interest related to the siting for critical facilities (Figure 15.10).

By considering a Cox process with a multivariate potential of volcanism, the assimilation of Quaternary geological information and geophysical data becomes operational for hazard calculations. Such a multivariate approach enhances the characterization of uncertainty and is expected to lead to uncertainty reduction when forecasting volcanic activity on the long term. In particular, the use of seismic tomographic data should improve hazard maps for regions located between clusters of past volcanic events.

The comparison of stochastic models with a deterministic potential (i.e. a deterministic intensity function; Connor and Connor, Chapter 14, this volume) shows that the integration of the uncertainty for the potential of volcanism causes the resulting probability estimates to

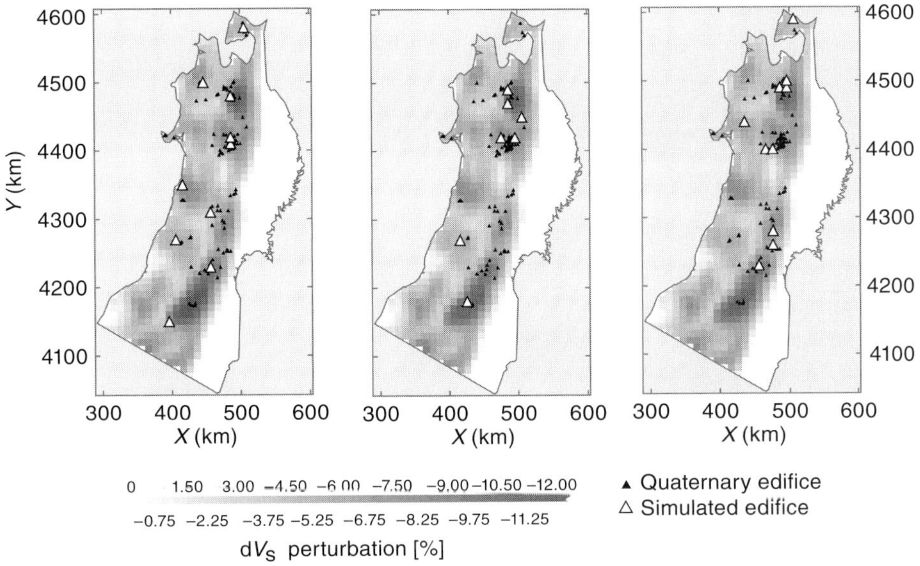

Fig. 15.9 Three Cox simulations with a multivariate potential of volcanism. The simulated edifices are likely to be located in zones with past activity as well as in zones with seismic anomalies (i.e. low dV_S values).

have dissemination effects: i.e. zones of high probability are likely to display lower values and zones of low probability have a tendency toward increased values. These results reveal the importance of accounting for the uncertainty associated with the potential of volcanism when assessing long-term volcanic hazard.

Concluding remarks

A stochastic model is proposed for the long-term assessment of volcanic hazard in relation to the siting for critical facilities. A Cox process, characterized by a multivariate potential of volcanism, was developed to allow for the assimilation of Quaternary geological information and seismic tomographic data. This model accounts for the observed spatial patterns which were characterized using variograms. Since future activity is more likely to occur near, or in the area of, past locations, the simulation method was made conditional on the sites of these events. Such a multivariate approach improves the description of uncertainty of future volcanic activity, in particular for regions located between clusters of past volcanic events.

This stochastic model is part of the development of a probabilistic methodology for the assessment of volcanic hazard for potential HLW repository sites in Japan; it provides a valuable contribution for the evaluation of uncertainties. In particular, as several probabilistic models have been developed for this siting issue in Japan (Jaquet *et al.*, 2008), they will contribute to the perception of the epistemic uncertainty (Woo, 1999). The comparison of hazard results relying on different concepts such as the non-homogeneous Poisson process

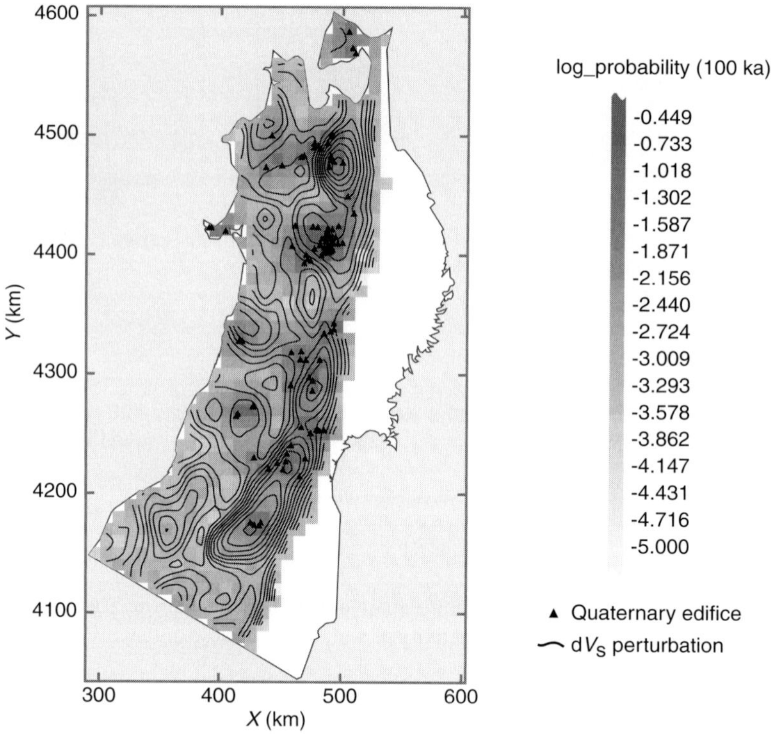

Fig. 15.10 Volcanic hazard map for the Tohoku region and for the next 100 ka (domain scale is 10 × 10 km), based on the cumulative results of the analysis.

(with a smooth deterministic potential) and the Cox model (using a random potential) should deliver some evidence with respect to the variability of this conceptual uncertainty and its possible reduction.

Non-homogeneous patterns for the recurrence rate of volcanic events (Smith and Keenan, 2005) need to be considered even when sufficient age data are lacking. Preliminary results have shown the existence of structured behaviors for the occurrence of volcanic events in time. The use of solely homogeneous Poisson models (in time) is by no means conservative when estimating volcanic hazard at long term. Further developments of the model will allow assimilation of additional information such as rock deformation data (e.g. fault location and strain) and geophysical data (e.g. gravity anomalies, high-resolution seismic tomography; Xia *et al.*, 2007) that also correlate with the spatial distribution of volcanic events. Ultimately, Cox processes with non-stationary potential are likely needed for complex volcanic regions.

Further reading

For an in-depth mathematical presentation of Poisson and Cox process models and their methods of simulation, the reader may refer to the book by Lantuéjoul (2002). And for an extensive and detailed presentation of the variogram and kriging methods, the book by Chilès and Delfiner (1999) offers the whole story.

Acknowledgments

We gratefully acknowledge the support of the Nuclear Waste Management Organization of Japan (NUMO) in making this work possible. Useful discussions with H. Tsuchi, K. Kitayama, H. Kawamura, H. Kondo, Y. Tamura, C. Connor, S. Sparks, N. Chapman and L. Connor improved this chapter.

References

Apted M., K. Berryman, N. Chapman *et al.* (2004). Locating a radioactive waste repository in the ring of fire. *Eos, Transactions of the American Geophysical Union*, **85**(45), 465, 480.

Bebbington, M. S. and C. D. Lai (1996). On nonhomogeneous models for volcanic eruptions. *Mathematical Geology*, **28**(5), 585–600.

Brown, G., C. Lantuéjoul and C. Prins (2005). Sample optimization and confidence assessment of marine diamond deposits using Cox simulations. In: Leuangthong, O. and C.V. Deutsch (eds.) *Geostatistics Banff*. Berlin: Springer Verlag, 315–324.

Chilès, J. P. and P. Delfiner (1999). *Geostatistics: Modeling Spatial Uncertainty*, Wiley Series in Probability and Mathematical Statistics. New York, NY: Wiley.

Connor, C. B. and B. E. Hill (1994). Estimating the probability of volcanic disruption of the candidate Yucca Mountain repository using spatially and temporally nonhomogeneous Poisson models. In: *Focus 'Ninety-Three: Site Characterization and Model Validation Proceedings, Las Vegas, NV, September 26–29, 1993*. La Grange Park, IL: American Nuclear Society, 174–181.

Connor, C. B. and B. E. Hill (1995). Three non-homogeneous Poisson models for the probability of basaltic volcanism: application to the Yucca Mountain region, Nevada. *Journal of Geophysical Research*, **100**(B6), 10 107–10 125.

Crowe, B. M., M. E. Johnson and R. J. Beckman (1982). Calculation of the probability of volcanic disruption of a high-level nuclear waste repository within southern Nevada, USA. *Radioactive Waste Management and the Nuclear Fuel Cycle*, **3**, 167–190.

Cox, D. R. (1955). Some statistical methods connected with series of events. *Journal of the Royal Statistical Society*, Series B (Methodological), **17**(2), 129–164.

Dietrich, C. and G. Newsam (1997). Fast and exact simulation of stationary Gaussian processes through the circulant embedding of the covariance matrix. *Siam Journal on Scientific Computing*, **18**(4), 1088–1107.

Geman, S. and D. Geman (1984). Stochastic relaxation, Gibbs distribution and the Bayesian restoration of images. *IEEE Transactions on Pattern Analysis and Machine Intelligence*, **6**, 721–741.

Hasegawa, A. and J. Nakajima (2004). Geophysical constraints on slab subduction and arc magmatism. In: Sparks, R. S. J. and C. J. Hawkesworth (eds.) *The State of the Planet: Frontiers and Challenges in Geophysics*, Geophysical Monograph Series 150. Washington DC: American Geophysical Union, 81–94.

Hasegawa, A., J. Nakajima, N. Umino and S. Miura (2005). Deep structure of the northeastern Japan arc and its implications for crustal deformation and shallow seismic activity. *Tectonophysics*, **403**, 59–75.

Ho, C.-H. (1991). Time trend analysis of basaltic volcanism at the Yucca Mountain site. *Journal of Volcanology and Geothermal Research*, **46**, 61–72.

Ho, C.-H. and E. I. Smith (1998). A spatial-temporal/3-D model for volcanic hazard assessment: application to the Yucca Mountain region. *Mathematical Geology*, **30**(5), 497–510.

Jaquet, O., C. B. Connor and L. J. Connor (2008). Probabilistic methodology for long-term assessment of volcanic hazards. *Journal of Nuclear Technology*, **163**, 180–189.

Lantuéjoul, C. (2002). *Geostatistical Simulation: Models and Algorithms*. Berlin: Springer Verlag.

Martin, A. J., K. Umeda, C. B. Connor *et al.* (2004). Modeling long-term volcanic hazard through Bayesian inference: an example from the Tohoku volcanic arc, Japan. *Journal of Geophysical Research*, **109**, B10208, doi:10.1029/2004JB003201.

Perry, F. V., B. M. Crowe and G. A. Valentine (2000). Analyzing volcanic hazards at Yucca Mountain. *Los Alamos Science*, **26**, 492–493.

Smith, E. I. and D. L. Keenan (2005). Yucca Mountain could face greater volcanic threat. *Eos, Transactions of the American Geophysical Union*, **86**(35), 317, 321.

Sparks, R. S. J. (2003). Forecasting volcanic eruptions. *Earth and Planetary Sciences Letters*, **210**, 1–15.

Sparks, R. S. J. and W. P. Aspinall (2004). Volcanic activity: frontier and challenges in forecasting, prediction and risk assessment. In: Sparks, R. S. J. and C. J. Hawkesworth (eds.) *The State of the Planet: Frontiers and Challenges in Geophysics*, Geophysical Monograph Series 150. Washington DC: American Geophysical Union, 359–373.

Tamura, Y., Y. Tatsumi, D. Zhao, Y. Kido and H. Shukuno (2002). Hot fingers in the mantle wedge: new insights into magma genesis in subduction zones. *Earth and Planetary Sciences Letters*, **197**, 105–116.

UNDRO (1979). Natural Disasters and Vulnerability Analysis, Report of Expert Group Meeting (9–12 July). Geneva: Office of the United Nations Disaster Relief Coordinator.

Wackernagel, H. (2003). *Multivariate Geostatistics*, 3rd edn. Berlin: Springer Verlag.

Wilkinson, J. H. (1965). *The Algebraic Eigenvalue Problem*. New York, NY: Oxford University Press.

Woo, G. (1999). *The Mathematics of Natural Catastrophes*. London: Imperial College Press.

Xia, S., D. Zhao, X. Qiu *et al.* (2007). Mapping the crustal structure under active volcanoes in central Tohoku, Japan using P and PmP data. *Geophysical Research Letters*, **34**, L10309, doi:1029/2007GL030026.

Zhao, D., A. Hasegawa and S. Horiuchi (1992). Tomographic imaging of P and S wave velocity structure beneath northeastern Japan. *Journal of Geophysical Research*, **97**, 19 909–19 928.

Zhao, D., A. Hasegawa and H. Kanamori (1994). Deep structure of Japan subduction zone as derived from local, regional and teleseismic events. *Journal of Geophysical Research*, **99**, 22 313–22 329.

Zhao, D., F. Ochi, A. Hasegawa and A. Yamamoto (2000). Evidence for the location and cause of large crustal earthquakes in Japan. *Journal of Geophysical Research*, **105**, 13 579–13 594.

16

Spatial distribution of eruptive centers about the Idaho National Laboratory

P. H. Wetmore, S. S. Hughes, L. J. Connor and M. L. Caplinger

Regional volcanic hazard investigations require an in-depth understanding of a region's spatial and temporal distribution of volcanic vents and variations in eruption rates. Usually assessments are based solely upon the distribution of vents and eruptive centers exposed at the surface. These assessments commonly assume relatively simple tectono-magmatic settings and evolutions (e.g. Connor *et al.*, 1992; Conway *et al.*, 1998). Hazard studies within tectonically complicated regions, such as the Basin and Range of the western USA (e.g. Yucca Mountain, Connor *et al.*, 2000; Valentine and Perry, Chapter 19, this volume), have demonstrated the need for more accurate knowledge of the regional volcanic stratigraphy. In this chapter, we describe an analysis of volcanic hazards that includes this more comprehensive view of volcano stratigraphy. Such detailed investigations, accounting for differential subsidence and the burial of older volcanic features, can vastly improve the accuracy of any volcanic hazard assessment.

The Idaho National Laboratory (INL) comprises several nuclear facilities, including the oldest power reactor in the world (see Chapman *et al.*, Chapter 1, this volume). The INL is located in a region of volcanic hazards stemming from its position on the eastern Snake River Plain (ESRP). The ESRP is one of the most volcanically active regions in North America. Recent volcanism on the plain is characterized by the effusion of very low viscosity lavas. The resulting lava flows are often < 10 m thick, but inundate vast areas, up to 1500 km^2. Volcanism on the ESRP is predominantly monogenetic, meaning that renewed volcanic activity and accompanying lava flows form from new batches of melt and issue from new volcanic vents (see Connor *et al.*, Chapter 3, this volume). Additionally, the ESRP transects, but continues to be structurally affected by, the northern Basin and Range Province. Thus, volcanic hazard assessments must consider not only the temporal rates of volcanic activity, the potential magnitudes of eruptions and the potential distribution of future volcanic vents from which lavas effuse, but also the potential complications associated with active normal faulting (e.g. differential vertical motions).

An accurate understanding of the spatial variability in recurrence and accumulation rate of volcanism in the central ESRP at and near the INL is important due to the presence of nuclear reactors and other highly sensitive facilities located within this US Department of Energy site. Most hazard assessments for the region have worked under the assumption that volcanism is focused into northwest-trending volcanic rift zones (VRZs; Hackett and Smith,

1994). However, a two-decade-long extensive study of the stratigraphy at depth beneath the INL (Anderson *et al.*, 1996, 1997; Champion *et al.*, 2002; Hughes *et al.*, 2002a) provides a unique opportunity to assess the spatial and temporal evolution of magmatism within one of the most productive basaltic volcanic fields in North America. These investigations reveal the locations of ≈ 50 buried eruptive centers/vents, and led to identification of differential subsidence of various parts of the plain. In this chapter we employ the results of these studies to illustrate how the distribution of exposed vents in the central portion of the ESRP results from the complex interplay between spatial variability in the recurrence of volcanic events, accumulation of lava flows, and differential subsidence and sedimentation.

In the following sections we describe the distribution of exposed and buried vents at and near the INL, the spatial variability of vent density and basalt accumulation, and the role and various scales of subsidence of the ESRP with potential connections to faulting in the adjacent Basin and Range Province. We then relate the distribution of volcanic vents and their preservation at the surface to the competing effects of spatial variations in magma productivity and subsidence. Finally, we demonstrate how including these vents, now buried in the subsurface, changes models of spatial density of vent distribution, and hence the volcanic hazard assessment for the INL.

16.1 Geology of the ESRP and the INL region

The ESRP of southeastern Idaho is a major tectonic depression, under-filled with late Cenozoic volcanic and sedimentary strata (Figure 16.1). The plain formed in the wake of the Yellowstone hot spot (Pierce and Morgan, 1992) as rhyolitic volcanism ceased and basaltic volcanism flared up. Basaltic volcanism of the plain is dominated by low-volume ($< 6\,\text{km}^3$) monogenetic eruptions (Kuntz *et al.*, 1986, 1992) with some evidence for isolated larger-volume ($8-20\,\text{km}^3$) flows during the late Pleistocene (Wetmore *et al.*, 1997; Wetmore, 1998; Scarberry, 2003).

Basaltic volcanism of the ESRP is traditionally inferred to be asymmetrically distributed into a series of narrow, curvilinear and northwest-trending VRZs, based on the distribution of exposed eruptive centers (Kuntz, 1977a, 1977b; Kuntz *et al.*, 1992). Boundaries of these VRZs are outlined in Figure 16.2. Many of these VRZs appear to be continuations of range-front faults of the Basin and Range Province to the north of the plain (e.g. Arco–Big Southern Butte VRZ and the Big Lost River fault zone). Similar to many other such basaltic volcanic fields of the Basin and Range Province, the VRZs of the ESRP are characterized by aligned vents, non-eruptive fissures, and small offset normal faults that are approximately parallel to the margins of the rift zones and the traces of the range-front faults in the adjacent Basin and Range (Kuntz *et al.*, 1992, 1994, 2002). Throughout the remainder of this chapter the VRZs will be utilized as geographic reference points. We will return to the issue of their viability as zones of focused magmatism in the discussion section.

The INL is located within the central portion of the ESRP (as shown in Figures 16.1 and 16.2) and is transected, from southwest to northeast, by four inferred VRZs, Arco–Big Southern Butte, Howe–East Butte, Lava Ridge–Hell's Half Acre and Circular

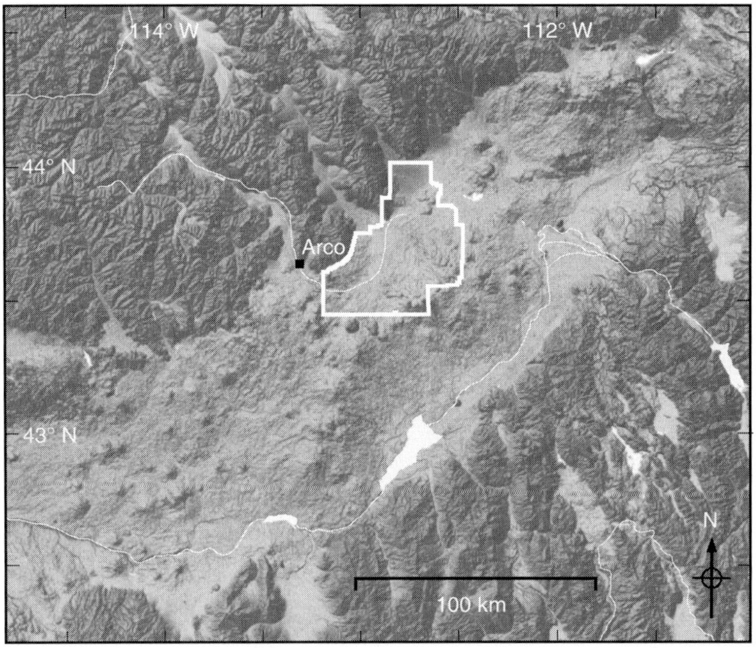

Fig. 16.1 This digital elevation model (DEM) of the ESRP shows its unique geology, with subdued topography compared with the surrounding fault-block mountain ranges. The ESRP itself is armored by low-viscosity lava flows that produce an overall smooth surface. The Idaho National Laboratory (INL) is centrally located and outlined in white. The Axial Volcanic Zone is a curvilinear string of large volcanoes that form a low ridge, trending northeast along the southest border of the INL. The town of Arco is indicated for reference.

Butte–Kettle Butte. Three of these VRZs, Arco–Big Southern Butte, Lava Ridge–Hell's Half Acre and Circular Butte–Kettle Butte, appear to be continuations of range-front faults. Additionally, the Axial Volcanic Zone approximately parallels and overlaps the southern and southeast boundary of the site.

Most of the INL occupies a low-relief Pleistocene–Holocene depositional basin known as the Big Lost Trough (Gianniny *et al.*, 1997, 2002). This basin (outlined in Figure 16.2) is bounded by the relatively high topography of the Arco–Big Southern Butte VRZ on the southwest, the Axial Volcanic Zone on the southeast, the Circular Butte–Kettle Butte VRZs to the northeast, and the Basin and Range Province to the northwest. It is transected by both the Howe–East Butte and Lava Ridge–Hell's Half Acre VRZs in its central and northeast portions, respectively. Relative to the bounding volcanic zones, sedimentation rates are as much as two to three times higher in the Big Lost Trough, as shown in Figure 16.3.

The Axial Volcanic Zone (Figure 16.2) is another region of the ESRP where volcanism is believed to be focused (Hackett and Smith, 1994). This feature parallels the axis of the ESRP and is most prominent near the central portion of the plain along the southeastern

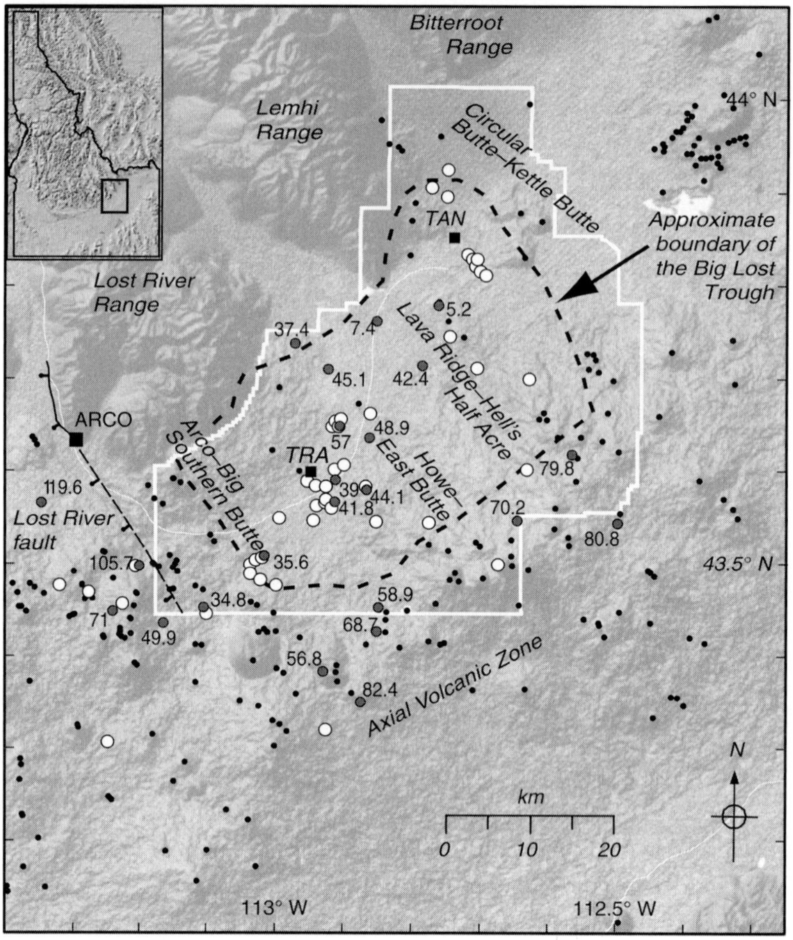

Fig. 16.2 This map of the central ESRP shows the INL border outlined in white, labeled boundaries of the VRZs and the Axial Volcanic Zone, exposed vents as black dots, inferred vents in the subsurface as white circles and the Big Lost Trough enclosed by a circular dashed line within the INL border. The distribution of exposed and inferred vents from Anderson and Liszewski (1997) shows that while many exposed vents lie within the VRZs described by Kuntz *et al.* (2002), many inferred vents lie between these zones. Gray circles are labeled with accumulation rates of basalt (mm a^{-1}) averaged over the last 600 ka that demonstrate some spatial variability in the accumulation of basalt, especially between the Axial Volcanic Zone and the Big Lost Trough. The black line tracing the Lost River Fault has barbs on the hanging wall. The dashed portion is the inferred position within the ESRP based on the location of small offset fault scarps in the Arco–Big Southern Butte VRZ (Kuntz *et al.*, 1994).

boundary of the INL. Eruptions of the Axial Volcanic Zone lavas have erected a low but massive topographic high, which is clearly visible on the DEM (Figure 16.1). Here the Axial Volcanic Zone serves as a drainage divide preventing rivers flowing south out of the Basin and Range to the north from crossing the plain. Rather, these rivers are diverted into

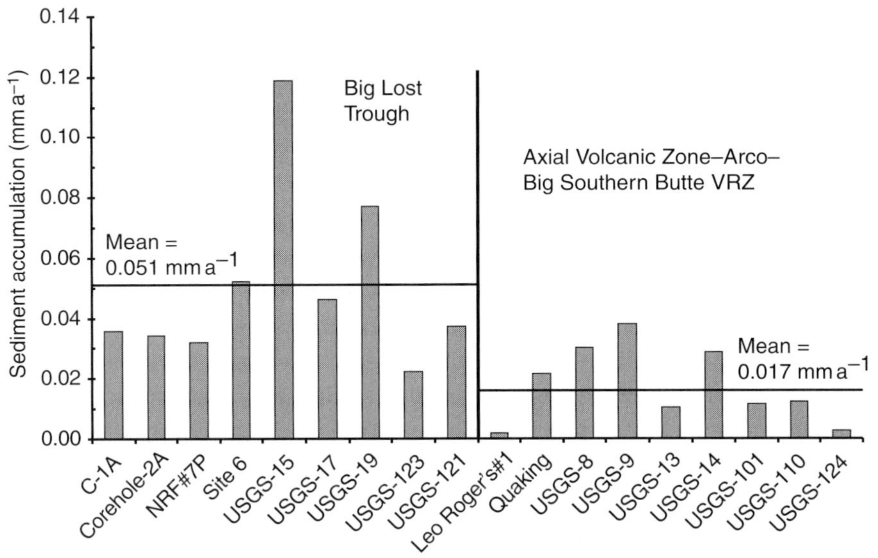

Fig. 16.3 Bars show sediment accumulation rates for selected wells within the Big Lost Trough and the Arco–Big Southern Butte VRZ and the Axial Volcanic Zone. Rates are calculated using stratigraphic distribution, thickness and age data from Anderson *et al.* (1996, 1997). The vertical bar separates wells from the two subprovinces.

large ephemeral lakes (e.g. Lost River Sinks) whose waters seep into the Snake River Plain aquifer (Bartholomay *et al.*, 2002). The Axial Volcanic Zone is also a region of the plain where highly differentiated magmas are extruded, such as those representing the high silica rhyolite domes and intermediate composition Cedar Butte (Hayden, 1992; McCurry *et al.*, 1999, 2008).

As a result of the need to remediate contamination of the Snake River Plain aquifer, geophysical, geochronological and geochemical/petrological data collected from several hundred wells and a few dozen coreholes help to define the subsurface distribution and stratigraphy of lava flows in the vadose zone and uppermost part of the Snake River Plain aquifer at and near the INL (Anderson and Lewis, 1989; Anderson, 1991; Anderson and Bartholomay, 1995; Anderson and Bowers, 1995; Anderson and Liszewski, 1997; Anderson *et al.*, 1996, 1997). The correlations of subsurface lava flows made by Anderson and co-workers were employed by Anderson and Liszewski (1997) and Wetmore (1998) to establish the locations of buried eruptive centers in the central portion of the ESRP. In general, this was accomplished through the construction of isopach and structural contour maps of upper and lower flow surfaces for each identified and regionally correlated flow package observed in the coreholes and wells. A full description of their methods for identifying the positions of buried eruptive centers can be found in those references. In addition to locating eruptive centers, the stratigraphic data of Anderson *et al.* (1996) were also used by Wetmore (1998) and Champion *et al.* (2002) to define regions of differential subsidence and uplift in the area of the INL. Similarly, Blair (2002) cited variation of the elevations of correlative sedimentary units between cores from the southern portion of the INL as evidence of post-depositional differential subsidence of the Big Lost Trough.

16.2 Spatial variation in basaltic magmatism

Within the central portion of the ESRP, 232 exposed eruptive centers have been located near the INL (Kuntz *et al.*, 1994). These are primarily confined to the Arco–Big Southern Butte VRZ, the Axial Volcanic Zone, and clusters of centers around the Test Area North (TAN) facility (locations are plotted in Figure 16.2). Within the central portion of the INL, eruptive centers and their encompassing shields are limited in number and prominence as they tend to be partially buried by younger flows and sediments. In the east-central and northern portions of the INL, a northwest-trending alignment of vents, west and southwest of the TAN facility, defines the Lava Ridge–Hell's Half Acre VRZ. East and southeast of the TAN facility are three vents (Antelope Butte, Circular Butte and an unnamed butte) that form the northern end of the Circular Butte–Kettle Butte VRZ.

Ages of these exposed eruptive centers reveal a pattern of younger (i.e. those less than ~ 200 ka) eruptive centers restricted to the Axial Volcanic Zone (Kuntz *et al.*, 1994). The oldest eruptive centers within the four VRZs are concentrated in the northwest portions of the zones (> 500 ka for the Arco–Big Southern Butte and > 730 ka for the other three VRZs). The youngest centers are present to the southeast where the VRZs intersect the Axial Volcanic Zone. The isolated and partially buried centers in the central part of the INL, especially those near INTEC, a facility in the middle of the site, are relatively old at ~ 650 ka (Anderson and Liszewski, 1997).

16.2.1 Inferred eruptive centers

Two recent studies of the basaltic flow field architecture in the subsurface of the central ESRP at and near the INL (Anderson and Liszewski, 1997; Wetmore, 1998) identify ≈ 47 additional eruptive centers concealed by younger lava flows and sediment (locations in Figure 16.2). While the methods for identifying the locations of vents concealed in the subsurface are not explicitly described by Anderson and Liszewski (1997), it is clear, based on comparison with well and corehole data from Anderson *et al.* (1996), that the vent location for any particular unit closely coincides with the region of thickest basalt accumulation where that unit is correlated in wells or coreholes. Wetmore (1998), using the Anderson *et al.* (1996) data, generated isopach and structural contour maps of the upper and lower surfaces of the flow units, to identify areas of greatest flow field thickness. These areas are coincident with the higher elevation of the upper surface to locate concealed eruptive centers.

The majority of the 47 vents identified by Anderson and Liszewski (1997) are located near facilities in the southwestern portion of the INL and south of the TAN facility (Figure 16.2).

Wetmore (1998) identified 32 vents within the southern INL area. In general, between the two studies, the differences in the locations of vents from the same basalt flow group are small (< 2 km). Therefore, those inferred by Anderson and Liszewski (1997) will be used in the analyses of vent distribution in the following sections, due to the broader area investigated in their study.

Throughout the central portion of the INL nearly all vents ranging in age from $\sim$650–247 ka are concealed in the subsurface. Only vents from the largest flow field (i.e. the *I* flow, $626 + 67$ ka; Anderson *et al.*, 1997) and the youngest extrusions (i.e. the *B* flow, $221 + 2$ ka; Anderson *et al.*, 1997) are exposed at the surface within the southern portion of the INL.

One way to visualize the distribution of vents near the INL is to generate a histogram plot relating the number of vents to distance across or along the ESRP (Figure 16.4). We overlaid a grid onto the distribution map shown in Figure 16.2 and tabulated the number of vents in each box as a function of distance from the margin of the plain (Figure 16.4a) and as a function of distance from the southwestern edge of the map (Figure 16.4b, c). Vents were counted in bins that were 5 km on a side. The vents shown in these plots lie between the northwest margin of the ESRP and $\approx$ 25 km southeast of the axis of the plain, and from $\approx$ 20 km southwest of the Lost River Fault/Arco–Big Southern Butte VRZ and $\approx$ 5 km northeast of the Beaverhead fault. The overlain grid is oriented with one axis approximately parallel to the axis of the ESRP, while the other axis is approximately parallel with the trends of the VRZs and the range-bounding faults of the Basin and Range Province to the north.

Figure 16.4a plots the number of vents with distance from the northwest margin of the ESRP, including the vents inferred to exist in the subsurface. In this plot there are relatively few vents in the northwest part of the ESRP as the grid extends $\approx$ 15 km beyond the actual margin in the southwest portion of the map area (Figure 16.2), so the plot also includes those vents near Arco and the TAN facility. Throughout the middle portion of the transect (i.e. 20–70 km) the average number of vents per bin is $\approx$ 20. Relatively few vents occur southeast of the Axial Volcanic Zone, which Kuntz (1992) attributes to a low-density barrier formed by the Taber caldera, a buried rhyolitic eruptive center associated with the passage of the Yellowstone hot spot (Pierce and Morgan, 1992).

Figure 16.4b and c plot the number of vents with distance from a southwest to northeast transect that crosses three faults and four VRZs. Figure 16.4b uses all vents, including those from the Axial Volcanic Zone, while Figure 16.4c excludes vents from the latter volcanic zone to emphasize the variability within and between VRZs and as a function of distance from faults. Figure 16.4c does not seem to support the inference that VRZs represent zones of focused volcanism since they are, in all cases, not coincident with the most prominent vent clusterings or surrounded by regions of reduced vent clustering. Similarly, the distribution of volcanoes relative to the location of faults is inconsistent. For example, the Lost River fault bisects a region with an average of 11.5 exposed vents per 5 km, while there are no exposed vents within 5 km southwest or northeast of the Lemhi fault. Between the two faults, however, is a region with relatively few exposed vents but an average of $\approx$ 5 vents per 5 km inferred to be present in the subsurface of the ESRP.

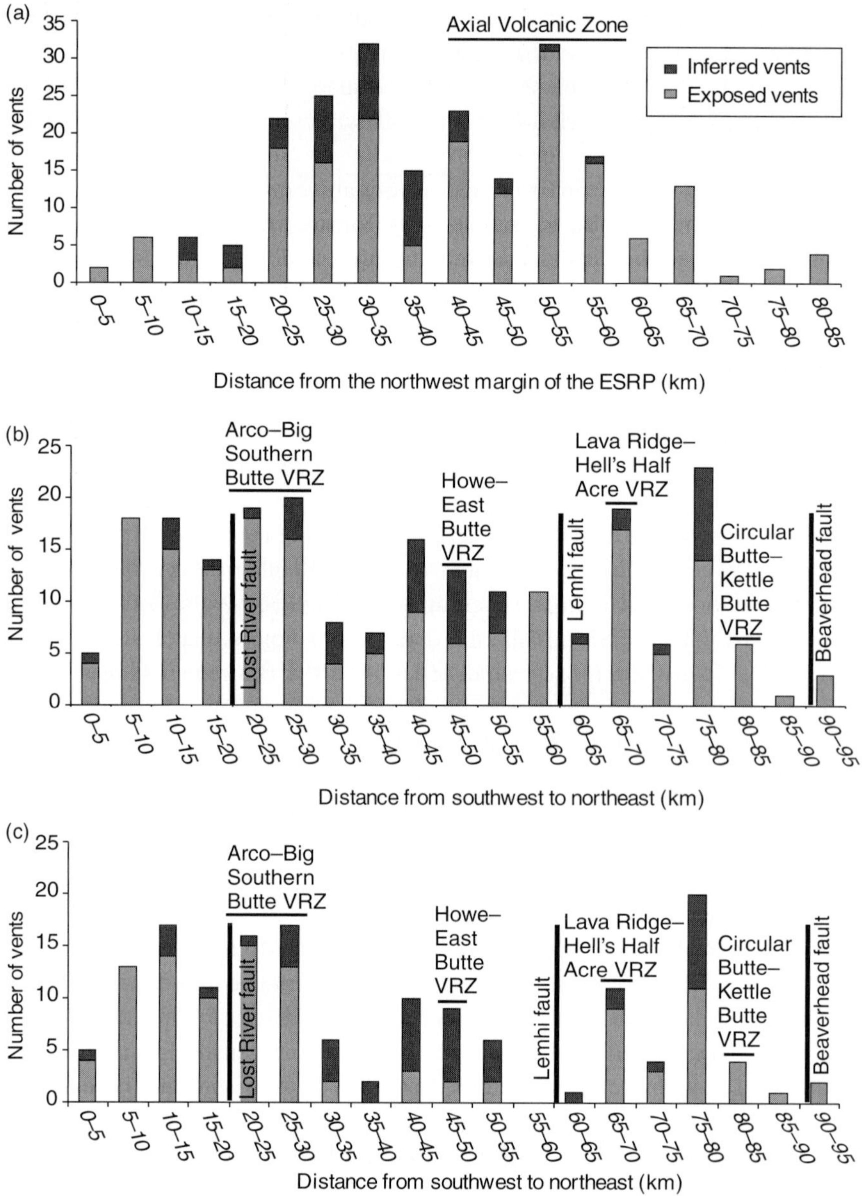

Fig. 16.4 The histograms show the distributions of exposed and inferred vents at and near the INL, inferred locations from Anderson and Liszewski (1997). Plot (a) shows the distribution of vents from the northwest margin of the plain to the axis and beyond. Plot (b) shows vent distributions from southwest of the Arco–Big Southern Butte VRZ to northeast of the Circular Butte–Kettle Butte VRZ. Plot (c) is identical to plot (b) but excludes vents within the Axial Volcanic Zone.

16.2.2 Regional density estimation using Gaussian smoothing

An alternative way to view the clustering and regional spatial distribution of volcanic vents located within the ESRP is by kernel density estimation based on vent locations. A robust and unbiased approach uses a Gaussian kernel function and an optimized bandwidth selector algorithm to calculate the spatial density across an area. Since spatial density is highly sensitive to the smoothing bandwidth used in the calculation, an optimal bandwidth is chosen by an unbiased algorithm that is based solely on the vent locations, in other words bandwidth selection is data-driven. This method is discussed in detail by Connor and Connor (Chapter 14, this volume).

For the exposed vents within the ESRP, the derived optimal bandwidth, **H**, is a 2 × 2 matrix of values given by:

$$\mathbf{H} = \begin{bmatrix} 122 & 79 \\ 79 & 87 \end{bmatrix}$$

which specifies the amount of spatial variation in the east–west and north–south directions and the overall trending direction of the data. The square root of the bandwidth matrix gives the following matrix:

$$\sqrt{\mathbf{H}} = \begin{bmatrix} 10.2 & 4.3 \\ 4.3 & 8.3 \end{bmatrix}$$

which corresponds to the amount of smoothing in km. Specifically, for exposed vents in the ESRP, the optimal smoothing distance is $\approx$ 10.2 km in an east–west direction, $\approx$ 8.3 km in a north–south direction, with a positive overall clustering trend from the southwest towards the northeast at $\approx 45°$ from north.

Similarly, the optimized bandwidth matrix, **H**, for calculating the spatial density of vents based on exposed and subsurface vents is given by:

$$\mathbf{H} = \begin{bmatrix} 130 & 82 \\ 82 & 88 \end{bmatrix}$$

and the square root of this matrix is given by:

$$\sqrt{\mathbf{H}} = \begin{bmatrix} 10.5 & 4.4 \\ 4.4 & 8.1 \end{bmatrix}$$

which corresponds to an optimal smoothing distance of $\approx$10.5 km in the east–west direction, $\approx$ 8.1 km in the north–south direction, with the overall clustering exhibiting the same southwest to northeast trend. Overall, the addition of subsurface vents to the exposed vent population does not significantly change the dimensions of the optimal smoothing bandwidth for this enlarged population. To see the unique smoothing effect of this bandwidth matrix, a spatial density map was calculated based on a single vent centrally located within the ESRP region. The results are shown as an x–y plot in Figure 16.5.

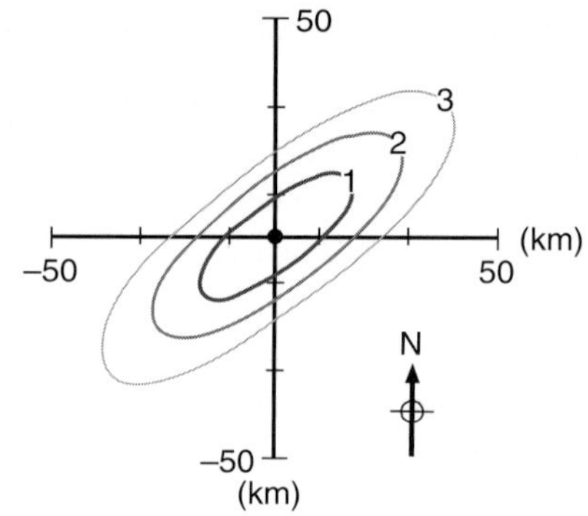

Fig. 16.5 This plot illustrates the unique Gaussian smoothing effect of the optimal bandwidth matrix centered around one vent. The area within the contours (from inside to outside) represent 68%, 95% and 99.7% of the total spatial density, that is 1, 2, and 3 standard deviations from the mean spatial density. A map is generated by summing this smoothing effect for each vent and then normalizing the grid to create a probability density surface (i.e. the surface integrates to one) which is then contoured as in Figures 16.6 and 16.7.

A spatial density estimation across the entire ESRP region, an approximate 325 km by 280 km area (which includes the INL), is calculated by summing the smoothing effect of the bandwidth around each event in the data set and then normalizing the entire grid so that the vent density integrates to 1. Figure 16.6 shows the density grid calculated for exposed vents within the ESRP contoured at quartile intervals (i.e. 25%, 50%, 75%) and intervals representing 95% and 99% of the entire spatial density (i.e. includes approximately two and three standard deviations from the mean spatial density). Notice the area of low density within the central region of the INL in Figure 16.6.

Figure 16.7 shows the spatial density variation across the region based on exposed and subsurface vents. The obvious difference between the two spatial density maps (Figures 16.6 and 16.7) is the disappearance of the low-density area that was centrally located within the INL boundary, obviously due to the inclusion of the subsurface vents in the analysis. This difference can be quantified by differencing the two spatial density grids, thereby creating a third grid which illustrates the relative differences in density (Figure 16.8). An interesting question is, does the spatial density map represent an accurate estimation of the probability of volcanic hazard to the INL facility and does the inclusion of subsurface (inferred) vents increase that hazard?

Based on these regional spatial density maps, contours of equal density are mostly elongate in a northeast direction, approximately parallel to the axis of the ESRP. Most prominent are three high-intensity zones located adjacent to the northwest margin of the ESRP. The Axial Volcanic Zone makes a fourth, less intense zone, off of the main trend. These regional spatial density maps do not appear to capture the pattern of discrete northwest trending zones of volcanism, perpendicular to the ESRP, as shown by the locations of the VRZs in Figure 16.2.

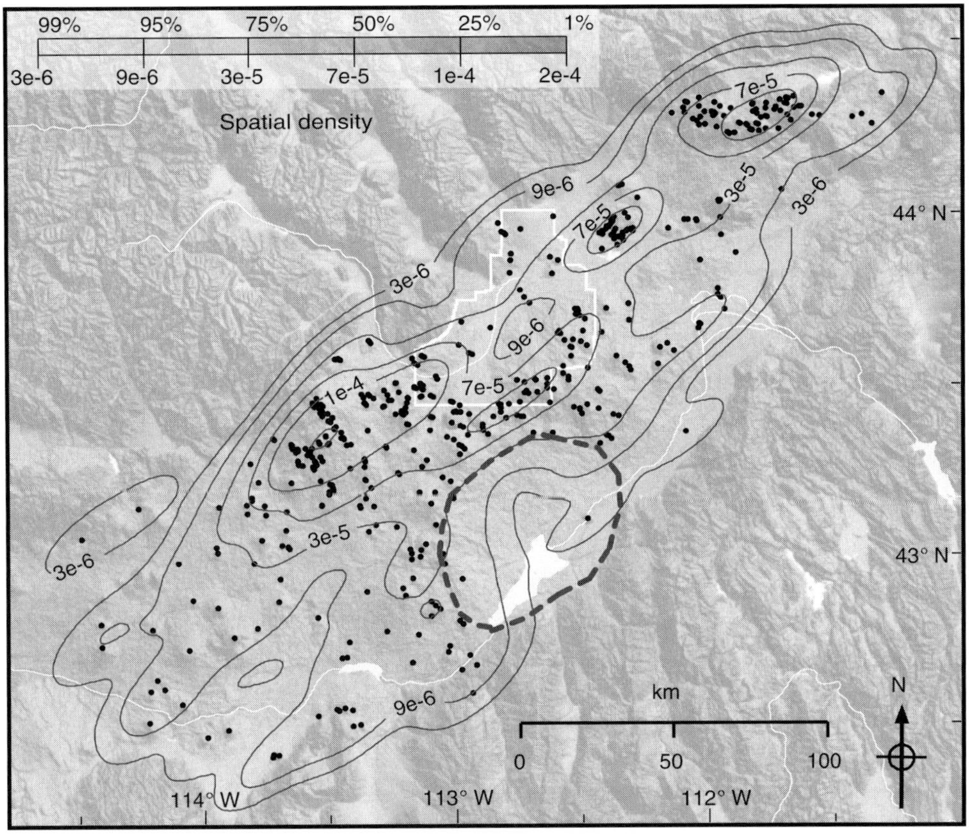

Fig. 16.6 This contour plot of the spatial density estimation of exposed vent locations within the ESRP is based on a grid that was calculated using a Gaussian kernel function and an optimal 2×2 smoothing bandwidth matrix. The contours indicate 25%, 50%, 75%, 95% and 99% of the total spatial density (e.g. 50% of the total spatial density falls within the 7×10^{-5} contour). Black dots represent the locations of exposed vents. The INL boundary is outlined in white. Notice the centrally located hole of lower spatial density within the center of the INL due to the absence of exposed vents. The dark gray (dashed) circle represents the approximate location of Taber caldera (Pierce and Morgan, 1992).

16.2.3 Spatial variations in the rate of basalt accumulation

Since ~ 650 ka the overall average accumulation rate of basalt at and near the INL has been $\sim 58\,\text{km}^3\,100\,\text{ka}^{-1}$ in volume and $\sim 49\,\text{m}\,100\,\text{ka}^{-1}$ in thickness (Wetmore *et al.*, 1997; Wetmore, 1998). However, the accumulation rate is highly variable throughout the INL as illustrated by the paleomagnetic interpretations of Champion *et al.* (2002). Within the boundaries of the INL, these authors observe dramatic gradients in the rate of basalt accumulation from corehole to corehole (e.g. a change of $47\,\text{m}\,100\,\text{ka}^{-1}$ over 13 km near the Test Reactors Area (TRA)). Perhaps most significant is their observation that accumulation

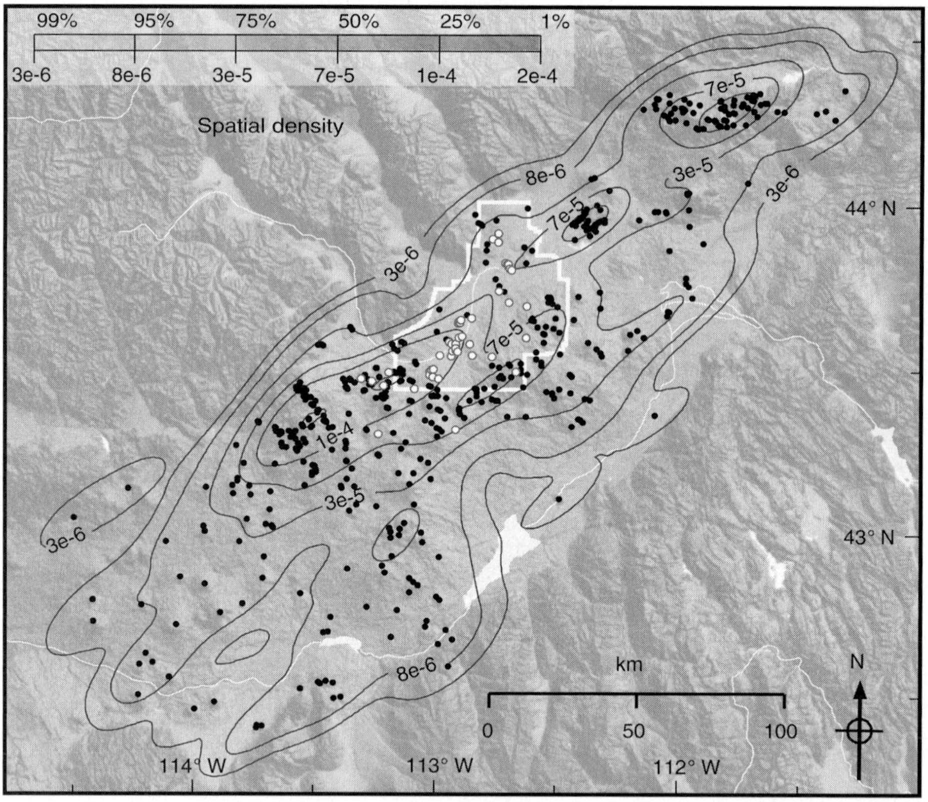

Fig. 16.7 This contour plot of spatial density was based the population of exposed (black dots) and sub-surface (inferred) vent locations (white circles) within the ESRP. This map was calculated and contoured as described in Figure 16.6. Notice the change in contours within the center of the INL (white lined boundary). The hole of lower spatial density, apparent in Figure 16.6, is no longer evident due to the inclusion of sub-surface (inferred) vent locations in the spatial density calculation. See color plate section.

rates increase toward the axis of the ESRP (i.e. with proximity to the Axial Volcanic Zone) while the recurrence interval between eruptive events decreases.

Figure 16.2 shows locations and rates of basalt accumulation (in mm a^{-1}) from select wells and coreholes about the INL. Spatial variability in the accumulation of basalt reveals an overall pattern that is generally consistent with that described by Champion *et al.* (2002). Specifically, an increase in the rate occurs toward the southeast, i.e. toward the Axial Volcanic zone; however, a general increase is evident toward the Arco–Big Southern Butte VRZ, with the highest rates observed southwest of the zone. Accumulation rates from wells within the Big Lost Trough are, on average, half as large as those from the two bounding volcanic zones (Figure 16.9).

It should be noted, however, that a substantial hiatus in volcanism occurred during the late Quaternary throughout much of the area shown in Figure 16.2, and certainly where the wells

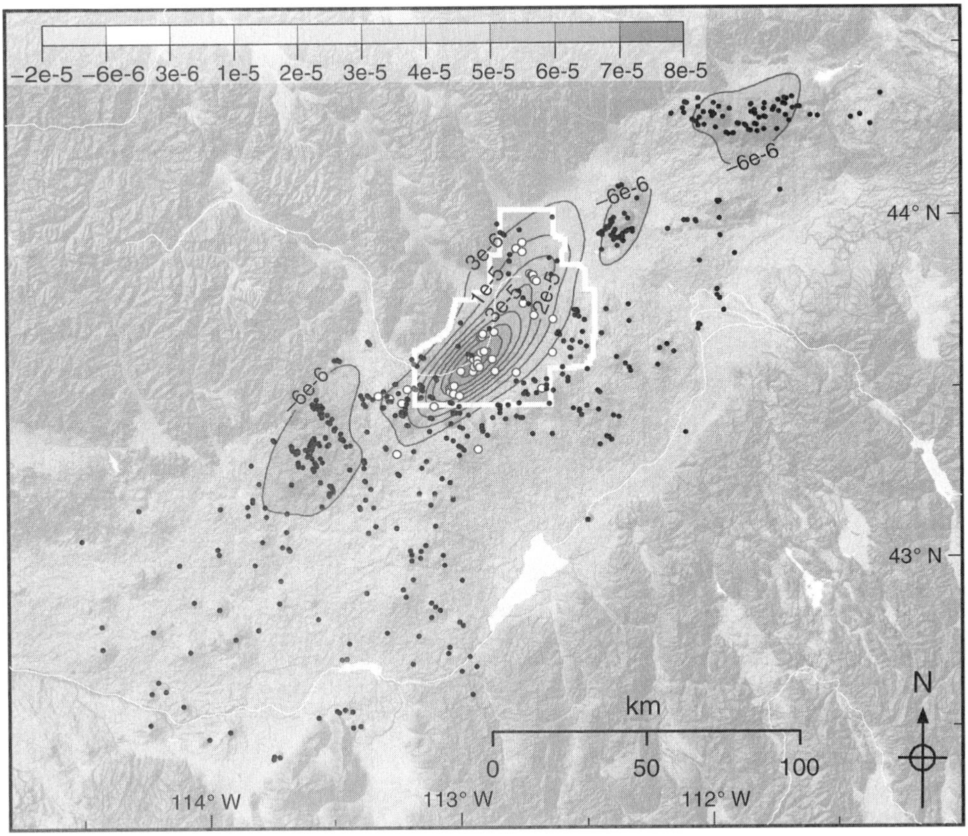

Fig. 16.8 This map shows the difference between the spatial density maps in Figures 16.6 and 16.7. The difference is contoured at a $1 \times 10^{-5}\,\mathrm{km}^2$ interval. The centrally located positive anomaly is caused by an increase in spatial density due to the inclusion of sub-surface vents. The smaller negative anomalies to either side of the "high" are a relative effect due to the centrally located positive anomaly. See color plate section.

were drilled from which accumulation data were derived. Within much of the southern Big Lost Trough, for example, no eruption has occurred for at least the last ~218 000 a. Most of the Arco–Big Southern Butte VRZ and the northern Big Lost Trough experienced a much longer hiatus, ranging from 300 ka to almost 500 ka (Kuntz *et al.*, 1994; Anderson *et al.*, 1997). Younger flows cover some portions of the Axial Volcanic Zone, but none are younger than ~95 ka in any of the wells (Anderson *et al.*, 1996, 1997).

16.3 Subsidence and the Big Lost Trough

Several lines of evidence demonstrate that the ESRP has been and probably still is undergoing subsidence relative to the adjacent Basin and Range Province. Additionally, a second-order subsidence is also observed locally within the plain. Whereas the former

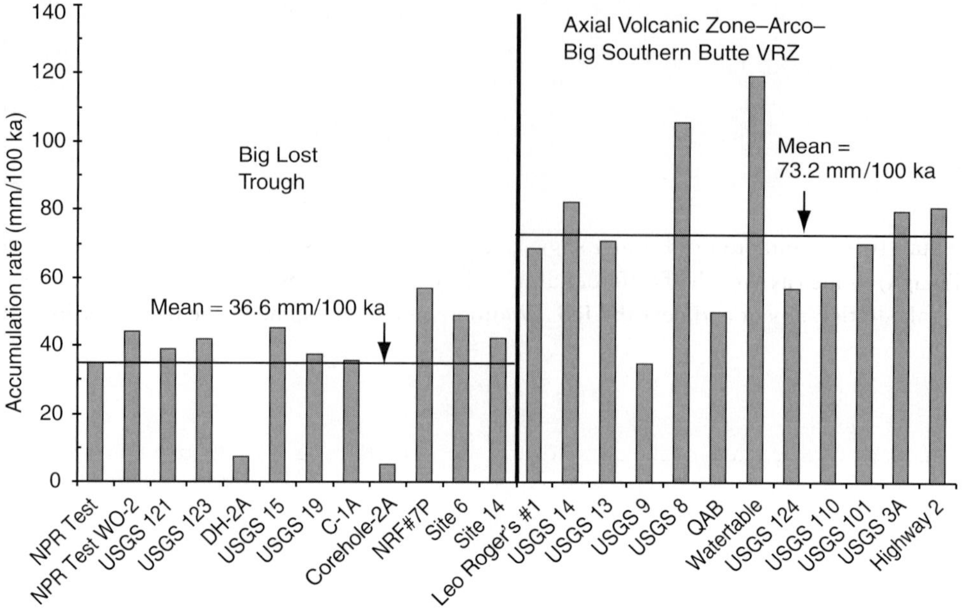

Fig. 16.9 This plot compares accumulation rates of basalt measured from select wells and coreholes located within two regions near the INL, the Big Lost Trough and the Arco–Big Southern Butte VRZ and Axial Volcanic Zone. Rates are calculated using data from Anderson *et al.* (1996, 1997).

subsidence is likely related to large-scale flexure due to the passage of the region over the Yellowstone hot spot and the consequential formation of a mafic mid-crustal sill, the latter appears to be the consequence of Basin and Range faulting continuing into the ESRP.

16.3.1 Subsidence relative to the surrounding Basin and Range

One of the most striking aspects of the ESRP is that it is a relatively low-standing region transecting the relatively high-standing northern Basin and Range Province. Rivers flow toward the plain from both the northwest and southeast sides, and the ridge lines of most ranges appear to rotate into the ESRP along its margins. McQuarrie and Rodgers (1998), studying the rotation of Mesozoic fold hinges from sub-horizontal orientations into steep southeast plunges along the northwest margin of the Plain, demonstrated as much as 8.5 km of subsidence of the ESRP relative to the adjacent Basin and Range Province. The amount of subsidence adjacent to the INL ranges from 4.5–8.5 km.

McQuarrie and Rodgers (1998) argue that the subsidence is related to flexure caused by a mid-crustal load, namely, a mafic sill or sill complex, between 17–25 km thick, that formed during the passage of the hot spot under this region of the ESRP. While most of the subsidence is inferred to have occurred prior to ~6.6 Ma, even before much of the rhyolitic volcanism in the area of the INL, the plain has subsided an additional 1.5 km since that time.

16.3.2 Local subsidence and basin formation

On a finer scale the ESRP is a mosaic of small basins and subtle topographic ridges (e.g. the Axial Volcanic Zone and Arco–Big Southern Butte VRZ as illustrated by the DEM in Figure 16.1). Wetmore (1998) first suggested that one of these small basins (the Big Lost Trough) formed as a result of differential subsidence based on observations of the subsurface basalt stratigraphy at and near the INL. Wetmore noted that some of the distal portions of lava flows observed in wells and cores within the Arco–Big Southern Butte VRZ were at elevations 120–200 m higher than those of their vent areas located to the east in the Big Lost Trough and the northern portion of the Axial Volcanic Zone. In contrast, there is no evidence for differential vertical movements between the Big Lost Trough and the Axial Volcanic zone. Blair (2002) also noted that the elevations of lake beds in coreholes from the INL exhibit between 120 and 220 m of elevation change, an observation that, Blair argues, requires post-depositional differential subsidence.

Wetmore (1998) suggested several possible causes for the observed subsidence of the Big Lost Trough relative to the Arco–Big Southern Butte VRZ, including the extension of faulting from the Basin and Range onto the ESRP and movement on ring faults associated with the calderas of the Heise volcanic field (a precursor to the Yellowstone caldera). Wetmore *et al.* (1999) and Blair (2002) both argued, however, that slip on the southern continuation of the Big Lost River fault onto the plain best explains the observed subsidence. Furthermore, Kuntz *et al.* (2002) suggested that a series of small offset faults (5–10 m) in the Arco–Big Southern Butte VRZ, which have been described as dike-induced features (e.g. Hackett and Smith, 1992), are most likely the on-plain expression of the Big Lost River fault.

16.4 Discussion

The spatial analyses of exposed and the inferred subsurface vents presented here indicate that, although northwest-trending vent alignments are apparent in some regions (Kuntz *et al.*, 1992; Hughes *et al.*, 2002a, 2002b) the concept of volcanic rift zones is a questionable characterization of ESRP magmatism. Rather, the data support an interpretation that magmatism is focused into northeast-trending elongate zones that approximately parallel the axis of the plain and are located adjacent to the northwest margin. Furthermore, the preservation of the vents at the surface is not solely related to spatial variations in the focus of volcanism (i.e. the volcanic-rift-zone concept), but rather is a function of the interplay between the spatial variability of volcanism, differential subsidence and burial by sediments and subsequent volcanic rocks. In this section we will use the descriptions of the vent distributions, regional densities, accumulation rates and differential subsidence to address two fundamental questions. How do the spatial variations in vent distributions, accumulation rates and subsidence result in the distribution of exposed vents seen at the surface today in the southern INL? What is driving the spatial variability in these aspects of central ESRP tectono-magmatism?

16.4.1 Spatially variable volcanism and differential subsidence

In the southern portion of the INL topography and the distribution of exposed vents clearly define the Arco–Big Southern Butte VRZ, Axial Volcanic Zone, and the Big Lost Trough (Figure 16.1). However, the assumption that vent distribution relates solely to the spatial distribution of volcanism between these three parts of the plain is unsupported by the preceding analysis. For example, when buried vents from the southern Big Lost Trough are taken into account the density of vents in that area is approximately as high as that to the west and south (Figure 16.7). The density of vents in the southern INL, in fact, does not exhibit significant gradients between these three regions of the ESRP.

The accumulation rates of basalt in the area of the southern INL do vary between the regions, but do not faithfully define their boundaries. Medium rates within the Big Lost Trough (~ 35 m in 100 ka) are also observed in wells in the eastern portion of the Arco–Big Southern Butte VRZ (Figure 16.2), but are much less compared to those west of the rift zone and within the Axial Volcanic Zone (~ 60 to more than 100 m in 100 ka).

In general, the descriptions of the subsidence characterizing the southern INL combined with the spatial variations in the distributions of vents and accumulation rates explains the ultimate distribution of exposed vents. Specifically, the subsidence of the Big Lost Trough has resulted in the burial of nearly all vents, although no major change in the vent distribution or rate of basalt accumulation between the Arco–Big Southern Butte VRZ and the Big Lost Trough can be documented. By contrast, vents within the volcanic rift zone have been uplifted and preserved without being buried by subsequent lavas or sediment. This is also supported by the observation that the oldest basaltic rocks at the tops of most wells and coreholes in the Big Lost Trough are ~ 200 ka younger than those at the surface in the Arco–Big Southern Butte VRZ (Kuntz *et al.*, 1994; Anderson and Liszewski, 1997). The lack of any significant topographic change between the Arco–Big Southern Butte VRZ and the Axial Volcanic Zone (elevations range between 1600–1650 m, also see Figure 16.1), although the latter too has subsided relative to the former, is explained by the fact that topography is maintained by accumulation rates that are twice as much as off-axis rates. This, too, is supported by the observation that many parts of the Axial Volcanic Zone are covered by flows much younger than the youngest flows in the Big Lost Trough (~ 221 ka; Anderson and Liszewski, 1997) and Arco–Big Southern Butte VRZ (Kuntz *et al.*, 1994).

West of the Arco–Big Southern Butte VRZ, well correlations reveal the highest accumulation rates, but this area is topographically equivalent to that of the rift zone and the Axial Volcanic Zone (~ 1600 m). This may be explained if the area west of the rift zone (Figure 16.2) experienced subsidence associated with slip on the continuation of the Lost River fault onto the ESRP, similar to the explanation for subsidence in the Big Lost Trough (Wetmore *et al.*, 1999; Blair, 2002). In this case, however, the subsided block is located on the hanging wall of the Lost River fault, and basalt accumulation was high enough to keep up with subsidence. The location of the trace of the Lost River fault likely coincides with the zone of small offset fault scarps (2–10 m) within the Arco–Big Southern Butte VRZ (Figure 16.2; Kuntz *et al.*, 1994, 2002). Elevated rates of basalt accumulation in this region of the ESRP may be due in part to the possibility that this region was subsiding, forming a relatively low area adjacent to an area of high lava output.

16.4.2 Underlying cause(s) of spatially variable volcanism

There are a few specific conclusions that can be made concerning the underlying causes of the spatial variability in ESRP volcanism based on data presented herein. The most significant conclusion is that upper crustal structures (e.g. range-bounding faults of the Basin and Range Province) do not appear to play a fundamental role in the distribution of basaltic volcanism.

The spatial intensity of basaltic volcanism on the ESRP reaches a maximum in three northeast-trending elongate zones adjacent to the northwest margin of the plain, as well as the Axial Volcanic Zone (Figure 16.7). Not only are the orientations of these high-intensity magmatism zones normal to the trends of the range-bounding faults, they appear to transect the extrapolations of those structures onto the plain without being affected. While the distribution of basaltic volcanism does not seem to be significantly affected by upper crustal structures, it is affected by the ambient regional stress field, as noted by the orientation/alignment of multiple vents or dikes in a given lava flow field. Magmas ascending through the crust as dikes are clearly oriented approximately parallel to other regional structures all forming in the same northeast–southwest oriented extensional stress regime (Kuntz *et al.*, 2002).

If the spatial distribution of basaltic volcanism on the ESRP is not strongly influenced by upper crustal structures then the geometry of melting in the source region may be the fundamental control. In fact, comparing the spatial density maps of vents on the ESRP (Figure 16.7) with mantle tomography shows a strong overlap between high vent intensity zones with thick (100–200 km) low-velocity zones (e.g. Figure 2 of Yuan and Dueker, 2005). Zones of lesser intensity, such as along the southeastern margin of the plain, also correspond with relatively high velocity, or colder regions of the upper mantle. Even the region of relatively low vent density southwest of the TAN facility and the Lava Ridge–Hell's Half Acre VRZ overlap in space with the a minor velocity increase in the upper mantle beneath that part of the ESRP. The Axial Volcanic Zone also overlies a region of low velocities in the upper mantle, as part of an ~250 km-long zone that extends southwest from the Yellowstone Plateau (Saltzer and Humphreys, 1997). These observed correlations are similar to those made of the distribution of Quaternary volcanoes in Japan where they are confined to regions overlying hot mantle "fingers" (Tamura *et al.*, 2002, Chapter 7, this volume; Kondo, Chapter 12, this volume) or low-velocity zones. This relationship strongly suggests to us, as it does to Tamura and others, that the geometry of melting in the source plays the fundamental role in the distribution of volcanism at the surface.

Concluding remarks

The distribution of volcanoes on the surface on the ESRP has traditionally led to the interpretation that volcanism is focused into narrow, northwest-trending zones known as volcanic rift zones. However, based on recent stratigraphic data from the subsurface of the plain within and immediately surrounding the INL, we can now test this view of ESRP volcanism, and more fully characterize the spatial variability in vent distributions, accumulation rates and differential subsidence.

The results of this exercise demonstrate that volcanism is not focused into northwest-trending volcanic rift zones, but rather is focused into a series of elongate, northeast-trending zones along the northwest margin and along the axis of the ESRP. The distribution of vents exposed at the surface of the plain, by contrast, results from a complex interplay between variations in the intensity of volcanism and accumulation rates, plus local differential subsidence related to slip on Basin and Range normal faults that extend onto the ESRP. For example, in the area of the southern INL the distribution of vents at the surface suggests that volcanism is focused into the Arco–Big Southern Butte VRZ and the Axial Volcanic Zone and limited in the Big Lost Trough. In part, this distribution results from the differential subsidence and burial vents within the Big Lost Trough, while those in the Arco–Big Southern Butte VRZ are uplifted and have avoided burial. No variation in the density of volcanism or rate of basalt accumulation exists between these zones. The Axial Volcanic Zone south of the Big Lost Trough also subsided relative to the Arco–Big Southern Butte VRZ, but its topography has been maintained by a much higher rate of accumulation relative to the other two zones.

Although the region surrounding the ESRP is seismically active, faults in the region are not spatially associated to any substantial degree with zones of high spatial density of volcanism on the plain. There is, however, a strong spatial correlation between low-velocity zones in the upper mantle and the zone of high spatial density of volcanism at the surface, suggesting that source geometry, and not near-surface structures, play the fundamental role in determining the spatial distribution of vents.

Acknowledgments

This chapter was improved by the editorial comments of Chuck Connor and Neil Chapman.

References

Anderson, S. R. (1991). Stratigraphy of the unsaturated zone and uppermost part of the Snake River Plain aquifer at the Idaho Chemical Processing Plant and Test Reactors Area, Idaho National Engineering Laboratory, Idaho, USGS Water-Resources Investigations Report 91-4010. Idaho Falls, ID: US Geological Survey INL Project Office.

Anderson, S. R. and R. C. Bartholomay (1995). Use of natural-gamma logs and cores for determining stratigraphic relations of basalt and sediment at the Radioactive Waste

Management Complex, Idaho National Engineering Laboratory, Idaho. *Journal of the Idaho Academy of Science*, **31**, 1–10.

Anderson, S. R. and B. Bowers (1995). Stratigraphy of the unsaturated zone and uppermost part of the Snake River Plain aquifer at test area north, Idaho National Engineering Laboratory, Idaho, USGS Water-Resources Investigations Report 95-4130. Idaho Falls, ID: US Geological Survey INL Project Office.

Anderson, S. R. and B. D. Lewis (1989). Stratigraphy of the unsaturated zone at the radioactive waste management complex, Idaho National Engineering Laboratory, Idaho, USGS Water-Resources Investigations Report 89-4065. Idaho Falls, ID: US Geological Survey INL Project Office.

Anderson, S. R. and M. J. Liszewski (1997). Stratigraphy of the unsaturated zone and the Snake River Plain aquifer at and near the Idaho National Engineering and Environmental Laboratory, Idaho, USGS Water-Resources Investigations Report 97-4183. Idaho Falls, ID: US Geological Survey INL Project Office.

Anderson, S. R., D. J. Ackerman, D. J. Liszewski and R. M. Feiburber (1996). Stratigraphic data for wells at and near the Idaho National Engineering Laboratory, Idaho, USGS Open-File Report 96-248. Idaho Falls, ID: US Geological Survey INL Project Office.

Anderson, S. R., M. J. Liszewski and L. D. Cecil (1997). Geologic ages and accumulation rates of basalt-flow groups and sedimentary interbeds in selected wells at the Idaho National Engineering Laboratory, Idaho, USGS Water-Resources Investigations Report 97-4010(doe/ID-22134). US Geological Survey INL Project Office: Idaho Falls, ID.

Bartholomay, R. C., L. C. Davis and P. K. Link (2002). Introduction to the hydrogeology of the eastern Snake River Plain. In: Link, P. K. and L. L. Mink (eds.) *Geology, Hydrogeology, and Environmental Remediation: Idaho National Engineering and Environmental Laboratory, Eastern Snake River Plain, Idaho*, Special Paper 353. Boulder, CO: Geological Society of America, 3–9.

Blair, J. J. (2002). Sedimentology and stratigraphy of sediments of the Big Lost Trough subsurface from coreholes at the Idaho National Engineering and Environmental Laboratory, Snake River Plain, Idaho. Unpublished M.S. thesis, Idaho State University.

Champion, D. E., M. A. Lanphere, S. R. Anderson and M. A. Kuntz (2002). Accumulation and subsidence of the Pleistocene basaltic lava flows of the eastern Snake River Plain, Idaho. In: Link, P. K. and L. L. Mink (eds.) *Geology, Hydrogeology, and Environmental Remediation: Idaho National Engineering and Environmental Laboratory, Eastern Snake River Plain, Idaho*, Special Paper 353. Boulder, CO: Geological Society of America, 175–192.

Connor, C. B., C. D. Condit, L. S. Crumpler and J. C. Aubele (1992). Evidence of regional structural controls on vent distribution: Springerville Volcanic Field, Arizona. *Journal of Geophysical Research*, **97**(B9), 12 349–12 359.

Connor, C. B., J. A. Stamatakos, D. A. Ferrill *et al.* (2000). Geologic factors controlling patterns of small-volume basaltic volcanism: application to a volcanic hazard assessment at Yucca Mountain, Nevada. *Journal of Geophysical Research*, **105**(1), 417–432.

Conway, F. M., C. B. Connor, B. E. Hill *et al.* (1998). Recurrence rates of basaltic volcanism in SP clusters, San Francisco volcanic field, Arizona. *Geology*, **26**, 655–658.

Gianniny, G. L., J. K. Geslin, J. W. Riesterer, P. K. Link and G. D. Thackray (1997). Quaternary surficial sediments near Test Area North (TAN), northeastern Snake River Plain: an actualistic guide to aquifer characterization. *Proceedings of the 32nd Annual Symposium on Engineering Geology and Geotechnical Engineering*. Moscow, ID: University of Idaho, 29–44.

Gianniny, G. L., G. D. Thackray, D. S. Kaufman *et al.* (2002). Late Quaternary highlands in the Mud Lake and Big Lost Trough sub-basins of Lake Terreton, Idaho. In: Link, P. K. and L. L. Mink (eds.) *Geology, Hydrogeology, and Environmental Remediation: Idaho National Engineering and Environmental Laboratory, Eastern Snake River Plain, Idaho*, Special Paper 353. Boulder, CO: Geological Society of America, 77–90.

Hackett, W. R. and R. P. Smith (1992). Quaternary volcanism, tectonics and sedimentation in the INEL area. In: Wilson, J. R. (ed.) *Field Guide to Geological Excursions in Utah and Adjacent Areas of Nevada, Idaho and Wyoming*. Salt Lake City, UT: Utah Geological Survey, 1–18.

Hackett, W. R. and R. P. Smith (1994). Volcanic hazards of the Idaho National Engineering Laboratory and adjacent areas, Technical Report INEL–94/0276. Idaho Falls, ID: Idaho National Engineering Lab.

Hayden, K. P. (1992). The geology and petrology of Cedar Butte, Bingham County, Idaho. Unpublished M.S. thesis, Idaho State University.

Hughes, S. S., M. McCurry and D. J. Geist (2002a). Geochemical correlations and implications for the magmatic evolution of basalt flow groups at the Idaho National Engineering and Environmental Laboratory. In: Link, P. K. and L. L. Mink (eds.) *Geology, Hydrogeology and Environmental Remediation: Idaho National Engineering and Environmental Laboratory, Eastern Snake River Plain*, Special Paper 353. Boulder, CO: Geological Society of America, 151–173.

Hughes S. S., P. H. Wetmore and J. L. Casper (2002b). Evolution of Quaternary tholeiitic basalt eruptive centers on the eastern Snake River Plain, Idaho. In: Bonnichsen, B., C. White and M. McCurry (eds.) *Tectonic and Magmatic Evolution of the Snake River Plain Volcanic Province*, B-30. Moscow, ID: Idaho Geological Survey, 363–385.

Kuntz, M. A. (1977a). Extensional faulting and volcanism along the Arco rift zone, eastern Snake River Plain, Idaho. *Geological Society of America Abstracts with Programs*, **9**, 740–741.

Kuntz, M. A. (1977b). Rift zones of the Snake River Plain, Idaho, as extensions of basin-range and older structures. *Geological Society of America Abstracts with Programs*, **9**, 1061–1062.

Kuntz, M. A. (1992). A model-based perspective of basaltic volcanism, eastern Snake River Plain, Idaho. In: Link, P. K., M. A. Kuntz and L. B. Platt (eds.) *Regional Geology of Eastern Idaho and Western Wyoming*. Boulder, CO: Geological Society of America, 289–304.

Kuntz, M. A., D. E. Champion, E. C. Spiker and R. H. Lefebvre (1986). Contrasting magma types and steady-state, volume-predictable basaltic volcanism along the Great Rift, Idaho. *Geological Society of America Bulletin*, **97**(5), 579–594.

Kuntz, M. A., H. R. Covington and L. J. Schorr (1992). An overview of basaltic volcanism of the eastern Snake River Plain, Idaho. In: Link, P. K. , M. A. Kuntz and L. B. Platt (eds.) *Regional Geology of Eastern Idaho and Western Wyoming*. Boulder, CO: Geological Society of America, 227–267.

Kuntz, M. A., B. Skipp, M. A. Lanphere *et al.* (1994). Geologic map of the Idaho National Engineering Laboratory and adjoining areas, eastern Idaho, I-2330. Moscow, ID: Idaho Geological Survey.

Kuntz, M. A., S. R. Anderson, D. E. Champion, M. A. Lanphere and D. J. Grunwald (2002). Tension cracks, eruptive fissures, dikes, and faults related to late Pleistocene–Holocene basaltic volcanism and implications for the distribution of hydraulic conductivity in the eastern Snake River Plain. In: Link, P. K. and L. L. Mink (eds.) *Geology, Hydrogeology, and Environmental Remediation: Idaho National Engineering and Environmental Laboratory, eastern Snake River Plain, Idaho*, Special Paper 353. Boulder, CO: Geological Society of America, 111–133.

McCurry, M., W. R. Hackett and K. P. Hayden (1999). Cedar Butte and cogenetic Quaternary rhyolite domes of the eastern Snake River Plain. In: Hughes, S. S. and G. D. Thackray (eds.) *Guidebook to the Geology of Eastern Idaho*. Pocatello, ID: Idaho Museum of Natural History and Idaho State University Press, 169–179.

McCurry, M., K. P. Hayden, L. H. Morse and S. Mertzman (2008). Genesis of post-hotspot, A-type rhyolite of the eastern Snake River Plain volcanic field by extreme fractional crystallization of olivine tholeiite. *Bulletin of Volcanology*, **70**, 361–383.

McQuarrie, N. and D. W. Rodgers (1998). Subsidence of a volcanic basin by flexure and lower crustal flow: eastern Snake River Plain, Idaho. *Tectonics*, **17**, 203–220.

Pierce, K. L. and L. A. Morgan (1992). The track of the Yellowstone hotspot: volcanism, faulting, and uplift. In: Link, P. K., M. A. Kuntz and L. B. Platt (eds.) *Regional Geology of Eastern Idaho and Western Wyoming*, Memoir 179. Boulder, CO: Geological Society of America, 1–53.

Saltzer, R. L. and E. D. Humphreys (1997). Upper mantle P wave velocity structure of the eastern Snake River Plain and its relationship to geodynamic models of the region. *Journal of Geophysical Research*, **102**(B6), 11 829–11 841.

Scarberry, K. C. (2003). Volcanology, geochemistry and stratigraphy of the F basalt flow group, eastern Snake River Plain, Idaho. Unpublished M.S. thesis, Idaho State University.

Tamura, Y., Y. Tatsumi, D. Zhao, Y. Kido and H. Shukuno (2002). Hot fingers in the mantle wedge: new insights into magma genesis in subduction zones. *Earth and Planetary Science Letters*, **197**, 105–116.

Wetmore, P. H. (1998). An assessment of physical volcanology and tectonics of the central eastern Snake River Plain based on the correlation of subsurface basalts at and near the Idaho National Engineering and Environmental Laboratory, Idaho. Unpublished M.S. thesis, Idaho State University.

Wetmore, P. H., S. S. Hughes and S. R. Anderson (1997). Model morphologies of subsurface Quaternary basalts as evidence for a decrease in the magnitude of basaltic magmatism at and near the Idaho National Engineering and Environmental Laboratory, Idaho. *Proceedings of the 32nd Symposium on Engineering Geology and Geotechnical Engineering*, 45–58.

Wetmore, P. H., S. S. Hughes, D. W. Rodgers and S. R. Anderson (1999). Constructional origin of the Axial Volcanic Zone of the eastern Snake River Plain, Idaho. *Geological Society of America Abstracts with Programs*, **31**(4), 61.

Yuan, H. and K. Dueker (2005). Teleseismic P-wave tomogram of the Yellowstone plume. *Geophysical Research Letters*, **32**, L07304, doi:10.1029/2004GL022056.

17

Modeling the flow of basaltic magma into subsurface nuclear facilities

T. Menand, J. C. Phillips, R. S. J. Sparks and A. W. Woods

Worldwide, a consensus is developing among countries using nuclear power that deep, geologic disposal of spent nuclear fuel and high-level radioactive waste is the safest long-term option (National Research Council, 1990, 2001; EPA, 2001). The geologic medium acts as a component of a multiple barrier system (including the waste form and engineering components) designed to isolate the waste from the biosphere. Regulations in many countries, therefore, require repository developers to consider various natural hazards when evaluating repository performance. Among the hazards considered is the potential for igneous activity at the site and surrounding area (Long and Ewing, 2004). For example, in the United States, regulations governing the geologic disposal of high-level radioactive waste at the potential Yucca Mountain, Nevada, repository require inclusion of risk (i.e. probability and consequence) in assessments of the safety of the repository system. Based on probabilities estimated for repository disruption by future basaltic volcanism (e.g. 1.8×10^{-8}: Bechtel SAIC Company, LLC, 2007; 1.0×10^{-6}: Smith and Keenan, 2005) and the potential risks for this natural hazard, performance assessments should evaluate the consequences of a basaltic volcano intersecting the drifts and tunnels of the potential repository, which might damage the emplaced waste packages and waste form, and could transport radioactive material to the biosphere (NRC, 2005).

There is almost no precedent for a volcanic eruption interacting with an underground storage facility of the kind envisaged for radioactive waste repositories. These facilities generally consist of a network of tunnels or drifts. Some designs require the drifts to remain empty apart from their inventory of radioactive waste containers, at least up to the time the repository is permanently closed (i.e. on the order of several hundred years in some cases). Thus, the generic processes that might occur if magma erupts into empty drifts have been a prominent topic of study. Because no such events have occurred and analogs such as eruptions into natural caves have not yet been identified, the assessment of igneous disruption will need to rely largely on non-empirical information. In general, such assessments may consider the limited empirical evidence of volcanic and intrusive processes; knowledge of the properties of erupting magmas that help constrain the dynamics of these processes; laboratory experiments designed to elucidate how multiphase fluids interact with drifts; and finally, development of models. Because of the complexity of the processes and current state of the art in representing those processes, a comprehensive approach to assessing igneous consequences that integrates knowledge from each of these sources is appropriate.

This chapter focuses on modeling with an emphasis on the use of analog laboratory experiments that are designed to either (i) test theories and numerical models or (ii) gain insights into processes in circumstances where numerical models are either poorly understood or too complex. To develop representative models, we considered results from volcanological studies of eruption behavior and products. We also incorporated the physical properties of magma when choosing relevant analog fluids and addressing scaling issues.

17.1 Magma properties and fluid dynamics relevant to magma–drift interaction

17.1.1 Physical properties of magmas

The magma properties that exert the strongest control on flow dynamics are the magma density ρ and viscosity μ, both of which decrease with increasing temperature. Typical basalt eruption temperatures range from 1000–1300 °C (Kilburn, 2000; Francis and Oppenheimer, 2004), and over this range of temperatures the density and viscosity of natural, dry basaltic melts range between 2600 and 2800 kg m^{-3} and between 1 and 1000 Pa s at atmospheric pressure, respectively (Murase and McBirney, 1973; Kilburn, 2000; Spera, 2000; Francis and Oppenheimer, 2004).

Magma typically consists of three phases: melt (liquid), crystals (solid) and bubbles (gas). The temperature dependence of the melt viscosity can be described by several models; the simplest is the Arrhenian model $\mu = \mu_0 \exp(E^*/RT)$, where μ_0 is the melt viscosity at infinite temperature, E^* is the melt activation energy, R is the universal gas constant and T is the temperature in Kelvin. According to this model, melt composition mainly affects the activation energy E^*, and Shaw (1972) estimated the activation energy from the partial molar coefficients of SiO_2. Although melts with high silica content do not exactly follow the Arrhenian temperature–viscosity relationship (Hess and Dingwell, 1996), it is usually necessary to employ the simpler Arrhenian model of Shaw (1972) for petrologic purposes (Spera, 2000), which is a good approximation for low-viscosity basalts (Giordano and Dingwell, 2003).

Small amounts of dissolved volatiles can have important effects on the density, viscosity and crystallization of melts and magmas, which will strongly influence the ability of magmas to flow. Of all the volatile species, water is the most abundant and accounts for the largest variations in density and, more importantly, viscosity. Basaltic magmas typically contain 1–4 wt.% volatiles, although dissolved water contents as high as 6 wt.% have been measured in arc basalts (Sisson and Layne, 1993); and water contents up to 4.6 wt.% have been estimated for the Lathrop Wells basalts near the proposed site for the high-level radioactive waste repository at Yucca Mountain, Nevada (Nicholis and Rutherford, 2004; Valentine *et al.*, 2007). Dissolved water contents of 3 wt.% will lower the density of basaltic melts by 5 wt.% (Lange, 1994; Wallace and Anderson, 2000) and lower the melt viscosity of basalts by two orders of magnitude (Shaw, 1972; Giordano and Dingwell, 2003). Water is not the only volatile species, however. Carbon dioxide is also present in magmas, but the amounts of dissolved CO_2 are typically one to two orders of magnitude less than those of water. Furthermore, the effect of CO_2 on melt density and viscosity is smaller than for water: Wallace and Anderson (2000) report that adding 3 wt.% of CO_2 to a basaltic melt will decrease its density by $\approx 3\%$. Contrary to water, dissolved CO_2 appears to have a minimal effect on melt viscosity. This effect depends on the speciation of CO_2, and dissolved CO_2 can increase melt viscosity slightly if the CO_2 is dissolved as carbonate (Lange, 1994). Dissolved CO_2 in melts can also have important indirect effects on melt viscosity because dissolved CO_2 lowers the solubility of water (Holloway and Blank, 1994).

The presence of exsolved gas bubbles and crystals also has a strong influence on magma density and viscosity. Crystals act to increase both magma viscosity and density. Estimating the speciation and volume fraction of different crystal phases present in the melt requires modeling the thermal and decompression history of the magma. This is a complex process using an incomplete understanding of the phase behavior, particularly the solubility of CO_2 in basalts. Therefore a more typical approach in recent studies has been to investigate a wide range of magma viscosities that will account for the ranges of temperatures, compositions and crystal contents that characterize basaltic magmas.

The presence of exsolved gas bubbles also significantly affects the density of magmas, which decreases linearly with the volumetric concentration c of the bubbles: $\rho \sim \rho_l(1-c)$ where ρ_l is the density of pure melt. The effect of bubbles on magma viscosity is more complex as it depends on the tendency of bubbles to deform under viscous stresses induced by flow, relative to their tendency to remain spherical as a result of interfacial stresses, and the rapidity of this response (Llewellin and Manga, 2005). For steady flows involving spherical bubbles, the commonly accepted empirical relationship at low volumetric fractions ($< 10\%$) for viscosity of the bubbly mixture, μ_b, as a function of bubble content and melt viscosity, μ_l, is $\mu_b = \mu_l/(1-c)$ (Llewellin and Manga, 2005; Menand and Phillips, 2007b). For higher bubble contents, viscosity appears to be better approximated by the relationship $\mu_b = \mu_l(1-c)^{-5/2}$ (Jaupart and Vergniolle, 1989; Menand and Phillips, 2007b).

An important consideration in the eruption of water-rich basalts is the crystallization that is principally related to the change in liquidus temperatures of the main stable mineral phases. Degassing-induced crystallization and the consequent rheological changes are key to understanding conduit flows and lava extrusions in andesite eruptions (Cashman, 1992; Melnik and Sparks, 1999). This is likely to be the case for wet basalt eruptions, too, although there is less supporting research. The viscosity increases dramatically as groundmass crystals form from degassing basalt, and the crystal content may become so high that the rheology can become non-Newtonian. For example, wet trachybasalt with a liquidus at 950–1000 °C (Nicholis and Rutherford, 2004) tends toward the solidus at one atmosphere pressure. The effects of degassing on crystallization and viscosity will be counteracted by the latent heat of crystallization (Blundy *et al.*, 2006) such that the temperature will be above the solidus in the fully degassed and decompressed state at one atmosphere. Fifty percent crystallization will increase the temperature by $\approx 100\,^{\circ}\mathrm{C}$ based on the latent heat of crystallization of plagioclase as the dominant groundmass mineral. Thus, the eruption temperature of trachybasalt should be $\sim 1050-1100\,^{\circ}\mathrm{C}$. The rheology of such magmas can be compared to the field rheological measurements of Etna trachybasalt lava (Pinkerton and Sparks, 1978), which is $\sim 10^5$ Pa s with $\approx 50\%$ total crystal content at $1070\,^{\circ}\mathrm{C}$.

At lower pressures, gases become less soluble in magmas, leading to an increase in gas exsolution and magma crystallinity. Additionally, gas bubbles expand as the magmatic pressure decreases, so the controls exerted by bubbles and crystals on magma properties become more significant at lower pressures, and these effects will be especially important at the typically shallow depths (~ 500 m) of radioactive waste repositories. The exsolved gas mass fraction n varies with pressure according to the solubility law (based on Henry's law),

$$n(P) = n_0 - sP^{1/2} \tag{17.1}$$

where n_0 is the total gas mass fraction, P is pressure and s is the solubility constant for water in basalt, with a value $3 \times 10^{-6}\ \mathrm{Pa}^{1/2}$ (Holloway and Blank, 1994). In general, the gas pressure will not be equal to the bulk flow pressure or to the surrounding rock lithostatic pressure. Exsolving gas bubbles are overpressured with respect to the surrounding fluid due to surface tension, viscous resistance and inertia, as gas bubbles expand in ascending magma due to diffusion and decompression (Sparks, 1978; Sparks *et al.*, 1994). In basalt magmas, overpressures due to surface tension and inertia are typically negligible but overpressures due to viscous resistance can be significant in very fast explosive flows. Additionally, the bulk flow pressure is initially determined by the pressure in the source chamber but decreases due to frictional losses in the magma flow. Thus, gas pressure evolves during magma ascent, which in turn determines volatile exsolution and depends on the detailed dynamics of the eruption (Massol *et al.*, 2001). A common approach is to assume that pressure is lithostatic, but magma pressures that deviate significantly from lithostatic are likely. For example, a dike typically requires internal pressure that exceeds lithostatic pressure and the tensile strength of the surrounding rock to propagate (Lister and Kerr, 1991), whereas an explosive eruption through an open conduit can result in large underpressures (e.g. Mason *et al.*, 2006). Significant disequilibrium is also possible for fast flows so that kinetics have to be taken into account. If flows are at equilibrium, however, the solubility law given by (17.1) can be used for any pressure assumption.

17.1.2 Magma flow dynamics

Magma ascends through the Earth's crust by means of dikes, which are sheet-like igneous intrusions typically several centimeters to meters or tens of meters (rarely several hundreds of meters) in thickness and several kilometers (rarely several hundreds of kilometers) in extent (Pollard, 1987). The present study considers a 1 m-width and a 1–10 km lateral extension as reasonable average dimensions for basaltic dikes (Lister and Kerr, 1991; Rubin, 1995). Magma fluxes can range from $1\ \mathrm{m}^3\ \mathrm{s}^{-1}$, an average replenishment rate for the summit reservoir at Kilauea volcano, Hawaii (Rubin and Pollard, 1987), to 10^6 m^3 s^{-1}, such as may be appropriate to very high volumetric flow flood basalt eruptions (Swanson *et al.*, 1975; Wilson and Head, 1981). An average magma flux of 10^3 m^3 s^{-1} would correspond to an average magma ascent rate of 1 m s^{-1} through a 1 m-wide, 1 km-long dike.

Magma ascent through dikes is mainly driven by magma buoyancy, initially determined by magma composition and ultimately controlled by volatile exsolution, which becomes the dominant control at shallower depths. A key fluid dynamical parameter for magma flow is the Reynolds number, $Re = \rho u L/\mu$, which represents the ratio of viscous to inertial forces (u is the flow velocity and L is a typical length scale such as dike thickness). Magma flow is laminar if $Re << 10$, and the flow is considered to become turbulent when the Reynolds number exceeds a critical value of ~ 1000 (Lister and Kerr, 1991). In most cases, magma flow is laminar with a Reynolds number of ~ 1 for basaltic magma with viscosity of 1000 Pa s and density of 2750 kg m^{-3} flowing at a rise velocity of 1 m s^{-1} in a 1 m-wide dike. However, flows that involve magmas of much lower viscosity or higher magma ascent rates, such as during flood basalt eruptions, may become turbulent (Huppert and Sparks, 1985).

An important question is whether exsolved volatile bubbles are uniformly distributed throughout the magma, forming a uniform bubbly mixture, or whether phase separation occurs, which would strongly modify the flow behavior. Two-phase flow regimes range, in order of increasing bubble content and flow explosivity, from: bubbly flows; to slug flows, where bubbles coalesce into larger gas pockets; to annular flows, where gas flows in the center of a dike or conduit while the fluid phase flows on its periphery; to dispersed flows, where fragmented magma is carried by gas flow (Wallis, 1969; Jaupart, 2000; Slezin, 2003; Figure 17.1). Although magma flow will evolve through these different regimes as the bubble content increases, how magma flows change from one regime to another is still not fully understood. The different two-phase flow regimes depend on various parameters that include, but are not restricted to, bubble contents, flow rates and flow geometries (Wallis, 1969). A reasonable assumption is to consider that deeper in a basaltic system, bubbles are well mixed due to low volumetric concentration, the relatively small size of the bubbles, the relatively low viscosity of basaltic melts and the effects of magma convection (Phillips

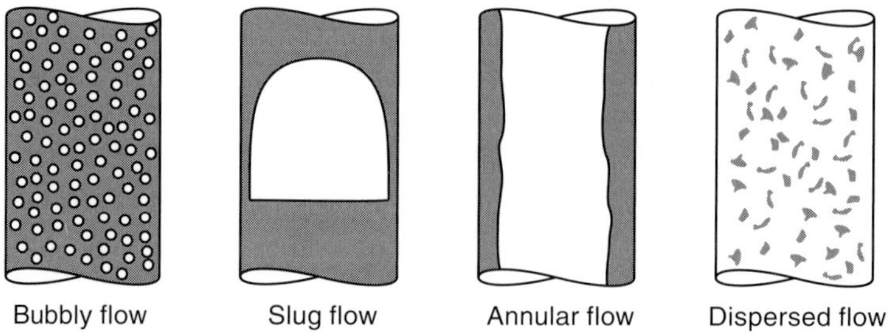

Fig. 17.1 The different flow regimes experienced by two-phase flows, going from bubbly flow to dispersed flow as both gas content and flow explosivity increase.

and Woods, 2001). As we shall show, the geometry of the magmatic system provides a strong control on bubble segregation from the melt, so an appropriate starting condition is to assume exsolved gas bubbles are uniformly distributed throughout the melt.

17.2 Modeling magma–repository interaction

17.2.1 Transient flows

The initial interaction with repository drifts involves the transient case where a magma-filled dike propagates and intersects a drift; however, it is unknown whether this magma will be degassed. Degassed lava emerges early in some basaltic eruptions and can be associated with simultaneous explosive activity. As mentioned previously, gas segregation processes are not understood well enough to determine whether the magma that first flows into a drift will be degassed. Thus, both end-member cases should be considered to bound possible interactions. Lejeune *et al.* (Chapter 18, this volume) consider the degassed case through laboratory experiments and theoretical analysis. Here we consider the explosive end-member where the magma and gas have not segregated.

An explosive flow is expected for the interaction of rapidly decompressing gas-rich magma rising in a dike with an underground drift structure. Repository drifts are usually proposed to be maintained at atmospheric pressure (Rosseau *et al.*, 1999), while at the potential Yucca Mountain repository depths of 200–300 m, the magma pressure just behind the tip of a dike is estimated to be typically 10–20 MPa, based on the lithostatic pressure and the fluid pressure required to drive a fracture at the dike tip (Pollard, 1987; Lister and Kerr, 1991; Woods *et al.*, 2002). When the dike intersects the drift, the magma will rapidly decompress, and at the relatively high water contents measured for Lathrop Wells basalts of up to 4.6 wt.% (Nicholis and Rutherford, 2004), this decompression will be explosive (Blackburn *et al.*, 1976), assuming that the gas has been retained during ascent.

A quantitative model of the process of magma decompression into a subsurface horizontal drift was proposed by Woods *et al.* (2002). On decompression, volatile exsolution within the magma in the dike and the drift will cause the magma to expand and accelerate, and if this process occurs sufficiently rapidly, the magma will fragment into a two-phase mixture of vesicular magma and gas. Woods *et al.* (2002) modeled this flow as a one-dimensional homogeneous mixture of magma and gas in a coordinate frame that was continuous for the flow from the dike into the drift (Figure 17.2). The cross-sectional area was assumed to vary smoothly between the dike and drift, and the flow was assumed to remain isothermal during volatile exsolution due to the high thermal inertia of the magma.

The motion of the magma–gas mixture can be described in terms of its averaged velocity, u, and averaged density, ρ, at position x, at pressure P and at time, t, leading to the equation for the conservation of momentum,

$$\rho\left(\frac{\partial u}{\partial t} + u\frac{\partial u}{\partial x}\right) = -\frac{\partial P}{\partial x} - fu - \rho g G(x) \qquad (17.2)$$

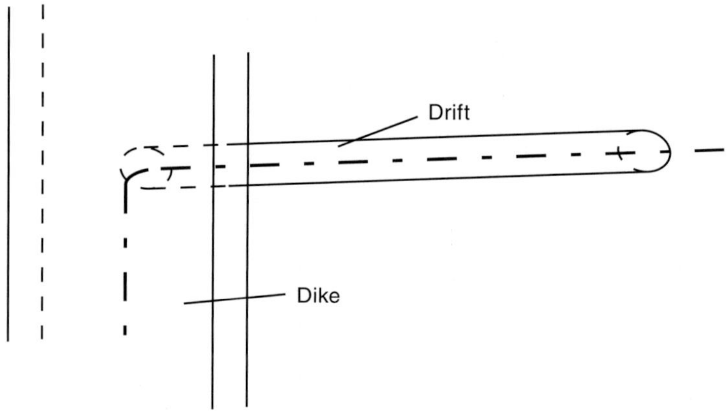

Fig. 17.2 The coordinate frame used in the one-dimensional simulations of Woods *et al.* (2002).

where f is a drag coefficient, g is the acceleration due to gravity and $G(x)$ has value 1 in the vertical dike and 0 in the horizontal drift. The terms on the left-hand side describe the inertia of the magma–gas mixture, and the terms on the right-hand side represent the pressure gradient in the flow, the resistance to motion due to flow against the walls of the dike or drift, and the buoyancy forces, respectively. The drag coefficient was parameterized as

$$f = \frac{\alpha\mu}{D^2} + \frac{2C\rho|\mu|}{D} \tag{17.3}$$

where α is a coefficient with value 12 for a two-dimensional dike and 8 for a cylindrical drift, D is dike width or drift diameter and C is the turbulent-drag coefficient. The first term on the right-hand side is the viscous drag, and the second term is the turbulent drag.

The equation of conservation of momentum was coupled with an equation for mass conservation,

$$\frac{\partial(A(x)\rho)}{\partial t} + \frac{\partial(A(x)\rho u)}{\partial x} = 0 \tag{17.4}$$

where $A(x)$ is the cross-sectional area of the dike or drift, and an equation for the bulk density of the magma–gas mixture (based on the perfect gas law),

$$\frac{1}{\rho} = \frac{n(P)RT}{P} + \frac{(1 - n(P))}{\rho_l} \tag{17.5}$$

where $R = 462\,\mathrm{J\,kg^{-1}\,K^{-1}}$ is the gas constant for H_2O, T is the (constant) temperature and $n(P)$ is the exsolved gas mass fraction given by the solubility law (17.1).

For their model simulations, Woods *et al.* (2002) assumed basaltic values $\mu = 10\,\mathrm{Pa\,s}$, $\rho = 2600\,\mathrm{kg/m^{-3}}$, a water content of 2 wt.% and $C = 0.01$. The results of a typical simulation for a dike intersecting a drift and remaining open are shown in Figure 1 in Woods *et al.* (2002). Initially, the magma–gas mixture rapidly expands as a rarefaction wave propagates back into the dike through the magma, and volatiles are exsolved. The expanding mixture accelerates along the drift, reaching speeds of tens to hundreds of meters per second, with the density decreasing as the pressure falls. Air is displaced and compressed ahead of the magma–gas mixture, and as a result, a shock forms in the air and moves down the drift at speeds of several hundreds of meters per second. If the drift is closed at its ends, then the shock is reflected when it reaches the end of the drift, increasing its amplitude by an order of magnitude. The reflected shock recompresses the magma–gas mixture, and a region of higher pressure, up to a few MPa, is formed in the drift. If the drift is open ended, the flow adjusts to a steady regime within seconds once the drift system is completely filled (Woods *et al.*, 2002).

Dartevelle and Valentine (2005) further investigated the eruption scenario proposed by Woods *et al.* (2002) using the GMFIX multiphase numerical model (Dartevelle, 2004). This model allowed the properties of each phase (pyroclasts and gas) to be determined in a two-dimensional Cartesian frame with full time-dependence, relaxing the assumption of homogeneous flow made by Woods *et al.* (2002). The results of a simulation corresponding to the intersection of an overpressurized dike containing basalt with 1 wt.% water with a drift at atmospheric pressure are shown in their Figure 1 (Dartevelle and Valentine, 2005). About 30–35% of the magma–gas mixture flows into the drift, forming a shock that propagates into the drift at speeds of about 200 m s^{-1}. The following flow forms a low-density current that flows along the drift roof at speeds of about 120 m s^{-1}. If the end of the drift is closed, the shock is reflected and weakens through interaction with the following gas flow before interacting with the current of pyroclasts and ash at a time of 1.10 s. On reaching the closed end of the drift, the current is concentrated in density and reflected to form a dense current that flows along the base of the drift. When the dense return flow reaches the dike, some of the material is entrained into the rising flow and reaches the surface, while some is recirculated back into the drift.

Both Woods *et al.* (2002) and Dartevelle and Valentine (2005) investigated scenarios where there are secondary openings in the drift due to the presence of a further dike and found that there is little difference to the flow patterns and velocities and pressures generated. Both studies show the generation of high-speed shocks due to the initial decompression of the magma–gas mixture into the drift, although the shock amplification on reflection from the closed end of the drift observed by Woods *et al.* (2002) is not recognized in the simulations of Dartevelle and Valentine (2005). The formation of high-pressure regions in the magma–gas flow has important implications for the potential disruption of waste containers and transport of small fragments of spent nuclear fuel, as discussed further in Section 17.2.3.

We are aware that other models have been presented in various reports on igneous consequences at Yucca Mountain by different panels and bodies. We have not referred to this work, which has not been subject to peer review. However, the results of such studies all confirm that fast explosive flows will occur if volatile-rich basaltic magma is rapidly decompressed into tunnels or drifts.

The only reported natural example of the interaction of basalt with an analogous man-made structure occurred as an eruption along a geothermal borehole during the 1977 Krafla eruption (Larsen *et al.*, 1979). The borehole was 10 cm in diameter and 1138 m in length. The eruption involved an explosive Strombolian jet, lasted 20 minutes and erupted a volume of 26 m^3. In the context of potential repository interactions, this example is important because it shows that basalt magma can flow along a hole that has a cross-sectional area two orders of magnitude smaller than a radioactive waste repository drift. This case shows that cooling during magma flow to form a quenched layer is not significant, so models that do not account for cooling reasonably simulate the pertinent processes.

17.2.2 Steady-state flows

Following the initial transient decompression of the gas-bearing magma into the drift, the flow will adjust to a steady state within seconds to hours depending on whether magma flow is diverted along the drift or is limited to the main dike if access drifts are backfilled with crushed rocks (Woods *et al.*, 2002; Dartevelle and Valentine, 2005). If steady-state magma flow is established in the drift, Woods *et al.* (2002) calculated that magma will flow past the waste containers with steady speeds of $\sim 10\,\mathrm{m\,s^{-1}}$. Waste containers will experience considerable thermal stress from the magma and gradually heat by thermal conduction and, for times greater than $\approx 1000\,\mathrm{s}$ (based on diffusion of heat into the waste containers), they will become deformable and may break open. If the end of the drift remains closed, magma flow will be limited to the drifts directly intersected by the main dike; but basaltic magma will nevertheless fill the drift. In both cases, magma pressure in the repository will ultimately decrease to be close to lithostatic (Woods *et al.*, 2002).

This latter result is consistent with observations of natural volcanic systems. It is commonly observed that many basaltic eruptions tend to become less explosive with time. Initial basaltic eruptions occur explosively along fissures, typically in Strombolian-style fire fountains. Within hours to a few days, activity focuses onto a progressively restricted number of vents along the fissures; tephra plumes form, along with subordinate volumes of lava (Thorarinsson, 1969; Fedotov and Markhinin, 1983; Macdonald *et al.*, 1983). There is a general tendency for such eruptions to become less explosive with time and for lava to become an increasingly dominant product. However, observations of eruptions such as Eldfell volcano (Iceland) in 1973 show that even at the very beginning of an eruption, explosive flow of gas-rich magma and discharge of degassed lava occur simultaneously. The interaction of a dike with an underground drift is therefore likely to involve magma flow of decreasing intensity. Basaltic magma can fill the drift, and subsequent magma circulation will depend on processes of gas segregation within the drift (Menand and Phillips, 2007a). Moreover, as the eruption proceeds, magma flow can be sustained for days to weeks in the vertical dike; for example, the great Tolbachik basaltic fissure eruption of 1975–1976 lasted for more than one and a half years (Fedotov and Markhinin, 1983). Over these time-frames, the dike may increase in size and change from a planar cross-section to a more circular cross-section owing to mechanical and thermal erosion as well as solidification of the dike in areas away from the focused flow (Macdonald *et al.*, 1983; Bruce and Huppert, 1990).

At shallow crustal depths (< 500 m) typical for repository drifts, magma volatiles are very likely to exist as exsolved bubbles. For instance, initial water contents of basalts that have erupted in the vicinity of Yucca Mountain range from 1.9 wt.%–4.6 wt.% (Nicholis and Rutherford, 2004) and at a depth of 300 m, a basaltic magma with 4.6 wt.% initial water will have exsolved 3.8 wt.% of its water (Holloway and Blank, 1994), which would correspond to volumetric gas fractions in the range of 70–90% at that depth in equilibrium (Menand *et al.*, 2008). The presence of exsolved bubbles as well as the amount of water that remains dissolved will strongly affect the density and viscosity of the basalts. Furthermore, there may be a range of bubble volumetric contents depending on exsolution at greater depths and gas loss during ascent. The amount of exsolved gas in magma within the drift will determine the nature and strength of magma circulation in the drift.

Menand and Phillips (2007a, 2007b) investigated gas segregation in a magma-filled drift intersected by a vertical dike using analog experiments. The apparatus consisted of a glass recirculating flow loop with a vertical mounting section (to simulate the vertical dike) connected to a horizontal section (to simulate the drift; Figure 1 in Menand and Phillips, 2007a). Electrolysis of the recirculating flow was used to simulate low volumetric gas fractions ($< 10\%$), producing micrometric bubbles in viscous mixtures of water and golden syrup. These low gas fractions correspond to the situation where magma has lost a large proportion of its gases at some depth greater than that of the repository, or during the latest waning stages of an eruption. To simulate higher volumetric gas contents, golden syrup was aerated before its injection into the recirculating flow loop, leading to volumetric gas fractions as high as 40%.

The experiments of Menand and Phillips (2007a) at low gas fractions ($< 10\%$) show that exsolved bubbles induce a buoyancy-driven exchange flow between the dike and the drift, whereby bubbly fluid flows from the dike into the drift as a viscous gravity current (Figure 17.3; Figure 3 in Menand and Phillips, 2007a). This exchange flow is slow enough that bubbles in the drift have time to rise, segregate from the fluid, and accumulate as foam at the top of the drift in conjunction with the accumulation of degassed fluid at the base of the drift. The maximum distance L_{max} that the gravity current can travel corresponds to the point where all bubbles have risen to the top of the side arm,

$$L_{max} = \frac{D^2}{d}\left(\frac{12c}{1-c}\right)^{\frac{1}{2}} F(c) \tag{17.6}$$

where D is the drift diameter, d is the average bubble diameter, c is the volumetric bubble concentration and F is a function of c which has a value ~ 0.1 (Menand *et al.*, 2008). Ultimately, a steady state is reached, whereby influx of bubbly fluid into the drift is balanced by outward flux of lighter foam and denser degassed fluid back into the dike. Moreover, gas segregation processes and rates appear to be independent of moderate changes in magma supply rates in the dike.

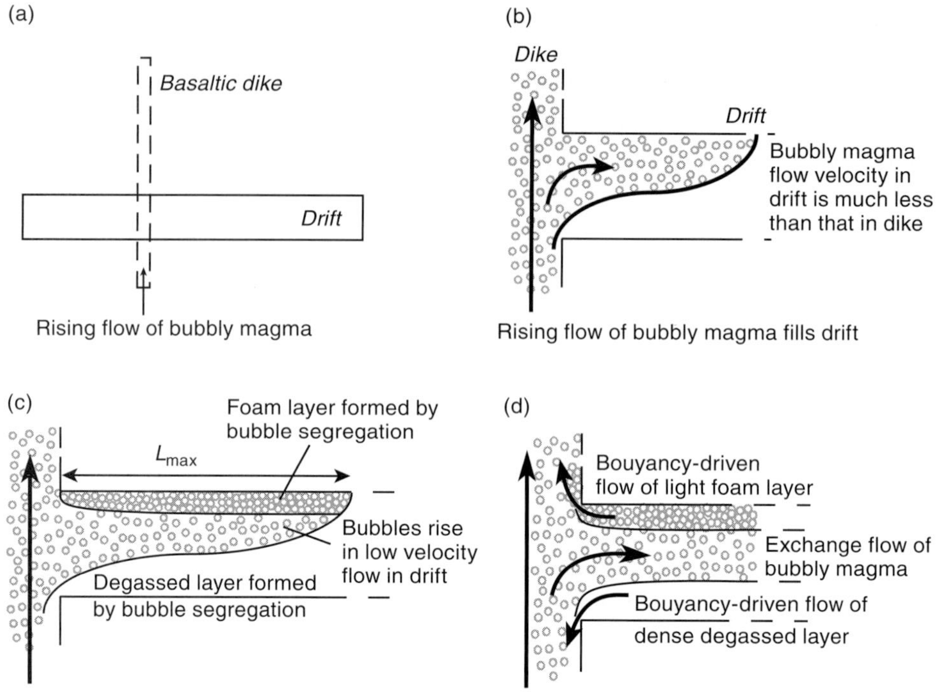

Fig. 17.3 Schematic illustration of magma flow in an interconnecting dike and drift. (a) System geometry. (b) Initial condition with the drift filling with bubbly magma from the dike. (c) Bubble segregation within the drift leads to the formation of a foam layer at the top of the drift and a degassed layer at its floor. (d) Steady-state exchange flows set up by bubble segregation.

The laboratory experiments at high gas fractions showed that the same processes occurred as for lower volumetric gas fractions, with the increased viscosity and reduced density contrast increasing the timescale for gas segregation (Menand and Phillips, 2007b; Menand *et al.*, 2008). The amount of foam that collects at the top of the drift is determined by the balance between the amount of bubbles rising from the bubbly fluid within the drift and the outward flux of foam that leaves it; the steady-state foam thickness $h(x)$ can be written as the product of a characteristic thickness H and a shape function $f(x)$, $h(x) = Hf(x)$, with

$$H = \left[\frac{c(1-c)^{\frac{7}{2}} d^2 L_{\max}^2}{\epsilon(1-\epsilon)^{\frac{5}{2}}(\epsilon - c)}\right]^{\frac{1}{4}} \qquad f(x) = \left(\frac{x}{L_{\max}} - \frac{x^2}{4L_{\max}^2}\right)^{\frac{1}{4}} \tag{17.7}$$

where ϵ is the volumetric gas fraction of the foam and x is the position along the foam (the origin is fixed at the dike–drift junction). The foam thickness is limited by the packing of the bubbles in the foam: as the foam thickness increases, bubbles deform and can coalesce, leading to the collapse of the foam. The maximum, critical thickness H_c the foam can sustain before collapsing is (Jaupart and Vergniolle, 1989; Menand *et al.*, 2008)

$$H_c = \frac{4\sigma}{\epsilon \rho_l g d} \tag{17.8}$$

where σ is the surface tension between the melt and the gas trapped in the bubbles of the foam. The steady-state foam thickness H described by (17.7) can develop only if it is smaller than the critical thickness H_c. If this is not the case, accumulation of bubbles at the top of the drift will lead to repeated collapse of the foam.

Two timescales are associated with this gas segregation process, and both are controlled by the rise of bubbles within the drift. The first one is the time needed by the bubbles to rise the diameter of the drift, D, and accumulate as the foam

$$T_b = \frac{12\mu_l D}{\rho_l g d^2 (1-c)^{\frac{7}{2}}} \tag{17.9}$$

The second timescale is the time needed for the steady-state foam to fully develop

$$T_f = \frac{12\mu_l L^{\frac{1}{2}}}{\rho_l g d^{\frac{3}{2}}} \left[\frac{\epsilon^3}{c^3 (1-c)^{\frac{21}{2}} (1-\epsilon)^{\frac{5}{2}} (\epsilon - c)} \right]^{\frac{1}{4}} \tag{17.10}$$

Gas segregation occurs in the drift if these two timescales are smaller than the timescale for cooling and solidification of the magma, T_s. Menand and Phillips (2007a, 2007b) based their calculations on cooling by pure conduction so that

$$T_s = \frac{D^2}{16\kappa\lambda^2} \tag{17.11}$$

where κ is the magma thermal diffusivity and the thermal constant, λ, depends on the temperature difference between the magma and surrounding rocks (Turcotte and Schubert, 1982).

Gas segregation leads to a steady-state recirculation of fluid in the drift with exchange of bubbly fluid with foam and degassed fluid, and this recirculation is characterized by a volumetric flux

$$Q = \frac{(1-c)^{\frac{7}{2}} \rho_l g d^2 D L}{12\mu_l \epsilon} \tag{17.12}$$

Up-scaled to the potential Yucca Mountain repository conditions, the results suggest that steady-state gas segregation would occur within hours to hundreds of years depending on the viscosity of the degassed magma, $10-10^5$ Pa s, and the average size of exsolved gas bubbles, 0.1–1 mm (right-hand plot of Figure 4 in Menand *et al.*, 2008). For comparison, Menand *et al.* (2008) give a solidification timescale by pure conduction of about three months for a 5 m-diameter drift; note that this estimate will be a lower bound due to circulation in the drift and replenishment with hotter magma as the eruption proceeds.

Using (17.12), Menand *et al.* (2008) calculated the fluxes that would be associated with gas segregation in a 5 m-diameter drift for different magma viscosities (right-hand plot of Figure 6 in Menand *et al.*, 2008). These range from 1 $m^3\,s^{-1}$ for the less viscous magmas to 10^{-8} $m^3\,s^{-1}$ for the most viscous degassed magmas. Gas segregation is likely to be in an unstable foam collapse regime, with the foam accumulated by gas segregation at the top of the drift reaching a few centimeters in thickness before its collapse due to bubble coalescence (Jaupart and Vergniolle, 1989; Menand *et al.*, 2008). The relative proportion of erupted degassed magma, which could potentially transport radioactive waste material towards the surface, depends on the value of the dike magma supply rate relative to the value of the gas segregation flux; with violent eruption of gas-rich as well as degassed magmas at relatively high magma supply rates, and eruption of mainly degassed magma by milder episodic Strombolian explosions at relatively lower supply rates (Menand and Phillips, 2007a). Menand *et al.* (2008) calculated that, depending on the average size of exsolved gas bubbles, the critical magma supply rate delimiting these two eruptive regimes would range from 10^{-4}–1 $m^3\,s^{-1}$ for magma viscosity of $\approx$ 10 Pa s to 10^{-8}–10^{-4} $m^3\,s^{-1}$ for magma viscosity of $\approx 10^5$ Pa s.

Menand and Phillips (2007a) also applied these general principles to degassing and eruption processes at Stromboli volcano, Italy. The results and their implications are consistent with a variety of independent field data. Gas segregation at Stromboli likely occurs in a shallow reservoir of sill-like geometry at a 3.5 km depth with bubbles of exsolved gas 0.1–1 mm in diameter. Menand and Phillips (2007a) also calculated that the transition between Strombolian activity, erupting gas-poor, highly porphyritic magmas, and violent explosions that also erupt gas-rich, low porphyritic magmas would correspond to a critical magma supply rate of $\sim$ 0.1–1 $m\,s^{-1}$.

If magma flows along the drift, either because the drift is open-ended or because magma pressure in the repository is able to drive open a new fracture in the surrounding rocks, steady-state flow will be characterized by speeds of $\sim$ 10 $m\,s^{-1}$ (Woods *et al.*, 2002; Dartevelle and Valentine, 2005), which corresponds to steady-state fluxes of $\sim$ 100 $m^3\,s^{-1}$ in a 5 m diameter drift. These fluxes are at least two orders of magnitude greater than fluxes induced by gas segregation processes. In this scenario, gas segregation processes are unlikely to affect the flow as they would occur on timescales much longer than those needed for magma to flow along the drift. Therefore, the steady-state flow pattern would depend on the magma supply rate from the dike.

17.2.3 Magma flow dynamics and cooling within repository drifts

As magma flows up the dike and along the drift, heat will be advected by the flowing magma and will simultaneously be lost by conduction into the colder surrounding rocks and waste containers. Competition between heat advection and conduction can affect magma dynamics, as investigated quantitatively by Bruce and Huppert (1989, 1990) and Petford *et al.* (1993). During the initial stage of magma flow up a dike, magma cools in response to the lower temperature of the surrounding rock walls. Subsequently, the continual supply of magma transfers heat into the solid walls. Magma cooling within the dike is typically confined to a thin thermal boundary layer adjacent to the dike walls, and the width of this thermal boundary layer increases with the length of the dike (Carrigan, 2000). Whether the dike becomes blocked or remains open is determined by the balance between the rate of solidification of the magma (dike closure) and that of melting of the walls (dike opening). These rates are in turn controlled by the magnitude of the latent heat (released during solidification and consumed during melting) as well as the difference between the heat supplied to the walls by the thermal boundary layer and that conducted into the surrounding rocks. Bruce and Huppert (1989, 1990) and Petford *et al.* (1993) showed that a critical width

$$w_c = 1.5\left[\frac{c_{\text{heat}}(T_w - T_\infty)^2}{L_{\text{heat}}(T_m - T_w)}\right]^{\frac{3}{4}}\left(\frac{\mu\kappa H_{\text{dike}}}{\Delta\rho g}\right)^{\frac{1}{4}}$$

exists, where c_{heat} is the specific heat, L_{heat} is the latent heat, T_m is the initial magma temperature, T_w is the temperature of the walls, T_∞ is the far field temperature of the rocks, H_{dike} is the dike length and $\Delta\rho$ is the density difference between magma and rocks. If the dike is thinner than this critical width, it will solidify before it can transport a significant volume of magma to the surface. When applied to basaltic dikes, these analyses show that dikes must be thicker than $\sim$ 0.5 m if they are to reach the surface before solidifying completely; this prediction agrees with field observations (Wada, 1994; Kerr and Lister, 1995; Wada, 1995). Three-dimensional analyses suggest that magma flowing through an initially long surface fissure will tend to localize to a number of isolated vents (Bruce and Huppert, 1989, 1990), as observed during basaltic fissure eruptions.

Insights about the cooling of magma as it flows within a drift can be obtained from studies of horizontal igneous intrusions (sills) and lava tubes. Holness and Humphreys (2003) observed that rocks surrounding the Traigh Bhàn na Sgùrra sill on the Isle of Mull, Scotland, displayed thermal aureoles up to 4 m thick around that sill, which taken in conjunction with the spatial distribution of crystals within the sill demonstrate that progressive focusing of magma flow into the wider parts of the sill was sustained for up to five months. Lava tubes, which form when the lava flow surface solidifies as a crust while hot lava continues to flow beneath, are a common feature of basaltic lava flow fields. If the flow rate is sufficiently high, lava can thermally erode its way into the surrounding solidified lavas (Francis and Oppenheimer, 2004). Lava tubes can extend significant distances because lava is well insulated by the tube crust and loses very little heat by conduction or radiation. Lava tubes up to 20 m in diameter and > 100 km in length have been observed in Queensland, Australia; these would have enabled the lava flow fields to develop over several months to years despite involving overall effusion rates perhaps as low as 10 $m^3\ s^{-1}$ (Stephenson *et al.*, 1998). The Krafla borehole eruption also shows that cooling may not be a major factor in the initial filling of a tunnel.

Precise assessment of magma cooling while flowing in a drift partially obstructed by waste containers is difficult because of the complex three-dimensional geometry of the flow field. Nevertheless, a conservative estimate can be made by assuming conductive cooling of the magma through the wall of the drift, which gives about three months for magma to solidify by conduction (Menand *et al.*, 2008). This estimate is comparable to the timescales associated with lava flows within sills and lava tubes and suggests that magma could remain fluid for several months, at least in some part of the drift. For comparison, based on the diffusion of heat into metal waste containers, Woods *et al.* (2002) calculated that the waste containers will become deformable and may break open for times $\gtrsim 1000\,\mathrm{s}$ (< 1 hr). Improved estimates for potential magma cooling rates within repository drifts and their effect on flow dynamics, if warranted, would require three-dimensional numerical simulations.

If transient or steady magma flow occurs through radioactive waste repository drifts, possible consequences include the generation of waste container motion due to drag exerted by the flowing magma, heating and possible disruption of the waste containers, and transport of the container contents. Woods *et al.* (2002) estimate the drag force acting on the waste containers

$$F_\mathrm{d} = C_\mathrm{d}\rho u^2 A \tag{17.13}$$

where A is the area of the face of the container perpendicular to the flow direction, and C_d is the drag coefficient, which is order unity for these flow conditions. For steady flow conditions, Woods *et al.* (2002) estimated the ratio of the drag force to container weight to be typically of order or smaller than 0.1–1.0, suggesting that the containers may be displaced down the drift. However, the flow is too weak to keep the containers in suspension, and any container motion is likely to be relatively slow. Later calculations of the flow conditions following the intersection of a basaltic dike with a repository drift conducted by Dartevelle and Valentine (2005) were made with smaller dike widths and resulted in much lower flow velocities than those calculated by Woods *et al.* (2002). In Dartevelle and Valentine's calculations, the drag force was insufficient to generate waste container motion; furthermore, because the waste containers will be placed in a line along the length of the repository drift, only the first container will feel the drag force estimated by Woods *et al.* (2002). Subsequent containers will feel a lower drag force in the wake flow behind the first container – the same effect exploited by the formations adopted by migrating birds and racing cyclists.

To the extent that waste containers can be disrupted by the combined effects of magma flow and heat transfer, waste container contents can possibly be transported in the magma flow through the drifts and to the surface. Although the exact contents of waste containers will vary, current models typically assume that the waste material is spent nuclear fuel that has been fragmented by disruptive processes in the size range 10–500 microns (CRWMS M&O, 2000) with a density of $\approx 10\,000\,\mathrm{kg\,m^{-3}}$. Erosive transport of small particles by turbulent stream flows has been widely studied in sedimentology, and the key parametric relationship between the Shields number θ (ratio of shear stress acting on a particle to its weight) and particle Reynolds number (the ratio of inertial to viscous forces acting on a particle in the flow) has been empirically determined for turbulent flow conditions (Figure 2 in Miller *et al.*, 1977). However, until recently, there has been little study of particle transport under viscous flow conditions, as would be appropriate for magma flow.

The experimental and theoretical studies of Charru *et al.* (2004) provide a framework for estimating the transport properties of small particles in a uniformly sheared viscous flow. The experiments were conducted in a rotating annular viscous flow, to achieve steady flow over long times and under conditions in which the secondary velocity generated by centrifugal forces was negligible compared to plane Couette flow in the channel. Direct observations showed that small particle motion took the form of a series of saltation "flights," whose duration τ was found to be independent of shear rate γ, $\tau \approx 15\, d_{\mathrm{p}}/v_{\mathrm{S}}$, where d_{p} is the particle diameter and v_{S} is the Stokes settling speed of the particle. The mean particle velocity $\bar{u}$ was found to depend linearly on the shear rate, $\bar{u} \approx 0.1\, \gamma d_{\mathrm{p}}$. The particle flow rate Q_{p} was found to have a quadratic dependence on shear rate,

$$Q_{\mathrm{p}} \approx 0.1\gamma(0.47/d_{\mathrm{p}})(\theta - 0.12) \qquad (17.14)$$

Erosive transport of small particles by fluid flow is a complex process that depends on interaction of the particles and fluid, and the particles with each other. Further work is required to understand the flow conditions representative of potential magma–waste repository interaction and particle transport under these conditions. However, the scaling arguments presented here form a fundamental framework for estimating transport properties of high-density particles in viscous magma flow. This can be illustrated with simple estimates for magma properties for basalts in the Yucca Mountain region (density of 2750 kg m^{-3}, viscosity of 10 Pa s, 70 vol.% of degassed bubbles 1 mm in diameter), assuming in this example that the magma flow pattern corresponds to flow through a circular cross-section drift. The presence of waste containers in the drift will complicate the flow patterns (e.g. create local eddies that can affect Couette flow dynamics), but the principles illustrated here should hold. For a maximum average velocity of degassed magma in a 5 m-diameter and 1 km-long drift of about 10^{-3} m s^{-1} (using (17.12) for the flux) and assuming a standard Poiseuille flow profile for viscous flow in a cylindrical cross-section (Schlichting, 1960), the maximum shear rate 0.1 m above the base of the drift is approximately 7.5 s^{-1}. Using the range of particle sizes and densities for fragments of spent nuclear fuel given previously, the maximum Shields number is $\sim$ 2 for this viscous flow (Charru *et al.*, 2004), corresponding to a particle flow rate of $\sim$ 1500 particles per unit drift width per second, (17.14). This particle flow rate is estimated for the largest fragments of spent nuclear fuel 500 μ in diameter. Particle transport from potentially disrupted waste containers will be limited by the volume of particles available to the flow.

Concluding remarks

The models discussed here provide some first-order constraints on the interaction of basaltic magma with an open-drift system that is the basis for some potential radioactive waste repositories. A first-order conclusion is that potentially intersected drifts rapidly will be filled by magma. This will be the case irrespective of whether the magma is explosive due to the exsolution and expansion of gases or degassed as a consequence of as yet poorly understood gas segregation processes during ascent. Lejeune *et al.* (Chapter 18, this volume) have presented the degassed end-member, and we present the explosive case. At typical eruption rates of monogenetic basaltic eruptions (10–1000 $m^3\,s^{-1}$), a drift can be filled in a few tens of seconds for the explosive case and a few tens of minutes for the degassed case. Although the magma will form a thin quench on contact with the drift walls and containers, estimates of cooling timescales indicate that these effects are small and will not inhibit the filling of the drifts. This view is verified by the observation that basalt flowed along a geothermal borehole for hundreds of meters despite having a volume-to-surface area ratio that is two orders of magnitude smaller than a repository drift.

Characterizing container disruption is complex. In addition to considering the state of magma upon entry into a drift, disruption is also dependent on the design and properties of container materials, particularly in relation to response to impacts and heating. Such considerations go well beyond the scope of this chapter, but some inferences can be made. Container failure could occur due to prolonged heating and pressure effects. Heating weakens the containers, making them more likely to fail, and the interior pressure of the container is expected to increase due to heating of the gases. Simultaneously, the pressure in the surrounding magma after a drift has been filled may increase substantially to lithostatic values or above. For example, if an eruption is occurring, the magma-static pressure at a 300 m-deep repository would be above 8 MPa for a column of degassed magma. In response to the combined thermal and mechanical stresses, the container might be deformed or broken open, and the contents of the affected containers might be released and transported to the surface by entrainment in the erupting column.

If a subvolcanic conduit developed through a drift, waste that is directly entrained in the erupting conduit could be transported to the surface. Two additional scenarios have been investigated that consider the ability of magma to entrain waste from drifts potentially intersected by a dike. Woods *et al.* (2002) described a "dogleg" scenario in which the dike intersects a drift and a secondary fracture develops in a new location along the drift so that once the magma breaches the surface, the magma flows along the drift to connect the inlet dike with the outlet dike. Here, a second scenario is considered where the original supply dike continues to the surface and there is convective exchange between the magma in the drift and the magma flowing up the dike.

In this alternative scenario, Menand and Phillips (2007a, 2007b) infer from their experiments that, independently of moderate changes in the dike magma supply rate, gas segregation processes will occur in the drift and lead to a convective exchange of gas-rich foam and degassed magma flowing out of the drift and back into the dike with bubbly magma from the dike. Using the potential Yucca Mountain repository geometry as an example, flows will likely be in an unstable collapse regime with the gas-rich foam experiencing repeated collapse as it accumulates at the top of the drift. The length and timescales of the gas segregation processes are controlled by the rise of bubbles in the drift. The time required for steady-state gas segregation is estimated to range from hours to hundreds of years, depending on the average size of exsolved gas bubbles; and on the viscosity of degassed magmas, which depends strongly on the degree of water exsolution, cooling and crystallization. The associated magma flux is estimated to range from $1\,m^3\,s^{-1}$–$10^{-8}\,m^3\,s^{-1}$, depending on the magma viscosity and the size of exsolved gas bubbles. The relative proportion of erupted degassed magma depends on the value of the dike magma supply rate relative to that of the gas segregation flux. If magma is supplied at a higher rate, then gas-rich as well as degassed magmas are expected to be violently erupted; if the supply rate is lower, then mainly degassed magma would be erupted by milder episodic Strombolian explosions generated by the repeated collapse of the foam accumulated at the top of drifts.

A related matter is whether a potential magma flow could transport very dense waste particles. Based on estimates of magma fluxes in the initial stages of basaltic volcanic eruptions, it is unlikely that significant transport of intact waste containers would occur, regardless of whether the flow is unidirectional along the drift (Woods *et al.*, 2002; Dartevelle and Valentine, 2005) or recirculates within the drift (Menand and Phillips, 2007a, 2007b). In the event of waste-container disruption, estimates suggest that small fragments of spent nuclear fuel of the size and density used in current design calculations might be transported along the drift in a viscous saltation regime. Two-phase flows of this type have not been widely studied, although recent interest has established the key principles and identified transport regimes (e.g. Charru *et al.*, 2004). As outlined in this chapter, precise estimates of potential magma transport dynamics for laminar Couette flows will not become available until the exact configuration of disrupted waste containers and drift geometry can be better constrained, but the first-order estimates presented here suggest that even modest magma shear rates can initiate erosive motion of dense waste fragments along the base of a drift, irrespective of the precise geometric details. For the scenario of volcanic activity potentially interacting with a repository drift, the calculations indicate magma flows are capable of transporting small waste fragments along the drift and into the erupting conduit, with the possibility that this material could be subsequently transported to the surface and dispersed in explosive eruptions or effusive flows.

Further lines of investigation

The presence of engineered barrier systems and their additional thermal mass will likely affect the flow of magma within drifts and thus the thermal evolution of the system. Assessing how waste containers and barrier systems may affect heat transfer into the containers and the cooling rate of magma, as well as gas segregation processes within drifts, is complex and may warrant three-dimensional computational modeling.

Confidence in models for potential magma–drift interaction processes would benefit from improved knowledge of the petrological evolution of magmas in the shallow subsurface during dynamic eruptions (see Spera and Fowler, Chapter 8, this volume). This evolution will affect magma viscosity and the size of exsolved gas bubbles, which in turn will exert a strong influence on the fluxes and average velocities associated with gas segregation and convective exchange flow. Constraining the petrological evolution of magmas will improve estimates of the heat transfer and cooling rates during and after gas segregation, and thus the duration of potential magma exchange flow within drifts.

Further reading

White (1979) provides a particularly clear introduction to fluid mechanics with applications to engineering problems including flow in ducts, flow around immersed bodies and calculations of drag coefficients for various flow configurations. Analytical techniques for the treatment of two-phase flow problems as well as practical applications can be found in Wallis (1969). Integrating observation, theory and experimental studies, Sparks *et al.* (1997) provide a technical and complete reference to physical volcanology using historical volcanologic events as case studies.

Acknowledgments

This work benefited from fruitful discussions with Chuck Connor, Brittain Hill, Andrew Hogg and Gary Matson. This chapter was prepared to document work performed by the Center for Nuclear Waste Regulatory Analyses (CNWRA) and its contractors for the US. Nuclear Regulatory Commission (NRC) under Contract No. NRC-02-07-006. The activities reported here were performed on behalf of the NRC Office of Nuclear Material Safety and Safeguards, Division of High-Level Waste Repository Safety. This chapter is an independent product of the CNWRA and does not necessarily reflect the view or the regulatory position of the NRC.

References

Bechtel SAIC Company, LLC (2007). Characterize framework for igneous activity at Yucca Mountain, Nevada, ANL-MGR-GS-000002, Rev. 3. Las Vegas, Nevada: Bechtel SAIC Company, LLC.

Blackburn, E. A., L. Wilson and R. S. J. Sparks (1976). Mechanism and dynamics of Strombolian activity. *Journal of the Geological Society of London*, **132**, 429–440.

Blundy, J., K. Cashman and M. Humphreys (2006). Magma heating by decompression-driven crystallization beneath andesitic volcanoes. *Nature*, **443**, 76–80.

Bruce, P. M. and H. E. Huppert (1989). Thermal control of basaltic fissure eruptions. *Nature*, **342**, 665–667.

Bruce, P. M. and H. E. Huppert. (1990). Solidification and melting along dikes by the laminar flow of basaltic magma. In: Ryan, M. P. (ed.) *Magma Transport and Storage*. New York, NY: John Wiley, 87–101.

Carrigan, C. R. (2000). Plumbing systems. In: Sigurdsson, H. *et al.* (eds.), *Encyclopedia of Volcanoes*. San Diego, CA: Academic Press, 219–235.

Cashman, K. V. (1992). Groundmass crystallization of Mount St Helens dacite, 1980–1986 – a tool for interpreting shallow magmatic processes. *Contributions in Mineralogy and Petrology*, **109**, 431–449.

Charru, F., H. Mouilleron and O. Eiff (2004). Erosion and deposition of particles on a bed sheared by a viscous flow. *Journal of Fluid Mechanics*, **519**, 55–80.

CRWMS M&O (2000). Miscellaneous waste-form FEPs, ANL-WIS-MD-000009 Rev 00 ICN 01. Las Vegas, NV: Civilian Radioactive Waste Management System, Management and Operating Contractor.

Dartevelle, S. (2004). Numerical modeling of geophysical granular flows: 1. A comprehensive approach to granular rheologies and geophysical multiphase flows. *Geochemistry, Geophysics, and Geosystems*, **5**, Q08003, doi:10.1029/2003GC000636.

Dartevelle, S. and G. A. Valentine (2005). Early-time multiphase interactions between basaltic magma and underground openings at the proposed Yucca Mountain radioactive waste repository. *Geophysical Research Letters*, **32**, L22311, doi:10.1029/2005GL024172.

EPA (2001). Background information document for 40 CFR 197: Public health and environmental radiation protection standards for Yucca Mountain, NV, EPA 402-R-01-004. US Environmental Protection Agency, Air and Radiation.

Fedotov, S. A. and Ye. K. Markhinin (eds.) (1983). *The Great Tolbachik Fissure Eruption: Geological and Geophysical Data 1975–1976*. Cambridge; New York: Cambridge University Press.

Francis, P. and C. Oppenheimer (2004). *Volcanoes*. Oxford: Oxford University Press.

Giordano, D. and D. B. Dingwell (2003). Viscosity of hydrous Etna basalt: implications for Plinian-style basaltic eruptions. *Bulletin of Volcanology*, **65**, 8–14.

Hess, K.-U. and D. B. Dingwell (1996). Viscosities of hydrous leucogranitic melts: a non-Arrhenian model. *American Mineralogist*, **81**, 1297–1300.

Holloway, J. R. and J. G. Blank (1994). Application of experimental results to C–O–H species in natural melts. In: Carroll, M. R. and J. R. Holloway (eds.) *Volatiles in Magmas*, Reviews in Mineralogy 30. Washington, DC: Mineralogical Society of America, 187–230.

Holness, M. B. and M. C. S. Humphreys (2003). The Traigh Bhàn na Sgùrra Sill, Isle of Mull: flow localization in a major magma conduit. *Journal of Petrology*, **44**, 1961–1976.

Huppert, H. E. and R. S. J. Sparks (1985). Cooling and contamination of mafic and ultramafic magmas during ascent through continental crust. *Earth and Planetary Science Letters*, **74**, 371–386.

Jaupart, C. (2000). Magma ascent at shallow levels. In: Sigurdsson, H. *et al.* (eds.), *Encyclopedia of Volcanoes*. San Diego, CA: Academic Press, 237–245.

Jaupart, C. and S. Vergniolle (1989). The generation and collapse of a foam layer at the roof of a basaltic magma chamber. *Journal of Fluid Mechanics*, **203**, 347–380.

Kerr, R. C. and J. R. Lister (1995). Comment on "On the relationship between dike width and magma viscosity" by Yutaka Wada. *Journal of Geophysical Research*, **100**(B8), 15 541.

Kilburn, C. R. F. (2000). Lava flows and flow fields. In: Sigurdsson, H. *et al.* (eds.), *Encyclopedia of Volcanoes*. San Diego, CA: Academic Press, 291–305.

Lange, R. A. (1994). Application of experimental results to C–O–H species in natural melts. In: Carroll, M. R. and J. R. Holloway (eds.), *Volatiles in Magmas*, Reviews in Mineralogy 30. Washington, DC: Mineralogical Society of America, 331–369.

Larsen, G., K. Grönvold and S. Thorarinsson (1979). Volcanic eruption through a geothermal borehole at Ná mafjall, Iceland. *Nature*, **278**, 707–710.

Lister, J. R. and R. C. Kerr (1991). Fluid-mechanical models of crack propagation and their application to magma transport in dikes. *Journal of Geophysical Research*, **96**, 10 049–10 077.

Llewellin, E. W. and M. Manga (2005). Bubble suspension rheology and implications for conduit flow. *Journal of Volcanology and Geothermal Research*, **143**, 205–217.

Long, J. C. S. and R. C. Ewing (2004). Yucca Mountain: Earth-science issues at a geologic repository for high-level nuclear waste. *Annual Review of Earth and Planetary Sciences*, **32**, 363–401.

Macdonald, G. A., A. T. Abbott and F. L. Peterson (1983). *Volcanoes in the Sea: The Geology of Hawaii*. Honolulu: University of Hawaii Press.

Mason, R. M., A. B. Starostin, O. E. Melnik and R. S. J. Sparks (2006). From Vulcanian explosions to sustained explosive eruptions: the role of diffusive mass transfer in conduit flow dynamics. *Journal of Volcanology and Geothermal Research*, **153**, 148–165.

Massol, H., C. Jaupart and D. W. Pepper (2001). Ascent and decompression of viscous vesicular magma in a volcanic conduit. *Journal of Geophysical Research*, **106**(B8), 16 223–16 240.

Melnik, O. and R. S. J. Sparks (1999). Nonlinear dynamics of lava extrusion. *Nature*, **402**, 37–41.

Menand, T. and J. C. Phillips (2007a). Gas segregation in dikes and sills. *Journal of Volcanology and Geothermal Research*, **159**, 393–408.

Menand, T. and J. C. Phillips (2007b). A note on gas segregation in dikes and sills at high gas fractions. *Journal of Volcanology and Geothermal Research*, **162**, 185–188.

Menand T., J. C. Phillips and R. S. J. Sparks (2008). Circulation of bubbly magma and gas segregation within tunnels of the potential Yucca Mountain repository. *Bulletin of Volcanology*, **70**, 947–960.

Miller, M. C., N. I. McCave and P. D. Komar (1977). Threshold of sediment motion under unidirectional currents. *Sedimentology*, **24**, 507–528.

Murase, T. and A. R. McBirney (1973). Properties of some common igneous rocks and their melts at high temperatures. *Geology Society of America Bulletin*, **84**(11), 3563–3592.

National Research Council (US) and F. L. Parker (1990). *Rethinking High-level Radioactive Waste Disposal: A Position Statement of the Board on Radioactive Waste Management, Commission on Geosciences, Environment, and Resources, National Research Council*. Washington, DC: National Academy Press.

National Research Council (US) (2001). *Disposition of High-level Waste and Spent Nuclear Fuel: The Continuing Societal and Technical Challenges*. Washington, DC: National Academy Press.

Nicholis, M. G. and M. J. Rutherford (2004). Experimental constraints on magma ascent rate for the Crater Flat volcanic zone hawaiite. *Geology*, **32**, 489–492, doi:10.1130/G20324.1.

NRC (2005). Integrated issue resolution status report NUREG-1762, 1(1). Washington, DC: US Nuclear Regulatory Commission.

Petford, N., R. C. Kerr and J. R. Lister (1993). Dike transport in granitoid magmas. *Geology*, **21**, 845–848.

Pinkerton, H. and R. S. J. Sparks (1978). Field-measurements of the rheology of lava. *Nature*, **276**, 383–385.

Pollard, D. D. (1987). Elementary fracture mechanics applied to the structural interpretation of dikes. In: Hall, H. C. and W. F. Fahrig (eds.) *Mafic Dyke Swarms*, Special Paper 34. St. Johns, Newfoundland: Geologic Association of Canada, 5–24.

Phillips, J. C. and A. W. Woods (2001). Bubble plumes generated during recharge of basaltic magma reservoirs. *Earth and Planetary Science Letters*, **186**, 297–309.

Rosseau, J. P., E. M. Kwicklis and D. C. Giles (1999). Hydrogeology of the undersaturated zone, North Ramp Area of the Exploratory Studies Facility, Yucca

Mountain, Nevada, Water-Resources Investigations Report 98B4050. Denver, CO: US Geological Survey.
Rubin, A. M. (1995). Propagation of magma-filled cracks. *Annual Reviews of Earth Planetary Sciences*, **23**, 287–336.
Rubin, A. M. and D. D. Pollard (1987). Origin of blade-like dikes in volcanic rift zones. In: Decker, R. W. *et al.* (eds.) *Volcanism in Hawaii*, Professional Paper 1350(2). US Geological Survey, Volcano Hazards Team, 1449–1470.
Schlichting, H. (1960). *Boundary Layer Theory*. New York, NY: McGraw-Hill.
Shaw, H. R. (1972). Viscosities of magmatic silicate liquids: an empirical method of prediction. *American Journal of Science*, **272**, 870–893.
Sisson, T. W. and G. D. Layne (1993). H_2O in basalt and basaltic andesite glass inclusions from four subduction-related volcanoes. *Earth and Planetary Science Letters*, **117**, 619–635.
Slezin, Yu. B. (2003). The mechanism of volcanic eruptions (a steady state approach). *Journal of Volcanology and Geothermal Research*, **122**, 7–50.
Smith, E. I. and D. L. Keenan (2005). Yucca Mountain could face greater volcanic threat. *Eos, Transactions of the American Geophysical Union*, **86**(35), doi:10.1029/2005EO350001.
Sparks, R. S. J. (1978). The dynamics of bubble formation and growth in magmas: a review and analysis. *Journal of Volcanology and Geothermal Research*, **3**, 1–37.
Sparks, R. S. J., J. Barclay, C. Jaupart, H. M. Mader and J. C. Phillips (1994). Physical aspects of magma degassing I. Experimental and theoretical constraints on vesiculation. In: Carroll, M. R. and J. R. Holloway (eds.) *Volatiles in Magmas*, Reviews in Mineralogy 30. Washington, DC: Mineralogical Society of America, 413–445.
Sparks, R. S. J., M. I. Bursik, S. N. Carey *et al.* (1997). *Volcanic Plumes*. Chichester: John Wiley and Sons.
Spera, F. J. (2000). Physical properties of magma. In: Sigurdsson, H. *et al.* (eds.) *Encyclopedia of Volcanoes*. San Diego, CA: Academic Press, 171–190.
Stephenson, P. J., A. T. Burch-Johnston, D. Stanton and P. W. Whitehead (1998). Three long lava flows in north Queensland. *Journal of Geophysical Research*, **103**(B11), 27 359–27 370.
Swanson, D. A., T. L. Wright and R. T. Helz (1975). Linear vent systems and estimated rates of magma production and eruption for the Yakima Basalt on the Columbia Plateau. *American Journal of Science*, **275**, 877–905.
Thorarinsson, S. (1969). The Lakagigar eruption of 1783. *Bulletin of Volcanology*, **33**, 910–927.
Turcotte, D. L. and G. Schubert (1982). *Geodynamics: Applications of Continuum Physics to Geological Problems*. New York, NY: John Wiley and Sons.
Valentine, G. A., D. J. Krier, F. V. Perry and G. Heiken (2007). Eruptive and geomorphic processes at the Lathrop Wells scoria cone volcano. *Journal of Volcanology and Geothermal Research*, **161**, 57–80.
Wada, Y. (1994). On the relationship between dike width and magma viscosity. *Journal of Geophysical Research*, **99**(B9), 17 743–17 755.
Wada, Y. (1995). Comment on, "On the relationship between dike width and magma viscosity". *Journal of Geophysical Research*, **100**(B8), 15 543–15 544.
Wallace, P. and A. Anderson (2000). Volatiles in magmas. In: Sigurdsson, H. *et al.* (eds.) *Encyclopedia of Volcanoes*. San Diego, CA: Academic Press, 149–170.
Wallis, G. B. (1969). *One-dimensional Two-phase Flow*. New York, NY: McGraw-Hill.

White, F. M. (1979). *Fluid Mechanics*, International Student Edition. Tokyo: McGraw-Hill Kogakusha.

Wilson, L. and J. W. Head (1981). Ascent and eruption of basaltic magma on the Earth and Moon. *Journal of Geophysical Research*, **86**, 2971–3001.

Woods, A. W., S. Sparks, O. Bokhove *et al.* (2002). Modeling magma-drift interaction at the proposed high-level radioactive waste repository at Yucca Mountain, Nevada, USA. *Geophysical Research Letters*, **29**(13), 1641, doi:10.1029/2002GL014665.

18

Intrusion dynamics for volatile-poor basaltic magma into subsurface nuclear installations

A.-M. Lejeune, B. E. Hill, A. W. Woods, R. S. J. Sparks and C. B. Connor

Igneous events create physical conditions that commonly are beyond the design basis of most engineered systems, with little data available for direct analysis of potentially hazardous scenarios. In addition, interactions with engineered systems can change the character of an igneous event in ways that never occur in nature. These potential changes in process may directly affect the impact of the resulting hazard. In this study, we will examine the potential changes in magma-flow processes that might occur if rising, volatile-poor magma intersects open, subsurface structures such as tunnels or drifts. Examination of decompression processes provides one end-member to the range of models that may need to be considered for potential subsurface hazards associated with basaltic igneous events. Although volatile-rich magma decompression may be viewed as a more likely scenario for some basaltic magma systems, examination of the volatile-poor scenario places important constraints on the extent and duration of potential magma flow into underground openings.

The United States has generated $\sim$ 50 000 000 kg of high-level radioactive waste from commercial and defense reactors. The current proposal is to dispose this waste in 300 m-deep tunnels beneath Yucca Mountain, Nevada, USA. The regulatory framework in the United States establishes limits on potential doses to the public for a period of at least 10 ka (NRC, 2001). This proposed site, however, is located in a geologically active basaltic volcanic field, where the probability of a new volcano forming at the potential repository site is generally calculated at 10^4–10^3 during the next 10 ka (e.g. Connor *et al.*, 2000; DOE, 2001). United States federal regulations require a detailed hazard analysis for natural events with a $> 10^4$ in 10 ka likelihood of occurrence (EPA, 2001). A component of this hazard analysis is how flow processes may be affected if ascending magma potentially intersects 5 m-diameter tunnels containing radioactive waste. The actual dike–tunnel interaction mechanism will depend on regional and local stress relationships between the tunnels and surrounding rock. A future dike rising beneath the repository, however, would likely intersect multiple tunnels located 20–80 m apart, as the tunnels are oriented roughly orthogonal to the direction of maximum horizontal compressive stress and thus orthogonal to the direction of dike propagation (e.g. Delaney *et al.*, 1986; Morris *et al.*, 1996). Repository tunnels are also laid out roughly perpendicular to faults that may influence magma ascent. Clearly, confined basaltic magma intersecting a tunnel at essentially atmospheric pressure would decompress and flow into that tunnel. The extent and duration of magma flow into the tunnels, however, determines how many radioactive waste canisters may be affected during a potential igneous event. If waste canisters are damaged by magma flow into the tunnels, radioactive waste may be released through hydrologic flow and transport processes that re-establish following the potential igneous event.

The goal of this work is to examine the extent and duration of magma flow into tunnels intersected by pressurized, volatile-poor basaltic magma ascending along a dike. Thus, we have developed a simple theoretical model of the time-dependent viscous flow based on an abstracted geometry of the dike–tunnel system and dimensional scalings of the different forces involved in the subsurface environment. We then designed an analog experimental model for this abstracted system. Results of our experiments examine how the flow rate depends on the pressure drop into the tunnel and describe the morphology of the flow front. We also compare these results with the theoretical model. We conclude by scaling the model to the subsurface repository, to gain insights on igneous hazard assessments for Yucca Mountain.

18.1 Background and conceptual model

We have developed a hierarchical program of numerical and experimental investigations to evaluate possible decompression-induced flow phenomena in the interaction of basaltic magma with a geological repository of radioactive waste with the potential design attributes of Yucca Mountain. Models and experiments in this chapter examine flow conditions appropriate for a volatile-poor basaltic magma. These models complement alternative models that consider shallow subsurface flow processes for volatile-rich basaltic magmas (Woods *et al.*, 2002; Bokhove *et al.*, 2005; Dartevelle and Valentine, 2005; Woods *et al.*, 2006). Although we recognize that basalt may contain appreciable quantities of magmatic volatiles, the present work helps establish the underlying principles that control decompression-induced flow without the complications of volatile exsolution and rapid fragmentation phenomena.

Pliocene–Quaternary-age trachybasalts in the Yucca Mountain region are mildly alkaline and contain phenocrysts of predominantly olivine and augite, with minor and variable amounts of plagioclase and occasionally titanian pargasitic amphibole (Vaniman *et al.*, 1982; Hill *et al.*, 1995). Hydrous crystallization experiments on these basalts (Nicholis and Rutherford, 2004) and glass inclusion analyses (Luhr and Housh, 2002) indicate initial magmatic water contents of approximately 4 wt.%. Such magmatic water contents are expected to produce a high-velocity fragmented flow following the rapid decompression at repository depths of 300 m (e.g. Woods *et al.*, 2002).

Although the early phases of a monogenetic basaltic eruption commonly are assumed to be predominantly explosive, there are in fact rather few well-documented eruptions or analyses of pre-historic sequences from scoria cones that can be cited to verify this assumption. Indeed, significant volumes of lava erupted in the first two days of the 1973 Eldfell eruption in Iceland, during a time when pyroclastic activity was relatively weak along an extended fissure (Thorarinsson *et al.*, 1973; Williams and Moore, 1976). The eruption was most extensive and intense on the second day when activity localized on the site where the scoria cone developed. Gutmann (1979) has shown that in the Pinacate volcanic field in Mexico, degassed lava is commonly the lowermost stratum in proximal section through scoria cone sequences. A detailed study of the scoria cone and lava of Lathrop Wells volcano in Nevada (Valentine *et al.*, 2007) also shows that early activity consisted of cone-forming Strombolian bursts and formation of a fan-like lava flow, with violent Strombolian explosive activity occurring after the early lava was emplaced. This evolution of eruption style may be related to degassing effects during initial ascent of the magma, suggesting that initial hazards from subsurface flow may be represented by decompression of a volatile-poor rather than volatile-rich magma. In addition, during the course of a basaltic eruption, degassed basaltic magma may intersect tunnels in association with syneruptive dike propagation. Thus, although initial explosive interaction is generally thought to be more likely, the case of initial interaction with partially degassed magma remains a plausible scenario.

Rising basaltic dikes typically have a fluid pressure $\sim$ 1–10 MPa in excess of local lithostatic pressure, which allows the magma to ascend from depth, fracture the surrounding rock and dilate fractures to a 1 m aperture (Pollard, 1973; Lister, 1991; Lister and Kerr, 1991). At a 300 m depth beneath Yucca Mountain, the tuff bedrock has a hydrofracture stress of 5.1–5.5 MPa (Stock *et al.*, 1985). In contrast, the potential repository tunnels would have atmospheric pressure (Rosseau *et al.*, 1999) and thereby provide the path of least resistance for the ascending magma. We therefore anticipate that on potential intersection of the dike with the tunnel, the magma flow will be diverted into the tunnel.

The design for the potential high-level radioactive waste repository at Yucca Mountain in Nevada is currently under development and may change significantly from conditions outlined in this chapter. Current plans are to locate approximately 100 horizontal tunnels at depths of 200–300 m below the surface of Yucca Mountain. The east-northeast-trending tunnels are $\approx$ 5 m in diameter, 600 m long and spaced 80 m apart. This tunnel orientation is approximately parallel to the direction of minimum horizontal *in situ* stress (e.g. Morris *et al.*, 1996). The tunnel system has a footprint of 1 $\times$ 5 km, and both ends of each tunnel are connected to the surface by access tunnels with a diameter of 7–10 m. Radioactive waste likely will be stored in chromium–nickel alloy canisters up to 2 m in diameter and up to 5 m in length, which can be covered by a titanium-alloy drip shield. Current plans are to seal tunnel ends and access tunnels after waste emplacement and to not emplace backfill around the drip shields. The tunnels containing waste will be at one atmosphere pressure and can effectively be considered as empty cavities, as the waste canisters occupy $\approx$ 20% of the volume of the tunnel.

As shallow dike lengths routinely exceed the 80 m spacing between waste emplacement tunnels, multiple tunnels likely would be intersected during a potential igneous event at Yucca Mountain. For typical magma ascent rates of 1 m s^{-1}, the ascending magma flux from each 80 m-long by 1 m-wide dike segment thus can be captured by a 20 m^2 tunnel if the flow accelerates to a speed of 4 m s^{-1} as it propagates into the tunnel. An understanding of the resulting hazard potential depends on the actual rate of flow, the effects of flow acceleration on magma supply in the feeder dike, and on the time needed to fill an intersected tunnel and re-establish magmatic pressures that allow continued ascent to the surface.

We now build a simplified model of this process and explore the flow regimes that may develop. We then describe an analog laboratory experiment that has been designed to simulate the flow, and test the model quantitatively. First, it is useful to examine the dimensionless parameters that describe the ratio of forces in the problem and to use these parameters to examine typical flow regimes in operation.

As the flow moves into the tunnel, it experiences a sudden decompression, ΔP, and an associated acceleration. We simplify our model by assuming that in this shallow part of the crust, the crust is sufficiently strong to withstand the decrease in pressure. We therefore assume the flow geometry to be fixed. This assumption simplifies the model, in that, following breakthrough, the rapid decompression of the magma may lead to partial closing of the dike and also some failure of the dike walls.

The typical flow speed in the basaltic dikes prior to breakthrough is $u \sim 1\,\mathrm{m\,s^{-1}}$, based on both observation and theory (Wilson and Head, 1981; Lister, 1991). For a dike width of $w = 1\,\mathrm{m}$, magma of viscosity $\mu = 300\,\mathrm{Pa\,s}$ and magma of density $\rho = 2500\,\mathrm{kg\,m^3}$, the typical Reynolds number of the flow, $\rho u w/\mu$, is ≈ 8. For such a flow, the turbulent drag exerted by the walls of the dike has a magnitude of $2fu^2\rho/w$, where f is the turbulent friction factor, ~ 0.01. The turbulent drag has a value of ≈ 50, whereas the viscous drag acting from the conduit walls has a value of $12\mu u/w^2$, which has a magnitude of ≈ 3600. Thus, for the typical flow regime in a dike, the flow is dominated by viscous drag, but the turbulent friction does contribute to the pressure losses experienced by the flow (cf. Wilson and Head, 1981). A rapid decompression of the flowing gas–magma mixture by $\approx$ 1–10 MPa as it breaks into a tunnel will lead to an acceleration of the flow. However, for an incompressible (i.e. volatile-free) magma, the ensuing magnitude of the flow is expected to be similar to the original value. This is because a decrease in pressure $\sim$ 1–10 MPa at the flow front, as the flow breaks through into the tunnel, increases the effective overpressure by an amount comparable to the original overpressure driving magma flow in the system.

Another important parameter concerns the hydrostatic pressure gradient in the nose of the flow as it spreads down a tunnel. In this nose region, of length L, the hydrostatic pressure gradient is represented as $rg\rho/L = 10^5/L$, where r is the radius of the tunnel. The free surface in the nose region can be maintained only if the hydrostatic pressure gradient is greater or equal to the driving pressure gradient, which is balanced by viscous resistance and turbulent drag. For a flow speed u along the tunnel $\sim 1\,\mathrm{m\,s^{-1}}$, the viscous resistance, which is represented as $8\mu u/r^2$, is $\sim 400\,\mathrm{kg\,m^{-2}\,s^{-2}}$. In contrast, the turbulent drag is represented as $2fu^2\rho/r$, which has a value of $\sim 20\,\mathrm{kg\,m^{-2}\,s^{-2}}$. The hydrostatic pressure gradient is therefore comparable to the net frictional resistance if the nose of the flow extends a distance of $L \sim$ 50–100 m. Behind this advancing gravity intrusion, the tunnel will be completely filled with magma.

These relationships establish the balance of forces and the leading order structure of the flow that may arise following breakthrough of degassed magma into a tunnel. We now use these results to develop the analog experimental study of the morphology of this advancing front. We compare the scaling for the extent of the nose of the flow with our experimental data, and we develop and test a model for the continuing flow behind the nose.

18.2 Analog experimental system

We developed an analog experimental apparatus to model the flow of magma up a dike and into a horizontal tunnel, with a fixed geometry along the flow path. This apparatus (Figure 18.1) consists of a vertical Hele–Shaw cell consisting of two parallel aluminum and glass plates that are 200 mm wide and 500 mm high. The 10 mm gap between these plates represents a fixed-geometry dike. The base of the cell is connected to a large cylindrical reservoir that contains pressurized liquid. This reservoir represents a deep magma source that drives magma ascent in the dike. The reservoir has an internal diameter of 153 mm and a height of 485 mm. Near the top of the Hele–Shaw cell there is a hydraulically operated gate. This gate, which opens in a fraction of a second, connects the cell to a horizontal glass tube of 40 mm radius and 450 mm length. The axis of the horizontal tube is normal to the plates of the cell. The whole system is sealed from the air, and the initial pressures at (i) the top of the reservoir above the layer of liquid, (ii) the top of the Hele–Shaw cell and (iii) the end of the horizontal tube are controlled independently by three vacuum regulators (SMC handle-operated vacuum regulators series T203) connected to a diaphragm vacuum pump (KNF

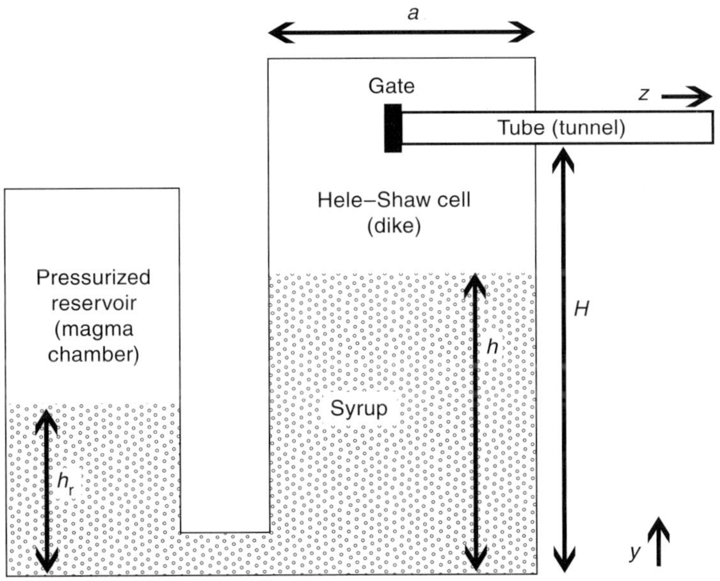

Fig. 18.1 Schematic diagram of the analog experimental apparatus, illustrating the key variables; w refers to the width of the cell (in and out of the page, not shown here); z refers to the position of the flow, measured from the cell; y refers to the position in the cell, measured upwards (see Appendix for a complete listing of all the variables used).

Neuberger model N180.3 FT.18). This configuration enables us to control the pressure difference between the reservoir and the top of the cell and also between the top of the cell and the horizontal tube. Pressures are measured with BOC Edwards active strain gauges (ASG NW16-1000 mbar). These gauges have been calibrated in the Calibration Laboratory of the Southwest Research Institute and have an accuracy better than ± 1 mbar. A series of miniature pressure sensors (113A21 sensors from PCB Piezotronics) were embedded at equal distances along the bottom of the horizontal tube, to record the changes of dynamic pressure during the flow propagation.

18.3 Experimental materials

We used three working fluids as analogs to volatile-poor basaltic magma. The primary fluid was pure golden syrup, a partially inverted refiners syrup manufactured by Tate and Lyle Company. The syrup contains 31–38 wt.% sucrose, 42–50 wt.% invert sugar and the remainder is dominantly water. This syrup has a Newtonian rheology with a strongly temperature-dependent viscosity (e.g. White, 1988; Davaille and Jaupart, 1993). To cover a wider range of Reynolds numbers, we also used less viscous liquids: golden syrup diluted with 5, 10 and 15 wt.% deionized water, called DGS5 and DGS15, respectively. Viscosities of the samples used in our experiments were measured over a range of shear rates and temperatures between 0 and 40 °C with a Haake RV20 Rotovisco rotating cylinder viscometer, using the sensor systems M5/MVI and M5/SVII. During these measurements, temperatures

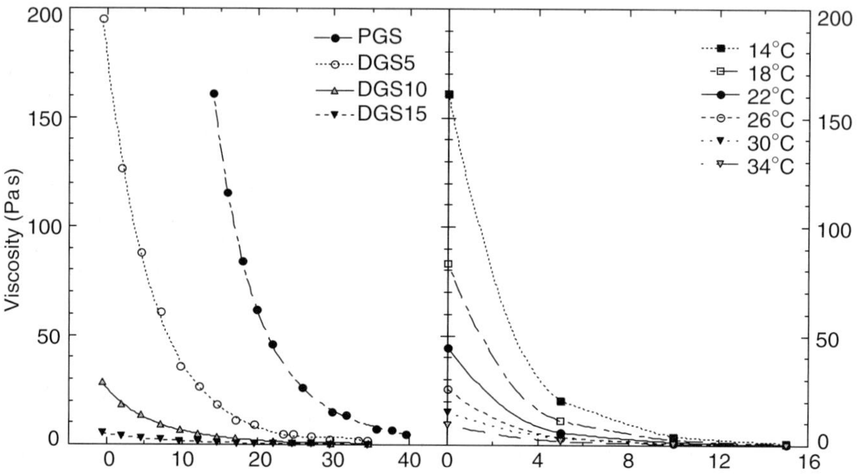

Fig. 18.2 Viscosity (μ) as a function of temperature and water content for pure golden syrup (PGS) and golden syrup diluted with 5, 10 and 15 wt.% deionized water (respectively, DGS5, DGS10 and DGS15). Temperatures are measured to within 0.1° C. Each viscosity value was measured and averaged over a range of shear rates.

were set and controlled by a temperature vessel connected to a thermal liquid circulator. Variations in syrup viscosity as functions of water content and temperature are shown in Figure 18.2. Densities were calculated by weighing different known volumes of each fluid at room temperature.

In the experiments, the flow regimes included slow viscous flows of very small Reynolds number, but also flows in which the Reynolds number was as large as 100. These latter experimental flow conditions are comparable to the range of likely basaltic flow conditions discussed in the previous section, in which Reynolds numbers for the tunnel, $\rho ur/\mu$, have values of $\sim 20-200$. In comparison, Reynolds numbers for the potential dike flow are $\sim 10-100$. Thus, the slow viscous flow experiments are useful as an analog model of the structure of potential flows into a tunnel, in that they illustrate the balance of applied pressure and dissipation and the slumping of the flow nose, even though the flow resistance in the tunnel is through viscous dissipation rather than turbulent drag.

18.4 Experimental results

Two series of experiments were performed with the experimental apparatus. First, a series of calibration experiments were conducted to examine the initial ascent of syrup up the Hele–Shaw cell as a function of the pressure difference between the reservoir and the top of the ascending layer of syrup. This procedure enabled us to test the model of flow resistance in the reservoir-cell part of the system, independently of the horizontal tube. The second series of experiments involved measuring the flow in the glass tube following the opening of the gate between the horizontal tube and the Hele–Shaw cell. In these experiments, the cell was initially filled with syrup while the gate was closed. The pressure in the tube was then lowered to a prescribed value below that in the cell, and the gate was opened. The rate of advance of the syrup was then measured, and the morphology of the flow front in the tube was recorded by high-speed video.

18.4.1 Initial calibration experiments

We performed a systematic series of experiments to examine the ascent of syrup in the Hele–Shaw cell as a function of the overpressure and the syrup viscosity. In Figure 18.3, we present data that illustrate the variation of the height of the syrup–air interface in the cell as a function of time. The different curves correspond to different pressure contrasts between the top of the reservoir and the top of the Hele–Shaw cell. In each case, the syrup gradually ascends through the cell to a final, static steady state in which the difference in pressures between the top of the cell and the top of the reservoir is accommodated by the difference in the head between the reservoir and the cell.

We now present a simple quantitative model of this experiment. For reference, the key parameters used in the model are illustrated in the schematic of the experimental apparatus shown in Figure 18.1. For the pure golden syrup experiments, with flow speeds $u = 0.01\,\mathrm{m\,s^{-1}}$, the Reynolds number $Re = \rho uw/\mu$ based on the width of the cell,

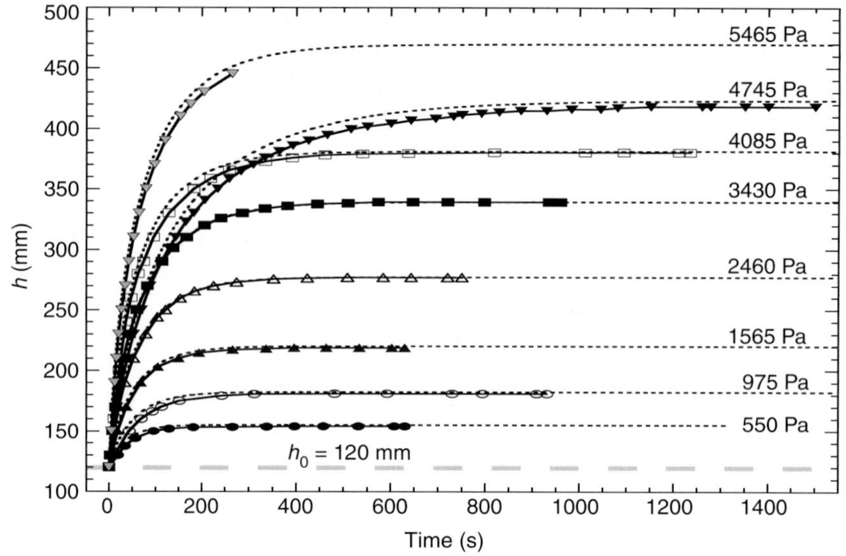

Fig. 18.3 Rise height of pure golden syrup in the Hele–Shaw cell as a function of time, for a series of different applied pressures ranging from 0.55 to 5.5 kPa. Dashed lines represent the expected value; connected symbols show the experimental data.

$w = 0.01$ m and the dynamic viscosity, μ. Given the dynamic viscosities shown in Figure 18.2, the Reynolds number is $Re \sim 10^3$–10^2. Thus, we expect inertia to be negligible. Therefore, the flow satisfies the approximate equation for the flow, averaged across the width of the cell

$$u = -\left(\frac{w^2}{12\mu}\right)\left(\frac{\mathrm{d}p}{\mathrm{d}y} + \rho g\right) \tag{18.1}$$

which applies in the limit of low Reynolds number (Batchelor, 1967), p represents the pressure, the y-axis is the direction of the flow w is the width of the cell, μ is the viscosity of the fluid, ρ the density of the fluid and g is the gravitational acceleration. The relation in (18.1) leads to

$$\frac{\mathrm{d}h^2}{\mathrm{d}t} = \left(\frac{w^2}{6\mu}\right)(\Delta p - \rho g h) \tag{18.2}$$

where $u = \mathrm{d}h/\mathrm{d}t, h = h(t)$ is the vertical position of the ascending flow front in the cell and Δp is the total pressure drop from the base of the cell to the top of the layer of syrup. This pressure decrease has a value given by

$$\Delta p = \Delta p_0 - \frac{\rho g A_C h}{A_R} \tag{18.3}$$

because the depth of syrup in the reservoir decreases by an amount $(A_C h)/A_R$, where A_C and A_R are the cross-sectional areas of the cell and the reservoir. The relationship between ascent height h and time t in (18.2) and (18.3) leads to

$$\left(\left(\frac{\Delta p_0}{(\lambda \rho g)^2}\right)\ln\left(\frac{\Delta p_0}{\Delta p_0 - \lambda \rho g h}\right)\right) - \frac{h}{\lambda \rho g} = \frac{w^2 t}{12\mu} \qquad (18.4)$$

where $\lambda = 1 + A_C/A_R$. Figure 18.3 compares the experimental data with the expected values calculated from (18.4), and we find very good agreement for each experiment. The early time rate of increase of the depth of the cell from (18.4) is given by

$$h^2 = \frac{\Delta p_0 w^2 t}{12\mu} \qquad (18.5)$$

which illustrates how the flow rate initially slows down with time as the head driving the flow decreases.

18.4.2 Experiments of flow into a horizontal tube

In Figure 18.4, we present a series of profiles of the flow front as captured from high-speed video, which illustrate the propagation of the pure golden syrup along the horizontal tube following the opening of the gate. Figure 18.4a shows an image from one representative experiment, with a pressure difference of 20 kPa driving the flow. Figure 18.4b shows the profiles obtained by tracing the shape of the flow front at successive times during four different experiments ranging from 10 to 80 kPa pressure difference between the reservoir and the end of the horizontal tube. With larger pressure differences and hence larger flow rates, the leading edge of the flow is nearly vertical. With a smaller pressure difference, the head becomes more inclined as the gravitational force at the nose of the flow becomes comparable to the pressure-driven flow, and the syrup slumps into the tube. In all cases, however, the shape of the leading edge is essentially invariant with time (Figure 18.5). In Figure 18.5, data are shown for a series of experiments in which different values of the pressure difference between the reservoir and the top of the cell were used. In each case, the nose and tail of the flow front move at constant speed, suggesting that a quasi steady state is established between the background pressure driving the flow and the gravitational pressure head driving the slumping of the nose. This result implies that the walls of the vertical Hele–Shaw cell exert the dominant resistance to flow, and so the flow rate remains constant.

We now model the dynamics of the flow into the horizontal tube, following the opening of the gate, assuming that the vertical cell is filled with syrup. At each height in the cell, the upward flux has value

$$Q = -\frac{w^2}{12\mu}\int \mathrm{d}x \left(\frac{\mathrm{d}p}{\mathrm{d}y} + \rho g\right) \qquad (18.6)$$

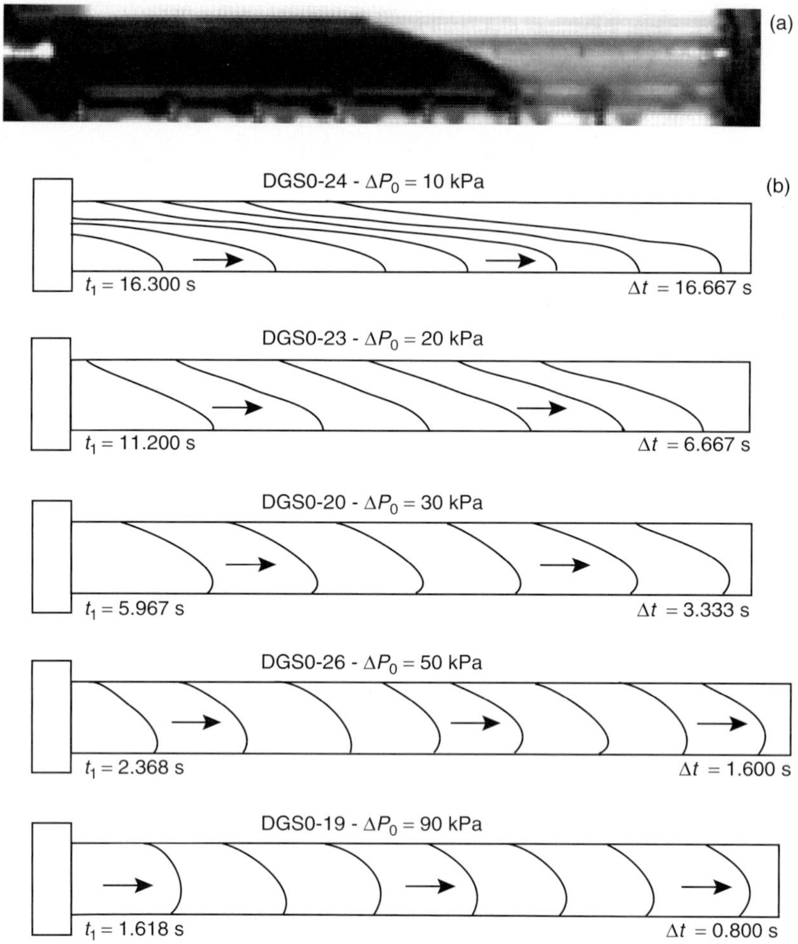

Fig. 18.4 Profile of the syrup–air interface as it flows down the horizontal tube. (a) Photograph of a typical experiment with a 20 kPa pressure difference driving the flow. (b) Tracings of the shape of the leading edge of the flow taken at a series of times in five experiments, with the pressure difference driving the flow being 10, 20, 30, 50 and 90 kPa. In this figure, Δt is the time elapsed between the two consecutive profiles and t_1 corresponds to the time of the first profile after the opening of the gate. To a good approximation, the shape of the head of the flow remains the same as it moves down the horizontal tube.

Integrating over the depth of the cell leads to the relation

$$Q = \left(\frac{a\beta w^3}{12\mu H}\right)(\Delta p_1 - \rho g h) \tag{18.7}$$

where Δp_1 is the pressure change between the base of the Hele–Shaw cell and the opening into the tube, a is the length of the cell and β is a dimensionless constant. The constant β

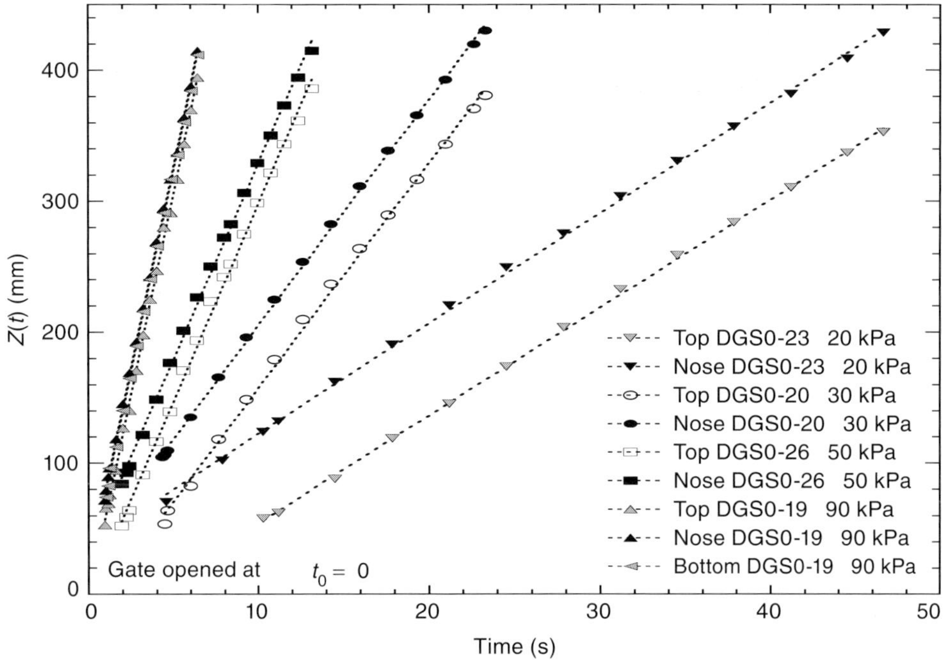

Fig. 18.5 Variation of the position of the leading and trailing edges of the nose of the flow as a function of time, for applied pressure differences of 20, 30, 50 and 90 kPa driving the flow into the horizontal tube. The speed of the trailing or top edge of the front is shown in Figure 18.6(a).

accounts for additional resistance that the flow experiences in the Hele–Shaw cell during a transition from a nearly parallel upward flow to a more focused radial flow in the vicinity of the horizontal tube. Detailed calculation of the value of β would require a full three-dimensional model of flow in the apparatus, particularly as flow migrates into the horizontal tube. These calculations are beyond the scope of this study. We therefore use (18.7) to determine the value of β for the experiments described below.

Parameter β depends primarily on the width of the cell compared to the width of the tube opening, which determines the degree of flow focusing. Parameter β also depends on the length-to-width ratio of the cell, which determines the fraction of the overall flow path over which the flow is focused. The focusing will occur over a length scale along the cell comparable to width, w. In this region, the resistance to flow will increase as the flow migrates through a progressively smaller area of the cell and hence at a greater speed. Because β may be interpreted as a measure of the reduction in the flow rate associated with this flow focusing, we expect it to have a value somewhat smaller than 1 in the experiments. Scaling the model to a possible dike–repository system, we expect that H is $\approx$ 10–30 km whereas a is $\approx$ 20–80 m. Thus, the effect of this focusing will be negligible, suggesting that $\beta = 1$ would be a good approximation.

Depending on the volume flux and the viscosity of the fluid, flow in the horizontal tube may be dominated by the applied pressure with little gravity slumping at the nose. Conversely, flow may include an extensive gravity-driven head that only partially fills the tube. In either case, in the region where the horizontal tube is completely filled with fluid, the pressure gradient is to a good approximation related to the flow according to the relation

$$Q = -\left(\frac{\pi r^4}{8\mu}\right)\frac{\mathrm{d}p}{\mathrm{d}y} \tag{18.8}$$

where r is the tube radius. Integrating along length L of the horizontal tube that is completely filled with fluid, we find the volume flux of fluid into the tube, Q, as a function of Δp_2, the pressure drop along the tube,

$$Q = \frac{\Delta p_2 \pi r^4}{8\mu L} \tag{18.9}$$

In the experiments, the total pressure drop

$$\Delta p = \Delta p_1 + \Delta p_2 \tag{18.10}$$

is controlled because ahead of the point $x = L$ in the horizontal tube the fluid has a free surface on which the pressure equals that in the far-field of the reservoir. Combining (18.7) and (18.9), we obtain the governing relation

$$\Delta p = \Delta P_{\mathrm{d}} + \rho g H = Q\mu\left(\frac{8L}{\pi r^4} + \frac{12H}{\beta a w^3}\right) + \rho g H \tag{18.11}$$

where ΔP_{d} is the overpressure that drives the flow. Here we have used the result that, by mass conservation, the net volume flux in the horizontal tube and in the Hele–Shaw cell is equal. The mean velocity of the fluid front that completely fills the area of the horizontal tube, $u = \mathrm{d}L/\mathrm{d}t$, is related to the volume flux in the tube according to the relation

$$\frac{\mathrm{d}L}{\mathrm{d}t} = \frac{Q}{\pi r^2} \tag{18.12}$$

Also, the overpressure driving the flow (18.11) decreases from the initial value ΔP_0 as fluid invades the tube and the level in the main reservoir decreases

$$\Delta P_{\mathrm{d}} = \Delta P_0 - \frac{\rho g L A_{\mathrm{T}}}{A_{\mathrm{R}}} \tag{18.13}$$

Combining (18.11) and (18.13), and integrating, we find that the horizontal extent of the fluid front that completely fills the tube, $L(t)$, is given by the relation

$$\left(\frac{8\mu A_{\mathrm{R}}}{r^2 \rho g A_{\mathrm{T}}}\right)\left(L + \left[\left(\frac{\Delta P_0 A_{\mathrm{R}}}{\rho g A_{\mathrm{T}}}\right) + \left(\frac{3\pi r^4 H}{2\beta a w^3}\right)\right]\ln\left(1 - \frac{L\rho g A_{\mathrm{T}}}{A_{\mathrm{R}}\Delta P_0}\right)\right) = -t \tag{18.14}$$

at early times, when L is small, or in the case that $\Delta P_0 >> L\rho g A_T/A_R$, so that there is very little change in liquid height in the reservoir and hence pressure at the base of the reservoir. As a result of the flow into the horizontal tube, (18.14) has the approximate form

$$L = \frac{\Delta P_0 \beta w^3 at}{12\mu\pi r^2 H} \tag{18.15}$$

corresponding to the situation in which the main frictional losses controlling the flow into the horizontal tube are dissipated in the Hele–Shaw cell, and hence in which the extent of the liquid-filled zone in the tube increases linearly.

The experimental data (Figure 18.5) show that to a very good approximation the flow advances along the tunnel with a constant speed. From (18.15), we expect the gradient of the lines, L/t, shown on Figure 18.5, to be proportional to the driving overpressure ΔP_0. Figure 18.6(a) illustrates the variation of the gradient of those lines, as measured from Figure 18.5, with overpressure ΔP_0. This variation confirms that dL/dt is indeed proportional to ΔP_0. According to (18.15), the constant of proportionality is given by $\beta w^3 a/12\mu r^2 H$. By measuring this constant of proportionality from Figure 18.6(a) and combining this with the dimensions of the experimental system, we find that $\beta = 0.95$. In comparison, the expected value for β is 1.0.

Although this model captures the propagation of the fluid-filled front along the horizontal tube, it does not account for the shape or extent of the gravitational slump zone at the nose of the flow. As described in Section 18.1, the extent of this slump zone, D, is determined by a balance between the gravitational pressure gradient, $\rho gr/D$, and the applied pressure

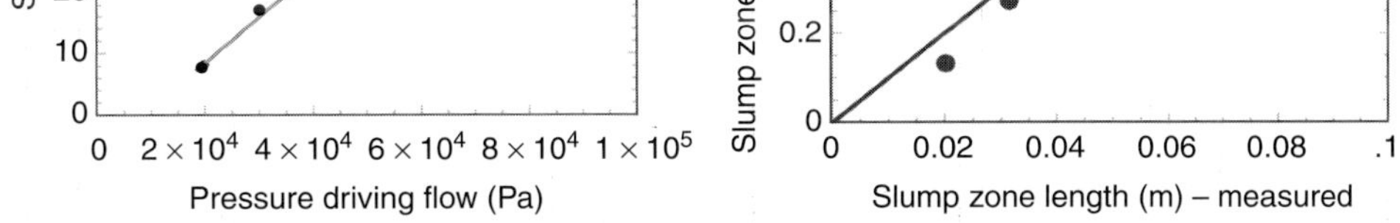

Fig. 18.6 (a) Variation of the speed of the flow front spreading into the horizontal tube as a function of overpressure driving the flow. The data collapse to a straight line to good approximation, as anticipated by (18.15). (b) Comparison of the idealized scaling law (solid line) for the length of the gravitationally induced slump with experimental observations (dots). In this figure, the calculated length of the slump is evaluated using the speed of the filling front as it advances along the horizontal tube, and the density contrast between fluid in the tube and air.

gradient, $\mathrm{d}p/\mathrm{d}z = \Delta p_2/L$, along the completely fluid-filled part of the horizontal tube. This relationship can be represented as

$$D \approx \frac{\rho g r}{\mathrm{d}p/\mathrm{d}z} \tag{18.16}$$

and may be re-expressed in terms of the speed of the flow along the tube, u_T, noting that the pressure gradient in the tube is given by $dp/\mathrm{d}z = 8\mu u_\mathrm{T}/r^2$, leading to the relation

$$D \approx \frac{\rho g r^3}{8\mu u_\mathrm{T}} \tag{18.17}$$

In Figure 18.6(b), we compare this expected scaling relationship for the extent of the slumping zone, as given by (18.17), and using measurements of the speed of the flow along the horizontal tube with the laboratory data on the width of the slumping zone. We find reasonable agreement between the expected order of magnitude scaling and the experimental data. Note that the speed u is related to the applied overpressure that drives the flow. In the limit that most of the frictional dissipation occurs in the Hele–Shaw cell rather than the horizontal tube, D can be expressed as

$$D = \frac{3H\rho g r^5 \pi}{2\beta a w^3 \Delta P_0} \tag{18.18}$$

which is equivalent to (18.17) because we estimated the value of β using (18.15). Equation (18.17) identifies how the length of the gravity intrusion decreases with the overpressure, the width of the cell and the length of the cell supplying fluid into the horizontal tube. In contrast, the slump zone increases with the radius of the tube and the length of the cell over which the applied overpressure is being dissipated.

18.4.3 Flows at higher Reynolds numbers

The experiments described above refer to low Reynolds number flows in which only the viscous resistance controls the flow dynamics. As discussed in Section 18.1, for a potential repository system, turbulent drag may also be an important process for flows having higher Reynolds numbers, $\sim$ 10–100. In Figure 18.7, a series of further experimental results are shown for flows using dilute golden syrup of lower viscosity. For these experiments using diluted syrup, Reynolds numbers are $\sim$ 10–100. In comparison, a possible magma flow within a subsurface tunnel will likely have a Reynolds number $\sim$ 20. In these experiments, an extensive slump zone always develops. The steady slump surface is of comparable length to the horizontal tube. Flow reaches a quasi steady state only as the fluid reaches the end of the tube. At this stage, fluid is reflected back from the end of the tube and develops a backward-propagating filling front. For higher pressures driving the flow, the front is sharper and more localized as expected, with gravity playing a smaller role in the overall flow.

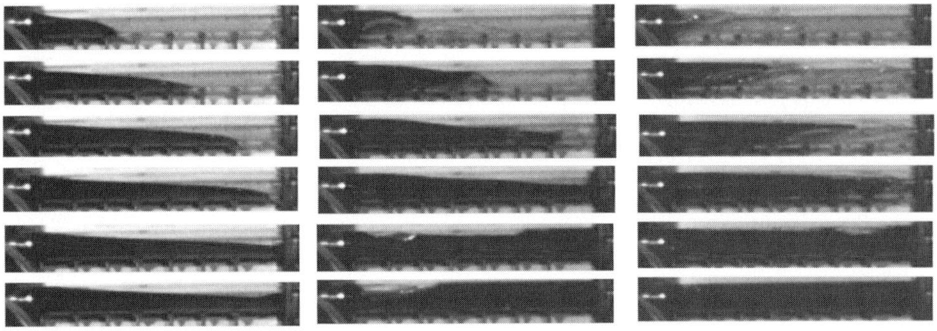

Fig. 18.7 Three sequences of photographs from experiments using a water–golden syrup mixture, with applied pressures of 20, 50 and 80 kPa. The Reynolds numbers of the flows in these experiments are more analogous to those expected for potential magma–repository interactions. For the higher flow rate, the flow front is somewhat irregular but quite localized. For the slower flow rate, the flow front shows more evidence of a gravitational slump region, as in the viscous syrup experiments (Figure 18.4(a)), except that the slump region is relatively more extensive.

18.5 Application to potential dike–repository conditions

We now scale our theoretical model for application to the potential repository, using representative parameters for magma viscosity and the possible dike–tunnel geometry. In developing the model for application to a geological repository, we draw from Section 18.1 in which we established that the viscous resistance is dominant, but that the turbulent drag, although small, is non-negligible; the Reynolds number of the flow has magnitude $\sim$ 10–1000. We consider the case of an open tunnel, in which the air has atmospheric pressure ahead of the magma; and a closed tunnel, in which the air gradually becomes compressed as the magma advances and decelerates the flow.

To model this flow regime, we extend the model in Section 18.4 for low Reynolds numbers flow to include the effects of flow inertia in the model. Flow motion in the dike thus is governed by (cf. Woods *et al.*, 2002)

$$\rho \frac{\mathrm{d}u}{\mathrm{d}t} = -\frac{\mathrm{d}p}{\mathrm{d}y} - \frac{12\mu u}{w^2} - \frac{c_D \rho u^2}{w} - \rho g \tag{18.19}$$

where c_D is the drag coefficient. Motion in the region of the repository tunnel that is filled with magma is governed by

$$\rho \frac{\mathrm{d}u_T}{\mathrm{d}t} = -\frac{\mathrm{d}p}{\mathrm{d}z} - \frac{8\mu u_T}{r^2} - \frac{c_D \rho u_T^2}{r} \tag{18.20}$$

where the speed in the tunnel, u_T is related to the speed in the dike according to the relation $u_T = uA/A_T$ where A and A_T are the cross-sectional areas of the dike and tunnel. If the dike extends a vertical distance H, and L is the extent of the tunnel that is fully filled, then (18.19) and (18.20) may be integrated and combined to give the relation

$$\rho(\lambda L + H)\frac{\mathrm{d}t^2}{\mathrm{d}^2 L} = \Delta p - 4\mu\frac{\mathrm{d}L}{\mathrm{d}t}\left(\frac{3H}{w^2} + \frac{2\lambda L}{r^2} - c_{\mathrm{D}}\rho\frac{\lambda L}{r} + \frac{H}{w}\right)\left(\frac{\mathrm{d}L}{\mathrm{d}t}\right)^2 - \rho g H \quad (18.21)$$

where $\Delta p - \rho g H$ is the effective overpressure driving the flow. Included in (18.21) are both the chamber overpressure and the release of the stress at the level of the repository as the magma flows into the tunnel.

18.5.1 Open-tunnel conditions

For potential magma flow into an open tunnel, the pressure ahead of the advancing magma is at atmospheric value. In contrast to the tunnel, the surrounding rock has stress associated with the overburden of a thickness L_H of rock, although the least principal stress σ_3 may be less than the lithostatic load. However, for simplicity, here we assume the stress is given by the lithostatic head $\rho g L_H$. Because the dike walls and thus the magma in the dike are subjected to this stress, the lithostatic head becomes available to drive the flow into the tunnel. As mentioned in Section 18.1, the model assumes that the walls of the dike are of fixed geometry. Although a fixed geometry is a simplification based on earlier models of conduit flow (Wilson and Head, 1981), it is a reasonable starting point for examining flow into a tunnel following possible intersection by an ascending dike. In practice, the flow may be initially weaker, as the rapid decompression of the magma may lead to temporary closure of the dike walls as a result of elastic strain effects. However, as magma continues to ascend from depth and recompresses the walls of the dike, magma will resume flowing into the tunnel. Our calculations are therefore an upper bound on the flow rates in the system, assuming that the dike wall geometry remains constant during the process.

We solved (18.21) numerically based on the assumption of a 5 m-diameter tunnel being fed by a 1 m-wide dike of 80 m lateral extent and supplied from a source reservoir below a brittle–ductile transition zone at $\approx$ 10 km. We assume that at this depth, the ascending magma has an overpressure ΔP, comparable to the strength of the wall-rock, of 5 MPa. We take L_H to have a value of 300 m, and the drag coefficient c_{D} to have a value of 0.01 (Wilson and Head, 1981). Finally, we take the value 300 Pa s for the viscosity of the basaltic magma. As in the syrup experiments, the drag in the tunnel is negligible compared to drag in the dike.

Figure 18.8a shows the time-dependent velocity calculated by this model as a function of the magma viscosity when the flow invades the tunnel, assuming that the flow has a vertical front. The calculation shows that if the flow enters the tunnel overpressured, and starting at rest, then the flow accelerates to steady state over a time $\sim$ 10–100 s, for magmas with a viscosity of $\sim$ 30–300 Pa s. As this initial acceleration develops, the slumping front of the flow may also grow. The scaling relationships in Section 18.4 and (18.16) suggest that for a viscosity of 300 Pa s, slumping at the flow front will be dominated by a balance of the viscous forces and gravitational acceleration terms. Given a characteristic magma speed of

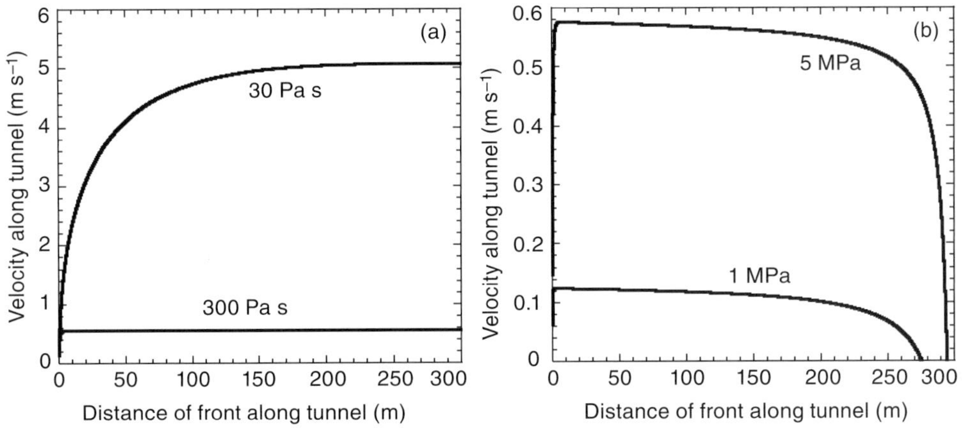

Fig. 18.8 (a) Calculated speed of the flow along a potential repository tunnel as a function of the magma viscosity, calculated using (18.21). The hydrofracture stress at 300 m below the surface at Yucca Mountain is $\approx$ 5 MPa, and the lithostatic load at this depth is $\approx$ 7 MPa. These calculations ignore the influence of any gravity slump zone at the leading edge of the flow. (b) Calculation of the rate of advance of a planar filling front into a closed tunnel. As the pressure in the tunnel increases, the flow decelerates and the model ceases to apply since the gravity slumping front will dominate the flow.

1 m s^{-1}, this balance implies a slump region of $\sim$ 50–100 m may grow ahead of the part of the flow that completely fills the tunnel.

18.5.2 Closed tunnel conditions

Tunnels that may contain high-level radioactive waste could be effectively sealed after waste emplacement. In contrast to an open tunnel, an advancing mass of magma will displace and compress the air between the magma and sealed end of the tunnel. Although rock units in the potential repository horizons have bulk permeabilities of $\sim 10^{-12}$–10^{-11} m^2 (Rosseau *et al.*, 1999), these permeabilities are small compared to the 20 m^2 cross-sectional area of the tunnel and thus are neglected in the following discussion. In addition, because the flow speeds are much less than the speed of sound in the air, the air pressure (P_a) essentially will remain spatially uniform. The effective pressure (P_e) will therefore increase inversely with the unfilled volume in the tunnel, according to

$$P_e = P_a \frac{X}{(X - L)} \qquad (18.22)$$

where X is the length of the tunnel and L is the length of the filled part of the tunnel. The effective pressure driving the flow therefore decreases with time, causing the flow to decelerate. As seen in Figure 18.8b, the effect of the closed tunnel is to cause the simple piston displacement flow to decelerate with time as the air is compressed. While this flow decelerates, the tunnel would continue to fill. However, as the magma slumps along the base of the reservoir, the flow evolves from the simple one-dimensional model presented herein.

18.6 Thermal effects on flow

Cooling with formation of a chilled margin against tunnel walls and canisters might inhibit the advance of magma down a repository tunnel. However, our results indicate that the formation of chilled margins would not be a significant impediment to magma flow. For magma viscosities ~30–300 Pa s, the time to fill a potentially intersected tunnel is ~ 100–1000 s. The thickness of chilled margins is ~ $(Kt)^{0.5}$, where K is the thermal diffusivity and t is the time. For a typical $K = 5 \times 10^{7}\ \mathrm{m^2\,s^{-1}}$, a chilled margin thickness will be ~ 1–2 cm. Even for the case of an end-member viscosity of 10^5 Pa s for fully degassed and partially crystallized trachybasalt magma, the modeled time to fill a tunnel is ~ 30 000 s and the resulting chilled margin thickness is only 10 cm. The reduction in cross-sectional area and increase of viscous resistance as a consequence of a progressive 1–10 cm chilled margin during flow emplacement thus will be small to negligible.

Concluding remarks

The modeling described in this chapter has examined the flow that may develop following breakthrough of a volatile-poor basaltic magma from a dike into a low-pressure subsurface tunnel. We developed both experimental and numerical models of the flow, assuming that the dike–tunnel geometry remains constant. Agreement of the experimental observations with the calculated results of this simplified model suggests that the model captures the first-order behavior of magma flow. A key result from our experimental study is the recognition of a gravitational slumping zone (Figure 18.4) ahead of the magma front into the tunnel. A simple scaling relationship developed and tested with the laboratory experiments indicates that the length scale of this zone depends on the ratio of the gravitational head across the tunnel compared to the pressure gradient along the tunnel associated with the frictional dissipation of the flow.

We have applied this model to the scale of a potential radioactive waste repository. For typical overpressures of 1–5 MPa that may be released as the magma flows from a dike into a single tunnel, we calculate a slumping zone develops with a lateral extent of ~ 50–100 m. The speed of the flow along the potentially intersected tunnel is expected to be ~ 1 m $\mathrm{s^{-1}}$.

The model represents a first-order abstraction of the complex processes anticipated for magma flow in a tunnel, and is based on a number of important simplifications. One key assumption is that the magma does not contain a volatile phase, and therefore the flow is not fragmented. This study thus provides a base case to evaluate the effects of volatiles in other models and experiments. For example, complementary studies described by Woods *et al.* (2002) and Dartevelle and Valentine (2005) explore some of the flow processes that might occur for volatile-rich magma, when a fragmented, high-velocity flow likely develops on interaction of a dike with a tunnel. From these studies, the speed of flow along the potentially intersected tunnel is expected to be ~ 100 m $\mathrm{s^{-1}}$.

A second key simplification is that the dike–tunnel geometry remains fixed following breakthrough into the tunnel. This simplification is important because as the pressure in the upper part of the dike decreases, the dike may partially close if the magma supply rate cannot match the rate of decompression (Bokhove *et al.*, 2005; Woods *et al.*, 2006). However, as magma continues to ascend from depth and the dike–tunnel system is repressurized, the dike would be expected to reopen. The present model is therefore likely to represent an upper bound on flow rate into a tunnel for volatile-poor magma.

The disposal of hazardous materials such as high-level radioactive waste poses great challenges to societies in trying to evaluate future risks. In the case of a repository for high-level radioactive waste, this evaluation considers potential hazards that could exist for longer than the span of recorded human history (EPA, 2001). Thus, scientists are required to assess the interactions of long-term geological processes with engineered systems to help evaluate risks to society for millennia to come. There are geological events of low probability, which have never been witnessed but can be clearly deduced from the geological record. This record also provides a basis to assess models that attempt to evaluate physical processes that are not well preserved.

Although the likelihood of igneous disruption is relatively small for the potential repository site at Yucca Mountain in Nevada, the possible radiological hazards from such disruption appears large relative to other higher likelihood events and processes (e.g. CRWMS M&O, 2000). Thus, models for this potential disruption are being developed, so that risks can be assessed appropriately. The abundance of a gas phase is a key component in evaluating potential magma–tunnel interactions. Relatively slow ascent of volatile-bearing magma may lead to volatile accumulation or loss in the leading regions of a dike, whereas rapid ascent may lead to disequilibrium effects and diffusion-limited bubble growth. An appropriate range of models can be considered and the range of potential hazards evaluated for that range.

The numerical models developed herein capture important, first-order processes of volatile-poor magma flow and are sufficiently general so that they can be tested against observations of natural or experimental flows. Experimental tests also support the fundamental results of these numerical models. Although the decompression of a volatile-poor basaltic magma may represent a less likely scenario than one for a volatile-rich magma, the volatile-poor scenario provides an appropriate lower bound to the range of conditions that appear appropriate to consider for potential magma–tunnel interactions.

Further reading

Wilson and Head (1981) provide an approachable overview of fundamental magma ascent and flow processes applicable to basaltic volcanism. Woods *et al.* (2002) applied these first-order principles to the modeling of initial magma–tunnel interaction processes. Additional insights on the mechanics of magma ascent in Lister and Kerr (1991) are used by Bokhove *et al.* (2005) and Woods *et al.* (2006) to evaluate couplings between magma pressure and wall-rock response, which affect the characteristics of openings in potential magma–tunnel interactions. Fully coupled two-phase flow models in Dartevelle and

Valentine (2005) provide additional insights on time-transient characteristics of potential magma–tunnel interaction processes. Although some important engineering characteristics of the potential Yucca Mountain repository may have changed since DOE (2001) was issued, this report provides a recent, publicly available overview of the potential geological repository system.

Acknowledgments

Detailed reviews by Stefan Mayer, Wes Patrick, Timothy McCartin and John Trapp greatly improved the content and readability of this chapter. The authors thank Mike J. Dury and Fred Wheeler, who helped design and build the experimental apparatus in the workshop of the Earth Sciences Department, University of Bristol, UK. Melissa Hill, Jerry Nixon and Alan Jenkinson also are thanked for their support and assistance in the laboratory, and we thank Rebecca Emmot for clerical support and Jim Pryor for editorial review. Documentation on the details of the model calculations is available from the authors. R. S. J. Sparks thanks the Natural Environment Research Council (NERC) for a NERC Professorship. This chapter was prepared to document work performed by the Center for Nuclear Waste Regulatory Analyses (CNWRA) for the US Nuclear Regulatory Commission (NRC) under Contract No. NRC-02-02-012. The activities reported here were performed on behalf of the NRC Office of Nuclear Material Safety and Safeguards, Division of High Level Waste Repository Safety. This chapter is an independent product of the CNWRA and does not necessarily reflect the view or regulatory position of the NRC.

Appendix

List of variables

A_C	cross-sectional area of cell
A_R	cross-sectional area of reservoir
A_T	cross-sectional area of tube
a	length of cell
c_D	drag coefficient
D	extent of the slump zone
g	acceleration due to gravity
H	vertical extent of flow pathway in cell
h	depth of syrup in cell
h_r	initial depth of syrup in reservoir
L	length of flow in filled part of horizontal tube
ΔP	pressure difference across flow path, including gravitational head
ΔP_0	initial driving overpressure
ΔP_d	overpressure driving the flow

List of variables

Δp	total pressure drop from base of cell to top of fluid layer
Δp_0	initial total pressure drop
Δp_1	pressure change between base of cell and horizontal tube opening
Δp_2	pressure drop along horizontal tube
Q	flow rate
r	radius of horizontal tube
u	speed in cell
u_T	speed in tube
w	width of cell
x	position across cell
y	position in cell, measured upwards
z	position along horizontal tube, measured from the cell
β	dimensionless scaling factor to account for diversion of flow in cell near horizontal tube
μ	viscosity of syrup
ρ	density of syrup/magma

References

Batchelor, G. K. (1967). *An Introduction to Fluid Dynamics*. Cambridge: Cambridge University Press.

Bokhove, O., A. W. Woods and A. de Boer (2005). Magma flow through elastic-walled dikes. *Theoretical and Computational Fluid Dynamics*, **22**, 261–286.

Connor, C. B., J. A. Stamatakos, D. A. Ferrill *et al.* (2000). Geologic factors controlling patterns of small-volume basaltic volcanism: application to a volcanic hazards assessment at Yucca Mountain, Nevada. *Journal of Geophysical Research*, **105**, 417–432.

CRWMS M&O (2000). Total system performance assessment for the site recommendation, TDR-WIS-PA-000001 Rev. 00 ICN 01. North Las Vegas, NV: DOE Yucca Mountain Site Characterization Office.

Dartevelle, S. and G. Valentine (2005). Early-time multiphase interactions between basaltic magma and underground openings at the proposed Yucca Mountain radioactive waste repository. *Geophysical Research Letters*, **32**, L22311, doi:10.1029/2005GL024172.

Davaille, A. and C. Jaupart (1993). Transient high-Rayleigh-number thermal convection with large viscosity variations. *Journal of Fluid Mechanics*, **253**, 141–166.

Delaney, P. T., D. D. Pollard, J. I. Ziony and E. H. McKee (1986). Field relations between dikes and joints: emplacement processes and paleostress analysis. *Journal of Geophysical Research*, **91**(B5), 4920–4938.

DOE (2001). Yucca Mountain science and engineering report DOE/RW-0539. North Las Vegas, NV: US Department of Energy, Office of Civilian Radioactive Waste Management.

EPA (2001). Public health and environmental radiation protection standards for Yucca Mountain, Nevada, Code of Federal Regulations, Title 40 ch. 1, p. 197. Washington, DC: US Environmental Protection Agency.

Gutmann, J. T. (1979). Structure and eruptive cycle of cinder cones in the Pinacate volcanic field and the controls of Strombolian activity. *Journal of Geology*, **87**, 448–454.

Hill, B. E., Lynton, S. J. and J. F. Luhr (1995). Amphibole in Quaternary basalts of the Yucca Mountain region: significance to volcanism models. *Proceedings, Sixth Annual Internationl High-level Radioactive Waste Management Conference*. La Grange Park: American Nuclear Society, 132–134.

Lister, J. (1991). Steady solutions for feeder dikes in a density stratified lithosphere. *Earth and Planetary Science Letters*, **107**, 233–242.

Lister, J. R. and R. C. Kerr (1991). Fluid-mechanical models of crack propagation and their application to magma transport in dykes. *Journal of Geophysical Research*, **96**(B6), 10 049–10 077.

Luhr, J. F. and T. B. Housh (2002). Melt volatile contents in basalts from Lathrop Wells and Red Cone, Yucca Mountain region (SW Nevada): insights from glass inclusions. *Eos, Transactions of the American Geophysical Union*, **83**, 1221.

Morris, A., D. A. Ferrill and D. B. Henderson (1996). Slip-tendency analysis and fault reactivation. *Geology*, **24**(3), 275–278.

Nicholis, M. G. and M. J. Rutherford (2004). Experimental constraints on magma ascent rate for the Crater Flat Volcanic Zone hawaiite. *Geology*, **32**, 489–492.

NRC (2001). Disposal of high-level radioactive wastes in a proposed geologic repository at Yucca Mountain, Nevada; final rule, Code of Federal Regulations, Title 10, p. 63. Washington, DC: US Nuclear Regulatory Commission.

Pollard, D. D. (1973). Equations for stress and displacement fields around pressurized elliptical holes in elastic solids. *Mathematical Geology*, **5**, 11–25.

Rosseau, J. P., E. M. Kwicklis and D.C. Giles (1999). Hydrogeology of the undersaturated zone, North Ramp Area of the Exploratory Studies Facility, Yucca Mountain, Nevada, Water-Resources Investigations Report 98-4050. Denver, CO: US Geological Survey.

Stock, J. M., J. H. Healy, S. H. Hickman and M. D. Zoback (1985). Hydraulic fracturing stress measurements at Yucca Mountain, Nevada, and relationship to the regional stress field. *Journal of Geophysical Research*, **90**(B10), 8691–8706.

Thorarinsson, S., S. Steinthórsson, Th. Einarsson, H. Kristmannsdóttir and N. Oskarsson (1973). The eruption on Heimaey, Iceland. *Nature*, **241**, 372–375.

Valentine, G. A., J. Donathan, D. J. Krier, F. V. Perry and G. Heiken (2007). Eruptive and geomorphic processes at the Lathrop Wells scoria cone volcano. *Journal of Volcanology and Geothermal Research*, **161**, 57–80.

Vaniman, D. T., B. M. Crowe and E. S. Gladney (1982). Petrology and geochemistry of hawaiite lavas from Crater Flat, Nevada. *Contributions to Mineralogy and Petrology*, **80**, 341–357.

White, D. B. (1988). The planforms and onset of convection with a temperature dependent viscosity fluid. *Journal of Fluid Mechanics*, **191**, 247–286.

Williams, R. S., Jr. and J. G. Moore (1976). Man against volcano: the eruption on Heimaey, Vestmann Islands, Iceland, USGS Report. Denver, CO: US Geological Survey, http://pubs.usgs.gov/gip/heimuey.

Wilson, L. and J. W. Head (1981). Ascent and eruption of basaltic magma on the Earth and Moon. *Journal of Geophysical Research*, **86**, 2971–3001.

Woods, A. W., R. S. J. Sparks, O. Bokhove *et al.* (2002). Modeling magma–drift interaction at the proposed high-level radioactive waste repository at Yucca Mountain, Nevada, USA. *Geophysical Research Letters*, **29**(13), 1–4.

Woods, A. W., O. Bokhove, A. deBoer and B. Hill (2006). Compressible magma flow in a two-dimensional elastic-walled conduit. *Earth and Planetary Science Letters*, **246**(3–4), 241–250.

19

Volcanic risk assessment at Yucca Mountain, NV, USA: integration of geophysics, geology and modeling

G. A. Valentine and F. V. Perry

The proposed Yucca Mountain repository license application, prepared by the US Department of Energy, was submitted to the US Nuclear Regulatory Commission in 2008, placing the project in the international lead in terms of potential implementation of permanent geologic disposal of high-level nuclear waste. Volcanic risk assessment is an important component of the license application for Yucca Mountain because the site is located in an area that has experienced sporadic, monogenetic basaltic volcanism for millions of years (Figure 19.1). This chapter summarizes the various issues that come into play in volcanic risk assessment at Yucca Mountain, and how a wide range of observational and theoretical approaches are integrated to support that risk assessment. The chapter is organized around four main themes: (i) definition of an igneous event that might affect the repository and the processes associated with such an event; (ii) estimation of event probability by integrating geology, geophysics and geochemistry of volcanism in the region and the use of expert elicitation; (iii) constraining the consequences of an event via analog and theoretical studies; and (iv) abstraction of the above knowledge and incorporation in probabilistic risk assessment. The focus of the chapter is on the underpinning science and approach that have been developed for the Yucca Mountain effort, which can be translated to other repository or long-term facility assessments in volcanic regions. Risk assessment for the license application is currently under way; therefore we do not provide specific, quantitative values of risk except for one historical example that is provided for illustration purposes only.

19.1 Background

Yucca Mountain is an ~ 20 km long, north–south-trending ridge located in southwestern Nevada. It is formed of tilted Miocene ignimbrites and related silicic pyroclastic rocks. The Miocene deposits are related to a large caldera complex that last erupted 11.45 Ma (Sawyer *et al.*, 1994). Most of the normal faulting and tilting that formed the topography of Yucca Mountain and its immediate surroundings occurred during caldera volcanism and prior to the 11.45 Ma Ammonia Tanks tuff eruption. The proposed geologic repository for high-level radioactive waste is located near the northern end of Yucca Mountain (Figure 19.1). The location was selected based upon several factors, including the remoteness of the area (~ 150 km from the city of Las Vegas), the arid climate that results

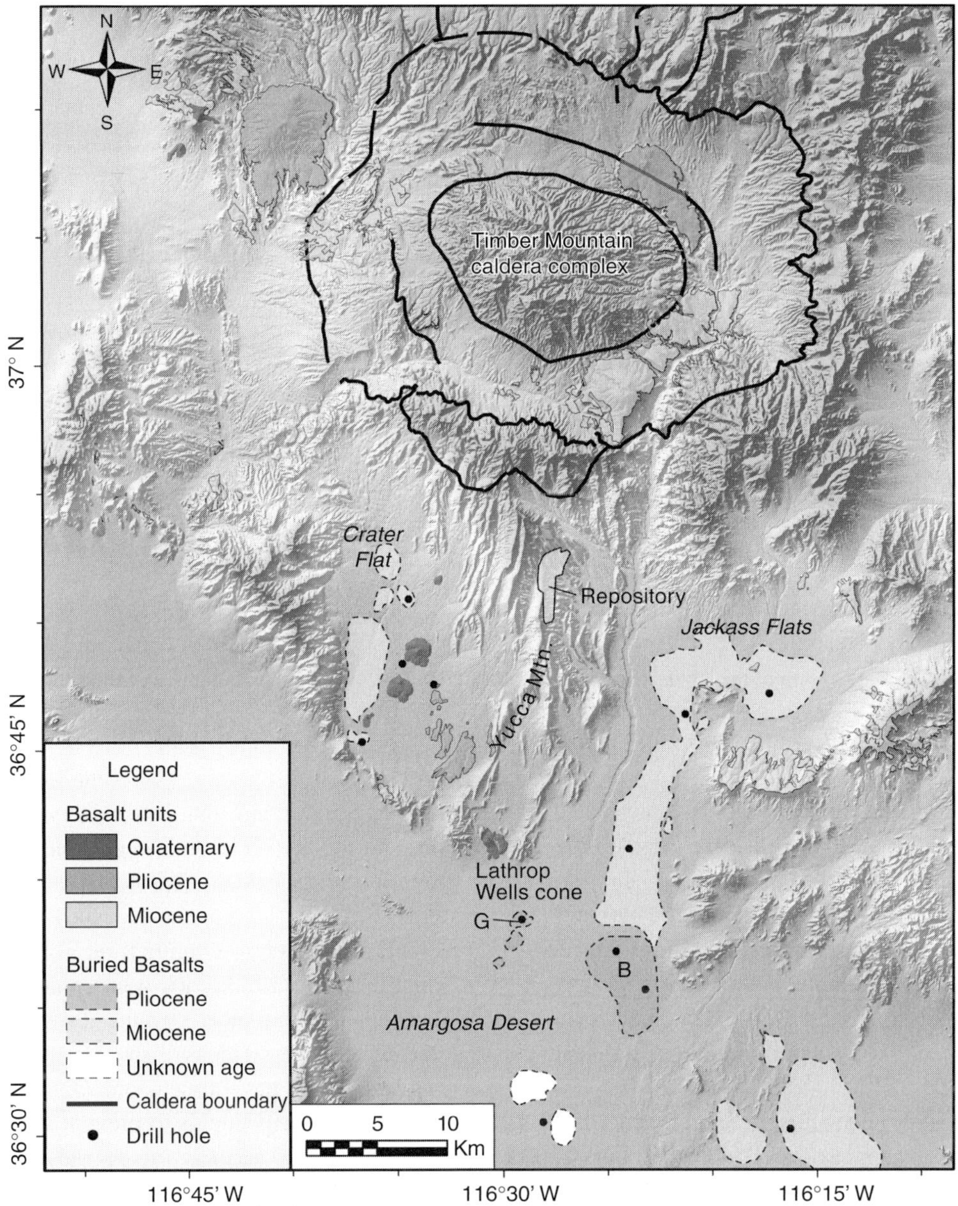

Fig. 19.1 Shaded relief map illustrates the southwest part of the southwest Nevada volcanic field showing Plio-Pleistocene volcanoes. Lathop Wells cone is the youngest volcano in the area (~ 77 ka). Buried basalts labeled **B** and **G** correspond to magnetic anomalies **B** and **G** in Figure 19.4, for comparison. Crater Flat and Amargosa Desert are major basins that host many of the Plio-Pleistocene basalts. Black lines are Miocene caldera boundaries from Wahl *et al.* (1997).

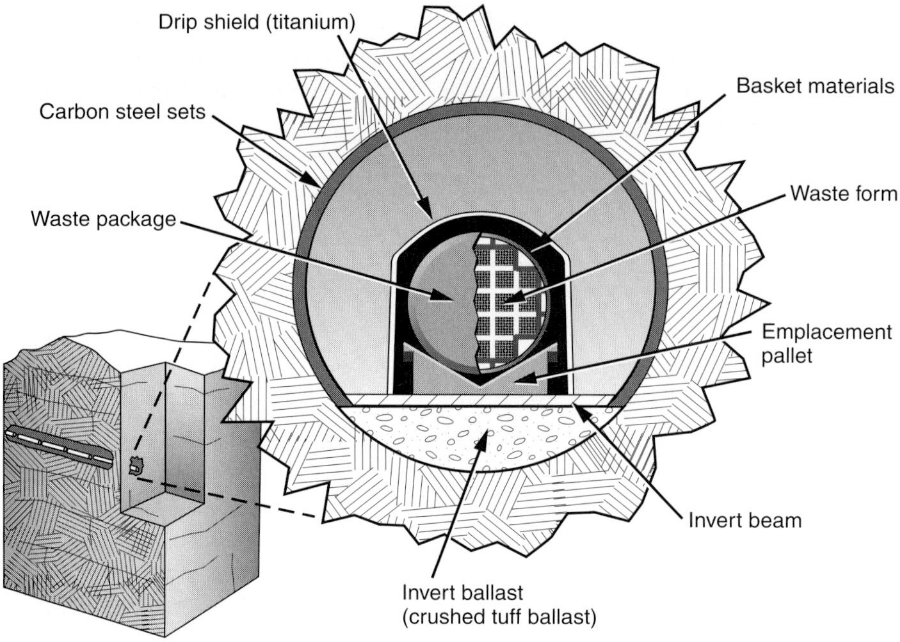

Fig. 19.2 Diagram of the inside of a waste emplacement drift shows various elements within the drift including the waste package and drip shield.

in relatively low-volume fluxes of water infiltrating the mountain, the deep water table ($\sim$ 600–700 m below the crest of Yucca Mountain) and the presence of clays and zeolites in the altered Miocene deposits that retard the migration of dissolved radionuclides; these factors comprise much of the natural barrier system. The proposed repository consists of a complex of horizontal, underground drifts (5.5 m diameter) at depths ranging from $\sim$ 250–350 m, variable due to the topography of the mountain (Figure 19.2). Each waste package will consist of a cylinder of corrosion-resistant, nickel–chrome–molybdenum alloy surrounding a stainless steel structural canister. The packages will hold racks of spent fuel rods and will be emplaced in the drifts via rail tracks. Packages will have dimensions of $\approx$ 6 m length and $\approx$ 2 m diameter, and will be covered by titanium drip shields to prevent seepage waters and rock fall from contacting the waste packages and accelerating degradation. Waste packages and drip shields are the main components of the engineered barrier system.

Repository performance is analyzed by an integrated total system performance assessment (TSPA), of which volcanic risk is but one part, for a period of 10 ka into the future (and potentially as long as 1000 ka, depending upon the fate of a new regulation that is under review). Compliance is defined in terms of probability-weighted dose of radiation to individuals (so-called "reasonably maximally exposed individual," or RMEI) in a hypothetical population located $\sim$ 20 km south of the repository on what is predicted to be the approximate axis of radionuclide transport in the groundwater. A comprehensive review of both the engineered and natural features that form the repository system can be found in US Department of Energy (DOE, 2001) as well as in the book edited by McFarlane and Ewing (2006).

After the eruption of the Ammonia Tanks tuff from the Timber Mountain caldera complex (Figure 19.1), silicic activity ended, while basaltic volcanism has persisted with a general pattern of decreasing volume flux on the scale of the volcanic field, as well as decreasing sizes of individual volcanoes (Crowe, 1986; Fleck *et al.*, 1996; Valentine and Perry, 2006, 2007). As described in Sections 19.3 and 19.4, basaltic activity in the area (the southwestern Nevada volcanic field, or SNVF (Sawyer *et al.*, 1994)) continued to as recently as the $\sim$ 77 ka Lathrop Wells volcano (Heizler *et al.*, 1999; Valentine *et al.*, 2007). Studies related to the risk of future volcanic activity began in the late 1970s, when Yucca Mountain was originally identified as a potential site for permanent underground disposal of high-level radioactive waste. A pioneering set of studies led by Bruce M. Crowe and co-workers comprised a rigorous probabilistic risk assessment for volcanic activity, with studies related to event probability (Crowe *et al.*, 1982; Crowe, 1986) and consequences (Link *et al.*, 1982; Crowe *et al.*, 1983). Volcanism studies continued into the mid 1990s as repository concepts and requirements evolved. During that time various research groups developed different models for volcanic activity in the Yucca Mountain region (e.g. Ho *et al.*, 1991; Perry and Crowe, 1992; Bradshaw and Smith, 1994; Connor and Hill, 1995), particularly with respect to event probability. A formal expert elicitation approach, probabilistic volcanic hazard assessment, PVHA, was used to account for varying interpretations and perspectives (CRWMS M&O, 1996; Bechtel-SAIC Company, 2004a) with the goal of obtaining a robust estimate of the annual probability of occurrence for basaltic activity intersecting the repository (Coppersmith *et al.*, Chapter 26, this volume). To our knowledge, this marked the first use of expert elicitation for volcanic hazards at a nuclear facility.

After a hiatus of about five years, Department-of-Energy-sponsored volcanic risk studies resumed in 2001 in response to new repository designs, regulatory requirements and new data that had emerged. This chapter focuses on the work conducted since 2001, which forms the basis for volcanic risk in the repository license application.

19.2 Risk assessment framework

In order to understand how scientific efforts are linked to risk assessments it is useful to have an overall framework such as shown in Figure 19.3. Ultimately, risk assessment studies culminate in a relatively simple result that can be compared to some criterion in order to make decisions and that incorporates some measure of uncertainty; this is represented by the apex of the triangle in Figure 19.3. For example, the regulatory criteria that are used to determine whether the proposed Yucca Mountain repository will perform satisfactorily boil down to probability-weighted radiation dose limits to a defined population, as described above. Thus our understanding and predictions about future volcanic activity are all captured in a plot of probability-weighted dose as a function of time into the future.

Risk assessment relies on a fundamental understanding of the events and processes of interest, as determined from field, experimental and theoretical research (base of the triangle

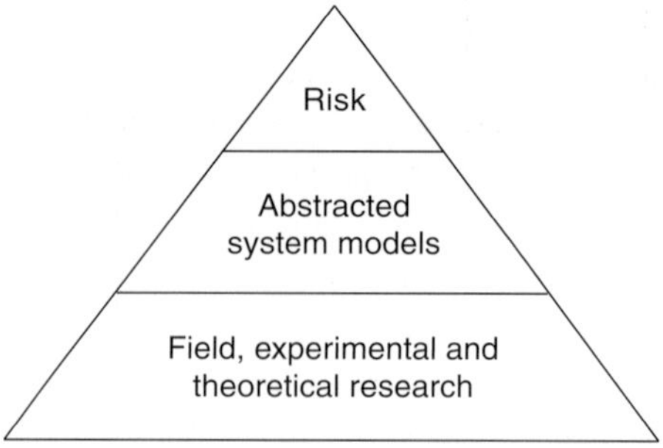

Fig. 19.3 This framework for volcanic risk assessment shows a foundation in basic understanding of the volcanic system and its interaction with a repository. Intermediate models are simplified (abstracted) such that they can capture the range of parameter values and uncertainties, and culminating risk value that can be used by decision makers.

in Figure 19.3). The event of interest must first be defined (Section 19.3). The probability of that event occurring (Section 19.4) is determined by a range of geological and geophysical data as well as detailed models for behavior of a volcanic field. Identification of processes that have adverse consequences likewise depends upon detailed field observations and models (theoretical and/or experimental) of volcanic intrusion and eruption dynamics and interactions with repository features, as discussed in Section 19.5. All of these factors have varying degrees of uncertainty, and furthermore there might be a wide range of possible volcano-repository scenarios. It is not practical to attempt first-principles predictions of all possible combinations of processes, but it is important to capture the range of possible outcomes in order to arrive at risk. This necessitates an intermediate level of modeling where the complexities of the system are represented by uncertainty distributions assigned to parameters used as inputs to simplified, or abstracted, process models (middle level of Figure 19.3). Monte Carlo sampling can then be conducted on the abstracted models in order to capture the range of possibilities. This abstraction level and the resulting decision-level results are the focus of Section 19.6. Iteration between the top and the base of the triangle (Figure 19.3) is necessary. A first cut at the decision-level risk assessment for a given problem can be made from a relatively simple understanding of the basic volcanic processes. However, this first cut is likely to have a large uncertainty. Specific processes can then be identified for which a better understanding will significantly reduce the uncertainty in the decision-level assessment, thus providing a mechanism for prioritizing research on the underlying volcanic processes. This iterative approach can be used until the uncertainty at the decision level is reduced to a satisfactory level, as defined by regulatory criteria (e.g. Hill *et al.*, Chapter 25, this volume).

19.3 Event definition and associated processes

The first step in risk assessment for any natural hazard is to define clearly the event(s) of interest. Based upon this definition it is possible to design a strategy to constrain the probability that the event(s) will occur. The second step is to describe the processes associated with that event and their interaction with and effects on assets of interest, in other words, the consequences of the event(s). Together the probability and consequences form the basis for risk assessment. In this section we describe the event definition and associated processes of interest for volcanic risk assessment at Yucca Mountain. Later sections describe more of the technical details and scientific bases for probability and consequence analyses within this framework.

A starting point for determining what types of volcanic events might occur at the repository in the 10^4–10^6 a regulatory timeframe is to look at the types of events that have happened in the most recent past over a similar timespan. During the past ~ 1.1 Ma, eight new monogenetic volcanoes have formed within a distance ~ 50 km from the repository site (Quaternary basalts on Figure 19.1). Assuming that the overall tectonic and volcanic framework for the region does not fundamentally change (an assumption that is contested by Smith and Keenan, 2005), it is reasonable to assume that these eight Quaternary volcanoes are representative of the type of event that could occur at the repository during its lifetime. There is evidence that the volcanic field waned between the Miocene and middle Pliocene and has been in a relatively steady state since ~ 3 Ma in terms of average eruptive flux (Valentine and Perry, 2007), which supports the assumption that the next 10^6 a will be similar. The Quaternary volcanoes are, in all but one case, each characterized by a single scoria cone with one or two main lava fields that typically extend ~ 1 km from each cone (the one exception is a very small volcano that might have been entirely fissure fed). The event of interest, therefore, is the formation of a new, small-volume, trachybasaltic scoria cone volcano and its shallow ($<$ 400 m depth) plumbing system that might interact with the repository.

19.4 Framework for basaltic volcanism and probability

Most volcanic hazard assessments depend on understanding the past patterns of volcanism at the volcano or the volcanic field. Data necessary to assess likely future behavior of a volcanic field for volcanic hazard assessment include the age, location, volume and composition of individual volcanoes, as well as information about volcano clusters, and how these parameters have evolved through time.

Significant basaltic volcanism in the SNVF began ~ 11.3 Ma, shortly after eruption of the last major ignimbrite eruption from the Timber Mountain caldera. In the central and southern parts of the volcanic field, episodes of basaltic volcanism (each involving from one to several spatially distinct volcanoes) occurred ~ 11.3, 10.5, 9.5, 4.6, 3.8, 2.9, 1.1, 0.35 and 0.077 Ma. Additional Miocene episodes occurred on the eastern, northern and western margins of the field ~ 9.1, 8.6 and 7.2 Ma (Fleck *et al.*, 1996; Perry *et al.*, 1998). Partly because of the long hiatus between Miocene and Pliocene activity (2.6 Ma in the field overall, ~ 5 Ma in the part of the field where Yucca Mountain is located), post-Miocene basaltic volcanism has been primarily emphasized in hazard studies. Post-Miocene activity is characterized by episodic formation of individual monogenetic volcanoes, and of clusters of such volcanoes, in scattered locations throughout the west-southwest portion of the SNVF

(Figure 19.1). Thus forecasting volcanic hazard requires constraining both the timing and location of potential future volcanoes (or clusters).

Two key characteristics of the volcanic history of the field are that the eruptive volume flux has decreased in the Quaternary compared to the Pliocene, and that the rate of formation of new volcanoes is relatively low compared to other fields in the Basin and Range Province. A total volume of $\sim$ 20–30 km^3 erupted in the Miocene, $\sim$ 5 km^3 in the Pliocene and $\sim$ 0.5 km^3 in the Pleistocene. The eruptive flux over the entire history of the volcanic field is $\sim$ 3 km^3 Ma^{-1}, but in the last 5 Ma the eruptive flux has decreased to $\sim$ 1 km^3 Ma^{-1} and during the most recent 3 Ma the flux has been $\sim$ 0.5 km^3 Ma^{-1} (Valentine and Perry, 2007). Average recurrence rates for new monogenetic volcanoes during the last 5 Ma (the time period that the geologic record can be most reliably interpreted) range from $\sim$ 3–4 volcanoes Ma^{-1}. This can be compared with more active fields such as Lunar Crater to the north and Cima to the south, which have recurrence rates of $\sim$ 30–50 volcanoes Ma^{-1}. Measured in terms of eruption rate and the volcano recurrence rate, basaltic volcanism near Yucca Mountain constitutes one of the least active, but longest lived, basaltic volcanic fields in the western United States.

Geophysical surveys and limited drilling carried out from the late 1970s to early 1990s indicate that buried Miocene and Pliocene basalt is present in the Crater Flat and the northern Amargosa Desert alluvial basins (Kane and Bracken, 1983; Carr and Parrish, 1985). Later ground and aeromagnetic data, combined with detailed modeling, highlighted the issue that a number of magnetic anomalies probably represented buried basalt and that drilling would be required to characterize the extent and age of buried basalt (Langenheim, 1995; O'Leary *et al.*, 2002). The youngest basalt drilled to this point was at anomaly B in the northern Amargosa Desert, encountered at depth of $\approx$ 100 m, with an age of 3.8 Ma (Figure 19.1).

It is important to characterize all of the buried basalts in order to have a full understanding of the volcanic history of the field. Beginning in 2004, the US Department of Energy sponsored an integrated aeromagnetic survey and drilling program to characterize buried basalts in the alluvial basins surrounding Yucca Mountain (Perry *et al.*, 2005). The high-resolution, helicopter-borne aeromagnetic survey was the first designed specifically to optimize detection of buried or intrusive volcanic features. Approximately 16 000 km of flight line data were collected in an area of $\sim$ 30 $\times$ 30 km; resolution over the entire survey area is nearly equivalent to detailed ground magnetic surveys (Figure 19.4). Attaining a uniformly high resolution within the survey area was critical in interpreting whether magnetic anomalies were due to buried basalt or magnetized Miocene tuff, interpretations that could be tested by drilling.

Seven drillholes were completed as part of the program, four of which encountered basalt flows. Three of these basalts were Miocene in age ($\sim$ 9–11 Ma) and the youngest, from drillhole VH-2, was 3.9 Ma. The combination of drilled basalt ages, depths of burial and aeromagnetic anomaly characteristics allowed undrilled anomalies to be categorized with a high level of confidence as to whether they represent basalts or fault blocks of Miocene silicic tuffs. The results of the aeromagnetic survey/drilling program have two major implications

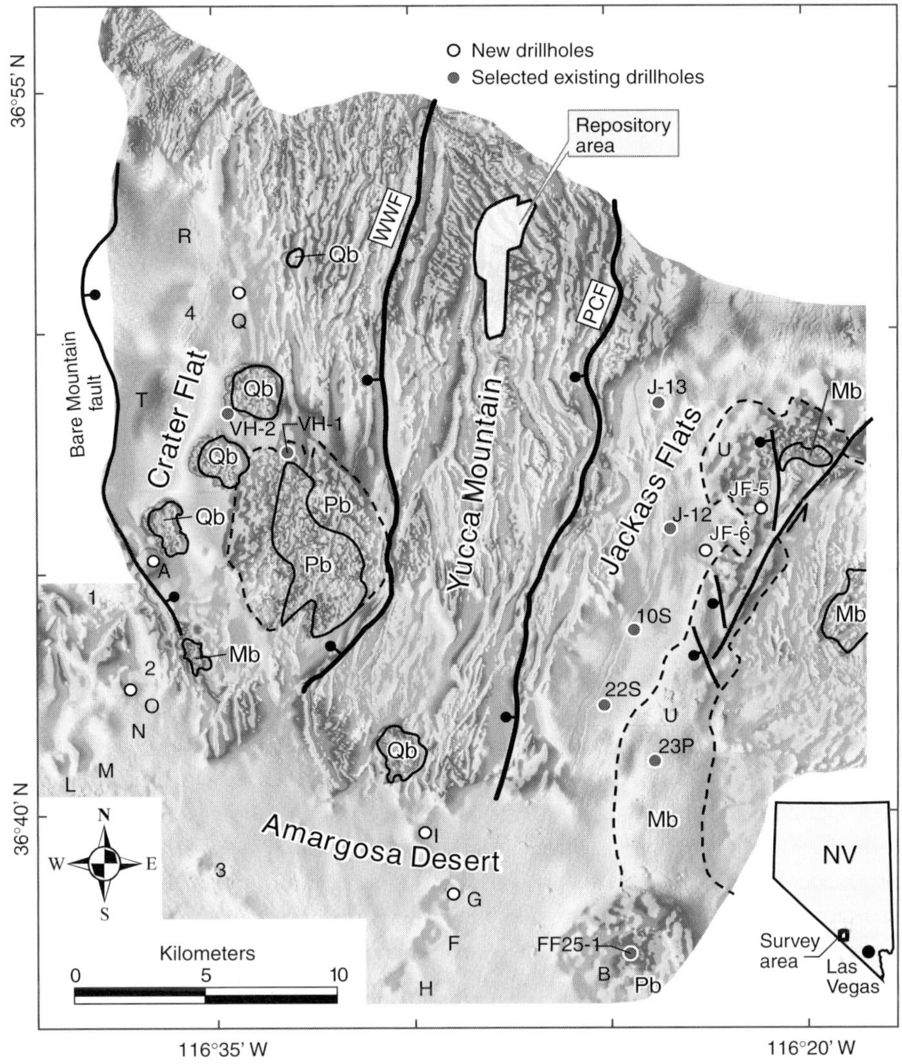

Fig. 19.4 This aeromagnetic map images the residual magnetic field (measured total field minus the international geomagnetic reference field) for Yucca Mountain and surrounding basins. Windy Wash fault (WWF) and Paintbrush Canyon fault (PCF) in the central part of the survey area define the approximate boundaries between uplifted Miocene tuffs of the Yucca Mountain range block and the Crater Flat and Jackass Flats basins. The Bare Mountain fault defines the western edge of the Crater Flat Basin. Single-character alphanumeric labels indicate anomalies suspected of representing buried basalt prior to drilling (cf. Figure 19.1). Solid lines enclose outcrops of Quaternary (Qb), Pliocene (Pb) and Miocene (Mb) basalt. The four Quaternary basalts in Crater Flat are 1.1 Ma scoria cone volcanoes. The Quaternary basalt south of Yucca Mountain is the $\sim$ 77 ka Lathrop Wells volcano. Dashed lines enclose areas of inferred buried basalt associated with outcrops of Pliocene and Miocene lava flows. Dark circles indicate selected previous drillholes, and white circles indicate drillholes from the recent drilling program. The bar and ball symbols on faults denote the downthrown side. The color version of the map shows the sense of polarity of the residual magnetic field; see color plate section. Map modified from Perry *et al.* (2005).

for hazard studies. First, the spatial distribution of Plio-Pleistocene volcanism primarily to the south and west of Yucca Mountain remains essentially the same as understood from locations of surface volcanoes; no Plio-Pleistocene volcanism was discovered in the terrain to the east of Yucca Mountain (Figure 19.1). Second, average temporal recurrence rates estimated for the past 5 Ma from surface expressions of volcanism are essentially unchanged in light of new data on buried basalts within the 30 × 30 km area. Additional magnetic anomalies outside the boundaries of the 2004 survey have not been drilled, but the possibility that they represent additional buried basalt centers is being accounted for through expert elicitation (see below). Finally, the aeromagnetic survey provided data that strongly suggest the feeder dikes for Pleistocene volcanoes coincide with pre-existing normal faults (see also Connor *et al.*, 1997). North to north-northwest-trending subsidiary faults are common in the Yucca Mountain region and serve to accommodate strain beneath major north to northeast-trending faults. While no Pleistocene vents are observed on major northeast-trending faults, the aeromagnetic data reveal that vents are commonly located on the north to north-northwest-trending faults, probably because these faults have significantly steeper dips that favor capture of ascending dikes in the shallow subsurface (Perry *et al.*, 2006; Gaffney *et al.*, 2007).

We base a petrogenetic model for the volcanic field on detailed geochemical, geochronology and physical volcanology studies, combined with studies of the tectonic history of the region. Fridrich *et al.* (1999) demonstrated a correlation between extension rate and eruption volume over the history of the SNVF. In both cases, activity peaked in the Miocene and has strongly declined into the Quaternary. The systematic decline in eruption volume flux over the history of the field is directly linked to the nature of the mantle source: ancient non-convecting lithospheric mantle that was relatively hot at the peak of volcanism and extension (due to thermal input within a subduction zone) and is now in an advanced state of thermal decline (Valentine and Perry, 2007). Instead of basaltic melt being generated by convective upwelling and decompression, melting within lithospheric mantle probably only occurs within localized zones rich in hydrous minerals that have a lower solidus temperature. If these zones contain a small melt fraction, they are weaker and deform preferentially under regional extensional strain. Deformation of the zones increases porosity, providing a mechanism for enhanced melt migration and accumulation within melt bands that further enhance shear localization. As melt is focused, buoyancy forces increase to a critical value resulting in upward propagation of dikes (Valentine and Perry, 2007). Over the timeframe of the volcanic field, the lithospheric source has conductively cooled with no further thermal input following the breakdown of subduction. Reduced extension rates still focus hydrous melts into melt bands, but at reduced rates and volumes that are reflected in declining eruption volumes and magmatic footprints (Valentine and Perry, 2006). In the Yucca Mountain region, a correlation between age, volume and trace-element composition indicates that the degree of partial melting in the mantle source has decreased systematically with time, consistent with lithospheric cooling and decreasing extension rates (Valentine and Perry, 2007). This conclusion places constraints on models of future volcanic activity in the volcanic field. There is no physical reason for eruptive volume fluxes to increase, for example, when the overall thermo-tectonic regime is decaying. Recently, a new expert elicitation (Coppersmith *et al.*, Chapter 26, this volume) has been convened for the primary purpose of assessing the potential impact of buried volcanoes on understanding the history of the volcanic field and the volcanic hazard (probability of an event intersecting the repository). Another emphasis has been an assessment of the likely characteristics of future volcanic events (event definition), which are now better informed by detailed physical volcanology studies in the Yucca Mountain region (Valentine *et al.*, 2006, 2007; Valentine and Keating, 2007; Keating *et al.*, 2008a). Results of the elicitation are expected in 2008.

19.5 Processes and effects

The regulatory criteria for the proposed Yucca Mountain repository focus on radiation dose to a hypothetical population $\approx$ 20 km south of the site. Therefore we must assess processes related to a volcanic event that could potentially transport radioactive contamination from the repository to that population. These transport processes define two scenario classes: (i) subsurface processes, where disruption of the repository by volcanic plumbing (dikes, conduits, sills) compromises the engineered barrier system (see Section 19.1) and, after the igneous event is over and the system has cooled to ambient temperatures, radioactive waste is contacted by seepage waters and potentially transported to the water table and the control population as dissolved species; and (ii) eruptive and surficial processes, where radioactive waste is erupted and transported to the control population in volcanic plumes or by post-eruptive surficial processes. Analyses of the two scenarios require basic knowledge of the expected plumbing (at depths of $\sim$ 250–350 m and less) of small-volume basaltic volcanoes, the interaction of that plumbing (i.e. dikes and magma flow) with repository elements, eruptive styles and surface transport processes. These issues are approached with a combination of field studies at analog sites and numerical and experimental modeling of specific scenarios.

19.5.1 Subsurface processes

Analog field studies

Quantifying the possible range of shallow intrusive geometries and orientations, based upon field studies at sites that are analogous (in part or in whole) to the most likely future event, is a first step in understanding potential subsurface effects. As a starting point for constraining shallow volcanic plumbing we use a conceptual model, illustrated in Figure 19.5, wherein magma is transported upward through the crust by a dike propagation mechanism. Where a dike system initially intersects the Earth's surface an eruptive fissure may form, and as activity continues the flow of magma focuses into one or more conduits that feed central vents, allowing the development of a scoria cone(s).

Because of the long history of basaltic volcanism in the Yucca Mountain region, erosional exposures range from relatively little (e.g. the $\sim$ 77 ka Lathrop Wells volcano; Valentine *et al.*, 2007) to as deep as $\sim$ 250 m below the base of a volcano (e.g. $\sim$ 8.8 Ma

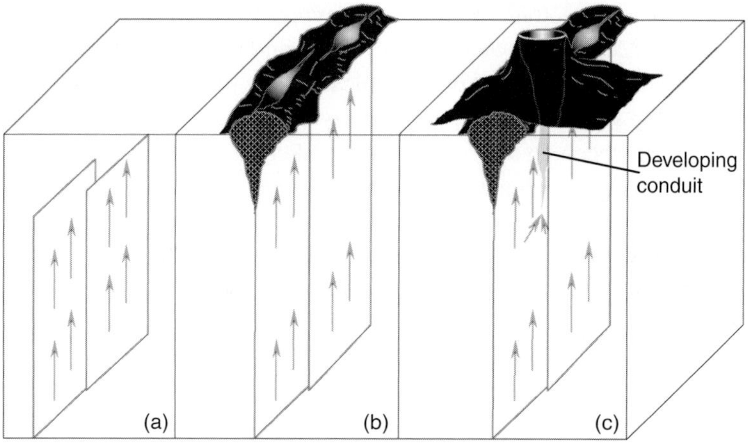

Fig. 19.5 Schematic diagram of three stages of development of a dike into a small-volume basaltic volcano. In panel (a), a pair of *en echelon* dikes approaches the surface. One or both of these dikes reaches the surface resulting in a fissure eruption, forming spatter ramparts and areas of localized magma flow (b). Within hours to days, the fissure eruption localizes into a single vent, an enlarged conduit forms in the upper ~ 100 m, and an edifice (cone) is constructed above (c), from Keating *et al.* (2008a).

East Basalt Ridge; Keating *et al.*, 2008a). Field data collected at sites where basalts erupted through tuff sequences similar to that at Yucca Mountain (Valentine and Krogh, 2006; Valentine and Perry, 2006; Keating *et al.*, 2008a) suggest that dike widths at repository depths could range between ~ 1–12 m, with a mean of ~ 8 m, and that likely dike lengths (along strike) would range between ~ 400 m and 8 km with a mean of ~ 2 km (Sandia National Laboratories, 2007a). In detail, these dike lengths may include *en echelon* segments of an identifiable main dike. Dikes at most eroded analogs occur in sub-parallel sets that can consist of 1–5 separate dikes, with spacing between dikes ranging up to ~ 1500 m (Sandia National Laboratories, 2007a). Typically one dike is thickest and is referred to as the master dike, while others are thinner and may be shorter. It is along the master dike of a given set that widened zones occur, which grade upward into pyroclastic vent facies, representing eruptive conduits (Keating *et al.*, 2008a), while other dikes in the set apparently often do not reach the surface. Based upon the Quaternary volcanic record it is most likely that for a potential individual volcano in the future, only one main conduit will form, resulting in one cone, although we allow for as many as three conduits along the master dike but with relatively low probability (Sandia National Laboratories, 2007a). This is based upon the typical spacing of conduits along individual dikes observed in the geologic record in the region (Keating *et al.*, 2008a), combined with the relatively short dike lengths and the single cone per volcano that characterize the Quaternary activity (Valentine and Perry, 2006; Valentine *et al.*, 2006, 2007; Valentine and Keating, 2007). Note that lava boccas or weak pyroclastic boccas might be fed by shallow lateral breakouts from the main conduit, most likely near the interface between the volcanic edifice and the pre-existing ground surface (Valentine *et al.*, 2006; 2007). Modeling studies by Gaffney and Damjanac (2006) suggest that conduits are most likely to form where a rising dike first intersects a topographically low area.

Dikes at eroded analogs (Miocene and Pliocene age) in the region are nearly ubiquitously coplanar with pre-existing normal faults (Valentine and Krogh, 2006; Keating *et al.*, 2008a). Where there is sufficient bedrock exposure, Quaternary volcanoes also seem to be located along steeply dipping normal faults (Valentine and Keating, 2007; Valentine *et al.*, 2007) and aeromagnetic data support this interpretation at volcanoes where bedrock is not exposed (Perry *et al.*, 2006).

Because many of the faults in the area have orientations that are related to late Miocene principle stresses or that reflect the faults' role as relays between larger structures (Perry *et al.*, 2006), the coplanar dikes often are not perpendicular to the Pliocene–Quaternary minimum principle stress as would be expected in an homogeneous elastic medium. Modeling by Gaffney *et al.* (2007) is consistent with field data (Perry *et al.*, 2006; Valentine and Krogh, 2006) showing that the capture of a vertically rising dike into a fault plane is most likely at shallow depths (uppermost few hundred meters) and with relatively steeply dipping faults (typically $< 60°$; Connor and Conway, 2000). Magma lubricates and opens the faults and can result in syn-intrusive fault slip (Parsons *et al.*, 2006; Valentine and Krogh, 2006). This inelastic process likely contributes to the relatively large width-to-length ratios of some observed dikes (Valentine and Krogh, 2006). In addition, syn-intrusive slip can result in local stress rotation along a fault and formation of small sills (Valentine and Krogh, 2006; Gaffney *et al.*, 2007). Observed sills at paleodepths of < 250 m extend laterally $\sim 30–500$ m from their parent dikes and range from ~ 3 to 50 m in thickness (Valentine and Krogh, 2006; Keating *et al.*, 2008a).

Modeling studies

The above features provide useful information on the geometry of a potential subsurface interaction between volcanic plumbing and a repository, but must be supplemented with modeling (numerical and experimental) studies that account for the effects of the repository on the plumbing and vice versa. Four main classes of problems have been considered: (i) effects of repository structures on vertical dike propagation at the beginning of a volcanic event; (ii) fluid dynamic processes of magma flow into and out of drifts and the dynamical conditions that result; (iii) magma flow and potential generation of secondary pathways (dikes) between the repository and the surface, during the months to years duration of a volcanic event; and (iv) hydrothermal mass transfer processes.

Extensive modeling of the interaction between a vertically ascending dike tip and the repository is reported in Sandia National Laboratories (2007b). A dike tip consists of the leading crack that is followed at some distance (depending upon rock and magma properties) by the magma front; between the leading crack and the magma front is a relatively fluid-free, low-pressure zone (dike tip cavity; e.g. see Rubin, 1995). Calculations indicate that the leading crack of an ascending dike at Yucca Mountain would accelerate ahead of the magma front as it approaches the Earth's free surface. Stress perturbations due to the presence of open drifts and heating of host rocks by radioactive decay of waste (during the first $\sim$ 2 ka of the repository lifetime) have little effect on the vertical propagation direction, although the perturbations might reduce the upward crack acceleration (Sandia National Laboratories, 2007b). When the rising magma front reaches repository level, magma that intersects or is near to open drifts may be diverted into the drifts and locally stall the ascending magma front. However, the 5.5 m-diameter drifts are planned to have $\approx$ 80 m spacings such that the bulk of the repository horizon is composed of intact country rock (referred to as pillars). Model results (Sandia National Laboratories, 2007b) indicate that between intersected drifts the magma front would continue to ascend vertically with little influence by the local diversion into drifts. From these results we expect that the trajectory of vertical or subvertical dikes will be relatively unaffected by the repository. This provides a level of confidence that the dike geometries observed at field sites are appropriate analogs for scenarios with the repository present. Modeling suggests that rising magma might first breach the Earth's surface at sites above the pillars.

The second class of problems focuses on the dynamics of magma flow into drifts just after initial intersection by a dike. Basaltic magmas at shallow depths can have a wide spectrum of properties depending upon temperature and upon the presence of crystals and volatiles. Two end-member types of behavior have been modeled: (i) incompressible (unfragmented) magma, and (ii) fragmented magma where compressible gas (mainly H_2O vapor) is the continuous phase transporting the silicate melt as pyroclasts. Analysis suggests that a drift could fill with incompressible magma over timescales of several tens to hundreds of seconds (Bechtel-SAIC Company, 2004b); these should be viewed as minimum timescales because of the simplifying assumptions in the analysis (for example, cooling effects with temperature-dependent viscosity were not included in the analysis, although the inferred timescales are short enough that heat loss to the relatively insulating country rocks would be limited). Additional studies of this non-explosive interaction between magma and drifts are presented by Lejeune *et al.* (Chapter 18, this volume).

Woods *et al.* (2002) considered the flow of fragmented magma from a dike into an initially low-pressure drift using a model similar to those used to predict flow in gas shock tubes, and suggested that initial interactions would be characterized by shock propagation into the drifts. They then applied a steady-state, one-dimensional, dusty gas model (i.e. particles and gas are assumed to be perfectly coupled in terms of heat and velocity), and considered three hypothetical scenarios: (i) intersecting flow fills drifts but then continues upward along the original dike path to vent at the surface; (ii) flow moves into intersected drifts but exits at secondary vertical dikes that extend to the surface at some distance away from the points of initial dike–drift intersection; and (iii) flow is established throughout the repository and exits to the surface up ventilation shafts. Dartevelle and Valentine (2005) and Bechtel-SAIC Company (2005) presented two-dimensional, time-dependent, multiphase calculations using similar boundary conditions to those of Woods *et al.* (2002) and showed that an initial shock propagation phase is followed by a period of several seconds during which a rapidly flowing gas–particle mixture circulates in an intersected drift. Recent work by Dartevelle and Valentine (submitted) applies more realistic boundary conditions based upon field analog data. Complex flow patterns arise as gas–particle flows are partially diverted into horizontal drifts, which eventually fill with a pyroclastic deposit. The multiphase calculations indicate that temperature varies little compared to the initial magma temperature during flow into drifts. Dynamic pressures, which are related to damage potential of pyroclastic flows (Valentine, 1998), range as high as 100–1000 kPa, but are most commonly in the range of 1–10 kPa.

The above modeling focuses on interactions as a dike tip approaches the repository from below, and when the magma front first arrives at the repository horizon; the third class of problems, on the other hand, explores processes that could occur during the lifetime of the volcanic event. Field studies of the Quaternary volcanoes in the Yucca Mountain region and comparison with historical scoria cone volcanoes suggest that each volcano might have been active for minimum periods of several months to a few years (Valentine *et al.*, 2006, 2007). Once a plumbing system is established beneath a volcano, it might experience transient variations in pressure that can be caused by temporary blockage of the vent and subsequent decompression when the blockage fails and is blown out. Blockage can be caused by avalanching of large quantities of cone scoria into the vent (e.g. McGetchin *et al.*, 1974), by collapse of conduit walls or by solidification of a magma plug in the upper conduit (e.g. Taddeucci *et al.*, 2004). Overpressures on the order of 5–10 MPa can develop within the plumbing system before the blockage or its surroundings fail and allow decompression and eruption (Bechtel-SAIC Company, 2005; Damjanac *et al.*, 2005; Gaffney *et al.*, 2005; Valentine and Krogh, 2006); these overpressures will be transmitted throughout the fluid-filled plumbing system, which might include magma-filled drifts that have not solidified. Dartevelle and Valentine (2005, 2008) modeled decompression of a dike connected to a pyroclast-gas-filled drift and showed that material streams out of the drift into and up the vertical dike, but a significant portion of material simply settles to the drift floor and forms a stagnant deposit. This suggests that while there is some potential that waste (fragmented and released from damaged packages) could be drawn from a drift into a dike under decompression scenarios, it is likely to be somewhat limited due to formation of a relatively stable deposit in the lower part of a drift (e.g. see Bechtel-SAIC Company, 2005). Magma overpressures might also result in propagation of new vertical fractures and dikes extending upward from intersected drifts, potentially establishing a secondary pathway for transport of waste to the surface. However, analyses suggest that magma in such secondary dikes would likely solidify before propagating an appreciable distance and that secondary crack-tip propagation rate lags significantly behind that of the main dike (Bechtel-SAIC Company, 2005; Sandia National Laboratories, 2007b).

An additional process that can occur once a dike–drift plumbing system is set up beneath a volcano is detailed by Menand and Phillips (2007a, 2000b) and Menand *et al.* (2007, Chapter 17, this volume). During phases of Strombolian activity, a relatively stable column of magma might develop in the volcanic conduit. Magma in the column and deeper source magmas degas and release bubbles, which in turn rise toward the surface and drag some melt with them. At the intersection between a vertical magma column and a horizontal drift, some bubbles and melt are diverted into the horizontal opening. As this mixture flows down the drift, bubbles continue to migrate upward, forming a bubble-rich mixture at the top of the drift and a gas-depleted, relatively dense fluid that sinks downward. A circulating process is set up that results in mass transfer between the drift and the vertical conduit. It is anticipated that such a process might prolong the cooling of magma in a drift as new material is advected in, although the temperature dependence of the process has not yet been fully explored.

Note that none of the above studies attempt to account for the presence of repository elements such as drift turnouts, emplacement pallets, waste packages, drip shields or rubble from local drift collapse. Rather, the work has focused on understanding the range of processes that might occur in magma–repository interactions.

A final mass transport process that must be assessed for its potential impacts on repository performance is hydrothermal flow. Analyses to date have included multiphase, reactive transport calculations focused on determining whether volatile species could migrate from a magma-filled drift to a neighboring drift that does not contain magma (Sandia National Laboratories, 2007b). Calculations indicate that there would be little or no migration of most corrosive species from one drift to another through host-rock pillars, mainly because of dissolution of species into pore water. Migration of volatiles through coarse backfill materials in the tunnels that connect drifts might occur under some conditions (Sandia National Laboratories, 2007b), but the impacts on non-magma-filled drifts are thought to be minor. These modeling results are consistent with field data where alteration of silicic tuff country rocks is limited to within $\sim$ 10 m of shallow basaltic dikes and necks in the vadose zone (Perry *et al.*, 1998; WoldeGabriel *et al.*, 1999), although more extensive alteration could occur directly above sills.

19.5.2 Eruptive and surficial processes

Analog field studies

Studies of volcanic products at the Quaternary volcanoes in the region indicate that fissure-fed eruptions, which are most likely during the very early stages of volcano formation, tend to produce lavas or coarse, localized accumulations of spatter and agglutinate indicative of relatively weak explosive activity (Valentine *et al.*, 2006, 2007; Valentine and Keating, 2007; Keating *et al.*, 2008a). From the perspective of a hypothetical population 20 km from an eruption site, there is little risk from direct radionuclide transport from such activity or from lava flows since these have extents limited to $\approx$ 2 km from their vents. Furthermore, it is unlikely that surficial processes would transport substantial quantities of such products to the control population over timescales of a few hundred thousand years.

Violent Strombolian activity that produces sustained tephra columns several kilometers high could, on the other hand, result in substantial dispersal of radioactive contamination (Jarzemba, 1997). Most of the Quaternary volcanoes in the region preserve some evidence of having produced violent Strombolian eruptions, and in each case that activity was associated with the main central conduit and cone (Valentine *et al.*, 2005, 2006, 2007; Valentine and Keating, 2007). Therefore it is important to constrain the size of such conduits at repository depths, where waste material could be directly entrained into a violent Strombolian eruption, assuming that intersected waste packages fail and the waste breaks up into fragments that can be transported in the conduit flow. Studies at eroded basaltic volcanoes that erupted through Yucca Mountain-like tuffs, and where eruptive facies are preserved but where erosional exposure is sufficiently deep, supplemented with numerical modeling, indicate that the conduits range around $\sim$ 15 m diameter at repository depths and flare rapidly in the uppermost $\sim$ 50 m to form vent complexes $\sim$ 100 m wide (in some cases with elongate shapes where the long axis is $\sim$ 200 m along the master dike; Keating *et al.*, 2008a). Minimum conduit diameter at repository depth is equal to the dike width, while maximum diameters may rarely exceed 20 m based upon indirect evidence from wall-rock lithic abundances in eruptive deposits at the Lathrop Wells volcano (Sandia National Laboratories, 2007a; Valentine *et al.*, 2007). Note that these conduit diameters are smaller than is commonly observed in diatremes formed from explosive magma–water interaction (e.g. Lorenz, 1986; Nemeth and White, 2003). There is little or no evidence for substantial hydrovolcanic activity in Quaternary volcanoes near Yucca Mountain (Valentine *et al.* 2006, 2007; Valentine and Keating, 2007).

A potential contaminated tephra deposit might extend 20 km or more from a volcano, as does the deposit associated with the Lathrop Wells volcano (Valentine *et al.*, 2007). If the winds are directed southward during an eruption at the repository site, tephra could therefore be directly deposited on the control population. If the winds are directed elsewhere, primary tephra deposition would not affect the population. However, the control population is located at the apex of a major alluvial fan that has accumulated from deposition of sediment from the Fortymile Wash drainage, which includes the eastern slopes of Yucca Mountain and the eastern moat of the Timber Mountain caldera to the north. Thus there is potential for contaminated tephra that is deposited in this drainage basin to be transported by surficial processes to the control population (Pelletier *et al.*, 2008).

The deposit from the $\sim$ 77 ka Lathrop Wells volcano provides some insight into the fate of a tephra sheet in this environment. In low-lying, relatively flat areas, the Lathrop Wells tephra was mostly buried by alluvial sediments. In relatively flat areas that are elevated above the surroundings (such as on top of lava platforms), the tephra deposits interact with eolian sediments to form relatively stable desert pavement surfaces. Valentine *et al.* (2007) show that on slopes of bedrock (in this case the bedrock is Miocene welded ignimbrite) the tephra sheet undergoes a multistage process of remobilization as small debris flows. (In cases where the tephra deposit is mainly coarse ash or lapilli, this remobilization occurs after an initial period of infiltration of eolian silt and fine sand into the top of the tephra deposit, providing a fine matrix that supports development of debris flows.) These small debris flows move tephra down slope some distance and then deposit as small sediment fans. The low points between these fans then become the nuclei for a second generation of small debris flows and down-slope transport. As tephra is gradually fed to the main fluvial channels (in this region these channels are only active during flash floods), it mixes with other channel sediments and is transported farther downstream (Pelletier *et al.*, 2008).

On longer timescales, scoria cones and lava fields are subject to erosion and the eroded material can be mobilized into drainage systems. The surface and erosional history of these features are strongly dependent upon their initial (volcanic) surface textures and morphologies (Valentine *et al.*, 2006, 2007).

Modeling approach

Potential dispersal of radioactive waste by violent Strombolian eruptions is modeled using a well-known plume dispersion and fallout deposition approach based upon that of Suzuki (1983) and Jarzemba (1997). Eruption column height is calculated based upon mass flux (or thermal power) at the vent. A range of mass fluxes is sampled, based upon historical violent Strombolian eruptions (Sandia National Laboratories 2007a, 2007c). Tephra is transported downwind using a simple advection–diffusion approach, and falls from the plume based upon particle settling velocities. Radioactive waste is assumed to break up into small fragments where the eruptive conduit intersects a drift, and the waste fragments are incorporated into pyroclasts using a simple model. The numerical implementation of the approach is the ASHPLUME code (Jarzemba, 1997), the output of which is a two-dimensional distribution of tephra thickness, grain size and radioactive waste mass on the ground. Although there are more sophisticated approaches to modeling tephra dispersal and fallout (e.g. Turner and Hurst, 2001; Bonadonna *et al.*, 2005; Folch and Falpeto, 2005; Volentik *et al.*, Chapter 9, this volume), the relatively simple ASHPLUME approach is considered to be appropriate for the Yucca Mountain problem due to the large uncertainties associated with potential eruption through a repository. Additionally, the most likely wind direction for a potential eruption is toward the northeast; any resulting contamination at the location of the control population would be transported there by surficial processes that would filter out many details of a tephra-fall prediction. Keating *et al.* (2008b) concluded that the potential refinements of a more sophisticated plume model would be swamped out by the inherent uncertainties and by the effects of surface redistribution.

Pelletier *et al.* (2008) describe the approach to modeling redistribution of contaminated tephra by surficial processes, implemented in the computational code FAR (Fortymile Ash Redistribution) and illustrated in Figure 19.6. The two-dimensional results of the ASHPLUME code are draped onto a digital elevation model of the Fortymile Wash drainage basin. Tephra deposited onto slopes steeper than a defined critical slope angle is moved into the drainage channel network. Once in the drainages, tephra is mixed with channel sediments according to an estimated scour depth (i.e. the depth of sediments in a channel bed that are mobilized and redeposited during flood events). The tephra–sediment mixture is transported to the head of the Fortymile Wash alluvial fan (the location of the control population) and distributed across surfaces in a manner that is determined by the timescale of interest. For a 10 ka timescale, only Holocene surfaces are considered, while for longer timescales (i.e. 1 Ma) Pleistocene portions of the fan surface may receive contaminated sediment. Once deposited, radionuclide migration into the soil profile is modeled as a diffusion process that is calibrated to measured radionuclide concentration profiles in desert soils (Pelletier *et al.*, 2005).

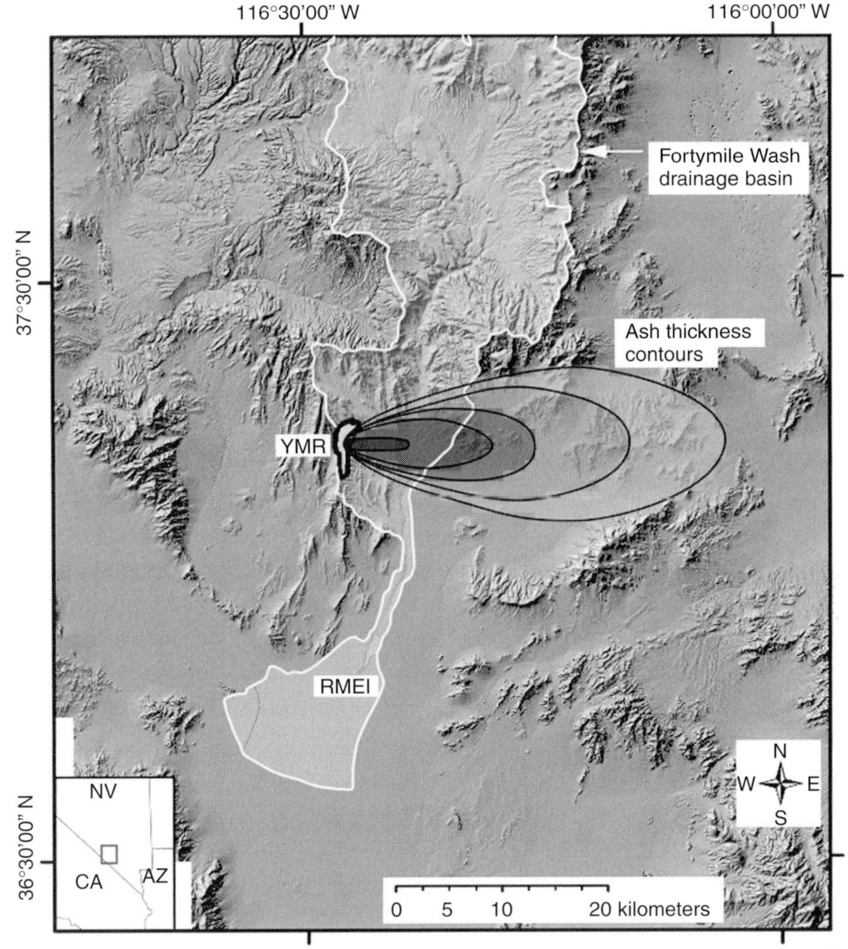

Fig. 19.6 Illustration shows the coupled tephra fallout and redistribution model on a digital elevation model of the Yucca Mountain area. The small area labeled YMR is the footprint of the proposed repository. Contours extending to the right (east) represent tephra fallout isopachs from a hypothetical violent Strombolian eruption at Yucca Mountain as might be simulated with the ASHPLUME code. Light-shaded area outlined in white is the Fortymile Wash drainage basin, which drains at its southern end in a broad alluvial fan. Contaminated tephra that falls in the drainage basin is remobilized, mixed with other channel sediments and transported toward the control population (labeled as RMEI, "reasonably maximally exposed individual"). Inset shows location relative to the states of California (CA), Arizona (AZ) and Nevada (NV). Map adapted from Keating *et al.* (2008b).

19.6 Abstraction and consequences

The detailed analog and modeling studies summarized above provide constraints on the geometries and dynamics that would likely characterize a potential future volcanic event at Yucca Mountain. These constraints then provide information for the second tier on

Figure 19.3, where abstracted or simplified models attempt to capture the range of parameter values and uncertainties and all their combinations. Here we briefly describe the abstracted models that capture subsurface disruption of the repository and subsequent groundwater-borne transport of radionuclides to the control population, transport of erupted material to the population, and the approach that is taken to convert radionuclide contamination (of groundwater or surface materials) to potential dose to the control population.

19.6.1 Subsurface igneous effects abstraction

The first step in the subsurface effects abstraction is to compute the number of waste packages that might be affected by an igneous event (i.e. those waste packages that might be contacted by magma). First, the probability that the subsurface plumbing of an igneous event intersecting the repository footprint is taken by sampling the probability distribution function for intersection probability derived from formal expert elicitation (Bechtel-SAIC Company, 2004a).

A given intersecting event, at repository depth, will have some combination of the following parameters: number of dikes, dike lengths, dike widths and dike orientations. Probability distribution functions are defined for each of these parameters based primarily upon field analog data such as that reported in Sandia National Laboratories (2007a). The parameters are sampled using a Monte Carlo or Latin hypercube sampling approach and combined to produce a realization of a potential event. This is overlain on the repository design and the intersected drifts are computed from geometry (e.g. Figure 19.7; Sandia National Laboratories, 2007d).

Given the number of intersected drifts in a realization, different approaches can be taken to estimate the number of waste packages that could be contacted by magma and potentially have their capability to contain radioactive waste compromised. The most conservative approach simply assumes that given any intersection of a dike with a drift, magma is able to flood all drifts and tunnels (Sandia National Laboratories, 2007d) and contact all waste packages. This is a somewhat extreme approach in that it does not account for viscosity variations due to heat loss and growth of groundmass crystals as magma flows through the repository, or potential flow blockage due to rock fall in drifts and tunnels that becomes increasingly likely as the repository ages. However, this simple approach recognizes the difficulty in predicting magma flow through a complex layout of drifts and tunnels and their contents. A second approach to computing number of waste packages affected assumes that only drifts that are directly intersected by dikes are flooded. This approach may be more reasonable for events that occur after some period of time, when access tunnels and drifts may have some blockage due to rock fall. Other approaches might assume that only part of an intersected drift is flooded, for example if there is a technical basis to show that magma might solidify or otherwise block flowage, or one could assume that only a fraction of intersected drifts are flooded. The simple approach to estimating intersected drifts therefore provides flexibility in testing sensitivity of risk to various models of magma–drift interaction, but it is important to note that in the final analysis any model that is selected must be technically defensible.

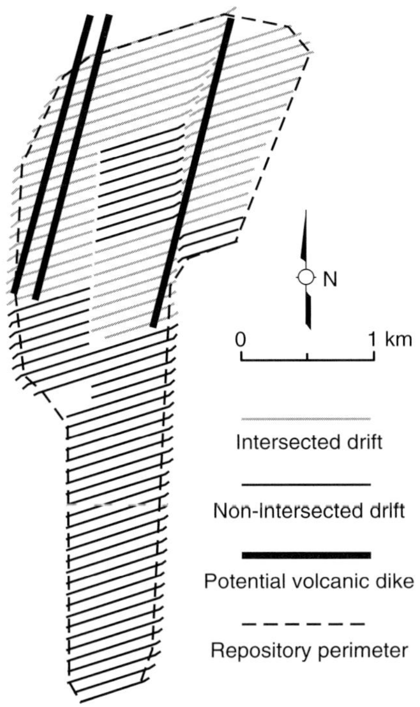

Fig. 19.7 Shown here is one realization of a dike set from a single volcano overlain on the repository layout. Dike-intersected drifts are light gray. One to three conduits could be located along one of the dikes. Any given realization samples from probability distributions (intended to represent the plumbing of a single monogenetic volcano) for dike length, number of dikes in the event, dike orientation, dike spacing, conduit size and location, and number of conduits (Sandia National Laboratories, 2007d).

For the purpose of consequence analysis it is assumed that any waste package that resides in a magma-flooded drift fails to a degree that essentially removes the capability of the waste package to act as a barrier to radionuclide migration. Failure can be due to a range of mechanisms including ductile deformation of waste-package walls as they heat to magmatic temperatures, differential pressures between the interior and exterior of a waste package that put stress on structural features such as welds, and metallurgical changes on package materials as they are exposed to magmatic conditions. During an igneous event and for some time afterwards, elevated temperatures will prevent liquid water from contacting waste. However, after the system has cooled it is expected that seepage will resume in the repository. Cooled basalt in magma-flooded drifts is likely to allow flow of these vadose zone waters through fractures such that the water can contact packages and any exposed waste (Sandia National Laboratories, 2007e) and carry dissolved radionuclides downward to the water table. The amount of contamination carried downward from the repository depends on the number of waste packages that are contacted by magma for a given realization, and the effects of basalt on seepage water chemistry. Once in the saturated zone, dissolved radionuclides may be transported as a contaminant plume within the regional groundwater regime toward the control population (Sandia National Laboratories, 2007f). If contaminated groundwater reaches the location of the control population it can be pumped to the surface for use as drinking and agricultural water (see below).

19.6.2 Eruptive effects and surficial transport abstraction

The most important mechanism through which waste could be dispersed onto the Earth's surface is by violent Strombolian activity that involves a central conduit(s). Therefore the probability of violent Strombolian waste dispersal is the probability that a conduit intersects a repository drift, given that the plumbing of a new volcano intersects the repository. For any realization this conditional probability is computed by sampling from probability distributions that describe the number of conduits in an event, conduit spacing, location of conduits along dikes and conduit dimensions at repository depth. These parameters are combined with the repository geometry to determine whether a given simulated conduit intersects a drift and, if so, how many waste packages it might intersect (Sandia National Laboratories, 2007d) and that might be entrained into violent Strombolian eruptive flows if waste packages break apart in the conduit flow.

Given an amount of waste entrained into an eruptive conduit for a realization, the next step is to compute its dispersal in an eruption plume and accumulation on the ground as fallout, and subsequent transport by fluvial processes to the control population (Figure 19.6). The ASHPLUME code samples historical wind data to determine wind velocity for a realization, and produces a simulated fallout deposit (which includes the mass of erupted waste as determined by the above steps). In the low-probability case where contaminated tephra is deposited directly by fallout on the area of the control population, the contaminant radionuclides then migrate into the underlying soil profile using a diffusion approximation (Pelletier *et al.*, 2005). In the more likely cases where contaminated tephra is deposited mainly to the north and east of the repository site, tephra remobilization, dilution (with other channel sediments) and transport to the control population area is computed with the FAR code (Section 19.5.2; Pelletier *et al.*, 2008). This code distributes contaminant mass along with other entrained sediments across some fraction of the area (e.g. the fraction covered by surfaces that have been active in the Holocene), and then the contaminants migrate into the soil profile by a diffusion approximation.

19.6.3 Biosphere dose analysis

Ultimately risk (or performance) assessment for the Yucca Mountain repository is measured in terms of radiation dose to individuals in the hypothetical control population. We have described three main transport mechanisms that might move radioactive contamination from the repository to the control population as a result of a volcanic event: (i) a groundwater contaminant plume due to subsurface disruption of the repository system; (ii) direct fallout on the control population area from a contaminated violent Strombolian eruption plume; and

(iii) deposition of reworked, contaminated fallout along with fluvial sediments in the control population area, which is on an alluvial fan. Each of these produces a concentration or mass of radionuclides at the control population either in underlying groundwater (i) or on the surface soils (ii and iii). Models referred to as biosphere dose conversion factors (BDCFs), which account for the pathways of radiation to individuals, are the final step in determining dose (Sandia National Laboratories, 2007g). For cases in which contaminated tephra is deposited directly or by fluvial processes at the control population, the BDCF accounts for inhalation of resuspended soil and radon decay products; ingestion of contaminated crops, animal products and soil; and from external exposure or "shine" from the soil. The BDCF for contaminated groundwater accounts for pathways associated with typical groundwater usages in analog modern populations (such as the farming community that currently exists to the south in the Amargosa Desert), including drinking the water; contamination of irrigated soils (with components due to "shine," inhalation of resuspended particulates and radon decay products, and direct ingestion), flora and fauna (which are then ingested by humans) through the use of the groundwater for irrigation; inhalation of aerosols associated with use of the water in evaporative coolers in homes and other structures. Development of BDCF models is a subject in the field of health physics and is a critical link in risk/performance assessment for a nuclear facility, but is beyond the scope of this chapter.

Concluding remarks

We have summarized a wide range of factors and scientific approaches to addressing volcanic risk for the proposed repository at Yucca Mountain. From the standpoint of defining event probabilities, spatial and temporal models for volcanism must be developed. These require data on tectonics of the region as well as on the geochemistry, petrology and geochronology of the volcanic field as a whole. The potential for young subsurface igneous features (either small volcanoes buried beneath alluvium in subsiding basins, or shallow intrusions) necessitated a program of aeromagnetic surveys and drilling. The potential for a range of interpretations, given the tectonic complexities of the region and the statistically sparse number of young volcanoes, has been accommodated by a process of formal expert elicitation in order to arrive at event probability. Because there is potential for a new volcano to form at the Yucca Mountain site, it is necessary to quantify the consequences of such an event. We have approached consequence assessments by field studies aimed at constraining the physical processes associated with volcanoes in the southwestern Nevada volcanic field during the past few million years; essentially, these are used as analogs for potential future events at the repository. Modeling approaches have been used to constrain processes that might specifically result from interaction of volcanic plumbing with repository structures, eruptive dispersal of contaminants and subsequent redistribution by surface processes. Finally, we described simplified, or abstracted, models that are used to capture volcano-repository interactions and the range of possible parameter combinations and uncertainties.

Section 19.2 describes an overall risk assessment framework that relies on a foundation of basic scientific understanding of the appropriate volcanic system and detailed process

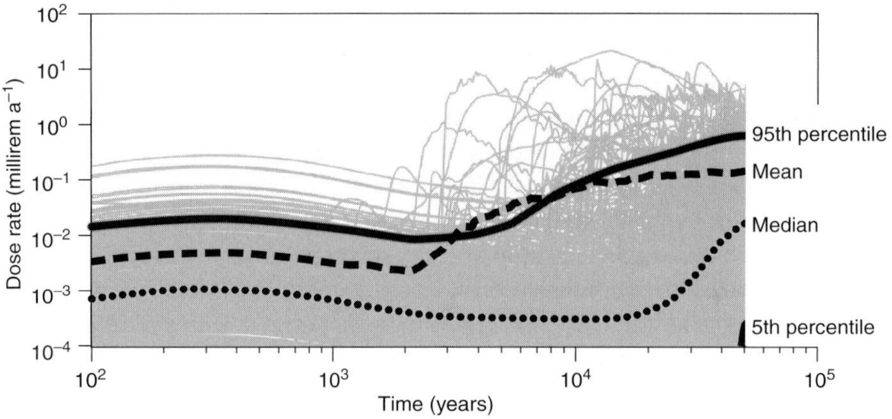

Fig. 19.8 Example of probability-weighted dose rate (millirem a^{-1}) due to igneous processes, plotted as a function of time, with results computed out to 500 ka in the future (from DOE, 2001). Each gray line represents a single realization, the dotted line is the median value, the dashed line is the mean and the solid black lines are 5th and 95th percentile values. The first maximum in probability-weighted dose rate (at $\sim$ 350 a) is due mainly to eruptive processes, while the later-time rise in dose rate reflects groundwater transport of radionuclides to the control population due to subsurface disruption of the repository.

models of how such a system might interact with a repository (base of Figure 19.3). After abstracting this understanding so that it can be used in a probabilistic framework, the assessment is boiled down into some relatively straightforward result that can easily be compared to some decision criteria (the apex of Figure 19.3). For the Yucca Mountain assessment, decisions are based upon regulatory criteria for the probability-weighted annual dose rate to individuals in the hypothetical control population. Figure 19.8 shows an example (DOE, 2001) of how all of the aspects we have described are compiled into a single result for use by decision makers. This figure shows an ensemble of Monte Carlo simulations of annual radiation dose rate to an individual from processes related to volcanic events at the repository, as a function of time in the future. It is important to note that the dose rate is weighted by the probability of the event, which in this case has a mean value of $\approx 1.7 \times 10^{-8}\,a^{-1}$ (5th and 95th percentile values of $7.4 \times 10^{-10}\,a^{-1}$ and $5.5 \times 10^{-8}\,a^{-1}$, respectively; Bechtel-SAIC Company, 2004a). Furthermore, consistent with the regulatory criteria, the mean dose history is an expected value that is the sum at all times of probability-weighted consequences of events that are equally likely to occur at any given time in the future; therefore the mean dose does not represent the conditional dose that any one individual might receive should a single volcanic event occur at Yucca Mountain. Calculation and interpretation of the probability-weighted mean annual dose is a subject in the field of probabilistic risk analysis, and, like the biosphere dose conversion factors discussed in the previous section, is outside the scope of this chapter. See Helton and Sallaberry (2007) for a more detailed discussion of the topic.

In the example shown here (Figure 19.8), which will be superseded by newer results for use in a license application, the first maximum in probability weighted dose rate, $\sim$ 350 a, is due mainly to eruptive processes, while the later-time rise in dose rate reflects groundwater transport of radionuclides to the control population due to subsurface disruption of the repository. In addition to the range of processes that are accounted for, as described in this chapter, some of the complexity of the result also reflects the changing radionuclide inventory with time due to radioactive decay.

Over the history of the Yucca Mountain effort, the type of result shown in Figure 19.8 has helped to focus additional research at the base of the risk framework, or upon refinements of abstracted models in the middle of the framework (Figure 19.3); such an approach can be used as a general framework for volcanology research and risk assessments in a range of problems, including hazards to nuclear facilities and to urban centers.

Acknowledgments

The work that we have summarized here has been a team effort that has involved many people over the years, beginning with the studies led by Bruce Crowe. The most recent efforts that we have emphasized are in many cases only reported in programmatic documents that do not provide attribution to the person(s) who did the work; we are taking the opportunity here to list those contributors whose products have been cited here but whose names do not appear in the citations. Bob Youngs provided outstanding expertise in development of probability models (Bechtel-SAIC Company, 2004a). Gordon Keating led the ASHPLUME modeling effort (reported in Sandia National Laboratories, 2007c), as well as the studies reported in Sandia National Laboratories (2007a) related to volcanic conduits. Donathan Krier has been a key participant in field analog studies and led preparation of Sandia National Laboratories (2007a). Edward Gaffney and Branco Damjanac have led most of the modeling work on dike propagation that is described in Sandia National Laboratories (2007b), with contributions from Eric Sonnenthal in the area of hydrothermal modeling. Stochastic and geometric modeling aimed at computing the number of waste packages affected by igneous processes, as reported in Sandia National Laboratories (2007d), has been led by Michael Wallace. Rick Kelley has provided valuable GIS expertise. Additional team members who have contributed in various key capacities over the past few years include Terry Crump, Jeff McCleary, Cheryl Hastings geographical information systems and Buddy Statham. Many others have participated on expert panels or as reviewers. The volcanic risk effort's recent progress could not have been accomplished without management support from Eric Smistad, Mike Cline and Thomas Pfeifle. Donathan Krier and Peter Swift provided very helpful reviews of the manuscript.

References

Bechtel-SAIC Company (2004a). Characterize framework for igneous activity at Yucca Mountain, NV, ANL-MGR-GS-000001(2). Las Vegas, NV: Bechtel-SAIC Company.

Bechtel-SAIC Company (2004b). Dike–drift interactions, MDL-MGR-GS-000005(1). Las Vegas, NV: Bechtel-SAIC Company.

Bechtel-SAIC Company (2005). Magma dynamics at Yucca Mountain, NV, ANL-MGR-GS-000005(00). Las Vegas, NV: Bechtel-SAIC Company.

Bonadonna C., C. B. Connor, B. F. Houghton *et al.* (2005). Probabilistic modeling of tephra dispersal: hazard assessment of a multiphase rhyolitic eruption at Tarawera, New Zealand. *Journal of Geophysical Research*, **110**, B03203, doi:10.1029/2003JB002896.

Bradshaw, T. K. and E. I. Smith (1994). Polygenetic Quaternary volcanism at Crater Flat, NV. *Journal of Volcanology and Geothermal Research*, **63**, 165–182.

Carr, W. J. and L. D. Parrish (1985). Geology of drill hole USW VH-2, and structure of Crater Flat, SW NV. US Geological Survey Open-File Report 87–475.

Connor, C. B. and F. M. Conway (2000). Basaltic volcanic fields. In: Sigurdsson, H. (ed.), *Encyclopedia of Volcanoes*. San Diego, CA: Academic Press, 331–343.

Connor, C. B. and B. E. Hill (1995). Three nonhomogeneous Poisson models for the probability of basaltic volcanism: application to the Yucca Mountain region, NV, USA. *Journal of Geophysical Research*, **100**, 10 107–10 125.

Connor, C. B., S. Lane-Magsino, J. A. Stamatakos *et al.* (1997). Magnetic surveys help reassess volcanic hazards at Yucca Mountain, NV. *Eos, Transactions of the American Geophysical Union*, **78**, 73, 77–78.

Crowe, B. M. (1986). Volcanic hazard assessment for disposal of high-level radioactive waste. In: Geophysics Research Forum – US (eds.) *Active Tectonics*. Washington, DC: National Academy Press, 247–260.

Crowe, B. M., M. E. Johnson and R. J. Beckman (1982). Calculation of the probability of volcanic disruption of a high-level radioactive waste repository within southern NV, USA. *Radioactive Waste Management and the Nuclear Fuel Cycle*, **3**, 167–190.

Crowe, B., S. Self, D. Vaniman, R. Amos and F. Perry (1983). Aspects of potential magmatic disruption of a high-level radioactive waste repository in southern NV. *Journal of Geology*, **91**, 259–276.

CRWMS M&O (1996). Probabilistic volcanic hazard analysis for Yucca Mountain, NV, BA0000000-01717-2200-00082(0). Las Vegas, NV: US Department of Energy.

Damjanac, B., Z. Radakovic-Guzina and E. S. Gaffney (2005). Conditions leading to sudden release of magma pressure. American Geophysical Union Fall Meeting, Abstract V33A-0664. San Francisco, CA.

Dartevelle, S. and G. A. Valentine (2005). Early-time multiphase interactions between basaltic magma and underground openings at the proposed Yucca Mountain radioactive waste repository. *Geophysical Research Letters*, **32**, L22311, doi:10.1029/2005GL024172.

Dartevelle, S. and G. A. Valentine (2008). Multiphase magmatic flows at Yucca Mountain, NV. *Journal of Geophysical Research*, **113**, B12209, doi:10.1029/2007JB005367.

DOE (2001). Yucca Mountain science and engineering report – technical information supporting site recommendation consideration, DOE/RW-0539. Washington, DC: US Department of Energy.

Fleck, R. J., B. D. Turrin, D. A. Sawyer *et al.* (1996). Age and character of basaltic rocks of the Yucca Mountain region, southern NV. *Journal of Geophysical Research*, **101**, 8205–8227, doi:10.1029/95JB03123.

Folch, A. and A. Falpeto (2005). A coupled model for dispersal of tephra during sustained explosive eruptions. *Journal of Volcanology and Geothermal Research*, **145**, 337–349.

Fridrich, C. J., J. W. Whitney, M. R. Hudson and B. M. Crowe (1999). Space–time patterns of late-Cenozoic extension, vertical axis rotation, and volcanism in the Crater Flat basin, SW NV. Geological Society of America, Special Paper 333, 197–212.

Gaffney, E. S. and B. Damjanac (2006). Localization of volcanic activity: topographic effects on dike propagation, eruption and conduit formation. *Geophysical Research Letters*, **33**, L14313, doi:10.1029/2006GL026852.

Gaffney, E. S., B. Damjanac, D. Krier and G. Valentine (2005). Deformation of scoria cone by conduit pressurization. American Geophysical Union Fall Meeting, Abstract V53B-1570. San Francisco, CA.

Gaffney, E. S., B. Damjanac and G. A. Valentine (2007). Localization of volcanic activity: 2. Effects of pre-existing structure. *Earth and Planetary Science Letters*, **263**, 323–338.

Heizler, M. T., F. V. Perry, B. M. Crowe, L. Peters and R. Appelt (1999). The age of the Lathrop Wells volcanic center: an $^{40}Ar/^{39}Ar$ dating investigation. *Journal of Geophysical Research*, **104**, 767–804.

Helton, J. C. and C. J. Sallaberry (2007). Illustration of sampling-based approaches to the calculation of expected dose in performance assessments for the proposed high-level radioactive waste repository at Yucca Mountain, NV, Report SAND2007-1353. Albuquerque, NM: Sandia National Laboratories.

Ho, C. -H., E. I. Smith, D. L. Feurbach and T. R. Naumann (1991). Eruptive probability calculation for the Yucca Mountain site, USA: statistical estimation of recurrence rates. *Bulletin of Volcanology*, **54**, 50–56.

Jarzemba, M. S. (1997). Stochastic radionuclide distributions after a basaltic eruption for performance assessments at Yucca Mountain. *Nuclear Technology*, **118**, 132–141.

Kane, M. F. and R. E. Bracken (1983). Aeromagnetic map of Yucca Mountain and surrounding regions, SW NV. US Geological Survey Open-File Report 83-616.

Keating, G. N., G. A. Valentine, D. J. Krier and F. V. Perry (2008a). Shallow plumbing systems for small-volume basaltic volcanoes. *Bulletin of Volcanology*, **70**, 563–582, doi:10.1007/s00445-007-0154-1.

Keating, G. N., J. Pelletier, G. A. Valentine and B. Statham (2008b). Evaluating suitability of a tephra dispersal model as part of a risk assessment framework. *Journal of Volcanology and Geothermal Research*, doi:10.1016/j.jvolgeores.2008.06.007.

Langenheim, V. E. (1995). Magnetic and gravity studies of buried volcanic centers in the Amargosa Desert and Crater Flat, SW NV. US Geological Survey Open-File Report 95-564.

Link, R. L., S. E. Logan, H. S. Ng, F. A. Rockenbach and K. -J. Hong (1982). Parametric studies of radiological consequences of basaltic volcanism, SAND 81-2375. Albuquerque, NM: Sandia National Laboratories.

Lorenz, V. (1986). On the growth of maars and diatremes and its relevance to the formation of tuff rings. *Bulletin of Volcanology*, **48**, 265–274.

McFarlane, A. M. and R. C. Ewing (eds.) (2006). *Uncertainty Underground: Yucca Mountain and the Nation's High-level Nuclear Waste*. Cambridge, MA: MIT Press.

McGetchin, T. R., M. Settle and B. A. Chouet (1974). Cinder cone growth modeled after Northeast Crater, Mount Etna, Sicily. *Journal of Geophysical Research*, **79**, 3257–3272.

Menand, T. and J. C. Phillips (2007a). Gas segregation in dykes and sills. *Journal of Volcanology and Geothermal Research*, **159**, 393–408.

Menand, T. and J. C. Phillips (2007b). A note on gas segregation in dykes and sills at high volumetric gas fractions. *Journal of Volcanology and Geothermal Research*, **162**, 185–188.

Menand, T., J. C. Phillips and R. S. J. Sparks (2007). Circulation of bubbly magma and gas segregation within tunnels of the potential Yucca Mountain repository. *Bulletin of Voclanology*, doi:10.1007/s00445-007-0179-5.

Nemeth, K. and J. D. L. White (2003). Reconstructing eruption processes of a Miocene monogenetic volcanic field from vent remnants: Waitpiata volcanic field, South Island, New Zealand. *Journal of Volcanology and Geothermal Research*, **124**, 1–21.

O'Leary, D. W., E. A. Mankinen, R. J. Blakely, V. E. Langenheim and D. A. Ponce (2002). Aeromagnetic expression of buried basaltic volcanoes near Yucca Mountain, NV. US Geological Survey Open-File Report 02-020.

Parsons, T., G. A. Thompson and A. H. Cogbill (2006). Earthquake and volcano clustering via stress transfer at Yucca Mountain, NV. *Geology* **34**, 785–788.

Pelletier, J. D., C. D. Harrington, J. W. Whitney *et al.* (2005). Geomorphic control of radionuclide diffusion in desert soils. *Geophysical Research Letters*, **32**, L23401, doi:10.1029/2005GL024347.

Pelletier, J. D., S. B. DeLong, M. L. Cline, C. D. Harrington and G. N. Keating (2008). Dispersion of channel-sediment contaminants in distributary fluvial systems: application to fluvial tephra and radionuclide redistribution following a potential volcanic eruption at Yucca Mountain. *Geomorphology*, **94**, 226–246, doi:10.1016/j.geomorph.2007.05.014.

Perry, F. V. and B. M. Crowe (1992). Geochemical evidence for waning magmatism and polycyclic volcanism at Crater Flat, NV. In: High-Level Radioactive Waste Management Program Committee (eds.) *High-level Radioactive Waste Management 1992*, Proceedings of the Third International Conference, Las Vega, NV, April 12–26. New York, NY: American Society of Civil Engineers, 2356–2365.

Perry, F. V., B. M. Crowe, G. A. Valentine and L .M. Bowker (1998). Volcanism studies: final report for the Yucca Mountain project, Report LA-13478-MS. Los Alamos, NM: Los Alamos National Laboratory.

Perry, F. V., A. H. Cogbill and R. E. Kelley (2005). Uncovering buried volcanoes at Yucca Mountain: new data for volcanic hazard assessment. *Eos, Transactions of the American Geophysical Union*, **86**, 485–488.

Perry, F. V., G. A. Valentine, A. H. Cogbill *et al.* (2006). Control of basaltic feeder dike orientation by fault capture near Yucca Mountain, USA. American Geophysical Union Fall Meeting, Abstract V11B-0572, San Francisco, CA.

Rubin, A. M. (1995). Propagation of magma-filled cracks. *Annual Reviews of Earth and Planetary Sciences*, **23**, 287–336.

Sandia National Laboratories (2007a). Characterize eruptive processes at Yucca Mountain, NV, ANL-MGR-GS-000002(3). Las Vegas, NV: Sandia National Laboratories.

Sandia National Laboratories (2007b). Dike–drift interactions, MDL-MGR-GS-000005(02). Las Vegas, NV: Sandia National Laboratories.

Sandia National Laboratories (2007c). Atmospheric dispersal and deposition of tephra from a potential volcanic eruption at Yucca Mountain, NV, MDL-MGR-GS-000002(03). Las Vegas, NV: Sandia National Laboratories.

Sandia National Laboratories (2007d). Number of waste packages hit by igneous events, ANL-MGR-GS-000003(03). Las Vegas, NV: Sandia National Laboratories.

Sandia National Laboratories (2007e). In-package chemistry abstraction, MDL-MGR-MD-000001(02). Las Vegas, NV: Sandia National Laboratories.
Sandia National Laboratories (2007f). Site-scale saturated zone transport, MDL-NBS-HS-000010(03). Las Vegas, NV: Sandia National Laboratories.
Sandia National Laboratories (2007g). Biosphere model report, MDL-MGR-MD-000001(02). Las Vegas, NV: Sandia National Laboratories.
Sawyer, D. A., R. J. Fleck, M. A. Lanphere *et al.* (1994). Episodic caldera volcanism in the Miocene southwestern Nevada volcanic field: revised stratigraphic framework, $^{40}Ar/^{39}Ar$ geochronology, and implications for magmatism and extension. *Geological Society of America Bulletin*, **106**, 1304–1318.
Smith, E. I. and D. L. Keenan (2005). Yucca Mountain could face greater volcanic threat. *Eos, Transactions of the American Geophysical Union*, **86**, 317.
Suzuki, T. (1983). A theoretical model for dispersion of tephra. In: Shimozura, D. and I. Yokoyama (eds.). *Volcanism – Physics and Tectonics*. Tokyo: Arc.
Taddeucci, J., O. Spieler, B. Kennedy *et al.* (2004). Experimental and analytical modeling of basaltic ash explosions at Mount Etna, Italy. *Journal of Geophysical Research*, **109**, B08203, doi:10.1029/2003JB002952.
Turner, R. and A. W. Hurst (2001). Factors influencing volcanic ash dispersal from the 1995 and 1996 eruptions of Mount Ruapehu, New Zealand. *Journal of Applied Meteorology*, **40**, 56–69.
Valentine, G. A. (1998). Damage to structures by pyroclastic flows and surges, inferred from nuclear weapons effects. *Journal of Volcanology and Geothermal Research*, **87**, 117–140.
Valentine, G. A. and G. N. Keating (2007). Eruptive styles and inferences about plumbing systems at Hidden cone and Little Black Peak scoria cone volcanoes (NV, USA). *Bulletin of Volcanology*, doi:10.1007/s00445-007-0123-8.
Valentine, G. A. and K. E. C. Krogh (2006). Emplacement of shallow dikes and sills beneath a small basaltic volcanic center – the role of pre-existing structure (Paiute Ridge, southern NV, USA). *Earth and Planetary Science Letters*, **246**, 217–230.
Valentine, G. A. and F. V. Perry (2006). Decreasing magmatic footprints of individual volcanoes in a waning basaltic field. *Geophysical Research Letters*, **33**, L14305, doi:10.1029/2006GL026743.
Valentine, G. A. and F. V. Perry (2007). Tectonically controlled, time-predictable basaltic volcanism from a lithospheric mantle source (central Basin and Range Province, USA). *Earth and Planetary Science Letters*, **261**(1–2), 201–216, doi: 10.1016/j.epsl.2007.06.029.
Valentine, G. A., D. J. Krier, F. V. Perry and G. Heiken (2005). Scoria cone construction mechanisms, Lathrop Wells volcano, southern NV, USA. *Geology*, **33**, 629–632.
Valentine, G. A., F. V. Perry, D. J. Krier *et al.* (2006). Small-volume basaltic volcanoes: eruptive products and processes, and posteruptive geomorphic evolution in Crater Flat (Pleistocene), southern NV. *Geological Society of America Bulletin*, **118**, 1313–1330, doi:10.1130/B25956.1.
Valentine, G. A., D. J. Krier, F. V. Perry and G. Heiken (2007). Eruptive and geomorphic processes at Lathrop Wells scoria cone volcano. *Journal of Volcanology and Geothermal Research*, **161**, 57–80, doi:10.1016/j.jvolgeores.2006.11.003.
Wahl, R. R., D. A. Sawyer, S. A. Minor *et al.* (1997). Digital geologic map database of the Nevada Test Site area, NV. US Geological Survey Open File Report 97-140.

WoldeGabriel, G., G. N. Keating and G. A. Valentine (1999). Effects of shallow basaltic intrusion into pyroclastic deposits, Grants Ridge, NM, USA. *Journal of Volcanology and Geothermal Research*, **92**, 389–411.

Woods, A. W., R. S. J. Sparks, O. Bokhove *et al.* (2002). Modeling magma–drift interaction at the proposed high-level radioactive waste repository at Yucca Mountain, NV, USA. *Geophysical Research Letters*, **29**, L1641, doi:10.1029/2002GL014665.

20

Geological issues in practice: experience in siting US nuclear facilities

L. Reiter

As a result of the civilian nuclear power program that began more than 45 years ago, many nuclear facilities have been sited in the United States. They include more than 100 nuclear power plants and several low-level radioactive waste disposal facilities. Extensive efforts are under way to site a high-level radioactive waste disposal facility (repository) at Yucca Mountain in the state of Nevada. These projects have resulted in the accumulation of an extraordinarily large and detailed geological database that has been useful to many outside the nuclear industry. These efforts, both successful and unsuccessful, have provided important insights into the role of geology, which often has played a critical or apparently critical role in siting and operating decisions. Some of these decisions have highlighted the interactions between scientific issues and political and social issues and the difficulty often faced in distinguishing between them. Geological uncertainties and the highly interpretive nature of the geologic sciences have made a particularly fertile ground for disputes between those in favor of and those opposed to the siting and operation of nuclear facilities.

Although this chapter discusses both nuclear power plants and nuclear waste disposal facilities, the role of geology in these two types of facilities is different. Nuclear power plants are highly complex, relatively short-lived (typically assumed to be 40 years, but see discussion by Chapman *et al.*, Chapter 1, this volume) facilities, requiring sophisticated control technologies, whose failure could result in the immediate release of large amounts of harmful radionuclides. The role of geology is to provide sufficiently stable locations so that these facilities can operate safely. Nuclear waste disposal facilities are typically simple, passive, long-lived (hundreds to hundreds of thousands of years) facilities, whose failure could result in the slow and sustained release of radionuclides well into the future. Geology, besides providing a stable environment for the engineered components of the facility, can be part of the disposal facility itself and play a critical role in the containment and isolation of the nuclear waste.

The approach taken here is to present some specific examples of geological issues at different facilities in the United States and to draw lessons from the examples. The discussion of nuclear power plants concentrates on seismic issues. Earthquakes have been particularly controversial because of the ability of fault rupture and vibratory ground motion (shaking) to threaten the sensitive pressurized systems in nuclear power plants. The discussion of nuclear waste repositories covers a broader range of issues emphasizing hydrogeology because groundwater is the primary means by which nuclear waste packages can be breached and radioactive waste transported to the environment.

20.1 Choosing a suitable site

The purpose of the siting process is to pick a suitable site, and geology has an important role in determining how "suitable" a site may be (McEwen and Andersson, Chapter 23, this volume). The geological attributes of suitability can include: the ability of the site to fulfill its assigned functions (stable environment and/or containment and isolation); ease of characterization (not too heterogeneous, key measurements easily made); ease of understanding (no need to rely on overly complex and uncertain models); and the absence of possible adverse disruptive factors, such as significant seismic or volcanic activity. No site is perfect, and trade-offs exist between these technical factors and between technical and non-technical issues, such as the willingness of the local population to host the facility. The discussion in this area focuses on two sites where serious technical concerns existed.

In the early 1980s, the US government enacted legislation encouraging states to form compacts and provide regional disposal facilities for commercial low-level radioactive waste. Such waste typically consists of metal components, resins, rags and protective clothing that have been exposed to radioactivity or contaminated with radioactive material. (The much more potent high-level waste discussed below consists of the spent fuel from nuclear power reactors and the byproducts of reprocessing the spent fuel.) According to a 1999 report of the US General Accounting Office (GAO, 1999; now renamed as the Government Accountability Office), almost $600 million had been spent over the prior 18 years by compacts or individual states on unsuccessful efforts to site and develop ten low-level waste disposal facilities. One such site, near the town of Martinsville, Illinois, was selected for detailed investigation in 1988. In 1992, the governor of Illinois appointed a Siting Commission to review the work done and provide its recommendations on the site. After 72 days of hearings, the Siting Commission found that the proposed Martinsville site was not suitable for the disposal of low-level radioactive waste (see NWTRB, 1995). Although the site "... could meet some of the statutory criteria with respect to the proposed design, it did not present a suitable geologic and hydrologic medium and it was not located so as to minimize radioactive releases into ground waters utilized as public water." According to the Siting Commission, "The underlying site geology and hydrology – how thick and tight is the Vandalia Till [a presumably low-permeability feature that was supposed to protect local aquifers directly beneath the site] and how long contaminated water would take to travel through to the Martinsville public wells – was never adequately explained."

The theme of too much uncertainty is a common thread in almost all disputes involving geological issues at nuclear facilities. In fact, as discussed later, depending on their background and motivation, different groups have different definitions of what constitutes an acceptable level of uncertainty. In the case of Martinsville, the Siting Commission found the uncertainty to be too large. Particularly damning was their conclusion that the site "... is entirely too small to provide an adequate buffer zone, given the hydrogeologic nature of the site . . . [and] . . . is virtually enveloped in water. The facility site itself has a nearby creek on one side and a stream bordering another. The [site] is upstream of the City [Martinsville] – precisely the wrong side of a populated area on which to locate a disposal facility." Proponents of the facility argued that the Siting Commission had gone beyond its mandate and had invoked its own criteria for the site and for how much uncertainty was acceptable. According to the Siting Commission, the site location was chosen to obtain local approval – the municipality wanted the facility, but the regional government did not. In the end, some $60 million had been spent on studies for a site that was, at best, marginally suitable as the location of a low-level radioactive waste facility.

An interesting example of how developing geological knowledge has changed perspectives on the suitability of a site is that of the Humboldt Bay nuclear power plant (collocated with two gas-fired non-nuclear units) on the northwest coast of California. Commissioned in 1963, the nuclear power plant operated without incident until 1975, when ground motions at the plant site from a nearby moderate earthquake slightly exceeded the original seismic design specifications (0.25 g peak ground acceleration). There was no damage to the plant. The plant subsequently was shut down for refueling and an evaluation of possible seismic retrofitting. Based on this evaluation the owner, in concurrence with the US Nuclear Regulatory Commission (NRC), decided to upgrade the seismic capacity of the plant to 0.5 g peak ground acceleration. In the meantime, local geologists were concerned that nearby faults appeared to be active. The plant operator decided not to restart the plant and a subsequent investigation showed that a nearby fault (within 1 km of the plant site) was indeed active. The plant operator decided that, although technically possible, it would be too costly to resolve the faulting issue and retrofit the plant. The plant was put into a safe condition (SAFSTORE) to await eventual decommissioning at an undetermined date. In 1980, a M 7.2 earthquake occurred 50 km offshore from the plant. The peak ground acceleration recorded at the site was 0.495 g. Although a nearby highway overpass did collapse, there was no structural damage at the plant site. In 1992, a M 7.1 earthquake occurred some 55 km south of the plant on the proposed Cascadia subduction zone, a major tectonic plate boundary. The peak ground acceleration recorded at the plant site was 0.22 g. Although some water did splash out of the pond holding spent nuclear fuel, there was no structural damage at the plant site. Analysis of the earthquake verified that the Cascadia subduction zone was indeed active and that an earthquake on this subduction zone could trigger ruptures on the faults near the plant. The plant operator decided to accelerate the decommissioning program. In 1994, a small M 5.4 earthquake some 8 km west of the plant gave rise to ground motions at the site that were even slightly larger (0.55 g) than those envisioned in the 1976 retrofit design. The earthquake was strongly felt at the plant site, but there was only minimal damage to the still-operating gas-fired units.

Recent studies have shown that the subduction zone fault ruptured during a truly great earthquake in 1700 associated with a devastating tsunami along the entire northern California–Oregon–Washington–British Columbia coast (see, for example, Penrose Conference, 2000). The Humboldt Bay nuclear power plant has now been partly dismantled. Spent fuel still remains in storage at the plant. However, because the fuel cannot practicably be moved to a completely different locality, efforts are under way to ensure that it is stored safely on-site in a dry-cask storage facility. To avoid future tsunami flooding, the dry-cask facility will be located in an underground bunker at a higher elevation on the same property. The operator has designed this facility (1.4 g peak ground acceleration, i.e. more than five times higher than the original plant design in 1963) to withstand the shaking on the hanging wall of a major thrust fault that lies directly beneath the site. This is the kind of seismo-tectonic setting that the 1999 Taiwan earthquake and other recent earthquakes have demonstrated to be among the most vulnerable to very intense earthquake shaking.

In retrospect, it is easy to say that the Humboldt Bay site was a poor location for a nuclear power plant. However the owner had no seismic siting guidance in the years prior to the issuance of Appendix A to Part 100 of Title I0 of the Code of Federal Regulations (Seismic and Geologic Siting Criteria for Nuclear Power Plants). But even had the local faults been recognized at the time of initial planning, the threat they posed would not have been evident. It would be decades before the seismic capability of the Cascadia subduction zone, the relationship of the local faults to the Cascadia subduction zone, and the nature of near-field strong ground shaking, particularly that associated with the hanging walls of thrust faults, would become clearly understood. It is the author's opinion that geological and seismological understandings of seismic hazards will continue to change and evolve in years to come, and that many existing nuclear facilities worldwide may have to undergo challenging safety reappraisals.

20.2 Estimating uncertainty

In a 1994 presentation to the US Nuclear Waste Technical Review Board (NWTRB), Wendell Weart, chief scientist for the Waste Isolation Pilot Plant (a repository for US defense-related transuranic radioactive waste in New Mexico), made an observation that is particularly applicable to geological investigations at many nuclear sites. He stated, "One is most confident of site and repository issues at the *beginning* [emphasis added] of detailed investigations." This has been particularly true for assumptions about hydrogeology at proposed nuclear waste repositories around the world. At Yucca Mountain, the original assumption was that the principal barrier to transport of harmful radionuclides from the repository would be provided by an underlying non-welded volcanic tuff, known as the Calico Hills formation. This optimism was based on the assumption of very slow flow of groundwater through the rock matrix of this formation and the presence of radionuclide transport-retarding zeolites within the rock matrix. Subsequent investigations revealed that much of this rock was fractured, allowing both fast flow and bypassing of many of the zeolites.

The experience at a number of nuclear sites shows an initial expectation that, as detailed investigations proceeded, uncertainties would be progressively reduced to the point where they were no longer significant. In practice, however, as site investigations progress, the amount of uncertainty about the site characteristics often appears to increase. In reality, what can occur is the gradual realization that the perceived uncertainty at the beginning of a site investigation is much less than the actual uncertainty. Eventually, if investigations are sufficiently thorough, the perceived uncertainty begins to match the actual uncertainty and a better assessment of the site's true properties can be made, along with the development of a firmer basis for assessing the site's ability to perform its assigned functions. This misperception can contribute to the frustration of management and legislative bodies with overly optimistic schedules and the apparent decrease in confidence associated with ongoing investigations at some sites.

20.3 Model uncertainty: the most difficult challenge

Although consensus exists on many important theories, such as plate tectonics, geologists are not united in their support for many of the conceptual models used to explain different phenomena. In the siting of nuclear facilities, disagreements among geologists over which is the "correct" model often surface as important issues. Although some models may be tested, others often require observations or tests that last long periods of time or involve parameters that are not readily measurable. If the uncertainty is such that one of the reasonably plausible models could result in a high probability of unacceptable consequences, the project may be doomed. If, on the other hand, such consequences are very unlikely, or if the models are constrained by the data in such a manner that the consequences are relatively bounded, model uncertainty may not be that significant.

For nuclear power plants, a source of great uncertainty was the occurrence of a large (M 7+) earthquake in Charleston, South Carolina, in 1886. Many nuclear power plants were sited under the assumption that such an earthquake could not occur elsewhere along the eastern seaboard of the United States. Repeated efforts to define the source of the earthquake resulted in a string of competing hypotheses whose lifetimes varied from months to years. This unresolved uncertainty caused some to question seismic safety at tens of operating nuclear power plants in the eastern United States. In the mid 1980s, emphasis shifted from finding the causative fault to finding other evidence of large earthquakes in the region. A detailed evaluation of earthquake-induced liquefaction (sand blows) identified five large, pre-1886 earthquakes that occurred in the Charleston region during the last several thousands of years (Obermeier *et al.*, 1987). This study also showed that the number and size of the sand blows decreased with increasing distance from Charleston. Although this study did not help in identifying the specific source of the 1886 earthquake, it did constrain the resulting seismic hazard analysis and limit the likelihood that locations away from Charleston would be affected by large earthquakes.

In many cases, the impact of conceptual model uncertainty can be quite significant. A particularly striking example of this can be found in National Research Council (2001). The report cites studies at the Idaho National Engineering and Environmental Laboratory conducted between 1965 and 1998. Over that time, as the understanding of unsaturated fracture flow and transport in volcanic rocks evolved, the estimated groundwater travel time from the surface through a 200 m-thick unsaturated zone to the water table has decreased by more than three orders of magnitude; from $\sim$ 50 ka to $\sim$ 20 a. In hydrogeology, surprises associated with conceptual models are so routine and well documented that they should be expected. As defined by Bredehoeft (2005), surprise comes in the form of new information that challenges the prevailing conceptual model and entails a complete paradigm shift. Overall, Bredehoeft (2005) found that surprises occurred in 20–30% of the 29 hydrogeological modeling studies he reviewed.

The estimation of volcanic consequences at the proposed high-level radioactive waste repository at Yucca Mountain, Nevada, provides an example of how poorly constrained model uncertainty can cause large fluctuations in risk estimates. Although there have been some variations over the years, the estimated mean probability of a volcanic event affecting the repository has remained relatively stable at about 10^{-8}–10^{-7} a^{-1}. There have been, however, remarkable changes in the estimates of the consequences of such an event. In 1991, analyses used through most of the 1990s by the Department of Energy's (DOE's) performance assessment (PA) contractors showed that even if one assumed the conditional case, i.e. that a volcanic eruption through the repository to the surface would occur (not taking into account its low probability), the repository could still meet the then current regulatory criteria (Barnard *et al.*, 1992). The risk analysts optimistically claimed that "... the models used are quite conservative, so any reevaluation of the parameters would adjust the results downward." In the past few years, dramatically different and more energetic models of volcanism–repository interaction have emerged. Since 2000, volcanism has dominated or been an important contributor to risk estimates of the proposed Yucca Mountain repository during the 10 ka regulatory period, even when the consequences are weighted by the very low (10^{-8}–10^{-7}) annual probability of occurrence. These estimates have been highly volatile (Reiter, 2003). In contrast to estimates in the 1990s, the 2002 PA results showed the conditional, non-probability-weighted dose from a volcanic eruption through the repository to the surface to be about 10 000 times greater than current regulatory criteria. Electric Power Research Institute studies (EPRI, 2004), on the other hand, maintain that a reasonable estimate of the conditional, non-probability-weighted dose from a volcanic eruption through Yucca Mountain is zero.

20.4 Dealing with uncertainties and surprises

Given that surprises do occur during geological investigations and that uncertainties exist and may even increase, developing appropriate coping strategies is important. Probabilistic analyses have often proven to be a particularly powerful means of dealing with uncertainty and surprises. Deterministic analyses, which rely on single scenarios and sets of parameters (e.g. in seismic hazard), do not reflect the uncertainty associated with that scenario or set of parameters, nor do they take into account the possibility of other less likely assumptions. The deterministic analysis could be based on a "worst case" scenario. Unfortunately, relying on the worst case when it is associated with a very low probability of being correct could, in many cases, lead to an "unwarranted" rejection of the site or increase in the cost of the facility. Judgment of what is a warranted or an unwarranted action is, of course, a societal decision, often referenced to an "acceptable level of risk," a criterion most easily used when explicitly defined in regulations. Uncertainty associated with the postulated locations where M 7+ earthquakes could occur in the future and possible concerns with the safety of existing nuclear power plants along the eastern seaboard of the United States (discussed above) led to a particularly powerful use of probabilistic seismic hazard analyses (PSHA). These analyses (Bernreuter *et al.*, 1989; EPRI, 1986) were able to take into account the different tectonic hypotheses and developing data sets, such as those using sand blows described above. The calculated probabilities of exceeding the seismic designs of existing nuclear power plants provided a rational basis for decision making. The analysis also led to the increased acceptance of PSHA in subsequent decisions by the US Nuclear Regulatory Commission.

However, the use of probabilistic analysis does not make all problems go away. Although probabilistic analyses can cast a broader net and take into account multiple hypotheses, there is always the possibility that such analyses have not captured the correct model or even given it sufficient weight. As stated by Oreskes and Belitz (2001), "Almost by definition, conceptual error cannot be quantified. We don't know what we don't know, and we can't measure errors that we don't know we've made." Probabilistic analysis can also force one to confront issues that are not otherwise considered. For example, one of the most extensive probabilistic seismic hazard analyses ever undertaken was carried out for both vibratory ground motion and fault displacement hazard at the proposed Yucca Mountain repository (Stepp *et al.*, 2001). Because of the proposed repository's very long operating life, the Yucca Mountain PSHA required extrapolation to very low probabilities ($< 10^{-6}\,a^{-1}$) and resulted in extremely high and, perhaps, unphysical ground motion (Reiter, 2004). Similar concerns were raised with respect to the PEGASOS seismic hazard project for nuclear power plants in Switzerland (Abrahamson, 2002). Efforts were, and are, being made to define upper bounds on earthquake ground motions. Different approaches ranging from looking for geologic evidence of very high ground motion in the deformation of local rocks to advanced numerical modeling techniques are currently being pursued to this end. An analysis and discussion of this issue can be found in Bommer *et al.* (2004). There is a difference between intuitively regarding a phenomenon as unphysical and demonstrating that it indeed is.

20.5 Sociopolitical considerations may override scientific ones

Sociopolitical considerations have been factors in siting and operating decisions for nuclear facilities. The significance of such considerations is particularly obvious in attempts to site low-level radioactive waste disposal facilities. The previous example of Martinsville shows how a technical evaluation by a state-appointed commission rejected a site primarily because of its hydrogeological characteristics. Following are two examples in which state-appointed reviews reached conclusions that accepted a site but were rejected or questioned at a higher state or federal level. Sociopolitical concerns may have played important, if not overriding, roles.

After a screening process that looked at a number of locations, the Southwestern Compact decided to site its low-level radioactive waste disposal facility in Ward Valley in the

Mojave Desert in California. Following site characterization and preparation of an application to build and operate the facility, the regulatory body responsible for approving the site, the California Department of Health Services, reviewed the application and granted approval in 1993 for the construction and operation of the facility. Because the site was on federal land, the US Department of Interior (DOI) had to transfer the land to the state of California. At this time, in an unofficial capacity, three scientists at the US Geological Survey (USGS) identified seven technical concerns having to do with the hydrological, geological, biological and engineering aspects of the site that they felt had not been adequately addressed in the application. The DOI referred the issue to the National Academy of Sciences' National Research Council (NAS). A 17-person panel was convened and in 1995 reached its conclusions (National Research Council, 1995). The panel dismissed many of the issues, and 15 of its 17 members felt that any remaining uncertainty (mostly hydrogeological) raised by the three USGS scientists could be addressed during the development and early operation of the site, and that work on the site could continue. The primary issue was the ability of radionuclides to migrate downward and eventually reach water resources. The NAS panel viewed this possibility as highly unlikely. Among other considerations, the panel pointed out that even if all of the plutonium in the proposed repository were to reach the Colorado River (a primary source of drinking water) over the life of the site, the level would be significantly lower than the levels of plutonium already in the river or permitted by accepted health-based standards. Environmental and local groups, along with some members of the California congressional delegation, argued against proceeding with site development. Eventually the DOI decided to hold up the land transfer until the uncertainties could be resolved. The 1999 GAO report (GAO, 1999) implies that this decision was due more to politics than to science. After much delay, a newly elected governor decided to try pursuing alternative ways of disposing of California's low-level radioactive waste.

At about the same time, efforts were under way to site a low-level radioactive waste disposal facility for the Texas Compact near the town of Sierra Blanca in west Texas. Site characterization was concluded, and an application was submitted to the appropriate regulatory body, the Texas Natural Resource Conservation Commission (TNRCC). The TNRCC technical staff conducted a review and issued a draft license approving the site. Hearings were held before administrative law judges (ALJs) in early 1998. The ALJs recommended that the license be denied, primarily because of concerns about a possible fault beneath the site that could rupture and cause earthquakes. In a detailed response (TNRCC, 1998), the technical reviewers pointed out that the supposed fault was overlain by almost 200 m of undeformed sediment, formed at least 780 ka ago. The ALJs based their rejection of the license application on the fact that very small earthquake-related deformations could not be detected in the overlying sediment and that there was not enough information on fault properties and possible connections to regional faults. The TNRCC technical reviewers argued that work sufficient to rule out such regional connections had been carried out and that the existing seismic design of the facility was already conservatively set at a very high ground acceleration level (0.7 g), based on the possibility that a M 6 earthquake could occur directly beneath the site and that a M 7 earthquake could occur 10 km away.

Coincidentally, this is almost the same scenario and design level used for the San Onofre nuclear power plant located in considerably more seismically active Southern California. The TNRCC technical reviewers argued further that even in the highly unlikely event of an earthquake-induced failure of the facility, analysis showed that the resultant dose would be significantly below regulatory limits. In October 1998, the TNRCC commissioners, citing the earthquake concerns, denied the license request. The decision to override the technical review has been largely attributed to opposition from Mexico (25 km from the site) and to political considerations before an upcoming statewide election (e.g. see Moore, 1998). Texas officials are currently considering a proposed low-level radioactive waste disposal site near the Texas–New Mexico Border, further away from Mexico.

Is it wrong to have political oversight of nuclear facility siting? Of course not! Science can characterize a site and estimate its risk to the public, but accepting this risk is a social and political decision, hopefully informed by relevant science. The problem with Ward Valley and Sierra Blanca is that, apparently, scientific concerns were used to hide political considerations. This doesn't have to be so. For example, in the 1998 environmental review of a proposed disposal concept for high-level nuclear waste in Canada, the review panel argued against approval of the proposed concept. The panel stated that although the case for the proposed design had been made technically, the public was not convinced (Canadian Environmental Assessment Agency, 1998).

Concluding remarks

Geology has played an important role in many siting and operating decisions for civilian nuclear facilities in the United States. Examples of geological issues at proposed low-level and high-level nuclear waste repository sites and operating nuclear power plants include seismic hazard in regions of high and low seismicity, volcanism and groundwater flow. Insights gained from past efforts emphasize the significance of choosing a suitable site, recognizing and dealing with uncertainty and avoiding undue complexity. A frequently cited concern has been the definition of acceptable levels of uncertainty and risk. Acceptable levels of uncertainty and risk are more often determined by unspecified mixes of social, political and technical factors than by technical considerations alone.

Further reading

Most of the basic information collected and methodologies used to license proposed nuclear facilities can be found, or are referenced, in Safety Analysis Reports issued by license applicants for each facility. The US Nuclear Regulatory Commission's evaluation of this material and its own supporting analyses can be found, or are referenced, in Safety Evaluation Reports. The general scientific literature does contain references to plant-specific investigations and methodologies.

Aside from this volume, general books on geological issues in siting nuclear facilities include Macfarlane and Ewing (2006), which discusses geologic issues at Yucca Mountain including volcanism and groundwater flow and transport. Alexander and McKinley (2007) provide a review of siting issues for repositories, including those outside of the USA. Reiter (1990) is a general book on earthquake hazard analysis that contains many examples related to nuclear power plants up to that time. The emphasis is on estimates of vibratory ground motion. An earlier book, Meehan (1984), provides an entertaining account of problems associated with estimating fault rupture at a number of proposed and actual nuclear facilities in California.

Acknowledgments

I would like to thank Clarence Allen who co-authored an earlier version of the material in this chapter and Stephen Etter for providing information on Sierra Blanca. This chapter benefited from very helpful comments and suggestions provided by Lloyd Cluff and David Diodato. The views presented are my own and not necessarily those of the reviewers or my former employers, the US Nuclear Regulatory Commission and the US Nuclear Waste Technical Review Board.

References

Abrahamson, N. A., P. Birkhauser, M. Koller *et al.* (2002). PEGASOS – a comprehensive probabilistic seismic safety hazard assessment for nuclear power plants in Switzerland. Presented at the 12th European Conference on Earthquake Engineering, September 9–13, London, UK, paper 633.

Alexander, W. R. and L. E. McKinley (2007). *Deep Geological Disposal of Radioactive Waste*, Radioactivity in the environment series, 9. Amsterdam: Elsevier Science.

Barnard, R. W., M. L. Wilson, H. A. Dockery *et al.* (1992). TSPA 1991: an initial total-system performance assessment for Yucca Mountain, Technical Report SAND91-2795. Albuquerque, NM: Sandia National Laboratories.

Bernreuter, D. L., J. B. Savy, R. W. Mensing and J. C. Chen (1989). Seismic hazard characterization of 69 nuclear power plant sites east of the Rocky Mountains, NUREG/5250. Washinton, DC: US Nuclear Regulatory Commission.

Bommer, J. J., N. A. Abrahamson, F. O. Strasser *et al.* (2004). The challenge of defining the upper limits on earthquake ground motion. *Seismological Research Letters*, **70**, 82–95.

Bredehoeft, J. (2005). The conceptualization model problem – surprise. *Hydrogeology Journal*, **13**, 37–46.

Canadian Environmental Assessment Agency (1998). Nuclear fuel waste management and disposal concept – report of the nuclear fuel waste management and disposal concept. Ottawa: Minister of Public Works and Services Canada.

EPRI (1986). Seismic hazard methodology for the central and eastern United States, EPRI Report NP-4726. Palo Alto, CA: Electric Power Research Institute.

EPRI (2004). Potential igneous processes relevant to the Yucca Mountain repository: extrusive release scenario, analysis and implications, EPRI Report 1008169. Palo Alto, CA: Electric Power Research Institute.

GAO (1999). Low level radioactive wastes: states are not developing disposal facilities, GAO/RCED 99-238. Washington, DC: US General Accounting Office.

Macfarlane, A. M. and R. C. Ewing (eds.) (2006). *Uncertainty Underground: Yucca Mountain and the Nation's High-level Nuclear Waste*. Cambridge, MA: MIT Press.

Meehan, R. L. (1984). *The Atom and the Fault*. Cambridge, MA: MIT Press.

Moore, R. M. (1998). Signs indicate license for nuclear dump to be denied. *El-Paso Times*, October 18.

National Research Council (1995). *Ward Valley: An Examination of Seven Issues in Earth Sciences and Ecology*. Washington, DC: National Academy Press.

National Research Council (2001). *Disposition of High-Level Waste and Spent Fuel: The Continuing Societal and Technical Challenges*. Washington, DC: National Academy Press.

NWTRB (1995). Report to the US Congress and the Secretary of Energy: 1994 findings and recommendations. Arlington, VA: US Nuclear Waste Technical Review Board.

Obermeier, S. F., B. E. Weems and R. B. Jacobson (1987). Earthquake-induced liquefaction features in the coastal South Carolina region. In: Jacob, K. H. (ed.) Proceedings from the Symposium on Seismic Hazards, Ground Motions, Soil-liquefaction and Engineering Practice in Eastern North America, Technical Report NCEER-87-0025. Buffalo, NY: MCEER Publications, 480–493.

Oreskes, N. K. and K. Belitz (2001). Philosophical issues in model assessment. In: Anderson, M. G. and P. D. Bates (eds.) *Model Validation Perspectives in Hydrological Sciences*. London: John Wiley and Sons, 23–41.

Penrose Conference (2000). Great Cascadia earthquake tricentennial, program summary and abstracts, Oregon Department of Geology and Mineral Industries, Special Paper 33, Geological Survey of Canada Open File Report 3938.

Reiter, L. (1990). *Earthquake Hazard Analysis: Issues and Insights*. New York, NY: Columbia University Press.

Reiter, L. (2003). A low probability–high consequence event. Presented at IUGG XXIII General Assembly, State of the Planet: Frontiers and Challenges, June 30–July 11, Sapporo, Japan.

Reiter, L. (2004). When are ground motion estimates too high? *Seismological Research Letters*, **74**, 282.

Stepp, J. C., I. Wong, J. Whitney *et al.* (2001). Probabilistic seismic hazard analyses for ground motions and fault displacements at Yucca Mountain, Nevada. *Earthquake Spectra*, **17**, 113–151.

TNRCC (1998). Application by Texas low level radioactive waste disposal authority for license to operate and dispose of radioactive waste for license RW-3100: Executive director's exceptions to the proposal for decision, Docket 96-1206-RAW. Austin, TX: Texas Natural Resource Conservation Commission.

21

Characterizing active tectonic structures for nuclear facilities in Japan

D. Inoue

The first commercial nuclear power plant (NPP) in Japan began operation in 1965. Since then 55 NPPs have been located and constructed in Japan, producing approximately 303 billion Kwh (Japan Nuclear Energy Safety Organization, 2007), one third of the total power generation for the entire nation.

The seismic design of operating NPPs in Japan is guided by regulatory documents. Basic design guidance is described in NSC (2006) and specific guidance for geological and geophysical surveys and seismic designs was provided by the Committee on Examination of Reactor Safety (1978).

Since 1966, several large-magnitude earthquakes have occurred in Japan (Figure 21.1, Table 21.1). While these earthquakes have not exerted any adverse influence on nuclear

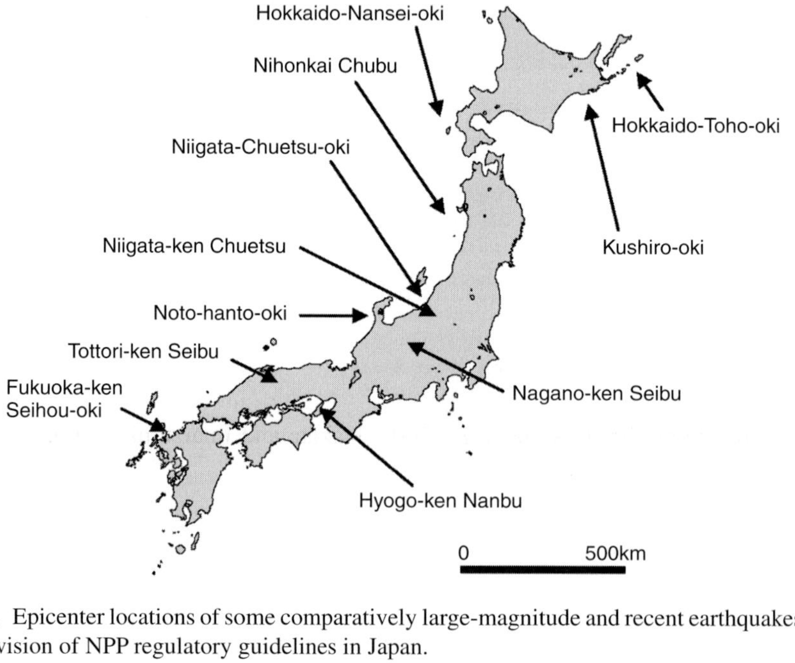

Fig. 21.1 Epicenter locations of some comparatively large-magnitude and recent earthquakes that led to the revision of NPP regulatory guidelines in Japan.

Table 21.1. Important recent earthquakes that have led to the revision of NPP regulation guidelines

Earthquake	Year	Mj	Description
Nihonkai Chubu	1983	7.7	Second largest earthquake since 1964 on the Japan Sea-side
Nagano-ken Seibu	1984	6.8	No surface fault
Kushiro-oki	1993	7.5	Deep (101 km) intraplate earthquake
Hokkaido-Nansei-oki	1993	7.8	Japan Sea-side with tsunami
Hokkaido-Toho-oki	1994	8.2	Largest plate-boundary earthquake in Japan in recent years
Hyogo-ken Nanbu (Kobe earthquake)	1995	7.3	Inland earthquake with more than 6300 casualties
Tottori-ken Seibu	2000	7.3	Inland earthquake with no surface fault shown on the standard active fault map (Inoue *et al.*, 2002)
Niigata-ken Chuetsu	2004	6.8	Inland earthquake in active folding area
Fukuoka-ken Seihou-oki	2005	7.0	Offshore earthquake without known surface faults
Noto-hanto-oki	2007	6.9	Offshore earthquake
Niigata-Chuetsu-oki	2007	6.8	Offshore earthquake in active folding area. Resulted in automatic shutdown and slight damage of Kashiwazaki-Kariwa Nuclear Power Plant

power facilities (one exception is the 2007 Niigata-Chuetsu-oki earthquake that resulted in automatic shutdown and slight damage at the Kashiwazaki-Kariwa NPP), valuable knowledge has been gained about patterns of seismic activity, properties of earthquake motion, and earthquake-resistant structures from their study.

Earthquake research has made remarkable progress in Japan during this forty-year time period. A coordinated, national effort has resulted in the production of a variety of active fault maps (e.g. Research Group for Active Faults of Japan, 1991; Nakata *et al.*, 1996; Okada and Imaizumi, 2000; Ikeda *et al.*, 2002; Nakata and Imaizumi, 2002). In addition to these studies, maps that provide geological and topographical information essential for active fault study have been published (e.g. neotectonics maps by the Geological Survey of Japan, Quaternary map of Japan by JAQUA, 1987; Machida and Arai, 1992; Koike and Machida, 2001). Furthermore, the 1995 Hyogo-ken Nanbu earthquake led the Japanese government to establish the Headquarters for Earthquake Research Promotion, to facilitate a series of studies on active faults, earthquake motion and prediction of earthquake damage.

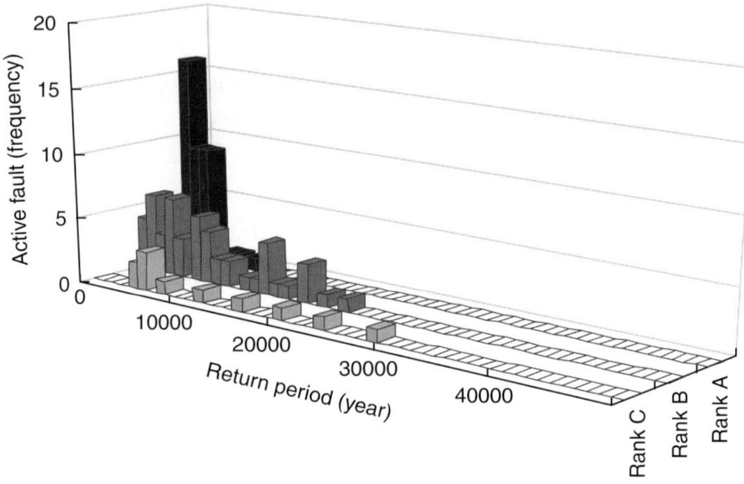

Fig. 21.2 A compilation of the return periods of 98 long active fault segments, in Japan, each studied by trench excavation. The longest return period among them is $\sim$ 30 ka. Faults are classified by slip rate, with ranks A, B and C corresponding to $> 1\,\mathrm{m\,ka^{-1}}$, $10\,\mathrm{cm}-1\,\mathrm{m\,ka^{-1}}$ and $< 10\,\mathrm{cm\,ka^{-1}}$, respectively.

As a result, ≈ 100 long fault zones with high rates of activity were chosen for additional detailed survey and research (Figure 21.2).

21.1 Revision process of regulatory guidelines

Based on the national reassessment of active faults, the Nuclear Safety Commission of Japan issued "*Order to revise the regulatory guideline by nuclear safety commission to the Special Committee on Nuclear Safety Standards and Guides*" in June 2001, and the "Formation of Subcommittee on Regulatory Guidelines for Aseismic Design" in July 2001, to initiate discussions for the revision of the "*Regulatory guide for seismic design of nuclear power reactor facilities.*" Subsequently, the final draft by the subcommittee was issued on April 28, 2006. The Nuclear Safety Commission invited the public to comment on the draft of the revised regulatory guide (hereafter referred to as the new regulatory guide) between May–June 2006, and enacted the new regulatory guide after a partial revision in September 2006 (NSC, 2006). The new regulatory guide will be applied to new NPPs, and will be used in the reevaluation of existing NPPs.

21.2 Major revisions to the regulatory guide for seismic design

The new regulatory guide acknowledges the existence of *residual risk* but demands that risk from seismic hazards be as small as rationally possible. The most notable point about the present revision is that the new regulatory guide accepts the concept of *residual risk*. In this context, *residual risk* means the risk that nuclear power facilities would be seriously damaged by earthquake motion exceeding the standard ground motion determined for seismic design, and that people near the nuclear power facility would be exposed to radiation as a result of this damage. In contrast, the former regulatory guide required nuclear facilities to be sufficiently earthquake-resistant to prevent "any" presumable earthquake motion from triggering serious damage. However, discussions during the revision process have led to a general consensus that the possibility of future earthquake motion exceeding the standard seismic design requirements cannot be denied from a seismological point of view.

The new regulatory guide defines the basic principle as, "The safety function of important nuclear facilities should be designed to remain undamaged by earthquake motion and the resultant seismic force, which are reasonably presumed not only to occur even at extremely rare intervals during the operating period of the facilities but also to inflict serious damage to the facilities from the seismological and earthquake-engineering viewpoint, based on the geology, geological structure and seismicity of the area." Although the basic policy of the new regulatory guide is equivalent to the former one, it assumes that basic earthquake ground motion can be "reasonably presumed" conceivably to exceed the standard seismic design requirements from the seismological and earthquake-engineering point of view, and requires that provisions be made for this possibility.

In addition to this philosophical change, the new guidelines provide specific suggestions about how seismic hazards should be assessed for NPPs. For example, the guidelines indicate that earthquake motion should be examined on the basis of fault models. The phrase, "the earthquake motion defined by specifying the hypocenters for the individual sites," in the new regulatory guide, means that several earthquakes expected to inflict serious damage to the site are selected and each of them is assessed in terms of earthquake motion on the basis of the response spectra and fault models. These results are used to determine the basic earthquake ground motion (Ss). Thus, the new regulatory guide supports the earthquake motion assessment based on fault models more definitively than the old guide. Furthermore, the new guide recommends applying an earthquake assessment method using fault models to assess the potential effects of earthquakes located close to the site (Figure 21.3).

Under the previous guidelines, active fault assessment was used in determining types of basic earthquake ground motion. Two types of basic earthquake ground motion were assumed to exist, S1 and S2. The S1 earthquakes include historical earthquakes and earthquakes associated with active faults that moved in the last 10 ka or faults that are classified by a slip rate; S2 earthquakes were associated with active faults that moved in the last 50 ka, seismo of $> 1\ \mathrm{m\,ka^{-1}}$. The tectonic structures, and near-field earthquakes (Mj = 6.5) presumed to occur beneath the NPP site. The new regulatory guide has adopted a new definition for basic earthquake ground motion, Ss, by unifying and enhancing the defining principles of S1 and S2 earthquakes. Two cases are identified: (i) "earthquake motion determined by specifying a hypocenter for each site," and (ii) "earthquake motion determined without specifying a hypocenter" (Figure 21.3).

In the first case, basic earthquake ground motion is determined for an NPP site by taking into account the properties of active faults, the occurrence of historical earthquakes and the

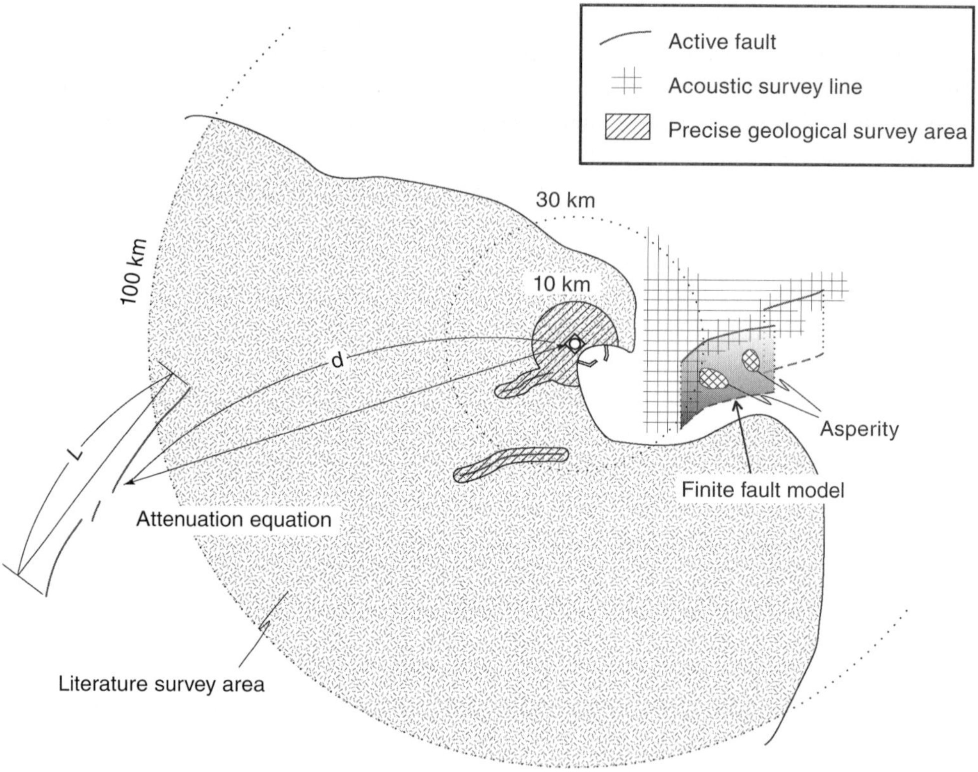

Fig. 21.3 Schematic image of an active fault investigation of an NPP in Japan. The area within 10 km of the NPP, both onshore and offshore, is subject to detailed investigations by field geological and geomorphological study, including geophysical exploration and trenching. In order not to overlook some active faults, the area within 30 km is further investigated by literature study and supplemented by additional investigations. The strong-motion on NPPs is calculated using two methods. First, by strong-motion evaluation based on the attenuation equation using the length, L, of the fault and the distance, d, between the fault center and the NPP. Second, by strong-motion evaluation based on finite fault models using parameters such as fault dimension and asperity. See color plate section.

types of earthquakes categorized by their mode of occurrence (e.g. inland crustal earthquake, interplate earthquake, oceanic intraplate earthquake). Concerning the active faults assessed for S2 earthquakes, the old regulatory guide states that "their activity should be 50 ka or later, or their recurrence period should be less than 50 ka." Regarding active faults triggering inland crustal earthquakes, the new regulatory guide states that "their activity in and after the Late Pleistocene should be undeniable, and their activation in the final interglacial epoch (80–135 ka); could be used for this identification" (Imbrie *et al.*, 1984; Chen *et al.*, 1995). In other words, the new guide expands the time range since slip last occurred on faults in order for these faults to be considered active. Specifically, faults that have experienced slip during or after marine isotope stage (MIS) 5 (80–125 ka) are now regarded as active (Figure 21.2).

In the second case, basic earthquake ground motion is determined for an NPP site without specifying a hypocenter for each site. In the old regulatory guide, this was represented by local earthquakes (Mj = 6.5) evenly applied to the whole country as a measure of conservatism. The new regulatory guide states "a response spectrum should be established by taking account of ground properties of the site and the collection of near-field observation records of inland crustal earthquakes for which it is difficult to relate hypocenters and active faults." The earthquake motion is expected to be determined by collecting near-field observation records of inland crustal earthquakes for which it is difficult to relate hypocenters with active faults, establishing a response spectrum that takes ground properties of the site into consideration, and by accounting for properties of the earthquake motion.

The new regulatory guide requires establishing the earthquake ground motion for elastic design, Sd, on the basis of Ss. This is done to verify that site facilities as a whole remain in the approximate range of elasticity. Establishment of Sd is expected to bear part of the role that the basic earthquake ground motion, S1, played in seismic design under the old regulatory guide.

The old regulatory guide required defining the static intensity of vertical seismic force as one half of the maximum acceleration amplitude of horizontal earthquake motion. In contrast, the new regulatory guide states that Ss should be determined as horizontal and vertical earthquake motion of the free bedrock surface for both cases of determined motion. Furthermore, the new guidelines request that the seismic force be "computed by properly combining the horizontal and vertical components using the basic earthquake ground motion, Ss."

Finally, the new guidelines introduce stochastic assessment into safety control. Concerning "earthquake motion determined by specifying a hypocenter for each site" and "earthquake motion determined without specifying a hypocenter for each site," it is desirable to estimate an approximate probability of exceedance suited to the response spectrum for each type of earthquake motion. Accordingly, the new regulatory guide requires referring to the probability of exceedance corresponding to the respective types of "earthquake motion in the safety examination."

21.3 An active fault survey for a nuclear power plant

The development of new regulatory guidelines for seismic hazard evaluation in Japan means that these regulations must be followed in practice. This section describes the work flow for assessing seismic hazards at NPPs, with particular emphasis on the identification and evaluation of active faults. Particularly for the evaluation of seismic hazard by active faults, it is very important to develop a comprehensive geoscientific model, considering not only results of topographic surveys on the surface, but also the results of geological and geophysical surveys beneath the surface to estimate the dimension and activity of the source faults.

21.3.1 Literature survey

When a potential site is initially selected for a possible NPP, a literature survey is conducted to check whether there is any description of active faults or the like in the area. All faults and lineaments with records or activity in the Quaternary are identified, including short lineaments that may not be geologically significant. These Quaternary faults and lineaments are described in terms of their location, trend, length, width, displacement, sense of slip, properties of lineament and nature of their activity. Based on this initial assessment, each fault is categorized, as described in detail in the following. The definition of the region itself depends on the length and scale of faults identified in the literature survey. In cases where an active fault is identified in the site region that is large in scale and activity, it is not uncommon to assess areas exceeding 30 km from the center of the site, as indicated by the Committee on Examination of Reactor Safety (1978).

This literature survey is extremely important because it guides much of the subsequent investigation. Literature surveys for assessing a general feature are followed by topographical surveys including aerial photo-interpretation, geological surveys and seismic surveys. Subsequently, drilling surveys, additional geophysical exploration and trenching surveys are conducted as the need arises. The latest active age of a fault is determined on the basis of the results of these surveys, and this information is used to decide whether the fault should be considered in seismic design.

21.3.2 Topographical surveys

Above all, potentially active faults identified by literature surveys must undergo a careful investigation. The next step is to investigate these features using topographical data. Generally, these data include topographical sheet maps (published at scales of 1:200 000, 1:50 000, 1:25 000, covering the whole country) and various types of aerial photographs (e.g. at scales of 1:40 000–1:10 000 issued mainly by the Ministry of Agriculture, Forestry and Fisheries and the Geographical Survey Institute; small-scale and large-scale air-photos are used for large and small tectonic structures, respectively; 1:40 000 air-photos taken by the US Airforce are used for surveying old topography before artificial changes). Photo-interpretation of lineaments suggesting the latest activity of faults in the site area is especially vital at this stage. Tectonic structures and faults are further characterized as a result of a large variety of topographical elements, displacement topography, such as linear topography (lineaments) and systematic bending of streams and ridges, and related features. The relationship of faults and related structures with the Quaternary geomorphic surface is particularly important.

Recently, new technology for acquiring detailed topographic digital data using airborne laser scanner methods has been developed. This method has an advantage that artificial structures and natural trees on the surface are removed from the data and, as a result, real topographic surfaces can be observed. This method is also advantageous because of the ease of comparison of topographic features with aerial photographs.

A primary goal of lineament photo-interpretation is to identify lineaments representing 2 km-long or more displacement topography. Because the length of fault zones is related to the potential maximum earthquake magnitude, the length of *en echelon* lineaments requires examination for evidence of continuity. Particularly, for a near-site area (within 5 km from the site center), it is vitally important not to overlook short lineaments, obscure lineaments or lineaments with suspicious signs of activity at this stage.

21.3.3 Geological and geophysical surveys

Geological surveys are implemented to understand rock distribution, tectonic structures and to confirm the presence of faults and their scale, properties and frequency of activity. The geologic survey area is usually limited to within 10 km of the NPP site. All of the exposures in the area are observed and described to make a geological map at a scale of ≈ 1:25 000. These data are then finalized to create a 1:50 000 scale geological map. In addition, surveys of faults and lineaments are conducted within the circular belt 10–30 km from the site (Figure 21.3). These surveys include examination of geological structures and rock distribution. As previously mentioned, since the new regulatory guide requires that detailed surveys not overlook any active fault in the near-site area, it is particularly important for geological surveys to recognize all potential faults, especially within 5 km of the site.

Like onshore surveys, offshore geological surveys aim to understand the distribution of rocks and strata; the distribution, scale and properties of faults; and the relationship of those faults with those on adjacent land areas. Offshore geological surveys are conducted along traverse lines by ship. The grid spacing and the length of traverse lines are determined such that active faults of a reasonable length cannot be overlooked. These offshore surveys are divided into two types: those that directly collect and observe sediments; and those that attempt to map large geological structures over broad areas using geophysical prospecting techniques. The former method adopts piston corers, dredges, submarine cameras and offshore borings.

Geophysical prospecting generally consists of 12 kHz precision echo sounding (for bathymetric survey), 3.5 kHz acoustic exploration (to obtain acoustic stratigraphy data indicating geological structures within a 10–100 m-thick stratum below the seafloor), the seismic reflection method (a method to investigate deeper structures than possible with 3.5 kHz acoustic exploration), the seismic refraction method (used to determine the velocity structure of seafloor strata), the sea-movies method (a method of planar precision echo sounding) and the sidescan-sonar method (a method to map the precise bottom-depth and sediment distribution).

Of these geophysical techniques, experience suggests that the seismic reflection method is the best for submarine topographical and geological surveys for the specific purpose of mapping of active faults for the siting and construction of NPPs. The seismic reflection method has a characteristic feature that the prospecting depth and resolution vary in response to the source frequency and with the number of receivers (multi-channel or single channel). Therefore, the seismic source and prospecting style must be properly selected according to the depth and prospecting precision of geological targets. For extensive seismic reflection mapping at sea, multi-source prospecting is adopted for high-precision exploration. For such surveys the traverse line interval in offshore areas ranges from 2–4 km on average.

Offshore geological investigation using geophysical methods has a distinct advantage over onshore investigation in that the geological structure can be interpreted using an image of the stratigraphic section as a whole. Onshore methods of geological mapping in Japan often involve significant interpolation between scarce outcrops. The biggest advantage for the evaluation of active faults in offshore areas is that it is easy to observe the accumulation of sediments adjacent to active faults. This gives a sense of slip rate, much as trenching methods do in onshore investigations. The fault activity is identified primarily by whether the fault deforms the last glacial unconformity that is usually found in continental-shelf sediments offshore Japan. Late Quaternary sediments are also checked to determine whether faults have deformed these older sediments.

21.3.4 Trenching studies

When faults are identified in the site region by topographical surveys, ground-surface surveys or geophysical prospecting, it is necessary to determine whether they are active faults, as all active faults need to be considered for seismic design of NPPs. In subaerial settings, trenching surveys are used to assess activity on specific faults. Trenching surveys have a tremendous advantage over natural exposures, in that trenches can be excavated at the ideal point on the fault to reveal its structure for detailed observation and recording (Figure 21.4). These observations include mapping geological formations exposed on the trench walls and along the bottoms of trenches. Mapping the dislocation, deformation and depositional condition of Quaternary formations helps to restore the stratigraphic section and quantify fault movement. In other words, trenching studies are used not only to determine if Quaternary slip has occurred, but to determine the history of activity and to detect the most recent activity.

Trench position is absolutely critical for fault characterization because active fault zones can be wide and complex structures. Conversely, trenches are generally short in length and expensive to dig. It is essential to narrow the position of faults down to within a few meters through detailed geological and geophysical investigations. The nature of sedimentation along the fault scarp is an equally important consideration in determining trench location. Continuous sedimentation along the fault scarp at the location of the trench is desirable, as this maximizes the possibility of defining a complete geological history of the fault and of identifying the most recent episode of activity. Maximum resolution is achieved where thin-bedded fine-grained sediments have accumulated. The scale and type of trench can be decided according to the types of faults in question (normal, reverse or strike-slip), and thickness of the Quaternary formations. When Quaternary formations are thin enough to permit excavating to the basement rock, properties of the crush zone and the relationship with Quaternary formations are observed in detail.

After excavating and smoothing trench walls, the displacement and deformation of strata is observed and sketched in detail to obtain the history of fault movement and the timing of earthquake occurrence. Repeated earthquakes often result in complicated overlapping of displacement and deformation structures. In these circumstances, evidence of older earthquakes is revealed by successively restoring the younger displacements observed in the geological section. If displacements are recognized, the absolute age of earthquakes can be constrained using ^{14}C radiometric age determinations of organic material, such as charcoal, preserved in the section. In Japan, tephra that is intercalated with sediment layers is

(a)

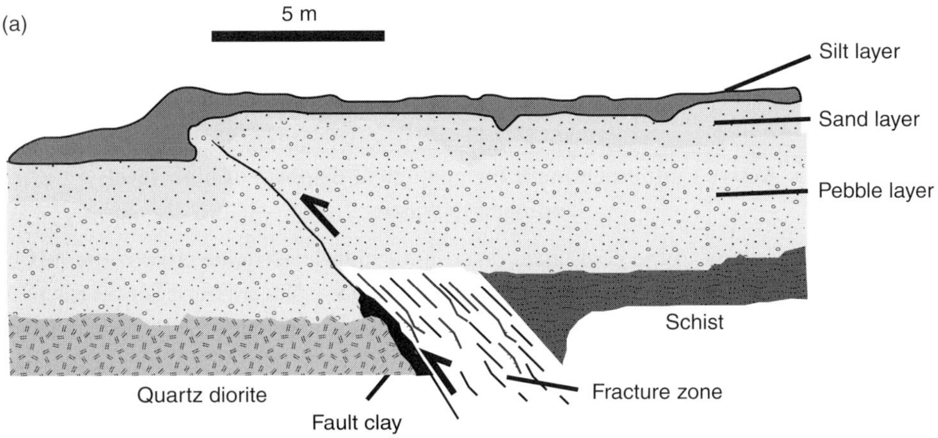

(b)

Fig. 21.4 Schematic geologic section (a) and photo (b) of a trench across the Fukouzu fault, Aichi Prefecture, Japan. This fault moved during the Mikawa earthquake, 1945. The earthquake fault continues from basement rocks of quartz porphyry and schist into Quaternary sediments. The crush zone is 4 m in width in the basement rocks. The vertical slip at the uppermost basement rocks is 1.3 m. The pebbles along the fault plane were rotated by the fault displacement so that the long-axes of pebbles are parallel to the fault. Borings adjacent to the trench indicate that the fault dip lessens with depth (Sone and Ueta,1990) (Photo: D. Inoue). See color plate section.

frequently very useful for determining the age of strata and hence constraining the timing of fault displacements (Machida and Arai, 1992). These detailed surveys are the only means of reconstructing a fault's history in sufficient detail for evaluation of seismic hazard.

21.4 Active fault assessment for geological disposal of high-level radioactive wastes

Active fault assessment in the geological disposal of high-level radioactive waste (HLW) differs from that for NPPs in two important ways. First, the performance period for HLW repositories is much longer than for NPPs. The performance period of NPPs never exceeds 100 a. For HLW facilities, the performance period is many tens of thousands of years, and the construction period alone for these facilities is many tens of years. Because Japan is a tectonically active island arc, it is well known that the return period of active faults is generally short compared to the performance period of HLW facilities. Consequently, in assessment of seismic hazards for HLW facilities, older faults are considered to be potentially active. For such facilities, the definition of active faults is extended to ones that have repeated activity during the latter half of the Quaternary period, and are also expected to move in the future. In light of this requirement, the Japan Society of Civil Engineers (2006) proposed an index of active faults "whether they repeated the movement throughout the latter half of the Quaternary period and their latest movement has displaced or deformed geological formations or topographical surfaces developed in and after the formation period of middle terraces" (MIS 5–6, 80–135 ka) (Imbrie *et al.*, 1984; Chen *et al.*, 1995). This time range is proposed because index layers, such as MIS 5 or MIS 6, are not consistently present in the stratigraphic section everywhere in Japan.

A second point is that the hazards associated with earthquake motion in geological repositories for HLW are not as alarming as those associated with NPPs. Japanese law states that a geological repository of HLW shall be constructed underground at the depth of 300 m or more. Since it turns out that the amplitude of earthquake motion under the ground ranges from one half to one third of that on the ground surface, it is not presumed that earthquake motion seriously affects the geological repository. Rather, the influence by active faults must be examined from two points of view. Over long periods of time, active faults may result in the direct destruction of the disposal facility and waste body in association with fractures, ruptures and displacements of the rock mass caused by fault movements. In addition, performance deterioration of the multi-barrier system may occur in response to repeated seismic events. This deterioration might include mechanical and hydrological changes in the surrounding rock mass generated by fractures, ruptures and displacements.

In light of the above, important problems to be solved as part of the active fault assessment for the geological disposal of HLW are divided into concerns with: (i) active fault assessment at the preliminary investigation stage; (ii) the examination of the limits of areas affected by active faults; and (iii) the possibility of the formation of new faults. Japan is much better equipped with excellent active fault maps than other countries. However, even these maps are insufficient for the active fault assessment for a HLW geological repository.

Rather, it is necessary to improve the precision of active fault surveys so as not to overlook any Quaternary fault. This is achieved by implementing more detailed offshore and onshore surveys, such as aerial photo-interpretation, geophysical prospecting, ground-surface surveys, drilling surveys and trenching surveys. Fortunately, if an active fault is encountered during this detailed investigation, the layout of the facility can be revised to avoid the fault.

Large-scale fault movements at shallow depths occur mainly along existing active faults or active fault zones, so long as the tectonic environment of the Japanese islands remains unchanged. Consequently, if the location of the repository is selected so as to avoid active fault zones, fault movements inflicting serious damage to the repository can be avoided. Earthquake source faults reach the ground surface and appear as earthquake faults. This process is repeated to accumulate displacement, resulting in active faults. However, earthquake faults do not always appear at the same location near the surface. Even earthquake source faults under the ground appear as parallel or *en echelon* faults on the surface to form an active fault zone. In areas with a thick sequence of Neogene and younger formations, the development of earthquake source faults results in a flexural deformation near the surface and flexural slip faults along bedding planes. Thus, it is desirable to identify an area expected to undergo large deformation over a long period of time, amounting to tens of thousands of years, and avoid it. Since the possibility that an active fault grows in length is undeniable, it is necessary to examine such a case in great detail.

The tectonics of the Japanese islands and their environs is controlled by plate movement. It is thought that a change in the configuration and dynamics of this plate system requires on the order of one million years. Consequently, it appears to be highly unlikely that a significant new active fault might develop on the timescale of the performance of a HLW geological repository. If a new fault were to form where there was no previous fault movement, it would take hundreds of thousands of years to millions of years for it to grow into a significant tectonic structure. Given these timescales, since the repository is constructed at a place devoid of faults, it is unnecessary to take the development of new faults into consideration for securing the safety of the repository system (JSCE, 2006). However, the activity and development of new faults should be carefully examined in the areas where the tectonic setting appears to be evolving (e.g. see Figures 21.5–21.7).

Concluding remarks

This chapter has provided an overview of the regulatory framework for assessment of active faults in Japan in the context of siting nuclear facilities, and provided background on the current methodologies used to assess active faults as part of seismic hazard investigations for NPPs and HLW geological repositories. Recent revisions of guidelines for assessment of active faults have realigned the regulatory framework with the state of the art in many aspects of seismic hazard assessment in Japan, including the geological and geophysical investigation of active faults and related structures. If properly implemented, this development should result in improved seismic hazard assessments.

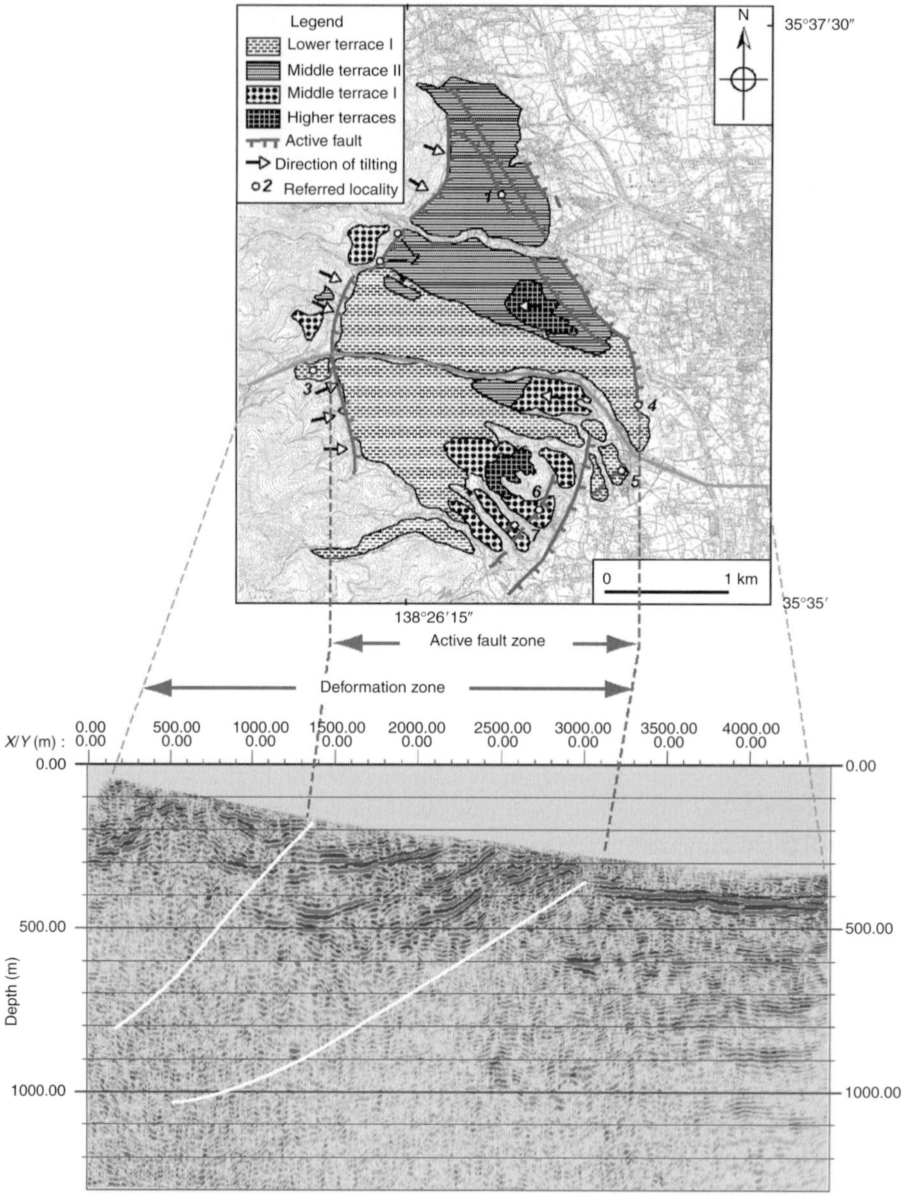

Fig. 21.5 Example of detailed investigations of a thrust fault, the Ichinose fault, that is part of the Itoigawa–Shizuoka tectonic line in Japan. The geological map of active faults and tilted marine terraces is used to define the active fault zone between the two active faults. The seismic profile was collected along the gray line shown on the geological map (horizontal scale shows meters from the west end of the profile) and shows that deformation of the basement layer extends beyond the western margin of the active fault zone. Addition trenching may indicate whether this part of the fault zone is active or not. The original active fault map is modified from Miura *et al.* (2002), the seismic profile is modified from JSCE (2004). See color plate section.

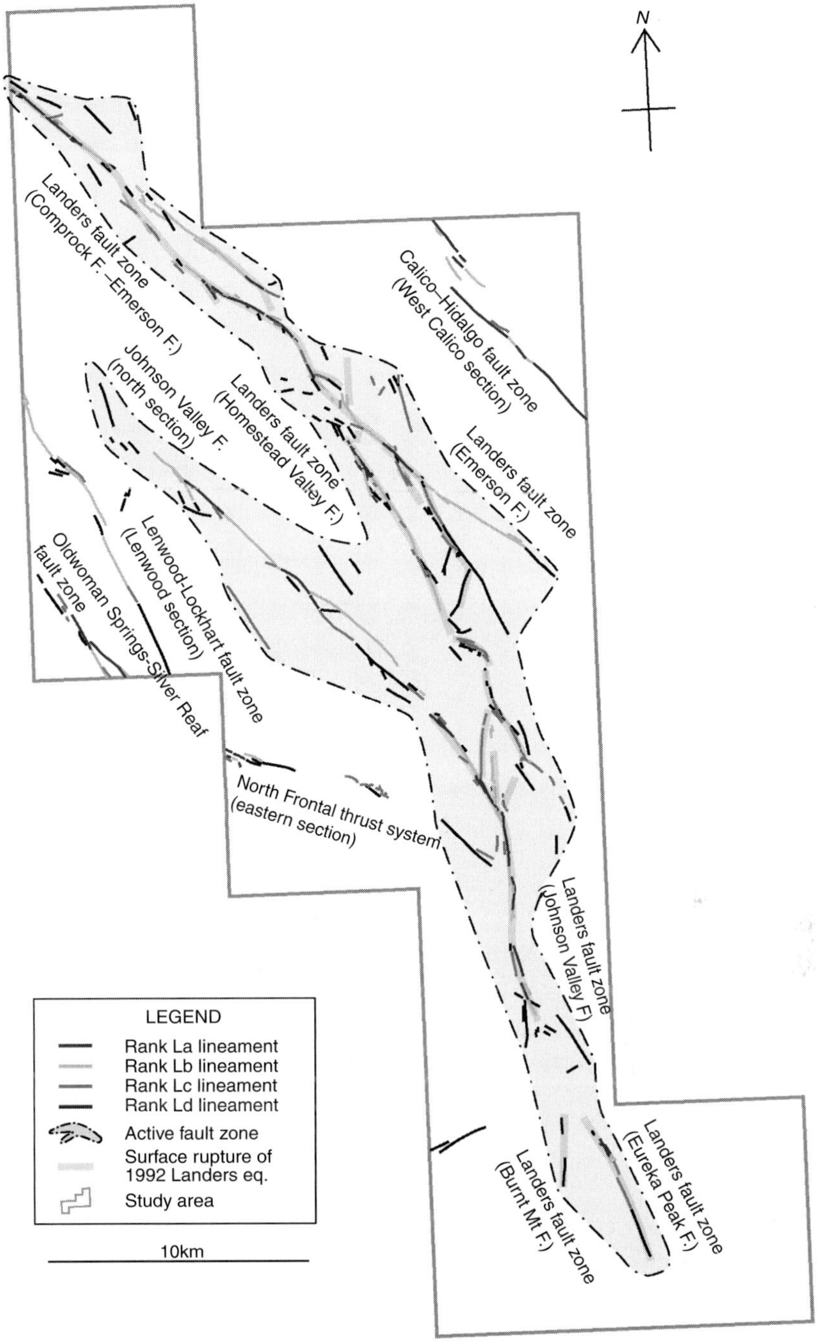

Fig. 21.6 An active strike-slip fault, the Landers fault system in California (Sieh *et al.*, 1993). The active fault zone is defined as a region between the eastern and western edge of the lineaments. The active fault zone widens where the lineaments merge. Ranks of lineaments are defined by Inoue *et al.* (2002). See color plate section.

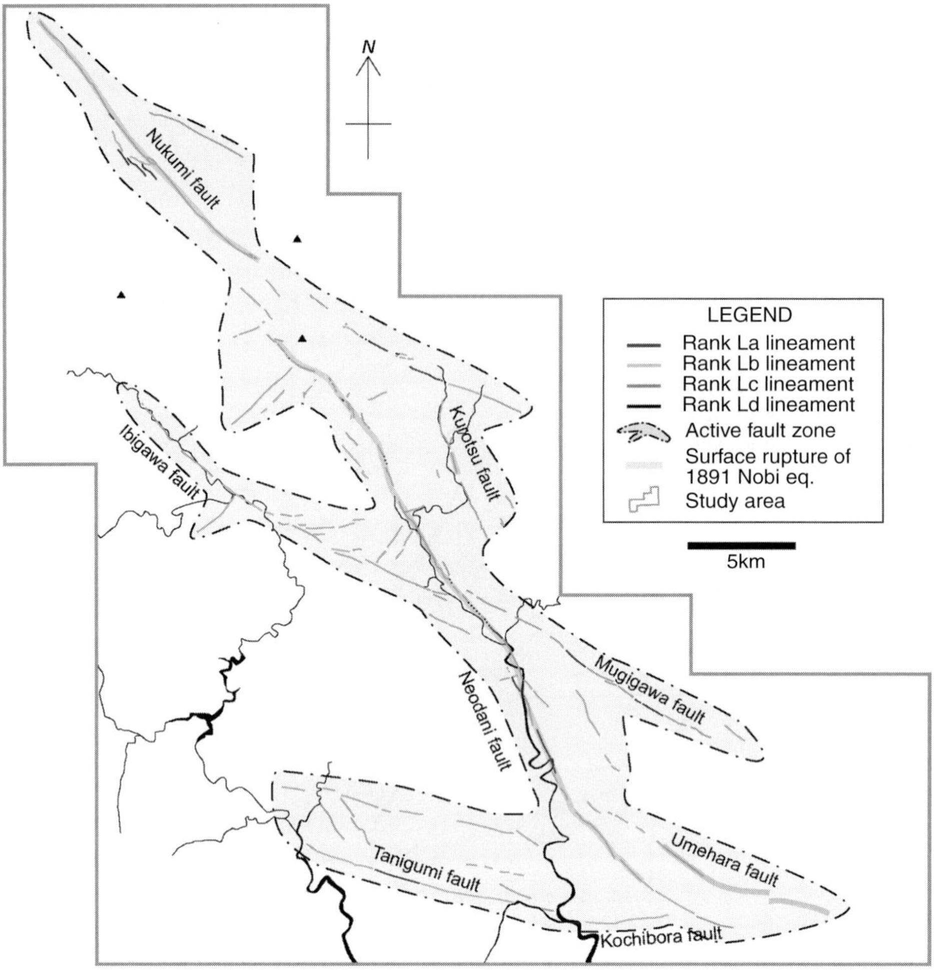

Fig. 21.7 An active strike-slip fault, the Neodani fault system in Japan, which experienced a Mj = 8 earthquake in 1891. The definition of active fault zone is the same as in Figure 21.6. The active fault zone widens where lineaments merge and at the end of lineaments. The lineaments should be checked in the field, as the active fault zone could be smaller than shown. See color plate section.

Further reading

Please refer to *Computer Graphics: Geology of Japanese Islands* edited by the Editorial Committee for the Geology of Japanese Islands (2002) for a complete discussion of the tectonic setting and earthquake seismology in Japan.

Acknowledgments

I gratefully acknowledge Mr. Katsuyoshi Miyakoshi, Mr. Keiichi Ueta, Dr. Shintaro Abe and Dr. Yasuhira Aoyagi of CRIEPI for useful discussions on these topics. I also thank

Mr. Toshinori Sasaki of CRIEPI for his contribution in the figure preparations. Figures 21.6 and 21.7 are the results of a study that was funded by NUMO. I acknowledge Dr. Laura Connor and Dr. Charles Connor for helping to refine the original manuscript.

References

Chen, J., J. W. Farrell, D. W. Murray and W. L. Prell (1995). Timescale and paleoceanographic implications of a 3.6 m.y. oxygen isotope record from the northeast Indian Ocean (Ocean Drilling Program site 758). *Paleoceanography*, **10**, 21–47.

Committee On Examination of Reactor Safety (1978). Guide for geological and ground safety examination of nuclear power plants, Report L-DS-II.OS. Tokyo: Nuclear Safety Commission.

Editorial Committee for the Geology of the Japanese Islands (2002). *Computer Graphics: Geology of Japanese Islands*. Tokyo: Maruzen Co. Ltd.

Geological Survey of Japan, http://www.gsj.jp/Map/EN/ntec.htm Neotectonic maps. Tsukuba: Geological Survey of Japan.

Ikeda Y. , T. Imaizumi, M. Togo *et al.* (2002). *Atlas of Quaternary Thrust Faults in Japan*. Tokyo: University of Tokyo Press.

Imbrie, J., J. D. Hays, D. G. Martinson *et al.* (1984). The orbital theory of Pleistocene climate, support from a revised chronology of the marine ^{18}O record. In: Berger, A. L. *et al.* (eds.) *Milankovitch and Climate: Understanding the Response to Astronomical Forcing*, NATO Science Series C. Berlin: Springer Verlag, 269–305.

Inoue D., K. Miyakoshi, K. Ueta, A. Miyawaki and K. Matsuura (2002). Active fault study in the 2000 Tottori-ken Seibu earthquake area. *Jishin*, **2**, 557–573 (in Japanese).

JAQUA (1987). *Quaternary Maps of Japan*. Tokyo: Japan Association for Quaternary Research, University of Tokyo Press.

JNES (2007). Nuclear facility operation control manual report. Tokyo: Japan Nuclear Energy Safety Organization.

JSCE (2004). The active fault evaluation technology of nuclear power plants, segmentation of long active fault system. Tokyo: Japan Society of Civil Engineers (in Japanese).

JSCE (2006). Fundamental views of geological environmental investigation and evaluation during the stage of selection of detailed investigation area of high-level radioactive disposal. Tokyo: Japan Society of Civil Engineers (in Japanese).

Koike K. and Y. Machida (2001). *Atlas of Quaternary Marine Terraces in the Japanese Islands*. Tokyo: University of Tokyo Press.

Machida Y. and F. Arai (1992). *Atlas of Tephra In and Around Japan*. Tokyo: University of Tokyo Press.

Miura D., R. Hataya, S. Abe *et al.* (2002). Recent faulting history at the Ichinose fault group, southern part of the Itoigawa–Shizuoka tectonic line, central Japan. *Jishin*, **2**, 23–45 (in Japanese).

Nakata, T. and T. Imaizumi (2002). Digital active fault map of Japan. Tokyo: University of Tokyo Press.

Nakata, T., A. Okada, Y. Suzuki, M. Watanabe and Y. Ikeda (1996). Osaka Seinanbu, Active Fault Map in Urban Area, scale 1:25 000. Tsukuba: Geographical Survey Institute (in Japanese).

NSC (2006). Regulatory guide for reviewing seismic design for nuclear power reactor facilities, Report NSCRG: L-DS.I.02. Tokyo: Nuclear Safety Commission of Japan.

Okada A. and T. Imaizumi (2000). *Active Faults in the Kinki Area, Central Japan*: Sheet Maps and Inventories. Tokyo: University of Tokyo Press.

Research Group for Active Faults of Japan (1980; revised in 1991) *Active Faults in Japan: Sheet Maps and Inventories* (revised edn.). Tokyo: University of Tokyo Press (in Japanese with English summary).

Sieh, K., L. Jones, E. Hauksson *et al.* (1993). Near-field investigations of the Landers earthquake sequence, April–July, 1992. *Science*, **260**, 171–176.

Sone K. and K. Ueta (1990). Evaluation of fault activity below alluvium – (1) deep groove fault trench investigation. CRIEPI Report U90029. Tokyo: Central Research Institute of Electric Power Industry (in Japanese).

22

Issues for coastal sites

I. G. McKinley and W. R. Alexander

A large number of nuclear facilities are situated directly on or very close to the coast. This is not coincidental. Such locations ease the transportation of bulky or sensitive materials by ship and offer an effectively limitless supply of cooling water. Nuclear facilities may also be less intrusive in coastal settings. For some countries there may be few alternatives to coastal sites, particularly those comprising islands with mountainous interiors, such as Japan (Figure 22.1).

Coastal sites also offer the potential to dilute releases, whether planned or inadvertent, from nuclear facilities. In remote locations, the dilution of released liquid waste has

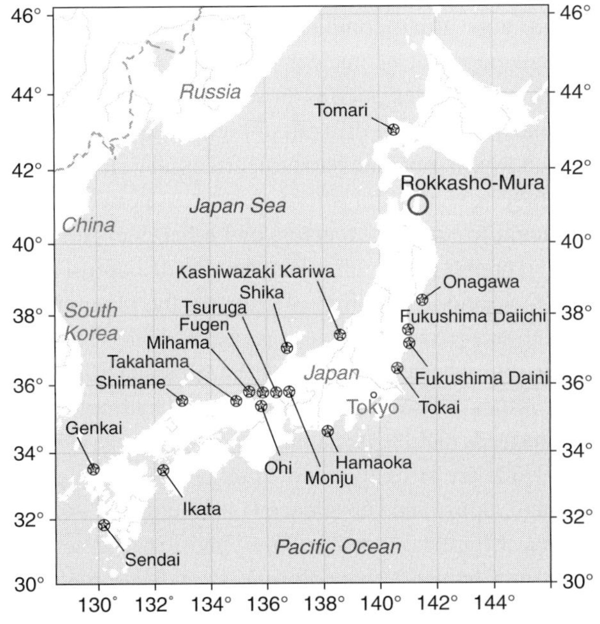

Fig. 22.1 Distribution of nuclear power plant sites in Japan. Rokkasho, indicated by the larger circle, is a center for nuclear fuel cycle activities, including enrichment, reprocessing, high-level waste storage and low-level waste disposal. Map data provided by FEPC (2008) and INSC (2008).

Table 22.1. The impact of tectonic phenomena and sea-level change on nuclear facilities[1]

Sea-level change (m)	Uplift	Subsidence	Tsunami	Mega-tsunami	Sedimentation
Maximum (+80)		S I D		S I	
Major increase (+10)		S	S	S	
Minor increase (+1)			R S		
Minor decrease (−1)					R
Major decrease (−20)	S I				S
Minimum (−150)	S I D				I

[1] Weighted by the probability of these events (see text for discussion). *Relevant facility*: R = reactor or surface nuclear facility; S = on- or near-surface repository; I = intermediate-depth repository; D = deep geological repository.

historically been considered an advantage. Such dilution can be very important for facilities that produce large volumes of low-toxicity waste, or facilities that may experience low releases in the future (i.e. radioactive waste repositories). However, there is increasing international pressure to limit, or even completely ban, such releases (cf. Mobbs *et al.*, 1989; DEFRA, 1999; Chapman and McCombie, 2003).

Demographically, the expanding global population is increasingly concentrated in coastal areas, predominantly in large urban complexes. Expanding demand for energy is likely to lead to the further development of nuclear facilities in such regions, with potentially increasing competition for prime sites, such as those with low risk from natural hazards. Nevertheless, for many countries it may be difficult to preclude a risk of some kind of dramatic natural phenomena, such as seismogenic tsunami, affecting potential sites. Indeed, natural hazards at coastal sites may be significant when risk is integrated over long timescales, many decades for the operational lifetimes of reactors and other conventional nuclear facilities, hundreds of years for the period of operation and active site control for near-surface waste disposal facilities, and thousands to millions of years for the required performance of deep geological repositories.

This chapter provides an overview of tectonic hazards specific to nuclear facilities in coastal settings and the issues which should be considered by implementors and regulators of such facilities. Such hazards include tsunami, inundation by sea-level change and erosion of coastlines. These hazards are strongly coupled to tectonic events, such as seismogenic and volcanogenic tsunami, uplift and subsidence (Litchfield *et al.*, Chapter 4, this volume; McKinley and Chapman, Chapter 24, this volume) and extreme sedimentation related to volcanism (Volentik *et al.*, Chapter 9, this volume). An indication of such coupling to sea-level change is given in Table 22.1.

Although the focus of this chapter is on marine coastal settings, some of the considerations are relevant to facilities situated beside large freshwater bodies. Even more than the ocean, these large freshwater bodies are vulnerable to very short-term changes as a result of both geological/climatic and anthropogenic processes. One example is the Aral Sea. In 1960, the Aral Sea was the world's fourth largest lake ($\sim$ 67 000 km^2). Since then its level has dropped by 16 m, reducing its volume by 75% (e.g. see Aral Sea at NASA, 2008). This indicates how quickly major environmental changes in such a location can occur; although, in this specific case, rapid anthropogenic effects only mirror slower natural changes. The Aral Sea has, in the past, dried up completely. Such perturbations are likely to be an even more acute problem as a consequence of climate change, especially if the expected major shortage of freshwater arises in coming decades (cf. UNEP, 2007).

22.1 Key issues at coastal locations

Coastal geography reflects the interaction between slow evolutionary processes and rapid tectonic events. Sea level is a key factor at coastal sites (Table 22.1, Figure 22.2), reflecting a balance between the volume of the oceans, influenced by climate change, and the local geodetic land surface, influenced by plate tectonic processes, isostatic adjustments and related processes (McKinley and Chapman, Chapter 24, this volume). Earthquakes, tephra falls and related high sedimentation events are some of the phenomena that cause abrupt changes in the morphology of coastal sites, and the topography of the coastline and adjacent offshore zone may be rapidly affected by coupled erosion and sedimentation processes.

Inundation is an obvious hazard at any coastal site. Rapid events that could lead to inundation include, in particular, tsunami. Estimated probabilities of tsunami directly associated with underwater fault movement/seabed displacement (Power and Downes, Chapter 11, this volume) are often based on historical and geological records, complemented by tectonic models of the type considered by Stirling *et al.* (Chapter 10, this volume), which can help delineate tsunami source characteristics. The resultant local impact is dependent on both large-scale wave propagation for distant sources and the detailed characteristics of the coastal region of interest. However, as was dramatically emphasized by the Indian Ocean

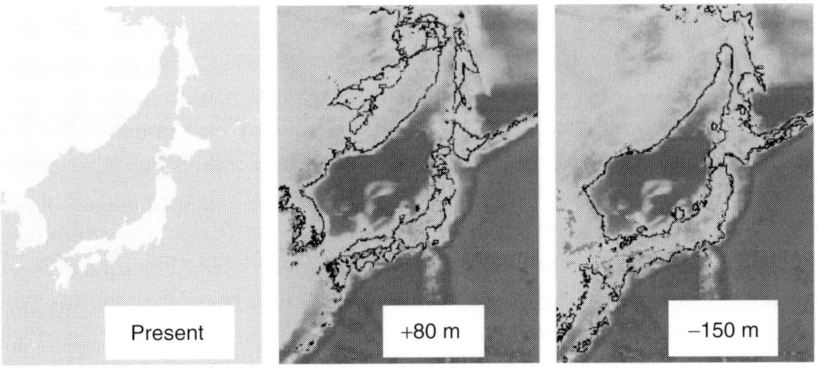

Fig. 22.2 Influence of extreme changes in sea level on the coastline of Japan. Map data from GEOID (2008).

tragedy of December 26, 2004, recognition of risks associated with a potential tsunami is very patchy, especially for large magnitude/low probability events.

Collapse of volcanoes, associated with eruptions or not, and similar landslips also result in tsunami. Again, historical and geological records, together with associated modeling, can provide background but, as emphasized by Power and Downes (Chapter 11, this volume), records associated with such events are less extensive and understanding of the fundamental water-displacement mechanisms rather rudimentary. Nevertheless, it is clear that, especially locally, such processes can give rise to giant tsunami, sometimes termed mega-tsunami as described by Power and Downes (Chapter 11, this volume). For completeness, it is noted that mega-tsunami also result from bolide impact and, indeed, over very long time periods this source produces the largest tsunami.

Giant tsunami are not usually explicitly considered in hazard assessments for nuclear facilities because the probability of occurrence of such events is small ($\leq 10^{-6}\,a^{-1}$). For particularly large source events (explosive volcanism or bolide impact), associated catastrophic non-nuclear consequences may be global (or, at least, regional), so reducing the relative significance of any impact on nuclear facilities. In some circumstances, such giant tsunami may have higher probabilities (e.g. localized giant tsunami caused by landslides into restricted bodies of water, including both freshwater lakes and coastal bays or fjords). Such special circumstances require focused hazard assessments.

Climate-related rapid events can cause inundation, including storm surges and seiches, resulting from extreme weather (e.g. hurricanes, typhoons), the consequences of which can be exacerbated by other damage caused by high winds and tides. The probability of such events is usually derived from historical records but, with increasing acknowledgment of the rising frequency of extreme weather as a consequence of global warming, these may actually be rather poor predictors of future conditions (UNEP, 2002, 2007).

Although usually not quite as dramatic, the position of the coastline can also alter relatively quickly. The most obvious process is erosion, which can occur particularly rapidly due to scouring associated with tsunami. This is especially evident where coastlines are maintained artificially (e.g. the Mississippi delta in the USA and the Rhine/Scheldt delta in the Netherlands).

Also important to be considered is coastline movement due to sedimentation. This can be coupled to intense erosion at one location being associated with re-deposition elsewhere. Additionally, major sedimentation can be associated with tephra falls and related volcanic phenomena (Volentik *et al.*, Chapter 9, this volume). High sedimentation rates can result in complete infilling of shallow bays, with rapid coastline displacement such as seen at Sakurajima volcano in southern Japan, where one eruption buried the strait between the island of Sakurajima and the Osumi Peninsula, turning the island into an extension of the peninsula (Fukuyama, 1978). Apart from the direct consequences of alteration of the coastline caused by erosion and sedimentation processes, such changes may also have the indirect effect of influencing the risk and potential consequences of the inundation processes considered above, which are very sensitive to local bathymetry (Power and Downes, Chapter 11, this volume).

The consequences of such events may be assessed at the present time based on probabilistic analyses, specified reference cases or worst conceivable scenarios. However, in the future, the coupled effects of slow processes also need to be considered. A particular concern is sea level, which is changing at present worldwide, predominantly as a result of global warming, and locally due to tectonics or glacial rebound (cf. Sasaki *et al.*, 2008). Rigorous predictions of the climate-related rate of such change are very difficult to make and this is currently an area of controversy (e.g. Intergovernmental Panel on Climate Change, 2007). There is a consensus amongst most modelers that, even for pessimistic global warming scenarios, maximum sea-level rise will be in the order of a few meters over the next couple of centuries (cf. Beckmann and Tetzlaff, 1998; Langenberg *et al.*, 1999; Church *et al.*, 2001; Woth *et al.*, 2006).

Nevertheless, anthropogenic increases in greenhouse gases are without precedent in the geological record and hence it is difficult to demonstrate conclusively that such scenarios are really worst cases. The maximum possible sea-level rise is calculated to be $\sim$ 80 m based on an inventory of all major ice masses (Table 22.1 and Figure 22.2). Indeed, scenarios have been proposed in which feedback mechanisms (e.g. release of undersea gas hydrates, methane release from melting permafrost, etc.) lead to runaway warming, in which all ice sheets melt within a few centuries. This event is devastating enough on its own, without considering the additional complication of further processes that would be associated with such a dramatic change to the global environment.

Concern presently focuses on anthropogenic global warming but, from a geological perspective, the present global climate is classified as interglacial and, within a period of tens of thousands of years, a return to ice-age conditions could be expected. In this case, based on the last ice age, global sea level could well drop by $\sim$120–150 m. As indicated in Figure 22.2 for the case of Japan, such a decrease in sea level would result in dramatic change in the location of coastal areas.

In coastal locations, a clear problem is the determination of the consequences of coupling between several different processes. Thus, rather than focusing on the processes themselves, the following sections concentrate on the timescales involved.

22.2 Perturbations occurring on short timescales

Nuclear facilities in coastal areas are vulnerable to a variety of phenomena on timescales of decades or less. Tectonism and volcanic activity, as previously described, result in tsunami and related phenomena on short timescales. In addition, storm surges and seiches present hazards to coastal sites on these short timescales, and the effects of these phenomena may change as a result of uplift or subsidence.

Tsunami

In a coastal environment, the threat of tsunami resulting from offshore earthquakes must be considered for specific sites. Earthquake hazard maps can now be combined with models of tsunami generation to produce reasonable coupled hazard maps, allowing facilities to be located away from regions of higher risk. It is probable, however, that comprehensive hazard assessment has not been carried out for many of the coastal nuclear facilities which could be subject to such threats, as national regulatory authorities are only slowly becoming aware of the degree of risk (e.g. DEFRA, 2005).

Even in areas where tsunami have never been considered in the past, safety cases increasingly require that arguments are developed to allow such events to be explicitly discounted. One recent example is the preliminary assessment by Alexander and Neall (2007) of the potential risk of tsunami inundation of the ONKALO facility (Posiva, 2003) on the Baltic coast of Finland. Here, conventional tsunami resulting from large offshore earthquakes could be excluded over the timescale of concern ($\sim$ 100 a) as being of "... negligibly low probability in the Baltic" due to the well-understood current tectonic quiescence of the region.

Nevertheless, there are many locations where the probability of a tsunami cannot be discounted. In such cases, depending on the evaluated risk, specific precautions or counter-measures can be adopted, tailored to the facility involved. The issues involved are highlighted by the analysis of perturbations to Indian nuclear power plants due to the tsunami of December 26, 2004 (see Appendix). It is noticeable that drop of sea level preceding the tsunami is here identified as a concern; a factor that is often forgotten if there is a focus on the hazard and consequences of inundation.

For reactors and other nuclear facilities requiring active control, if the risk is significant then engineering counter-measures can be considered (e.g. structured breakwaters, locating key infrastructure in strengthened buildings or underground, siting cooling water intakes in deeper water). In such cases, it would also be prudent to have established operational procedures to put the facility into a safe mode following the warnings that can be provided by existing (and future) earthquake and tsunami monitoring networks. In the case of locations where potential tsunami sources are nearby, such procedures may need to be automated as the warning time may be rather short. In any case, it is important to ensure that key services are protected from potential disruptions or have sufficiently robust backup provided.

For passively controlled facilities (e.g. waste stores or repositories) all access points should be located above likely surge reaches of maximum expectable tsunami. Although not relevant to modern designs of heavily engineered facilities, the direct disruption of surface coastal dumps (as experienced for toxic waste in Somalia as a result of the tsunami on December 26, 2004) may need to be considered for old facilities in vulnerable locations. In particular for geological repositories, any shafts or tunnels that have any possibility of being flooded should have bulkheads that can be sealed rapidly. As was the case for reactors, any vulnerable services should have appropriate levels of backup (e.g. drainage and ventilation systems), which should be dimensioned to allow for the likely period of institutional disruption that would inevitably result from such an event.

Very large, regional-scale tsunami can probably be discounted on such short timescales on the basis of low probability. Nevertheless, given the high sensitivity of nuclear facilities, siting of facilities should consider any local features that could give rise to specific risks (e.g. due to major landslips into restricted volumes of water). An additional concern could be areas that may be subject to tsunami caused by local volcanic activity, which should be explicitly considered in the analysis of the potential consequences of undersea volcanism.

Storm surges and seiches

The direct consequences of storm surges due to extreme weather will depend on both the coastal geography and sea level. To date, although there is clearly an awareness of the potential impacts (e.g. Sumerling, 2005; Kim *et al.*, 2006; Gallani, 2007), few formal assessments of the potential effects of storm/tidal surges and seiches on a coastal facility have been carried out.

A recent preliminary evaluation of the potential impact of storm surges on the ONKALO facility on the Baltic Sea coast of Finland by Alexander and Neall (2007) concluded that, in the short term, they are unlikely to impact the facility significantly. Although situated very close to the coast on an island, the design of the facility has taken advantage of the topography and is well above any currently foreseen surges. However, if some of the more extreme increases in sea level which have been proposed are realized, storm surges may impact the facility within its currently planned 100 a operational lifetime. Such a vulnerability could be even greater in areas showing high subsidence rates, although, to be significant on this timescale (several millimeters per year), tectonic processes would be a less likely mechanism than local water, gas or oil extraction (Browitt *et al.*, 2007).

Anomalous sedimentation

Regional-scale processes associated with volcanism could significantly disrupt nuclear facilities and those on the coast could be particularly vulnerable. An example would be major tephra falls which, depending on the volcanic event and prevailing winds, could have significant effects over distances from tens to hundreds of kilometers. Structural collapse due to the weight of tephra would not be expected to be a problem for sensitive nuclear facilities, due to the very conservative building specifications normally utilized, but such falls could not only disrupt services, but also clog intake for water-cooled facilities, which would be particularly problematic for facilities situated on shallow bays. Furthermore, these fall deposits normally remobilize quickly and sedimentation rates in coastal areas can change abruptly as a result. Unfortunately, such coupled processes are rarely considered by current hazard assessments. In case of a risk of tephra fall, procedures for rapidly placing operations in a safe mode would be needed.

A critical aspect of disruptive events occurring over these short time periods is that active responses can be assumed. Siting, design, layout and operational procedures should be specified to avoid any risk of catastrophic disruption due to any credible scenario; unlikely major perturbations could, however, lead to requirements for extensive repair or remediation. For example, the old disposal shaft at Dounreay has been shown not to reach current standards for waste isolation and, indeed, could be vulnerable to the active coastal erosion at this location, which could well increase as a result of global warming (Figure 22.3). Extensive and costly remediation is thus ongoing (e.g. Highland Council, 2006). The decision on

Fig. 22.3 The shaft at Dounreay, UK, which was used for disposal of solid radioactive waste from 1959 to 1977, when use was discontinued following an explosion. Apart from problems due to waste leakage, the shaft is clearly threatened by coastal erosion; ongoing remediation includes both waste removal and grouting the surrounding area (figure provided courtesy of UKAEA, 2004).

how much to invest in precautions against less likely events that could lead to expensive responses must be made, however, on the basis of a cost–benefit analysis.

22.3 Perturbations occurring on long timescales

When looking at time periods of centuries to millennia, the focus is entirely on repositories (or very long-term monitored stores) for radioactive waste. Here it is important to consider not only the integrated risk of low-probability events, but also the coupled influences of slow processes (see Table 22.1). In terms of consequences, the existence of the resources and infrastructure to support active response to disruptions cannot be assumed in the case of closed repositories. Hence, either consequences have to be shown to be acceptable or passive design measures implemented to reduce these consequences.

Key slow processes to be considered in a coastal setting are the net effects of sea-level change and regional uplift/subsidence. For long timescales, near-surface or surface disposal sites may be particularly vulnerable, especially if located relatively short distances from coastlines and/or only a little above sea level (e.g. Drigg, UK (Figure 22.4); La Hague, France; Rokkasho, Japan; Dounreay, UK).

Fig. 22.4 Drigg, the UK national site for low-level wastes, sits on the coast of northwestern England; note the Sellafield nuclear reprocessing facility in the background (photo from NEI, 2005).

Tectonic subsidence and uplift

Subsidence, combined with rising sea levels, could give rise to a risk of flooding or inundation of facilities, particularly if coupled with the possible increase in extreme weather associated with climate change. This is clearly a greater concern for open storage facilities; nevertheless such a change in boundary conditions would need to be specifically addressed even for closed repositories, taking account of any infrastructure that may be present for active institutional control (e.g. drainage and monitoring systems).

Over the timescale of centuries, uplift is likely to be minor compared to the potential for sea-level rise as a result of anthropogenic warming. Locally, there are exceptions, however. For example, in the Baltic region post-glacial uplift of Olkiluoto island is expected to mitigate the expected sea-level rise due to climate change (Alexander and Neall, 2007). Uplift may often reduce the risk of flooding, and therefore be considered a positive effect. At coastal locations, however, the main concern with uplift may be erosive displacement of the coastline. This has been highlighted during the recent review of the Drigg, UK, safety case (EA, 2005), where a scenario involving erosion of the coastline was estimated to give rise to radiological risks two orders of magnitude above regulatory guidelines. It is notable that many of the present coastal low-level waste disposal sites were originally licensed before recent concerns about global warming emerged. As a result, this is an issue that may be open for many such repositories.

When a repository is covered by the sea, there is the potential for massive dilution of any releases of radioactivity. Conversely, when the coastline recedes during uplift, releases may take place on land, with consequent higher potential exposures. This is currently an issue for the SFR repository in Sweden (SKB, 2001). Although currently located at 50 m depth in rock below the Baltic seabed, it will lie under land within 2–3 ka owing to ongoing post-glacial uplift. The change in biosphere receptors for any potential releases has to be accounted for in safety analyses of this facility and is identified by regulators to be a weakness in the current safety case (Klos and Wilmot, 2002).

Sedimentation

Although less obviously disruptive, displacement of the coastline caused by sedimentation (either as a slow process or resulting, for example, from tephra fall) can also significantly influence the safety case for near-surface disposal facilities. Slow leaching of waste is to be expected as soon as a repository is closed and a particular advantage of coastal locations is the extent of dilution if releases occur to a marine (or brackish) environment. Loss of such dilution could significantly increase calculated doses and hence would need to be considered by alternative biosphere scenarios in the safety case.

Sea-level change

For deep geological repositories (several hundred meters below surface; see Alexander and McKinley, 2007), the effect of sea-level change on timescales of centuries to millennia would be expected to be relatively small. A borderline case would be intermediate-depth repositories, such as the L1 facility planned at Rokkasho (to be between 50 and 100 m below surface; Karigome, 2005) In such cases, changes in sea level relative to the ground surface/repository horizon (Figure 22.2) could cause significant changes in both the regional hydrogeology and geochemistry (Figure 22.5), which would need to be assessed by appropriate scenarios (e.g. Sasaki *et al.*, 2007).

Apart from sea level, the evolving shape of the coastline may, over this timescale, change the vulnerability of specific locations to tsunami and storm surges, which may need to be explicitly considered for surface/near-surface facilities.

Increased probability of extreme events

Finally, the issue of low-probability/high-consequence scenarios may be particularly important over long timescales and need explicit consideration. There is considerable uncertainty over the probability and consequences of very large tsunami but, integrated over a timescale of $\sim$ 1 ka, it may be difficult to argue that such events are improbable. For completeness, it can be noted that, in some locations, the development of new volcanic centers might be possible, also potentially impacting coastal environments. Analysis of the probability of new vent formation, say in nearby shallow-water environments, and potential consequences

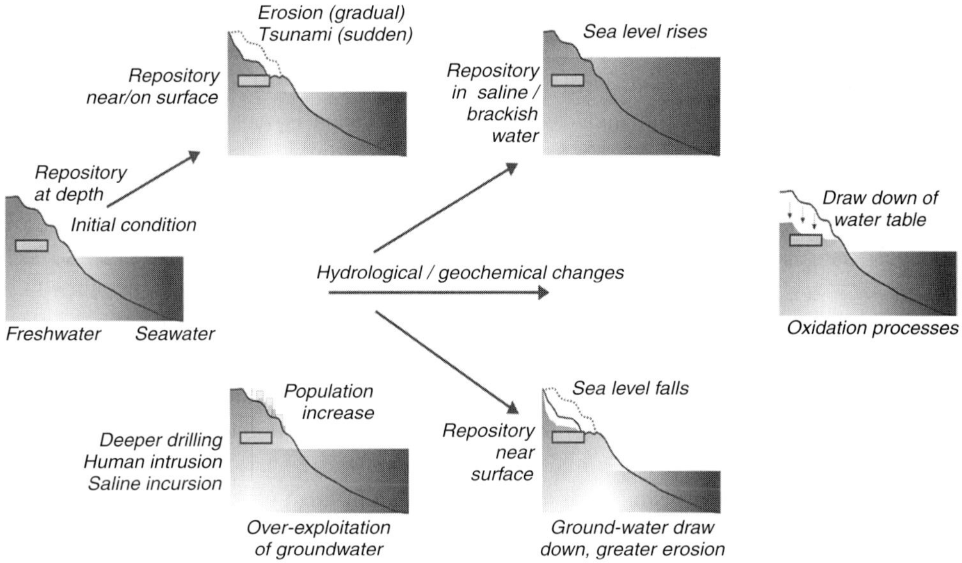

Fig. 22.5 Sea level may vary between 80 m above to 150 m below the present level (Figure 22.2), with potentially significant implications on the local groundwater chemistry and hydrogeology.

of this activity in terms of change in tsunami potential and sedimentation would have to be included in the safety case (e.g. Hill *et al.*, Chapter 25, this volume; US DOE, 2002).

Overall, for near-surface or intermediate-depth facilities near the coast, this time period of centuries to millenia may be critical. Scenarios that give rise to direct exposure of waste could represent significant hazards, even for wastes classified as low- or intermediate-level. Such isolation-failure scenarios can, for new facilities, be avoided by siting (e.g. hard rock, sheltered coastlines where erosion is not a problem) or design (predominantly emplacing the waste deeper and/or further from vulnerable coastlines).

For deep repositories, the inherent buffering of the geological and engineered barriers would probably minimize any significant degradation in performance due to such perturbations of the surface environment. In cases where first releases of more mobile radionuclides may occur, potential alterations to the biosphere and the geosphere/biosphere interface may, however, need to be explicitly considered within safety assessments.

22.4 Perturbations occurring on geological timescales

Over timescales of tens of thousands to hundreds of thousands of years, concerns generally involve only geological repositories. Nevertheless, although near-surface or interim-depth facilities are intended for disposal of shorter-lived wastes, the consequences of their disruption can still be significant over such very long timescales (McKinley and Chapman, Chapter 24, this volume).

Major variations in sea level can be assumed to be driven by natural Milankovitch cycles which could, potentially, cover the full range from maximum ice cover to an effectively ice-free state over such a time period (cf. Table 22.1). This would be coupled to the integrated effects of slow tectonic uplift or subsidence. The net effect is that submergence and/or erosion would be difficult to preclude for almost any coastal near-surface facility and, unless excluded by legislation, should be carefully assessed to determine potential consequences (using either dose, risk or, possibly, alternative indicators of safety).

Major changes in sea level will clearly have practical impacts on coastal dwellers, with the sites of cities, harbors and other facilities moving with the migrating coastline. It is not reasonable to assume that knowledge of the existence of a repository can be preserved in the light of such disruptions. Especially for near-surface or intermediate-depth facilities, the additional risk of anthropogenic perturbations may need to be addressed.

Even for deep repositories, a coastal location can increase the potential for disruption. Thus, for an initial depth of 300 m (the minimum allowed by law for a Japanese high-level waste repository), a decrease of sea level of about 120 m combined with a tectonic uplift rate of 1 mm a^{-1} could give rise to potential exposure of the repository within about 180 ka. This would certainly be long enough to considerably reduce the radiotoxicity of high-level waste, but calculated doses may still be significant. If the repository contained long-lived intermediate-level waste and/or spent fuel, resultant doses might be much higher than those generally considered acceptable (e.g. McKinley and Chapman, Chapter 24, this volume; Mathieson *et al.*, 2006).

Extreme erosion of coastal sites due to sea-level variation would be a worst-case scenario but, even if this does not occur over relevant timescales, the effects of hydrological and geochemical perturbations of both the near-field and far-field environments would need to be considered in performance assessments. For example, the change in the groundwater salinity as the coastline migrates (Figure 22.5) may change the efficiency of the repository engineered barriers; higher-salinity groundwater significantly reduces bentonite buffer swelling pressure (Karnland *et al.*, 2003), although recent work by Savage (2005) suggests that it may be possible to "engineer around" this by means of increasing the bentonite density. In addition, changes in salinity may also increase overpack corrosion, reduce retardation of radionuclides in the geosphere and influence elemental solubility, although these effects are expected to be relatively small (Alexander and McKinley, 1999).

Apart from absolute uplift/subsidence rates, differential movements could be important if they could significantly alter the direction and/or rate of groundwater flow. As noted above, the major changes in the biosphere into which any releases occurs are also very important, particularly in terms of the extent of dilution that can be assumed, but also due to inherent differences in the concentration and dose pathways in terrestrial and marine systems.

The significance of these effects is likely to be very site- and repository-concept-specific. In all cases, however, the cyclic variation in sea level is likely to introduce an additional level of complexity to the already great challenge of adequately simulating the long-term evolution of repository systems in their specific geological and biosphere settings.

22.5 Synthesis: issues to be considered by implementers and regulators

For reactors and other similar nuclear facilities, experience gained applying current regulations (Inoue, Chapter 21, this volume; Hill *et al.*, Chapter 25, this volume; IAEA, 1993) has already been extended to coastal sites. Coupling of perturbations is noted in IAEA (1993), e.g. "... the damage to the plant due to the combination of seismic and flood effects may be greater than that arising from either occurring separately" and "... the combination of fire and structural motion may together produce more damage in other safety systems than if they had occurred separately. This aspect of indirect seismic failures is obviously quite complex and has yet to be fully modeled." In the period since publication of IAEA (1993) there have certainly been moves to bring these aspects together (e.g. Hill *et al.*, Chapter 25, this volume; IAEA, 2003a, 2003b). For high seismic risk areas such as Japan, such analyses for reactor sites and fuel-cycle facilities are now commonplace. Such considerations are readily extended to include both direct and indirect effects of volcanic activity.

Large tsunami are events associated with low-probability/high-consequence perturbations. Although probabilities may be "low," the uncertainties associated with submarine initiating events are larger than with sources on (or under) land, which are inherently easier to explore. Emphasis in risk assessment has been focused on utilizing the history of past occurrences of similar perturbations to develop models of future events, but scarcity of records and lack of understanding of the mechanisms involved for volcanic and landslip sources makes quantification of probabilities very difficult. This is further complicated if future occurrences could be influenced by processes such as global warming or sea-level change, where extrapolations can be based only on models based on expert judgments of how future anthropogenic drivers will develop. Due to the sensitivity of nuclear facilities, however, it may be prudent in such cases to attempt to minimize risks to new plants by appropriate siting and/or design of structures and operational procedures.

Indeed, for facilities such as reactors, assessment of risk should also take into account public concerns. Major damage to a coastal reactor anywhere in the world due to an earthquake, volcano or a tsunami could cause concern about any plants located in similar sites. Such a risk, for new plants, can be reduced by careful siting but, in many countries, efforts to reduce both probability and consequences of disturbance by design should, at least, be considered. In particular, constructing plants underground could be a very cost-effective option, which may also make them effectively invulnerable to other low-probability risks that worry the public (e.g. terrorist attacks).

For long-term waste stores, operational repositories and repositories that are under active institutional control, similar issues to those above for nuclear power plants arise. Here, however, the timescales may be longer. Monitoring for three hundred years is planned for the Rokkasho R1 and R2 repositories in Japan and even longer periods have been considered for the La Manche waste site in France. These are extremely costly programs. La Manche has a proposed budget of FF 12–13 million ($\sim$ 2 million euros) per year until the year 2300. This provoked the then director of ANDRA (French National Nuclear Waste Management Agency), Mr. Kaluzny, to ask, "Do financial instruments guaranteeing such revenues for three hundred years actually exist?" (Le Monde, 1995; IAEA, 2006). Because of the nature of these sites, both slow processes and disruptive events need to be included in the associated safety case. Although active site control can be expected, allowing remediation in the event of major disturbances, the high cost and potentially negative impact on public confidence of such remediation should be carefully balanced against the costs of potential counter-measures.

Over longer time periods, assuring safety of surface or intermediate-depth disposal sites in coastal locations is particularly tricky due to the combination of rising sea level together with other potentially disruptive events. For wastes with significant inventories of long-lived radionuclides, risks of major disruption might be reduced by locating new facilities at greater depths and further from the coast (this should be explicitly considered during the siting and design process).

For deeper facilities, perturbations caused by surface processes may be buffered for many thousands of years by the geological and engineered barriers. Eventually, however, some or all of the barrier system is likely to be modified by cycles of sea-level change that act in combination with tectonically driven processes (uplift/subsidence) and events (e.g. volcanic eruptions). Safety assessment needs to consider explicitly all relevant processes which influence release of activity from the waste package, its transfer through the geosphere and the doses resulting from its release into the biosphere. In a coastal setting, the treatment of the geosphere/biosphere interface and the associated model of human activities and food chains could be very sensitive and have a large effect on calculated hazards. In some cases, possible ranges of processes can be simplified by "conservative assumptions" that err on the side of caution.

In conclusion, coastal sites have many positive features which make them attractive locations for siting nuclear facilities. Some sites, however, are associated with particular environmental risks, or an assessment of their safety reveals particular complications, which do not occur in inland settings. Therefore, it is important that all pros and cons of specific settings are realistically but comprehensively assessed and, if needed, specific design counter-measures introduced to minimize any unacceptable consequences of credible future scenarios.

Further reading

There is no other publication known to us that specifically examines the sensitivity of nuclear (or any other important industrial) facilities located in coastal areas to the full spectrum of geohazards examined here. Basic background material on hazards relevant to coastal locations can, however, be derived from McGuire *et al.* (2004) and some specific hazards have been examined, albeit in a more general framework (e.g. DEFRA, 2005). In addition, some specific nuclear facilities have undergone a preliminary hazard assessment (e.g. Alexander and Neall, 2007), but this remains the exception, rather than the rule. Interpreting such risks for nuclear reactors and fuel-cycle facilities can be examined using the methodology covered in Fullwood (2000). For radioactive waste disposal facilities, an overview of designs and both operational and post-closure safety issues is provided by Alexander and McKinley (2007).

Appendix

Impact of a tsunami on nuclear reactor sites at Kalpakkam (edited from AERB NEWS, 2008)

The tsunami caused by the December 26, 2004 earthquake hit the east coast of India and, at the time of the incident, Unit-2 of the Madras Atomic Power Station (MAPS) was operating (Unit-1 of the MAPS has been shut down since August 2003). Seawater entered the pump house through the intake tunnel and resulted in tripping of the cooling pumps. The control-room operator tripped the turbine and consequently the reactor tripped. Cool down of the system was initiated and the reactor was shut down safely. There was seawater inundation ($\sim$ 0.5 m) over the ground/road up to the east periphery of the turbine building, but there was no entry of seawater in the reactor building, turbine building and the service building. The low-water-level alarm is triggered at 3.4 m in the pump house fore bay. As the water level did not drop to this level, the alarm was not activated. The lowest water level during the incident is not known. To assess the impact of the tsunami, a team of senior inspectors from the Atomic Energy Regulatory Board (AERB) visited the nuclear reactors at Kalpakkam on December 29, 2004. The AERB team inspected all the important areas of the plant. All radiological conditions in the plant were normal and there was no release or discharge of radioactivity from the plant. The AERB team also noted that the operator response to the event was satisfactory. Based on the AERB inspection, Unit-2 of MAPS was restarted on January 2, 2005, and remains operational.

References

AERB NEWS (2008). Impact of Tsunami waves on nuclear reactors at Kalpakkam. AERB News article, http://www.aerb.gov.in/cgi-bin/News/AERBNews/detail.asp?ID=19. India: Atomic Energy Regulatory Board.

Alexander, W. R. and I. G. McKinley (1999). The chemical basis of near-field containment in the Swiss high-level radioactive waste disposal concept. In: Metcalfe, R. and C. A. Rochelle (eds.) *Chemical Containment of Wastes in the Geosphere*, Special Publication 157. London, UK: Geological Society, 47–69.

Alexander, W. R. and L. E. McKinley (eds.) (2007). *Deep Geological Disposal of Radioactive Wastes*, Radioactivity in the environment, 9. Amsterdam: Elsevier, http://www.sciencedirect.com/science/publication?issn=15694860&volume=9.

Alexander, W. R. and F. B. Neall (2007). Assessment of potential perturbations to Posiva's SF repository at Olkiluoto caused by construction and operation of the ONKALO facility, Working Report 2007-35. Olkiluoto, Finland: Posiva Oy.

Beckmann, B.-R. and G. Tetzlaff (1998). Modifications of the frequency of storm surges at the Baltic coast of Mecklenburg–Vorpommern. *Global Atmosphere and Ocean System*, **6**, 177–192.

Browitt, C., A. Walker, P. Farina *et al.* (2007). Terra not so firma. *Geoscientist Online*, **17.6**, http://www.geolsoc.org.uk/gsl/geoscientist/features/page3694.html.

Chapman, N. A. and C. McCombie (2003). *Principles and Standards for the Disposal of Long-lived Radioactive Wastes*, Waste Management Series, 3. Amsterdam: Pergamon.

Church, J. A., J. M. Gregory *et al.* (2001). Sea level changes. In: Houghton, J. T., Y. Ding and D. J. Griggs *et al.* (eds.) *Climate Change 2001: The Scientific Basis: Contribution of Working Group I to the Third Assessment Report of the Intergovernmental Panel on Climate Change*. Cambridge: Cambridge University Press, Ch. 1.

DEFRA (1999). Report by the United Kingdom on intentions for action at the national level to implement the OSPAR strategy with regard to radioactive substances. London: Department for Environment, Food and Rural Affairs.

DEFRA (2005). The threat posed by tsunami to the UK. London, UK: Department for the Environment, Food and Rural Affairs.

EA (2005). The Environment Agency's assessment of BNFL's 2002 environmental safety cases for the low-level radioactive waste repository at Drigg, NWAT/Drigg/05/001 (Version: 1.0). Bristol: The Environment Agency for England and Wales.

FEPC (2008). Location of nuclear power plants in Japan. Washington, DC: Washington Office of the Federation of Electric Power Companies of Japan, http://www.japannuclear.com/nuclearpower/program/location.html.

Fullwood, R. R. (2000). *Probabilistic Safety Assessment in the Chemical and Nuclear Industries*. Boston, MA: Butterworth–Heinemann.

Fukuyama, H. (1978). Geology of Sakurajima volcano, southern Kyushu. *Journal of the Geological Society of Japan*, **84**, 309–316 (in Japanese with English abstract).

Gallani, M. L. (2007). Review of medium to long-term coastal risks associated with British Energy sites: climate change effects – final report, CPM/2006/C001C/P176/1. Exeter: Meteorological Office for British Energy.

GEOID (2008). Building the digital earth. Geoscience Interactive Databases. Ithica, NY: Institute for the Study of the Continents (INSTOC), Cornell University. http://atlas.geo.cornell.edu/.

Highland Council (2006). The socio-economic impact of the Dounreay decommissioning programme: a strategy for Caithness and North Sutherland, report by Caithness Area Manager, September 7, http://www.highland.gov.uk/NR/rdonlyres/1AECB73A-FC10-4DDC-A828-CE6F5B4059FE/0/ItemNo10hc2406.pdf

IAEA (1993). Probabilistic safety assessment for seismic events, IAEA-TECDOC-724. Vienna, Austria: International Atomic Energy Agency.

IAEA (2003a). Site evaluation for nuclear installations, IAEA Safety Standards Series, NS-R-3. Vienna, Austria: International Atomic Energy Agency.

IAEA (2003b). Extreme external events in the design and assessment of nuclear power plants, IAEA-TECDOC-1341. Vienna, Austria: International Atomic Energy Agency.

IAEA (2006). Integrated regulatory review service (IRRS) - full scope - to France. Paris, France, IAEA-NS-IRRS-2006/01. Vienna, Austria: International Atomic Energy Agency.

INSC (2008). Reactor Maps. International Nuclear Safety Center at Argonne National Laboratory, http://www.insc.anl.gov/.

Intergovernmental Panel on Climate Change (2007). *Climate Change 2007: The Physical Science Basis: Working Group I Contribution to the Fourth Assessment Report of the IPCC*. Cambridge: Cambridge University Press.

Karigome, S. (2005). Policy on management and disposal of low level waste in Japan. In: *Disposal of Low Activity Radioactive Waste: Proceedings of an International Symposium on Disposal of Low Activity Radioactive Waste, Held in Cordoba, Spain, 13-17 December*. Vienna, Austria: International Atomic Energy Agency, 87–94.

Karnland, O., A. Muurinen and F. Karlsson (2003). Bentonite swelling pressure in NaCl solutions – experimentally determined data and model calculations. In: Alonso, E. E.

and A. Ledesma (eds.) *Advances in Understanding Engineered Clay Barriers: Proceedings of the International Symposium on Large Scale Field Tests in Granite, Sitges, Barcelona, Spain, 12–14th November*. Leiden: Balkema, 241–256.

Kim, K., H. Lee, M. Haggag and T. Yamashita (2006). Storm surge simulation on Hurricane Katrina using air–wave–sea coupling model. *Kaigan Kogaku Rombunshu*, **53**, 416–420 (in Japanese).

Klos, R. and R. Wilmot (2002). Review of project SAFE: comments on biosphere conceptual model description and assessment methodology, SSI Report 2002:17. Stockholm, Sweden: Swedish Radiation Protection Authority.

Langenberg, H., A. Pfizenmayer, H. von Storch and J. Suendermann (1999). Storm-related sea level variations along the North Sea coast: natural variability and anthropogenic change. *Continental Shelf Research*, **19**, 821–842.

Le Monde (1995). Interview with Mr Kaluzny, director of ANDRA, November 1, 1995. Paris, France: Le Monde.

McGuire, B., C. Kilburn, P. Burton and O. Willets (2004). *World Atlas of Natural Hazards*. London: Hodder Arnold.

Mathieson, J., A. Hooper, W. R. Alexander, M. Shiotsuki and G. Kamei (2006). International progress in developing cases for long-term safety of repositories for transuranic and long-lived intermediate level wastes: summary of the third international workshop. Presented at: WM'06 Conference, February 26–March 2, Tucson, AZ.

Mobbs, S. F., D. Charles, C. E. Delow and N. P. McColl (1989). Assessment of subseabed disposal of vitrified high level waste for the PAGIS Project, NRPB Report R218. London: Health Protection Agency, National Radiological Protection Board.

NASA (2008). NASA Visible Earth, a catalog of NASA images and animations of our home planet, http://visibleearth.nasa.gov/.

NEI (2005). Carry on at CoRWM. *Nuclear Engineering International*, feature article, March 29.

Posiva (2003). ONKALO Underground characterisation and research programme (UCRP), Report POSIVA 2003-03. Olkiluoto, Finland: Posiva Oy.

Sasaki, T., T. Moritomo, H. Ikeda, T. Shiraishi and S. Sugi (2007). Groundwater flow prediction method in consideration of long-term topographic changes in uplift and erosion. Presented at: IGSC Workshop on the Stability and Buffering Capacity of the Geosphere for Long-term Isolation of Radioactive Waste: Application to Crystalline Rock, November 13–15, Manchester, UK.

Savage, D. (2005). The effects of high salinity groundwater on the performance of clay barriers, SKI Report 2005-54. Stockholm, Sweden: Swedish Nuclear Power Inspectorate.

SKB (2001). Project SAFE: scenario and system analysis, R-report SKB R-01-13. Stockholm, Sweden: Swedish Nuclear Fuel and Waste Management Co.

Sumerling, T. (2005). Control, loss of control, causes, putative scenarios and option performance, support to CoRWM – Task TS108/4, Report SAM-J114-TN1, Version 2. Reading, UK: Safety Assessment Management.

UKAEA (2004). Project Update – September 2004, Decommissioning the Dounreay waste shaft. The United Kingdom Atomic Energy Authority, http://www.ukaea.org.uk/downloads/dounreay/Shaft.

UNEP (2002). *Global Environment Outlook 3: Past, Present and Future Perspectives*. Nairobi, Kenya: United Nations Environment Programme.

UNEP (2007). *Global Environment Outlook: Environment for Development, GEO 4*. Nairobi, Kenya: United Nations Environment Programme.

US DOE (2002). Yucca Mountain science and engineering report, DOE/RW-0539-1. Washington, DC: US Department of Energy, Office of Civilian Radioactive Waste Management.

Woth K., Weisse R. and H. von Storch (2006). Climate change and North Sea storm surge extremes: an ensemble study of storm surge extremes in a changed climate projected by four regional climate models. *Ocean Dynamics*, **56**, 3–15.

23

Stable tectonic settings: designing site investigations to establish the tectonic basis for design and safety evaluation of geological repositories in Scandinavia

T. McEwen and J. Andersson

Knowledge of the tectonic setting of the site for a nuclear facility is required in order to develop a design for the facility and to evaluate its safety, both during its operational phase and, in the case of a radioactive waste repository, following its closure. Understanding the tectonic evolution of the site and of the area in which it lies permits estimation of, for example, future seismic or volcanic activity. In this chapter, we focus specifically on the investigation of sites proposed for deep geological repositories for radioactive waste, although the structured approach and many of the general principles discussed are more widely applicable to the siting of other nuclear facilities. This chapter also focuses on the situation of a relatively stable tectonic environment, which exists in much of Scandinavia, where volcanic activity is not a consideration and where the maximum level of seismic activity is low, at least as far into the future as the next glaciation. Olkiluoto, in Finland, is used as an example of such a stable tectonic environment to illustrate the approach that is being taken in its characterization. Neotectonic activity is still of interest, however, in a Scandinavian context, and reference is also made to the situation of sites in less tectonically stable environments.

An understanding of the tectonic setting of a site, even where the tectonic activity is relatively benign, is an integral part of developing an understanding of the overall geological environment, so that a convincing and well-founded geological model of the site can be developed. This geological model will, in turn, be used to develop other models (e.g. hydrogeological, hydrogeochemical and rock mechanics models) of the site and eventually a fully integrated "site descriptive model." Developing such an understanding could be a major part of the site investigation program.

The site descriptive model is an important factor in deciding on the most suitable repository design concept, in determining the most suitable location and depth for the repository and in determining its layout. Such decisions need to be based on a good understanding of the geology, hydrogeology, hydrogeochemistry and rock mechanics of the site. It is also important in developing a safety case for the repository, both during its operational stage and for tens of thousands of years after it is closed.

23.1 Strategy of site investigation

The tectonic setting of a site and the area in which it lies needs to be established using a phased approach. Initially, and almost certainly before the site is selected, a literature search will have been carried out. The level of information at this stage may vary considerably between different sites, but any assessment of the tectonic setting will be made easier if the site has been selected in an area that is geologically less complex. A site with a relatively simple geological environment is also more likely to be easier to characterize. The ease of site characterization has been considered by Nirex, a UK organization dedicated to radioactive waste management issues, as a scoring factor when comparing alternative sites during site selection programs (e.g. Chapman and McEwen, 1991; Nagra, 1994). It has been termed the site's *explorability* (NEA, 1991). This aspect of the site selection process is obviously very dependent on the geological history of the area in which the site lies.

At the earliest possible stage in the site investigation it is necessary to develop sufficient confidence to devote large resources to a fuller investigation, in particular, by establishing whether there are any obvious geological features that could make the site clearly unsuitable. This confidence must be derived from an evaluation of the literature review results plus the data and their interpretation gathered during the initial investigations, which are likely to be limited in duration and scope. Parenthetically, it is common to divide the investigation of a potential radioactive waste disposal site into stages, or phases, with the initial phase lasting perhaps three to five years. This, for example, has been the situation in Finland and Sweden and is the proposed approach in Japan. There is a natural break at the end of the initial phase before any underground construction can commence, after which subsurface investigations can take place from excavations.

In considering how the tectonic regime of a site can be established, it is assumed (for the purpose of this discussion) that the investigations can be divided broadly into initial investigations (INI) (which are assumed to last perhaps three to five years and to take place on and above the surface, including deep boreholes) and detailed investigations (DEI) (which include investigations underground, but are likely to include additional surface-based investigations). Both the INI and DEI are likely to be subdivided into several shorter phases related to specific elements of the program, such as early-phase and later-phase boreholes. This terminology, *initial* and *detailed*, is also used in the following to describe the other studies that will need to be carried out at the site, related to repository design and safety studies. National repository programs have used a range of terminology for these phases, and the timeframe of three to five years is introduced here as a practical example that is similar to investigations that have been taking place at two potential repository sites in Sweden (e.g. SKB, 2005a, 2005b, 2005c). This terminology is also consistent with the Nuclear Waste Management Organization of Japan (NUMO) plans for their preliminary investigations in Japan at potential geological repository sites for high-level waste (NUMO, 2004a, 2004b).

The confidence referred to above will be built upon two main foundations. First, there must be a very low probability of finding any problematic geological features, characteristics or properties (subsequently referred to as problematic features) at any stage during the investigations (but especially when such investigations take place underground during the DEI stage) that would have a major effect on the long-term safety of the facility. Examples of these problematic features in a Scandinavian context include the presence of extensive shear zones and ore minerals, as discussed in Andersson *et al.* (2000). In order to establish confidence, either the investigation program is able to demonstrate that no such problematic features exist, or is able to estimate the likelihood that they will not be found later (and that this likelihood is low). Second, a good understanding of the factors that will control the feasibility and costs of constructing/operating a repository and developing a strong safety case must be established.

The problematic features that are related to the tectonic setting of a site, and applicable to any level of tectonic activity, include the presence of recent volcanism, active faulting, excessive uplift rates or mineral reserves. The probability of finding any of these is obviously related to the level of tectonic activity in the area of the site and much can be done to limit the likelihood of their occurrence by selecting sites where evidence for such activity is lacking. In addition to these, there are other potentially important features and properties of the geological environment, related to the tectonic setting of the site. These include rock mass properties that do not allow the safe construction of the repository; *in situ* stresses that are abnormally high and result in potential unstable conditions underground; volumes of the rock mass that are insufficiently large to allow for repository development; high groundwater fluxes at repository depth; and obvious fast groundwater pathways at depth (included here as these are, in part, often tectonically controlled). Determining their existence or absence can be considered the goals of the investigation program.

In contrast to the problematic features, it is more difficult to reduce the likelihood of finding such features and properties of the geological environment by any form of screening as part of the site selection process, as there is likely to be limited information on the geological conditions at depth. The strategy to be developed for the initial site investigations depends on four main factors: (i) the extent to which it is considered possible to provide answers to the key questions discussed above within the time allocated for the initial investigations (perhaps three to five years); (ii) the parameters that will need to be measured to answer these key questions and satisfy these goals; (iii) the process that is to be used to decide whether the area is suitable for further, more detailed investigations; and (iv) the process that is to be used to decide the boundaries of these further investigations and the type of geological environment that is being investigated. The methodology applied in developing a suitable investigation program, based on these strategic factors, is discussed in Section 23.2.

The goals, stages and principal components of the geoscientific work during the initial phase, therefore, need to be defined, together with the structure of the investigations and modeling based around the various disciplines that are of interest in developing a repository: geology, rock mechanics, thermal properties, hydrogeology, hydrogeochemistry, transport properties of the rock and surface ecosystems or biosphere. These include additional disciplines to those normally included in the assessment of a site, in particular thermal properties and transport properties. The thermal properties are of interest, as the waste produces a considerable amount of heat, and transport properties refer to the properties of the rock mass that determine the rate of radionuclide transport away from the repository.

The assessment of a site can also be divided into the principal technical areas of investigations, initial design study, initial safety judgment, and the investigations designed

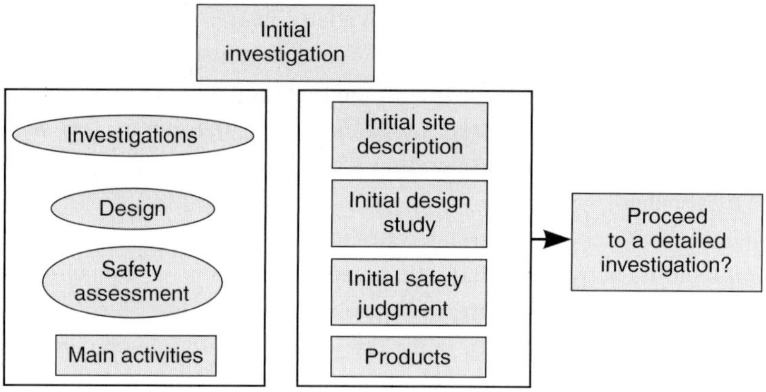

Fig. 23.1 Illustration of the main activities (ovals) and products (rectangles) which are expected during an initial investigation (INI).

accordingly (Figure 23.1). The information will be evaluated and assembled and the required technical supporting documents produced in the form of an *initial site description*, an *initial facility description* and an *initial safety report* (waste management organizations around the world use a variety of terminology to refer to these three documents, although all do produce, or intend to produce, documents with similar contents). A major component of the initial site description will be the initial site descriptive model (SDM), an integrated model that incorporates sub-models from the various disciplines listed above. Such integration between modeling in different disciplines is essential, since all the models describe the same volumes of rock. Good examples of such integration are shown by the work at Olkiluoto, Finland in the development of the SDM (Andersson *et al.*, 2005, 2007), in the model developed as part of the study of the Opalinus Clay in the Zürcher Weinland, Switzerland by Nagra (Nagra, 2002a, 2002b) and by SKB in their investigations of Forsmark and Laxemar, Sweden (e.g. SKB, 2005a).

Integration of understanding across scientific disciplines is enhanced by requiring that the presentation of information from each discipline follow a pre-set outline. This outline needs to include: the presentation of the conceptual model, an evaluation of the available information (in terms of its sufficiency, quality, whether it is representative, etc.), an assessment of the interactions with other disciplines, a presentation of the descriptive modeling and an evaluation of uncertainties.

The process of undertaking prediction/outcome studies (i.e. studies in which predictions of geological properties are made, which are later tested against outcomes of field measurements and tests), and the need to complete an overall confidence assessment, also enhance integration. Although the SDM should be an integrated description, encompassing all different disciplines, it is a necessary and practical requirement to divide such a model into the different discipline areas: geology, hydrogeology, rock mechanics, etc.

The investigations result in primary data, both measurement values and directly calculated values, which are collected in a database. In order for the collected (measured) information

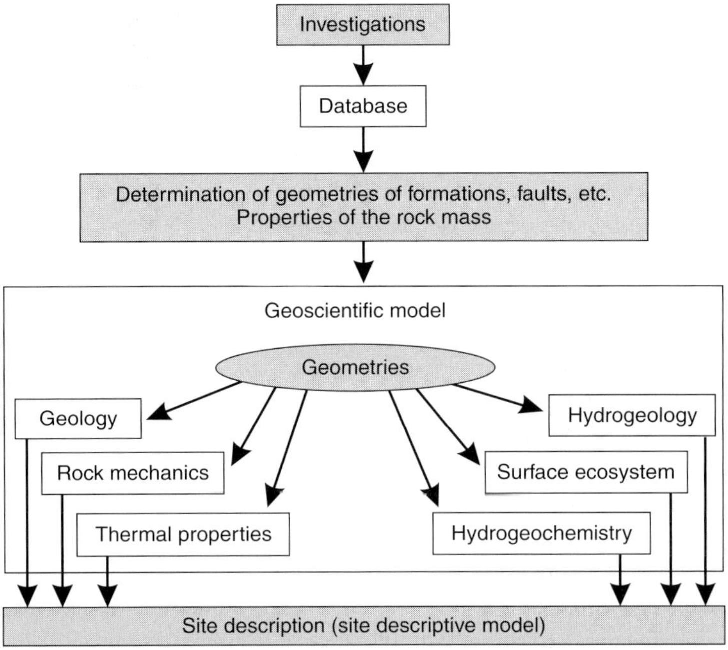

Fig. 23.2 The primary data from the investigations are collected in a database. Data are interpreted and presented in an SDM, which consists of a description of the geometry and different properties of the site.

to be used for the design study and the safety assessment and to enable the reliability of the information to be judged, it will need to be interpreted and presented in the (mainly geoscientific) initial SDM (Figure 23.2), which consists of a description of the geometry and different properties of the site and comprises, together with databases, the backbone of the site description. There may be more than one version of this model during the INI (and further versions during a subsequent DEI). An example of the development of an SDM for Olkiluoto, in Finland, is described in Section 23.3 (SKB, 2005a).

The way in which the main activities, the investigations and the initial design and safety assessment, are to be integrated needs to be defined. When the investigations have gathered and interpreted new data, which could be incorporated into a revised site model, the repository layout can first be modified and then the safety assessment can be revised in light of the new model. The interdependence between the different principal activities requires coordination to ensure that the initial safety assessment and the repository layout are based on consistent versions of the site model. Unnecessary overlap of modeling work is, thereby, avoided, and judgments that are made with reference to one principal activity can be applied in the other activities in a consistent manner.

The investigation program should also determine parameters that require undisturbed conditions and initiate monitoring of parameters where long time series are essential. These matters are most significant in relation to the definition of the baseline conditions. Parenthetically, these are the conditions that pertain at a site before the investigations commence. It is important that they are defined with sufficient accuracy before any underground construction starts, as this construction will irrevocably perturb, for example, the groundwater flow and *in situ* stress conditions. Variation in the amount and quality of information available in advance of an INI may be considerable and, together with the type of geological environment under consideration, will affect the design of the investigation and the area considered necessary for the INI. The design of such an INI may, therefore, differ considerably between sites.

23.2 Methodology

A logical and well-structured investigation program is needed in order for the activities to produce the necessary information with sufficient precision, level of detail and accuracy, and with an efficient utilization of resources. Integration is required in two main areas: between the different disciplines and with the main users of the information from the investigations (the repository design and safety assessment teams). The integration between disciplines is of primary importance during the field operations in order to make best use of boreholes, time and resources, as well as in developing the SDM. Planning the INI so that it meets its objectives requires careful consideration. Experience from existing investigations, such as the SKB site investigation planning (SKB, 2000), suggests this is best achieved by using the following methodology.

Step 1. Assess existing information and develop a Version 0 SDM. Before the actual planning of the INI, an assessment of the information that is already available is needed. Several questions need to be asked, including: what is the quality of the available data and what reliance can be placed on it? Are more data likely to become available? What are the geological characteristics of the site and its environment? Can a potential repository host-rock formation be identified? What are the main uncertainties associated with reponses to these questions? The overall understanding of the main characteristics of the site is then formulated into the first (Version 0) SDM. In order to set INI targets and needs, it is also necessary to establish the range of possible repository design concepts that could apply to a specific site.

Step 2. Establish the main targets. The actual INI planning starts by establishing the investigation targets, which should be set by the end-users of the information (i.e. safety assessment, engineering design, environmental group/surface facility designers). The INI must also provide feedback to the planning of any subsequent characterization of the site (i.e. planning the DEI). These targets would generally concern the problematic features and other potentially important features and properties of the geosphere, related to the tectonic setting of the site, as discussed in Section 23.1; and it is important that such generic targets be adapted to the existing knowledge of the site and to the main uncertainties. Furthermore, it is not likely that the scope of the INI will allow all the specified targets to be met and priorities will therefore need to be established.

Step 3. Identify and select investigation methods. When selecting methods, such as specific geophysical techniques used to characterize target geology, it is necessary also to identify the constraints on the practicalities of the investigations, such as climate, infrastructure, environmental impacts, budget and time.

Step 4. Prepare a geoscientific database. A common geoscientific database for storing and retrieving all investigation results is needed. Where data are derived from the interpretation of the raw data, both the raw data and the interpretation of these data should be entered into the database, and a clear distinction must be made between the two types. Strict quality assurance and version control procedures are needed when storing and retrieving these data.

Step 5. Plan the evaluation modeling. In order to be able to make use of the collected (measured) information, it must be interpreted and presented in a geoscientific SDM. Detailed plans for this modeling can be developed once the targets and planned investigations have been identified.

Step 6. Establish the integrated INI program. Finally an integrated INI program is established by setting out the sequence of investigations and characterization activities, a process

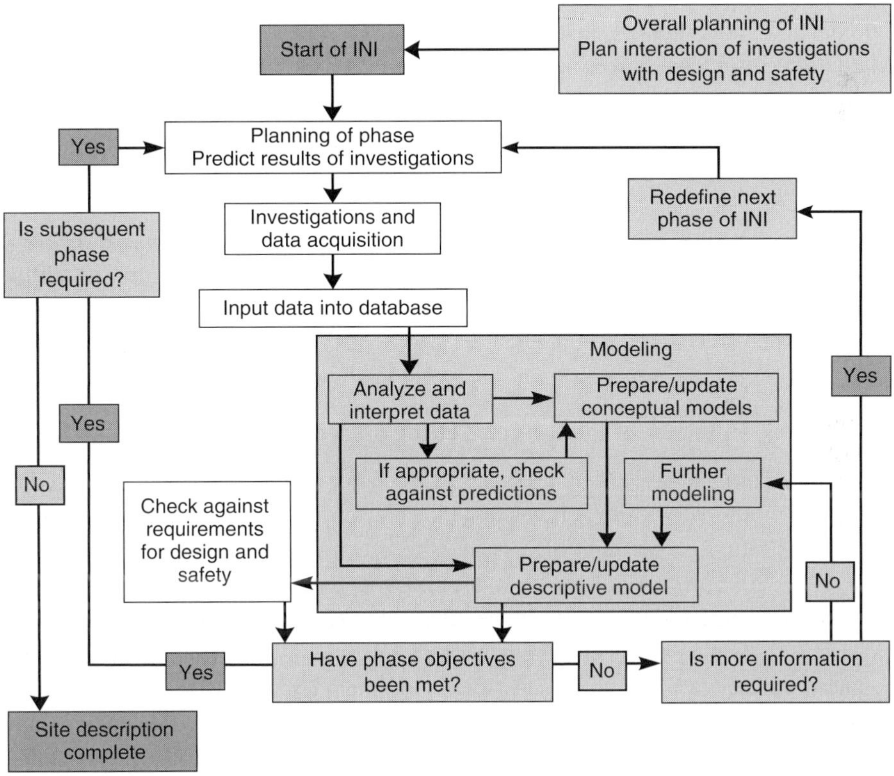

Fig. 23.3 The phasing of investigations, as applied to the INI, illustrate how such phasing is applied in the evaluation of a site. The site investigation data acquired in any one phase are analyzed and interpreted and used as input to the development of conceptual and descriptive models. When the objectives of any phase have been met, either the next phase can commence or, after sufficient phases, the SDM can be considered complete.

that needs to consider the priorities of the different investigation targets, combined with the constraints on the different investigation methods.

The INI also needs to be subdivided into phases or steps in order to allow the investigation methodology to be adapted to the site and for a more efficient feedback from the evaluation of the investigation data as they are collected (Figure 23.3). Each new phase or step entails confirming or rejecting the main results of the preceding step of the investigations, answering the questions that have arisen and carrying out the tasks set for the particular step, with each step building further on the description that emerged from the preceding investigation step. An important element of each step is verifying whether the outcome of the investigation step agrees with predictions made in advance (referred to as *prediction/outcome studies* by Posiva, the organization responsible for radioactive waste management in Finland, (Andersson *et al.*, 2005, 2007)).

At the end of each phase or step a decision has to be made as whether to continue with the investigations, or whether sufficient data have been obtained, and also whether the objectives of the step have been met. If they have not, there are three possibilities: (i) it may still be decided, nevertheless, that sufficient data have been collected, and a preliminary SDM could be produced, based on these data (even if this situation is not ideal); (ii) it may be decided that there is no more time available for the investigations (this may imply that the site is unsuitable); or (iii) it may be decided that the subsequent phase of the investigations will need to be redefined (i.e. perhaps some of the objectives of the investigations may need to be modified).

If the objectives of that phase of the investigations have been met then a decision needs to be made as to whether a subsequent phase of investigations is required; or if no additional information is required, then the preliminary SDM, based on these data and the associated modeling, can be considered complete.

23.3 Influence of the tectonic setting on repository development

The tectonic setting of a site has a considerable influence on a repository development program and on the way a site is investigated. This subject is considered below using the example of Olkiluoto, Finland which is the site of a proposed repository for spent fuel (Posiva, 2003, 2006; Andersson *et al.*, 2005, 2007). The relative importance of the various factors discussed here for Olkiluoto will be different in other geological environments, such as those in sedimentary rocks and those in more tectonically active areas, but the general principles are likely to be applicable to any environment and site.

Olkiluoto is located in a tectonically stable part of Europe, where neotectonic activity is limited to relatively small-scale movements on large brittle deformation zones, and is in an area with a low level of seismic activity. The current regional trend of the maximum principal stress is approximately southeast-northwest and is considered to be due to the combined effect of mid-Atlantic ridge push and European–African plate collision. However, the stress state at Olkiluoto may also be affected by the larger-scale lineaments and brittle deformation zones that define and cut the island. The most recent major modification to the current stress field is the glacial loading and unloading that occurred during the most recent ice age. The magnitudes of the associated stress changes were probably insufficient to create new fractures, but geological evidence exists that demonstrates that reactivation of some fractures has occurred and that the ice loading and current postglacial uplift (of order $5\,\mathrm{mm\,a^{-1}}$ in the Olkiluoto area) has further locally perturbed the regional stress state (Hudson and Cosgrove, 2006). A survey of glacio-isostatic (postglacial) faults in the Olkiluoto area has revealed a small number of such faults, but none in the vicinity of Olkiluoto Island was considered to have an indisputable postglacial component (Lindberg, 2007).

Posiva has operated a local micro-seismic network at Olkiluoto since February 2002 and the low level of seismic activity is demonstrated by the monitoring results for 2006, during which only two microearthquakes were detected (Mattila, 2007). Evidence of seismic activity that would affect either the current underground construction or the safety of the future repository has not been observed. The observed earthquakes in 2006 had magnitudes of $\mathrm{ML} = -0.6$ and $\mathrm{ML} = -0.9$ and relate to small movements on brittle deformation zones. All other micro-seismic activity was related to blasting associated with the construction of the ONKALO (the name given to the initial part of the underground phase of the DEI).

Global positioning system-based deformation studies have also been carried out in the Olkiluoto area since 1995, when a network of ten GPS pillars was established, with 22 GPS measurement programs having been carried out. According to the time series of the GPS results, $\approx 30\%$ of the baselines at Olkiluoto have statistically significant rates of change; however, the observed movements are $> 0.22\,\mathrm{mm\,a^{-1}}$ (Mattila, 2007).

Of greater interest is the impact of past tectonic activity, which has left a legacy of complex relationships between the various lithologies, large brittle deformation zones and a well-developed foliation, which has an important effect on the rock mechanics properties (Paulamäki *et al.*, 2002, 2006; Hudson and Cosgrove, 2006; Milnes *et al.*, 2006).

23.3.1 Investigations at Olkiluoto, Finland

An extensive investigation has taken place at Olkiluoto, including the construction of the ONKALO (Figure 23.4), which is located on the coast of western Finland in an area of very subdued topography. It lies in the western part of the Svecofennian area, part of the extensive Precambrian Fennoscandian shield, in which dominantly supracrustal rocks, mainly consisting of migmatized mica gneisses, are present. After the Fennian and Svecobaltian orogenies, 1910–1800 Ma, the deformation of the Fennoscandian shield was dominated by a series of tectonic events, which resulted in extensive brittle deformation (Paulamäki *et al.*, 2002; Andersson *et al.*, 2007). The relationship between these main tectonic events and the brittle deformation zones observed at Olkiluoto is not well understood, and is the subject of current study, but these events do provide a framework for constraining the timing and development of the brittle structures.

The larger of these brittle deformation zones can be seen from topography and airborne geophysics (mainly magnetics) as lineaments. However, many lineaments identified by

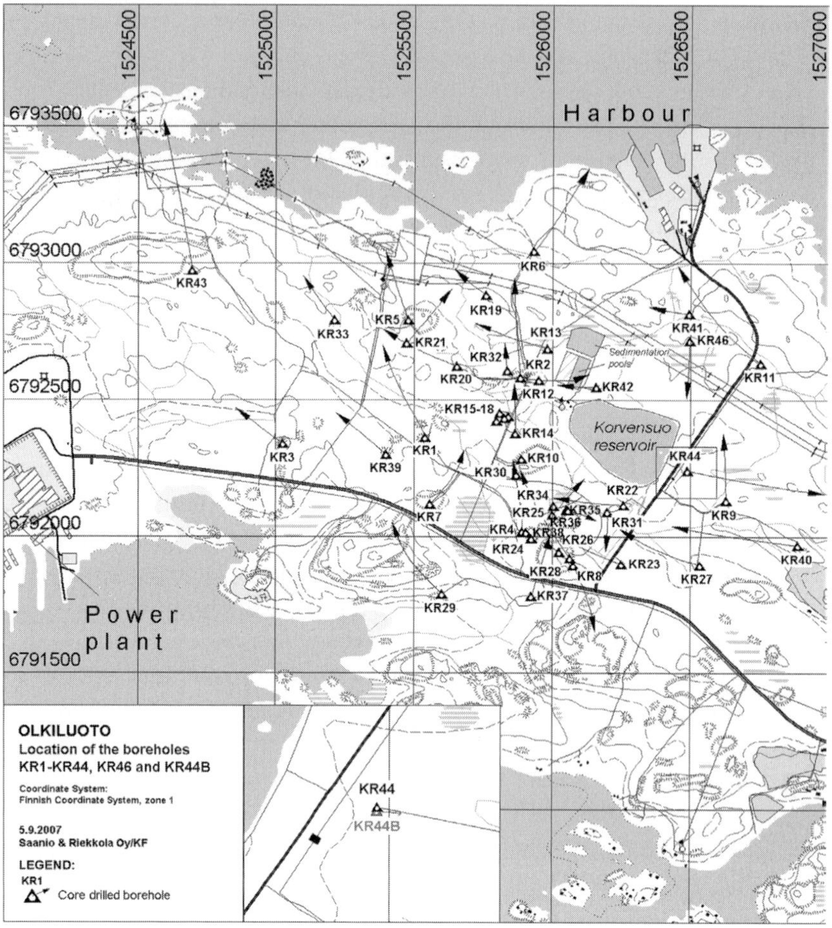

Fig. 23.4 Deep boreholes at Olkiluoto. The ONKALO is currently under construction to the south of the Korvensuo reservoir. The size of the grid squares is 500 × 500 m. Copyright (2006) Posiva Oy. Reproduced by permission of Posiva Oy.

geophysics may actually not relate to brittle deformation zones, but instead to other geophysical anomalies, such as intruding dikes; the identification of topographic lineaments is uncertain due to variations in the overburden thickness (mainly glacial deposits). Nevertheless, the larger deformation zones are believed to be where any substantial future movement and seismic activity could take place. There is a considerable separation of the proposed repository from such zones; however any seismic activity on such zones (which is likely to be significant only during the early phase of any future glacial retreat) will result in displacements on smaller-scale zones and on single fractures, and this has to be taken into account in the location of disposal tunnels and, in particular, in the location of the waste canisters within tunnels.

23.3.2 Development of the SDM

The main product of the investigations and the associated modeling is the SDM, a model that includes a description of the interacting processes and mechanisms that are relevant for understanding the evolution of the site to the present day and the potential for future radionuclide migration. This SDM is comprised of several discipline-specific models: the geological, hydrogeological, hydrogeochemical and rock mechanics (which includes thermal aspects) models. Although the surface conditions are included in the description of the site (Andersson *et al.*, 2007), they are not included within the SDM itself, nor are the transport properties, as both are considered as part of the development of the safety case.

The main approach in the construction of the current geological model has been the development of an understanding of the geological processes that have created the observed geological features at Olkiluoto (Paulamäki *et al.*, 2006). A thematic approach has been taken when describing the model, which is composed of four sub-models (Figure 23.5): (i) the ductile deformation model, (ii) the lithological model, (iii) the alteration model, and (iv) the brittle deformation model. It is emphasized that these sub-models are not independent entities, as the geological processes are closely interrelated. For example, the ductile deformation is an important precursor to the subsequent brittle deformation. This subdivision is, therefore, for the convenience of handling different types of data and should not be regarded as of fundamental significance.

The *ductile deformation model* describes and models the products of polyphase ductile deformation, which makes it possible to define the dimensions and geometrical forms of individual lithological units determined in the lithological model (e.g. Figure 23.6) and also to assess the orientation and effects of the lithological anisotropy. The ductile deformation has resulted in the formation of a pervasive foliation which imparts a marked anisotropy to the rock mass, and which is of considerable importance when considering its geotechnical and thermal properties.

The *lithological model* provides a general view of the lithological properties of definite rock volumes or units that can be defined on the basis of a set of parameters. The goal of the

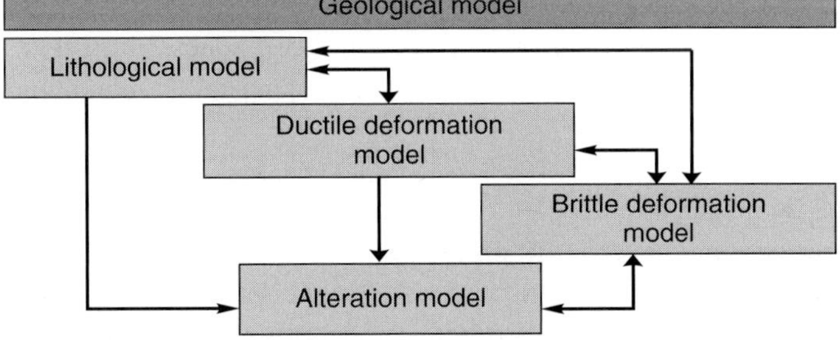

Fig. 23.5 Thematic sub-models of the geological model developed for Olkiluoto showing the links between the sub-models. Copyright (2006) Posiva Oy. Reproduced by permission of Posiva Oy.

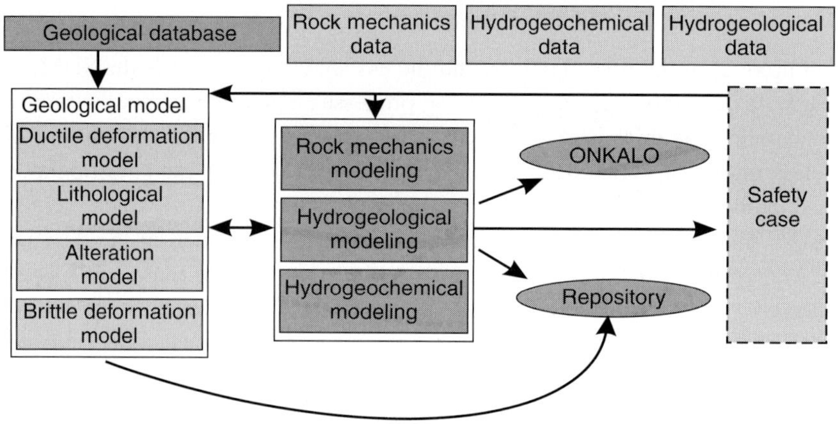

Fig. 23.6 Flow chart illustrating the interactions between the geological model, the other modeling activities, the design of the ONKALO and the repository, and the safety case. Copyright (2006) Posiva Oy. Reproduced by permission of Posiva Oy.

model is to represent the spatial distribution of lithologically fixed and genetically related bedrock units, which, from the perspective of underground construction, have uniform properties (Figure 23.6).

The *alteration model* deals mainly with the products of hydrothermal alteration, but retrograde metamorphism and subsequent low-temperature weathering, which have also affected the lithological units in the site area, are also considered as a part of the long-term alteration history. The goal of the alteration model is to present the shapes and volumes of altered bedrock units as well as the types of altered rocks.

The *brittle deformation model* describes the large-scale structures produced during the long history of brittle deformation (i.e. the fault zones and joint cluster zones). Brittle deformation products may have important implications for construction and long-term safety, and an initial attempt at evaluating their properties is presented in Andersson *et al.* (2007).

This information is then used as an important reference for the development of the hydrogeological and geomechanical models, for the construction and layout design both of the ONKALO and the repository, and as input to the safety assessment (Figure 23.6).

Lithological model

The rocks at Olkiluoto can be placed into two main classes on the basis of texture, migmatite structure and major mineral composition and, in total, five different lithological units are defined (an example of the three-dimensional distribution of one of these is shown in Figure 23.7). There are inevitable uncertainties concerning the geometry and extension of the various rock units at depth, even with data from the 33 boreholes that are used to develop the model, as the three-dimensional form of these units can be complex.

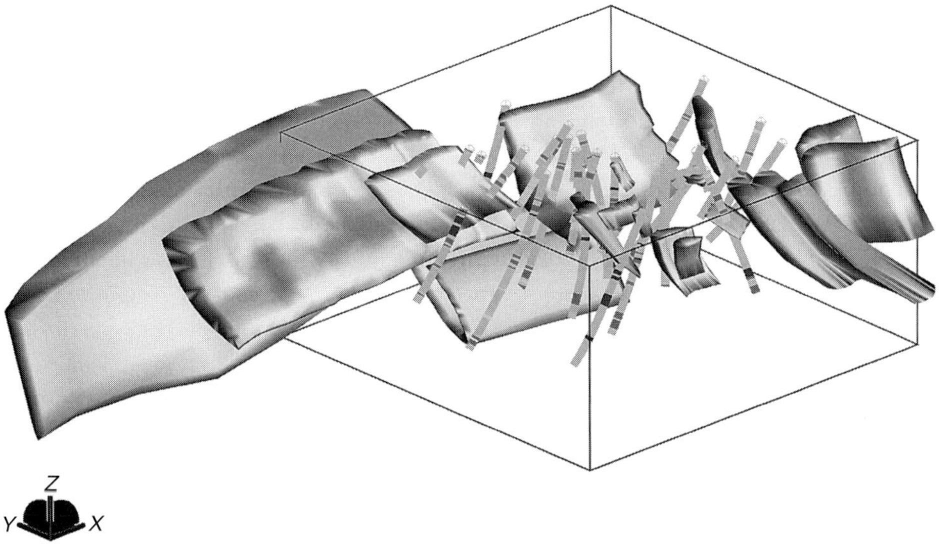

Fig. 23.7 Modeled tonalitic–granodioritic–granitic gneiss units are one of the lithological units defined in the lithological model. The black frame indicates the Olkiluoto site volume and borehole traces are also shown. Copyright (2006) Posiva Oy. Reproduced by permission of Posiva Oy.

Alteration model

Three different alteration episodes can be identified at Olkiluoto (Paulamäki *et al.*, 2006), with the hydrothermal alteration being the most important. Three types of hydrothermal alteration can be distinguished, with the addition of calcitic fracture fillings and sets of calcite stockworks; and all of these have been separately included in the alteration model. These processes have transformed the physical and chemical properties of rock material, and altered rocks may have physical properties that are notably different from those of primary, fresh rocks. Thus the degree and type of secondary alteration and retrogressive metamorphism are important parameters in evaluating, for example, the mechanical strength of the rocks. An example of the distribution of kaolinitization is shown in Figure 23.8.

Ductile deformation model

An understanding of the ductile deformation history is obviously important in demonstrating sufficient knowledge of the site, but the characterization of the resulting foliation, which is considered the most important element of the ductile deformation at Olkiluoto, is essential for an understanding of both rock mass properties and for rock engineering. Foliation planes represent incipient planes of weakness, resulting in anisotropic rock mass properties; consequently, the orientation of the foliation can have implications for the constructional properties of the rock mass. Different rock types show variations in the foliation type and the intensity of foliation, so that the significance of the foliation varies with rock type.

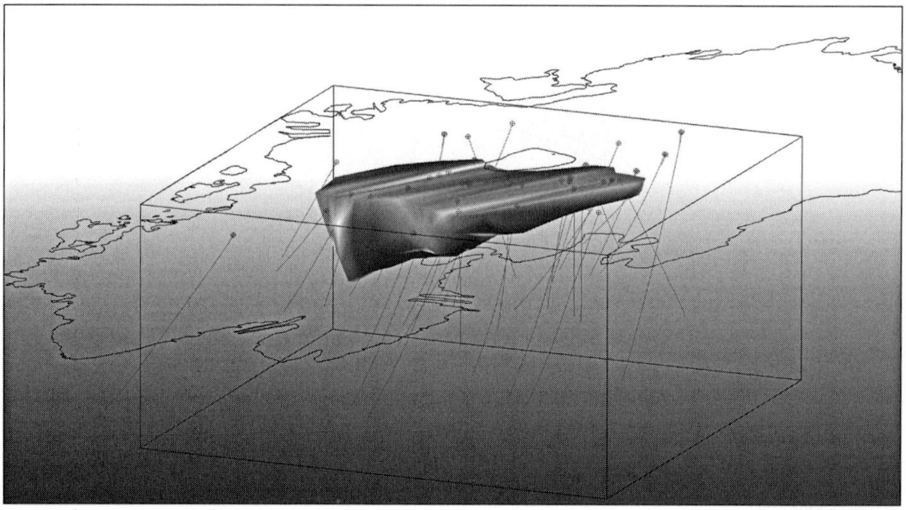

Fig. 23.8 Kaolinized domain (gray-shaded volume) is located in the uppermost part of the site volume (indicated by the black frame). Also shown are the borehole traces (light-gray lines) and the coastline of Olkiluoto Island. Copyright (2006) Posiva Oy. Reproduced by permission of Posiva Oy.

Brittle deformation model

The first step in developing the brittle deformation model involves the detection of all fractures that carry the imprints of tectonic movements (e.g. slickensides). These are studied in order to obtain the characteristics of the fracture surfaces, such as their shape, indicative traces of movement, orientation and kinematics (fault direction vector, direction of slip). The second step includes studying the slickensides, and dividing the deformation-zone intersections (within the drillholes, trenches and the ONKALO tunnel) into five classes: high-grade ductile deformation zone, low-grade ductile shear zone, semi-brittle fault zone, brittle joint zone and brittle fault zone. The locations and orientations of these zones have important links with the hydrogeological and rock mechanics models.

From a total of ≈ 1700 fault-plane and fault-vector directions that were analyzed, more than 20 groups of kinematically similar brittle faults were detected, based on the result of a simplified and generalized classification of fault-slip data (Andersson *et al.*, 2007). Tentatively, five groups of brittle faults can be defined from the total number of brittle faults; the distribution of one is shown in Figure 23.9.

Studies on the structural evolution of the bedrock have shown the obvious statistical similarity between the orientation of the regional, composite, pervasive foliation and the slickensided fracture surfaces, measured from the drillholes. One explanation for this association is that the faults appear to have exploited the planes of weakness imposed by the pervasive foliation. The products of ductile deformation seem, therefore, to be important precursors for subsequent brittle deformation; however, many brittle fault structures are not controlled by any older deformation elements.

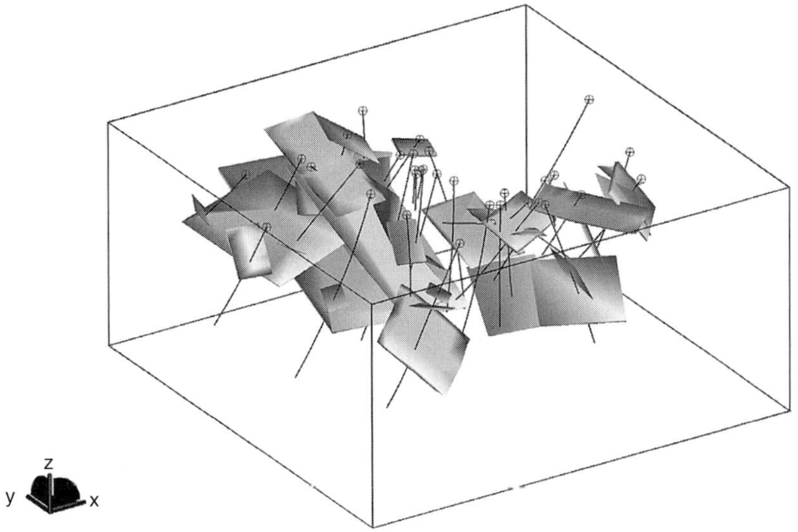

Fig. 23.9 Fault zones at Olkiluoto in one of the fault groups, borehole traces are also shown. Copyright (2006) Posiva Oy. Reproduced by permission of Posiva Oy. See color plate section.

The *hydrogeological model* consists of the *hydrostructure model* and the *flow model*. The *flow model* consists of the groundwater simulation model with its boundary conditions and assigned hydraulic properties. The *hydrostructure model* is the geometrical distribution of the permeable features of the rock. It is closely related to the geometrical structure of the *geological model*, but is not identical, and one of the modeling tasks is to describe the relationship of the *hydrogeological model* with the geological model. Hydrogeological zones with at least one measured transmissivity greater than $T = 10^{-7}\ m^2\ s^{-1}$ are explicitly incorporated into the *hydrogeological model*. The brittle deformation model was reconciled with the measurements of transmissivity and the hydrogeological responses to various field activities (pumping tests, water sampling, etc.), and strongly suggests the existence of three gently dipping, hydrogeologically dominant zone systems. Figure 23.10 shows hydrogeological zone HZ20, its relationship to three brittle fracture zones (BFZs), multiple boreholes and the tunnel.

The *rock mechanics model* consists of the geometrical, mechanical and thermal descriptions of the rock mass. The geometrical description is based on the bedrock geological model, as described above, and also has obvious relationships with the hydrogeological model. The mechanical description includes: the *in situ* stress state and the deformation and strength properties of the intact rock; the fractures; the rock mass between deformation zones and the deformation zones themselves; and their interaction with the underground excavations (Figure 23.11). Data for the model are derived from the drillholes and, eventually, from the excavations themselves. Any underground openings have to be located and designed with such interactions in mind, and the safety case for the repository needs to consider such interactions in both the short and long term.

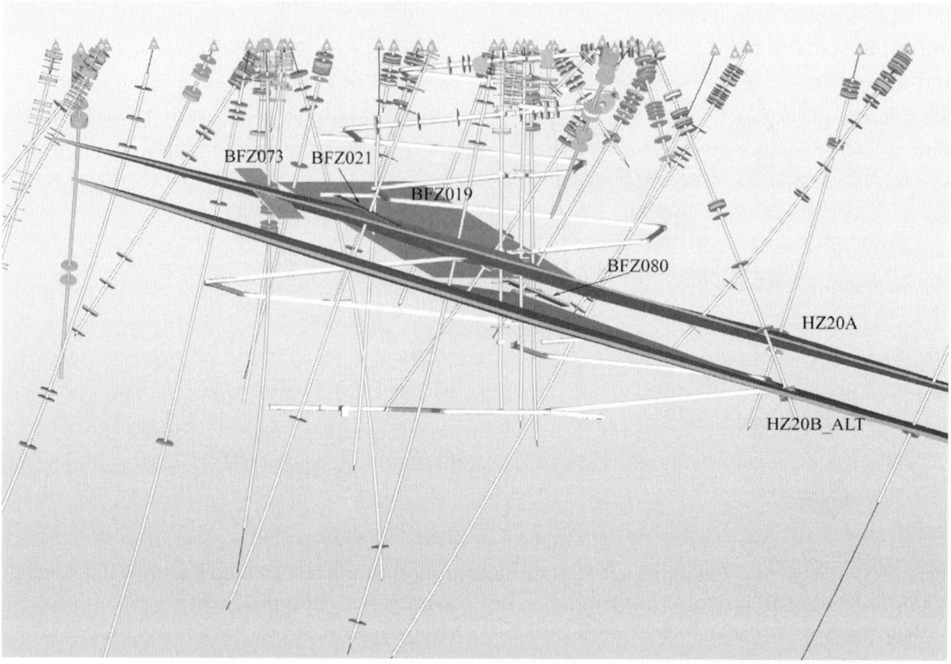

Fig. 23.10 Hydrogeological zones (HZ20A and HZ20B-ALT) and brittle fracture zones (BFZs) with approximately the same location and orientation are shown with boreholes (essentially vertical), used to characterize these zones and tunnels (essentially horizontal). Copyright (2006) Posiva Oy. Reproduced by permission of Posiva Oy. See color plate section.

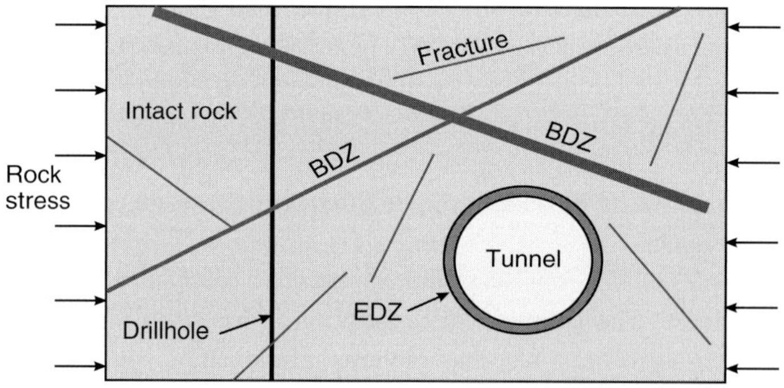

Fig. 23.11 Conceptual rock mechanics model, illustrating the relationships between the excavation damage zone (EDZ) around underground openings, the brittle deformation zones (BDZs), the intact rock lying between the BDZs and the applied rock stress. Copyright (2006) Posiva Oy. Reproduced by permission of Posiva Oy.

The *hydrogeochemical model* consists of a description of the groundwater composition and an assessment of the processes controlling the evolution of the groundwater composition in time and space. There are obvious links between the hydrogeochemical evolution of the site and the geological structure, as the hydrogeological zones (which have many affinities with the brittle deformation zones) tend be considerably more transmissive than the rock lying between these zones.

23.4 Predictions of properties in the ONKALO and their impact on construction

The adequacy of the site model is tested by making different kinds of predictions as the tunnel work on the ONKALO progresses. Type A predictions use only the latest version of the overall site model. Type B predictions also use data that were not available at the time of producing the site model, such as results from recently completed tunnel mapping and pilot holes. Type C predictions are made after excavation and involve establishing whether the modeling method could indeed have predicted the known outcome by adjustment of the input parameters.

Predictions concern both what will be encountered during construction of the ONKALO, such as the lithology of the rock mass and the rock mechanics properties; and the disturbance caused by the construction, such as the draw down and upconing of groundwater. Clearly, strong deviations between the A, B and C predictions are reasons to update the overall site model; however, it may not be feasible to formulate strict criteria for determining the level of "success" or "failure" for all parameters. Instead, a set of four principles for the prediction/outcome studies have been established. The first of these is that the main objective of the prediction-outcome studies is to contribute to the updating and revision of the descriptive models, such that they can be used in the safety case and for the appropriate detailed design of the repository. The second principle is that the uncertainties and spatial resolutions of the predictions should be provided, so that it is acceptable to make a prediction in terms of a statistical description, such as a range, a distribution or just an average. The third principle is that the results of the different types of prediction should be compared and the findings documented (the comparison will generally indicate a successful prediction if the outcome lies within the uncertainty range of the predicted properties). The fourth principle is that the deviations between predictions and outcomes should be investigated so that the reasons for the deviation can be identified; for example, was the deviation due to lack of information or improper modeling?

The prediction/outcome work is still in a development stage; not all of the above principles have yet been fully applied and more principles may be added. The analysis of this work is also preliminary in nature, but the prediction/outcome studies are showing good promise.

Concluding remarks

The example from Olkiluoto illustrates the approach that can be taken at an actual site to understand its tectonic setting and produce an integrated set of models, which, together, make up the SDM for the site. These models can, in turn, be used to help locate and design the repository and provide information necessary for the development of a safety case. There is an increasing level of confidence in the Olkiluoto site description; the main remaining challenges of the site characterization work are to assess properly the confidence in the description outside the well-characterized volume of the ONKALO (i.e. to extend it into the part of the rock mass where the repository will be located) and to ensure that the rock can be adequately characterized at the detailed scale. The latter will be necessary, amongst other things, for locating precise deposition locations for each container of spent nuclear fuel.

Some outstanding issues that are related to the tectonic setting of Olkiluoto are the geological understanding of the deformation history and the distribution and orientation of stress in relation to the geological structures. The first of these may require the development of a *deformation history model*, which relates the geological history to the ductile and brittle deformation zones. For the second, a certain level of understanding is required of the pre-existing rock stress state and how the geological structures at Olkiluoto (the lineaments and brittle deformation zones) can locally cause perturbations; knowledge of this stress state is desirable, as rock mechanics analyses are required to support repository design. These subjects are being considered at the moment, for example in the area of rock mechanics, where the impact of the geological history of the area on the rock mechanics properties is under investigation (Hudson and Cosgrove, 2006; Milnes *et al.*, 2006).

More generally, the results from Olkiluoto are applicable to other similar tectonic regimes and are, therefore, directly applicable to sites in Sweden (e.g. SKB, 2005a, 2005b, 2005c) and also to waste management organizations in other countries, which may consider the development of a repository in crystalline basement rocks, perhaps in areas of more active tectonics.

In stable geological environments there is less interest in the level of tectonic activity and more interest in how past tectonic activity has resulted in the geological structures that determine the design, layout and depth of the repository, and may have important controls on the groundwater flow system. The location and orientation of any larger, regional brittle deformation zones are, however, still of interest, as they have some influence on the stress state at the site; but, more importantly, are the locus of possible future seismic activity. In the case of sites in Finland and Sweden, and other countries which may be glaciated in the future, the only large seismic events are expected to be associated with movement on these zones during the early stages of glacial retreat (Lambeck and Purcell, 2003). Considerable work has been carried out in modeling the effect of such seismic activity on the long-term safety of a repository in both Sweden and Finland (e.g. La Pointe and Hermanson, 2002; La Pointe *et al.*, 2002) and on the control that such activity has on the design of the repository and, in particular, on the location of waste canisters within it (e.g. Munier, 2006). Munier and Hökmark (2004) have shown that it is necessary to have a separation distance (known as a *respect distance*) from a regional brittle deformation zone of only 100 m for a case of a magnitude M 6 event (which might be expected during glacial retreat) for there to be acceptably small displacement on a fracture within the rock mass, so that the integrity of the waste canister and the engineered barrier system of the repository is not impaired.

In more tectonically active regimes there is greater interest in the current tectonic activity, both during the operational phase of any repository (i.e. perhaps 100 a or more) and during the long term with regards to the effect of this activity on long-term safety. This interest can be seen in NUMO's exclusion factors in selecting areas for their preliminary investigations in Japan. Such investigations will be carried out in preliminary investigation areas (PIAs), which are to be selected by a rigorous process, in which the various national and site-specific evaluation factors, known as the evaluation factors for qualification (EFQ), and referred to as nationwide evaluation factors (NEF) and site-specific evaluation factors (SSEF), respectively (described in NUMO, 2004b), have been applied. These EFQs relate to the presence of Quaternary volcanism, active faults, rock deformation and uplift, and mineral resources, and are considered to be unacceptable features of the geological environment. The hope is that it will be possible to detect the presence of these EFQs during the literature review phase and not to investigate sites where such geological features exist. However, after selecting a site, it will still be necessary to design the site investigation program with their presence in mind. These EFQs are the problematic features referred to earlier in this chapter and, in areas where tectonic activity is considerably more significant than in Finland and Sweden, any site investigation program may well be driven by determining their presence or absence. In the Japanese context, for example, any evidence for active faulting within the PIA obtained from the site investigation would eliminate the site from further consideration. A site investigation program in Japan in a similar rock type to that being investigated at Olkiluoto would almost certainly, therefore, place considerably greater emphasis on detecting any such activity. This is likely to involve investigating each brittle deformation zone to examine it for signs of recent movement, perhaps using a combination of inclined drilling, geophysics and more direct methods, such as trenching.

Further reading

The majority of relevant reading in this field is associated with the disposal of radioactive waste, as it provides good examples of the integration of many disciplines. Examples of such work can be found in the description of these programs which are taking place in France (e.g. ANDRA, 2005a, 2005b, 2005c; potential repository host rock: Jurassic clay), Finland (e.g. McEwen and Äikäs, 2000; Posiva, 2003, 2006; Andersson *et al.*, 2005, 2007), Sweden (e.g. SKB, 2005a, 2005b, 2005c; similar geological environments to that at Olkiluoto), Switzerland (Nagra, 2002a, 2002b; potential repository host rock: Jurassic clay), the USA (e.g. Whitney and Keefer, 2000; potential repository host rock: unsaturated volcanic tuffs) and Belgium (Ondraf/Niras, 2001; potential repository host rock: Oligocene clay). The sites referred to in France, Finland, Sweden and Belgium are in areas with subdued tectonic activity, the site in Switzerland is in a slightly more tectonically active area and that in the USA is in an area with a considerably greater level of tectonic activity.

References

Andersson, J., A. Ström, C. Svemar, K.-E. Almén and L. Ericsson (2000). What requirements does the KBS-3 repository make of the host rock? Geoscientific suitability indicators and criteria for siting and site evaluation, Technical Report TR-00-12. Stockholm, Sweden: Swedish Nuclear Fuel and Waste Management Co.

Andersson, J., H. Ahokas, J. A. Hudson *et al.* (2005). Olkiluoto site description 2004, Report POSIVA 2005-03. Olkiluoto, Finland: Posiva Oy.

Andersson, J., H. Ahokas, J. A. Hudson *et al.* (2007). Olkiluoto site description 2006, Report POSIVA 2007-03. Olkiluoto, Finland: Posiva Oy.

ANDRA (2005a). Dossier 2005 Argile: architecture and management of a geological repository. Châtenay-Malabry, France: French National Agency for Radioactive Waste Management (ANDRA).

ANDRA (2005b). Dossier 2005 Argile: phenomenological evolution of a geological repository. Châtenay-Malabry, France: French National Agency for Radioactive Waste Management.

ANDRA (2005c). Dossier 2005 Argile: safety evaluation of a geological repository. Châtenay-Malabry, France: French National Agency for Radioactive Waste Management.

Chapman, N. A. and T. J. McEwen (1991). Geological aspects of the British programme for the deep disposal of nuclear wastes. In: Downing R. A. and W. B. Wilkinson (eds.) *Applied Groundwater Hydrology – A British Perspective*. Oxford: Clarendon Press, 143–164.

Hudson, J. and J. Cosgrove (2006). Geological history and its impact on the rock mechanics properties of the Olkiluoto site, Working Report 2006-14. Olkiluoto, Finland: Posiva Oy.

La Pointe, P. and J. Hermanson (2002). Estimation of rock movements due to future earthquakes at four candidate sites for a spent fuel repository in Finland, Report POSIVA 2002-02. Olkiluoto, Finland: Posiva Oy.

La Pointe, P., T. Cladouhos and T. Follin (2002). Development, application and evaluation of a methodology to estimate distributed slip of fractures due to future earthquakes for nuclear waste repository performance assessment. *Bulletin of the Seismological Society of America*, **92**(3), 923–944.

Lambeck, K. and A. Purcell (2003). Glacial rebound and crustal stress in Finland, Report POSIVA 2003-10. Olkiluoto, Finland: Posiva Oy.

Lindberg, A. (2007). Search for glacio-isostatic faults in the vicinity of Olkiluoto, Working Report 2007-05. Olkiluoto, Finland: Posiva Oy.

Mattila, J. (ed.) (2007). Results of monitoring at Olkiluoto in 2006 – Rock Mechanics, Working report 2007-53. Olkiluoto, Finland: Posiva Oy.

McEwen, T. and T. Äikäs (2000). The site selection process for a spent fuel repository in Finland – Summary report, POSIVA 2000-15. Olkiluoto, Finland: Posiva Oy.

Milnes, G., J. Hudson, L. Wikström and I. Aaltonen (2006). Foliation: geological background, rock mechanics significance, and preliminary investigations at Olkiluoto, Working Report 2006-03. Olkiluoto, Finland: Posiva Oy.

Munier, R. (2006). Using observations in deposition tunnels to avoid intersections with critical fractures in deposition holes, Report R-06-54. Stockholm, Sweden: Swedish Nuclear Fuel and Waste Management Co. (SKB).

Munier, R. and H. Hökmark. (2004). Respect distances. Rationale and means of computation, Report R-04-17. Stockholm, Sweden: Swedish Nuclear Fuel and Waste Management Co. (SKB).

Nagra (1994). Kristallin-1 safety assessment report, Technical Report 93-22. Wettingen, Switzerland: Nagra.
Nagra (2002a). Project Opalinuston – Synthese der geowissenschaftlichen Untersuchungsergebnisse. Entsorgungsnachweis für abgebrannte Brennelemente, verglaste hochaktive sowie langlebige mittelaktive Abfälle, Report NTB 02-03. Wettingen, Switzerland: Nagra.
Nagra (2002b). Project Opalinus Clay – Safety Report, NTB 02-05. Wettingen, Switzerland: Nagra.
NEA (1991). Disposal of high level radioactive wastes: radioactive protection and safety criteria. Proceeding of an NEA workshop, November 1990. Paris: OECD Nuclear Energy Agency.
NUMO (2004a). Development of repository concepts for volunteer siting environments, Report NUMO-TR-04-03. Tokyo: Nuclear Waste Management Organization of Japan (NUMO).
NUMO (2004b). Evaluating site suitability for a HLW repository site: scientific background and practical application of NUMO's siting factors, Report NUMO-TR-04-04. Tokyo: Nuclear Waste Management Organization of Japan (NUMO).
Ondraf/Niras (2001). SAFIR 2: Safety assessment and feasibility interim report 2, NIROND 2001-06E. Brussels, Belgium: Ondraf/Niras.
Paulamäki, S., M. Paananen and E. Seppo (2002). Structure and geological evolution of the bedrock of southern Satakunta, SW Finland, Report POSIVA 2002-04. Olkiluoto, Finland: Posiva Oy.
Paulamäki, S., M. Paananen, S. Gehör *et al.* (2006). Geological model of the Olkiluoto site – Version 0, Working Report 2006-37. Olkiluoto, Finland: Posiva Oy.
Posiva (2003). Nuclear waste management of the Olkiluoto and Loviisa power plants. Programme for research, development and technical design for 2004–2006, Report TKS-2003. Olkiluoto, Finland: Posiva Oy.
Posiva (2006). Nuclear waste management of the Olkiluoto and Loviisa power plants. Programme for research, development and technical design for 2007–2009, Report TKS-2006. Olkiluoto, Finland: Posiva Oy.
SKB (2000). Geoscientific programme for investigation and evaluation of sites for the deep repository, Report SKB TR-00-20. Stockholm, Sweden: Swedish Nuclear Fuel and Waste Management Co.
SKB (2005a). Preliminary safety evaluation for the Forsmark area. Based on data and site descriptions after the initial site investigation stage, Report SKB TR-05-16. Stockholm, Sweden: Swedish Nuclear Fuel and Waste Management Co.
SKB (2005b). Preliminary site description. Forsmark area – version 1.2, Report SKB R-05-18. Stockholm, Sweden: Swedish Nuclear Fuel and Waste Management Co.
SKB (2005c). Preliminary site description. Simpevarp subarea version 1.2, Report SKB R-05-08, updated November 9. Stockholm, Sweden: Swedish Nuclear Fuel and Waste Management Co.
Whitney, J. W. and W. R. Keefer (eds.) (2000). Geologic and geophysical characterization studies of Yucca Mountain, Nevada, a potential high-level radioactive-waste repository. *USGS Digital Data Series*, **58**(1), CD-ROM. Denver, CO: US Geological Survey.

24

The impact of subsidence, uplift and erosion on geological repositories for radioactive wastes

I. G. McKinley and N. A. Chapman

Earth is a dynamic planet and all points on the land surface are subject to processes that cause either uplift or subsidence. These types of vertical movements can vary considerably in magnitude and can occur on a wide range of timescales. Short-term cyclic motions, caused by planetary orbits, range from minute daily movements due to gravitational effects (Earth tides) through larger, millennial-scale effects of glacial loading or unloading and sea-level variations caused by climate changes driven by Milankovitch variations in Earth's movement around the Sun, or other mechanisms. Over a longer timescale, tectonic processes resulting from plate-motion-driven crustal deformation and the emplacement and evolution of magma bodies cause more significant movements. For completeness, we should note that a further series of processes may need to be considered, where land surfaces rise due to the pressurization of underground reservoirs and sink due to depressurization of fluids or material removal, caused by either anthropogenic or natural processes.

Uplift is generally accompanied by erosion of the uplifted surface, with erosion rates dependent on the mechanical and chemical properties of the rocks, climate, altitude and uplift rate; high uplift rates generally correlate with high erosion rates. The corollary is that subsidence is generally accompanied by sedimentation onto the sinking surface, often into basinal structures, with erosion on the flanks of the subsiding area.

The fundamental background to the identification and characterization of tectonically-driven uplift and erosion is presented by Litchfield *et al.* (Chapter 4, this volume) and the special case of the influence of such processes on coastal sites is discussed by McKinley and Alexander (Chapter 22, this volume). This chapter examines the consequences of uplift and erosion, and to a lesser extent, subsidence and burial on geological repositories in inland locations. We focus on the effects on radioactive waste repositories as, in general, the rates of uplift and subsidence are so small that the effects will be negligible for reactors and other nuclear plants with design lifetimes in the order of several decades to perhaps one hundred years (Chapman *et al.*, Chapter 1, this volume). An exception might be facilities such as the currently planned large linear accelerator (ILC, 2008), which, with a total length of more than 30 km and a high spatial and geometrical sensitivity, may be unusually vulnerable to differential crustal movement.

Safety assessments of geological repositories for radioactive waste usually concentrate on fluid flow processes (groundwater or gas) that could mobilize radionuclides from the waste and transport them back to the accessible environment. In particular cases, however, it is possible that uplift and erosion may be sufficiently fast to affect the process and rate of mobilization and provide an alternative transport mechanism. Although sometimes dismissed by technical groups, erosion is often a concern of the general public, especially in regions where this process is obviously active today (e.g. mountainous areas).

On the other hand, subsidence and burial would usually tend to increase isolation from the biosphere and, thus, contribute to safety. In terms of predictability, if not necessarily performance, an area with negligible change in surface over hundreds of thousands to millions of years has advantages. Arid regions in the center of stable cratonic blocks can display ancient landforms that have been little affected by erosion for millions of years. Such conditions were a consideration when looking for highly stable and "undynamic" locations worldwide as possible sites for an international repository. For example, the Pangea project (Black and Chapman, 2001) considered parts of central western Australia that have these characteristics.

This chapter introduces some of the key concerns for repositories associated with subsidence and uplift and illustrates these by case studies from Switzerland, where uplift and erosion have been seen to be important in the safety assessment of repositories for different types of waste. The possibility of using studies of eroding ore bodies to put repository analyses in context is examined and key issues and open questions are summarized.

24.1 Subsidence and burial

The potentially positive effects of subsidence are so evident that disposal in active subduction zones has, indeed, been proposed as a strategy for long-lived radioactive (or other highly toxic) wastes periodically, since the earliest days of geological disposal (e.g. summary in NNC, 2004). Despite being firmly ruled out on a number of occasions (for reasons outlined below), it is still sometimes suggested as a viable option.

The difficulties are that subduction zones are complex, unstable and essentially unpredictable in behavior at the timescale of relevance to waste disposal. Even if waste packages could be emplaced securely into ocean-floor sediment or shallow ocean crust in the lower reaches of the downwarping plate on the deep flanks of an ocean trench, there would still be enormous uncertainties about what would happen to them. First, there are serious logistical problems in terms of guaranteeing safe emplacement of the waste in a condition that will prevent releases before subduction has contributed to its removal from contact with the biosphere. Second, such zones are seismically highly active and difficult to characterize in detail, with the eventual fate of any particular block of rock or sediment (subducting, folding, over-riding, being scraped off the subducting plate, etc.) very uncertain. It would be hard to develop strong arguments that disruptive damage to waste packages and subsequent release of radionuclides to the ocean floor would not occur over the first few thousands or tens of thousands of years. Even if immense dilution in ocean waters would probably make the radiological consequences of such releases imperceptible, the waste management community has always been reticent to accept such levels of uncertainty. The rate of subduction also means that, for high-level waste (HLW) for example, the radiotoxicity would have decayed to ore-body levels (Chapman *et al.*, Chapter 1, this volume) before it had been significantly buried. In any case, the concept is academic, as sub-sea disposal of radioactive waste from ships or platforms (as would presumably be required here) is currently forbidden by international law.

In less tectonically dynamic conditions, locations where slow, uniform regional subsidence and sedimentation are occurring could be considered favorable, particularly for long-lived wastes; but, again, the rate of burial will be slow compared to the decay of long-lived wastes, so any added isolation benefits are likely to be marginal. In addition, significant subsidence rates may introduce their own problems in terms of demonstration of long-term safety. These would include the need to assess the consequences of differential subsidence on extended repository structures (e.g. tunnel liners, caverns for cemented wastes) and the impact of this process on regional flow, geochemistry and resultant radionuclide transport properties. This might lead to the conclusion that a site with slow, large-scale subsidence might make it easier to make a safety case than one with (empirically better) rapid subsidence.

24.2 Uplift and erosion

Uniform uplift over a broad geographical area around a repository may appear unlikely to have a direct influence on repository performance, as the mechanical and hydrogeological state of the deep host rock would be expected to be little perturbed by this process. However, in areas where uplift results from isostatic rebound after ice loading, with relatively rapid ($\sim$ mm a^{-1}) uplift associated with changing sea level (as in the Baltic Sea region over the last 20 ka) there may be significant impacts on repositories situated near the coast or under the sea (accessed by land). This is discussed further by McKinley and Alexander (Chapter 22, this volume).

A more general concern is where uplift will be associated with increased erosion, resulting in a reduction in the thickness of the geological barrier. It is, of course, true to say that, over very long periods of geological time ($\sim$ 100s Ma), almost any repository will be destroyed and its contents distributed by large-scale geological processes of crustal cycling. As discussed by Chapman *et al.* (Chapter 1, this volume), however, in evaluating long-term safety of radioactive wastes, our chief concern is with periods $\lesssim$ 1 Ma. Consequently, uplift and erosion become potentially important when, over timescales of the order of a few millions of years at most, the remnants of the repository could be significantly perturbed, especially if it could be exposed on the surface and components distributed directly into the biosphere. Indeed, this is considered sufficiently significant a scenario in some countries that regulatory guidelines forbid siting of geological repositories in areas with high uplift rates. In Japan, for example, the siting guidelines for the deep repository for HLW state that the implementing agency, NUMO, will only consider areas where there is no clear literature evidence of uplift amounting to $>$ 300 m during the last one hundred thousand years (NUMO, 2004).

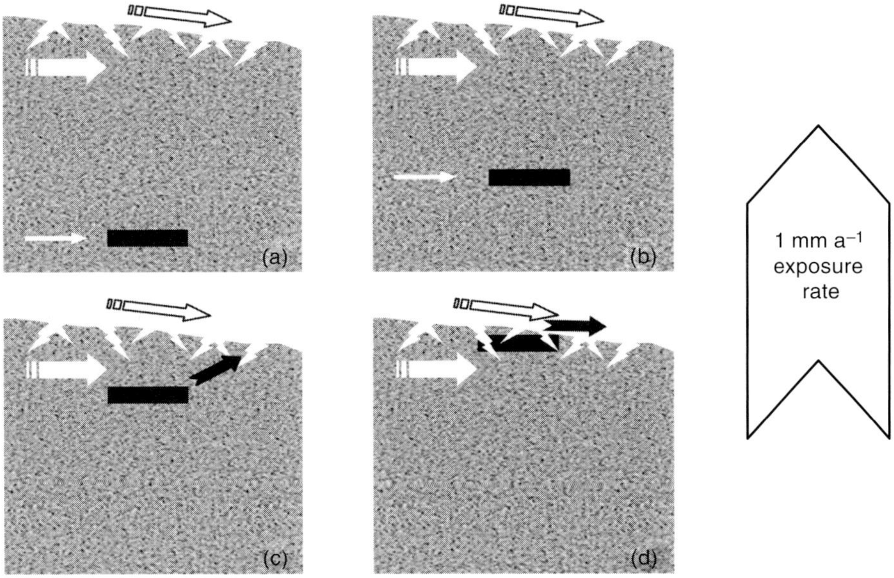

Fig. 24.1 Aspects of possible uplift and erosion of a deep geological repository for spent fuel (SF) for a hypothetical denudation rate (combined result of uplift and erosion) of ~ 1 mm^{-1}a. (a) The repository lies in an undynamic, low groundwater flux, deep environment at a depth of several hundred meters. The relatively more dynamic near-surface flow regime and erosion of the uplifting surface are shown. (b) After ~ 200 ka the radiotoxicity of the SF is equivalent to that of uranium ore and the situation of the repository may be little changed. Fluxes may be a little higher, but it might be assumed still that the engineered barrier system performs as designed. (c) After a further few hundred thousand years the repository lies in a more dynamic region of higher water flux leading to increased mobilization and dispersion of radionuclides. (d) Eventually the remains of the repository are exposed in eroding surficial rocks.

First, we look at the stages by which the wastes in a repository might be exposed to erosion and dispersion as a result of long-term uplift. Figure 24.1 illustrates schematically the stages that may need to be considered in regions where significant uplift and erosion occur. A critical aspect highlighted in Figure 24.1 is the extent to which material is dispersed in stage (c). If the upper ~ 100 m of the rock has high permeability and high water fluxes, then the radionuclides (by this time, equivalent in toxicity to uranium ore) would be expected to be progressively mobilized and dispersed over tens of thousands of years. As discussed later, if the immediate host rock surrounding the repository retains substantial hydrogeological and hydrochemical integrity, even after uplift into the upper ~ 100 m or so of the crust, or if the containment properties of the repository engineered barriers have exceptional longevity, then such dispersion may be less effective.

The consequences of uplift and erosion on long-term safety will depend on the inventory of radionuclides; the nature of the engineering barriers in the repository; the average rates of uplift and erosion; and the extent of localized uplift/erosion.

Although the consequences of uplift and erosion could be significant in many geographical settings, these processes have, in the past, rarely received much consideration in total system performance assessments. Nevertheless, two case studies from Switzerland provide examples of the issues raised for the particular cases of geological repositories for low- and intermediate-level radioactive waste (L/ILW) in a mountainous terrain (with lateral access into a mountainside) and for HLW, spent fuel (SF) and long-lived intermediate level radioactive waste (ILW) in an area with considerably less topographic relief where the repository would be accessed by shaft or incline. The erosion of near-surface radioactive waste repositories and other nuclear facilities is examined for coastal settings by McKinley and Alexander (Chapter 22, this volume).

24.2.1 Case 1: Oberbauenstock and Wellenberg, Switzerland

During the 1980s and 1990s, two locations in the fore-Alps of Switzerland were under detailed consideration as potential sites for geological repositories for lower activity classes of radioactive waste. We consider these two repository projects together as their settings are very similar and they highlight the same issues. Initial site studies led to selection of Wellenberg as the favored repository site but, as a result of lack of political and societal support at the regional level, this project had to be abandoned following a Cantonal referendum in 2002.

In both cases, the intention was to emplace L/ILW in large caverns within mountains with an overburden of $\sim$ 500 m. The host rock comprises intensively folded sediments (marls); a conceptual design for Wellenberg is shown in Figure 24.2. The siting region is undergoing continuing uplift as a consequence of the Alpine orogeny and the erosion rate is high, owing to the considerable topographic relief.

Oberbauenstock was the focus for the Project Gewähr analysis to demonstrate the fundamental feasibility of radioactive waste disposal in Switzerland (Nagra, 1985a) and the acceptance of this project by the regulatory authorities was a major milestone in the Swiss nuclear program. The safety assessment for this repository (described in detail in German in Nagra, 1985b; summarized in English in Nagra, 1985a) clearly identified uplift and erosion as a concern and hence an "erosion scenario" was defined for quantitative analysis. At this early stage of the program, when site information was rather limited, the net rate of erosion was based on consideration of local denudation, river erosion, uplift and glacial erosion, resulting in an estimated range of 0.2–2 mm a^{-1}. Taking a "best estimate" of 1 mm a^{-1} and a depth of 500 m results in an expected erosion time of $\sim$ 500 ka.

For the quantitative assessment of a potential repository at Oberbauenstock, it was conservatively assumed that the repository would be exposed after 100 ka. The repository content would eventually become mixed with eroded rock and form soils and sub-soils for land that could be used for agriculture and also serve as a reservoir for groundwater supply. Again conservatively assuming a small slope of the topographic surface, eroded repository material would become mixed with rock in a ratio of 1: 2500. A conventional biosphere model was used to calculate doses for ingestion, inhalation and direct radiation. Even though this

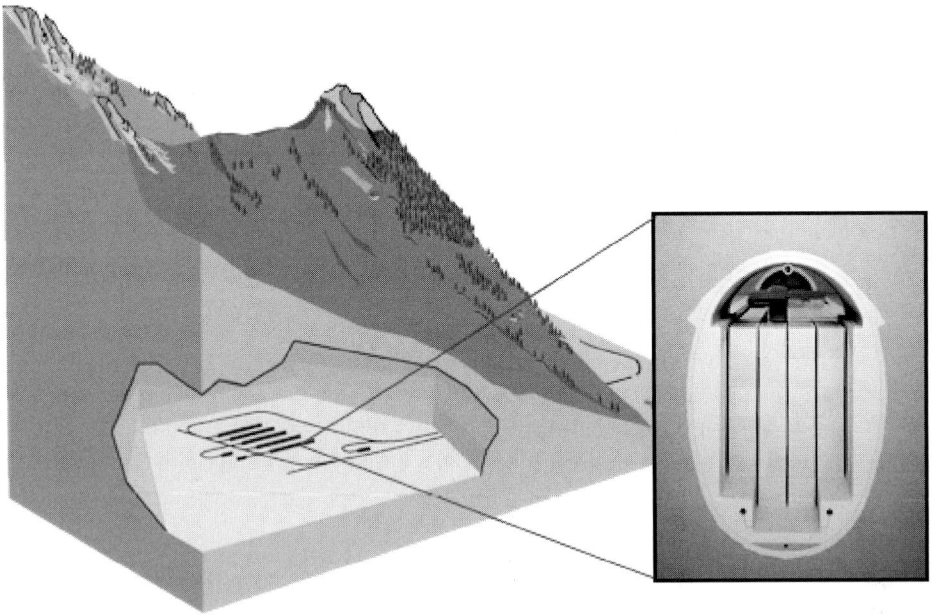

Fig. 24.2 Conceptual layout for a repository at Wellenberg. Disposal caverns for packaged L/ILW are located under a mountain, providing geological isolation with the convenience of horizontal access. See color plate section.

repository was for waste containing predominantly short-lived radionuclides, these always contain traces of longer-lived species. Thus, even at 100 ka, the resultant doses are relatively high, calculated to be 0.13 mSv a^{-1} and slightly above the regulatory guideline of 0.1 mSv a^{-1}. Indeed, this was the highest dose calculated for any of the credible scenarios for radioactivity releases from the repository that were assessed quantitatively, and the only one to exceed regulatory guidelines.

The setting and repository design were very similar for Wellenberg, situated about 10 km to the west of Oberbauenstock. This site was, however, subject to much more intensive site characterization, allowing both the geological structure and the regional setting of uplift and erosion to be better defined (Nagra, 1997). This resulted in a body of evidence suggesting that, even with conservative assumptions, a repository would still be covered by $\geq$ 60 m of overburden after 100 ka. The hydrogeological properties of the site were, however, acknowledged to depend on the overburden; alternative hydrogeological data sets were thus derived to correspond to erosion after 50 and 100 ka. Within the safety assessment (Nagra, 1994a), even such decreased performance of the natural barrier was calculated to result in doses that were still well under the regulatory guideline. For direct erosion, four different scenarios were defined and assumed to occur at 100 ka, reflecting either general erosion or localized erosion by streams. In each scenario (Table 24.1) the doses were below regulatory guidelines although, in the most extreme case of stream erosion, the safety reserve was rather

Table 24.1. Calculated doses from erosion scenarios summed for all waste categories[1]

Scenario	Description	Maximum dose (μSv a^{-1})
Base case	General erosion	0.7
Variant 1	General erosion, enhanced removal in repository areas	7
Variant 2	Stream erosion	7
Variant 3	Stream erosion under extreme conditions	70

[1] Summary based on Nagra (1994a).

small (less than a factor of two). For this case, however, it was also noted that the affected group would probably be very small (~80 people, based on current population levels).

24.2.2 Case 2: deep repository for HLW, SF and long-lived ILW in northern Switzerland

For the Project Gewähr (Nagra, 1985a) study of disposal of HLW, SF and long-lived ILW (which focused principally on the HLW component), the repository was assumed to be situated at a depth of 1300 m within crystalline basement rocks overlain by a sedimentary cover in northern Switzerland. The maximum expected denudation rate in the area of interest was calculated to be ~ 100 m Ma^{-1}, so uplift and erosion were not considered to be of concern. In the later "Kristallin-I" study, based on a more extensive geological data set (Thury *et al.*, 1994), the potential for alteration of regional hydrogeology due to down-cutting of the Rhine river into the northernmost part of the Swiss crystalline basement was acknowledged, leading to variant parameter sets for geosphere transport and biosphere models of radionuclide behavior (Nagra, 1994b). The type and extent of erosion estimated was highly dependent on the assumed climatic conditions, but the most significant influence on repository safety involved a potential southwards shift of the course of the Rhine river, resulting in abandonment of the current bed (filled with gravels) with the new course potentially lying directly on crystalline basement. Release of radionuclides directly into the Rhine resulted in very low radiation doses (10^{-7} mSv a^{-1}), but a conservative case of release into shallow soils when the gravel aquifer was not present gave much higher doses (0.002 mSv a^{-1}), although still well below the regulatory guideline and a minute fraction of natural background levels. Direct exposure by erosion was not considered, however.

More recently, the "Entsorgungsnachweis" study (Nagra, 2002b) extended earlier work, aiming to provide the demonstration of siting feasibility for such wastes as a required extension to Project Gewähr. In this project, the repository was planned to be located at a depth of ~ 800 m in a gently dipping layer of the Opalinus Clay (a consolidated siltstone) in the Zürcher Weinland area of northern Switzerland. The repository concept is illustrated in Figure 24.3. High-level waste and SF are emplaced axially in tunnels backfilled with

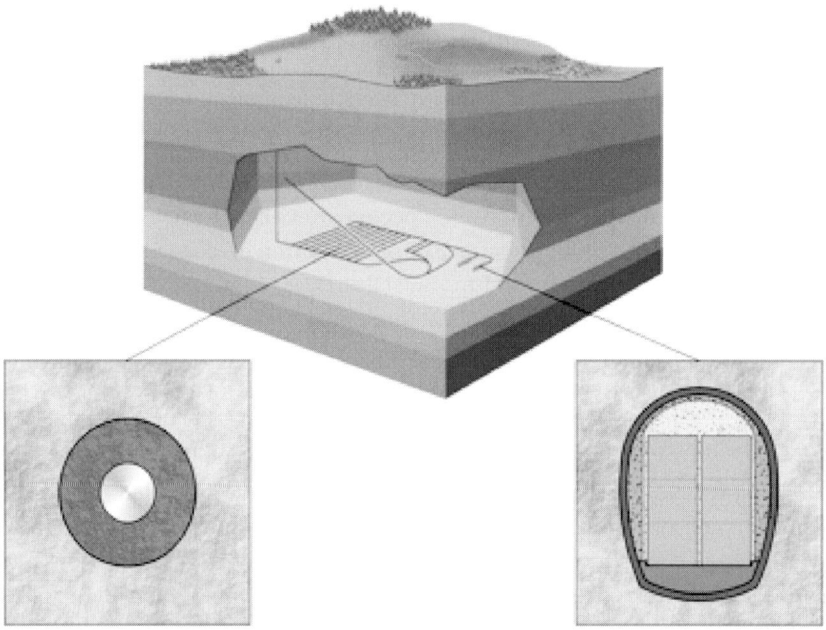

Fig. 24.3 Concept for co-disposal of HLW, SF and long-lived ILW in Opalinus Clay in northern Switzerland. The HLW and SF within massive steel overpacks are emplaced axially in small-diameter tunnels while ILW is emplaced in larger caverns in a separate part of the repository. See color plate section.

compacted bentonite while the ILW is emplaced in caverns that are backfilled with cement. The Opalinus Clay has an extremely low hydraulic conductivity, which leads to very high performance for the reference case of radionuclide releases caused by groundwater transport.

Uplift and erosion were discussed extensively in the integrated geological study supporting this project (Nagra, 2002a). The conclusion was that, for the shallowest parts of the repository (original depth, 650 m below surface), $\geq$ 450 m of overburden would remain after 1 Ma. The safety assessment (Nagra, 2002b) focused on the effects of such uplift and erosion on the local hydrogeology and the geosphere–biosphere interface but, given the high isolation capacity of the host rock, such effects did not give rise to any significant increase in calculated doses over the 1 Ma period of concern. Further examination of evolution beyond 1 Ma was restricted to qualitative discussion (Nagra, 2002b):

After in excess of 5 million years, erosion can be expected to expose some tunnels, but the large area of the repository and the development of topographic relief due to erosion would lead to only localized exposure of repository tunnels. Chemical conditions would remain reducing in the vicinity of the decayed wastes until erosion reduces the overburden to $\sim$ 10–20 m. The radiotoxicity of the SF, the most hazardous of the wastes, will have declined by this time to a value similar to that of the uranium ore from which the fuel was produced. No rigorous assessment of doses for times beyond several million years has been carried out, but some illustrative calculations of releases have been performed that indicate that sufficient safety is maintained even after millions of years.

24.3 Repository erosion in context: analogy with eroding uranium ore bodies

The models utilized in the Swiss case studies are acknowledged to be simplistic and are very sensitive to the assumed properties of the surface environment (particularly the geosphere–biosphere interface, which influences the extent of dilution of deep groundwater, and the biosphere characteristics that are used to translate releases into doses). Compared to the engineered barriers and the deep geosphere, such properties are inherently unpredictable over timescales of millions of years and hence must be represented by idealized representations. If uplift and erosion rates are high enough to lead to significant removal of repository overburden, such that more toxic wastes such as SF or HLW might become more easily mobilized into near-surface waters, it might be useful to consider natural analogs that can better indicate the levels of exposure to which people might be subjected.

As already noted, the general similarity in radiotoxicity levels of HLW and SF with those of uranium ore bodies after a few thousand years and a few hundred thousand years, respectively, suggests that study of eroding uranium deposits may shed some light on the uplift/erosion hazard potential. It is estimated that worldwide, 27 000–32 000 tons a^{-1} of uranium are released from rocks by weathering and natural erosion (CCME, 2007), which can be compared to commercial uranium production levels of $\sim$ 42 000 tons a^{-1} (EWG, 2006). Much of this release, however, is eroding from rocks with very low concentrations of uranium, rather than from ore deposits.

In order to determine particular situations that would be of relevance, it is important to consider the stages of erosion shown in Figure 24.1, as the relative importance of each will vary considerably depending on the characteristics of the repository and its associated geological setting. The first challenge is determining the transitions between phases (a), (b) and (c) shown in this figure. The Oklo (Gabon) and Cigar Lake (Canada) deposits are examples of very old, stable uranium ore bodies showing that minimal disruption may occur over timescales of many millions of years at depths of several hundred meters below surface (see Miller *et al.*, 2001 for descriptions of these "natural analogs" and lists of supporting references). Indeed, Oklo indicates that slow uplift has resulted in little disturbance of high-uranium-content ores unless they are brought into the zone of near-surface weathering (as is the case at the nearby Bangombé site). This is, naturally, very site specific, but observation of a number of different ore bodies studied as analogs indicates that the more dynamic zone, in which mobilization is greatly encouraged by higher fluxes of more oxidizing groundwaters, might typically occur at depths from several tens to about a hundred or so meters below surface.

Application of conventional approaches to repository performance assessment to this enhanced mobilization period would tend to predict rather high rates of release of uranium due to its relatively high solubility under more oxidizing conditions. This, however, contrasts with the observed low concentrations in the surface waters found around uranium ores that are actually in the process of being weathered (see Appendix for a commentary on some

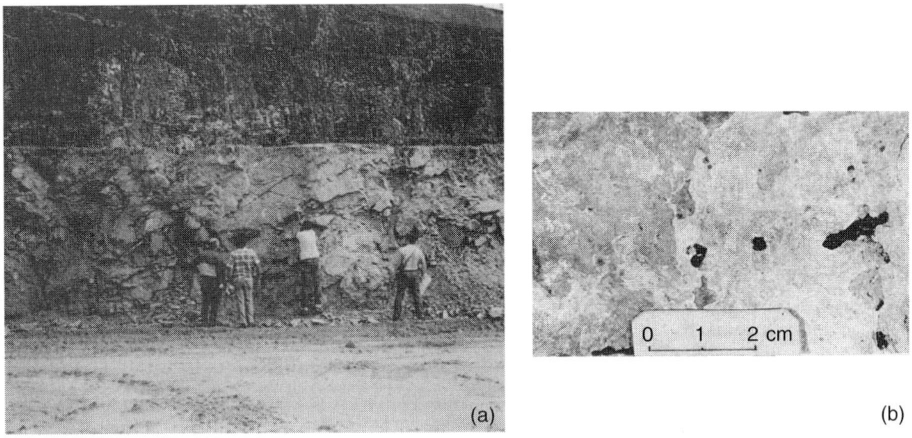

Fig. 24.4 (a) A marked redox front at the Osamu Utsumi uranium mine, Brazil. (b) On a small scale it can be seen that uranium is mobilized on the oxidizing (left) side of the front and concentrated as black nodules on the reduced (right) side. See color plate section.

of the available analog information from ore bodies). One reason for this is that, although they are often not contained in repository performance assessment databases, there are a number of U(VI) minerals that can give rise to secondary immobilization of uranium, even under oxidizing conditions. In addition, as dramatically illustrated in the Osamu Utsumi uranium mine near Poços de Caldas, Brazil (Miller *et al.*, 2001), slow downward penetration of a weathering front (as would result from uplift coupled to erosion) can result in uranium being concentrated at the resulting redox front rather than being released to flowing groundwater (Figure 24.4).

In cases where the transition from reducing to oxidizing conditions occurs very close to the surface, the engineered barriers are very strong or significant lateral or localized erosion occurs, the possibility of phase (d) in Figure 24.1, when significant quantities of repository materials are directly exposed in eroding surface rocks, has to be considered.

At the outset of this book we noted that the hazard potential of SF in concentrated form and/or in large masses will remain relatively high, even far into the future, and this includes hazard from the viewpoint of external exposure. The same can be said of rich uranium ores (e.g. the $\sim$ 40% UO_2 content of some parts of the Cigar Lake ore body make standard mining practices that bring mine-workers into close proximity to large volumes of ore unacceptable from the viewpoint of radiological dose). If a scenario could be envisaged where SF (or secondary minerals resulting from the alteration of HLW) were brought to the surface in large masses, before it could be dispersed by natural processes of chemical weathering and physical erosion, the fuel would constitute a significant external radiation dose hazard. Hedin (1997) provides data on the long-term hazard potential of SF, showing that, even after 100 ka, a mass of one metric ton would give external gamma doses of around 10 mSv hr^{-1} at a distance of one meter. The comparable dose from a point source of one ton of natural uranium is about 300 times less ($\sim$ 30 μSv hr^{-1}, using WISE, 2006).

We emphasize, however, that such numbers are indicative only and are very dependent on assumptions made on the geometry of the source. The Appendix to this chapter provides analog evidence of external exposure levels from near-surface ore bodies.

In addition to external radiation exposure, other exposure pathways are potentially more important contributors to the hazard potential of exposed or weathering ore or waste, in particular the possibility of inhalation of radon and ingestion of uranium and daughter products (particularly ^{226}Ra and its decay chain). Over the first millions of years following disposal of SF, the inventory of such radionuclides increases significantly as they grow-in at a rate determined by the half-life of ^{234}U (0.25 Ma). In passing, it may be noted that this same process will increase the hazard from depleted uranium (DU, a byproduct of nuclear fuel manufacture) if this is considered as waste, indicating that it is not only the conventional high activity wastes like HLW and SF that need to be carefully considered for erosion scenarios.

For repository evolution phases (a), (b) and (c), direct release of radon from either repositories or analogous ore bodies will not need to be considered, as the half-life of the longest-lived isotope (^{222}Rn) is less than four days and thus trivial compared to likely geosphere transit times. Concern is thus focused on transport of its longer-lived parent (^{226}Ra; half-life 1.6 ka), which, in many settings, is likely to be constrained by its limited release and mobilization rates, at least during phases (a) and (b). Most concern, therefore, is likely to be associated with ^{226}Ra release, transport and reconcentration near the surface during phases (c) and (d) and direct gaseous release of ^{222}Rn during phase (d). Although they certainly represent a worst case, uranium mine tailings indicate the possible health hazard from such processes, which can give rise to hazardous radiation doses, exceeding International Atomic Energy Agency intervention levels.

Although limited, available information on undisturbed (un-mined) ore bodies (see Appendix) indicates that doses well above normal background could result from erosion scenarios for some types of repository (e.g. for SF or DU). The question is then, how hazardous would such levels actually be?

There are regions of Earth's surface where radiation exposures are considerably greater than those discussed above. In some parts of the city of Ramsar in northern Iran, people receive an annual radiation absorbed dose from background radiation of up to 260 mSv a^{-1}, $\sim$ 100 times global average exposures to natural background. Ghiassi-nejad *et al.* (2002) note that the inhabitants of Ramsar have lived for many generations in these high background areas and cytogenetic studies show no significant differences between people in the high background compared to people in normal background areas.

24.3.1 Issues raised

Simple scoping calculations demonstrate that erosion scenarios can give rise to doses above regulatory guidelines for individual exposure of members of the public. This is inevitable, as such guidelines are conservatively set and are at levels that are between one and two orders of magnitude below natural radiation background exposures. A significant component of such natural radiation background results from the direct and indirect dose from the rocks and soils we live on. As pointed out more than twenty years ago in Project Gewähr (Nagra, 1985a), it is unreasonable for radiation dose resulting from an eroding mountain containing a repository to be less than the dose arising from erosion of the mountain without the repository. For such cases, rather than the regulatory guidelines, comparison of radionuclide concentrations from the waste with those in ambient rock may be a more appropriate safety measure. Although it was not explicitly discussed at the time, the fact that the demonstration of safety for Oberbauenstock was accepted by regulators, despite this scenario exceeding the formal guidelines, implies tacit acceptance of the argument (at least for cases where calculated exposures are not multiples of the regulatory standard).

For the case of the scenario considered in Entsorgungsnachweis, erosional exposure is likely to give rise to estimated doses significantly higher than regulatory standards for radioactive waste repositories. For SF, even if exposure occurs after several millions of years, this is equivalent to the erosion of a particularly rich uranium ore body. Indeed, given the high isolation capacity of the Opalinus Clay, it is difficult to argue that much dispersion would occur before such exposure. The analogs provided by exposed and eroding ore deposits suggest that such doses (external, radon and ingestion pathways) are likely to be in the order of typical natural background worldwide and within the broad global range of natural background variability. It is certainly debatable to what extent any consequences at such levels occurring in the very distant future need to be addressed (see below) but, formally, Swiss legislation requires a demonstration of safety "for all time." Nevertheless, Entsorgungsnachweis has been accepted by the regulatory authorities, without explicit discussion of this point.

How realistic are such exposure scenarios? Figure 24.1 indicates that direct exposure is only likely to arise if the uplift and erosion mechanisms are inefficient in degrading rock containment properties in the upper hundred meters or so of the geological profile above a repository. For SF to be exposed at the surface in large masses would mean that the chemical and hydrogeological isolation properties of the repository host rock would have to be just slightly perturbed by uplift and erosion until within 10–20 m of the surface. While there are indications that this could be the case for the Opalinus Clay, in many geological environments it is difficult to conceive, especially in regions likely to be affected by repeated glaciation over the next hundreds of thousands of years. While glacial erosion typically has only removed a few tens of meters of hard rock during the Quaternary (Glasser and Hall, 1997), the associated impacts on the fracture network in the rock of ice loading/unloading and permafrost development have increased hydraulic conductivity and consequent groundwater fluxes to depths of more than 100 m, thus increasing the likelihood of progressive dispersion before exposure occurs. The indication is that, if a SF repository were to be eroded progressively, it would not present an unusual hazard in terms of Earth's surface rocks and waters (see Appendix). Exposures associated with uranium ore bodies, whilst elevated, are equivalent to or a few times larger than global average exposures from natural background and appear to be generally well below maximum known natural exposures (e.g. Ramsar, Iran) where no obvious radiological effects are observed (Karam, 2002).

There are important ethical issues that have to be considered here. For some wastes (particularly SF, but also possibly DU), calculated doses using conventional assessment approaches may be high from the perspectives of radiological protection (above IAEA intervention levels) even if they do occur at times > 1 Ma and even if they do lie within the band of global natural background. How should this be handled? If we wish to dispose of these materials, either the system has to be redesigned to reduce the resultant far-future doses or we need to consider how much weight to place on the significance of the hazard that they represent. Mitigation by design would argue against the use of very long-lived engineered containment of SF (e.g. copper, titanium containers) in areas when sufficient performance to meet regulatory guidelines during phases (a) and (b) can be achieved with shorter-lived materials that would encourage more dispersion over this period and hence less concentration of materials in phases (c) and (d).

It is not really a technical issue to decide whether slightly increased (but still well below background, and hence with no hazard potential) releases over the first million years resulting from such mitigation measures should be balanced against reduction of potentially hazardous exposure in the distant future. Nor is it a purely technical issue to decide whether secure isolation today, when wastes such as SF represent a real and present hazard, should be balanced against possible doses in the range of natural background in the remote and uncertain future. These are topics that involve value judgments, which should really be discussed openly by regulators and other key societal stakeholders before safety cases for such repositories are developed and evaluated. In any case, as noted in the Appendix, better understanding would be gained from investigation of actual radiological exposures and health impacts of undisturbed, near-surface uranium ore bodies in relevant settings and such projects should be encouraged.

Concluding remarks

In many potential repository sites, uplift or subsidence processes can be significant over the period during which wastes remain potentially hazardous and will hence require quantification within associated safety cases. The procedures for measuring and modeling uplift and subsidence and hence estimating associated erosion and burial were discussed previously by Litchfield *et al.* (Chapter 4, this volume). Conventional approaches to repository safety assessment can be utilized to evaluate the consequences of gradual changes to the performance of the geological barrier but, especially for longer-lived wastes, there is the potential for high doses to result from direct erosion and exposure of waste material. This problem is particularly acute for disposal of SF, but may also be important for other wastes (e.g. depleted uranium). The issues involved in handling such high probability processes, which give high consequences, but only in the distant future, need to be discussed and agreed by stakeholders before safety cases are developed. Several points should be borne in mind:

(i) The timescales under discussion are orders of magnitude longer than recorded human history ($\sim$ 5 ka) and of a similar order to the estimated age of emergence of modern human beings from an African source. There is no universally accepted basis for assuming an ethical responsibility by the current generation for any people who might be living at such distant times in the future. At one extreme it could be accepted that current generations have discharged their responsibility by providing strong, safe isolation for future generations whilst the wastes are at their most hazardous, leaving only a residual hazard that is unexceptional by global natural standards. Essentially, we cannot expect to do a better job than nature at very long times. Having done this, it is more responsible to devote any additional resources to solving other present-day problems with high current impacts. At the other extreme, it could be accepted that all future generations, no matter how remote and tenuous their existence, should be protected to the same level as our present generation and that we should devote resources to doing this (in fact, the approach tacitly accepted by many regulatory agencies). Clearly, a choice between these two end-points involves personal value judgments informed by variable levels of optimism about the future of the human race and pessimism about its capability to look after itself. Such considerations should, rightfully, be colored by what we accept as current disposal practices for other long-lived toxic materials, which, if analyzed in the same way that we do radioactive waste repositories, would give much larger consequences in much shorter periods of time.

(ii) The growth with time of many forms of uncertainty about disposal system behavior leaves the results of detailed quantitative modeling of the far future open to endless challenge and argument, making them untenable as serious decision tools, owing to their diminished levels of credibility. If quantitative, far-future consequences are to be assessed, then evaluation of appropriate natural analogs and comparison of potential doses with natural background variations provide a sounder and more defensible basis for such analysis.

(iii) The source of the problem in the cases that give potentially high hazard can be attributed to over-designed repository systems. Rather than concentrating on very long-term confinement, overall performance might be optimized if credit was taken for the advantages of very slow dilution and dispersion. We believe that a repository providing assured isolation such that releases could be confidently estimated to be an order of magnitude below extremely strict regulatory guidelines can be considered completely safe. Measures to reduce such base-case releases yet further (e.g. by use of very long-lived containers) are not only cost-ineffective, they can lead to creation of this type of problem.

This is not an easy issue to resolve and does involve the difficult task of comparing long-standing regulatory approaches against a fresh look at the realism of long timescale assessments in the light of value judgments that many find uncomfortable to discuss. The very fact that issues of this type are seriously considered by the waste management community indicates how seriously the concerns of the public are taken and is an indication of the extremely high level of safety that deep geological disposal can provide.

Appendix

Analogs in uranium ore bodies and uranium mining areas

Both individual uranium ore bodies and areas with extensive uranium mineralization and associated mining provide useful analogs for the radiation impacts of an eroding waste repository. There is an extensive literature on the environmental impacts of uranium mining, especially in Australia, Canada and Germany. From the perspective of this chapter, however, most of this information is of limited use, as studies have concentrated on the impacts of perturbing the natural system by mining. The mobilization of uranium and other natural series radionuclides from ore bodies is considerably accelerated by mining operations, with rock spoil, mine tailings, processing residues and wastewater being primary sources of radioactivity entering the surface environment. Relatively limited information is available on the natural radiation environment around ore deposits before mining, although several studies contain baseline data for comparative purposes. Below, we look at some of the available information.

Extensive environmental studies have been carried out over many years in the uranium mining districts of northern Australia. Martin and Ryan (2004) studied the diet of Aboriginal groups living in the uranium-mining region of Australia's Northern Territories and estimated that the contribution to annual dose from naturally occurring radionuclides (rather than those mobilized by mining) was 1.6 $mSv\,a^{-1}$ for an adult. This value, $\sim$ 60% of the average worldwide natural background exposure, seems a useful yardstick for people living off the land in a region with many important, near-surface uranium ore bodies.

When a new mine was proposed at Jabiluka in Australia's Northern Territories, a risk assessment of release of mine waters into local surface waters estimated maximum doses to people living downstream of the mine at 20 $\mu Sv\,a^{-1}$. It was concluded that the proposed water management system would give rise to an insignificant radiological risk to people living in the vicinity of the mine and consuming traditional foods obtained from water bodies downstream from the mine (Johnston and Milnes, 2007). The migration of uranium from mine tailings was also modeled as part of an environmental impact assessment for the proposed Jabiluka mine. It was estimated that migration would be limited to tens to hundreds of meters over times up to 10 ka, with concentrations of uranium and radium at these distances being negligible compared with naturally occurring concentrations. The northern Australian ore deposits are generally shallow (with open pit mines exploiting the upper tens of meters of rock) in a region with high seasonal rainfall, characterized by relatively intense weathering. The Jabiluka work suggests that, even with relatively rapid weathering of near-surface ore, there is limited bulk migration of materials, with water passing through the ore deposit region giving low radiation doses downstream, indicating that any health impacts would tend to be localized to those living on top of the ore body.

The Canadian government (CCME, 2007) looked into concentrations of uranium in soils and waters around ore bodies when developing soil quality concentration guidelines for uranium in agricultural or residential areas. Canada possesses some of the richest uranium ore bodies in the world, with those developed earliest in its mining program being open-pit mines exploiting relatively shallow regions of deposits that can extend from near-surface to depths of hundreds of meters. The CCME proposes a guideline upper concentration for uranium of 23 $mg\,kg^{-1}$ and cites typical concentrations in residential areas of 1–2 $mg\,kg^{-1}$, whilst background concentrations in the vicinity of uranium ore bodies in Canada vary from $\sim$ 2.0–3.5 $mg\,kg^{-1}$ (Key Lake, South March, Prairie Flats) up to 33 $mg\,kg^{-1}$ (Port Hope).

In glacial till and alluvium deposits (typical erosion products) concentrations of some tens of $mg\,kg^{-1}$ are typical and values can reach $\sim 500\,mg\,kg^{-1}$. The postglacial, surficial Prairie Flats deposit, which lies in the upper few meters of an organic-rich fluviatile basin, has a maximum recorded value of $572\,mg\,kg^{-1}$, but this is in a region of apparently active uranium deposition and ore formation.

Uranium concentrations in surface waters around uranium ore deposits in Canada (CCME, 2007) are typically ~ 0.1–$5.0\,\mu g\,l^{-1}$, which is unexceptional when compared to nationwide figures of ~ 0.1–$2.0\,\mu g\,l^{-1}$. Around the Bancroft uranium deposit, concentrations ranged from < 1–$700\,\mu g\,l^{-1}$, with an overall average of $1.6\,\mu g\,l^{-1}$. A background uranium concentration range of 1–$2\,\mu g\,l^{-1}$ was reported for surface water in this region (Gordon 1992). In surface water samples from control lakes not impacted by mining activity near Elliot Lake, Ontario, mean total uranium concentrations ranged from < 1 to $1\,\mu g\,l^{-1}$ (Clulow *et al.*, 1998).

Groundwater from shallow (1.5 m) wells in peat and clay at the Prairie Flats surficial uranium deposit has uranium concentrations of $10-4000\,\mu g\,l^{-1}$. The groundwater in 3 m-deep wells in sand and gravel contained $12-740\,\mu g\,l^{-1}$, with vertical hydraulic gradients indicating a significant upward discharge of groundwater into the peat and clay unit (Tixier and Beckie, 2001). Again, it should be noted that this is an area of active postglacial uranium ore deposition.

As noted above, the activity of mining and processing uranium is a major contributor to mobilizing radioactivity into surface waters and sediments and this is often difficult to disentangle form the natural signal. Consequently, it is not always clear from the above data which of the higher values might be due to mining activities. Nevertheless, it is noticeable that many of the average and background levels of uranium concentrations in both soils and waters are little different from regional or nationwide averages.

On the assumption that direct exposure to large masses of decayed SF is unlikely, it is also useful to look at external exposures from weathering ore bodies where uranium is more widely dispersed in rocks at the ground surface. First, it is not always evident that regions of Earth's crust hosting uranium deposits will have characteristically high surface radiation exposures. Darnley *et al.* (2003) report two large-scale airborne gamma surveys over the whole of the USA and much of Canada, noting that the Athabasca Basin in Saskatchewan, although containing the world's largest known uranium deposits, is an area of very low surface radioactivity.

Where the Radium Hill ore body (Australia) intersects the ground surface, dose rates up to $14\,\mu Gy\,hr^{-1}$ have been measured, although the general background around the ore body and the mine is 0.5–$1\,\mu Gy\,hr^{-1}$ (McCleary, 2004). For someone spending 12 hours per day in this area, these exposures correspond to an average of $\sim 2.5-5.0\,mSv\,a^{-1}$ up to $70\,mSv\,a^{-1}$ (from slightly above to $\sim$ 30 times global average exposures to natural background). External (gamma) doses fall to normal background levels at a typical distance of $\sim$ 50 m from waste rock and uranium tailings piles.

Mudd (2003) cites compiled data on gamma dose rates taken by ground surveys, pre-mining, from a range of uranium ore deposits in the Alligator Rivers region of northern

Australia, noting that, for most deposits, there is no significant or elevated gamma radiation dose rate noticeable, although some sites have small and localized areas of high dose (e.g. the Ranger deposit, with 30–250 times background).

It is clear from the above discussion that further, more directly useful data could be obtained from focused studies of eroding, undisturbed uranium ore bodies. Combined studies of mobilization processes, uptake processes, actual population exposures and epidemiological studies of health impacts (such as those that have been performed on some global high natural background areas) would be particularly valuable.

References

Black, J. H. and N. A. Chapman (2001). Siting a high-isolation radioactive waste repository, Technical Report PTR-01-01. Baden, Switzerland: Pangea Resources International.

CCME (2007). Canadian soil quality guidelines for uranium: environmental and human health, Scientific Supporting Document PN 1371, ISBN 978-1-896997-64-3 PDF. Canada: Canadian Council of Ministers of the Environment.

Clulow, F., N. Dave, T. Lim and R. Avadhanula (1998). Radionuclides (lead-210, polonium-210, thorium-230, and -232) and thorium and uranium in water, sediments, and fish from lakes near the city of Elliot Lake, Ontario, Canada. *Environmental Pollution*, **99**, 199–213.

Darnley, A. G., J. S. Duval and J. M. Carson (2003). The surface distribution of natural radioelements across the USA and parts of Canada: a contribution to Global Geochemical Baselines. Natural Resources Canada, http://gsc.nrcan.gc.ca/gamma/dist/index_e.php.

EWG (2006). Uranium resources and nuclear energy, EWG-series 1/2006. Energy Watch Group, http://www.wise-uranium.org/clit.html.

Ghiassi-nejad, M., S. M. J. Mortazavi, J. R. Cameron, A. Niroomand-rad and P. A. Karam (2002). Very high background radiation areas of Ramsar, Iran: preliminary biological studies. *Health Physics*, **82**(1), 87–93.

Glasser, N. F. and A. M. Hall (1997). Calculating Quaternary glacial erosion rates in north east Scotland. *Geomorphology*, **20**, 29–48.

Gordon, S. (1992). Link between ore bodies and biosphere concentrations of uranium, AERCB Project 5.140.1. Ottawa, Canada: Atomic Energy Control Board.

Hedin, A. (1997). Spent nuclear fuel – how dangerous is it? Report SKB TR-97-13. Stockholm, Sweden: Swedish Nuclear Fuel and Waste Management Co.

ILC (2008). International Linear Collider, http://www.linearcollider.org/cms/.

Johnston, A. and A. R. Milnes (2007). Review of mine-related research in the Alligator Rivers Region 1978–2002, Supervising Scientist Report 186. Prepared for ARRTC9 meeting, February 25–27, 2002. Darwin, NT, Australia: Supervising Scientist.

Karam, P. A. (2002). The high background radiation area in Ramsar, Iran: geology, NORM, biology, LNT and possible regulatory fun. Presented at: Waste Management 2002 Symposium, February 24–28, Tucson, AZ.

Martin, P. and B. Ryan (2004). Natural-series radionuclides in traditional Aboriginal foods in tropical Northern Australia: a review. *The Scientific World Journal*, **4**, 77–95.

McCleary, M. (2004). Radium Hill uranium mine and low level radioactive waste repository, management plan – phase 1 – preliminary investigation, Report Book 2004/9. Adelaide, SA: Primary Industries and Resources South Austrailia.

Miller, W. M., W. R. Alexander, N. A. Chapman, I. G. McKinley and J. A. T. Smellie (2001). *Geological Disposal of Radioactive Wastes and Natural Analogues*. Oxford: Pergamon Press.

Mudd, G. M. (2003). Uranium mining in Australia: environmental impact, radiation releases and rehabilitation. Presented at: The 3rd International Symposium on the Protection of the Environment from Ionising Radiation (SPEIR 3), July 22–26, Darwin, Australia.

Nagra (1985a). Project Gewähr – nuclear waste management in Switzerland: feasibility studies and safety analyses, Project Report NGB 85-09. Baden, Switzerland: National Cooperative for the Storage of Radioactive Waste.

Nagra (1985b). Project Gewähr – Endlager für schwach- und mittelaktive Abfälle: Sicherheitsbericht, Project Report NGB 85-08. Baden, Switzerland: National Cooperative for the Storage of Radioactive Waste.

Nagra (1994a). Bericht zur Langzeitsicherheit des Endlagers SMA am Standort Wellenberg, Technical Report NTB 94-06. Wettingen, Switzerland: National Cooperative for the Storage of Radioactive Waste.

Nagra (1994b). Kristallin-I Safety Assessment Report, Technical Report NTB 93-22. Wettingen, Switzerland: National Cooperative for the Storage of Radioactive Waste.

Nagra (1997). Geosynthese Wellenberg 1996, Technical Report NTB 96-01. Wettingen, Switzerland: National Cooperative for the Storage of Radioactive Waste.

Nagra (2002a). Project Opalinus Clay: Synthese der geowissenschaftlichen Untersuchungsergebnisse, Technical Report NTB 02-03. Wettingen, Switzerland: National Cooperative for the Storage of Radioactive Waste (in German).

Nagra (2002b). Project Opalinus Clay safety report, Technical Report NTB 02-05. Wettingen, Switzerland: National Cooperative for the Storage of Radioactive Waste.

NNC (2004). Disposal in subduction zones, CoRWM Document 627. London: Committee on Radioactive Waste Management.

NUMO (2004). Evaluating site suitability for a HLW repository site, scientific background and practical application of NUMO's siting factors, Technical Report NUMO-TR-04-04. Tokyo, Japan: Nuclear Waste Management Organization.

Tixier, K. and R. Beckie (2001). Uranium depositional controls at the Prairie Flats surficial uranium deposit, Summerland, British Columbia. *Environmental Geology*, **40**, 1242–1251.

Thury, M., A. Gautschi, M. Mazurek *et al.* (1994). Geologie und Hydrogeologie des Kristallins der Nordschweiz, Technical Report NTB 93-01. Wettingen, Switzerland: National Cooperative for the Storage of Radioactive Waste.

WISE (2006). Uranium project calculators. World Information Service on Energy, http://www.wise-uranium.org.

25

Recommendations for assessing volcanic hazards at sites of nuclear installations

B. E. Hill, W. P. Aspinall, C. B. Connor, A. R. Godoy, J.-C. Komorowski and S. Nakada

Volcanic events are, at best, a parenthesis in current regulations or guidance for determining site suitability and for licensing decisions for most nuclear installations. This condition is understandable, as volcanic eruptions are rare natural events that have not created a significantly adverse condition at an operating nuclear installation. Nevertheless, unlike most geologic hazards, generally acceptable methodologies have not been established to assess volcanic hazards at a surface site or to determine if future volcanic events could be withstood by an appropriately designed nuclear installation. To address some of these challenges, the International Atomic Energy Agency (IAEA) has commissioned a multinational panel of consultants to revise preliminary guidance for assessing volcanic hazards at surface sites for nuclear installations. This chapter represents a summary of the consultants' recommendations, which are being considered for adoption within an IAEA Safety Guide for volcanic hazards assessment.

The goal of this chapter is to formulate a systematic approach for evaluating volcanic hazards at any candidate surface site. The approach must be flexible enough to assess a broad range of complex, often interrelated volcanic phenomena, yet still provide a transparent methodology to support decision making. Two fundamental outcomes need to be supported by the volcanic hazards assessment. If the assessment determines that volcanic hazards are credible external events at a site, the results of the assessment will need to provide sufficient technical detail to support development of design bases or operational criteria to mitigate the effects of potential future events on safety (e.g. IAEA, 2003a). This chapter represents a summary of the panel's recommendations. These recommendations will guide the formation of an IAEA Safety Guide for volcanic hazard assessment.

Although most nuclear installations are located at surface sites, some installations such as geologic waste repositories may be located in the subsurface. Clearly, most hazards from surface volcanic processes have limited potential to affect the safety of a subsurface installation. Direct intrusion of magma, or other igneous processes, may be credible hazards for some subsurface sites (e.g. Menand *et al.*, Chapter 17, this volume). Nevertheless, the methods and recommendations herein are developed for a comprehensive volcanic hazard assessment at a surface site.

Volcanic hazards arise from phenomena that have broad ranges, scales and magnitudes of physical characteristics. These processes may occur in isolation, or in combination with other phenomena, even during a single volcanic eruption. Some of these phenomena can occur long before or long after an eruption. Thus, the term "volcanic event" is adopted in this chapter to indicate a set of potentially hazardous phenomena that may occur before, during and after volcanic eruptions.

Both deterministic and probabilistic approaches are currently used to assess volcanic hazards, but with different degrees of formalism. Simply stated, deterministic methods use thresholds to screen specific phenomena from further consideration. Conversely, probabilistic methods use probability density functions to estimate the likelihood of specific volcanic phenomena. Although a deterministic approach may provide a transparent basis for decision making, screening criteria are often difficult to develop and defend, because the geologic record often contains only poorly preserved examples of a limited number of past events. Accommodation of large uncertainties in the number and character of past events can drive a deterministic approach to use extreme events as the basis for decision making. Reliance on extreme events can result in the rejection of a potentially acceptable site, or require design bases that are not commensurate with safety. A probabilistic approach, however, can readily incorporate uncertainties that arise from an incomplete geologic record, account for an appropriate range of natural variability in volcanic phenomena and consider uncertainties in scientific knowledge of processes that control volcanic phenomena. Probabilistic approaches can also result in quantitative assessments that allow for direct comparisons of hazard, or risk, between geologic events and other external events. Nevertheless, a probabilistic approach relies on numerical models that may be complex and sometime difficult to test, and can result in a more complex basis for decision making. Neither approach currently presents a clear advantage over the other in assessing volcanic hazards at sites for nuclear installations. Thus, either, or both, of these approaches appear suitable for consideration in a volcanic hazard assessment.

25.1 Principles of volcanic hazard assessment

Volcanic events are infrequent, relative to most other natural events that can affect the performance of surface nuclear installations. Some volcanoes have erupted after lying dormant for thousands of years, or even longer. As a general guide, volcanoes that have erupted during the last 10 ka (i.e. the Holocene) are usually considered active (e.g. Simkin and Siebert, 1994). Around the world, there are more than 1500 volcanoes that can be considered active on this basis. Holocene volcanoes may experience eruptions after long periods of inactivity. However, some volcanoes have reactivated after periods of inactivity longer than 10 ka. Therefore, consideration of volcanic hazards should not be limited only to Holocene volcanoes.

Within a geographic region, volcanic activity can persist for longer timescales than associated with individual volcanoes. For example, many volcanic arcs exhibit recurring volcanic activity for longer than 10 Ma, although individual volcanoes within the arc itself may remain active only for around 1 Ma. Because such distributed activity can persist for many millions of years, volcanic regions that have had activity during the past 10 Ma should be considered to have at least the potential for future activity.

Episodes of eruptive activity at individual volcanoes can last from hours to decades and, in rare cases, for even longer periods of time. The intensity of volcanic eruptions can vary from low-energy events, which may produce small lava flows and limited-range ballistic projectiles, to high-energy events that bury the countryside in tens of meters of hot ash. Even volcanoes located hundreds of kilometers from a site can produce hazardous phenomena such as tephra fallout or tsunami, which may adversely affect the performance of a nuclear installation. A summary of volcanic phenomena and primary hazards associated with these phenomena is presented in Table 25.1. Additional information on the physical characteristics of these potentially hazardous phenomena is presented in, for example, Connor *et al.* (Chapter 3, this volume).

Volcanic events rarely produce just a single hazardous phenomenon. Eruptions usually initiate a complex sequence of events that produce a wide range of volcanic phenomena. The occurrence of some volcanic phenomena may change the likelihood of occurrence for other phenomena. A volcanic hazard assessment should use a systematic methodology to evaluate credible, interrelated phenomena and ensure that all relevant hazards are integrated into the analysis.

Non-eruptive phenomena at volcanoes can also produce hazards for nuclear installations. Volcanoes are often unstable landforms. Even after long periods of repose, portions of volcanoes may suddenly collapse to form landslides and debris flows. Such events can impact areas of thousands of square kilometers around the volcano. Some volcanoes are closely linked to tectonic faults or geothermal activity. In such instances, seismic activity related to fault movement also may cause collapse of the volcano edifice. Volcanic hazard assessments for a nuclear installation should consider the influence of hydrologic and tectonic processes on the likelihood and characteristics of future volcanic events.

25.1.1 Data requirements

The cogency and robustness of any volcanic hazard assessment are dependent on a sound understanding of (i) the character of each individual volcanic source within the appropriate geographic region; (ii) the wider volcanological, geological and tectonic context of such volcanic sources; and (iii) the types and magnitudes of volcanic phenomena potentially produced by each of these sources. To achieve an appropriate level of transparency in the assessment, detailed information for each of the volcanic sources and their context in the region should be established or acquired, and compiled in a database.

The database should incorporate all the information that is needed to support decisions at each stage of the volcanic hazard assessment. The database structure should be flexible enough to accommodate increasing levels of information, completeness and integration as the assessment progresses through advancing stages of complexity. Initially, the database may be based upon, or include, information from existing international and national compilations of volcanological data. As site characterization progresses, additional data collected specifically for the assessment should be incorporated into the database.

Table 25.1. Volcanic phenomena and associated hazards for nuclear installations, with implications for installation design and siting

Phenomenon	Primary hazards	Design	Siting
Tephra fall	Static physical loads, abrasive and corrosive particles in air and water	Yes	No
Pyroclastic flows, surges and blasts	Dynamic physical loads, atmospheric overpressures, projectile impacts, temperatures > 300 °C, abrasive particles and toxic gases	No	Yes
Lava flows	Temperatures > 700 °C, dynamic physical loads, water impoundments and floods	Yes	Yes
Debris avalanches and slope failures	Dynamic physical loads, atmospheric overpressures, projectile impacts, water impoundments and floods	No	Yes
Debris flows and lahars	Dynamic physical loads, water impoundments and floods and suspended particulates in water	Yes	No
Opening of new vents	Dynamic physical loads, ground deformation and continuous seismic tremor	No	Yes
Ballistic projectiles	Projectile impacts, static physical loads and abrasive particles in water	Yes	No
Volcanic gases	Toxic and corrosive gases, water contamination and gas-charged lakes	Yes	No
Tsunami and seiches	Water inundation	Yes	Yes
Atmospheric phenomena	Dynamic overpressures, lightning strikes and downburst winds	Yes	No
Ground deformation	Ground displacements > 1 m landslides and volcanic gases	No	Yes
Volcanic earthquakes	Continuous tremor and multiple shocks (usually < M 5)	Yes	No
Geothermal fluids	Thermal water > 50 °C, adverse chemical compositions and water inundation or upwelling	No	Yes

Design represents the general practicality of mitigating potential hazard from this phenomenon by either facility design or operational planning. *Siting* indicates the presence of a credible hazard from this phenomenon generally constitutes a site suitability criterion. A *Yes* in both categories indicates that although a design basis may be achievable, sites with this hazard usually are avoided.

Partitioning the data collection requirements by distance from the site (e.g. IAEA, 2002a) is not necessarily the most effective way to approach a volcanic hazard assessment. This is because volcanic hazards, although they are associated often with a single, easily identified point source, can (i) occur as a range of phenomena with widely varying magnitudes and intensities that are less attenuated with distance than, for example, earthquake effects; and (ii) affect widely varying areas, depending on their individual characteristics as well as local variations in topography and the meteorological conditions in the region.

In addition to serving as an information resource, the database should also provide a structure that documents the treatment of data during the volcanic hazard assessment. This structure will serve to record the evidence and interpretations on which scientific decisions are made, as well as providing a basis for data quality assurance. For instance, all data used to formulate screening criteria and any consequent decisions should be contained in the database. Data considered in the assessment but rejected or otherwise not used should also be retained in the database and identified as such. Additional guidance on database goals and considerations is given in Section 25.2.

25.1.2 Geologic record and data uncertainty

The representative characteristics and frequencies of past events are critical data for any volcanic hazard assessment. The geologic record, however, is usually an incomplete source of these data. Large-magnitude volcanic events are much more likely to be preserved in the geologic record than small events. Yet such unrecorded small events may represent credible hazards to nuclear installations. Events missing from the geologic record, and interpretation of this record, create uncertainties that should be included in the hazard assessment.

The geologic record of an individual volcano does not necessarily encompass the potential characteristics and extent of future activity. Hazard assessments should consider that volcanic systems evolve, and that the characteristics of their hazards may change over time, sometimes quite rapidly. Information from analogous volcanoes can help both to constrain and reduce uncertainties arising from interpretations of an incomplete geologic record and also to further characterize potential changes in volcanic hazards through time.

The frequency and timing of past events is incompletely understood and relatively uncertain at most volcanoes. For example, ages of the most recent volcanic eruptions can be difficult to determine at volcanoes lacking a record of historical activity. Criteria to decide whether a volcano is dormant or extinct are often subjective and difficult to defend.

At most volcanoes, there is less uncertainty about the physical characteristics of past events than there is about the ages of these events. Thus, a volcanic hazard assessment that focuses on determining the geological characteristics of volcanic phenomena and their spatial extent will usually be less uncertain than one focusing on estimating the likelihood of occurrence for hazardous phenomena. Consequently, we develop an approach that emphasizes the initial screening of volcanic hazards based on their physical characteristics, rather than on their exact likelihood of occurrence. The concept of a *capable volcano* is introduced to define the potential for a volcano or volcanic field to produce hazardous phenomena that may affect a site. A capable volcano or volcanic field is one for which both (i) a future eruption or related volcanic event is credible; and (ii) such an event has the potential to produce phenomena that may affect a site. This definition is modified from McBirney and Godoy (2003) to more fully reflect the site-specific character of hazard investigations for nuclear installations. Identification of one or more capable volcanoes should result in development of a detailed, site-specific volcanic hazard assessment. The detailed hazard assessment, if warranted, should then consider the likelihood of occurrence and associated uncertainties for volcanic phenomena that may reach a site.

25.2 Volcanic hazard assessment methodology

A successful outcome of a volcanic hazard assessment is a transparent and traceable basis for making decisions about site suitability or facility design. A graded approach for information is warranted. Indeed, a graded approach for data analysis will allow the assessment to focus on volcanic phenomena that represent credible hazards to a site, rather than require an equivalent level of investigation and support for all possible types of hazards. The assessment approach advocated herein recognizes the need for increasing levels of information for increasing levels of potential hazard at the site. This approach also recognizes that sites located far from potentially active volcanoes may need to consider only a limited subset of potential hazards (i.e. distant tephra falls, volcanogenic tsunami), whereas sites located closer to potentially active volcanoes may need to consider the full range of potential hazards.

During the initial stage of the site selection and evaluation process, relevant data should be collected from available sources (e.g. publications, technical reports and related material) in order to identify volcanic phenomena with the potential for hazardous effects at the site. At each stage of the assessment, a determination should be made whether sufficient information is available to evaluate adequately the issue of volcanic hazards at the site. In some cases, available information could be sufficient to screen specific volcanic phenomena from further consideration. In other cases, additional information should be acquired in order to estimate volcanic hazards and determine site suitability, including consideration of volcanic hazards as design basis events (e.g. design for tephra loading).

The general goal for the volcanic hazard assessment is to determine the capability of a volcano or volcanic field to produce potentially hazardous phenomena that may reach the site of the nuclear installation, culminating in a comprehensive volcanic hazard model for the site, if deemed necessary. This goal can be accomplished in four stages, which are outlined in Figure 25.1 and presented in the remainder of this section.

25.2.1 Stage 1: initial scoping of past volcanism in a region

Approach

Stage 1 of the assessment focuses on two primary considerations: definition of an appropriate geographic region for the initial scoping of volcanic hazards and collection of evidence of

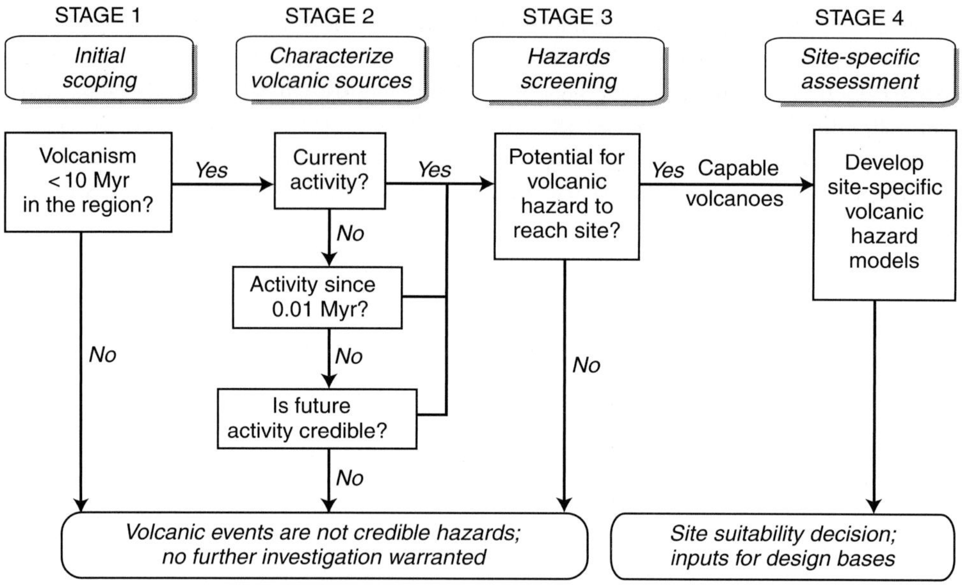

Fig. 25.1 Flow chart showing an approach to determine volcano capability and subsequent needs for a more detailed volcanic hazard assessment.

volcanic activity occurring within the last 10 Ma. Stage 1 includes a detailed review of available information sources for an appropriate geographic region around the site. This detailed review would typically include geologic maps, results from previous geologic investigations and other information.

The geographic region for the assessment does not have predetermined, symmetrical dimensions, but should consider the types of potentially hazardous phenomena that may have occurred at volcanoes younger than 10 Ma. For tephra fall and other atmospheric hazards related to volcanoes, this region can extend for hundreds of kilometers from the site, giving due consideration to regional wind field patterns (e.g. Hoblitt *et al.*, 1987). Assessment of volcanogenic tsunami may need to consider an entire ocean basin for some coastal sites. Other volcanic phenomena likely extend for shorter distances around a volcano. The region considered for such potential hazards might only extend for tens of kilometers away from the site.

For surface flow phenomena, consideration should be given to the topography between the site and potential volcanic sources. Areas with low elevation topography or broad, shallow drainages may be ineffective in diverting high-energy surface flows, even from volcanoes located more than 100 km from the site. Conversely, areas with steep topography and deep drainages may effectively capture and divert high-energy surface flows from volcanoes located much closer to the site. The definition of the appropriate region should be justified, to ensure that potentially hazardous volcanoes have been duly considered in the assessment.

Initial scoping studies should evaluate the evidence of volcanic activity occurring within the last 10 Ma. Because regions of volcanic activity can persist for millions of years or longer, a period of 10 Ma encompasses the geologic processes that could possibly affect an understanding of the potential for future volcanic activity within a region. Furthermore, a simplistic estimate of a regional volcanic recurrence rate of less than 1 event in 10 Ma would imply a probability of future activity $< 10^{-7}\ a^{-1}$, which is a commonly used screening probability level for external events in hazard analyses for nuclear installations (e.g. IAEA, 2002b).

Data requirements

A hierarchy of geological maps and volcanological data is needed for initial scoping in Stage 1. Available geological maps may be adequate if they provide appropriate data at various scales. For example, a 1:500 000 scale map may serve for the full area of study, moving down to 1:50 000 for nearby detailing. Geologic maps of volcanoes at a scale of 1:50 000 or larger will normally be required for initial scoping. Relevant information likely includes international and national compilations of volcanological data, especially for Holocene and Quaternary volcanoes.

Volcanism should be characterized in terms of the types of volcanoes concerned (cf. Connor *et al.*, Chapter 3, this volume). In Stage 1, past volcanic activity should be considered in terms of age, overall spatio-temporal trends, morphology, eruptive products and associated range of eruptive behaviors, and tectonic setting. At some sites, offshore data, such as bathymetry or drill-core logs or descriptions, may be important to consider in identification of potential volcanic sources during initial scoping. This characterization provides the groundwork for determination of the appropriate geographic region for the volcanic hazard assessment.

Age determinations are fundamental information for the initial scoping assessment. Such age determinations may include historical information, stratigraphic relationships, radiometric dating and morphological considerations. The level of information should be critically assessed for assurance that all relevant volcanic sources have been identified and have age determinations of suitable quality. If reliable age determinations are available they may provide an adequate basis for initial scoping.

For some cases, however, available information for initial scoping may not be sufficient for a robust appraisal at this stage of a site evaluation. In these circumstances, additional geological and volcanological data may need to be sought out, collected or commissioned. For instance, further age determination sampling may be needed in order to ascertain the age of volcanic products in the geographic region.

25.2.2 Stage 2: characterize sources of volcanic activity

Approach

If the outcome of the initial scoping in Stage 1 indicates that volcanoes or volcanic fields younger than 10 Ma are present in the selected geographic region, then these volcanic sources should be further characterized by additional investigations. If there is evidence of current or historical volcanic activity, then future eruptions should be assumed credible and the hazard assessment should proceed to Stage 3. Evidence of current or historical volcanic activity includes records of volcanic eruptions, ongoing volcanic unrest, an active hydrothermal system (e.g. presence of fumaroles) and related phenomena.

Evidence of an eruption during the last 10 ka (i.e. the Holocene) is a widely accepted indicator (e.g. Simkin and Siebert, 1994) that future eruptions are credible. Information for determining if Holocene volcanic activity has occurred may come from multiple sources. Radiometric dating of volcanic products, however, provides the most direct evidence that volcanic eruptions occurred within the Holocene.

In some circumstances, especially in the early stages of site investigations, the exact age of the most recent products may be difficult to determine. In such circumstances additional criteria may be used to consider a volcano as Holocene, including: (i) volcanic products overlying latest Pleistocene glacial debris; (ii) youthful volcanic landforms in areas where erosion should have been pronounced after many thousands of years; and (iii) vegetation patterns that would have been far more developed if the volcanic substrates were more than a few thousand or hundred years old.

Nevertheless, reliable sources may disagree over the evidence of Holocene volcanism, or there may be significant uncertainty about the most reliable age estimate of the most recent eruption. In this case, such volcanoes should be classified as Holocene(?), which is consistent with established volcanological terminology (Simkin and Siebert, 1994). From a safety perspective, future eruptions should be considered credible for Holocene(?) volcanoes (i.e. those with an uncertain record of eruptions in the Holocene) and the analysis should proceed to Stage 3.

If evidence of current or Holocene activity does not exist, additional consideration should be given to assess the timing of older activity in the region. Evidence of an eruption during the last 2 Ma generally indicates future activity remains possible. Furthermore, for some volcanic systems such as distributed volcanic fields or infrequently active calderas, activity during the last 5 Ma or so may also indicate some potential for future activity. To ensure an adequate evaluation, the geologic data should be assessed to determine if any of the volcanoes or volcanic fields in the region as old as 10 Ma has the potential for a future eruption.

A probabilistic analysis of the potential for future volcanic events can provide useful information for this stage of the analysis. Probabilistic methods for this assessment can include frequentist approaches based on the recurrence of past volcanic eruptions, Bayesian methods that can incorporate additional volcanological information, or process-level models, such as those based on time–volume relationships. Expert elicitation might be used to help inform an assessment of the probability of future activity (e.g. US Nuclear Regulatory Commission, 1996; Aspinall, 2006).

When designing new facilities, some countries use the value of 10^{-7} per year of anticipated nuclear power plant lifetime as one acceptable limit on the probability value for interacting events having serious radiological consequences (IAEA, 2002b). As volcanism is an external hazard with potentially adverse consequences for safe facility operation, an annual probability of renewed volcanism at or below $10^{-7}\,a^{-1}$ could be considered a criterion for screening future events in the absence of additional information regarding potential volcanic hazards at a site.

Alternatively, a deterministic approach can be used. For example, analogous volcanoes might be investigated to determine the maximum duration of gaps in eruptive activity. For a volcano with an ongoing period of quiescence, the possibility of return to activity could be compared with the maximum duration of such gaps in activity at analogous volcanoes. An additional deterministic approach might invoke time–volume or petrologic trends in the volcanic system. For example, a time–volume relationship may show an obvious waning trend and demonstrable cessation of volcanic activity in the early Pleistocene or older periods. In this situation, renewed volcanism can be considered unlikely. In cases where a resolution based on these other criteria is not achieved, a deterministic approach could simply assume that future eruptions are possible for any volcano younger than 10 Ma.

The analyses in Stage 2 may determine that future volcanic activity in the geographic region is considered not possible. If sufficient information is available to support this conclusion, no further analysis is required and volcanic hazards do not need further investigation for this site. Conversely, in the absence of sufficient evidence, or if future volcanic events in the region of interest appear to be credible, additional analyses are warranted and the hazard assessment should proceed to Stage 3.

Data requirements

An expanded scope and level of detail is required in the information needed for source characterization, hazards screening and a site-specific assessment (i.e. Stages 2–4). Decisions resulting from Stages 2–4 of the hazard assessment rely on the information about the timing and magnitude of activity at potential volcanic sources. Therefore, the database should document (i) spatial distribution of volcanic sources and geologic controls on the distribution of these volcanic sources; (ii) number and timing of eruptions at each source; (iii) the repose intervals between eruptions, and durations of eruptive episodes at each source, where it is possible to determine; (iv) the range of eruption magnitudes, dynamic processes such as eruption intensity and style, eruptive products, and associated phenomena such as seismicity, ground deformation and hydrothermal activity; and (v) the information about trends in eruptive activity, such as spatial migration of volcanic sources or temporal evolution of geochemical variations and changes in the volume of eruption products.

For volcanic sources with any documented historical activity, the database should contain information relevant to an understanding of the scale and timing of this activity. Possible volcanological information taken from historical sources should include (i) dates and durations of eruptions; (ii) description of the types of eruptive products, including areal extent, mass and composition; (iii) areal extent and magnitude of associated seismic activity, ground deformation, and other geophysical and hydrological activity or anomalies; and (iv) description of current activity at the volcano including monitoring programs and review of monitored data, if any.

The database should include descriptions of any volcanic products younger than 10 Ma. For Holocene and younger volcanoes, including those that are currently active, the geologic history of the volcano should be investigated, not only the period of most recent volcanic activity. An evaluation of the uncertainty in age determinations should be included in this assessment. For example, the stratigraphy of pyroclastic units is often complex and incomplete. Assessment of the completeness of the geologic record should be attempted, even if all volcanic deposits cannot be mapped. The ages of volcanic deposits should be numerically expressed and correlated to provide a complete description of the history of volcanic activity.

Information in the database will form the substantive basis on which to assess the potential for specific phenomena to affect the site, and will be used to develop screening distance values for these phenomena (i.e. Stage 3). Therefore, data should be compiled on volcanic products that could reach the site from each potential source. Deposits younger than 10 Ma in the site vicinity should be identified and evaluated to provide information on (i) the type and distribution of the deposits, and identification of the likely source or sources; (ii) the ages and volcanological characteristics of the associated eruptions; and (iii) the chemical and lithological compositions and physical properties, including areal extents, thicknesses, densities and particle size distribution.

The viability and usefulness of this type of information is highly dependent on the age of the deposits and completeness of the geologic record. Wherever possible, an appropriate range of volcanological information should be collected, in order to characterize individual phenomena and to evaluate long-term trends in the volcanic system. Care should be taken to appropriately characterize deposits that might credibly reach the site in the future. For example, a tephra fall from a nearby volcano that did not deposit at the site itself, perhaps only because of meteorological conditions during the eruption, should also be included in the database if future meteorological conditions could potentially direct tephra falls toward the site. Conversely, a pyroclastic flow deposit that was diverted from the site by a large topographic barrier should not be included, if that barrier is likely to exist during future eruptions and if larger or more energetic flows are not expected from future eruptions.

Geophysical and geochemical survey data collected at individual volcanoes within the region of interest can improve the overall hazard assessment. There are several reasons to survey such volcanoes: (i) to help reduce the level of uncertainty in the understanding of particular volcanic phenomena; (ii) to provide an objective basis for detecting changes in the level of activity of the volcano and prospects for future eruptive phenomena; (iii) to take advantage of new emerging or improved technologies or techniques to strengthen information about a specific volcano; and (iv) for capable volcanoes, to comply with safety requirements for dedicated monitoring (e.g. IAEA, 2003a). Conversely, there may be certain capable volcanoes within the geographic region around the site where surveys will not enhance a site-specific hazard assessment, depending on the nature of the volcanic phenomena concerned.

The type and extent of geophysical and geochemical surveys should be evaluated based on information needs for the volcanic hazard assessment. In the case of a new site evaluation, surveys should be considered at the earliest stages of the site characterization process. Survey data should be interpreted and integrated with other data that contribute to the site evaluation process, and included in the database. Close cooperation should be sought with existing monitoring systems, such as those implemented by national programs for prediction of volcanic eruptions and mitigation of disasters. Exchange of observational data and consultation with experts in volcanology working in such programs is generally beneficial.

25.2.3 Stage 3: screening volcanic hazards

In cases where future volcanic activity appears possible in the appropriate region around a site, the potential for hazardous phenomena to affect the site should be analyzed. This analysis should be performed for each of the phenomena associated with volcanic activity (e.g. tephra fallout, pyroclastic flows, lahars). In some cases, specific hazardous phenomena may be screened from further consideration, if there is negligible likelihood of these phenomena reaching the site. Screening decisions also should consider whether such phenomena might result from secondary processes or a complex scenario of volcanic events.

A deterministic approach to assessing hazards at Stage 3 can be based on establishing screening distance values for specific phenomena. Screening distance values can be defined in terms of the maximum known extent of a particular eruptive product, considering the characteristics of the source volcano and possibly the nature of topographic controls between the source volcano and the site. For example, most basaltic lava flows are known to travel no more than 10 to 100 km from source vents. A generic screening distance value of 100 km for basaltic lava flows appears justified for most basaltic volcanoes in most terrains. A shorter screening value distance may be justified, however, based on data gathered at analogous volcanoes or where topography prevents the phenomenon from reaching the site. In general, justification for the use of specific screening distance values for all types of volcanic phenomena should be consistent with representative examples from analogous volcanoes.

If the site falls outside the screening distance for a specific volcanic phenomenon, then no further analysis is needed for that phenomenon. Alternatively, if future volcanic activity appears possible and the site falls within the screening distance for a specific volcanic phenomenon, then the volcano or volcanic field should be considered capable and a comprehensive hazard assessment should be undertaken (i.e. Stage 4). An analysis for capability should be completed for each volcanic phenomenon that is associated with each potential source volcano, as each of these phenomena may have a different screening distance value.

An alternative approach to assessing hazards at Stage 3 is to estimate the conditional probability of a specific volcanic phenomenon reaching the site, given an eruption at the source volcano. Multiple methods are available to estimate this probability. These methods are considered further in the discussion on Stage 4. In most circumstances, site characterization data alone will likely be insufficient to determine a robust estimate of this probability, because the geologic record incompletely preserves past activity from volcanoes and because past activity may not have encompassed the appropriate range of phenomena potentially resulting from a future volcanic event.

Using a conditional probability estimate for a specific volcanic phenomenon with accompanying uncertainties can produce a range of likelihood values that can be used in the site assessment. If the potential for a volcanic event to produce any phenomenon that may reach the site is negligibly low, no further analysis is required and volcanic hazards do not represent credible design basis events for this site. If this potential is sufficiently high, the volcano or volcanic field concerned should be considered *capable* and a comprehensive site-specific volcanic hazard analysis should be undertaken (Stage 4).

Complications often arise in the use of both deterministic and probabilistic approaches to screening hazards because some volcanic phenomena may involve coupled processes. For example, tephra fallout on distant topographic slopes sometimes creates new source regions for debris flows and lahars. Water impoundments can be created by debris flows and lava flows. Screening decisions should consider secondary sources of hazards that result from such complexities.

25.2.4 Stage 4: site-specific volcanic hazard assessment

Stage 4 seeks to provide a clear technical basis for evaluating site-specific volcanic hazards when considering nuclear installation site selection and design. If one or more *capable volcanoes* are identified, a site-specific volcanic hazard assessment for the nuclear installation should be conducted. Specific outcomes of this hazard assessment should provide sufficient information to determine whether each volcanic hazard is a credible external event. If an event is credible, then the hazard assessment should also provide sufficient information to determine if a design basis or other practicable solution for this event can be established. If a design basis or other practicable solution for this credible external event cannot be resolved, the site will likely be considered unsuitable (e.g. IAEA, 2003a).

A combination of deterministic and probabilistic approaches can be used to decide whether or not an acceptability issue exists for the site due to volcanic hazard. Each hazard that is included in the design basis should be associated with a quantified parameter or set of parameters so that its value can be compared with the design basis values of other external events to the extent possible. For some of the hazards it may be possible to demonstrate that design basis parameters derived for other external events envelop those derived for volcanic hazards. Recommendations are provided in this section for volcanic phenomena that should be considered as part of a site-specific volcanic hazard assessment. Relevant volcanological information that should be considered for each of these phenomena is discussed in Connor *et al.* (Chapter 3, this volume).

Tephra fallout

Tephra fallout is the most widespread hazardous phenomena from volcanoes, including the opening of new vents. Hazards associated with tephra fallout include static load on structures, particle impact, potential blockage and abrasion of water circulation systems, mechanical and chemical effects on ventilation and electrical systems, and particle load in the atmosphere. Water can significantly increase the static load of a tephra deposit. Tephra fallout hazard assessments should consider (i) potential sources of tephra; (ii) magnitudes of potential tephra producing volcanic eruptions and the physical characteristics of these eruptions; (iii) frequency of tephra-producing eruptions; (iv) meteorological conditions between source regions and the site that will affect tephra transport and deposition; and (v) secondary effects of tephra eruptions.

A deterministic approach should consider the maximum credible thickness for tephra fallout deposits at the site. For example, actual deposits from analogous volcanoes could define the maximal thickness of accumulation at the site from a *capable volcano*. Particle-size characteristics (e.g. size distribution and maximum size) could be estimated from these deposits. Analog deposits or eruptions can also provide information about soluble ions that form corrosive, acidic condensates, which often accompany tephra falls.

A probabilistic approach should use a numerical simulation of tephra fallout at the site. In such an analysis, Monte Carlo simulation of tephra fallout from each capable volcano should be conducted, accounting for variation in eruption volume, eruption column height, total grain size distribution, wind velocity distribution in the region as a function of altitude and related parameters. Such models produce a frequency distribution of tephra accumulation, commonly presented as a hazard probability of exceedance curve (Hill *et al.*, 1998; Connor *et al.*, 2001; Bonadonna *et al.*, 2005; Volentik *et al.*, Chapter 9, this volume). The uncertainty in the resulting hazard curve can be expressed by confidence bounds, with a stated basis for selection of the reported confidence levels.

To support the potential development of design bases, results of deterministic or probabilistic assessments of tephra fallout for each capable volcano should be expressed in terms of parameters, such as mass accumulation, accumulation rate and grain size distribution. In order to estimate potential static loads, the contribution for each capable volcano should be integrated into a single, site-specific maximum credible value or single tephra fallout hazard curve (Hobblit *et al.*, 1987; Volentik *et al.*, Chapter 9, this volume). This information may also be used to assess particle size distribution and potential for remobilization of tephra deposits to create atmospheric mass loads of particles or debris flows and lahars.

Pyroclastic flows, surges and blasts

Pyroclastic flows, surges and blasts, known collectively as pyroclastic density currents, accompany both explosive volcanic eruptions and effusive volcanic eruptions that generally form lava domes or thick lava flows. Impacts of pyroclastic density currents are severe for obstacles in their flow paths, because these flows move at high velocities and commonly have high temperatures (e.g. more than 300 °C). In addition, these flows are destructive due to the momentum of the massive ground-hugging mixture of hot lava blocks, ash and volcanic gas; and due to their transport of projectiles. Although their main flowage is controlled topographically, surges and blasts are less constrained by topography than pyroclastic flows. All types of pyroclastic density currents are known to surmount topographic obstacles in certain circumstances or to flow across bodies of water.

A deterministic approach should consider the volume and energy of the pyroclastic flow or surge resulting from an eruption and hence its potential maximum travel distance (i.e. runout). Screening distance for these phenomena could be determined based on the volume and nature of pyroclastic flow or surge deposits exposed within the geographic region of concern, or by referring to flow events identified at analogous volcanoes. Potential runout can also be estimated using physics-based numerical models. Pyroclastic surges are generated directly from the vent by a fountain-like collapse of the eruption column, by blast, or from collapse of large domes, and may travel more than 10 km from the vent. Surges or blasts associated with pyroclastic flows may extend several kilometers more beyond the pyroclastic flow front. Thus, a deterministic approach for a pyroclastic surge or blast will be based on a screening distance value generally greater than that for pyroclastic flows.

Probability of pyroclastic flows should be calculated as a conditional probability of an eruption of given intensity, multiplied by conditional probability distributions for (i) the occurrences of flow and surge; (ii) runouts of these phenomena; and (iii) directivity effects. The value for conditional probability of pyroclastic surge should be representative of the magma's physical properties; the geometry and structure of the volcano; the dynamics of the eruption; and the physics of flow spreading and diffusion. The uncertainty in the resulting probability level can be expressed by confidence bounds, with a stated basis for selection of the reported confidence levels.

Several additional factors should be considered in deriving design basis and in making site acceptability judgments related to hazards from pyroclastic density currents. Hazards related to most pyroclastic density currents can be evaluated empirically and approximately by using the energy-cone model (Sheridan, 1979; Malin and Sheridan, 1982) to estimate potential runout distances. However, more sophisticated numerical models (e.g. Wadge *et al.*, 1998; Woods, 2000; Patra *et al.*, 2005; Neri *et al.*, 2007) coupled with Monte Carlo simulations can generate probabilistic assessments of runout and destructive effects. Although this is an area of intense research in volcanology, comprehensive dynamic models of pyroclastic flows, surges and blasts are not yet fully established. Consequently, both deterministic and probabilistic approaches should be considered. Results of analyses of pyroclastic flow, surge and blast impacts may be presented, for example, in terms of dynamic pressure, temperature and velocity. Some pyroclastic density currents can give rise to secondary hazards, such as tephra fallout, debris flows and tsunami.

Lava flows

Lava flows essentially cause total destruction on their path. The impact of lava flows will depend on the physical characteristics of the lava, the discharge rate, the duration of the eruption, the morphology at the vent and the topography. Lava flows have direct impacts due to their dynamic and static loads, flow thickness and temperature up to $\sim$ 1200 °C. In order to evaluate hazards associated with lava flows for each capable volcano, estimates are needed for (i) potential magnitude (e.g. mass discharge rate, areal extent, velocity, thickness) of lava flows; (ii) frequency of future effusive volcanic eruptions; (iii) eruptive scenario, such as individual lava flows, lava tubes and flow fields; and (iv) physical properties of erupted lava.

A deterministic assessment should first address the locations of vents and the potential formation of new volcanic vents. Subsequently, assessment for potential lava flow inundation should determine the maximum credible length, areal extent, thickness, temperature and potential speed of lava flows that could reach the site. This assessment can be achieved using data from other volcanoes from the region of concern, from analog volcanoes; or from empirical lava flow emplacement models. Topography along the path and at the site should be considered. A screening distance value can thus be defined for lava flows beyond which lava incursion is not thought to be a credible event.

A probabilistic approach should also address plausible variations for vent locations, and the potential formation of new volcanic vents. The probabilistic approach should entail numerical modeling of lava flows and proceed with numerical simulations from each capable volcano to account for a range of values for parameters that control flow length and thickness, using stochastic methods. Lava flow hazard curves should then be determined and combined to express the probability of exceedance of lava flow incursion and thickness at the site. Uncertainty in the resulting hazard curves can be expressed by confidence bounds, with a stated basis for selection of the reported confidence levels.

There are empirical correlations between flow length and effusion rate for many lavas (Walker, 1973), whereas others are volume-limited (Malin, 1980). Assessment of the potential for lava flow inundation usually involves numerical models of maximum lava flow length, area of inundation, speed and thickness of the flows (Barca *et al.*, 1994; Miyamoto and Sasaki, 1997; Vicari *et al.*, 2007). In these numerical simulations, topography, discharge rate, viscosity of the flow and duration of the eruption are key parameters that control modeled lava flow emplacement. Probabilistic assessments use numerical models of lava flow emplacement coupled with Monte Carlo simulations. Probabilistic or deterministic approaches should result in estimates of the potential for any lava flows to reach the site, their likely thicknesses, as well as their thermal properties. This assessment should include the effects of phenomena associated with lava flows such as tephra fall, generation of floods following interaction with ice and snow fields, water impoundments, and generation of pyroclastic flows from the collapse of viscous lava domes and flows.

Debris avalanches, landslides and slope failures

Debris avalanches resulting from edifice collapse should be considered separately from other slope failures because of the potentially large volumes involved (e.g. up to several km^3), high velocities and the considerable distances that can be reached (e.g. 150 km). Other, smaller-scale slope failures can be treated within the scope of non-volcanic geotechnical hazards (e.g. IAEA, 2004). The effects of volcanic debris avalanches are predominantly mechanical due to the mass of material involved and associated high velocities. A hazard assessment for debris avalanches, landslides and slope failures for each capable volcano should consider (i) potential source regions of these events; (ii) potential magnitude (volume, aerial extent, thickness) of these events; (iii) frequency of such events; and (iv) their potential flow paths. These assessments should identify potential source regions and areas of potential instability. Modifications of the flow properties along the path, as well as the topography from the source region to the site, should also be considered.

A deterministic approach should determine the maximum credible runout distance and thickness of avalanche deposits at the site using information collected from actual deposits from analogous volcanoes, and empirical avalanche flow emplacement models. A screening distance value can thus be defined for debris avalanches and other associated mass flows beyond which they are not credible events.

A probabilistic approach should extend the numerical modeling of these flows and proceed with numerical simulations for each capable volcano accounting for a range of values for parameters that control flow length, velocity and thickness using stochastic methods. Hazard curves should then be determined and combined to express the probability of incursion at the site. Uncertainty in the resulting hazard curves can be expressed by confidence bounds, with a stated basis for selection of the reported confidence levels.

Several additional factors should be considered in deriving design basis and in making site acceptability judgments related to debris avalanches, landslides and slope failures. The results of probabilistic or deterministic approaches should include parameter estimates of potential for incursion of the site, as well as flow thickness and velocity. This assessment should consider the other indirect phenomena associated with debris avalanches, landslides and slope failures such as tephra fall, projectiles, pressure waves, debris flows, floods and tsunami. Large slope failures are potential non-eruptive volcanic events and may be triggered by rainfall or tectonic earthquakes.

Volcanic debris flows, lahars and floods

Debris flows, lahars and floods of volcanic origin should be considered separately from other ordinary floods (e.g. IAEA, 2003b) mainly because of the short warning time available after the onset of the flow, high flow velocities and discharge rates, high flow volumes, and the considerable distances that can be reached (e.g. 150 km from the source). Their impact is mechanical due to the mass and velocity of material involved, erosive power and other effects related to flooding by water with a high sediment load. Modifications of the flow properties along the path, the sources of water and topography from the source region to the site should be considered. The hazard assessment for debris flows should also consider the fact that potentially adverse effects can persist over a time period that greatly exceeds the duration of an eruption. A hazard assessment for lahars, debris flows and floods of volcanic origin for each capable volcano should (i) identify regions of potential source for volcanic debris and for water; (ii) estimate the potential magnitude and flow characteristics; (iii) determine the frequency of such events in the past; and (iv) acquire meteorological data at the source region and along the potential path of such potential flows.

A deterministic approach should consider the maximum credible distance for debris flows and lahar deposits at the site using information from capable and analogous volcanoes, and empirical debris flow emplacement models. A screening distance value should be defined for debris flows, lahars and other associated floods, beyond which they are not credible events.

A probabilistic approach should use numerical modeling of these flows (e.g. Iverson *et al.*, 1998; Pitman *et al.*, 2003) and proceed with numerical simulations for each capable volcano to account for a range of values for parameters that control flow geometry and discharge rate, using stochastic methods. Hazard curves should then be derived that express the probability of exceedance for flow incursion and discharge at the site. The uncertainty in the resulting hazard curves can be expressed by confidence bounds, with a stated basis for selection of the reported confidence levels.

Several additional factors should be considered in deriving design basis and in making site acceptability judgments related to debris flows, lahars and floods. Probabilistic or deterministic approaches should result in estimates of the potential for these phenomena to reach a site as well as their likely flow geometry and discharge. Indirect event sequences, such as tephra fall on neighboring non-capable, snow-clad volcanoes, could act as sources for debris flows. Debris flows can also occur from floods generated by eruption under ice or snow and from the sudden release of water and debris from breakage of volcanic dams in craters or valleys filled with volcanic debris. Other, smaller-scale floods can be treated within the scope of floods of non-volcanic origin (IAEA, 2003b).

Opening of new vents

The opening of new vents is a geologically rare phenomenon but one that can produce significant flow, ballistic and ground-deformation hazards for a nuclear installation located close to the site of a new volcano (e.g. scoria cone). Vents generally form clusters within volcanic fields, or are closely associated with large volcanic systems, such as shield volcanoes and calderas. Assessment of the likelihood of formation of new vents requires information about the distribution, type and age of volcanic vents in the region. Additional information, such as geophysical surveys of the region, often is used to identify vents buried by subsequent activity or that are otherwise obscured. In addition, geological and geophysical models of the site region often provide important information about geological controls on vent distribution, such as the relationship between vents and faults or similar tectonic features.

A deterministic assessment of the possibility of new vent formation should determine a screening distance value for the site, beyond which the formation of a new vent is not thought to be a credible event. Additional information, such as significant changes in tectonic regime with distance from an existing volcanic field, should also be considered in a deterministic analysis.

Modern analyses of volcanic hazards associated with new vent formation normally involve probabilistic assessment (Connor and Connor, Chapter 14, this volume).

Probabilistic assessments should estimate a spatial probability density function describing the spatial, or spatio-temporal, intensity of volcanism in the region. Additional geological or geophysical information should be incorporated into the analysis. Uncertainties in the resulting probability density functions can be expressed by confidence bounds, with a stated basis for selection of the reported confidence levels.

Several additional factors should be considered in deriving design basis and in making site acceptability judgments related to the opening of new volcanic vents. Probabilistic and deterministic approaches may be used together. Results of this analysis could be expressed as the probability of a specific type of new vent forming within a specified time period (e.g. one year) and specific area (e.g. the area of the site vicinity). The potential for new vent formation should be considered as part of the hazard assessment of potential sources of other volcanic-coupled phenomena, such as lava flows, ballistics, tephra fallout and surges. In the case of opening of new vents, ground deformation of large magnitude (e.g. meters displacement), volcanic seismicity and gas flux may occur in the site vicinity.

Ballistic projectiles

Ballistic projectiles can be compared with impacts due to tornado-borne missiles, but the potential number of volcanic projectiles that may fall on a site within 5 km of a volcano can be very high. At the vent, ballistic projectiles have velocities in the range of 50 to 300 m s^{-1}, and distance traveled by ballistics is a function of their size and drag, which can be reduced behind shock waves in large eruptions. These factors mean that even large ballistics, 1 m in diameter, can travel kilometers from the volcanic vent. Hazard estimates for ballistics from each capable volcano need to consider the source locations, potential magnitude and frequency of future explosive eruptions.

A deterministic approach should consider the definition of a screening distance using information from the maximum distance and size of ballistics in previous explosive eruptions from analogous volcanoes. Empirical explosion models could also be used to determine a screening distance as a function of the exit speed, density of ballistics, exit angle and wind-field parameters. The analysis should consider the effect of topographic barriers between the site and the vent.

A probabilistic approach should consider a numerical simulation of ballistic trajectories at the site. In such an analysis, a stochastic analysis of ballistic trajectories from each capable volcano should be conducted, accounting for variation in explosion pressure, estimated drag coefficient for ballistics, exit angle and related parameters. Such models produce a frequency distribution of ballistic accumulation, commonly presented as a hazard curve. Uncertainty in the resulting hazard curve can be expressed by confidence bounds, with a stated basis for selection of the reported confidence levels.

Several additional factors should be considered in deriving design basis and in making site acceptability judgments related to ballistic projectiles. Probabilistic and deterministic approaches may be used together. Results of this analysis could be expressed as the probability of potential ballistic impacts beyond a screening distance. The potential for ballistics should be considered as part of the hazard assessment of the potential for the opening of new vents; and as impacts related to tephra fallout. Temperature effects from ballistic fragments also may need to be considered. Results of the analysis should be consistent with similar external hazards, such as tornado-borne missiles (cf. IAEA, 2002b).

Volcanic gases

Volcanic gases can be released in very large quantities during explosive volcanic eruptions, but can also be released from some volcanoes even during periods of non-eruptive activity and can diffuse through soils and along fracture systems on and adjacent to volcanoes. Adverse effects of volcanic gases include toxicity and corrosion, often associated with the condensation of acids from volcanic gases, dry deposition and heavy acid loading. Estimation of hazards due to volcanic gases relies on accurate estimation of the potential flux of such gases in volcanic systems, and the meteorological and topographical data used to model the dispersion, flow and concentration of gases in the atmosphere.

A deterministic approach should consider using information from analogous volcanoes or gas concentration measurements at the capable volcano to define an offset distance between potential volcanic gas sources and the site. Alternatively, assuming that degassing will occur from a capable volcano, a deterministic approach could estimate the impact of this degassing using an atmospheric dispersion model, assuming a conservative value for the mass flux of volcanic gases. This modeling should provide some indication of the extreme gas concentrations and acid loading that might occur at the site.

A probabilistic approach should consider the expected variation in mass flux from the volcano, including the possibility of degassing pulses at otherwise quiescent volcanoes, and the variability of meteorological conditions at the site. These probability distributions would be used as input into a gas dispersion model to estimate acid loading and related factors. Uncertainty in the models can be expressed by confidence bounds, with a stated basis for selection of the reported confidence levels.

Several additional factors should be considered in deriving design basis and in making site acceptability judgments related to gases. Probabilistic and deterministic approaches may be used together. Results of this analysis are generally expressed in terms of the expected atmospheric concentration of volcanic gases and expected dry deposition in the site vicinity. This analysis should consider hazards from direct degassing from volcanic vents and eruptive plumes as well as from indirect passive degassing of erupted products, through the ground, the hydrothermal system and crater lakes. The analysis also should evaluate the potential for catastrophic degassing of gas-charged (e.g. CO_2, CH_4) water bodies (e.g. crater or fault-bounded lakes) to affect the site.

Tsunami and seiches

Massive amounts of rock can abruptly enter large bodies of water during an eruption. Furthermore, volcano slopes can become unstable and collapse without warning or eruptive activity. Underwater volcanic eruptions also can displace large volumes of water, from both slope collapse and the release of volcanic gases, and should be considered in site-specific hazard assessments. Coastal sites, or sites located near large bodies of water, normally consider tsunami and seiche hazards as part of the site assessment (e.g. IAEA, 2003b). Nevertheless, specialist knowledge will be needed to evaluate fully the likelihood and source characteristics of potential volcanogenic tsunami. The effects from volcanically induced tsunami and seiches on sites are the same as those from seismically induced tsunami and seiches.

Currently, tsunami and seiche hazards are evaluated using deterministic numerical models that consider the locations of potential sources; volume and rate of mass flow; the source and characteristics of water displacement; and the resulting propagation of waves based on location-specific bathymetry (e.g. IAEA, 2003b). For sites located in areas potentially affected by volcanically induced tsunami or seiches, consideration should be given to the potential for large volumes of rock from volcanic eruptions or unstable volcanic slopes to enter water bodies, as part of the analysis of the potential distribution of tsunami sources.

Atmospheric phenomena

Explosive volcanic eruptions can produce atmospheric phenomena that have potentially hazardous characteristics. Overpressures from air shocks can often extend for kilometers beyond the projection of volcanic material. Eruptions that produce tephra columns and plumes commonly are associated with frequent lightning and occasionally with strong downburst winds. Because explosive volcanic eruptions would be considered rare events for atmospheric phenomena (e.g. IAEA, 2003c) and involve exceptional conditions, hazard assessments should consider a deterministic approach to model the potential maximum hazard for each phenomena associated with a potential volcanic eruption.

Volcanoes can be considered as stationary sources of explosions when considering air shocks in the hazard analysis (e.g. IAEA, 2002b). Hazard analyses described in, for example, IAEA (2002b) for stationary sources of explosions, are generally applicable to the analysis of air shocks from explosive volcanic eruptions. The air-shock analysis should focus on determining the potential maximum explosion for the volcanic source and a simplified analysis for shock attenuation with distance from that source.

Volcanically induced lightning has the same hazardous characteristics as lightning from other meteorological phenomena but is a widespread feature associated with tephra columns formed by explosive volcanic eruption. The likelihood for ground strikes is high and may exceed the strike rate for extreme meteorological conditions (e.g. IAEA, 2003c). A deterministic hazard assessment for volcanically induced lightning strikes should consider the screening criteria used in hazard assessment of rare atmospheric phenomena (IAEA, 2003c); but consider that there is a potential for a large number of column-to-ground lightning strikes during an explosive eruption.

Ground deformation

Ground deformation typically occurs prior to, during and following volcanic activity. Hazards associated with ground deformation take several forms. In the case of ground deformation at an existing capable volcano, ground deformation associated with intrusion of magma may have indirect effects, such as an increased potential of landslide, debris flow or related phenomena, and increased potential for volcanic gas flow. The potential magnitude of ground deformation should be estimated in terms of displacement, and results should be superimposed on topographic maps or digital elevation models in order to assess the potential for secondary impacts.

In a deterministic assessment, the potential magnitude of ground deformation at the site should be estimated using analytical solutions for deformation associated with magma movement of various geometries and from various source regions. Probabilistic assessment of potential ground deformation may simply link the magnitude of ground deformation estimated using models to the likelihood of such events, and a range of potential intrusion geometries.

Results of this analysis should include the estimation of the potential ground displacement to occur at the site as a result of volcanic activity, such as the opening of new vents. The most significant impact of the ground-deformation analysis, however, should involve coupling this analysis with analysis of potential for other volcanic phenomena. In particular, it is critical to assess the potential of ground deformation in landslide and volcanic debris avalanche source regions, as ground deformation in these zones may greatly change the potential volume of such surface flows and consequently their potential for reaching the site of the nuclear installation. Volcanic activity or subsurface intrusions of magma may change groundwater flow patterns or cause fluctuations in the depth of the water table. The potential hazards associated with such changes should be considered as part of the flood hazards assessment (e.g. IAEA, 2003b).

Volcanic earthquakes and seismic events

Volcanic earthquakes and seismic events normally occur as a result of stress and strain changes associated with the rise of magma toward the surface. The characteristics of volcano-seismic events may differ considerably from tectonic earthquakes, and volcanic earthquakes can be large enough or numerous enough (i.e. hundreds to thousands per day) collectively to represent a potential hazard. Volcano-seismic events may result in an increased possibility of slope failure and may compound loads on stressed structures (e.g. in tandem with tephra loading). Thus, a specific volcano-seismic hazard assessment should be undertaken using similar methods to those set out in IAEA (2002a).

In line with the approach to tectonic earthquake (i.e. seismic) hazard assessment, a deterministic method for assessing volcano-seismic ground motions should evaluate the combination of volcano-seismic event magnitude, depth of focus and distance from site that produces maximal ground motion at the site, with account taken of local ground conditions at the site. The analysis may need to consider that a volcano-seismogenic source structure cannot be construed as a capable fault (e.g. IAEA, 2002a). Suitable relationships for volcano-tectonic earthquakes should be derived for alternative ground-motion parametrization, such as peak acceleration, duration of shaking or spectral content, because specific ground-motion characteristics of volcano-tectonic earthquakes may differ from those considered in other seismic hazard assessments (e.g. IAEA, 2002a).

A probabilistic assessment of volcano-seismic hazard at a site should follow similar principles as those outlined in, for example, IAEA (2002a). Allowance should be made for uncertainties in the parameters as well as alternative interpretations. Application of the probabilistic method should include steps for (i) construction and parameterization of a volcano-seismic source model, including uncertainty in source locations; (ii) evaluation of event-magnitude frequency distributions for all such sources, together with uncertainties; and (iii) estimation of the attenuation of seismic ground motion for the site region and its stochastic variability. With these steps, the results of a probabilistic ground-motion hazard computation should be expressed in terms of the probability of exceedance of different levels of relevant ground-motion parameters (e.g. peak acceleration and an appropriate range of response spectral accelerations), for both horizontal and vertical motions. The uncertainty in the resulting probability level can be expressed by confidence bounds, with a stated basis for selection of the reported confidence levels.

In many cases, a site close to a capable volcano will also lie in a region of significant seismic hazard from tectonism. Simple scoping calculations may demonstrate that volcano-seismic hazards at a site are significantly lower than those associated with other sources of seismic activity. When such an analysis does not provide a clear margin of difference, a deterministic or probabilistic volcano-seismic hazard assessment should be undertaken.

Hydrothermal systems and groundwater anomalies

Hydrothermal systems can generate steam explosions, which eject rock fragments to a distance of several kilometers and can create craters up to hundreds of meters in diameter. Hydrothermal systems also alter rock to clays and other minerals, which creates generally unstable ground that can be highly susceptible to landslides. Currently, it is not possible to determine the likelihood for steam explosions to occur in most hydrothermal systems. Hazard evaluations for these systems are deterministic, and should consider evaluating the potential maximum ballistic or air-shock hazard for the hydrothermal source zone.

25.3 A comprehensive model

A comprehensive, site-specific volcanic hazard model is almost certainly complex. Such models will depend on assistance from informed volcanological experts, preferably through a formal expert elicitation process designed to consider all aspects of volcanic hazard at the site (e.g. US Nuclear Regulatory Commission, 1996; Aspinall, 2006). Furthermore, external peer review of the technical basis and application of the hazard model should be undertaken to increase confidence that an appropriate range of models and data has been considered in the assessment.

Volcanic events can give rise to multiple hazardous phenomena (e.g. tephra loading and seismic loading). In combination, these hazards can exacerbate the risk at an installation, even though the risk stemming from each hazard may be relatively minor on its own. A comprehensive model of volcanic hazard phenomena should therefore account for combined effects of volcanic phenomena.

Non-volcanic events such as regional earthquakes or tropical storms can initiate the occurrence of hazardous phenomena at a volcano. A comprehensive model for volcanic hazards should consider the likelihood of such hazards, which are coupled to non-eruptive initiating events. Additionally, in comparison to many external hazards, volcanic activity may persist for longer periods of time and may affect larger areas around a nuclear installation. For example, debris flows may not damage a nuclear installation directly, but may render normal operation of the installation temporarily impossible due to extensive or devastating impacts on the population and infrastructure of the surrounding region.

Overall, development of a site-specific volcanic hazard model should inform decisions about site suitability and installation design. In reaching these decisions, the potential for future volcanism and assessment of its potential effects should be considered from the perspectives of the impact on (i) the site, resulting in uncontrolled release of radionuclides into the biosphere; (ii) the site, resulting in controlled shutdown or other emergency response; and (iii) the surrounding communities, as volcanic impacts on communities may adversely affect both safe operation of the installation and the capability of the installation to deliver energy to the community, especially in a time of adverse circumstances.

Concluding remarks

Although volcanic events rarely occur, they can create a range of phenomena that could present potentially significant hazards to surface nuclear installations. The proposed approach to assess volcanic hazards provides a transparent technical basis to support risk-informed decision making for surface site suitability and consideration of design bases. The initial volcanic hazard assessment first considers the possibility of future eruptions from sites of past eruptions during the last 10 Ma. For volcanoes with the potential for future eruptions, the hazard assessment then evaluates the ability of future eruptions to produce phenomena that could reach the site of a nuclear installation. Identification of such capable volcanoes warrants the development of a site-specific volcanic hazard assessment, which more explicitly evaluates the likelihood of future eruptions and the specific characteristics of hazardous phenomena. Although deterministic methods can successfully support this evaluation, probabilistic methods provide a more transparent basis to consider data and model uncertainties and determine a range of potential hazards in addition to maximum credible events. Probabilistic methods also permit direct comparison of risks from volcanic hazards to risks from other external natural hazards, which allows for straightforward development of design bases for external and internal hazardous events at an installation. Incompleteness in the geologic record generally requires the use of numerical and statistical modeling to ensure that an appropriate range of potentially hazardous volcanic phenomena have been considered in the assessment. Although such process-level models are available for most volcanic phenomena, no model currently represents an explicitly validated methodology. Thus, models that are used to support public health and safety decisions will need to be supported by objective comparisons to well-studied analogous volcanoes, field observations, detailed process models and laboratory experiments (e.g. ASTM Standard C1174-07, 2007).

Further reading

Hoblitt *et al.* (1987) and Chung *et al.* (1990) provide good examples of comprehensive volcanic hazards assessments for several nuclear installations in the western United States. Karakhanian *et al.* (2003) and McBirney *et al.* (2003) discussed volcanic and associated hazards at an existing nuclear power plant in Armenia and a proposed site in Indonesia, respectively. These early studies relied primarily on deterministic methods to develop screening arguments for volcanic hazards. The assessment approach developed herein represents an enhancement of concepts originally expressed in McBirney and Godoy (2003), which resulted from an earlier IAEA project to develop a safety guide for volcanic hazards.

Acknowledgments

The NRC staff views expressed herein are preliminary and do not constitute a final judgment or determination of the matters addressed or of the acceptability of a license application for a geologic repository at Yucca Mountain. Discussions with Timothy McCartin, Giorgio Pasquaré, Akira Chigama and Yuichi Uchiyama helped focus and refine the concepts developed in this chapter.

References

Aspinall, W. P. (2006). Structured elicitation of expert judgment and its use for probabilistic hazard and risk assessment in volcanic eruptions. In: Mader, H. M., S. G. Coles, C. B. Connor and L. J. Connor (eds.) *Statistics in Volcanology*, London: Geological Society of London, 15–30.

ASTM Standard C1174-07 (2007). Standard practice for prediction of the long term behavior of materials, including waste forms, used in engineered barrier systems (EBS) for geological disposal of high level radioactive waste. West Conshohocken, PA: ASTM International.

Barca, D., G. M. Crisci, S. Di Gregorio and F. P. Nicoletta (1994). Cellular automata for simulating lave flows: a method and examples of the Etnear eruptions. *Transport Theory and Statistical Physics*, **23**, 195–232.

Bonadonna, C., C. B. Connor, B. F. Houghton *et al.* (2005). Probabilistic modeling of tephra dispersal: hazard assessment of a multiphase rhyolitic eruption at Tarawera, New Zealand. *Journal of Geophysical Research*, **110**, doi:10.1029/2003JB002896.

Chung, D. H., D. W. Carpenter, B. M. Crowe *et al.* (1990). Assessment of potential volcanic hazards for new production reactor site at the Idaho National Engineering Laboratory, UCRL-ID-104722. University of California: Lawrence Livermore National Laboratory.

Connor, C. B., B. E. Hill, B. Winfrey, N. M. Franklin and P. C. La Femina (2001). Estimation of volcanic hazards from tephra fallout. *Natural Hazards Review*, **2**, 33–42.

Hill, B. E., C. B. Connor, M. S. Jarzemba *et al.* (1998). 1995 eruptions of Cerro Negro volcano, Nicaragua, and risk assessment for future eruptions. *Geological Society of America Bulletin*, **110**, 1231–1241.
Hoblitt, R. P., C. D. Miller and W. E. Scott (1987). Volcanic hazards with regard to siting nuclear-power plants in the Pacific Northwest, USGS Open-File Report 87-297. Reston, VA: US Geological Survey.
IAEA (2002a). Evaluation of seismic hazards for nuclear power plants, Safety Standards Series NS G 3.3. Vienna: International Atomic Energy Agency.
IAEA (2002b). External human induced events in site evaluation for nuclear power plants, Safety Standards Series NS G 3.1. Vienna: International Atomic Energy Agency.
IAEA (2003a). Site evaluation for nuclear installations, Safety Standards Series NS R 3. Vienna: International Atomic Energy Agency.
IAEA (2003b). Flood hazard for nuclear power plants on coastal and river sites, Safety Standards Series NS G 3.5. Vienna: International Atomic Energy Agency.
IAEA (2003c). Metrological events in site evaluation for nuclear power plants, Safety Standards Series NS G 3.4. Vienna: International Atomic Energy Agency.
IAEA (2004). Geotechnical aspects of site evaluation and foundations for nuclear power plants, Safety Standards Series NS G 3.6. Vienna: International Atomic Energy Agency.
Iverson, R. M., S. P. Schilling and J. W. Vallance (1998). Objective delineation of lahar-inundation hazard zones. *Geological Society of America Bulletin*, **110**, 972–984.
Karakhanian, A., R. Jrbashyan, V. Trifonov *et al.* (2003). Volcanic hazards in the region of the Armenian Nuclear Power Plant and adjacent area. *Journal of Volcanology and Geothermal Research*, **125**, doi:10.1016/S0377-0273(03)00115-X.
McBirney, A. R. and A. Godoy (2003). Notes on the IAEA guidelines for assessing volcanic hazards at nuclear facilities. *Journal of Volcanology and Geothermal Research*, **126**, 1–9.
McBirney, A. R., L., Serva , M., Guerra and C. B. Connor (2003). Volcanic and seismic hazards at a proposed nuclear power site in central Java. *Journal of Volcanology and Geothermal Research*, **126**, 11–30.
Malin, M. C. (1980). Lengths of Hawaiian lava flows. *Geology*, **8**, 306–308.
Malin, M. C. and M. F. Sheridan (1982). Computer assisted mapping of pyroclastic surges. *Science*, **217**, 637–640.
Miyamoto, H. and S. Sasaki (1997). Simulating lava flows by an improved cellular automata method. *Computers in the Geosciences*, **23**, 283–292.
Neri, A., T. E. Ongaro, G. Menconi *et al.* (2007). 4-D simulation of explosive eruption dynamics at Vesuvius. *Geophysical Research Letters*, **34**, L04309, doi:10.1029/2006GL028597.
Patra, A. K., A. C. Bauer, C. C. Nichita *et al.* (2005). Parallel adaptive numerical simulation of dry avalanches over natural terrain. *Journal of Volcanology and Geothermal Research*, **139**, 1–21.
Pitman, B. E., C. C. Nichita, A. K. Patra *et al.* (2003). Computing granular avalanches and landslides. *Physics of Fluids*, **15**, 3638–3646.
Sheridan, M. F. (1979). Emplacement of pyroclastic flows: a review. In: Chapin, C. E. and W. E. Elston (eds.) *Ash-flow Tuffs*, Geological Society of America, Special Paper 180 125–136.
Simkin, T. and L. Siebert (1994). *Volcanoes of the World*. Tucson, AZ: Geoscience Press.

US Nuclear Regulatory Commission (1996). Branch technical position on the use of expert elicitation in the high-level radioactive waste program, Report NUREG-1563. Washington, DC: US Nuclear Regulatory Commission.

Vicari, A., H. Alexis, C. Del Negro *et al.* (2007). Modeling of the 2001 lava flow at Etna volcano by a cellular automata approach. *Environmental Modelling and Software*, **22**, 1465–1471.

Wadge, G., P. Jackson, S. M. Bower, A. W. Woods and E. Calder (1998). Computer simulations of pyroclastic flows from dome collapse. *Geophysical Research Letters*, **25**, 3677–3680.

Walker, G. P. L. (1973). Explosive volcanic eruptions – a new classification scheme. *Geologische Rundsche*, **65**, 431–446.

Woods, A. W. (2000). Dynamics of hazardous volcanic flows. *Philosophical Transactions of the Royal Society of London – Series A*, **358**, 1705–1724.

26

Formal expert assessment in probabilistic seismic and volcanic hazard analysis

K. J. Coppersmith, K. E. Jenni, R. C. Perman and R. R. Youngs

Quantification of uncertainties is an essential component of probabilistic seismic and volcanic hazard assessments. Formal expert assessment is a structured and documented process for identifying and quantifying uncertainties. Expert judgment is used in any technical assessment, but often is implicit and undocumented. Formal expert assessment explicitly includes judgments of multiple experts to represent the range of views within the scientific community. We use the term "formal expert assessment" rather than the more common term "expert elicitation" to emphasize how this approach differs from the traditional view both in philosophy and in practice.

In a typical *expert elicitation*, subject-matter experts are asked narrowly defined questions about specific uncertain quantities within their area of expertise, and they provide their judgments in the form of probability estimates or distributions. For example a climate scientist might be asked to provide an estimate of "the equilibrium change in global average surface temperature" given a specific set of circumstances (Morgan and Keith, 1995). In this approach experts are treated as independent point estimators of an uncertain quantity, and the elicitation "problem" is viewed primarily in terms of determining how to ask the right questions as clearly as possible of the most knowledgeable experts. Under this perspective, the elicitors may focus significant effort on ensuring that they have well-calibrated and informative experts; that is, experts who give both accurate estimates and a narrow range of uncertainty in estimates of quantities similar to those of interest but for which a "true" value can be determined (e.g. Chapter 10 of Bedford and Cooke, 2001). For narrow assessment tasks, the elicitor may focus on designing elicitation questions to motivate "honest" responses through the use of proper scoring rules (Gneiting and Raferty, 2007). All of these tools and approaches reflect the general philosophy that probabilities are something that exist in the experts' minds, and the job of the elicitor is to extract, or elicit, those probabilities.

Formal expert assessments, in contrast, take a broader view of the process: subject-matter experts are asked to participate in an interactive process of ongoing data evaluation, learning, model building and, ultimately, quantification of uncertainty. Experts are explicitly tasked with considering the opinions of the broader technical community of which they are part, and to ensure that those views are appropriately represented in the resulting analyses. Interactions among experts during the assessment process up to and including discussion of preliminary assessments of specific uncertain quantities are strongly encouraged and, in most cases, are built into the structure of a formal expert assessment project. The formal expert assessment approach reflects a view that expressing uncertainty with probabilities is not something that comes naturally to most people and thus cannot be "elicited" by the use of clever questions. Rather, probabilities are constructed through careful consideration of the technical questions, available data, models and so forth. The job of the assessor is to assist the expert in developing and expressing those probabilities in a way that is consistent and that accurately reflects their knowledge and beliefs.

The goals of this chapter are to provide an overview of the formal expert assessment approach used in probabilistic seismic and volcanic hazard analyses through a brief review of some case histories where the approach has been used to support hazard analyses for nuclear facilities; and to provide some lessons learned regarding process elements that will lead to successful applications. Formal expert assessment provides a means for properly and fully incorporating the uncertainties represented by diverse technical interpretations, as well as providing transparency through the complete documentation of the process and results. For these reasons, formal expert assessment has gained increasing acceptance within the regulatory community for dealing with seismic hazard analyses for nuclear facilities. The process promotes the direct involvement of technical experts and provides regulators with assurance that the hazard analysis reflects the current understanding of the technical issues and that all competing hypotheses have been considered. It is hoped that the reader will gain an understanding of formal expert assessment processes and an appreciation for the specific issues that are important to implementation of the approach to seismic and volcanic hazard analyses. The conclusions drawn in this chapter are based on experience gained over the past two decades on seismic and volcanic hazard analyses conducted for nuclear facilities in the eastern and western USA and Europe.

For probabilistic seismic hazard analyses (PSHA) the locations of future earthquakes, their recurrence rates and maximum size, and the ground motions that may result at a site of interest are all quantities that are uncertain and require careful consideration. The earliest PSHA model developed by Cornell (1968) provided for a single interpretation of seismic source geometries and characteristics, a single ground-motion attenuation relationship, and resulted in a single seismic hazard curve. With time, we have come to recognize more fully the uncertainties in alternative models to explain seismic-source and ground-motion predictions, and uncertainties in the parameter values for those models. The quantification and incorporation of those uncertainties led to a probability distribution defined by a family of hazard curves. Similarly, the basic components of probabilistic volcanic hazard analyses (PVHA) are the locations of future events, the characteristics of those events (defined by their hazard-significant parameters such as eruptive volume, duration and magnitude), and the recurrence rates of those events.

Two major PSHA projects conducted in mid 1980s represent significant landmarks in the development of formal expert assessment and the systematic, explicit incorporation of the diversity of expert interpretations in hazard assessments. Conducted by the Lawrence Livermore National Laboratory (LLNL; see Bernreuter *et al.*, 1989) and the Electric Power

Research Institute (EPRI; see EPRI, 1989), both projects focused on developing seismic hazard curves for the large region of the central and eastern United States to develop a basis for assessing hazard at the 69 nuclear power plant sites east of the Rocky Mountains. Although the two studies each utilized large numbers of technical experts and were similar in many technical and procedural ways, the results of the studies, specifically the mean seismic hazard curves, differed significantly for most sites in the eastern USA.

Based on the differing results obtained from these two large PSHA studies, it eventually became clear that the *process* used to conduct a hazard assessment using a diversity of technical experts, opinions and models was equally as important as the technical content of the interpretations. Accordingly, the US Department of Energy (DOE), the US Nuclear Regulatory Commission (NRC) and EPRI sponsored a project to study and offer recommendations for how such hazard analyses should be conducted in the future. The Senior Seismic Hazard Analysis Committee (SSHAC) developed a process (the "SSHAC process") that we now refer to as formal expert assessment. The final guidelines resulting from the study were published in 1997 in the US NRC's NUREG/CR-6372 (SSHAC, 1997) and have been used subsequently for numerous studies, including the case studies described in this chapter. Although the SSHAC report uses *seismic* hazard analysis as a case study, the procedural guidance is also applicable to volcanic and other types of hazard analyses. The descriptions in this chapter of the key elements required for successful probabilistic hazard assessments are based on both the SSHAC guidance and the experience of the authors, who participated in developing and implementing the formal expert assessment processes in each of the case studies described.

26.1 A brief overview of the SSHAC guidance

The objective of the SSHAC project was to provide methodological guidance on how to perform a probabilistic hazard analysis, with particular emphasis on approaches to dealing with uncertainties in PSHA. The latter is an analytical methodology that estimates the likelihood that various levels of earthquake-caused ground motion will be exceeded during a given future time period for a given location. Such estimates, however, can be attained only with significant uncertainty as there are major gaps in our understanding of the mechanisms that cause earthquakes and the effects of earthquakes at specific locations. Significantly, there are often wide differences of legitimate scientific opinion on key inputs into a PSHA. The SSHAC probabilistic formulation for dealing with seismic hazards specifically embeds uncertainties in the core of the methodology. Two different classes of uncertainties are identified. *Epistemic* uncertainties arise from lack of complete knowledge of the processes and/or quantities of interest: that lack of knowledge means it is not possible to make precise assessments of those processes or quantities, and they are treated probabilistically. In principle, epistemic uncertainties are reducible through further research and additional data collection. In seismic and volcanic hazard analyses, the randomness in a physical process results in *aleatory* uncertainty, or more accurately, *aleatory variability*. This randomness represents variability that cannot be known in detail nor reduced through further research. For example, recurrence rates for various-magnitude earthquakes reflect the degree of epistemic uncertainty, but the exact location and magnitude of a future large magnitude earthquake is an aleatory variability. Athough there are different schools of thought on the importance of this distinction between epistemic uncertainty and aleatory variability (Panel on Seismic Hazard Evaluation *et al.*, 1997), we have found it to be very useful throughout the formal expert assessment process for geologic hazards. The distinction between variability and lack of knowledge appears to be straightforward and helpful in this context.

The SSHAC guidance provides advice for four "study levels," which are differentiated as a function of the importance, complexity, diversity of views and contentiousness of an issue. The level of study required for a hazard analysis is related to factors such as the regulatory framework, the resources (money and time) available to conduct the study, perceptions of the importance of the project and scheduling constraints. Regardless of the level of study, the goal is the same: to provide a representation of the informed scientific community's view of the important components and issues, and, finally, the hazard. The term "informed" in this sense assumes, hypothetically perhaps, that the entire scientific community of experts was provided with the same data and level of interaction as that of the members of the formal expert assessment process. In more modest studies (SSHAC Study Levels 1–3), a Technical Integrator (TI) utilizes interpretations found in the published literature supplemented by conversations or workshops with individuals conducting relevant research. The TI then evaluates the viability and credibility of various hypotheses with an objective of capturing the range of alternative interpretations and their uncertainties to provide an overall assessment that represents the informed scientific community's view of the subject. In these studies, the TI is the "owner" of the hazard results, in the sense that the TI is expected to be able to document, explain and defend those results within the technical community. When resources are available and the level of sophistication of the required analyses are high (typically for issues that are highly contentious, significant to hazard and highly complex) the assessments needed for a hazard analysis can be made by multiple experts (called "evaluator experts"), who evaluate alternative models and parameters in a process that involves a series of workshops and individual assessment meetings. This is called SSHAC Study Level 4 and is the subject of this chapter. Consistent with the SSHAC guidance, a Technical Facilitator/Integrator (TFI) (this can be a single individual or a team) is responsible for facilitating the interactions among the experts and integrating the judgments of the expert panel to develop a composite distribution that reflects the informed technical community. For SSHAC Study Level 4, the ownership of the hazard results is shared between the TFI and the experts.

26.2 Hazard analysis case studies

Formal expert assessment has been used to conduct hazard analyses over the past twenty years in a wide variety of seismo-tectonic environments. Although sharing some common attributes, each study is unique in its implementation. Furthermore, each study has led to an evolution of methods and approaches that have been successful. The salient elements of some of these studies are summarized in this section, followed by a discussion of the key procedural elements of the lessons learned that should be considered in future formal expert assessments. The hazard results of these case studies is not the subject of this chapter, but the interested reader can refer to the cited references for those results.

26.2.1 *EPRI seismic hazard analysis*

The central and eastern United States is characterized generally by a 200 a historical record of low to moderate levels of seismicity punctuated by infrequent large-magnitude earthquakes (e.g. the M $\sim$ 7–8 New Madrid earthquakes of 1811 and 1812; the M $\sim$ 7 Charleston earthquake of 1886) and virtually no geologic information on active faulting. These conditions present significant challenges for assessments of seismic hazard because of the considerable uncertainties associated with key inputs. The large EPRI project (EPRI, 1986, 1989), in which seismic hazard was assessed for 69 nuclear power plant sites in the central and eastern USA, was important for developing methodologies and procedures on the conduct of a formal expert assessment utilizing multiple experts; and, as described above, was a direct predecessor to the SSHAC study. This hazard analysis focused on developing a methodology for PSHA that included a highly structured procedure for interpreting the tectonics of an area to define the seismic source zones and utilized statistical analyses of a historical earthquake catalog to develop earthquake size and rate parameters. An initial part of the study involved compilations of comprehensive geophysical and seismological databases. These databases were then distributed to six earth-science teams, composed of individuals representing the fields of seismology, geology and geophysics. The teams independently developed seismic source zones and associated seismicity parameters for the area of focus, explicitly accounting for uncertainties in the evaluations using alternative, weighted interpretations for individual zones or features. To implement the methodology numerous and extensive workshops and meetings with project participants were convened, and the methodology team worked with the participants to assess their scientific judgments and to format those judgments to be suitable for hazard calculations.

26.2.2 *Satsop seismic hazard analysis*

Considerable uncertainty exists regarding the earthquake potential of the Cascadia subduction zone in the Pacific northwest of the United States due to the historically aseismic nature of the interface between the Juan de Fuca and North American plates. A PSHA involving formal expert assessment of multiple experts was conducted in this region for the Satsop nuclear power plant site in western Washington state (Coppersmith and Youngs, 1990). To develop a complete seismic source characterization spanning the range of interpretations regarding the earthquake potential of Cascadia, a group of 14 experts was selected based on their experience in the region and at convergent margins worldwide. These experts assessed source characteristics, including subduction zone geometry of the plate interface, intraslab and crustal seismic sources; probability that each potential source is seismogenic;

expected locations and dimensions of rupture; maximum earthquake magnitude; earthquake recurrence models; paleoseismic recurrence intervals; plate convergence rate; and seismic coupling. Important sources of uncertainty included alternative conceptual models regarding the nature and rate of seismogenic convergence across a plate interface that has not experienced moderate to large earthquakes during the historical period of about 100 a. Prior to the elicitation, new geologic evidence of episodic coastal subsidence had become available, which was interpreted by some as evidence for pre-historical seismogenic coupling along the plate interface. The results of the seismic hazard analysis were submitted by the electric utility to the NRC as part of licensing activities.

26.2.3 Yucca Mountain probabilistic seismic hazard analysis

Assessing hazard for a nuclear waste repository presents unique technical challenges and significant uncertainties. Probabilistic seismic hazard analyses were conducted to estimate both ground-motion and fault-displacement hazards at the proposed geologic repository for spent nuclear fuel and high-level radioactive waste at Yucca Mountain, Nevada (CRWMS M&O, 1998; Stepp *et al.*, 2001). The methodology followed was consistent with SSHAC recommendations for a Level 4 PSHA as well as guidelines established by the NRC for expert elicitations (Kotra *et al.*, 1996). Six teams of three experts performed the seismic-source and fault-displacement evaluations, and seven individual experts provided ground-motion evaluations. State-of-the-practice formal expert assessment processes were implemented including dissemination of a common database, structured workshops, field trips to visit sites of paleoseismic investigations, participation of "resource" experts to share data gathered at analog locations, and open exchanges about alternative conceptual models. Site-specific ground motions were characterized for a given site condition in order to allow for subsequent use in site response studies. The major emphasis of the study was on quantification of epistemic uncertainty. The results of the PSHA are being used to establish seismic design bases for surface facilities and to evaluate the performance of the repository during the regulatory period following closure (10–1000 ka). The NRC, other oversight groups and participatory peer reviewers were closely involved in the review of the project throughout its course.

26.2.4 Switzerland seismic hazard analysis (PEGASOS project)

Although Switzerland is generally considered to have a low to moderate level of seismicity, the Swiss Federal Nuclear Safety Inspectorate (HSK) identified seismic hazard as a potentially significant contributor to the risk at four nuclear power plant sites (Mühleberg, Gösgen, Beznau and Leibstadt). The HSK also identified the need to update the seismic hazard analyses at the sites and requested that the Swiss electric utilities conduct a PSHA following SSHAC Level 4 methodologies. Under the direction of the National Cooperative for the Disposal of Radioactive Waste (NAGRA), a PSHA was conducted for Swiss nuclear power plant sites. The study has since become known under the name of the "PEGASOS project" (Probabilistische Erdbeben-Gefährdungs-Analyse für KKW-StandOrte in der Schweiz) (NAGRA, 2004). The objective of the project was to assess the relevant earthquake-induced ground motions at the building foundation levels of the four sites, which would be used subsequently for probabilistic safety analyses. A full-scope formal expert assessment process was used, including dissemination of a comprehensive database, multiple workshops for identification and discussion of alternative models and interpretations, assessment interviews, feedback to provide the experts with the implications of their preliminary assessments and full documentation of the assessments. The study brought together experts from all over Europe. Four teams consisting of three experts conducted the seismic source characterization, five individual experts addressed ground-motion characterization and four experts characterized the site response. The entire study was subject to participatory peer review by an HSK Review Team, which monitored and provided feedback on the procedural and technical aspects of the project, as well as provided a review of the final report.

26.2.5 Yucca Mountain volcanic hazard analysis and update

The performance of the high-level radioactive waste repository must be evaluated for future time periods of over 10 ka and must include evaluations of the potential for disruption due to volcanic events. The Department of Energy (DOE) conducted the PVHA following a formal expert assessment process and consistent with NRC guidance (Kotra *et al.*,1996). This original Yucca Mountain PVHA (CRWMS-M&O, 1996) was conducted with a panel of ten experts and included the following steps: identification and selection of experts; training of experts in probability and probability assessment; trial assessment to identify significant issues; workshops on data requirements, alternative models, feedback and results; field trips to visit key localities; and individual assessments with each panel member. The PVHA expert assessments focused on spatial models defining the future locations of volcanic events and temporal models defining the rate of occurrence of events. To make their assessments and to quantify the associated uncertainties, the experts considered the geologic data in the Yucca Mountain region as well as their own experience at analog basaltic volcanic fields.

In 2005–2007, the DOE conducted the probabilistic volcanic hazard analysis update (PVHA-U) involving a panel of eight experts and a SSHAC Level 4 process. This first-of-a-kind update of a major formal expert assessment was motivated by the availability of new data subsequent to the 1996 study, and by advances in approaches for modeling volcanic hazard. Newly available data included new high-resolution aeromagnetic data and drilling/geochronology/geochemical data. Several aeromagnetic anomalies were identified in alluvial basins in the region that were postulated to represent buried basaltic bodies. The drilling data provided information on the composition, depth and age of the igneous features giving rise to the anomalies. At their request, the experts were also provided with information regarding characteristics of analog volcanic events in the region. Conceptual models developed by the experts for the future spatial distribution of volcanic events included spatial smoothing of applicable past events, consideration of parametric field shapes and identification of alternative tectonic zones defined by variations in the rate of volcanism. Temporal models developed by the experts included models of recurrence rate as a Poisson process, defined by observed events of specified age, temporally clustered models and time-volume models that account for the decline in eruptive volumes over the past several million years. In the PVHA-U, particular emphasis was placed on defining the characteristics of future volcanic events, including the number and dimensions of dikes and eruptive conduits in an event, the geometry of these features and the expected eruptive type.

26.3 Key elements of formal expert assessment

Based on experience gained from the formal expert assessments outlined above, there are several lessons learned regarding key process elements. These elements should be considered when planning a formal expert assessment.

Experts should be trained in probability theory, uncertainty quantification and ways to avoid common cognitive biases

Many experts who have knowledge relevant to seismic hazard assessments do not necessarily have experience with probabilistic modeling or development of probability distributions that reflect their state of knowledge. Accordingly, training in the language of probability and quantifying uncertainties (including recognizing the distinction between aleatory variability and epistemic uncertainty) should be provided. Possible biases may also unknowingly be expressed by the experts unless they have been educated to recognize and minimize such biases. These include cognitive biases such as overconfidence, anchoring, availability and the reporting of narrower-than-justified probability distributions. Motivational biases may occur if an expert is rewarded for being a strong proponent of a particular viewpoint. The TFI must be aware of this possibility and attempt to eliminate such bias by stressing the importance of having each expert act as an evaluator representing the larger technical community.

Typically, training of the experts in these areas occurs during the first workshop on the project and is followed by reminders and facilitation throughout the project. Support from the TFI may need to be provided to assist earth scientists with statistical and probabilistic calculations, as long as the expert is fundamentally responsible for the expression of uncertainty. A careful expert selection process (SSHAC, 1997, p. 42), as well as the structured interactions between experts during workshops and field trips, help to mitigate the potential for motivational bias in the assessments. Simple awareness of potential cognitive biases and reminders throughout the project are typically adequate for mitigating this source of bias.

Comprehensive and user-friendly databases should be provided to the expert

From the time of the first large formal expert assessments in the 1980s, there has been an explosion in availability of tools to compile, display and evaluate complex geologic and geophysical data sets. Further, the range of information that is used to characterize the spatial and temporal elements of hazard analyses has expanded. It is important that all experts on a panel have access to data that they find most pertinent to the assessments that need to be made. Relevant data, as defined by the experts themselves, should be provided to all experts in the form of a comprehensive and uniform database. Any processing of "raw" data that is made at the request of one expert should also be provided to other experts who wish to review the processed data. The effort to compile and disseminate the data can be significant. For example, in the PEGASOS project, a dedicated database contractor was responsible for compiling geologic maps, geophysical data, seismo-tectonic data and other information in various formats specified by the experts. In many cases, multiple geographic information system (GIS) layers were combined to form maps for use by the experts (Figure 26.1). Likewise, on the Yucca Mountain PSHA, site-specific data gathered over many years were distributed to the experts, including seismicity catalogs, paleoseismic trench logs, Quaternary geologic maps and other data (Figure 26.2).

The PVHA-U for Yucca Mountain included the development of a database of maps, references and pertinent geophysical, geochemical and geochronology data. The database was structured around requests made by members of the expert panel. For example, to

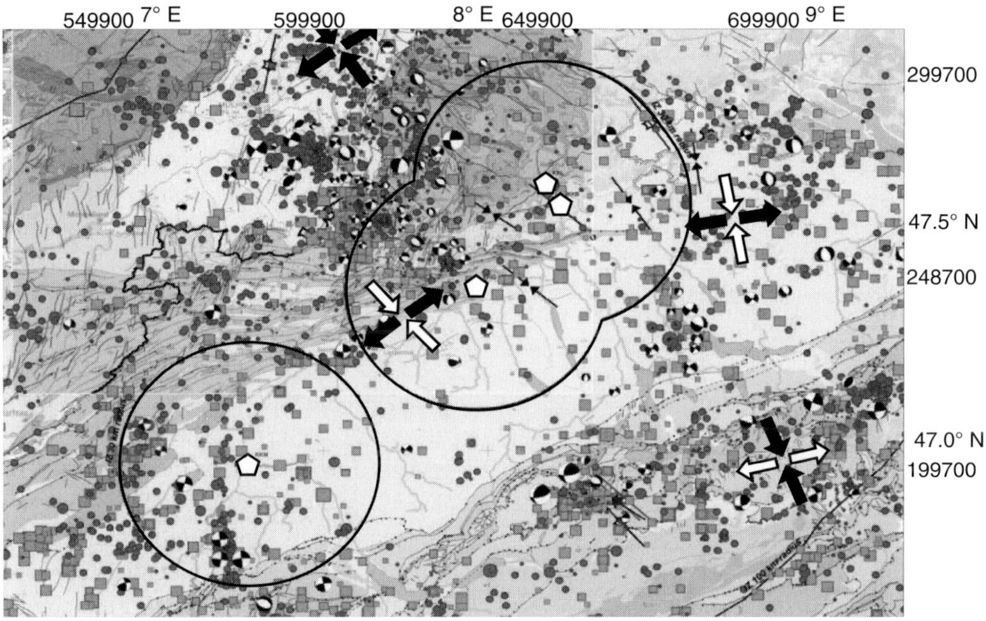

Fig. 26.1 Example of a map developed for the PEGASOS seismic source characterization experts. The GIS layers include seismicity (instrumental shown by small circles; historical by small boxes), regional stress orientations (large arrows), focal mechanisms and regional geology. The four Swiss nuclear power plant sites are marked as pentagons. Circular black lines indicate 25 km distance ranges from the sites (NAGRA, 2004). See color plate section.

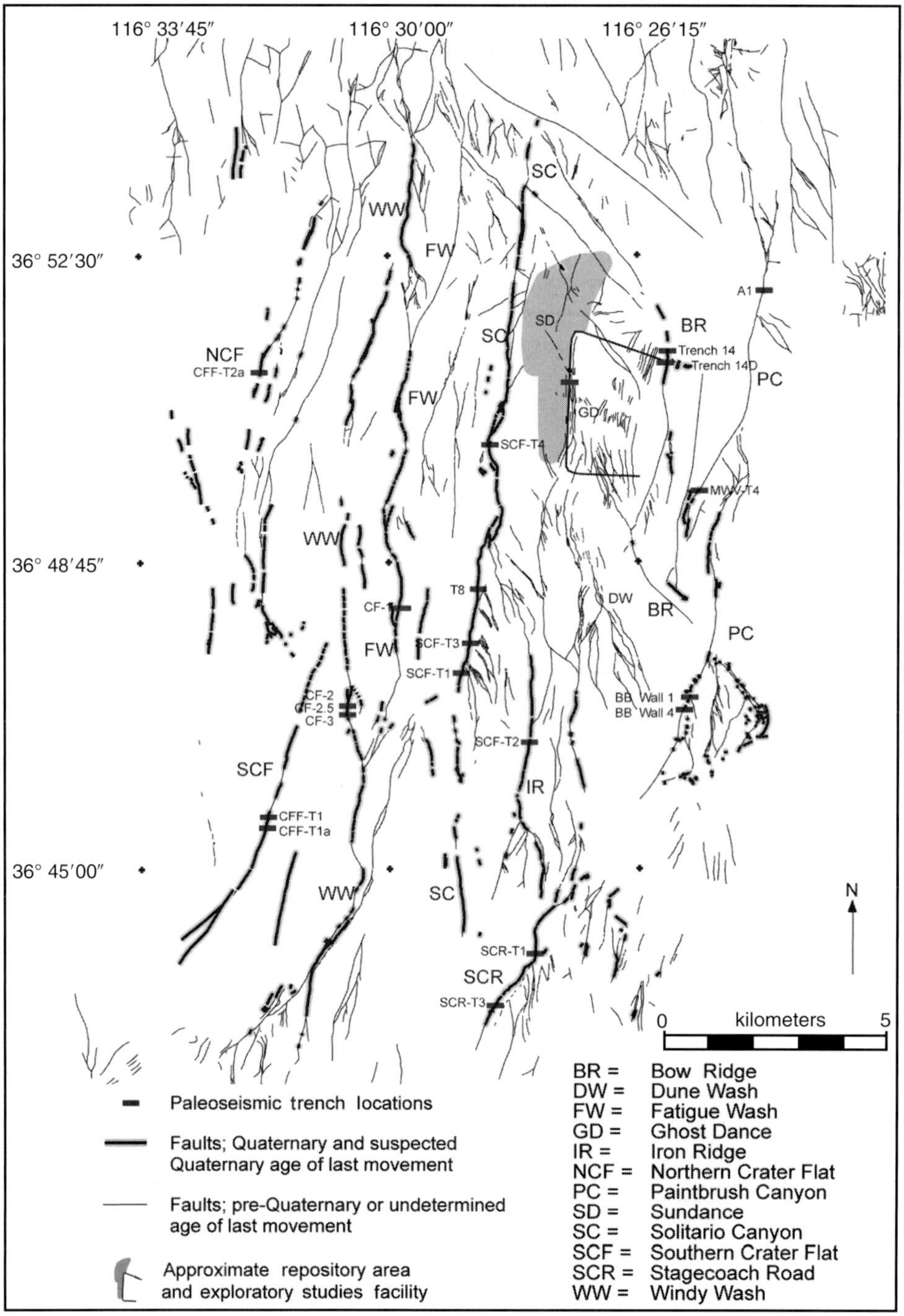

Fig. 26.2 Example of a map provided to the experts of the Yucca Mountain probabilistic seismic hazard analysis. Map shows faults and paleoseismic trench locations in the Yucca Mountain region. Fault map is modified from Simonds *et al.* (1995) and Pezzopane *et al.* (1996). See color plate section.

assist in modeling the future spatial distribution of volcanic events in the region, maps were developed that incorporated multiple GIS layers. To assist with assessments of the characteristics of volcanic events that could affect the Yucca Mountain site, data were compiled at a number of analog sites within the southern Great Basin. Interpretations of possible buried volcanic centers beneath the alluvial basins in the region were assisted by a detailed aeromagnetic survey and drilling, together with geochronology and geochemical analyses (Valentine and Perry, Chapter 19, this volume).

Experts should be required to evaluate all potentially credible hypotheses

For a formal expert assessment, members of an expert panel are specifically required to be "evaluators." In other words, they are expected to evaluate all potential hypotheses and the bases for the hypotheses from available information. This role is distinctly different from the "proponent" role, which is common for scientists. Proponents advocate particular hypotheses and points of view, based on their interpretation of the data. However, "evaluators" are charged with considering alternative interpretations in addition to those that they may favor, and to arrive at their representation of a community distribution. As such, experts should assign a relative weight or credibility, including zero weight, to hypotheses considered to be non-credible-to alternative hypotheses, recognizing that these are current representations of knowledge and that no single hypothesis is likely to be the "ultimate truth."

A common tool for expressing the relative credibility of alternative conceptual models is a logic tree (Figure 26.3). The branches of the tree represent alternative credible models or hypotheses and the weights on the branches indicate the relative credibility of the alternatives. For example, the experts in the Satsop PSHA evaluated alternative hypotheses regarding the seismogenic potential of the plate interface of the Cascadia subduction zone. These were given as branches of a logic tree, expressed as the probabilities that the plate interface and the intraslab source are seismogenic (defined as capable of generating a $M > 5$ earthquake). The aggregate distribution across all experts on the panel is shown in Figure 26.4.

Workshops and other interactions among the experts and proponents of alternative viewpoints should be encouraged

Interaction among experts, particularly in facilitated workshops, is a fundamentally important aspect of a formal expert assessment process. This is a natural extension of how earth scientists work together to formulate their ideas. Typically, they rely on common data sets and interact frequently as part of their professional activities. In workshops, the technical issues of greatest importance to hazard can be identified and the data needed to address those issues specified. In addition, the methods and procedures to characterize the fundamental inputs to the hazard analyses can be discussed among the experts to ensure that all are aware of the tools that are available. Presentations on available data by resource experts may be made and, importantly, proponents of alternative viewpoints can provide their arguments

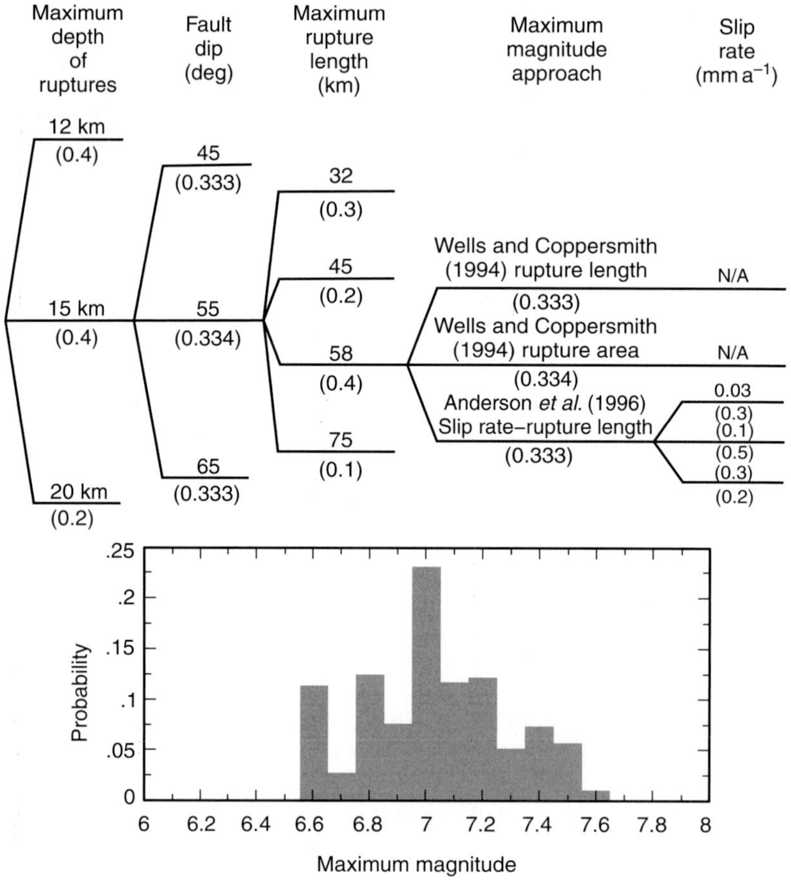

Fig. 26.3 A sample logic tree for maximum-magnitude assessment for a fault-specific seismic source using relationships between rupture dimensions and earthquake magnitude. At each node, alternative branches represent alternative conceptual models or parameter values. These alternatives are each assigned relative weights that express the expert's degree of belief that the particular model or parameter is the correct value. The relative weights are treated as conditional probabilities. The maximum-magnitude values associated with the various branches of the tree are multiplied by the weights associated with each branch to arrive at the probability density function shown at the bottom of the figure.

to the expert panel. Such workshops need to be carefully facilitated by the TFI to ensure that the viewpoints are presented and discussed in a balanced manner. Review, technical challenge, and defense of hypotheses and interpretations are important objectives for these workshops. In studies evaluated by SSHAC it was found that unless experts interact and discuss alternative technical interpretations, unintentional disagreements can arise because of reliance on different data sets, unstated assumptions or a lack of understanding of the basis for alternative positions.

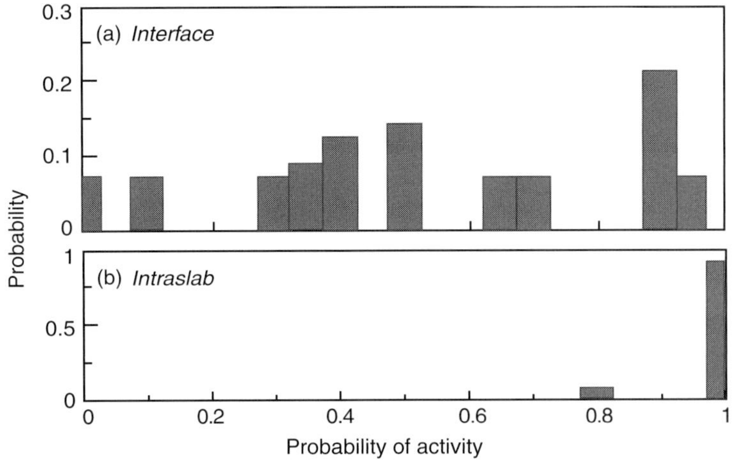

Fig. 26.4 Aggregate distribution of 14 experts' assessments of the probability of activity for the following elements of the Cascadia subduction zone: (a) the plate interface and (b) the intraslab source (Coppersmith and Youngs, 1990). "Activity" was defined as capable of generating a moderate-to-large earthquake ($M > 5$). Copyright (1990) Geological Society of America. Reproduced by permission of the Geological Society of America.

Assessment interviews should start with conceptual models and develop towards parameter assessments

After the experts have received training in probability assessment procedures and after they have been exposed to a full range of data and interpretations in workshops, it is recommended that individual experts or small teams of experts be elicited in small interview sessions. The interview should be conducted by the TFI (a TFI team may consist of a technical expert and a normative expert with experience in subjective probability assessment; others who can provide specialized knowledge may also attend the interview sessions as a resource for the expert). Commonly, the initial parts of the interview deal with the overall structure of the assessment and the general identification of basic hazard model elements. This session may include development of draft logic trees (described above) or influence diagrams (e.g. Section 2.3.3.5 in CRWMS-M&O, 1996), which illustrate the relationships between the various model components and help to identify key uncertainties that will be part of the assessment.

The experts should then consider conceptual models that address the hazard model elements. For example, alternative conceptual models regarding the future spatial distribution of volcanic events might be: (i) future volcanic events will be proximal to past events of a certain age, (ii) they will lie within certain structural settings, or (iii) they will occur randomly within certain tectonic domains. Each conceptual model can then be defined by modeling approaches that implement the conceptual model for purposes of the hazard analysis. For example, the notion that future volcanic events will be proximal to past events of a certain age might be implemented by spatial smoothing around existing volcanic events (e.g. Connor and Connor, Chapter 14, this volume). From there, more detailed assessments of model specifics and parameter values can be made. It is important to encourage the expert to consider the technical merits of all hypotheses and assess the relative credibility of each, keeping in mind the diversity of views within the larger informed technical community.

Feedback should be provided to the experts to give them insight into the significance of their assessments to the hazard results

Experience has shown that at least two rounds of assessment interviews, separated by feedback, can be valuable to the experts in developing their final assessment. After the first interview, the preliminary expert assessments can be used in hazard calculations and sensitivity analyses, which can then be discussed with all experts in a workshop setting. The purpose is to allow each expert to see the preliminary interpretations made by the other experts, to understand the implications that their own and others' assessments have to the calculated hazard results and to identify those aspects of their assessments that are most important to the hazard results. This allows each expert to focus on the important elements of their assessment during the second round of interviews. Possible problems or inconsistencies in the first round of assessments can also be identified using this approach. For example, correlations between some parameter values, as in the assessment of recurrence parameters, can lead to some unintended combinations and resulting rates. Providing feedback on the recurrence implications of the assessed models and parameters can allow these to be clearly identified and discussed.

A feedback workshop provides each expert the opportunity to discuss interpretations and evaluations with other experts while focusing on the technical bases for the assessments. Constructive scientific debate, while seeking areas of consensus and resolution of misunderstandings or identification of different assumptions, is a valuable component of a feedback workshop. Examples of the feedback provided to the experts on the Yucca Mountain PVHA-U project are given in Figures 26.5–26.7.

The assessments of multiple experts should be aggregated using equal weights

As discussed in the introduction, we view probabilistic seismic and volcanic hazard analyses not as elicitations of numbers and values that exist in experts' minds, but as scientific assessments that require consideration of data sets, learning and interactions. The interactions shown to be successful in a SSHAC process (e.g. workshops and field trips that foster discussions among the experts) is counter to the notion of independent point estimators of "the truth." Furthermore, the SSHAC process insists that experts act as evaluators, not proponents, and consider the views of the larger informed technical community.

This type of expert assessment process requires an aggregation method that acknowledges the roles that experts play as evaluators: the assessment of all experts should be equally weighted. The formal expert assessment process develops a defensible basis for equal weighting, defined in the SSHAC as requiring the following elements: a careful

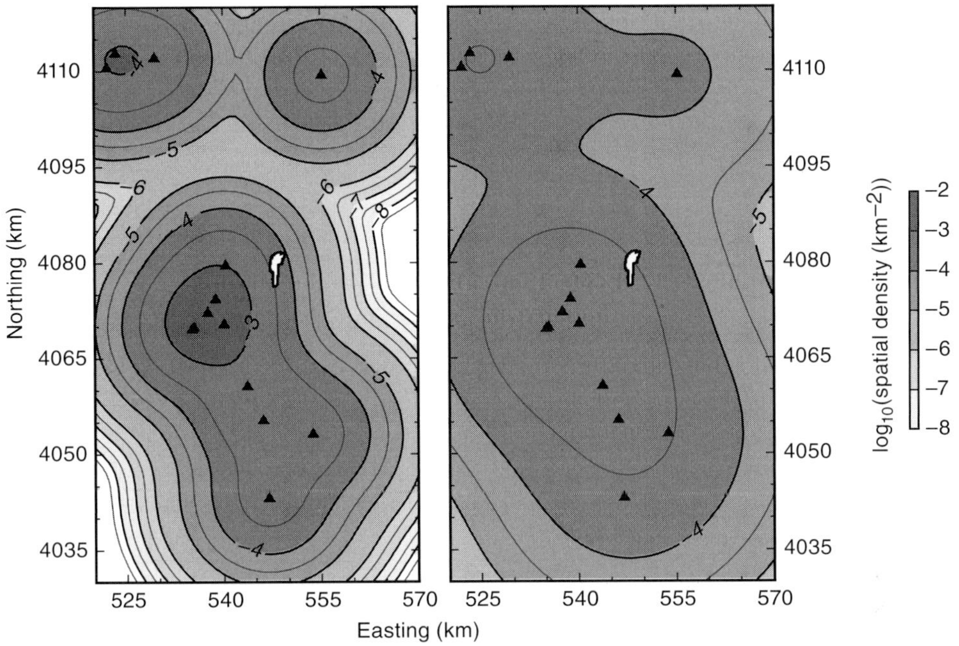

Fig. 26.5 Example of feedback provided to an expert on the PVHA-U at Yucca Mountain. Two alternative smoothing (h) distances are considered: $h = 5$ km on the left, and $h = 10$ km on the right. Contours are $\log_{10}$ of the conditional spatial density (number of events per square kilometer) calculated using a Gaussian kernel function. Observed volcanic events are shown as black triangles; proposed repository location indicated by the white polygon.

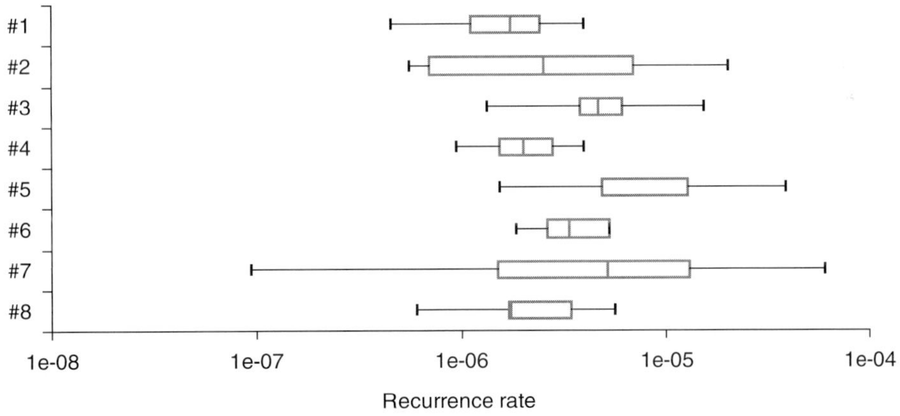

Fig. 26.6 Example of feedback provided to the experts on the PVHA-U for Yucca Mountain. Shown are preliminary assessments of the recurrence rate for volcanic events in the Yucca Mountain region made by each of the eight experts. The bars represent the 5th–95th percentiles of the assessment, boxes show the 25th–75th percentiles, and the vertical line in the box marks the median of the assessment. Each assessment is specific to the region of interest defined by each expert and thus the assessed rates must be compared cautiously.

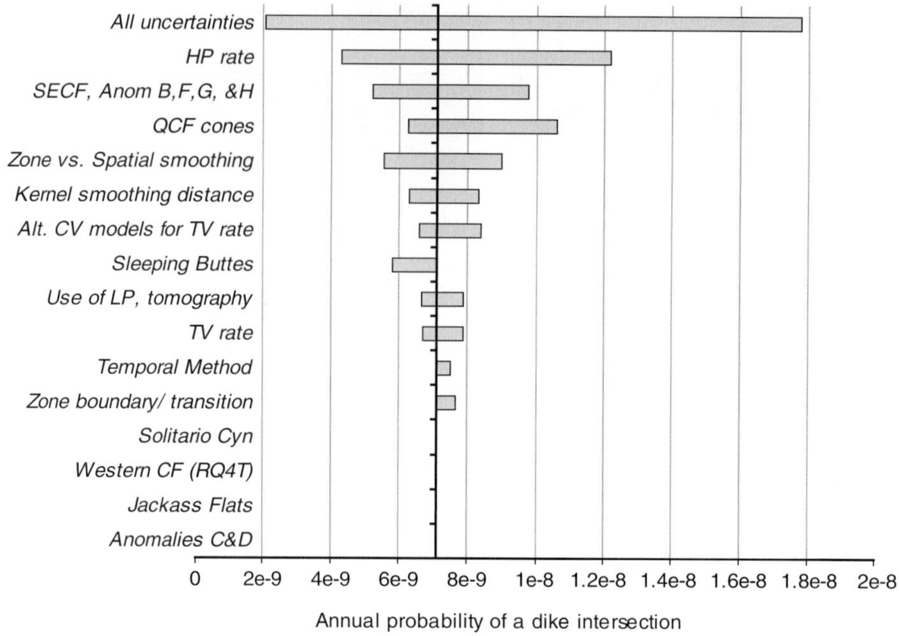

Fig. 26.7 Example of feedback provided to an expert on the PVHA-U for Yucca Mountain. The y-axis crosses the x-axis at the "most likely" or nominal value, which represents the annual probability of dike intersection at the repository with all model inputs set at their 50th percentile or most likely value. The length of the top bar represents the 5th–95th percentiles of the full distribution on the probability of intersection across all model inputs. Subsequent bars illustrate the uncertainty that results from uncertainty in a single specified input. These are calculated by setting all parameters equal to their most likely values, and then varying one from its lowest to highest value. This type of plot helps the expert to understand the relative contribution that uncertainties in various elements of their models makes to the total uncertainty in the hazard assessment.

expert selection process is followed; all experts provide the required effort and commitment; a comprehensive database is developed and is made accessible to all experts; experts are trained in probability and uncertainty quantification; the experts serve as expert evaluators; facilitated interactions are provided to foster a free exchange of data, including scientific debate with respect to alternative interpretations; and feedback and sensitivity analyses are provided to the experts such that all experts were aware of the implications of their assessments prior to finalization. The aggregate hazard distribution developed for the Yucca Mountain PVHA (CRWMS M&O, 1996) based on equal weighting of the expert assessments is given in Figure 26.8.

Complete documentation should include the technical basis for all assessments

Formal expert assessments are commonly conducted when there is a high level of regulatory and/or public scrutiny. Accordingly, the inputs and results of the hazard assessment will be extensively reviewed, so complete documentation of the elicitation process and the technical

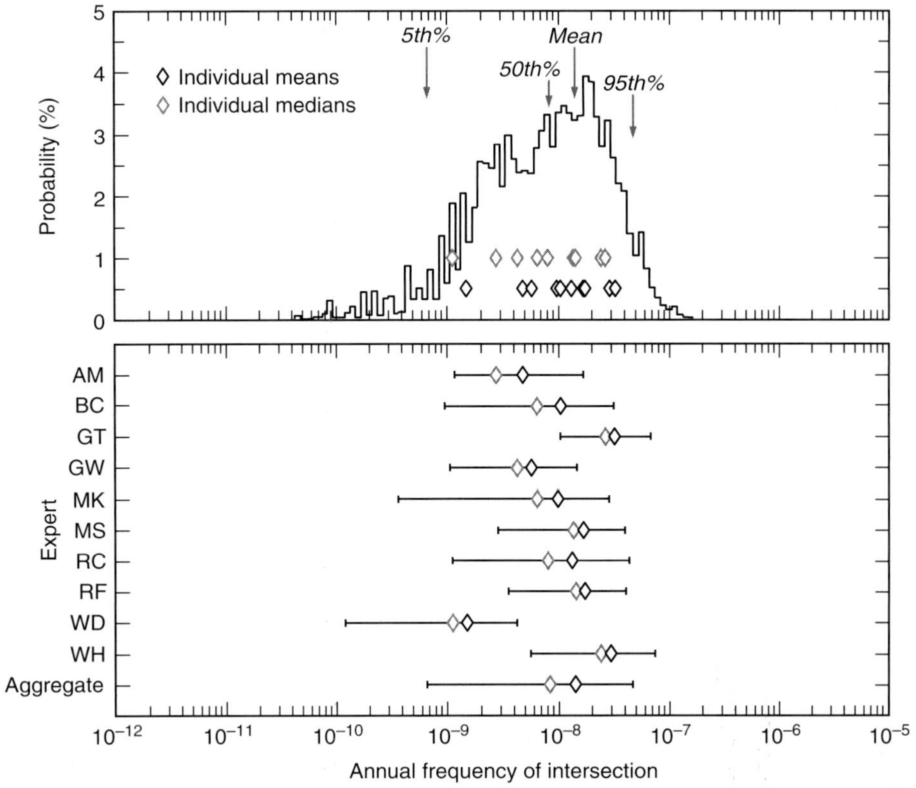

Fig. 26.8 Hazard result, expressed as the annual frequency of dike intersection with the repository, for the PVHA conducted for Yucca Mountain (CRWMS M&O, 1996). Individual expert hazard results are shown in the bottom plot: the bar extends from the 5th–95th percentile of the expert's assessment, and the median and mean of the distributions are shown with open and closed diamonds, respectively. The total distribution of hazard is shown on the top plot based on aggregation of the expert assessments using equal weights; percentiles and the mean of the total distribution are indicated.

bases for all judgments are essential. The elicitation interviews in particular must be carefully documented, as well as the proceedings of the workshops, the methodologies used, and the hazard results and sensitivity analyses. Both the interpretations and uncertainties expressed by the experts and the technical bases for the interpretations must be documented. A formal expert assessment for a hazard analysis occurs at a particular point in time and it is important to capture in the documentation the state of knowledge at that time. This allows reviewers, including those who may examine the study many years hence, a basis for understanding the data, models and thought processes that drove the assessments.

Concluding remarks

Over the past two decades, formal expert assessment has emerged as an effective tool for capturing the important uncertainties associated with seismic and volcanic hazard analyses.

Because of the resources required to carry out these projects, they have been reserved for projects under regulatory scrutiny where uncertainties are considerable and significant. The conduct of several elicitations in a variety of tectonic and regulatory environments has provided insights into the key process elements that will help lead to successful projects. These elements relate to the training of the experts, the data and information provided to the experts, the manner in which experts are allowed and encouraged to interact, the feedback that is provided to inform their final assessments, the aggregation of the expert judgments, and the documentation that is needed to allow for a review by others.

Further reading

Those interested in processes for eliciting expert judgment for use in safety assessments for nuclear facilities will find Meyer and Booker (1991, 2001) and Budnitz *et al.* (1998) to be useful. Procedural guidance for structured formal expert assessments is given in Kotra *et al.* (1996) and SSHAC (1997). Formal expert assessment is part of the larger field of decision analysis and information related to the elicitation and use of expert judgments can be found in papers such as Keeney and von Winterfeldt (1991).

References

Anderson, J. G., S. G. Wesnousky and M. W. Stirling (1996). Earthquake size as a function of fault slip rate. *Bulletin of the Seismological Society of America*, **86**, 683–690.

Bedford, T. and R. Cooke (2001). *Probabilistic Risk Analysis: Foundations and Methods*. Cambridge: Cambridge University Press.

Bernreuter, D. L., J. B. Savy, R. W. Mensing and J. C. Chen (1989). Seismic hazard characterization of 69 nuclear power plant sites east of the Rocky Mountains, NUREG/CR-5250(1–8). Washington, DC: US Nuclear Regulatory Commission.

Budnitz, R. J., G. Apostolakis, D. M. Boore *et al.* (1998). Use of technical expert panels: applications to probabilistic seismic hazard analysis. *Risk Analysis*, **18**(4), 463–469, doi:10.1111/j.1539-6924.1998.tb00361.x.

Coppersmith, K. J. and R. R. Youngs (1990). Probabilistic seismic hazard analysis using expert opinion: an example from the Pacific northwest. In: Krinitsky, E. L. and D. B. Slemmons (eds.) *Neotectonics in Earthquake Evaluation*, Reviews in Engineering Geology, 8. Boulder, CO: Geological Society of America, 29-46.

Cornell, C. A. (1968). Engineering seismic risk analysis. *Bulletin of the Seismological Society of America*, **58**, 1583–1606.

CRWMS M&O (1996). Probabilistic volcanic hazard analysis for Yucca Mountain, Nevada, BA0000000-01717-2200-00082(0). Las Vegas, NV: US Department of Energy.

CRWMS M&O (1998). Probabilistic seismic hazard analyses for ground motions and fault displacement at Yucca Mountain, Nevada, Report EPRI NR-4726(1–10). Las Vegas, NV: US Department of Energy.

EPRI (1986). Seismic hazard methodology for the central and eastern United States: Vol. 1: Part 2, Methodology (Revision 1): Final Report EPRI-NP-4726-A-1(1). Palo Alto, CA: Electric Power Research Institute.

EPRI (1989). Probabilistic seismic hazard evaluations at nuclear plant sites in the central and eastern United States: resolution of the Charleston earthquake issue, Report EPRI NP-6395-D. Palo Alto, CA: Electric Power Research Institute.

Gneiting, T. and A. Raferty (2007). Strictly proper scoring rules, prediction, and estimation. *Journal of the American Statistical Association*, **102**(477), 359–378.

Keeney, R. L. and D. von Winterfeldt (1991). Eliciting probabilities from experts in complex technical problems, engineering management. *IEEE Transactions*, **38**(3), 191–201.

Kotra, J. P., M. P. Lee, N. A. Eisenberg and A. R. DeWispelare (1996). Branch technical position on the use of expert elicitation in the high-level radioactive waste program, Report NUREG-1563. Washington, DC: US Nuclear Regulatory Commission.

Meyer, M. A. and J. M. Booker (1991). *Eliciting and Analyzing Expert Judgment: A Practical Guide*. San Diego, CA: Academic Press.

Meyer, M. A. and J. M. Booker (2001). *Eliciting and Analyzing Expert Judgment: A Practical Guide*, ASA-SIAM Series on Statistics and Applied Probability. Philadelphia, PA: SIAM, ISBN: 0-89871-474-5.

Morgan M. G. and D. W. Keith (1995). Subjective judgments by climate experts. *Environmental Science and Technology*, **29**, A468–A476.

NAGRA (2004). Probabilistic seismic hazard analysis for Swiss nuclear power plant sites (PEGASOS project), Final Report, 1. Wettingen: prepared for Unterausschuss Kernenergie (UAK) der Ueberlandwerke (UeW) by Nationale Genossenschaft für die Lagerung radioaktiver Abfälle.

Panel on Seismic Hazard Evaluation *et al.* (1997). *Review of Recommendations for Probabilistic Seismic Hazard Analysis: Guidance on Uncertainty and Use of Experts*. Washington, DC: National Academy Press.

Pezzopane, S. K., J. W. Whitney and T. E. Dawson (1996). Models of earthquake recurrence and preliminary paleoearthquake magnitudes at Yucca Mountain. Seismotectonic framework and characterization of faulting at Yucca Mountain, Nevada, Milestone 3GSH100M. Denver, CO: US Geological Survey.

Simonds, F. W., J. W. Whitney, K. F. Fox *et al.* (1995). Map showing fault activity in the Yucca Mountain area, Nye County, Nevada. Miscellaneous Investigations Series Map I-2520. Denver, CO: US Geological Survey.

SSHAC (1997). Recommendations for probabilistic seismic hazard analysis: guidance on uncertainty and use of experts, NUREG/CR-6372. US Nuclear Regulatory Commission.

Stepp, J. C., I. Wong, J. Whitney *et al.* (2001). Probabilistic seismic hazard analyses for ground motions and fault displacement at Yucca Mountain, Nevada. *Earthquake Spectra*, **17**(1), 113–151.

Wells, D. L. and K. J. Coppersmith (1994). New empirical relationships among magnitude rupture length, rupture width, rupture area, and surface displacement. *Bulletin of the Seismological Society of America*, **84**(4), 974–1002.

Winkler, R. L. (1969). Scoring rules and the evaluation of probability assessors. *Journal of the American Statistical Association*, **64**(327), 1073–1078.

Index